1st International Conference on Maritime Sciences and Advanced Technology 2017

Ocean Science and Technology Toward a Global Maritime Axis

IOP Conference Series: Earth and Environmental Science
Volume 162

Denpaser, Indonesia
3 - 5 August 2017

Editor:

R. Dwi Susanto

ISBN: 978-1-5108-6580-8
ISSN: 1755-1307

Printed from e-media with permission by:

Curran Associates, Inc.
57 Morehouse Lane
Red Hook, NY 12571

proceedings.com

Some format issues inherent in the e-media version may also appear in this print version.

Printed by Curran Associates, Inc. (2018)

For permission requests, please contact the Institute of Physics
at the address below.

Institute of Physics
Dirac House, Temple Back
Bristol BS1 6BE UK

Phone: 44 1 17 929 7481
Fax: 44 1 17 920 0979

techtracking@iop.org

Additional copies of this publication are available from:

Curran Associates, Inc.
57 Morehouse Lane
Red Hook, NY 12571 USA
Phone: 845-758-0400
Fax: 845-758-2633
Email: curran@proceedings.com
Web: www.proceedings.com

Table of contents

Volume 162

1st International Conference on Maritime Sciences and Advanced Technology "Ocean Science and Technology Toward a Global Maritime Axis"

3–5 August 2017, Denpasar, Bali, Indonesia

Accepted papers received: 5 June 2018
Published online: 6 July 2018

Preface

Papers

Ocean and Coastal Dynamics

Air-Sea Interaction and Climate Variability

Marine Disaster and Hazards & Marine Degredation

Living Marine Resources (Sub-theme: Coastal Conservation, Biodiversity, Marine Pollutant, Marine Food Security and Aquaculture)

Non-living Marine Resources (Sub-theme: Marine Geology and Marine Renewable Energy)

Satellite and Ocean Remote Sensing

Marine Technology (Sub-theme: Offshore Technology, Underwater Acoustic and Ocean Observation)

IOP Conf. Series: Earth and Environmental Science **162** (2018) 011001 doi:10.1088/1755-1315/162/1/011001

Preface

The Institut Teknologi Bandung (ITB) will celebrate 100 years anniversary in 2020. One of the faculty, the Faculty of Earth Science and Technology (FEST) celebrated the 10th anniversary by November 2017. The Department Oceanography is within this Faculty. In accordance to government vision to put the nation "Toward Maritime Axis", we decided to take advantage of this momentum by organizing *The 1st Maritime Sciences and Advanced Technology (MSAT)* conference on August 3-5, 2017 in Denpasar, Bali. This conference is one of the implementations of our Memorandum of Agreement between the Faculty of Earth Science and Technology - ITB, Geospatial Information Agency of Indonesia, and the Ocean College Zhejiang University.

About 70% of Indonesia is covered by ocean. Indonesian seas with complex coastline geometry, topography, and passages provide the only direct communication between the tropical Pacific and Indian Ocean known as Indonesian throughflow (ITF). Indonesia as one of the largest archipelago on Earth, plays a pivotal role in global ocean circulation and weather climate system. It hosts the strongest equatorial convective center that drives the global tropical atmospheric circulation. Along the route within the Southeast Asian seas, the water undergoes strong tidal mixing and air-sea interactions and other oceanic/atmospheric climatic processes associated with upwelling, Madden-Julian Oscillation (MJO), monsoon, El Nino Southern Oscillation (ENSO) and possibly the Indian Ocean Dipole (IOD). Hence, Indonesian seas are the centre for Tuna fishing ground, world-most diverse marine biodiversity, and the same time prone to natural hazard associated with extreme climate events associated with ENSO and IOD. Fisheries and marine resources serve as main income generating sector for people in coastal areas. Because of strong tidal currents and many narrow passages with, many straits provide an ideal place for tidal current renewable energy. Therefore, our first MSAT conference theme was "Ocean Science and Technology Toward a Global Maritime Axis" covered broad topics such as ocean-atmosphere dynamics, marine biogeochemistry, air-sea interactions, fisheries, food security, marine environment and pollution to marine technology and renewable energy. The Conference had been a good opportunity for participants coming from China (38 scientists), Japan (2 scientists), Australia (2 scientists), USA (2 scientist), Bangladesh (2 scientists), UK (1 scientist), India (1 scientist), Germany (1 scientists), Malaysia (1 scientist), Turkey (1 scientist) and Singapore (1 scientist) to present and discuss topics in their respective research areas.

This Conference Proceeding provides an opportunity for readers to engage with refereed papers presented during the 1st MSAT. The papers published in this proceeding were selected from a total of 110 oral presented and 52 flash presented papers. This proceeding is divided into eight (8) mini-symposia: (1) Ocean and Coastal Dynamics, (2) Climate Change and Biogeochemical Processes, (3) Air-Sea Interaction and Climate Variability, (4) Marine Disasters and Hazards & Marine Degradation, (5) Living Marine Resources (Sub-theme : Coastal Conservation, Biodiversity, Marine Pollutant, Marine Food Security and Aquaculture, (6) Non-living Marine Resources (Sub-theme : Marine Geology and Marine Renewable Energy), (7) Satellite and Ocean Remote Sensing, (8) Marine Technology

(Sub-theme: Offshore Technology, Underwater Acoustic and Ocean Observation). Therefore, in this Conference Proceeding, readers might discover the recent issue and results of research in the broad topics of maritime science and advanced technology.

We would like to sincerely appreciate all parties for the successful of the 1st MSAT. We express special acknowledgement for the Head of Geospatial Information Agency of Indonesia for their generous financial and technical supports for this conference, the Ministry of Research and Technology – Higher Education of Indonesia for the financial support, and Udayana University for the venue support and hospitality, which allowed all foreign participants to enjoy the conference. We deeply express our gratitude for the organizing committee, keynote and invited speakers, reviewers, editors, and editing staffs for fully dedication, tireless efforts, and continuing hard work along the conference events and the process of this proceeding publication. We also deeply appreciate all of participants and authors for the excellent participating and taking the best opportunity for disseminating, discussing, and publishing the papers. We deliver appreciation to many participants from Zhejiang University, together with their honourable Professors and Presidential boards. Finally, we hope this proceeding provides the readers up to date and prominent information in maritime science and advanced technology from various points of view.

Chief editor,

R. Dwi Susanto, PhD.

Adjunct Professor, Faculty of Earth Science and Technology - ITB, Indonesia.

Department of Atmospheric and Oceanic Science, University of Maryland, College Park, USA.

IOP Conf. Series: Earth and Environmental Science **162** (2018) 011001 doi:10.1088/1755-1315/162/1/011001

IOP Conf. Series: Earth and Environmental Science **162** (2018) 011001 doi:10.1088/1755-1315/162/1/011001

IOP Conf. Series: Earth and Environmental Science **162** (2018) 011001 doi:10.1088/1755-1315/162/1/011001

IOP Conf. Series: Earth and Environmental Science **162** (2018) 011001 doi:10.1088/1755-1315/162/1/011001

IOP Conf. Series: Earth and Environmental Science **162** (2018) 011001 doi:10.1088/1755-1315/162/1/011001

IOP Conf. Series: Earth and Environmental Science **162** (2018) 011001 doi:10.1088/1755-1315/162/1/011001

IOP Conf. Series: Earth and Environmental Science **162** (2018) 011002 doi:10.1088/1755-1315/162/1/011002

Peer review statement

All papers published in this volume of *IOP Conference Series: Earth and Environmental Science* have been peer reviewed through processes administered by the proceedings Editors. Reviews were conducted by expert referees to the professional and scientific standards expected of a proceedings journal published by IOP Publishing.

Ocean numerical model experiment on estimating the variation of volume and heat transport in Karimata strait

I P Anwar[1], M R Putri[1], and A Setiawan[2]

[1] Department of Earth Science, Faculty of Earth Sciences and Technology, Bandung Institute of Technology, Jalan Ganesha 10, Bandung 40132, Indonesia
[2] Marine Research Centre, Agency for Marine & Fisheries Research & Human Resources, Ministry of Marine Affairs & Fisheries, Jalan Pasir Putih 1, Ancol Timur, Jakarta Utara 14430, Indonesia

E-mail: iwan.p.anwar@gmail.com

Abstract. Volume and heat transport variations through Karimata Strait during 2010-2014 were studied using three-dimensional numerical model of Hamburg Shelf Ocean Model (HAMSOM). Simulated area is located in 1°0'0" N - 4°30'0" N and 104°0'15"E - 110°30'0"E with a horizontal resolution 0.5' (approximately 0.94 km). The average of root mean square error of current velocity between model and observation are 0.17 for zonal (v) and 0.20 for meridional (v) components. The results of signal analysis showed that the volume and heat transport is strongly influenced by the seasonal period. There was also weak and medium phase of El Niño and La Niña during 2010 - 2014 that affected volume transport. The average volume (heat) transport during 2010-2014 through Karimata Strait is -2.75 Sv (-0.031 PW), meanwhile the average volume (heat) transport during boreal winter and boreal summer are -0.66 Sv (-0.06 PW) and -0.38 Sv (-0.03 PW), respectively (1 Sv = 10^6 m^3/s; 1 PW=10^{15} W). Southward total transport indicates that there are parts of South Chinese Sea-Indonesia seas flow through Karimata Strait.

1. Introduction

Indonesian Throughflow (ITF) is the flow of water from the Pacific Ocean to the Indian Ocean through Indonesian waters. The ITF's main line is the Makassar Strait [1]. Thus, that statement makes the other line of ITF not getting more attention, for example is the current in the Karimata Strait. According to previous study, Karimata Strait throughflow is part of the Indonesian Throughflow [2]. The waters flow from the Pacific through the Luzon Strait into the South China Sea (SCS) and flowing into the Karimata Strait and Indonesian waters. The transport variation in the Luzon Strait is influenced by ENSO, strengthened to the west and cools during El Niño. Transport in the Luzon Strait ranges from almost zero to 6 Sv with an average of 4.5 Sv [3]. Meanwhile, It show the transport average is 2.9 Sv. The transport of the Luzon Strait is a source of water mass for the South China Sea [3].

The number of rivers that contribute fresh water into the South China Sea, Natuna Sea, and Karimata Strait to the Java Sea causes the water mass through the Karimata Strait as main line of ITF to have a relatively low density. The results is Karimata Throughflow disrupts the surface flow in the southern part of the Strait of Makassar in the northwest monsoon [3]. Based on the estimation results from observation data when peak of northwest monsoon the transport volume reach -3.6 ± 0.8 Sv and the heat transport is -0.36 ± 0.08 PW [4]. Meanwhile other research calculated volume and heat transport in the Karimata Strait (without Gaspar Strait) recorded the maximum value of - 2.565 Sv and -0.258 PW in the west season and 1.233 Sv and 0.131 PW in the southeast monsoon [5]. However, the measurements were discontinues.

Due to lack of data measurements, many researchers conducted the numerical simulation to estimate volume transport. So, this research used HAMburg Shelf Ocean Model (HAMSOM) to simulate water dynamics in Indonesian seas. The results from hydrodynamics model such as current velocity and water temperature used to calculate volume and heat transport.

2. Method

2.1. Numerical and verification experiments

HAMburg Shelf Ocean Model (HAMSOM) was built by Backhaus on 1984 [6]. This model applied Arakawa-C for horizontal grid and *z*-level types for depth. HAMSOM has been used several times to conduct studies in Indonesian waters [7]. HAMSOM is used to study the volume transport in Karimata Strait with nesting techniques. The main model derived from Mayer [6] has a horizontal resolution of 6 'and a depth of 39 levels. The large model area is 11°N-12°S and 92°E-132°E. Then Putri performs a higher resolution model for the area 1°0'0"S-3°0'0"S and 104°0'15"E-110°30'0"E. It has horizontal resolution of 0.5' and a depth of 23 levels (3 m for each level) [6]. Higher resolution model add input of tidal and river flow from southeast Sumatra and western Kalimantan. It used a nested technique and runs from 2009 to 2014 [6].

The current velocity obtained from model is verified with data from South Chinese Sea Indonesian Sea Transportation/Exchange (SITE) project (26 June 2012-10 October 2012) [5]. The average of root mean square error of current velocity between model and observation are 0.17 for zonal (v) and 0.20 for meridional (v) components (Figure 1). Thus, the model results can be used to estimate transport in the Karimata Strait with continuous data.

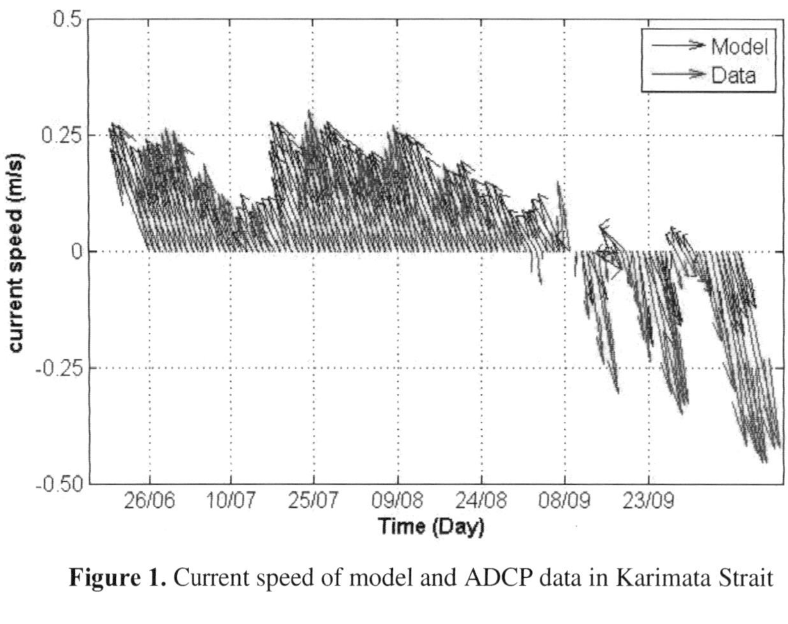

Figure 1. Current speed of model and ADCP data in Karimata Strait

The volume and heat transport are calculated using equation (1) and (2) [2][4][5].

$$F_v = \int_A udA \tag{1}$$

$$F_H = \rho C_p \int_A (T - T_o)udA \tag{2}$$

With, F_v is the transport volume (m^3/s), u is the current velocity (m/s), dA is the cross-sectional area perpendicular to the velocity vector (m^2). F_H is a heat transport (W), ρ is the density of sea water (kg/m^3), C_p is the heat capacity per unit volume ($4{,}1 \times 10^6$ J m^{-3} C^{-1}), T is temperature (^{0}C), T_o is the reference temperature (3.72 ^{0}C).

3. Results and Discussions

3.1. Volume and Heat Transport

Figure 2. show the variations of volume and heat transport every month during 2010-2014. The time-series data of volume and heat transport are overlaid by NINO 3.4 index to see the correlation with El Nino Southern Oscillation (ENSO) phenomenon. The previous study stated that there is no correlation between ENSO with water transport in Karimata Strait [3]. However in this research ENSO influenced on transportation in the Karimata Strait.

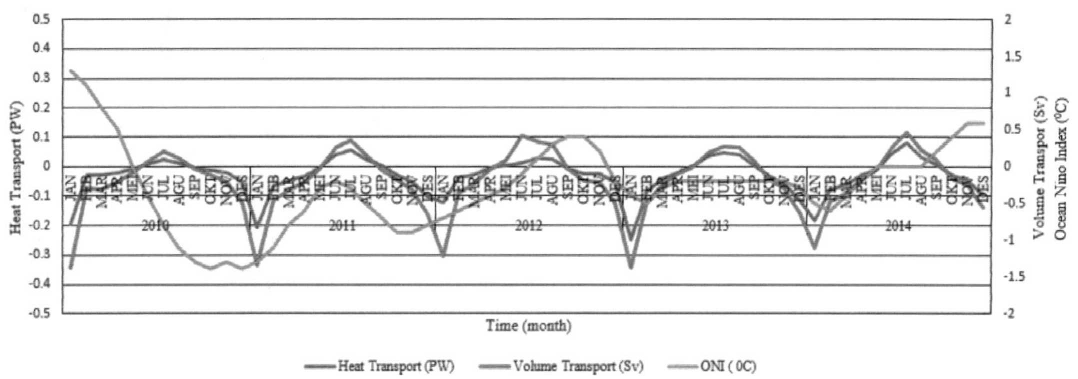

Figure 2. (a) Volume and heat transport in Karimata Strait (b) Ocean Nino Index at NINO 3.4

The monthly variation of volume and heat transport during 2010-2014 are shown in Figure 2. In the peak of northwest monsoon (January), the transport is always move southward. On the other hand, in the peak of southeast monsoon (July) the transport is move to north. The maximum value of volume transport is found in January (northeast monsoon) 2013 by -1.38 Sv (southward) and 0.47 Sv (northward) in July 2014 (southeast monsoon). Direction of wind and transport show no difference. It is because the average of Karimata Strait depth only 45 m.

Volume and heat transport in the northwest monsoon which occurred with medium phase El Niño in January of 2010 has a value -1.38 Sv and -0.19 PW. While at the time of La Nina phase occurred in July 2011, the volume transport is -1, 35 Sv and - 0.21 PW. In the northwest monsoon, the currents generally move from SCS to Indonesian waters [2][4][5]. However El Nino and La Nina affect the amount of volume transport. At the time of El Nino, volume transport in the Luzon Strait weakened near zero [3]. The water transport in Luzon Strait is a stream from the Pacific to SCS then to the Karimata Strait.

Volume and heat transport variations in southeast monsoon and La Nina occurred in July of 2010 which have a value of 0.21 Sv and 0.02 PW. This is due to the current transports of southeast monsoon in general to the north but when it coincides with La Nina the transport strengthens to the south. So when compared with southeast monsoon and La Nina does not occur then the value is greater 0.27 Sv and 0.05 PW.

The volume and heat transport system in the Karimata Strait is closely related to the monsoon and ENSO phenomena. This is because of the Karimata Strait has a shallow water depth (up to 45 m), so that it is strongly influenced by the wind. Furthermore, ENSO affects volumes and heat transport in the Karimata Strait because its water flow comes from SCS originating from the Luzon Strait and is strongly influenced by ENSO [3].

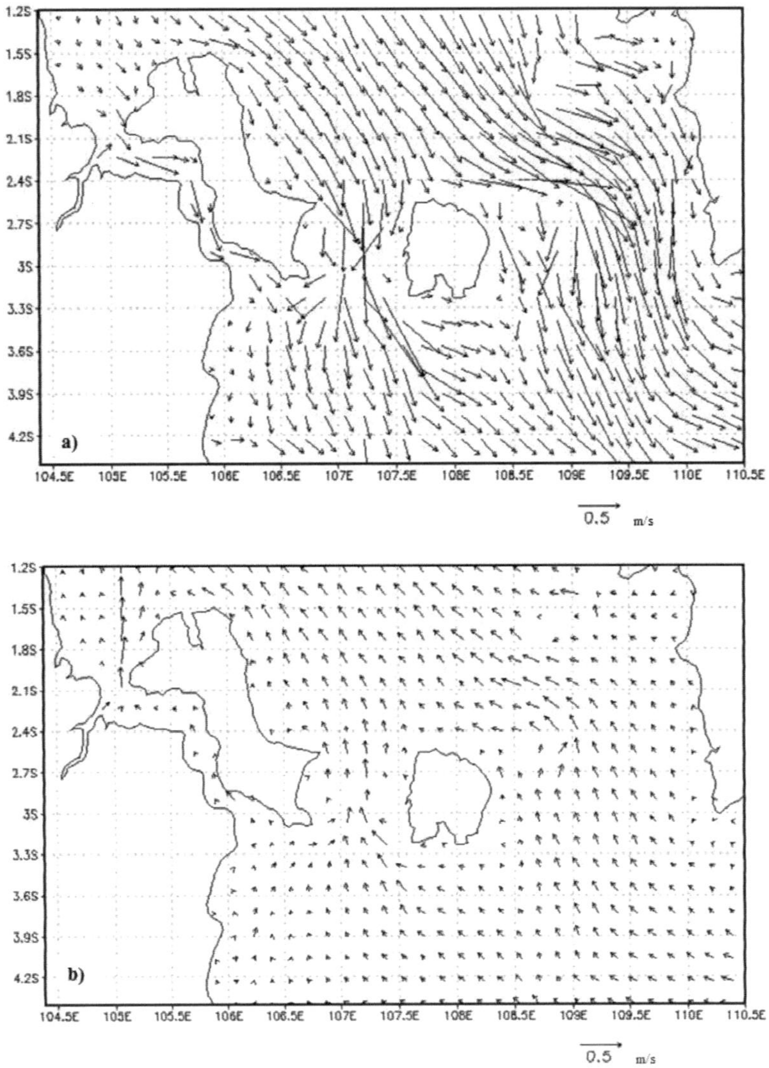

Figure 3. Average current (a) Januari and (b) July

Average current velocity and water heights in January and July during 2010-2014 indicate conformity with seasonal wind movements (Figure 3). January is the peak of the northwest monsoon, thus the current moves from SCS to the Java Sea, while July is the peak of the southeast monsoon so that the current velocity moves otherwise.

4. Conclusions

The average of root mean square error of current velocity between model and observation are 0.17 for zonal (v) and 0.20 for meridional (v). The maximum volume (Heat) transport during the northwest monsoon is -1.38 Sv (-0.25 PW) and the southeast monsoon is 0.47 Sv (0.08 PW). The mean volume (Heat) transport during the northwest monsoon is -0.66 Sv (-0.06 PW) and the southeast monsoon is 0.36 Sv (0.03 PW). Volume (Heat) transport during the northwest monsoon and El Nino is -1.38 Sv (-0.19 PW) and La Nina is -1.35 Sv (0.21 PW). Volume (heat) transport during southeast and La Nina monsoons is 0.21 Sv (0.02 PW) and normal 0.27 Sv (0.05 PW). For a more comprehensive study of volume and heat transport variations, the research should be conducted with data assimilation approach.

References

[1] Gordon A L 2005 *Oceanography* **18** **4** 1427
[2] Susanto R D, W Zexun, T R Adi, F Bin, L Shujiang and F Guohong 2013 *Acta Oceanol. Sin.* **32** **5** 16
[3] Gordon A L, B A Huber, E J Metzger, R D Susanto, H E Hurlburt and T R Adi 2012 *Geophys. Res. Lett.* **39** L11602
[4] Fang G, R D Susanto, S Wirasantosa, F Qiao, A Supangat, B Fan, Z Wei, B Sulistiyo, and S Li 2010 *J. Geophys. Res.* **115** C12020
[5] Setiawan A, M A Rahman, M R Putri, T Agustiadi, M Trenggono, B Priyono, Z Wei, S Li and R D Susanto 2015 *AOGS 12th Annual Meeting (Singapore)*
[6] Putri M R 2015 *DAAD Report*
[7] Mayer B and P E Damm 2012 *J. Geophys. Res.* **117** C07020

Evaluation of a global ocean general circulation model; The Lat-Lon-Cap (LLC90) configuration of the MITgcm

Bernawis Lamona[1], Judith Hauck[2] and Christoph Völker[2]

[1]Department of Oceanography, Institut Teknologi Bandung, Indonesia
[2]Alfred Wegener Institute, Germany

E-mail: lamona@fitb.itb.ac.id

Abstract. An evaluation of a general circulation model from Massachusetts Institute of Technology (MITgcm) with Lat – Lon – Cap (LLC90) configuration was done. Simulation of 100 years yields the annual means of potential temperature, salinity, meridional overturning stream function and transport of several throughflows. A reference run with widened Davis strait from the model was used to compare with observational WOA09 data, and the MITgcm has reproduced reasonably reliable data. The patterns of potential temperature can resemble the WOA09, however not so for the salinity fields. An experiment to simulate the model in 5 different tuning set were then proceeded. For the potential temperature and salinity fields, the North Pacific and the Southern Ocean still has significant difference to the WOA09. The difference was in the range of -6 to 5 ^{0}C for the potential temperature and -0.5 to 3 psu for the salinity field. The meridional overturning circulation stream function was still lower than the observational (± 17 Sv); the MITgcm simulates it 10 ± 1Sv in the experiment. The root mean square (RMS) of the difference were calculated, the surface layer has the largest magnitude of difference due to the forcing dynamics. The best simulation, which has the least difference to WOA09, was the one with the original bathymetry and smallest vertical diffusivity coefficient, 1 x 10^{-5} m^2 s^{-1}.

1. Introduction

MITgcm is widely used by oceanographers in wide spectrum of expertise; physics, biology, chemistry, and ocean-atmosphere interactions among others. The uniqueness of this general circulation model is that it employs the non-hydrostatic equation and fluid isomorphism to simulate fluid phenomena of both ocean and atmosphere, from the small scale of convection with 100m range to the general circulation of the world ocean on a scale of thousands of kilometers. In our simulations we used a resolution of 1^0.

Marshall [1] elaborated ocean models based on hydrostatic primitive equations (HPE), quasi hydrostatic (QH) and non-hydrostatic (NH) equations. In HPE, the vertical momentum equation is reduced to hydrostatic balance and an approximation is made with approximated Coriolis force and a shallow atmosphere approximation is made. With this, in large scale, the terms omitted in HPE are negligible, but on a small scale they become large enough to become a problem. For the application to global circulation, an HPE model is very accurate but it will break down somewhere between 10 to 1 km as the horizontal scale of the motion becomes comparable with its vertical scale.

In QH, the precise balance between gravity and pressure gradient forces is relaxed. The Coriolis force is treated exactly by including the cosine of latitude and becomes significant when approaching the equator; this provides the model a complete angular momentum. And in NH, the incompressible Navier Stokes equations were used.

Marshall [1] simulated small, meso and large scale with the three models and drew conclusions that there are equations that are more accurate than HPE which can be formulated and implemented e.g. QH and NH. These two are more versatile because they can simulate both small and large scale.
MIT scientists then developed milestones on the development of this general circulation model; [2],[3],[4] in aspects of efficiency of ocean modelling using the non-hydrostatic equation, new treatment of Coriolis terms in C-grid models, the using of parallel processor to run models, and construction of the adjoint MIT ocean general circulation model , respectively.

Currently, as in MITgcm manual by Adcroft, [5], the NH equation that is employed by the model is as below. Density,

$$\rho = \rho_0 + \rho' \tag{1}$$

Variations with depth ρ_0 are negligible and compressible in ρ' are:

$$\rho_0 = \rho_c$$

$$\rho' = \rho(\theta, S, p_0(z)) - \rho_0 \tag{2}$$

This leads to "semi-compressible" Boussinesq equations:

$$\frac{D\vec{v}_h}{Dt} + f\hat{k} \times \vec{v}_h + \frac{1}{\rho_c} \nabla_z p' = \vec{F} \tag{3}$$

$$\epsilon_{nh} \frac{Dw}{Dt} + \frac{g\rho'}{\rho_c} + \frac{1}{\rho_c} + \frac{1}{\rho_c} \frac{\partial \rho'}{\partial z} = \epsilon_{nh} F_w \tag{4}$$

$$\nabla_z . \vec{v}_h + \frac{\partial w}{\partial z} = 0 \tag{5}$$

$$\rho' = \rho(\theta, S, p_o(z)) - \rho_c \tag{6}$$

$$\frac{D\theta}{Dt} = Q_\theta \tag{7}$$

$$\frac{DS}{Dt} = Q_s \tag{8}$$

Where $\vec{v}_h$ (u,v,0) is horizontal component of velocity (on pressure surface), f is Coriolis force; $2\Omega \sin \varphi$, ρc is constant reference density of water, F is force, $w = \frac{Dp}{Dt}$ is vertical velocity in p-coordinate, ϵ_{nh} is coefficient of non-hydrostatic to make eq.(4) not negligible, θ is potential temperature, S is salinity, and Q is heat rate per unit mass.

Griffies, [6] have done comparative study by using seven models i.e. NCAR-POP, FSU-HYCOM, GFDL-MOM, GFDL-HIM, Kiel-ORCA, KNMI-MICOM, and MPI (see section 5 for acronyms). The study was within the frame of Coordinated Ocean-ice Reference Experiments (COREs). They aimed the COREs to become a tool to learn the behaviour of global ocean-ice models forced by common atmospheric dataset. In their work they evaluated the simulation results from the seven models, which were:

- the annual mean temperature, salinity
- anomaly of SST and SSS
- monthly values of heat content
- annual mean sea-ice area
- sea-ice concentration
- upper ocean temperature on the equator in the Pacific
- upper ocean zonal velocity component on the equator in the Pacific
- maximum mixed layer depth
- anomalous zonal-mean decadal mean of potential temperature and salinity
- Atlantic Basin meridional overturning stream function
- global meridional overturning stream function
- annual mean of volume transports.

The second phase of COREs (COREs-II) was conducted by Danabasoglu, [7] focusing in the North Atlantic, more specifically in the AMOC (Atlantic Meridional Overturning Circulation). AMOC has large heat and salt transport that influence the North Atlantic climate and finally affected the global climate through the ocean-atmosphere interaction. AMOC plays an important role in decadal and longer time scale of the climate variability, hence it will be important aspect to develop model to predict the earth's future climate in that time scale. The COREs-II experiment was a hindcast simulation to provide reconstruction of the AMOC behavior the past decades/centuries. They evaluated more or less the same parameters as Griffies, [6] have done, but with 13 different models, namely : MOM4p1, MICOM, NEMO 3.1.1, 3.2, 3.3, 3.4, HYCOM 2.2, GOLD, GISS Model E2-R, INMOM, MITgcm, MOVE/MRI.COM 3, POP2 simulated by 18 research institutions. Some institutions used the same model with different configurations.

Hauck, [8] used MITgcm to investigate seasonally different carbon flux changes in the Southern Ocean in response to the Southern Annular Mode (SAM). They did not include the Arctic in that work. In 2015, they modified a global configuration, namely LLC90 (lat lon cap) provided by Forget, [9] to include the Arctic into the model. This paper will discuss the evaluation process of the simulation result and the tuning of some parameters to reproduce the closest possible salinity and potential temperature to the observational data.

2. Model

The Massachusetts Institute of Technology general circulation model (MITgcm) was set up globally with a 1^0 resolution. The ocean water column was set to 50 layers, with 10m increment from 5 m to 115 m, and larger increment down to 5906m depth. A thermodynamic and dynamic sea-ice model [10] was used. One bathymetry product from LLC90 configuration provided by Forget, [9] was applied. The Davis Strait was widened on purpose to avoid the model break down because of ice flood.

The reference run was set up as follow. The diffusive rate coefficient was $2x10^{-5}$ m^2 s^{-1}. The Davis Strait was widened on purpose to avoid numerical problems with too strong ice formation. Some other changes at the Arctic and Anarctic coast were also applied to prevent too strong build-up of sea ice. The model then simulated for 100 years, the first 90 years is considered as the model spin-up. The last ten years were then calculated to get the annual mean for the analysis. The analysis was mostly done by using the gcmfaces, a Matlab package to analyze MITgcm provided by [11]. This reference run was then used to evaluate by comparing the data it reproduced to the observational World Ocean Atlas 2009 (WOA09).

3. Evalution of the reference run

In this project we only evaluate the potential temperature and salinity field in global scale and in Atlantic basin, the zonal mean of those fields, the meridional overturning stream function and the transports volume in several throughflows.

We observe the difference between the global salinity and potential temperature fields of WOA09 and the simulated field by the model from the reference run as in Figure 2 and 3.

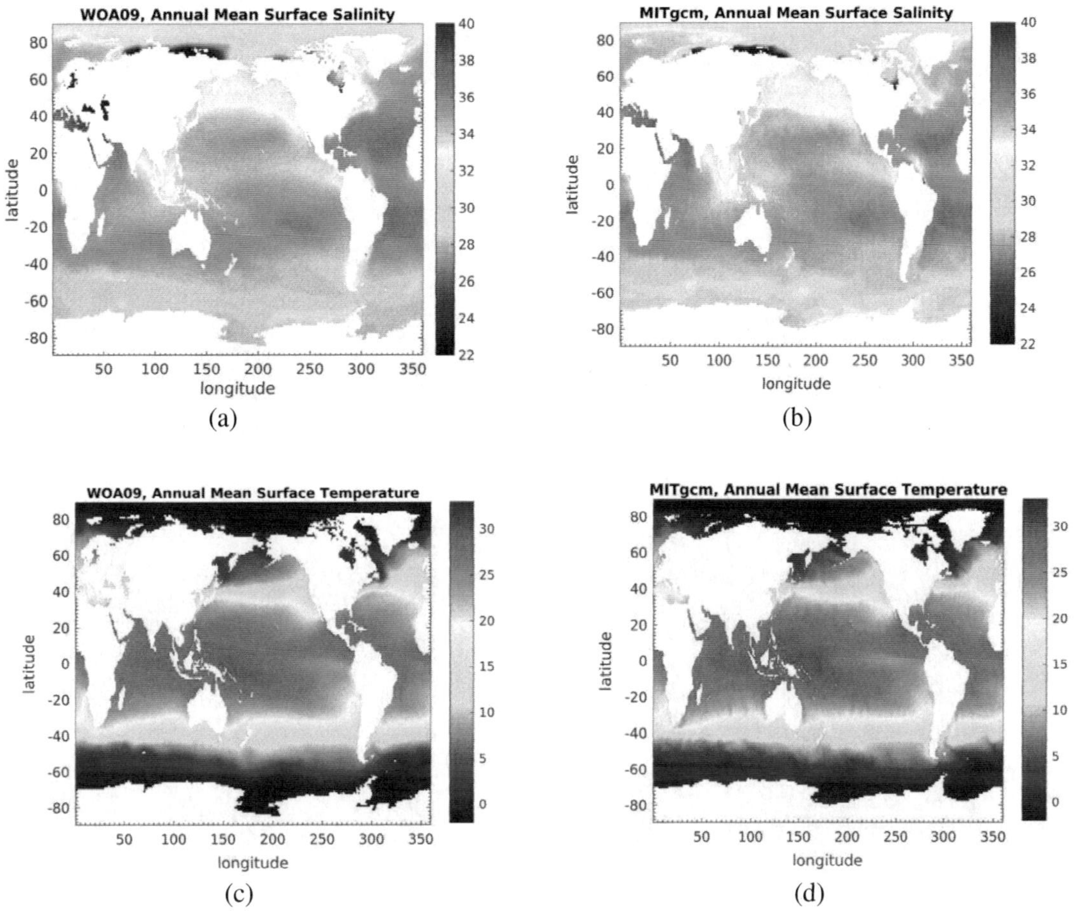

Figure 1. Surface salinity and potential temperature fields from WOA09 (a,c), and from MITgcm reference run result (b,d).

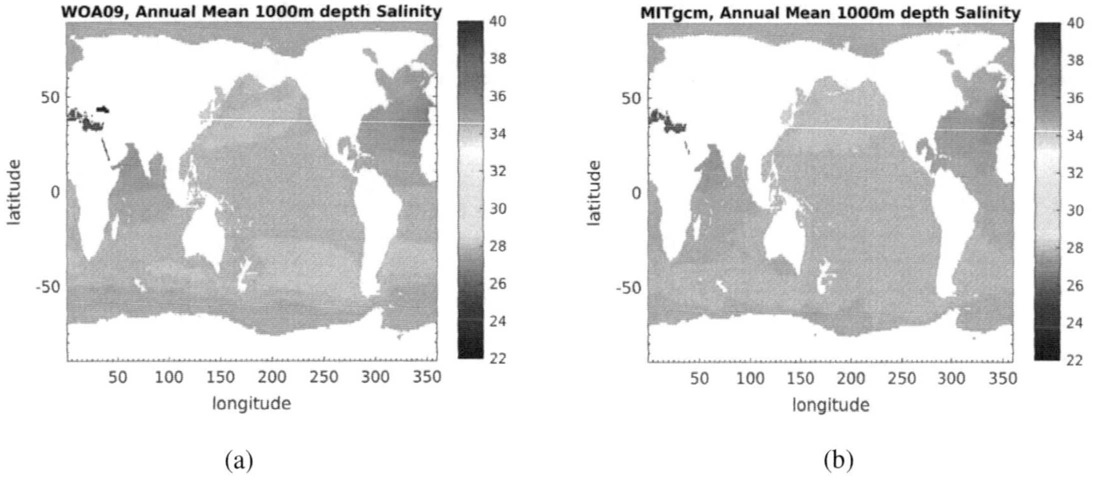

IOP Conf. Series: Earth and Environmental Science **162** (2018) 012002 doi:10.1088/1755-1315/162/1/012002

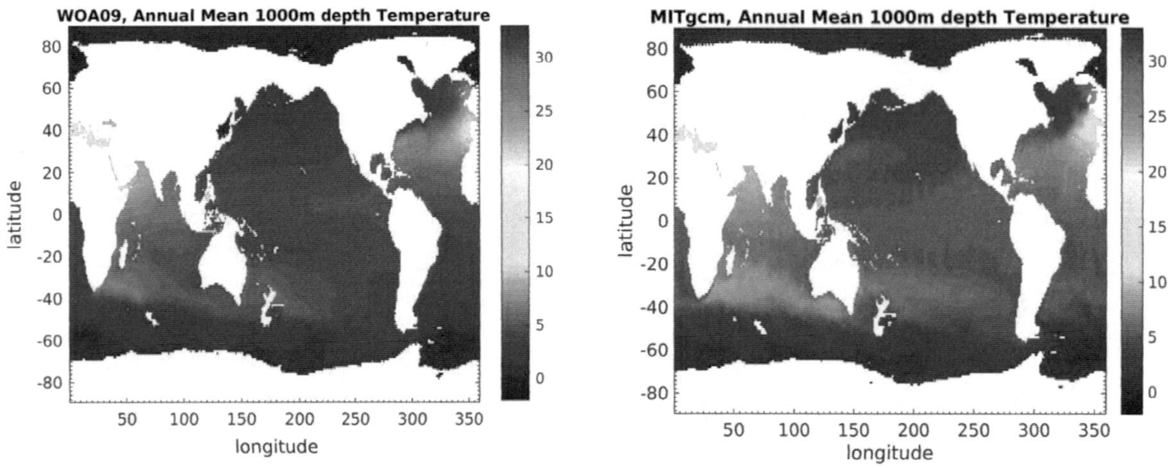

(c) (d)

Figure 2. Salinity and potential temperature fields of 1000m depth from WOA09 (a,c) and MITgcm reference run result (b,d).

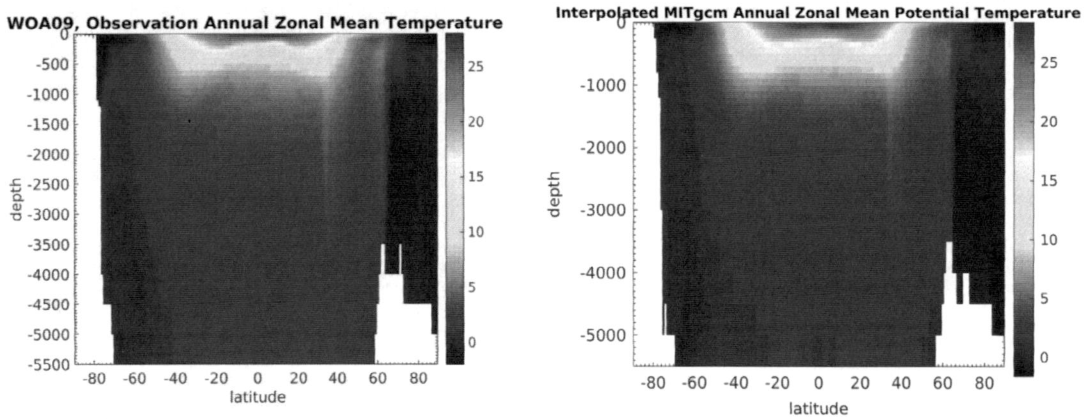

(a) (b)

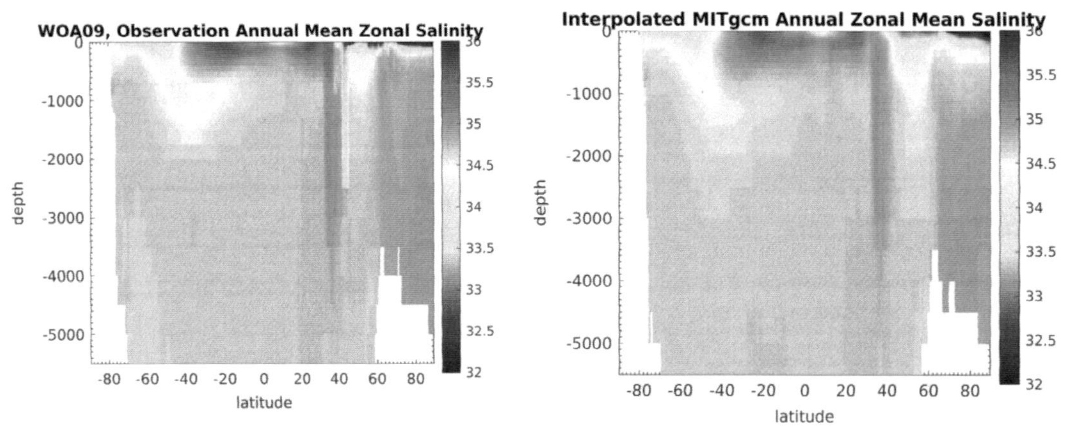

(c) (d)

Figure 3. Annual zonal mean potential temperature and salinity (a,c) from WOA09 and (b,d) from MIT gcm reference run.

Figure 3 shows the comparison between zonal mean of potential temperature and salinity from WOA09 and from the reference run. We needed to interpolate the model results because WOA09 has 33 depth layers while MITgcm has 50 depth layers. Therefore, we interpolated the MITgcm depth layers into WOA09's to make those comparable. We can say from these Figures that MIT gcm is able to reproduce the potential temperature and salinity reasonably well. Further on we would like to calculate the difference between model and data in a plot to get a more detailed look.

Figure 4(a) is showing some transports calculated from the reference run. We can see that the value for Davis Strait is out of range of -3 ~ 1 Sv; probably due to the widened bathymetry of it in the model configuration to prevent the piling-up of ice, and 4(b) the volume of meridional overturning. More to discuss in section 5.

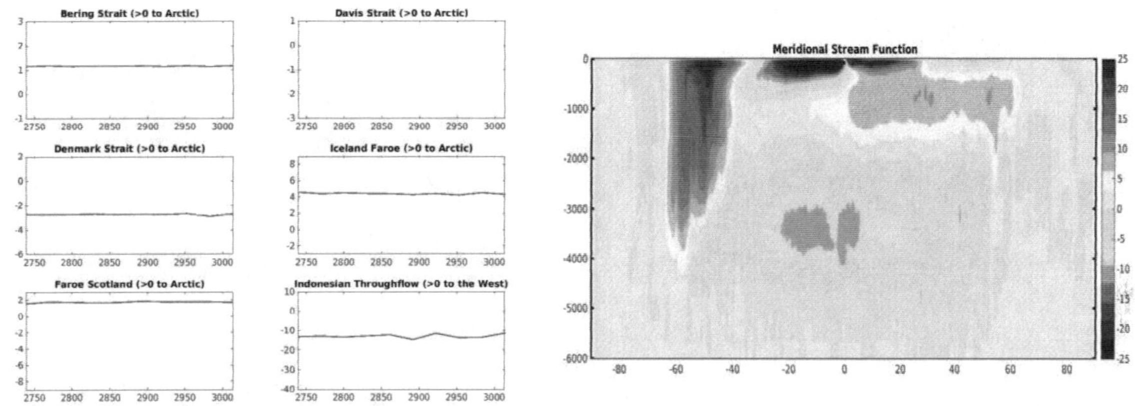

(a) (b)

Figure 4. (a) the transport in several throughflows of (from left to right); Bering Strait, Davis Strait, Denmark Strait, Iceland Faroe, Faroe Scotland, and Indonesia Throughflow, respectively. (b) the meridional stream function.

4. Experiment

We continued to the next step from model evaluation i.e. to conduct some tuning experiment. We did five more simulations with different sets of parameters. One change was to change the bathymetry to the original bathymetry that was provided in the LLC-90 configuration as in [9]. Other parameter that was altered was the vertical diffusivity and delta Z of sea-ice growth calculation. We named the runs as run5,6,7,8, and 9, respectively.

The following table is the summary of the tuning parameter set in the runs of our experiment. We refer to as "original bathymetry" for that of ecco configuration (bathy_ecco_llc_90x50_min2pts.bin in the table). For the newly set up bathymetry with widen Davis Strait and some changes in Arctic and Antarctic coast, we refer to as "new bathymetry" (bathymetry.llc in the table). BL79 in run 8 is a vertical diffusivity scheme suggested by Bryan, [12]. In run9, we enabled a flag in SEAICE_OPTIONS.h and altered the code in seaice growth.f to limit the effect of sea-ice on the non-linear free surface to a maximum of one full grid box. In all other runs this flag was not set, hence the effect of sea-ice on the non-linear free surface was not limited.

Table 1. Summary of the tuning parameter set in simulation experiment

Run	Bathymetry	Initial T/S	Vertical diffusivity ($m^2\ s^{-1}$)	Delta Z in seaice_growth.F
reference	bathymetry_llc.bin	llc90_12months.bin	2×10^{-5}	not set
5	bathy_ecco_llc_90x50_min2pts.bin	llc90_12months.bin	1×10^{-5}	not set

6	bathy_ecco_llc_90x50_min2pts.bin	llc90_12months.bin	2 x 10^{-5}	not set
7	bathy_ecco_llc_90x50_min2pts.bin	llc90_12months.bin	3 x 10^{-5}	not set
8	bathy_ecco_llc_90x50_min2pts.bin	llc90_12months.bin	BL79	not set
9	bathymetry_llc.bin	llc90_12months.bin	2 x 10^{-5}	(1)/1

4.1 Global

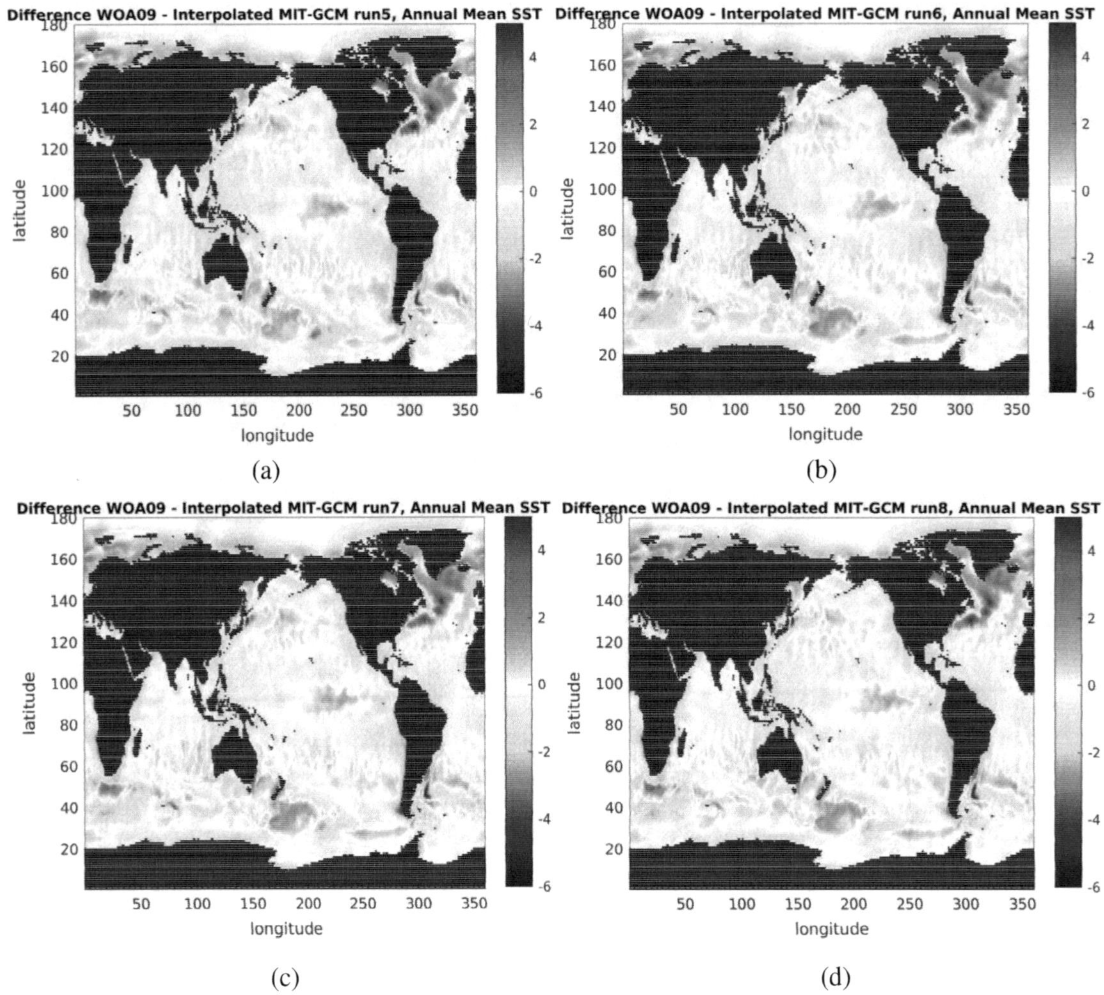

Figure 5. The difference (^{0}C) between observational SST WOA09 and MITgcm for run 5, 6, 7, 8.

We can see in Figure 5, the value difference of observational SST data from WOA09 and MITgcm from run 5 to 8. We can see some high difference mostly in the North Atlantic. The difference is 2 to 4^{0}C, where the MITgcm is colder than WOA09. We also get a notable difference at the Southern Ocean, where the MITgcm is warmer than WOA09. However it is not so clear here, with some areas are too warm and some areas are too cold. The largest difference is in the North Atlantic and also in the Arctic where WOA09 is consistently warmer.

IOP Conf. Series: Earth and Environmental Science **162** (2018) 012002 doi:10.1088/1755-1315/162/1/012002

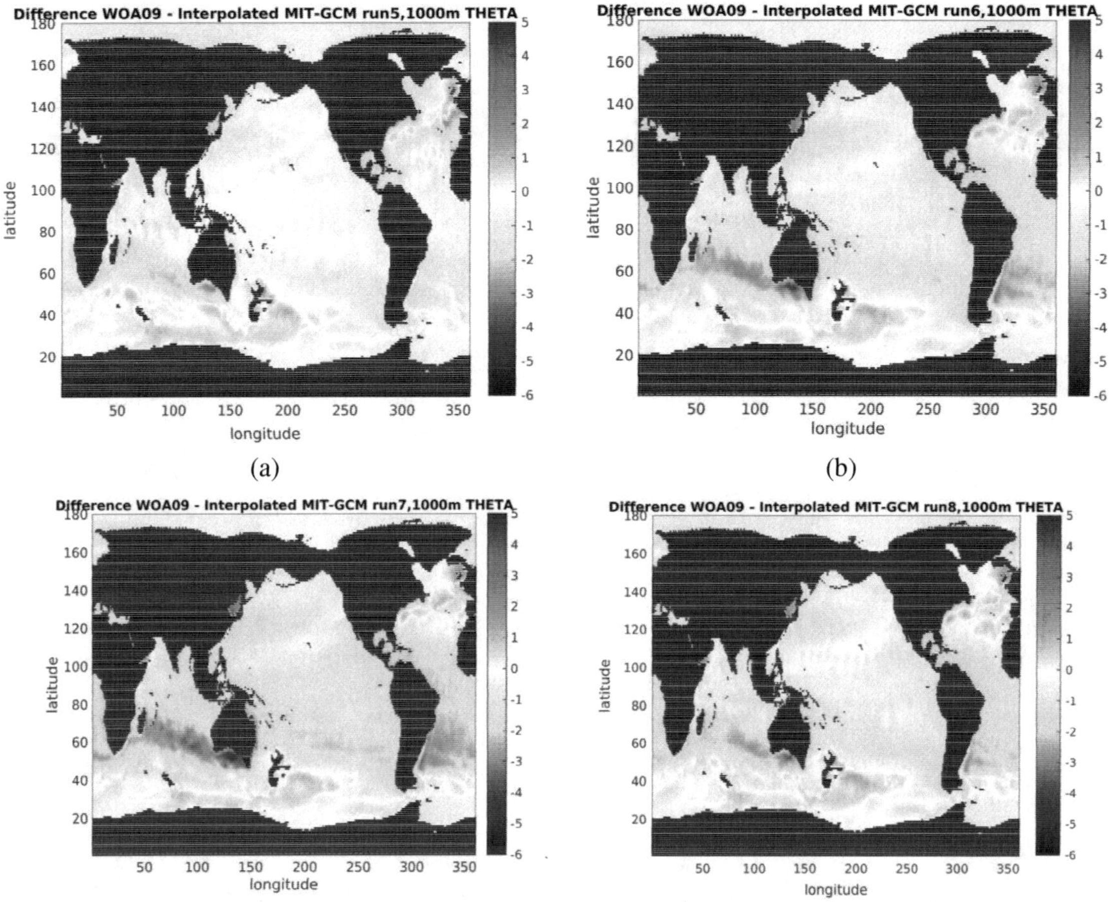

(a) (b)

(c) (d)

Figure 6. The difference between observational potential temperature at 1000m depth WOA09 and MITgcm for run 5, 6, 7, 8.

At 1000m depth as in Figure 6, in general, -we have the MITgcm data about 2^0C warmer than WOA09. We can see in Figure 6 that run 5 is the best run with the least difference to WOA09 among other runs. At this depth the model is less affected from the surface forcing, as expected for a 100 year run.

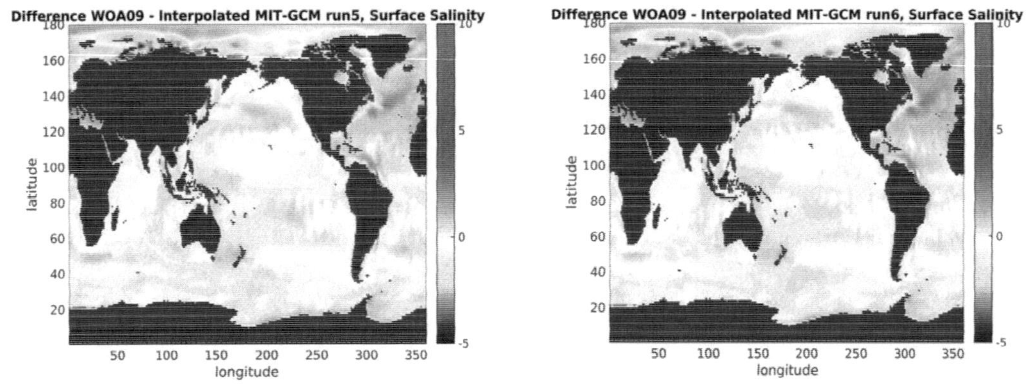

(a) (b)

IOP Conf. Series: Earth and Environmental Science **162** (2018) 012002 doi:10.1088/1755-1315/162/1/012002

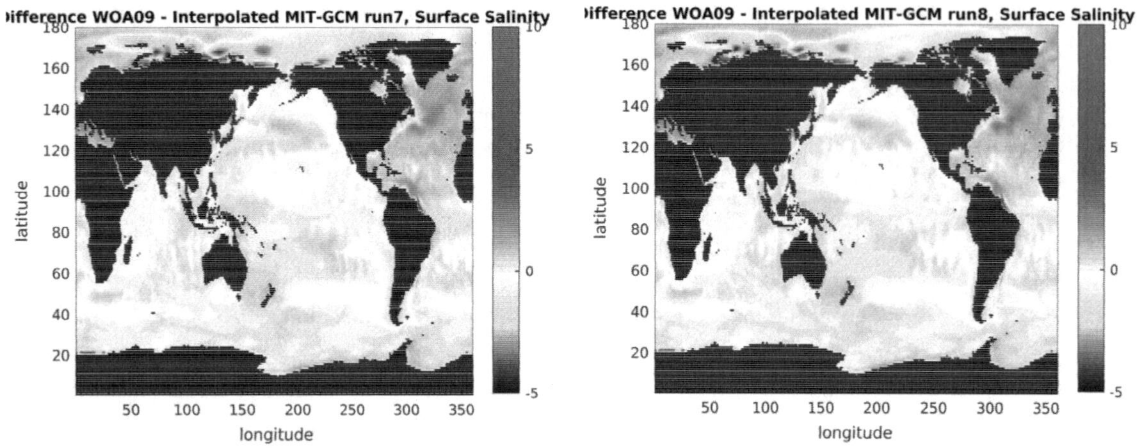

(c) (d)

Figure 7. The difference (psu) between observational surface salinity WOA09 and MITgcm for run 5, 6, 7, 8.

As for the surface salinity, we have the most contrast difference (still) in the North Atlantic; the MITgcm is less saline than WOA09 up to 5 psu. In the Arctic Ocean in contrary, the MITgcm is more saline than WOA09 up tu 5 psu.

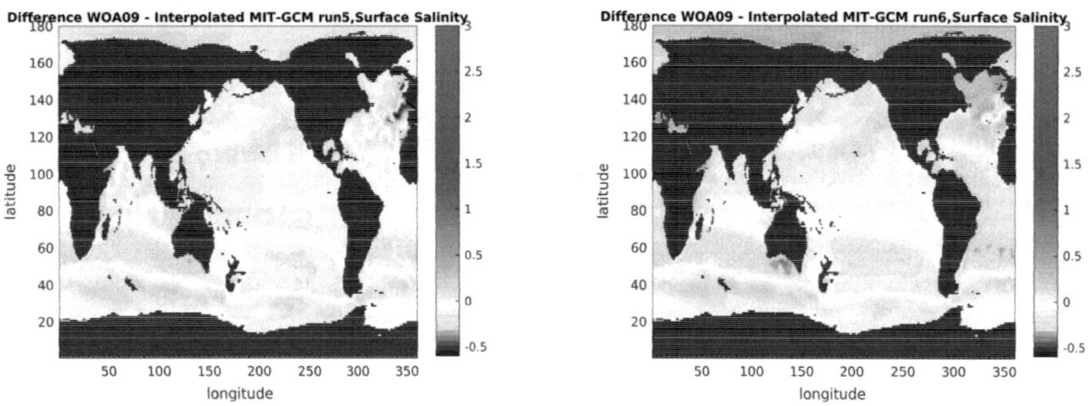

(a) (b)

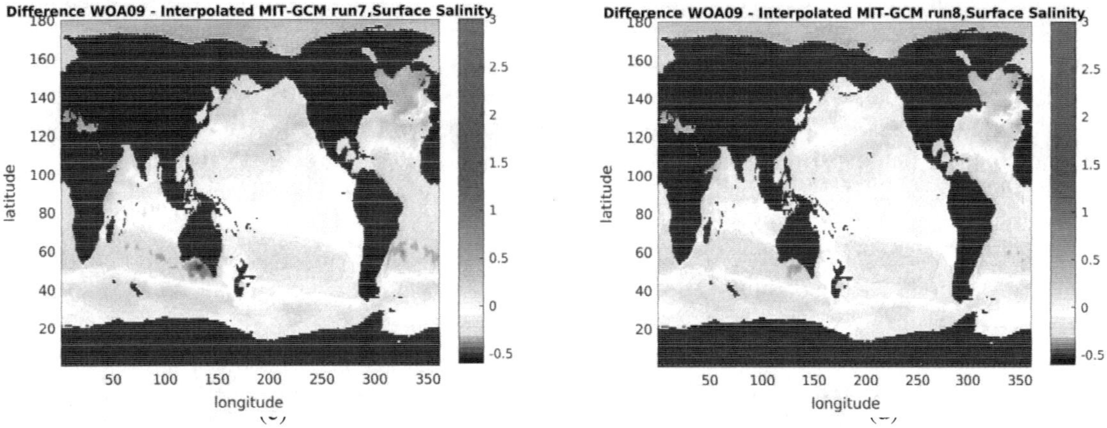

Figure 8. The difference between observational 1000m depth salinity WOA09 and MITgcm for run 5, 6, 7, 8.

Opposite to the surface where the MITgcm is more saline in the Arctic Ocean, at 1000m depth the MITgcm is more saline at Southern Ocean. Run 5 still hold the least difference to WOA09. From these results we would like to give more focus to North Atlantic Ocean because there is constant significant difference.

4.2 Atlantic basin

The zonal mean of potential temperature from all runs almost look identical (Figure 9). The model can resemble the WOA09 quite well, however the surface to 1000 m depth layer isotherms in the South Atlantic are deeper than WOA09. In contrary, the isotherms in the North Atlantic seem shallower. At the surface of low latitude, the MITgcm is warmer than WOA09 –and of high latitude it is colder. The isotherms of 2-3^{0}C pattern from the models are also different. Those stay at depth of 2000-1000m through the equator, while in the WOA09 those stay at much lower, below 3000m depth.

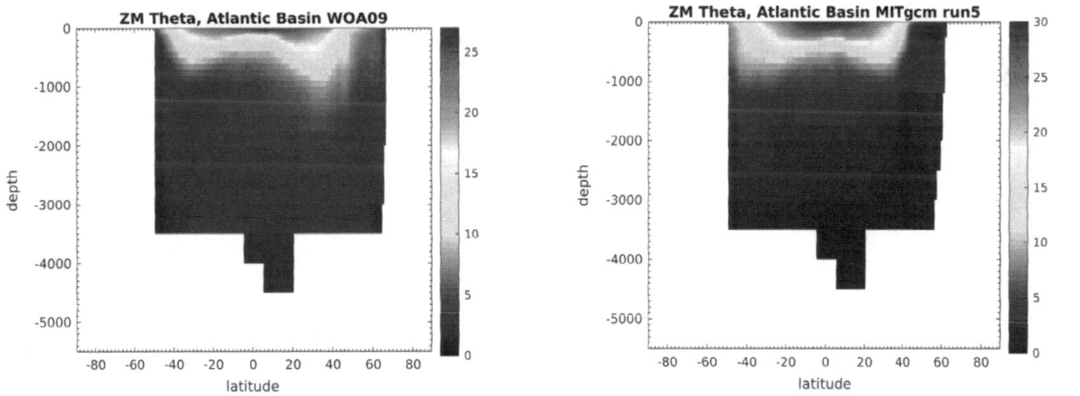

(a) (b)

Figure 9. The zonal mean of potential temperature from (a). The WOA09 observational data. (b) MITgcm run 5.

In Figure 10 we show the difference of zonal mean potential temperature between the WOA09 and MITgcm. This Figure sustains that MITgcm is warmer WOA09 at the low latitude surface layer and

IOP Conf. Series: Earth and Environmental Science **162** (2018) 012002 doi:10.1088/1755-1315/162/1/012002

colder the high latitude surface layer. It takes longer time for the deep layer to experience the dynamic forcing from the surface layer.

We continue to look at the zonal mean salinity vertical distribution of WOA09 as in Figure 11 (a). Here MITgcm does not seem to reproduce the data very well. From surface to 1000 m depth we have two high salinity cores of 36-37 psu from 40^{0}S to equator, less saline in equator due to high precipitation, then continues from near equator toward 40^{0}N. The Northern core is almost 300% larger than the Southern one. MITgcm reproduced these cores at least 2 psu less than WOA09, with the Southern core is much larger than the Northern one. The WOA09 also has tongue of 34-34.5 psu at the Sorthern part, which MITgcm failed to reproduce. Rather it produced the tongue at the Northern part of Atlantic basin. The rest of salinity gradient of MITgcm is not as smooth as in WOA09 with more significant difference.

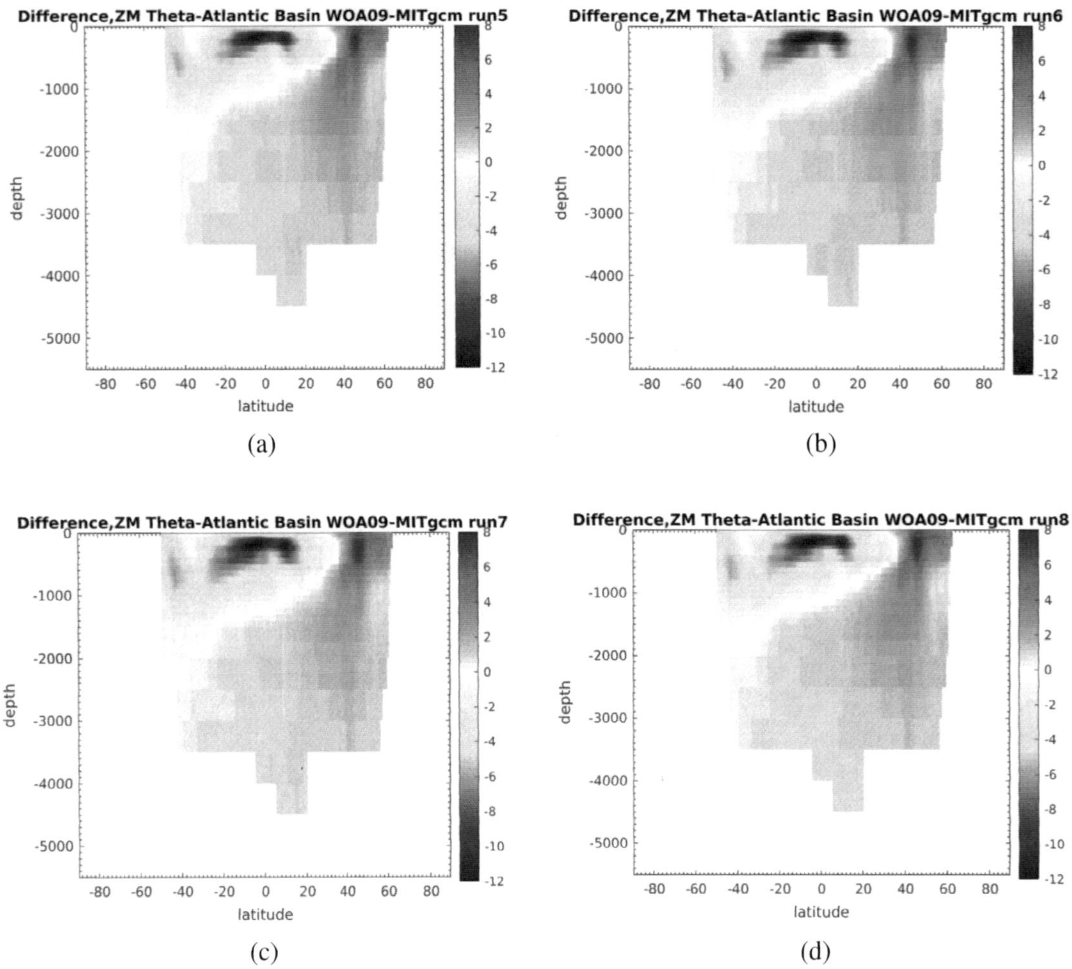

Figure 10. The difference of zonal mean potential temperature between WOA09 and MITgcm from (a), (b), (c), (d) run 5, 6, 7, and 8, respectively.

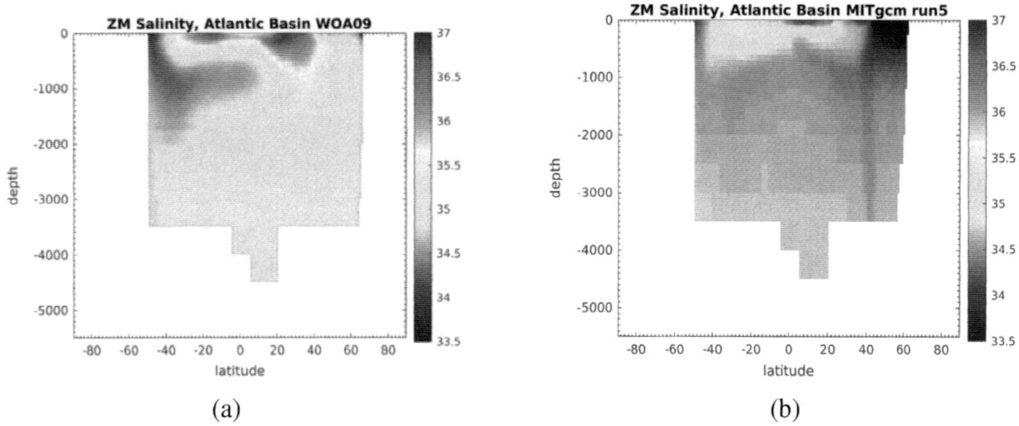

Figure 11. (a). The zonal mean salinity in Atlantic basin of WOA09. (b). The same for MITgcm run 5.

Coming from these result, we checked on the difference between WOA09 and MITgcm (Figure 12). We found two contrast cores of highest difference from surface to 1500m depth. In the southern part the higher difference takes place at the 34-34.5 psu salinity tongue. MITgcm is more saline than WOA09. In the northern part it takes place at the high salinity core of WOA09. This extended up to 60^0N. The difference value is at least 1.5 to 2.5 psu, the MITgcm is less saline than WOA09.

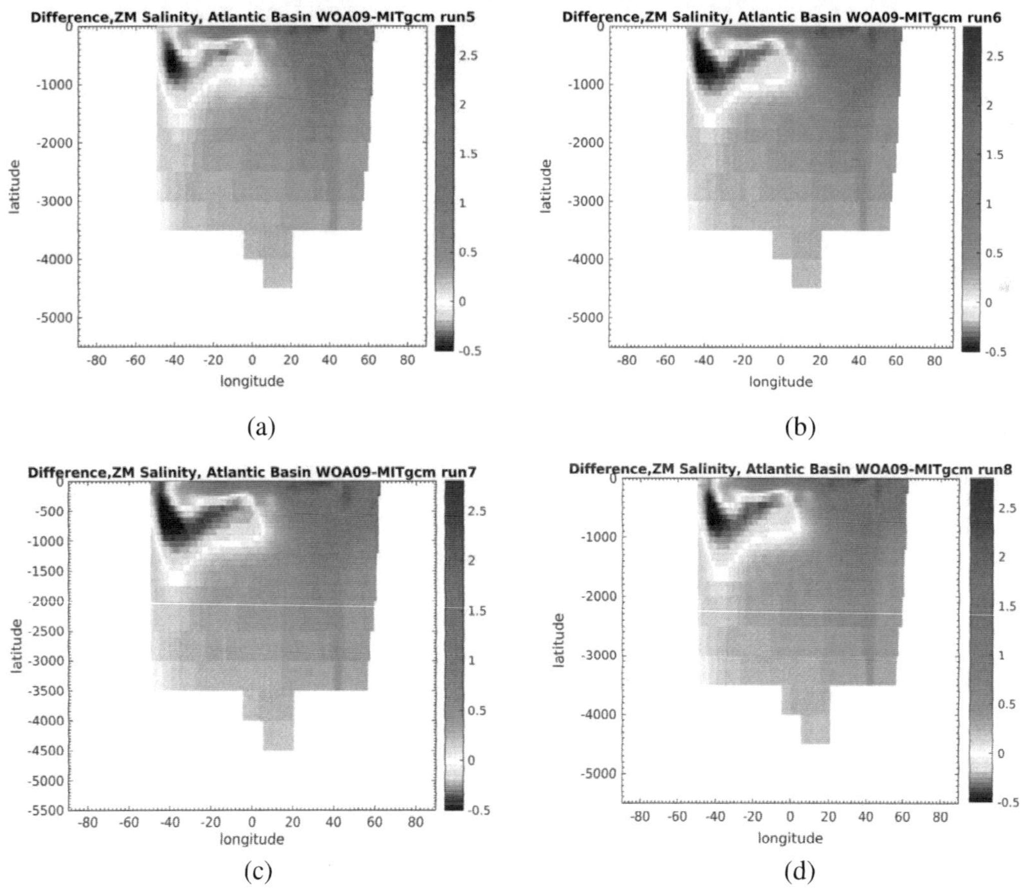

Figure 12. The zonal mean salinity difference between WOA09 and MITgcm from (a), run 5 (b) run 6, (c), run 7 and (d) run 8.

IOP Conf. Series: Earth and Environmental Science **162** (2018) 012002 doi:10.1088/1755-1315/162/1/012002

4.3 Meridional overturning circulation

Our reference run did not reproduce well the North Atlantic meridional overturning (see section 5. Discussion). Run 5 gave almost 10 Sv of the overturning and all the rest runs; 6,7,8,9, resulted in smaller volume than observational value of 17 Sv [1] of the overturning. Run 8 and 9 are not shown due to insignificant difference with run 6 and 7.

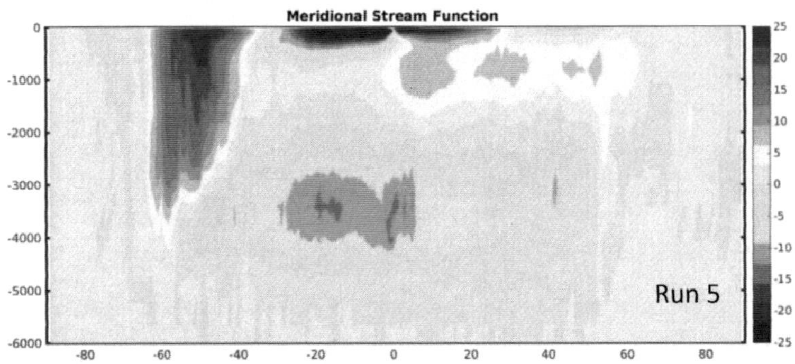

(a)

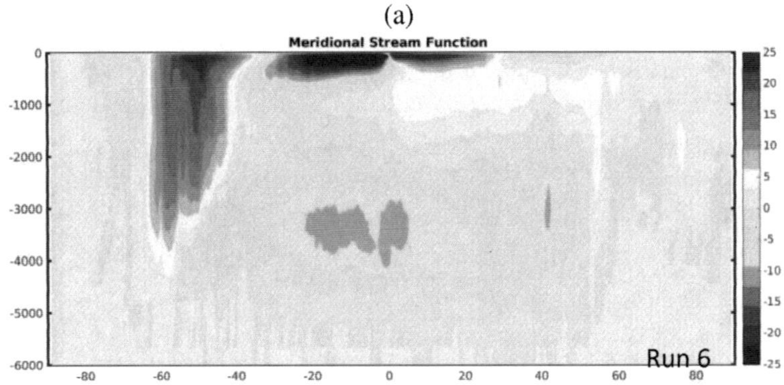

(b)

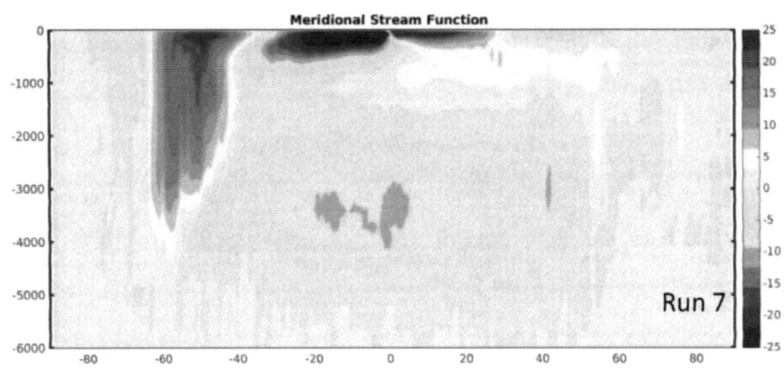

(c)

Figure 13. Meridional overturning from MITgcm (a). run 5, (b). run 6, (c). run 7. Other runs are not shown due to similarity to run 6 and 7. In unit of Sverdup ($10^6 m^3 s^{-1}$).

All the simulations of experiment return the value for the Davis Strait back in the scale range, because we used the original bathymetry again with more realistic Davis Strait width. The transport volumes among runs are not very much different; hence we only show result of run 5 in Figure 14. However these transports are still not showing seasonality, due to the values come from the annual mean.

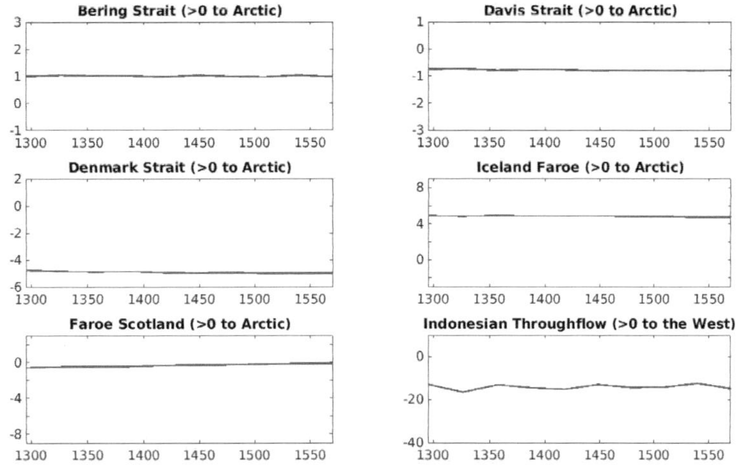

Figure 14. Volume transport of throughflows (Sv).

4.4 Root Mean Square

To further analyze the magnitude of the difference, we calculated the root mean square (RMS) of it. Figure 15 are the RMS of the potential temperature and salinity difference for each depth layers of Atlantic basin. The RMS peak of potential temperature is at 10m depth, and from 0 to 5m depth for salinity. All runs are almost identical for the RMS as in Figure 15(a) and (b). To check the magnitude along the latitude, we chose surface, 1000 m and 3000 m depth layers. For run 5 (Figure 15 (c)) the low latitudes have the smallest RMS difference in all layers and the difference are larger in the north and south of the Atlantic basin. Layer 3000m depth is the most stable RMS and only peaks in the north Atlantic. Run 6,7,8,9 have similar RMS for the difference, hence we only show run 8 in Figure 15(d). The magnitude of difference is larger than in run 5. At surface, the RMS is peaked higher in the north than in the south of Atlantic. In contrary at 1000m depth, it peaks higher in south than north. For 3000m depth it is the same as run 5, less difference with peak only in north of Atlantic. Surprisingly, the RMS of salinity difference is only large enough at the surface layer as in Figure 15(e).

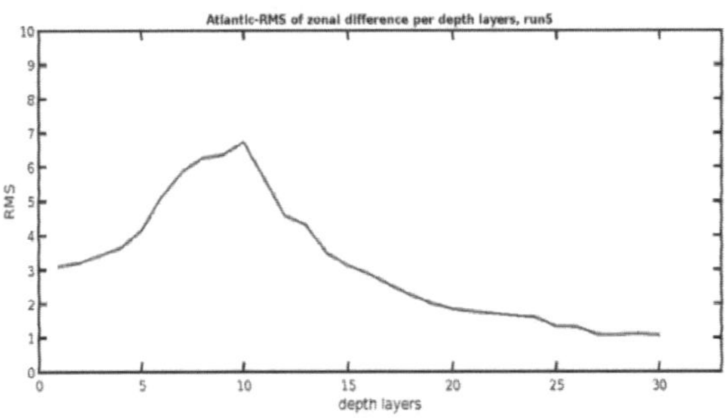

(a)

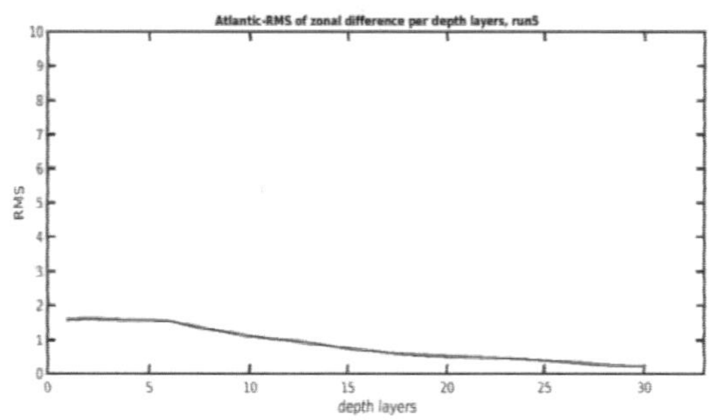

(b)

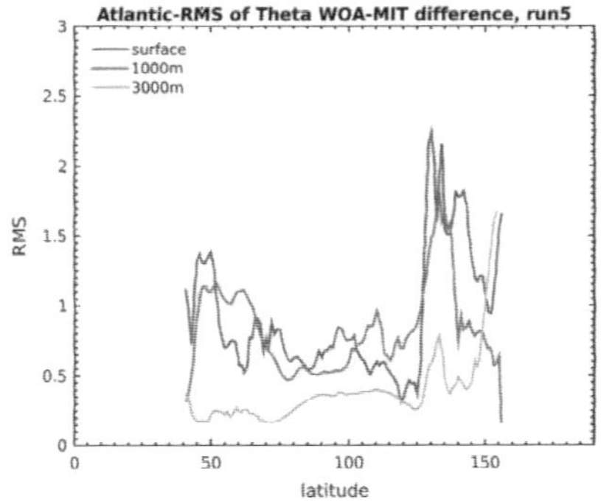

(c)

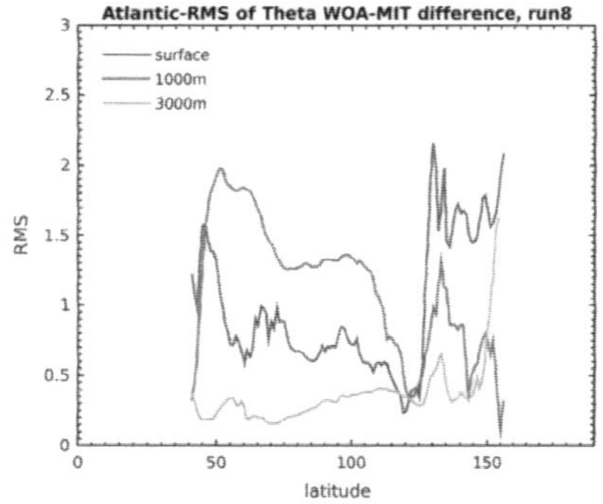

(d)

IOP Conf. Series: Earth and Environmental Science **162** (2018) 012002 doi:10.1088/1755-1315/162/1/012002

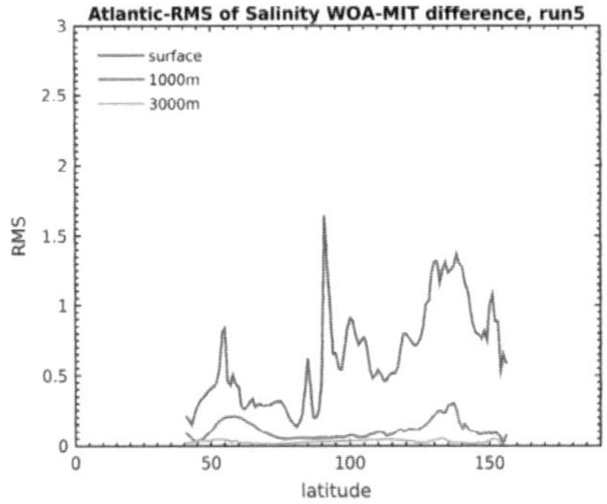

(e)

Figure 15. Root Mean Square of (a) Atlantic basin potential temperature difference and (b) salinity. (c), (d). RMS of surface, 1000m, 3000m depth layers of potential temperature difference along the latitude from run 5 and 8. Other runs are quite similar to run 8, hence not shown. (e). the same for salinity difference, run 5.

For the global scale, Figure (16) shows us that the potential temperature difference magnitude is 0 to 1.5 ^{0}C, peaked at near 0^0 and 300^0 longitude for the surface layer in run 5. In run 6, the 1000 m is mostly with larger difference than the surface. The deep layer of 5000m gives the least difference. Other runs are relatively similar to run 6, hence not shown here.

As for the salinity, the surface layer is always the highest difference through all longitude. At 1000 m depth the model is less different than WOA09 except at longitude of 0^0 to 40^0E. All runs are mostly having similar pattern of RMS value, and will be discussed in section 5.

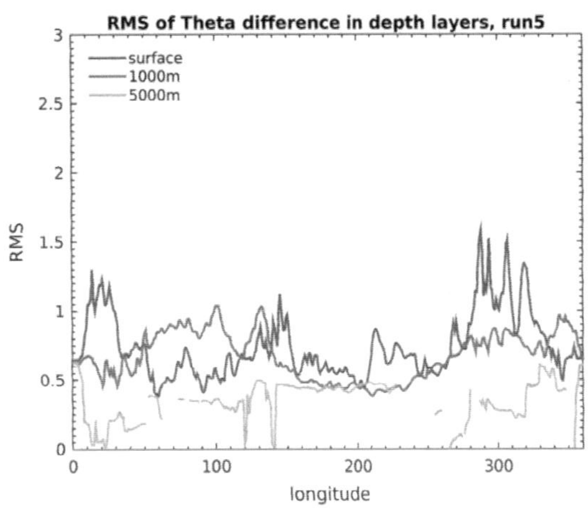

(a)

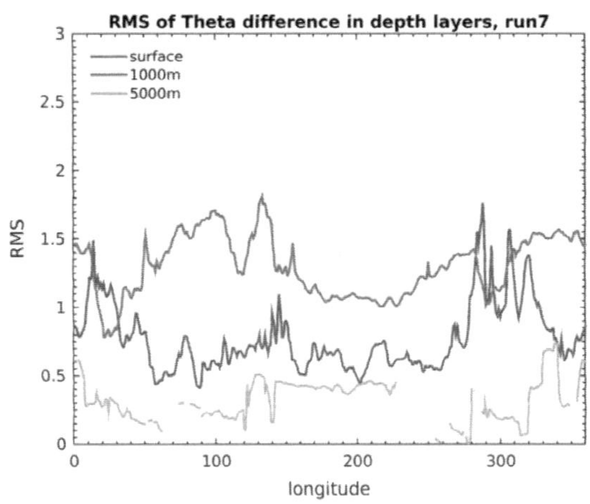

(b)

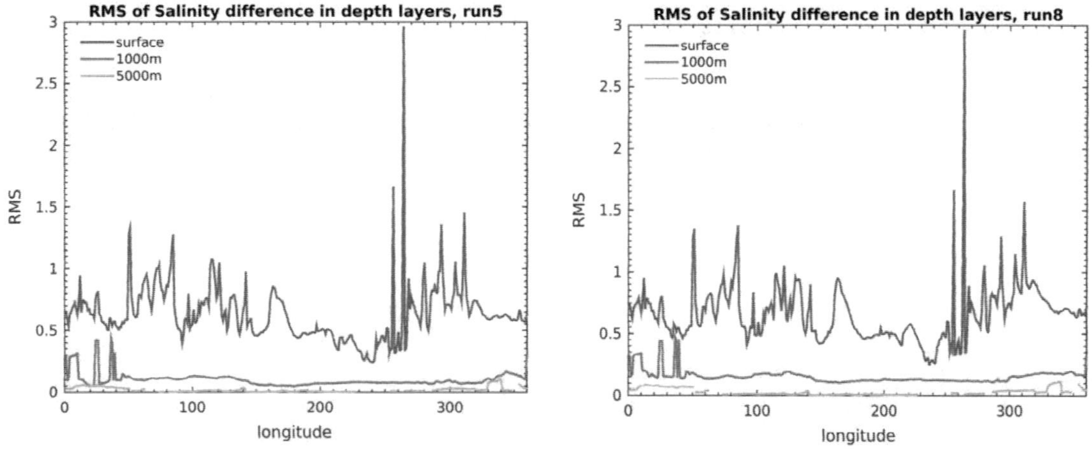

(c) (d)

Figure 16. (a), (b). RMS of global potential temperature difference (WOA09-MITgcm) of surface 1000m, 5000m in run 5 and 7, respectively. Run 6 and 8 are not shown due to quiet similarity to run 8. (c),(d). The same for salinity difference.

5. Discussion and concluding remarks

The model in general, from all the 7 experiments is having more difficulties in the North Atlantic and in the fronts of Southern Ocean as seen in Figure 5,6,7. [13] mentioned at least the separation of the Gulf stream, the strength and structure of meridional transport, the mode water formation and subduction, the nonlinear dynamics of eddies and interbasin exchange with the Mediterranean and Nordic seas are difficult to model. They suggested that direct simulation of wind-driven and the MOC are not very successful yet due to high uncertainty being the sensitivity of the models. As for the Southern Ocean, is location of several fronts i.e the Subtropical Front (STF), South Subtropical Fronts (SSTF), Subantarctic Fronts (SAF), and Polar Fronts (PF) according to Belkin and Gordon [14]. These fronts' dynamics are also difficult to simulate.

From the reference run, the volume transport of Davis Strait was out of range, with mean value -4.66 Sv. Davis Strait is the main conduit from the North Atlantic to Arctic Sea, exchanging not only sea water but fresh water as well from the Canadian glacial and West Greenland. Cuny et al. [15] found the net volume of -2.6 ± 1Sv, and fresh water of -92 ± 34 mSv from three consecutive years of moorings. Some moorings and gliders have been deployed to measure the transport volume within 2004 - 2010 and resulted in -1.6 ± 0.5 Sv, with -93 ± 6 mSv of liquid freshwater and sea ice export -10 ± 1mSv [16]. Our range was -3 to 1 Sv, hence with the new bathymetry in which Davis Strait was widened with some change in North Atlantic and Arctic coast we get the volume exceeding this range. In our experiment we used the original bathymetry and the transport is back in range, -0.2 to -0.5 out of the reproduced data. Apparently this number is still too low compared to the observational data.

Roemmich and Wunsch [17] calculated the net northward meridional transport of North Atlantic from transatlantic hydrographic sections in 1981 along latitudes 24.5° and 36.25°N. The data collected was obtained from 215 stations of CTD probe and water samples for other tracers. They came up with the number of 17 Sv of surface and intermediate water above $\sigma_2 = 36.82$. Ganachaud [18] conducted inverse study on observational (WOCE, World Ocean Circulation data) estimates and obtained number of 16 ± 2 Sv transport at 48^{0}N. Other inverse studies are from Ganachaud and Wunsch [19]; 15 ± 2 Sv at 42^{0}N and Lumpkin, Speer [20]; 16 ± 2 Sv at 48^{0}N. Lumpkin, Speer [21] and Lumpkin, Speer [20] had 13 ± 2 Sv at 42^{0}N. Most of our experiment simulations resulted with less than 10 Sv for this transport. If we compare to other models as in Griffies et al. [6], in which they run the simulation for

500 years; we would like to show out some numbers of their models' annual mean Atlantic meridional overturning stream function values at year 100 as follows :

- NCAR-POP (National Center Atmospheric Research – Parallel Ocean Program) : ±10 SV
- FSU-HYCOM (Florida State University—Hybrid Coordinate Ocean Model) : ±8.5 SV
- GFDL-MOM (Geophysical Fluid Dynamic Laboratory - Modular Ocean Model) : ±12.5 Sv
- GFDL-HIM (GFDL - Hallberg Isopycnal Model) : ±9.5 Sv
- Kiel-ORCA (Kiel University – Ocean model configuration of the NEMO ocean code) : ±10.5 SV
- MPI (Max Planck Institute) : ±12.5 SV
- KNMI-MICOM (Royal Neterlands Meteorological Institute – Miami Isopycnal Coordinate Ocean Model) : ±10 SV

The pattern of Root Mean Square tends to be larger at the surface layer due the dynamics of the forcing at this layer. RMS is also larger at 0-40^{0}E longitude comes from the Mediterranean Sea, even in depth layers of 1000m and 5000m. It is more difficult for the model to reproduce data from this sea because of its semi enclosed position gives it unique range of temperature and salinity values and different forcing.

Out of the 5 simulation experiment, run 5 holds the best run as it has the least difference to WOA09.

Acknowledgement
This short research was funded by the Nippon Foundation – Partnership for Observation of Global Ocean – Alfred Wegener Institute Center of Excellence Program, 2015.

References

[1] Marshall J, Jones H, and Hill C (1998), Efficient ocean modeling using non-hydrostatic algorithms, *Journal of Marine Systems*, ***18*(1–3)**, 115-134, doi:http://dx.doi.org/10.1016/S0924-7963(98)00008-6.

[2] Adcroft C. Hill N, and Marshall J C (1999), A New Treatment of the Coriolis Terms in C-Grid Models at Both High and Low Resolutions, *Monthly Weather Review*, ***127*(8)**, 1928-1936, doi:10.1175/1520-0493(1999)127<1928:ANTOTC>2.0.CO;2.

[3] Hill C N, Adcroft A J, Jamous D, and Marshall J (1999), A Strategy for Terascale of the Eight ECMWF Workshop on the Use of Parallel Processors in Meteorology.

[4] Marotzke J, Giering R, Zhang K Q, Stammer D, Hill C, and Lee T (1999), Construction of the adjoint MIT ocean general circulation model and application to Atlantic heat transport sensitivity, *Journal of Geophysical Research: Oceans*, ***104*(C12)**, 29529-29547, doi:10.1029/1999JC900236.

[5] Adcroft *et al* (2015), *MITGCM User Manual.*

[6] Griffies S M *et al* (2009), Coordinated Ocean-ice Reference Experiments (COREs), *Ocean Modelling*, ***26*(1–2)**, 1-46, doi:http://dx.doi.org/10.1016/j.ocemod.2008.08.007.

[7] Danabasoglu G *et al* (2014), North Atlantic simulations in Coordinated Ocean-ice Reference Experiments phase II (CORE-II). Part I: Mean states, *Ocean Modelling*, ***73***, 76-107, doi:http://dx.doi.org/10.1016/j.ocemod.2013.10.005.

[8] Hauck J, Völker C, Wang T, Hoppema M, Losch M, and Wolf-Gladrow D A (2013), Seasonally different carbon flux changes in the Southern Ocean in response to the southern annular mode, *Global Biogeochemical Cycles*, ***27*(4)**, 1236-1245, doi:10.1002/2013GB004600.

[9] Forget G, Campin J M, Heimbach P, Hill C N, Ponte R M, and Wunsch C (2015), ECCO version 4: an integrated framework for non-linear inverse modeling and global ocean state estimation, *Geosci. Model Dev. Discuss.*, ***8*(5)**, 3653-3743, doi:10.5194/gmdd-8-3653-2015.

[10] Losch M, Menemenlis D, Campin J M, Heimbach P, and Hill C (2010), On the formulation of sea-ice models. Part 1: Effects of different solver implementations and parameterizations, *Ocean Modelling*, ***33*(1–2)**, 129-144, doi:http://dx.doi.org/10.1016/j.ocemod.2009.12.008.

[11] Forget G (2015), gcmfaces a matlab framework for the analysis of gridded earth variables.

[12] Bryan K and Lewis L J (1979), A water mass model of the World Ocean, *Journal of Geophysical Research: Oceans*, ***84*(C5)**, 2503-2517, doi:10.1029/JC084iC05p02503.

[13] Haidvogel D B, Arango H G, Hedstrom K, Beckmann A, Malanotte-Rizzoli P, and Shchepetkin A F (2000), Model evaluation experiments in the North Atlantic Basin: simulations in nonlinear terrain-following coordinates, *Dynamics of Atmospheres and Oceans*, ***32*(3–4)**, 239-281, doi:http://dx.doi.org/10.1016/S0377-0265(00)00049-X.

[14] Belkin I M and Gordon A L (1996), Southern Ocean fronts from the Greenwich meridian to Tasmania, *Journal of Geophysical Research: Oceans*, ***101*(C2)**, 3675-3696.

[15] Cuny J, Rhines P B, and Ron K (2005), Davis Strait volume, freshwater and heat fluxes, *Deep Sea Research Part I: Oceanographic Research Papers*, ***52*(3)**, 519-542, doi:http://dx.doi.org/10.1016/j.dsr.2004.10.006.

[16] Curry B, Lee C M, Petrie B, Moritz R E, and Kwok R (2013), Multiyear Volume, Liquid Freshwater, and Sea Ice Transports through Davis Strait, 2004–10*, *Journal of Physical Oceanography*, ***44*(4)**, 1244-1266, doi:10.1175/JPO-D-13-0177.1.

[17] Roemmich D, and Wunsch C (1985), Two transatlantic sections: meridional circulation and heat flux in the subtropical North Atlantic Ocean, *Deep Sea Research Part A. Oceanographic Research Papers*, ***32*(6)**, 619-664, doi:http://dx.doi.org/10.1016/0198-0149(85)90070-6.

[18] Ganachaud A (2003), Large-scale mass transports, water mass formation, and diffusivities estimated from World Ocean Experiment (WOCE) hydrographic data *Journal of Geophysical Research*, ***108***, doi:10.1029/2002JC001565.

[19] Ganachaud A and Wunsch C (2000), Improved estimates of global ocean circulation, heat transport and mixing from hydrographic data, *Nature*, ***408***, 453-456.

[20] Lumpkin R, Speer K G, and Koltermann K P (2008), Transport across 48°N in the Atlantic Ocean, *Journal of Physical Oceanography*, ***38*(4)**, 733-752, doi:10.1175/2007JPO3636.1.

[21] Lumpkin R, and Speer K G (2003), Large-Scale Vertical and Horizontal Circulation in the North Atlantic Ocean, *Journal of Physical Oceanography*, ***33*(9)**, 1902-1920, doi:10.1175/1520-0485(2003)033<1902:LVAHCI>2.0.CO;2.

The roles of mixed layer depth in affecting chemical and biological characteristics of Makassar straits, Indonesia

I C Dewi[1], M R Putri[1] and A Setiawan[2]

[1] Department of Earth Science, Faculty of Earth Sciences and Technology, Bandung Institute of Technology, Jalan Ganesha 10, Bandung 40132, Indonesia
[2] Research and Development Agency of Marine Affairs and Fisheries, Jakarta Utara, Indonesia

E-mail: intan.cade@gmail.com

Abstract. Mixed Layer Depth (MLD) is able to determine the amount of light and nutrients abundance for the growth of phytoplankton, its known that phytoplakton pigments are the main absorbers and light scatterers in the ocean. Location wich has deeper MLD and large seasonal variations are the most productive areas, also could have blooming or the maximum growth of large phytoplankton occurring in the existence of a limited number nutrients for several days. The aims of this study are to understand the characteristics of the MLD and it roles to the abundance of Chlorophyll and nutrient in Indonesia Through-Flow/ITF (Indonesia: "ARLINDO"), specifically the area along Makassar Strait. The Chlorophyll and oceanographic parameters were obtained based on measuring by CTD in TIMIT 2016 expedition. The MLD was deep (~43 m) in the northern part of the transect (around Central Sulawesi Island) and shoaled progressively until about 13 m in the southern transect (the Southern Makassar Strait). The Chlorophyll-a (Chl-a) concentration show a decline pattern from the north to the south transects with the maximal Chl-a is 0.37 mg/m^3. Nutrient parameters in southern Makassar Strait studied were NODC (National Oceanographic Data Center) data from WOA (World Ocean Atlas), it can indicated a correlation between nutrient (Nitrate) with Chlorophyll. The mean correlation in mixing layer for whole area is 0.78, showing that Nitrate concentration is highly related proportional to Chlorophyll abundance, while below the mixing layer, correlation value is -0.52, indicating the inverse relation between Chlorophyll and nutrient. Based on those correlation, and assumption that the source of Chlorophyll is from phytoplankton, it can be concluded that phytoplankton needs Nitrate in its metabolism process as well as other productivities reason in mixing layer, while in depth below the mixing layer, growth of phytoplankton is no longer depends on concentration of Nitrate nutrient.

1. Introduction

Ecosystem condition in a region can be estimated by the physical and biological processes in it. The physical processes which involved are temperature, salinity, and light. While the biological process is due to energy cycle in food chain, which involved nutrient, phytoplankton, and zooplankton [1]. In physical processes, variations on values of temperature, salinity, and light distribution, are closely related with air-sea interactions. Wind stress on the sea surface, is capable to generate water mass movement in turbulent flow. The different of those turbulence movement could induce fluid mixing processes on a large scale. This mixing processes generally occurs in Mixed Layer Depth (MLD), with depth starts from below ten meters of sea surface [2],[3].

Mixing layer is an interface layer between the atmosphere and the ocean, and it has an important role to determaine the average conditions and variability of the sea [4]. The forces that occur in atmosphere (e.g. wind, solar heating, evaporation, and precipitation) along with waves from the sea will caused turbulence in mixing layer. The turbulence will give very high vertical mixing uniformity [5]. When solar radiation and wind gust weakened, the depth of mixing layer will become shallower, as it obtained more stratified ocean and blocking the vertical mixing process [6]. In addition, turbulent mixing can also be caused by internal wave breaking events due to the topography of the waters, such as the narrowing area and/or the existence of seamounts [7]. The differences of water depth are capable to generated different currents in surface and below layers, furthermore this type of currents can generate the formation of water patterns with relatively stable temperature changes in mixing layer [8].

The mass mixing in MLD could affect Chlorophyll, which is one of the parameter that necessary to determine primary productivity in the ocean [9]. Vertical and horizontal distribution of Chlorophyll depend on temperature, salinity, and light. However in the tropics region horizontal distribution has a little differences. Generally, Chlorophyll could be found in plants, photosynthetic bacteria, and algae (phytoplankton) [10]. Mass mixing phenomenon and vertical fluid movement will greatly affect the phytoplankton life for photosynthesis needs, which also affect the nutrient distribution in water column [11].

Makassar Strait is main pathway of Indonesia Through-Flow (ITF), this water mass flow formed as a result from sea level differences between Pacific Ocean and Indian Ocean, and lead the water mass moves through the straits in Indonesia [12]. The transport of mass which has been reviewed in Januari 2004 up to November 2006 is up to 11.6±3.3Sv (Sv=10^6 m^3/s), carrying out 80% ITF mass [13]. Previous research showed that there are a vertical diffusivity and heat flux release from wind stress, and lead the intensity of vertically mixing in Makassar Strait to increase [14]. The vertical change in both of water mass mixing and fluid movement will affect nutrient distribution in water column, and also photosynthesis process of phytoplankton [11].

The aims of this study are to understand the characteristics of MLD and its role to the abundance of Chlorophyll and nutrient in the main pathway of ITF, especially along the Makassar Strait. The assumption used in this study are; (a) Chlorophyll assumed as a Phytoplankton because we already know that most of Chlorophyll in the ocean can be found in the form of algae (phytoplankton), (b) Nitrate can be assumed as a Nutrient because in of the effect of Nitrate concentration on Chlorophyll-a has a greater value than the other nutrients in Makassar Strait [3].

2. Research methods

The data used in this study were obtained from The Transport, Internal Wave and Mixing in The Indonesia Through Flow Region and Its Impact in Marine Ecosystem (TIMIT) expedition, which is a collaboration between *Badan Penelitian dan Pengembangan Kelautan dan Perikanan (Balitbang KP)* and *Pusat Penelitian Oseanografi Lembaga Ilmu Pengetahuan Indonesia (P2O LIPI)* by using Baruna Jaya VIII research vessel in September 2016. Data were retrieved by using Conductivity Temperature Depth (CTD) multi-cylinder water sampler. Locations of data retrieval were divided into 12 observation stations, which spread across Makassar Strait to Lombok Strait. The study area starts from $0^0$47'9.60 N to $8^0$25'19.20 S and $119^0$46'19.20 E to $115^0$57'57.59 E (Figure 1). Depth ranges of CTD were between 300-2242 meter, depending on the depth of the water column. Meanwhile, Nitrate data were obtained from National Oceanographic Data Center (NODC) collection data by World Ocean Atlas (WOA) in September 2016. To find out the depth of MLD and Chlorophyll abundance value, it requires temperature and fluoresecence data along the depths that obtained by CTD.

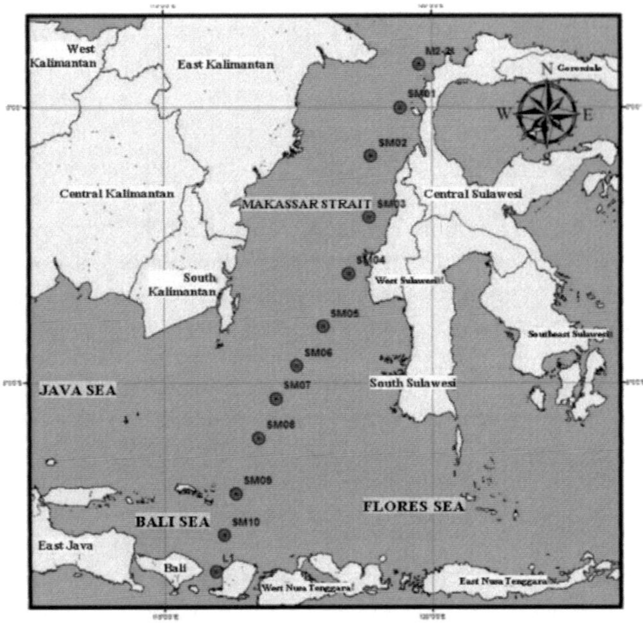

Figure 1. Location of Data Retrieval

2.1. Estimation of Mixed Layer Depth (MLD)

The determination of MLD criteria based on temperature data can be estimated by using threshold method which has previously been used in Pacific Ocean [15]. The implementation of this method is based on the similarity of water criteria as the water mass movement in Makassar Strait and Lombok Strait are sourced from Pacific Ocean. The source of water brought by ITF is from northern and southern Pacific Ocean while Makassar Strait is more influenced by the northern Pacific Ocean water [16]. The threshold method was implemented by estimating based on the depth of reference which is started from 10 meters while the $\Delta T = 0.1^0C$.

2.2. Estimation of Chlorophyll Abundance and Nitrate

Chlorophyll has physical properties such as it could receive and reflect light and fluorescence with different spectrum [17]. So, the fluorescence data recorded in CTD could be used as an input data for abundance Chlorophyll estimation at each depth. Meanwhile, the Nitrate abundance can be estimated from NODC WOA data collection at each depth.

2.3. Estimation of Correlation and Magnitude of Influences

The relations of Nitrate-Chlorophyll, MLD-Nitrate and MLD-Chlorophyll can be estimated by using the analysis methods of linear regression and correlation. The linear regression method was implemented by using scatter graph to produce result as in the form of r-square value. R-squared or determination coefficient can be used to estimate how far a variable is able to provide information related to other variables [18]. The range of R-squared value is between 0 and 1. If the r-squared value is close to zero then it shows that the ability of a variable to gives information about the other variable is restricted. Meanwhile, if r-squared value close to one, it shows that the variable is able to give various information of the other variables.

Correlation value consists of negative and positive values as both of the correlation values can show the relation between each variables. Positive correlation shows that the tested variables are directly proportional while negative correlation the opposite relationship [3].

3. Result and discussion

3.1. Mixed Layer Depth (MLD) Conditions

The MLD condition in Makassar Strait tends to shallowing from north to south as it is showed in Figure 2. The deepest MLD is up to 42.76 m located in SM02 station near Central Sulawesi. Meawhile, the shallowest MLD is up to 12.93 m located in SM09 station in the southern Makassar Strait. The data that was retrieved on September represents the southeast monsoon period (July-September) as the wind was blown from Flores Sea to Java Sea. In general, the sea surface temperature (SST) distribution shows that souther stations tend to have cooler SST as a result from rapturing deep water mass by Ekman Upwelling during the southeast monsoon period [19],[20].

Figure 2 shows that the deeper MLD are located in Central Makassar Strait, i.e. SM01, SM02, SM03, and SM04 stations with the MLD value respectively are 33.31, 42.76, 25.36, and 36.79 m. In fact, the four stations have deep water-depth as the depths are 1002 m for SM01, SM02 and SM03 while SM04 has a water depth of 2002 m. The deepest depth is located in SM04 station that is located in Labani Channel which has the narrowing topography characteristic. The characteristic causes internal waves breaking which is one of the factors of turbulent mixing [7].

Meanwhile, M2-2 station that is located in northern Makassar Strait and SM06 to L1 stations thatare located in southern Makassar Strait, both of them are have a shallow MLD with respectively values are 13.43 m in M2-2, and in the southern Makassar Strait the range values are about 12.93 m up to 18.90 m. The shallow MLD dominantly influenced by the location of station where adjacent to Sulawesi Sea and Java Sea. The condition of a location lose to Java Sea can be influenced by a direct water mass mixing i.e. the movement of the water mass upward rapidly that will cause a relatively small homogenous conditions [11].

3.2. Chlorophyll and Nitrate Distribution

The thinning event of MLD thickness can be seen drastically starting from SM04 station located in central Makassar Strait, headed to SM05 in the south with the change of 21.88 meter. This change is the starting point of the shallowing of MLD and will continue to SM09 station in the depth of 12.93 m. It shows that the phenomenon on stations located between SM04 and SM09 are intriguing, so the study on how big the influence of MLD to ecosystem condition—represented by Chlorophyll and nitrate distribution—needs to be specifically-conducted.

Chlorophyll distribution value can be known by grouping the maximum value of

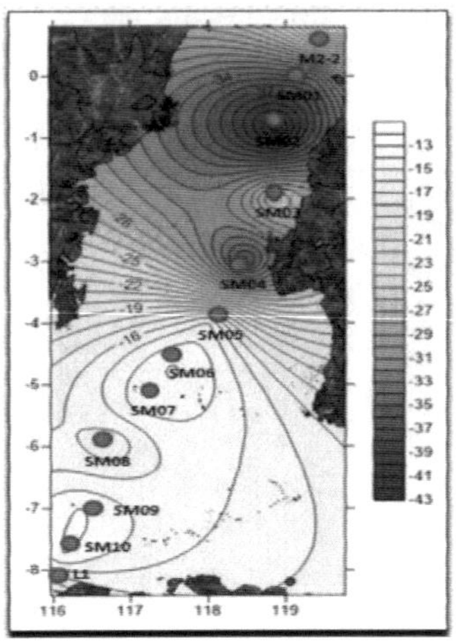

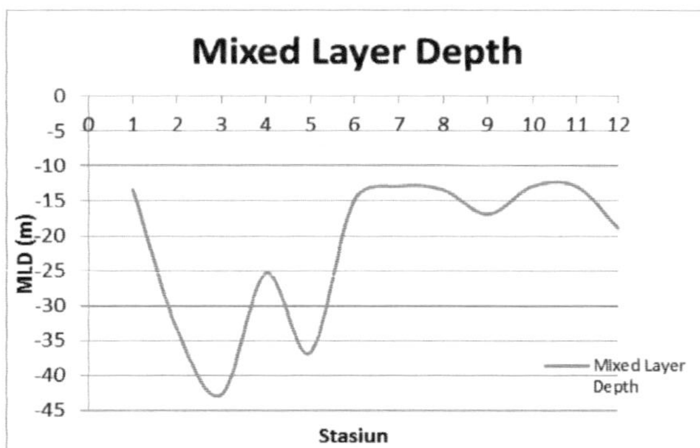

Numb.	1	2	3	4	5	6	7	8	9	10	11	12
Station	M2-2	SM01	SM02	SM03	SM04	SM05	SM06	SM07	SM08	SM09	SM10	L1

Figure 2. MLD Chararcteristics on each station

Chlorophyll in MLD and below MLD. The fluorescence data that has been measured with CTD showing that the Chlorophyll in MLD tends to decrease from north to south. The highest

IOP Conf. Series: Earth and Environmental Science **162** (2018) 012003 doi:10.1088/1755-1315/162/1/012003

concentration is up to 0.37 mg/m^3 located in SM04 while the lowest concentration is up to 0.12 mg/m^3 located in SM08. Meanwhile, the chorophyll abundance below MLD tends to increase from north to south as the highest concentration is up to 1.28 mg/m^3 located in SM09 and the lowest concentration is up to 0.61 mg/m^{3}located in SM04.

Nitrate distribution also can be known by grouping the maximum value in MLD and below MLD. Based on NODC WOA data collection, nitrate concentration in MLD from north to south (SM06 station) tends to decrease, then increase again to SM09. The highest concentration is up to 1.16 μmol/l located in SM04 and the lowest concentration up to 0.12 μmol/l located in SM06. Meanwhile below the MLD, nitrate concentration tends to decrease from north to south, although there is an increase at SM08.

Based on the Chlorophyll and nitrate distribution patterns that are showed in Figure 3 and Figure 4, and also in Table 1, it is known that in the MLD, the highest Chlorophyll and nitrate concentration is located in SM04 station which is the nearest station to equator. In this location, there is a considerable warming in sea surface that will induce upwelling to decrease the temperature of the surface layers. This upwelling process causes an increase of nitrate concentration on the surface, so the abundance of Chlorophyll become higher. There is an increase of Chlorophyll concentration and primary productivity in equatorial area that is caused by upwelling in divergence area of equator [21].

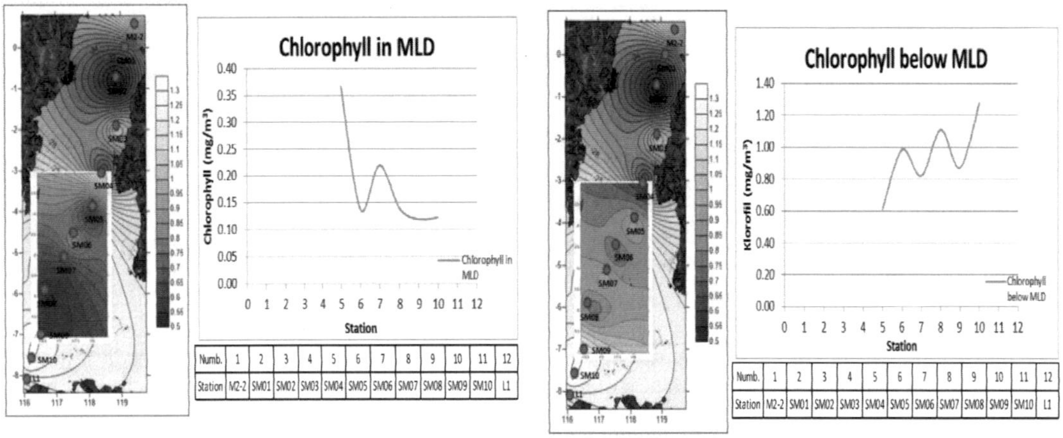

Figure 3. Chlorophyll horizontal distribution

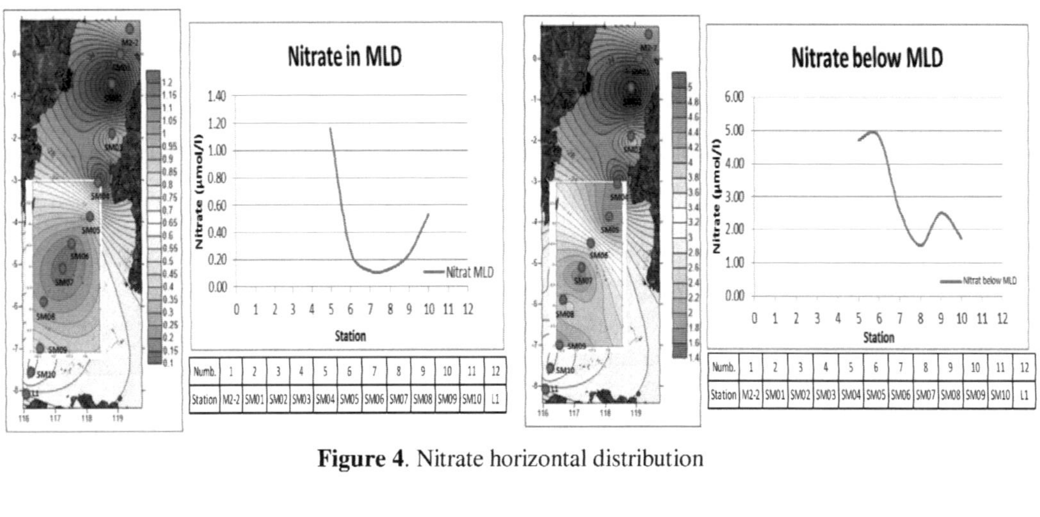

Figure 4. Nitrate horizontal distribution

Table 1. Chlorophyll and Nitrate Value

Station	MLD	MLD		Below MLD	
		Chlorophyll	Nitrate	Chlorophyll	Nitrate
SMO 4	-36.79	0.37	1.16	0.61	4.71
SMO 5	-14.92	0.14	0.27	0.98	4.82
SMO 6	-12.93	0.22	0.12	0.82	2.57
SMO 7	-13.43	0.14	0.13	1.11	1.53
SMO 8	-16.91	0.12	0.24	0.87	2.51
SMO 9	-12.93	0.12	0.53	1.28	1.74

3.3. The Influence of MLD and Correlation between parameters

The influence of MLD on the abundance of Chlorophyll and nitrate concentration can be known by the correlation value between each parameters and the range of depth waters. The correlation value for Chlorophyll and Nitrate in MLD, respectively are 0.88 and 0.91. Meanwhile in the depth below MLD, the correlation value respectively are -0.74 and 0.63. Based on these results, it shows that on Chlorophyll case, as the MLD going deeper, the Chlorophyll concentration will be higher due to the presence of homogeneous layer and uniformity lighting which support the Chlorophyll growth. Meanwhile in the depth below MLD, the correlation value classified as a strong negative correlation showing that Chlorophyll concentration will be decreased along with the increasing depth. This event can be caused by the light that can't reach the deep waters.

In another case, both nitrate parameter are having positive correlation, with strong classified in MLD and medium classified on below MLD depth. Based on those correlation, it shows that as the MLD becomes deeper, nitrate concentration will be higher. That event also happens on the depth below MLD as the nitrate concentration will be higher along with the increasing depth. The high concentration of nitrate in deeper MLD is due to the presence of mixing process from upwelling, while in the depth below MLD case, the concentration have higher value because of there already are a lot of settled nitrate on the sea floor.

The correlation between Clorophyll and nitrate can be seen by using Pearson correlation method because both parameters are having scale depth interval. Based on the results, it shows that Chlorophyll and nitrate correlation has different value in MLD and below MLD. Chlorophyll and nitrate correlation value in MLD is 0.78, classified as strong positive correlation. This correlation shows that as the nitrate concentration increases, the Chlorophyll abundance also increases in MLD. It is caused by the presence of nutrient (nitrate) abundance and lighting uniformity which supports the Chlorophyll growth. Meanwhile in below MLD depth, the correlation has been classified as a medium negative correlation with the value of -0.52. This correlation shows that Chlorophyll is inversely proportional to nitrate concentration, although nitrat concentration have higher value at deeper waters. The decreased Chlorophyll concentration can be caused by the minimum light penetration in deep waters.

The determination of correlation criteria is in accordance with the statement explaining that correlation criteria (r) value in range of 0.40 – 0.59 are classified as medium correlation, and (r) value with range of 0.60 – 0.79 classified as strong correlation [22]. R-square value of the three scatter graphs that have been calculated show that in MLD the r-square have higher values than on below MLD. The influence—quantified as r-square value—of MLD to Chlorophyll and nitrate abundance are 0.78 and 0.83, respectively, while the influence of nitrate to Chlorophyll in MLD is 0.60. Meanwhile on below MLD, the influences for same cases are 0.55, 0.40, and 0.28, respectively.

Based on correlation value and assumption of Chlorophyll as phytoplankton, it can be concluded that in MLD, phytoplankton community depends on nitrate concentration for metabolism and other activities, while on below MLD phytoplankton community growth is not depend on nitrate concentration. This result shows that the growth rate of phytoplankton is directly proportional to the increase of nutrient concentration until it reaches the nitrate saturation limit, after that, phytoplankton growth no longer depend on nitrate concentration [23].

4. Conclusion

Based on the measurements on TIMIT expedition, it can be shown that the MLD from northern Makassar Strait to Lombok Strait on the southeast monsoon period is divided into three waters with different characteristics as the characteristics from north to south tends to be shallowing. The deepest MLD is up to 42.76 m at SM02 station near the Central Sulawesi and the shallowest MLD is up to 12.93 m at SM09 station, southern of Makassar Strait .

In MLD, the Chlorophyll and Nitrate concentration will be higher on deeper MLD. Meanwhile on the depth below MLD, Chlorophyll concentration will be decreased, while nitrate concentration is getting higher along with increasing water depth. Phytoplankton metabolism in MLD depends on nitrate concentration, while on depth below MLD nitrate no longer have big influences to phytoplankton growth.

Acknowledgments

Thanks to the Research Team "The Transport, Internal Wave and Mixing in The Indonesia ThroughFlow and It's Impact in Marine Ecosystem (TIMIT)", Badan Penelitian dan Pengembangan Kelautan Perikanan (Balitbang KP) for data usage from TIMIT Expedition 2016.

References

[1] Kemili P 2012 *Estimasi Distribusi Horisontal dan Vertikal Produktivitas Primer Di Perairan Laut Banda* (Bandung: ITB)

[2] Kantha L and Clayson C A 2003 On The Effect Of Surface Gravity Waves On Mixing In An Oceanic Mixed Layer *Ocean Modelling*. **6.** 101-124

[3] Kaharuddin 2012 *Kajian Fluks Nutrien Dan Kandungan Klorofil-A serta Keterkaitannya dengan Proses Percampuran Di Selatan Selat Makassar* (Bogor: IPB)

[4] Keerthi M G, Lengaigne M, Vialard J, Montegut C B, and Muraleedharan P M 2012 Interannual Variability of The Tropical Indian Ocean Mixed Layer Depth *Clim Dyn.* **40**:743-759

[5] Kumar S P and Narvekar J 2005 Seasonal Variability of Mixed Layer In The Central Arabian Sea And Its Implication on Nutrients and Primary Productivity *J. Deep Sea Res.* **52**:1848-1861

[6] Narvekar J and Kumar S P 2006 Seasonal Variability of Mixed Layer In The Cemtral Bay of Benfal and Associated changes in nutrients and Chlorophyll *J. Deep Sea Res.* **53**:820-835

[7] Naulita Y 2016 Proses Percampuran Turbulen Di Kanal Labani, Selat Makassar *J. Ilmu dan Teknologi Kelautan Tropis.* **8**(1):345-355

[8] Wijesekera H W and Gregg M C 1996 Surface Layer Response to Weak Winds, Westerly Bursts and Rain Squalls in the Western Pacific Warm Pool *J. Geophys Res.* **101**: 977-997

[9] Narvekar J and Kumar S P 2014 Mixed Layer Variability and Chlorophyll-a Biomass in The Bay of Bengal *Biogeosciences*, **11** 3819-3843

[10] Hutabarat S and Evans S M 1984 *Pengantar Oseonografi* (Jakarta: UI Press)

[11] Ryandhini N A, Zainuri M, and Kuswardani R T D 2014 Karakteristik Mixed Layer Depth dan Pengaruhnya Terhadap Konsentrasi Klorofil-a *J. Ilmu Kelautan* **19**(4):219-225

[12] Ilahude A G dan Nontji A 1999 *Oseanografi Indonesia Dan Perubahan Iklim Global (El Nino Dan La Nina)* (Serpong: Lokakarya AIPI)

[13] Gordon A L, Susanto R D, Ffield A, Huber B A, Pranowo W and Wirasantosa S 2008 Makassar Strait throughflow, 2004 to 2006 *Geophysical Research Letters*, **35**

[14] Atmadipoera A S and Widyastuti P 2014 A Numerical Modelling Study on Upwelling Mechanism in Southern Makassar Strait *J. Ilmu dan Teknologi Kelautan Tropis* **6**(2):355-371

[15] Montegut C B, Madec G, Fischer A S, Lazar A and Iudicone D 2004 Mixed Layer Depth over The Global Ocean: An Examination of Profile Data and A Profile-based Climatology *J. Geophysical Research*, **109**

[16] Wyrtki K 1987 Indonesian Throughflow and The Associated Pressure Gradient. *J. Geophysics Research*. **92**:12941-12946

[17] Ai N S and Banyo Y 2011 Konsentrasi Klorofil Daun Sebagai Indikator Kekurangan Air Pada Tanaman *J. Ilmiah Sains* **11**(2)

[18] Ghozali I 2005 *Aplikasi Analisis Multivarians dengan Program SPSS* (Semarang: Badan Penerbit Universitas Diponegoro)

[19] Tubalawony S, Kusmanto E, and Muhadjirin 2012 Suhu dan Salinitas Permukaan Merupakan Indikator Upwelling Sebagai Respon Terhadap Angin Muson Tenggara di Perairan Bagian Utara Laut Sawu *J. Ilmu Kelautan* **17**(4): 226-239

[20] Gordon A L 2005 Oceanography of the Indonesian seas and their throughflow *Oceanography* **18**(4), 14– 27

[21] Rasyid A 2009 Distribusi Klorofil-a pada Musim Peralihan Barat-Timur Di Perairan Spermonde Propinsi Sulawesi Selatan *J. Sains dan Teknologi* **9**(2): 125 -132

[22] Beldjazia A, and Alatou D 2016 Precipitation Variability On The Massif Forest of Mahouna (North Eastern-Algeria) from 1986 to 2010 *I.J. Management Sciences and Business Research* **5**(3)

[23] Tubalowony S 2007 *Kajian Klorofil-a dan Nutrien serta Interelasinya dengan Dinamika Massa Air di Perairan Barat Sumatera dan Selatan Jawa-Sumbawa* (Bogor: IPB)

IOP Conf. Series: Earth and Environmental Science **162** (2018) 012004 doi:10.1088/1755-1315/162/1/012004

The study of MJO impact on wave height and wind speed in Indonesian Seas

I Hilmi[1], N S Ningsih[1], I Sofian[2], F Hanifah[1], A M Rizal[1]

[1] Research Group of Oceanography, Faculty of Earth Sciences and Technology, Bandung Institute of Technology (ITB), Jl.Ganesha No 10, Bandung 40312, Indonesia
[2] Geospatial Information Agency (BIG), Jl. Raya Jakarta-Bogor, Cibinong 16911, Indonesia

E-mail: irfan.hilmiu@gmail.com

Absract. Indonesia is passed by an atmospheric phenomenon, called the Madden-Julian Oscillation (MJO), which has an impact on the wave height in the Indonesian Seas. The significant wave height is simulated using WAVEWATCH-III (WW3) numerical model in Indonesian region (90 °E-150 °E, 20 °N-20 °S) forced by surface winds from Cross-Calibrated Multi-Platform (CCMP), Navy Global Environmental Model (NAVGEM), and Navy Operational Global Atmospheric Prediction System (NOGAPS). This simulation is concentrated on MJO phase 3, 4, and 5 which passed through Indonesia and its adjacent waters that occurred in particular time between 1990-2015. In this study, the impact of MJO was analyzed during every monsoon season. In addition, wind speed analysis was carried out to further enrich the analysis of the MJO impact. The simulation result shows that MJO exerts the highest impact during phase 5 and DJF, which contributes to the increase of wind speed (WS) and significant wave height (SWH) in Indonesian inner seas by 6 m/s and 30 cm, respectively, and in southern Lesser Sunda Island by 8 m/s and 1.2 m, respectively. MJO can also contribute to decreasing of the WS and SWH, when it occurred during DJF and MAM phase 3, and JJA phase 4. There is no noticabe change of WS and SWH during SON.

1. Introduction

Madden-Julian Oscillation (MJO) is a planetary-scale eastward propagation of convection and precipitation zone along the equator with 40-50 days of oscillation period (Madden and Julian, 1972). The enhanced convection is mainly caused by a formation of low-pressure area associated with high sea surface temperature (SST) that starts around east of Africa. It propagates with average speed of 5 m/s from western Indian Ocean to the eastern Pacific Ocean, bringing with it a number of impacts on some areas such as weather, climate, and ocean condition. Zhang (2005; 2013), in his study, briefly summarized that MJO has influences on precipitation, sea surface temperature, tropical cyclone activities, flood, Indonesian forest fires, and even global and lightning electromagnetic field. Among the various influences brought by MJO, SST is one of the notable impact on the ocean condition along the equator. The change in SST will influence the wind condition and furthermore, will also have an effect on wave condition. Duwaliser et al (2013) on his study using 6 years (2008-2013) of National Oceanic and Atmospheric Administration (NOAA) WAVEWATCH-III hindcast data forced by National Centers for Environmental Prediction (NCEP) wind data, found that MJO, depending on its phase, may cause both positive and negative anomaly globally. Moreover, As studied by Marshall et al (2015), MJO is able to cause a large fetch (~10.000 km) over a period of weeks. Using WAVEWATCH-III wave hindcast model from Centre for Australian Weather and Climate Research (CAWCR) forced by Climate Forecast System Reanalysis (CFSR) 10-m wind data for 1979-2009 period, this study by Marshall et al (2015) also found that MJO have different impact, both positive and negative, on wind speed (WS) globaly ranging from 1-5 m/s, and on significant wave height (SWH) ranging from 20-50 cm based on its phase and location. In addition, Marshall et al (2015) also found that MJO can cause an increase in wave period, indicating that the wave generation is more influenced by remote wind forcing caused by MJO rather than by swell. This result furthermore shows that the wind anomalies caused by MJO is able to cause both positive and negative anomalies on SWH condition.

This purpose of this paper is to further expand the study about MJO effect on wave, especially within the Indonesian Archipelago where MJO propagate through regularly. To meet this goal, the study will focus mainly on region adjacent to Indonesia's waters to investigate how MJO affects the wind and wave condition around Indonesia. Only MJO phase 3, 4, 5 that happened between 1991-2015 will be included in this study as they occurred near Indonesia, with phase 3 occurring west of Indonesia, phase 4 adjacent to Java and Lesser Sunda Islands, and phase 5 east Indonesia. MJO occurrence in every season (DJF, 1st transitional period of monsoon, JJA, and 2nd transitional period of monsoon) will be considered. The time of occurrence has been chosen to not coincide with the event of El-Niño Southern Oscillation (ENSO) and Indian Ocean Dipole (IOD) to exclude their effect on winds and wave condition around Indonesia . It is also well-known that MJO tends to be more active in the ENSO-neutral event compared to one with ENSO occurrence. This approach will allow us understand how the phase and season play an important role on the effect of MJO on WS and SWH around Indonesian regions, without interference from ENSO and IOD, the most dominant interannual phenomena in the region.

The assessment of the MJO impact is done by using output result from wave hindcast model, WAVEWATCH-III (WW3) forced with 25 years (1991-2015) of surface wind reanalysis. The method and data used in wave hindcasting will be described in Section 2. Section 3 will reveal the impact of MJO on WS and SWH around Indonesia region, and study is summarized in Section 4.

2. Materials and Method

Wave hindcast in this study is produced using WAVEWATCH-III with structured 0.125° x 0125° grid resolution over the Indonesian region (90°E-150°E, 20°N-20°S). This domain is large enough to cover South China Sea (SCS), western Pacific Ocean, and eastern Indian Ocean as to see the MJO impact on those areas as well. Cross-Calibrated Multi-Platform (CCMP) is used as the wind forcing with resolution of 0.25° x 0.25° and interval of 6 hours. As for the bathymetry, the ETOPO data with 30-arc second (0.0833°) resolution is being used with unchanged depth over time. The time of each MJO phase in this study is obtained from Australian Bureau of Meteorology website (bom.gov.au) that use RMM method and MJO diagram from Wheeler and Hendon (2014) to determine the MJO phase. The MJO phase is then compared to monthly Ocean Nino Index (ONI) data from NOAA and Indian Ocean Dipole (IOD) index from Japan Agency for Marine-Earth Science and Technology (JAMSTEC) to eliminate MJO occurences that coincide with any ENSO and IOD event.

Rather than using composite data of MJO over years, this study will focus on one MJO occurrence on each phase and season that occur in ENSO and IOD neutral year from 1991-2015. Although there is a strong El-Niño event on 1997, ONI data from NOAA does not show any El-Niño event on March, thus, March 1997 is considered ENSO-neutral month. Moreover, the choosen MJO should satisfy the following requirements. According to study by Ramdhani (2015), only the strong MJO (index>1) is used in this study, as the strength of MJO is proportionate to the effect it caused on WS and SWH. MJO with index more than 1 indicates that the low-pressured area created is able to cause significant change in the state of wind and wave. On contrary, with index less than 1, MJO will be classified as weak or even considered inactive. Thus, giving insignificant change in the state of wind and wave. In addition to that, the MJO choosen must have a minimum duration of 5 days to be able to induce a significant anomaly (Ramdhani, 2015). With this duration, wind anomalies caused by MJO will have enough time to create large fetch and transfering energy to the sea and create SWH higher than the normal state. The selected MJO occurrence will then be separated into 4 seasons; northeast monsoon (DJF), first monsoon's transitional period (MAM), southwest monsoon (JJA), an second monsoon's transitional period (SON) to observe which season gives the biggest effect on MJO's impact. It is hoped that by adhering to these requirements, this study will provide the result of highest anomaly an MJO can cause on WS and SWH around Indonesia region on each season. MJO choosen in this study is shown in the Table 1.

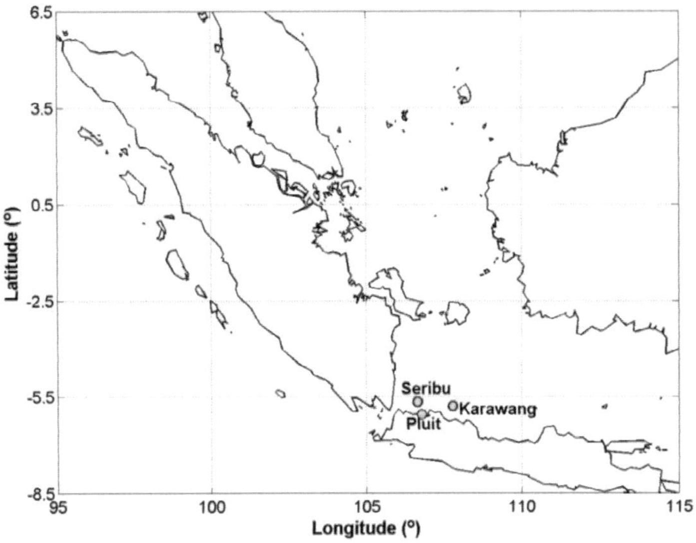

Figure 1. Three of buoy locations used in verification

Table 1. Chosen time of MJO occurrence for each phase and season

	Phase 3	Phase 4	Phase 5	No MJO
DJF	16 – 21 January 1993	12 – 20 February 1997	3-9 February 2009	December 2008
MAM	4 – 10 March 1993	1 – 6 March 1990	7-13 March 1990	March 1997
JJA	21 – 28 June 1998	17 – 22 June 2007	10 – 15 August 2001	August 1993
SON	15 – 21 October 2008	22 – 27 October 2008	19 – 24 November 2004	October 1990

IOP Conf. Series: Earth and Environmental Science **162** (2018) 012004 doi:10.1088/1755-1315/162/1/012004

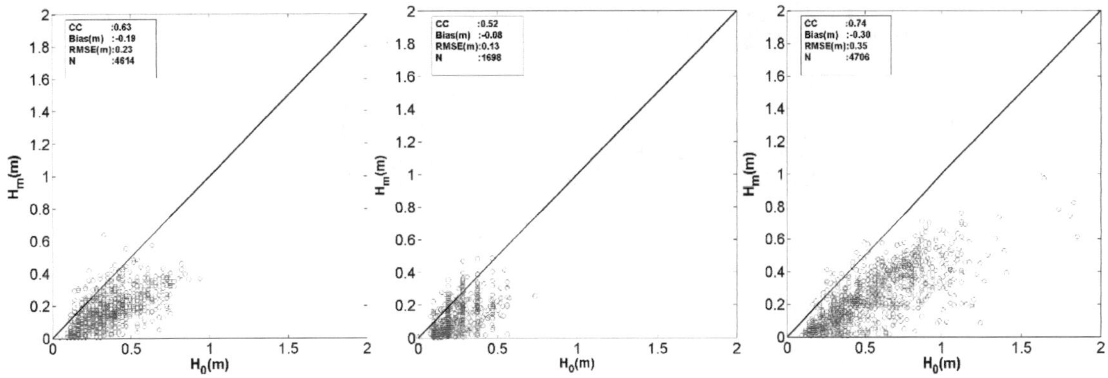

Figure 2. Verification result (from left to right) for Kepulauan Seribu, Pluit, and Karawang with period of 1998-1999. H_o is buoy significant wave height and H_m is model significant wave height. The diagonal line indicates y=x.

Result of the hindcast is validated using 1 year (1998-1999) SEAWATCH buoy data from BPPT (Badan Pengkajian dan Penerapan Teknologi) in 3 locations shown in figure 1; Kepulauan Seribu, Pluit, and Karawang. All of the locations are located in shallow water area in Indonesian inner seas. Generally, the validation result from figure 2 shown that the 3 locations give good verification result with each Correlation Coefficient (CC) being 0.63, 0.52, and 0.74 from left to right, respectively. In every location, almost all of the SWH give an underestimate result, with buoy's SWH being higher than model's SWH, shown by dot on the picture that mainly located below the line. It is possible because the location of the buoy is in the shallow water area which does not exactly coincide with the WAVEWATCH-III native grid, resulting in a slightly lower SWH compared to the buoy.Statistical parameter here is also including bias to see the difference between model's SWH and buoy's SWH. All 3 locations give low bias and RMSE result, meanings that there is a small difference between the buoy's SWH and model's SWH, which is to be expected from the result in shallow region.

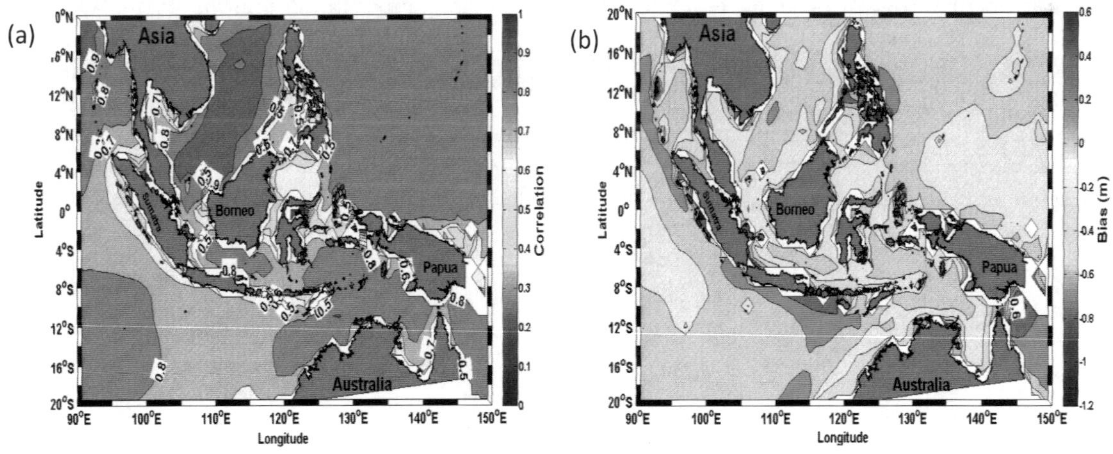

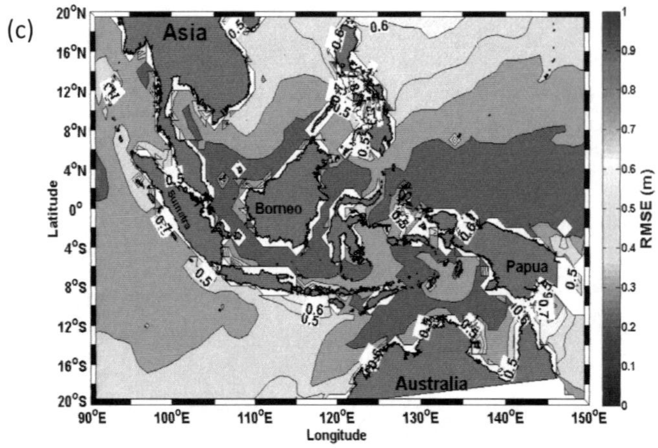

Figure 3 Spatial statistical parameter validated using AVISO satellite data consist of (a) correlation; (b) bias; and RMSE (c).

Aside from using buoy, verification is also conducted using 7 years (2009-2015) of AVISO satellite data on a larger area covering all Indonesia region. Compared to the verification using buoy data, this verification is able to show the statistical parameter spatially over the Indonesia region. However, despite being able to provide SWH data covering a wide area, AVISO satellite comes with a disadvantage of poor ability to differentiate between land and sea, resulting in unavailability of some SWH data near land and unreliable result in some of the near coastal area. In addition to that, the location of verification is chosen on every point on model that coincides with the point of AVISO to mitigate the difference in grid between model and AVISO. The result of CC, bias, and RMSE of this method is shown in figure 3, with the white area being devoid of SWH data from AVISO. Spatial correlation gives quite satisfactory result in a large part of the model's domain. The two data sets are most consistent around SCS with CC close to 1, followed by Indonesian inner seas, Banda Sea, and Pacific Ocean with the result ranging from 0.7 to 0.8. CC in the Indian Ocean are similar to those in the Pacific Ocean. Generally, the bias result is also quite good with result ranging from 0 m to -0.2 m around SCS, Pacific Ocean, and Indonesian inner seas, and ranging from -0.2 m in 0.2 m in Indian Ocean. The more accurate depiction of deviation between AVISO and WW3 model is provided in the spatial RMSE. In Indonesian inner seas, RMSE result is close to 0 m while in Banda Sea it is ranging from 0 m to 0.2 m, as expected of the result in a closed seas that have quite small SWH. In the oceans, however, the result is larger, 0 m to 0.6 m in Pacific Ocean and 0.3 m to 0.4 m in Indian Ocean. This large result is to be expected and proportional to the high SWH occurs in these areas.

IOP Conf. Series: Earth and Environmental Science **162** (2018) 012004 doi:10.1088/1755-1315/162/1/012004

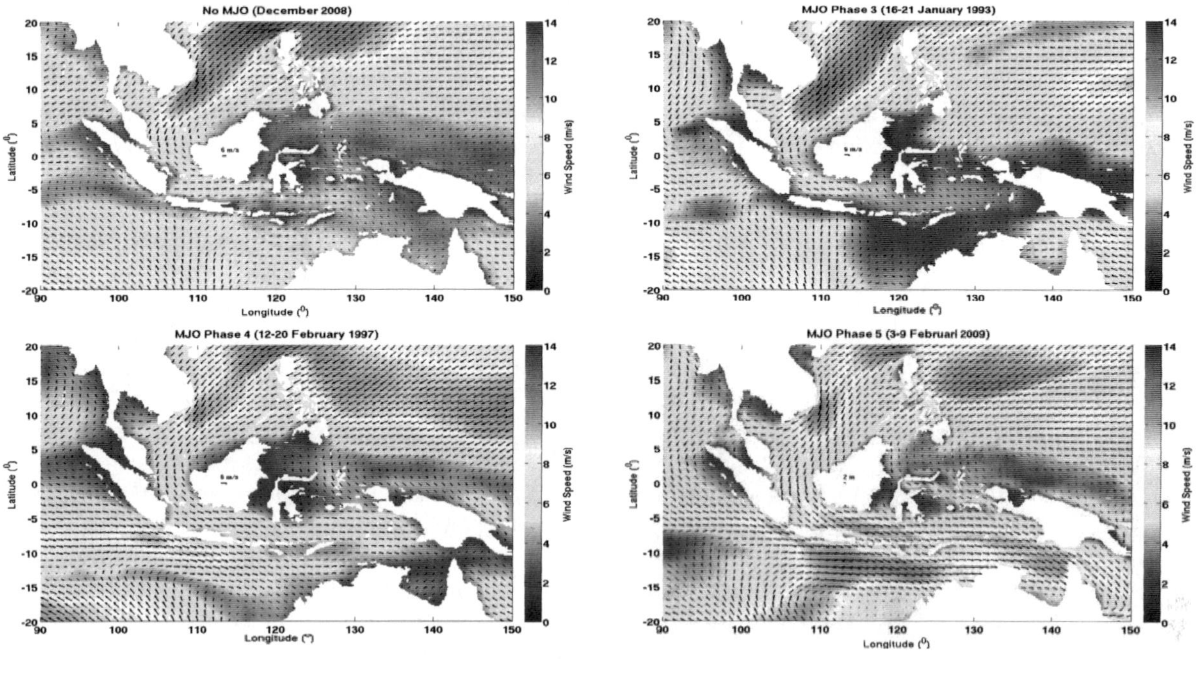

Figure 4. Averaged wind speed anomalies of each MJO phase in DJF season

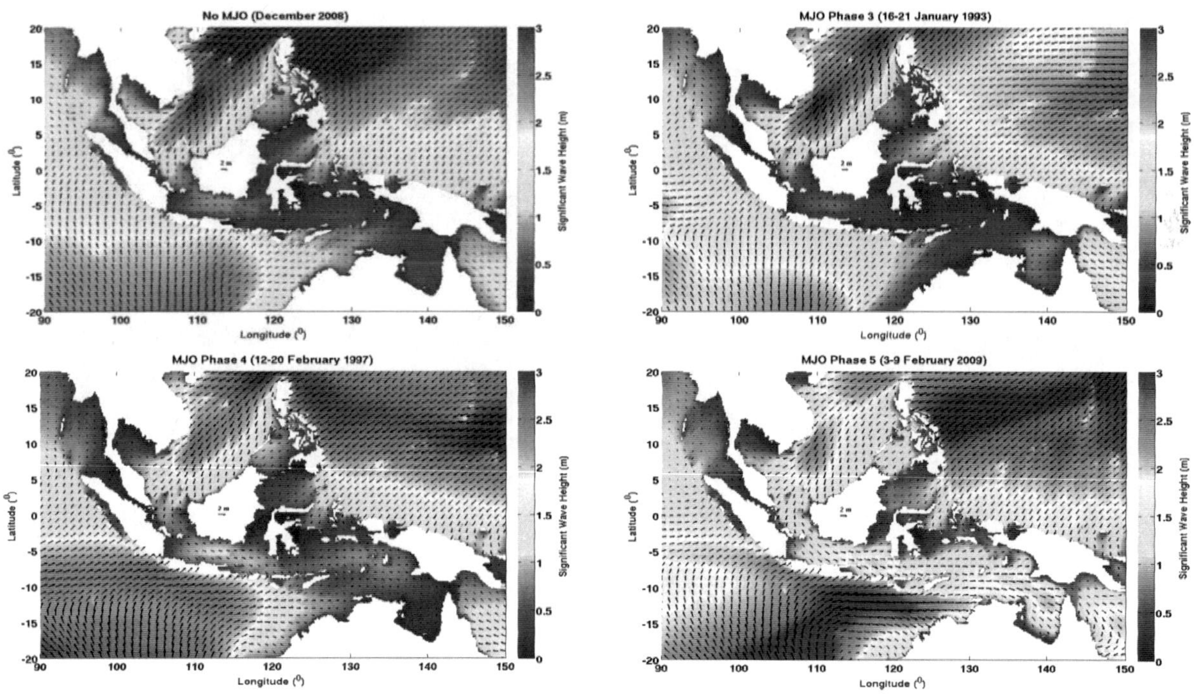

Figure 5. Averaged significant wave height anomalies of each MJO phase in DJF season

IOP Conf. Series: Earth and Environmental Science **162** (2018) 012004 doi:10.1088/1755-1315/162/1/012004

3. Results and Discussion

3.1. DJF and MAM season

As shown in figure 4, anomalies on wind speed caused by MJO in DJF season can be seen clearly on each picture. Compared to the condition of no MJO, each phase of the MJO shows quite different change. The wind speed in MJO phase 3 is seen as the lowest among the others, indicating that MJO may cause a decrease in wind speed. It may have occured because on phase 3, MJO is located around western of Indonesia. It means, the low-pressured area in the west of Indonesia will force wind over the Java Sea to blow westward. This direction is the opposite of the northwest monsoon that blows over Indonesia region, resulting in a net decrease in the wind speed over the Java Sea. On the other hand, there is an increase of WS shown in phase 4 and 5, with phase 5 being the phase with the most impact on the WS. On phase 4, the WS seems to increase for about 2 m/s in inner seas and 5 m/s in the southern Java, raising the WS to 8 m/s from 6 m/s and to 10 m/s from 5 m/s respectively. The strongest impact shown by phase 5 increase the WS by 6 m/s (from 6 m/s to 12 m/s) in inner seas, in the southern Java and Lesser Sunda Islands by 8 m/s (from 5 m/s to 13 m/s) and in Banda Sea by 5 m/s (from 5 m/s to 10 m/s). This increase is to be expected as some of the study on MJO found that the strongest impact of MJO around Indonesia region is observed in phase 5 in DJF season (Duwaliser et al (2013); Marshall et al (2015); Ramdhani (2015)). Strongest impact on phase occurs because in this phase, the MJO-driven low-pressured areais situated in the eastern part of Indonesia, forcing the wind to blow eastward to the low-pressured area. When this happens during the northwest monsoon, MJO-forced wind will blow in the same direction with the northwest monsoon and resulted in the amplification of the wind speed and larger fetch, hence the signifcant increase in wind speed across inner seas and southern Java.

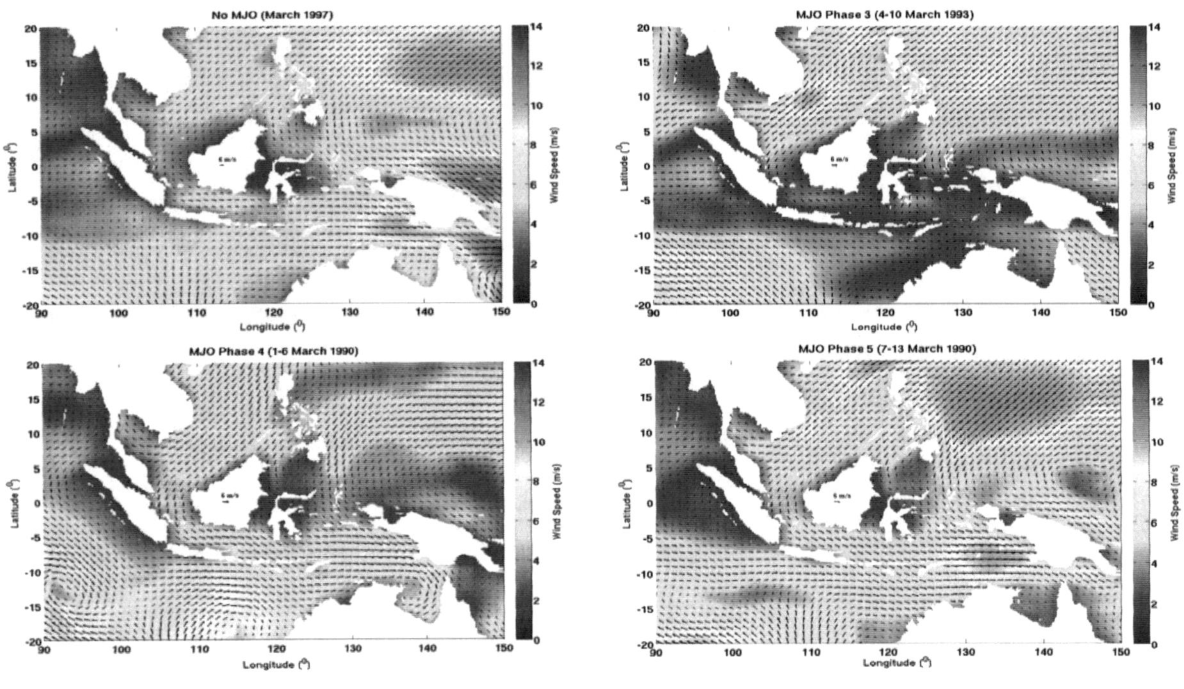

Figure 6. Averaged wind speed anomalies of each MJO phase in MAM season.

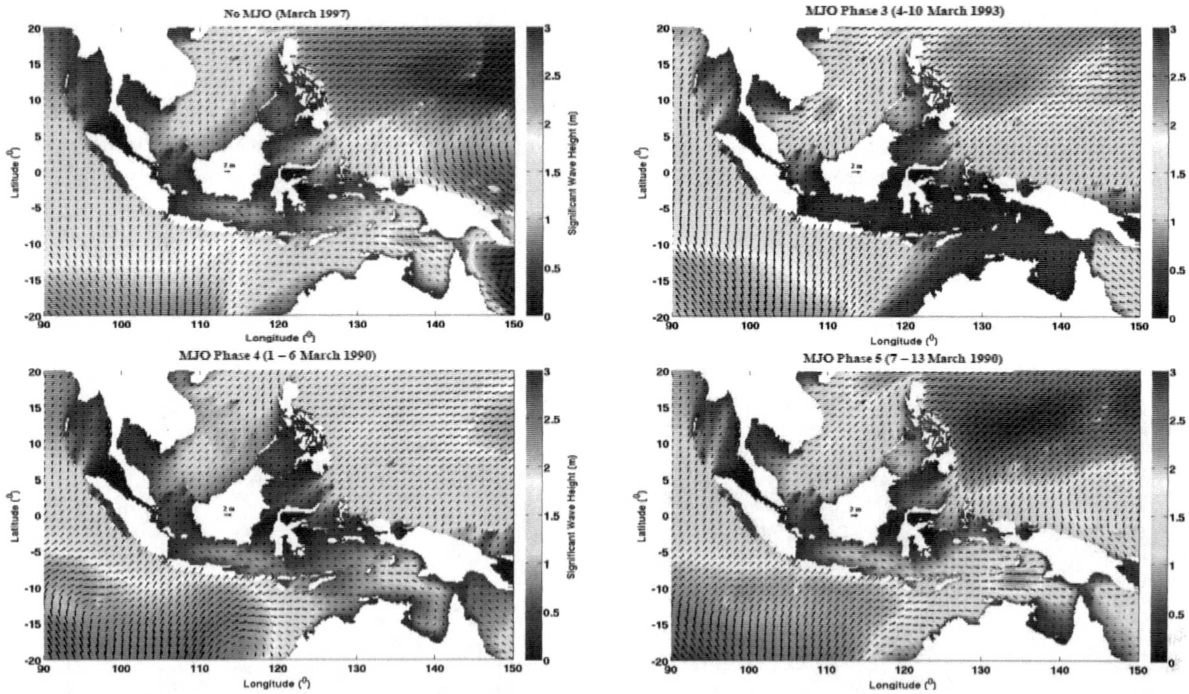

Figure 7. Averaged significant wave height anomalies of each MJO phase in MAM season

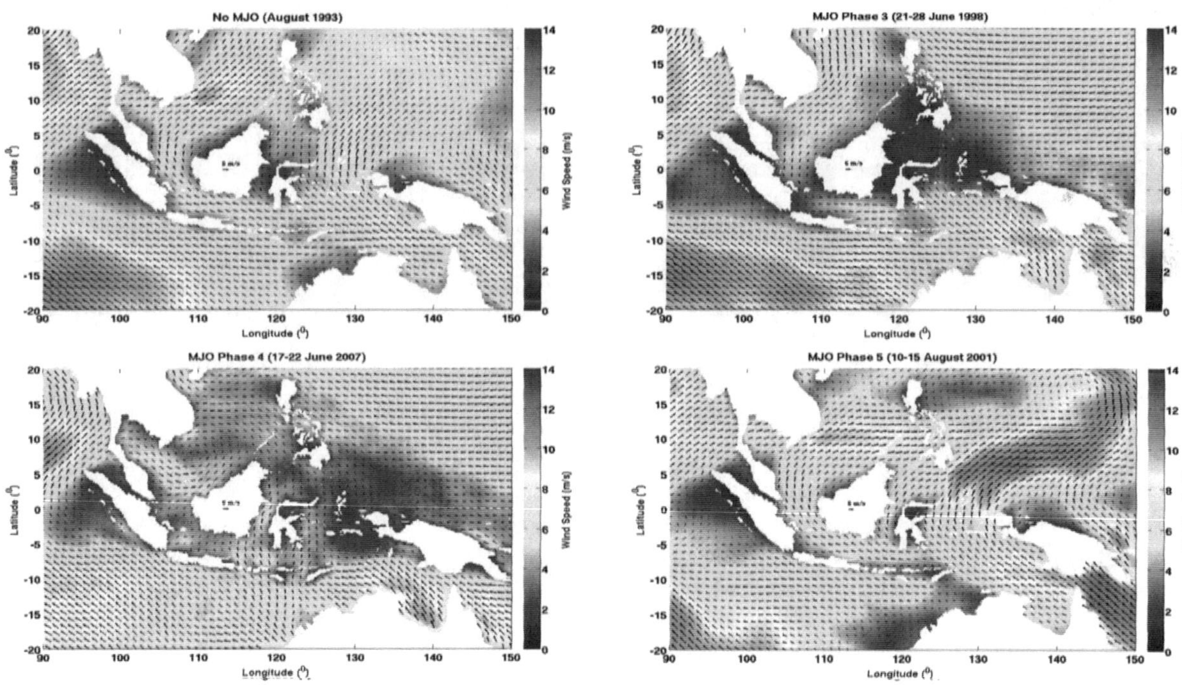

Figure 8. Averaged wind speed anomalies of each MJO phase in JJA season

The impact on SWH is also in accordance with the WS. In figure 5, it can be observed that there is a small decrease of SWH in phase 3 compared to No-MJO state. SWH in Java Sea is decreased by 0.1 m (from 0.5 m to 0.4 m), while the most significant decrease in this phase occurred in southern Lesser Sunda Islands with a decrease by 0.5 m (from 1.4 m to 0.9 m). Oppositely, the raise of SWH is shown by MJO phase 4 and 5 as the result of increase in WS in these phases. Wind is one of the most important wave generator in Indonesian inner seas and southern Java, thus, the stronger the wind blows over those areas, the higher the wave will be generated. Wind amplification on the event of MJO phase 5 transfers much more energy into the sea over the MJO duration, resulted in much higher wave compared to the other phase. In phase 4, the SWH is increased by 30 cm (from 0.5 m to 0.8 m) in inner seas and increased by 0.5 m in southern Java (from 1.5 m to 2 m). The most significant change in SWH can be found in phase 5, which raise the SWH in the inner seas by 1 m (from 0.5 m to 1.5 m), in southern Java by 0.7 m (from 1.5 m to 2.2 m), 0.8 m in Banda Sea (from 0.4 m to 1.2 m), and 1.2 m in southern Lesser Sunda Islands (from 1.4 m to 2.6 m). This large increases is may be due to the strong wind force created by MJO's low-pressured area in eastern part of Indonesia.

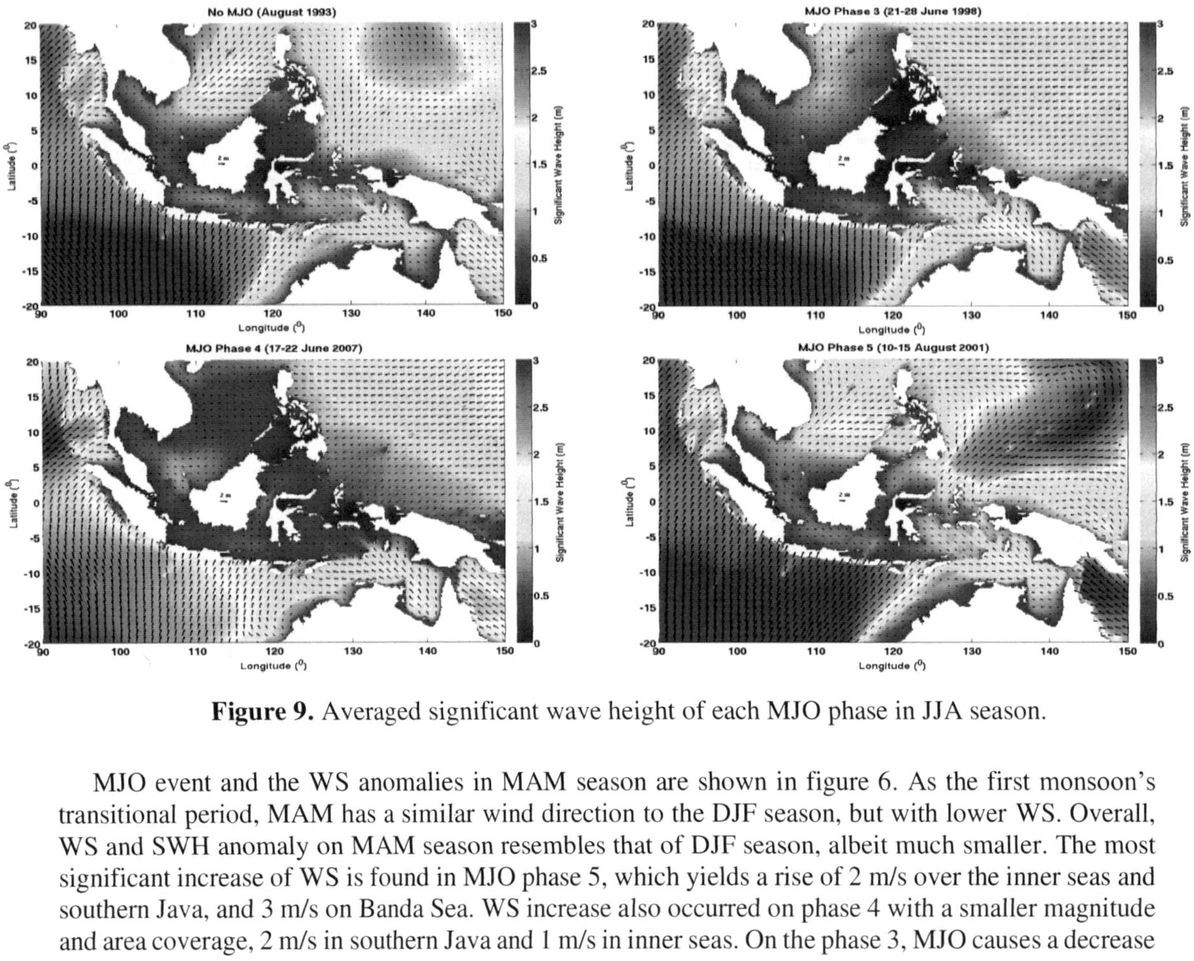

Figure 9. Averaged significant wave height of each MJO phase in JJA season.

MJO event and the WS anomalies in MAM season are shown in figure 6. As the first monsoon's transitional period, MAM has a similar wind direction to the DJF season, but with lower WS. Overall, WS and SWH anomaly on MAM season resembles that of DJF season, albeit much smaller. The most significant increase of WS is found in MJO phase 5, which yields a rise of 2 m/s over the inner seas and southern Java, and 3 m/s on Banda Sea. WS increase also occurred on phase 4 with a smaller magnitude and area coverage, 2 m/s in southern Java and 1 m/s in inner seas. On the phase 3, MJO causes a decrease in the WS over the inner seas and southern Java by 2 m/s (from 5 m/s to 3 m/s). This occurrence of increasing and decreasing WS happened because of the same mechanism as the change that happened in DJF season. Increase on MJO phase 4 and 5 occurred because of amplification in WS by the low-pressured area created by MJO in east of Indonesia, while a decrease in phase 3 happened because of wind collision caused by low-pressured area in west of Indonesia. SWH anomalies shown in figure 7 also follow the pattern of WS anomalies. The highest wave height occurrence is found in phase 5, with an increase of 0.3 m in inner seas and 0.5 m in southern Java and Lesser Sunda Islands. MJO phase 4 does not pose significant increase in inner seas (0.2 m) but it causes 0.4 m increase of wave height in southern Java and Lesser Sunda Islands. MJO also causes the SWH to decrease in phase 3, but the change is insignificant (0.2 m over the inner seas and 0.3 m in southern Java).

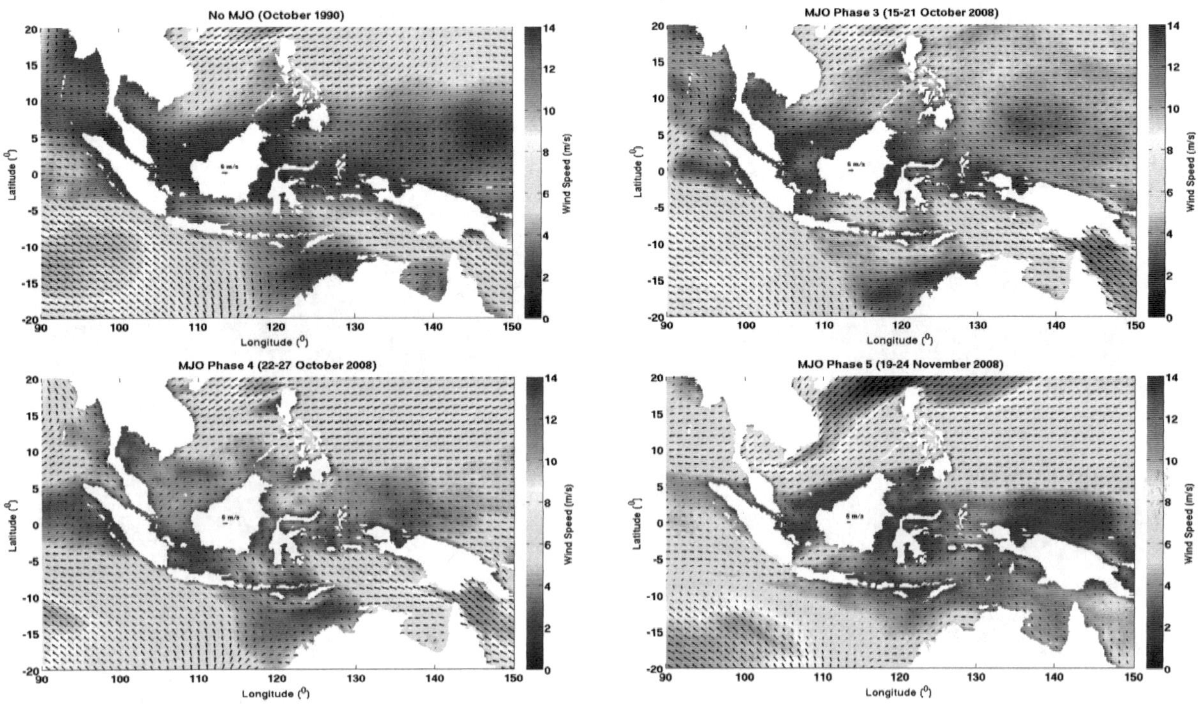

Figure 10. Averaged wind speed anomalies of each MJO phase in SON season.

3.2. JJA and SON

Being the opposite of the DJF season, JJA season has a reversed monsoon wind direction that flows from southern hemisphere to northern hemisphere across Indonesia, creating a southeastern monsoon. Low-pressured area created by MJO in this season also tends to be located farther north compared to DJF and MAM season (Wheeler and Hendon, 2004; Gotschalck, 2014), causing the change in wind speed and wave height mainly happened in northern part of Indonesia. This anomaly happens because in JJA season, the sun is located closer to the northern hemisphere (NH), resulting in NH area becomes warmer than the souther hemisphere (SH). In this season, MJO which happened mainly because of the warm SST can be easily formed in the warm NH rather than SH due to lack of sunlight heating in that area.

There is no significant change in WS except on MJO phase 5. As can be inferred from figure 8, Compared to the event of no MJO, WS during MJO phase 3 and 4 seems to only decreased in SCS, 3 m/s in phase 3 and 5 m/s in phase 4. The change is hypothesized as the effect of low-pressured area around middle and western of Indonesia caused by MJO. The low-pressured area located in southern of SCS attracts wind over the SCS to blow southward. This direction is the opposite of the southwestern monsoon that blows northward in SCS, resulting in decrease of WS over the SCS. The significant change in WS occurred on MJO phase 5 in SCS and southeastern of Philippine. Low-pressured area created by MJO in this phase is located around east of Phillipines, forcing the wind over the SCS to blows eastward, as well as the wind over western and eastern Phillipine, North Maluku, and Sulu Sea to blow even stronger. It can be seen from figure 6 that the wind over the SCS in phase 5 tends to shift eastward as the result of wind forcing by low-pressured area east of Phillipine. This change is followed by an increase of WS in Sulu Sea by 2 m/s, and in both eastern and western Philippine by 5 m/s and 3 m/s respectively. This change in WS is also followed by change in SWH, which significantly happened in MJO phase 5 that cause an increase of 0.5 m in Sulu Sea and 1 m in eastern Philippine. Conversely, the significant decrease in wave height can be found over the SCS in phase 3 and 4, where the SWH is decreased by around 0.5 m and 0.8 m respectively.

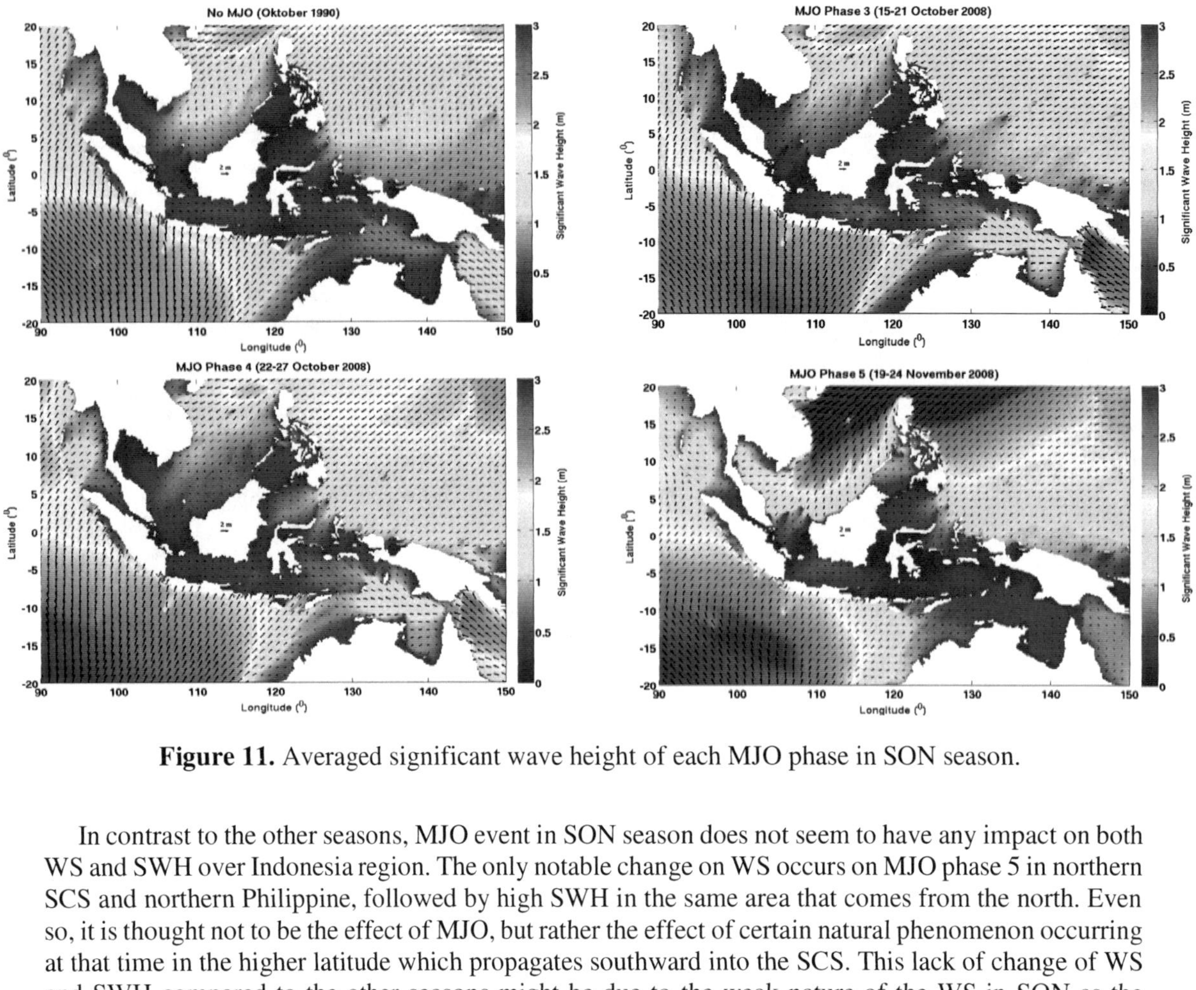

Figure 11. Averaged significant wave height of each MJO phase in SON season.

In contrast to the other seasons, MJO event in SON season does not seem to have any impact on both WS and SWH over Indonesia region. The only notable change on WS occurs on MJO phase 5 in northern SCS and northern Philippine, followed by high SWH in the same area that comes from the north. Even so, it is thought not to be the effect of MJO, but rather the effect of certain natural phenomenon occurring at that time in the higher latitude which propagates southward into the SCS. This lack of change of WS and SWH compared to the other seasons might be due to the weak nature of the WS in SON as the second monsoon transitional period, so the MJO cannot cause significant change despite having enough power to do so. The uncertainty of the wind direction in this phase is also thought to play a role in minimizing the MJO effect, depending on whether it blows upwind or downwind the low-pressured area.

4. Conclusion

The low-pressured area created by MJO in phase 3, 4, and 5 that occurred adjacent to Indonesia is able to cause anomalies in WS and eventually influencing the SWH. Increases in WS and SWH occur because the location of MJO's low-pressured area is situated downwind of the monsoon, regardless of its season and phase. It forces the wind to blow to the low-pressured area along with the moonson and creates an amplification of the WS. MJO phase 5 on DJF season is the prime example on how MJO is able to cause significant change of WS in southern Java by 8 m/s, inner seas by 6 m/s, and Banda Sea by 5 m/s, and cause significant change in SWH by 0.7 m in southern Java, 1 m in inner seas, 0.8 m in Banda Sea, and 1.2 m in southern Lesser Sunda Islands, respectively. On MAM, MJO phase 5 also has the same effect, but with lower value of increase, which is 2 m/s in wind speed over inner seas and southern Java, 0.3 m of wave height in inner seas, and 0.5m in Southern Lesser Sunda Islands. In lieu of increasing anomalies, MJO can also causes decreasing anomalies as occuring in MJO phase 3 on both season if its location is upwind of the monsoon. In DJF and MAM season, due to the heating in southern hemisphere, MJO tends to occur southward, resulting in WS and SWH anomalies mainly occur in inner seas, southern Java, Banda Sea, and around Lesser Sunda Straits. On the contrary, MJO tends to occur northward in JJA and SON seasons, thus creating WS and SWH anomaly in the northern part of Indonesia such as SCS, Sulu Sea, and western Pacific Ocean. The most significant impact on this season can be seen on MJO phase 5 which increases WS by 2 m/s in Sulu Sea, 3 m/s on western Philippine, and 5 m/s on eastern Philippine, followed by increase in SWH by 0.5 m in Sulu Sea and western Philippine. In SON, MJO does not seem to be creating any WS and SWH anomalies during any of its phases.

Acknowledgement

This work is supported by a grant of Ministry of Research, Technology, and Higher Education of the Republic of Indonesia (Project: Excellent Research of University). Authors are also thankful to the Agency for the Assessment and Application of Technology for given buoy data for model verification.

References

[1] A G Marshall et al., Madden julian oscillation impacts on global ocean surface waves, Ocean Modelling (2015), http:// dx.doi.org/10.016/j.ocemod.2015.06.002
[2] Gottschaclk J, Kousky V, Higgins W and L'Heureux M 2014 *MJO Summary* (National Oceanic and Atmospheric Administration)
[3] Madden R A and Julian P R 1972 Description of global-scale circulation cells in the tropics with 40-50 day period. *J. Atmos. Sci.* **29** 1109-1123.
[4] Ramdhani A 2015 *Pengaruh Siklon Tropis dan Madden-Julian Oscillation (MJO) Terhadap Kejadian Gelombang Tinggi di Perairan Indonesia Bagian Dalam* (Doctoral Dissertation. Institut Teknologi Bandung).
[5] Waliser D E, Wu B, Yung Y L, Guan B, Webb F and Kedar S 2013 The Impact of the MJO on Global Ocean Surface Wave Heights, *Bull. Am. Meteorol. Soc.* **94**.
[6] Wheeler M C and Hendon H H 2004 An All-Season Real-Time Multivariate MJO Index: Development of an Index for Monitoring and Prediction, *Monthly Weather Rev* 132 (**8**), 1917-1932.
[7] Zhang C 2005 Madden-Julian Oscillation *Rev. of. Geophysics.* **43** (2).
[8] Zhang C 2013 Madden-Julian Oscillation: Bridging Weather and Climate *Bull. Am Meteorol. Soc.* **94**, 1849-1870.

IOP Conf. Series: Earth and Environmental Science **162** (2018) 012005 doi:10.1088/1755-1315/162/1/012005

Integrated 3D numerical modeling during La Niña and El Niño events using Regional Ocean Modeling System (ROMS) in Makassar Strait: preliminary study

S Mubarrok, M Riza, A A Nur, and I Mandang

Physical Oceanography and Modeling Laboratory, Program Study of Physics, Faculty of Mathematic and Natural Sciences, Mulawarman University, Samarinda, 75123, Indonesia

E-mail: mubarrok@fmipa.unmul.ac.id

Abstract. This research showed the difference of thermocline layer depth during La Niña and El Niño events in the Makassar Strait, the main passage connecting the Pacific and Indian Oceans. La Niña event represented by 2011 and El Niño event represented by 2015. Three-dimensional numerical modeling using Regional Ocean Modeling System (ROMS) is used to determine ocean dynamic characteristics. The tides and atmospheric factor used as generating forces for coupled Ocean-Atmosphere-Wave-Sediment simulation model. Tidal elevation verification results from January to February 2017 showed a high confidence with RMSE = 0.148 m and MAPE = 13.51 %. Ocean circulation showed that water flows from north to south of the Makassar Strait in both conditions, El Niño and La Niña, indicating a transport from the Pacific to the Indian Oceans during those time. Two-dimensional seasonal sea surface temperature (SST) was used to determine the variability of thermocline layer depth. In general, based on simulation results, the depth of thermocline during La Niña was deeper than that during El Niño. The upper threshold of thermocline layer during El Niño phenomenon was shallower (average 27.29 m) compare to the threshold during La Niña (average 39.25 m). So do the lower threshold, during El Niño was shallower (average 160.47 m) compare to the threshold during La Niña (average 165.70 m). The thermocline thickness during El Niño (133.19 m) was found to be thicker compared to La Niña (126.43 m).

1. Introduction

Makassar Strait, geographically, is water that connects the Pacific Ocean and the Indian Ocean. Many unique phenomena, such as Indonesian Troughflow (ITF) [1, 2], tidal mixing caused by steep slope bathymetry [3], internal Kelvin wave propagated from Lombok Strait [4], and effects of ENSO events to water characteristics [5, 6]. Makassar Strait is also affected with seasonal trade wind that helped mixing in surface layer between sea water and fresh water from river [7].

To understand its complexity, a wind-wave-tide integrated numerical model needed to describe physical phenomenon in Makassar Strait. One of high level model that many others used in previous study is the Coupled Ocean-Atmosphere–Wave–Sediment Transport (COAWST) [8, 9, 10]. Figure 1 showed a schematic high level model COAWST which is comprised of the Model Coupling Toolkit to exchange data fields between the ocean model ROMS, the atmosphere model WRF, the wave model SWAN, and the sediment capabilities of the Community Sediment Transport Model. This formulation builds upon previous developments by coupling the atmospheric model to the ocean and wave models (figure 2), providing one-way grid refinement in the ocean model, one-way grid refinement in the wave model, and coupling on refined levels [9]. This study only focused to apply an integrated hydrodynamic numerical model using Regional Ocean Model System (ROMS) during ENSO event as main interest of the study.

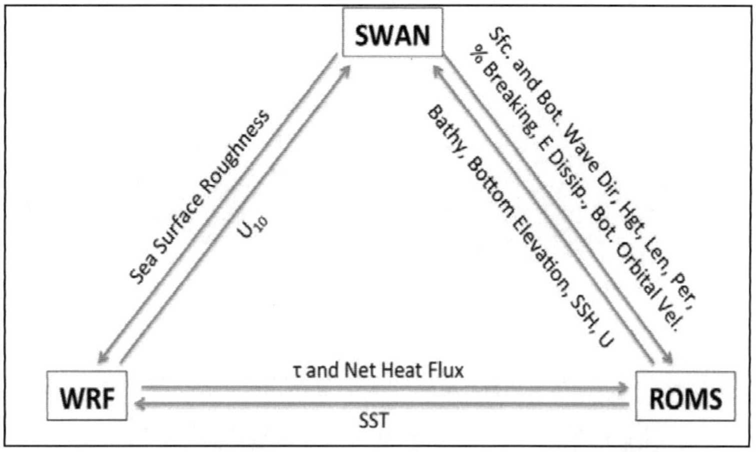

Figure 1. Basic schematic of the COAWST modeling system, include their 'link' between each other, ROMS as ocean model, SWAN as wave model, and WRF as atmospheric model [10].

Objective of this study is to understand the variability of ENSO phenomena, for both El Niño and La Niña conditions, to understand the characteristic water masses changes in Makassar Strait. To know its difference between two conditions, two variables were used as determinant factors; they are thermocline depth and sea surface temperature (SST). We compared these variables among three conditions, the La Niña in year of 2015, normal condition in year of 2013, and El Niño in year of 2011. The interest of this research area showed using figure 3. We compared the model elevation result with tide observation as verification

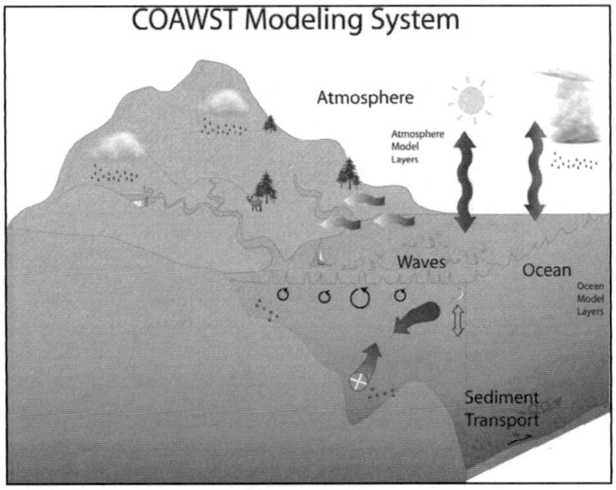

Figure 2. An illustration of the COAWST Modeling System that provides exchange between an ocean model, an atmosphere model, a waves model, and a sediment transport model [9].

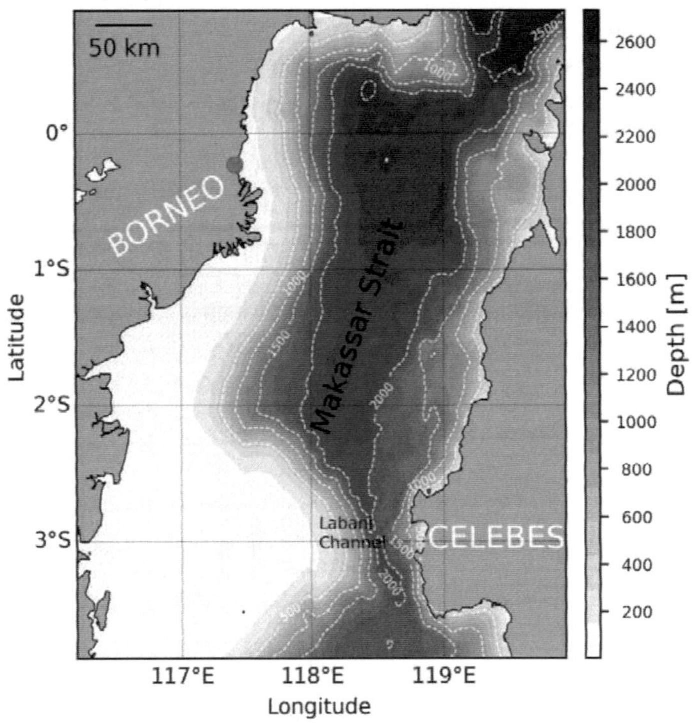

Figure 3. Study area of ROMS Model, Makassar Strait, and bathymetry isobaths interval is 500 m, verification point showed by red dot symbol.

2. Data

2.1 Tides

Forcing from tides is put to ROMS, and represented with eight tidal constituents of the M2, S2, N2, K2, K1, O1, P1, and Q1. They were imposed on the lateral open boundaries with constituents obtained from global model of ocean tides TPXO8-atlas TOPEX/Poseidon (http://volkov.oce.orst.edu/tides/global.html). TPXO8-atlas is current version of tides model with complex amplitudes harmonic constituents of earth-relative sea-surface elevation with ¼ degree resolution global grid and used a least-squares sense, the Laplace Tidal Equations and along track averaged data from TOPEX/Poseidon and Jason (on TOPEX/Poseidon tracks since 2002) [11, 12]. As a model verification, we used sea surface elevation, field observation data from Compact TD instrument, and compared to model result of 2017. We verified model elevation at latitude of 00°13' 26,8" S and longitude of 117° 25' 11,2" E with date of verification is from January 29th to February 19th, 2017.

2.2 Temperature, Salinity, and Current

Global reanalysis assimilation data sets from Hybrid Coordinate Ocean Model (HYCOM) and Navy Coupled Ocean Data Assimilation (NCODA Global 1/12°), e.g sea water salinity, sea water potential temperature, sea water velocity (*u* and *v*), and sea surface elevation, used as initial condition to the model (http://hycom.coaps.fsu.edu/thredds/catalog.html). During the simulation, the interior temperature and salinity were nudged to the tracer fields from HYCOM with a time scale of 1 day for each year simulation.

2.3 Atmospheric Forcing

Atmospheric forcing such as surface wind, air pressure, air temperature, air humidity, net fresh water flux, rainfall rate, net long-wave radiation flux, and solar shortwave radiation flux extracted from European Centre for Medium-Range Weather Forecasts (ECMWF) (http://apps.ecmwf.int/datasets/) and applied to model for each scenario. Surface wind is velocity of wind 10 meters above sea surface elevation and used as generating force for surface current circulation. Surface wind was imposed every 3 hours for a-year long simulation. Atmospheric forcing were computed by ROMS using bulk-fluxes formulation internally and turbulent fluxes for wind, heat, and moisture are computed using Monin-Obukhov similarity theory [13]. These atmospheric forcing were imposed to the model every 3 hours time step.

2.4 Bathymetry

Figure 3 showed bathymetry that was used for the model area. It was extracted from global topography data fusion of NASA Shuttle Radar Topography Mission (SRTM) [14] land topography with measured and estimated seafloor topography (SRTM15_PLUS) (ftp://topex.ucsd.edu/pub/srtm15_plus/). This data is corrected by sounding [15] and gravity data [16] and modified from SRTM30 product distributed by USGS EROS data center. The grid resolution is 30 second which is roughly one kilometer. Land data are based on the 1-km averages of topography derived from the USGS SRTM30 gridded DEM data product created with data from the NASA Shuttle Radar Topography Mission. GTOPO30 data are used for high latitudes where SRTM data are not available. Version 1 of SRTM15 is based on V11 of the SRTM30_PLUS grid. Basically the land data is completely new based on a combination of SRTM, ASTER, and CryoSat-2 ice sheet topography. The ocean data are the same as SRTM30_PLUS but required more extensive editing to remove the bad points mostly along edges of the swath data. Ocean data are based on the global seafloor topography from satellite altimetry and ship depth soundings [17] global 1-minute grid between latitudes +/- 81 degrees. Higher resolution grids have been added from the LDEO Ridge Multibeam Synthesis Project, the JAMSTEC Data Site for Research Cruises, and the NGDC Coastal Relief Model. Arctic bathymetry is from the International Bathymetric Chart of the Oceans (IBCAO) [18]. For model, depth minimum settled to -5 m and deepest depth is -2734 m.

3. Methods

As a part of future goal research, in this study we focused to apply ROMS model to determine characteristic water mass in the Makassar Strait. We forced this model using ECMWF data and HYCOM data as initial condition to the model. We did yet included WRF model and SWAN model to see the good quality of our ROMS model as the main hydrodynamic model. To determine ENSO effect in Makassar Strait, we compiled three different scenarios hence represented El Niño, Normal, La Niña year. La Niña acted by 2011, Normal scenarios by 2013, and El Niño scenario by 2015. These scenarios were taken based on Ocean Niño Index (ONI) Extended Reconstructed Sea Surface Temperature (ERSST) v4 (https://data.nodc.noaa.gov/cgi-bin/iso?id=gov.noaa.ncdc:C00884). The Extended Reconstructed Sea Surface Temperature (ERSST) dataset is a global monthly sea surface temperature dataset derived from the International Comprehensive Ocean–Atmosphere Dataset (ICOADS). It is produced on a 2° × 2° grid with spatial completeness enhanced using statistical methods. This monthly analysis begins in January 1854 continuing to the present and includes anomalies computed with respect to a 1971–2000 monthly climatology. The newest version of ERSST, version 5, is based on optimally tuned parameters using the latest datasets and improved analysis methods [19, 20, 21].

We compared model results from year 2017 model using field observation data and used the setting to be applied to the three scenarios of model. We compared 3-month average temperature profiles (SST and vertical temperature profiles) between each scenario. We collected 7 random points in the model from north part to south part to see the vertical profiles and compared them between La Niña, El Niño, and normal scenario. Thus thermocline layer calculate using gradient criterion 0.05 ºC/m to determine the depth of upper and lower threshold of thermocline layer [22].

To see the correlation between model and observation data, we used Root Mean Square Error (RMSE) and Mean Percentage Absolute Error (MAPE) formula that shown by equation (1) and (2) [23] below

$$RMSE = \left[\frac{\sum_t \left(A_{i,t} - B_{i,t} \right)^2}{n} \right]^{1/2} \tag{1}$$

$$MAPE = \frac{1}{n} \sum_{t=i}^{t=imax} \left| \frac{A_{i,t} - B_{i,t}}{A_{i,t}} \right| \times 100\% \tag{2}$$

with t is time step model in hour, n is total data, A is actual value of observation data, and B is model result.

3.1 Hydrodynamic Model ROMS

The simulation of hydrodynamic process in domain area were conducted by using the numerical model Regional Ocean Modeling System (ROMS). This model was quite popular among modeler, scientist and researcher who wanted to study the coastal application and developed from Princeton Ocean Model (POM). Many literature described this model capability especially in the regional ocean domain. ROMS is a three-dimensional, free surface, terrain-following numerical model that solves finite difference approximation of the Reynolds-averaged Navier-Stokes (RANS) equation using the hydrostatic and Boussinesq assumption with a split-explicit time stepping algorithm [24, 25, 26]. It uses a horizontal curvilinear Arakawa "C" grid and vertical stretched terrain-following coordinates [26]. This model also can be configured depending of users application which has several choices for advection schemes, pressure gradient algorithms, turbulent closure, and many types of boundary conditions.

IOP Conf. Series: Earth and Environmental Science **162** (2018) 012005 doi:10.1088/1755-1315/162/1/012005

The governing equations used in ROMS were presented in flux form on the Cartesian horizontal coordinates and sigma vertical coordinates. For the momentum equations on the x- and y- axis (equation 3 and 4) directions are:

$$\frac{\partial(H_z u)}{\partial t}+\frac{\partial(uH_z u)}{\partial x}+\frac{\partial(\upsilon H_z u)}{\partial y}+\frac{\partial(\Omega H_z u)}{\partial s}-fH_z\upsilon=-\frac{H_z}{\rho_0}\frac{\partial p}{\partial x}-H_z g\frac{\partial \eta}{\partial x}$$
$$-\frac{\partial}{\partial s}\left(\overline{u'w'}-\frac{\nu}{H_z}\frac{\partial u}{\partial s}\right)-\frac{\partial(H_z S_{xx})}{\partial x}-\frac{\partial(H_z S_{xy})}{\partial y}+\frac{\partial S_{px}}{\partial s} \quad (3)$$

$$\frac{\partial(H_z \upsilon)}{\partial t}+\frac{\partial(uH_z \upsilon)}{\partial x}+\frac{\partial(\upsilon H_z \upsilon)}{\partial y}+\frac{\partial(\Omega H_z \upsilon)}{\partial s}+fH_z u=-\frac{H_z}{\rho_0}\frac{\partial p}{\partial y}-H_z g\frac{\partial \eta}{\partial y}$$
$$-\frac{\partial}{\partial s}\left(\overline{\upsilon'w'}-\frac{\nu}{H_z}\frac{\partial v}{\partial s}\right)-\frac{\partial(H_z S_{yx})}{\partial x}-\frac{\partial(H_z S_{yy})}{\partial y}+\frac{\partial S_{py}}{\partial s} \quad (4)$$

$$0=-\frac{1}{\rho_0}\frac{\partial p}{\partial s}-\frac{g}{\rho_0}H_z\rho \quad (5)$$

with the continuity equation:

$$\frac{\partial \eta}{\partial t}+\frac{\partial(H_z u)}{\partial x}+\frac{\partial(H_z \upsilon)}{\partial y}+\frac{\partial(H_z \Omega)}{\partial s}=0 \quad (6)$$

and scalar transport:

$$\frac{\partial(H_z C)}{\partial t}+\frac{\partial(uH_z C)}{\partial x}+\frac{\partial(\upsilon H_z C)}{\partial y}+\frac{\partial(\Omega H_z C)}{\partial s}=-\frac{\partial}{\partial s}\left(\overline{c'w'}-\frac{\nu_\theta}{H_z}\frac{\partial C}{\partial s}\right)+C_{source} \quad (7)$$

These equations are closed by parameterizing the Reynolds stresses and turbulent tracer fluxes as

$$\overline{u'w'}=-K_M\frac{\partial u}{\partial z},\quad \overline{v'w'}=-K_M\frac{\partial v}{\partial z},\quad \overline{\rho'w'}=-K_M\frac{\partial \rho}{\partial z} \quad (8)$$

where K_M is the eddy viscosity for momentum and K_H is the eddy diffusivity. Eddy viscosities and eddy diffusivities are calculated using one of five options for turbulence-closure models in ROMS: (i) Brunt-Väisälä frequency mixing in which mixing is based on the stability frequency; (ii) a user-provided analytical expression such as a constant or parabolic shape; (iii) the *K*-profile parameterization [27], expanded to include both surface and bottom-boundary layers [28]; (iv) Mellor-Yamada level 2.5 (MY2.5) method [29]; and (v) the generic length-scale (GLS) method [30] as implemented in [31] that also includes the option for surface fluxes of turbulent kinetic energy due to wave breaking. and for this study, we applied option (v) to

IOP Conf. Series: Earth and Environmental Science **162** (2018) 012005 doi:10.1088/1755-1315/162/1/012005

calculated eddy viscosities and eddy diffusivities. For the details of each variables on the equations are listed in table 1.

3.2 Model Design

This model have 160 east-west grid cells and 208 north-south grid cells with 2.6 km grid spacing uniformly. This model have 413.4 km wide and 538.2 km long with width area is about 222.49 10^3 km^2. We used masking to determine which one is land and which one is ocean using 1 and 0, respectively. This model used sigma vertical coordinate with 30 layers and the shallowest depth is -5 m and deepest is -2734 m. Vertical depth references is mean sea level and no wet and dry scenarios used. We used 3 momentum open boundary (north, south, and east, see figure 4) with tidal elevation as major forcing input and coupled with atmospherics data from ECMWF as mention in Section 2. No nesting model used and no river discharge in domain model used. Time step was calculated using Courant–Friedrichs–Lewy condition (CFL) [32] and obtained 123.32 second. We plotted ocean current pattern 3-months averages based on ONI for 3 scenarios (El Niño, La Niña, and Normal) and so do for SST. Also we plotted temperatures vertical profiles and compared them to see upper threshold and lower threshold for thermocline layer.

Table 1. List of variables

Variables	Description	Dimensions
u	Velocity$x$-direction	m s^{-1}
υ	Velocity y-direction	m s^{-1}
Ω	Velocity s-direction	m s^{-1}
s	Vertical sigma coordinate	–
z	Vertical elevation	m
η	Wave averaged free surface elevation	m
D	Total water depth ($D = h + \eta$)	m
h	Depth below mean sea level of the sea floor	m
H_z	Grid cell thickness	m
f	Coriolis parameter	s^{-1}
u'	Turbulent velocity x-direction	m s^{-1}
v'	Turbulent velocity y-direction	m s^{-1}
w'	Turbulent velocity s-direction	m s^{-1}
c'	Turbulent concentration (temperature, salt, or suspended-sediment concentration)	°C, salinity, or kg m^{-3}
p	Pressure	N m^{-2}
ρ	Total density of seawater	kg m^{-3}
ρ_0	Reference density of seawater	kg m^{-3}
g	Gravity	m s^{-2}
v	Tracer kinematic viscosity	m^2 s^{-1}
v_θ	Tracer kinematic diffusivity	m^2 s^{-1}
C	Tracer (temperature, salt, or suspended-sediment concentration)	°C, salinity, or kg m^{-3}
C_{source}	Tracer source/sink term	C units m s^{-1}

4. Results

4.1 Verification Model

Figure 4 showed a tidal verification between observation data (red line) and ROMS model results (blue line). It showed that model still under-estimated with observation data have higher value than output model. This verification is still good with pattern of tidal is match and RMSE value is about 0.148 m and MAPE is around 12.22 %.

4.2 Surface Current Circulation in Makassar Strait

Ocean pattern circulation in Makassar Strait was influenced mostly by wind, tidal forcing, and gradient pressure difference between West Pacific Ocean and East Indian Ocean. Figure 5 (a), (b), (c), and (d) showed ocean current pattern seasonal depth-averaged, January February March (JFM), April May June (AMJ), July August September (JAS), October November December (OND), respectively, for 2011 La Niña condition. Figure 6 (a), (b), (c), and (d) showed ocean current pattern seasonal depth-averaged, JFM, AMJ, JAS, OND, respectively, for 2013 Normal condition. Figure 7 (a), (b), (c), and (d) showed ocean current pattern seasonal depth-averaged, JFM, AMJ, JAS, OND, respectively, for 2015 El Niño condition. And figure 8 (a), (b), and (c) is ocean current pattern for yearly averaged for 2011, 2013, 2015, respectively.

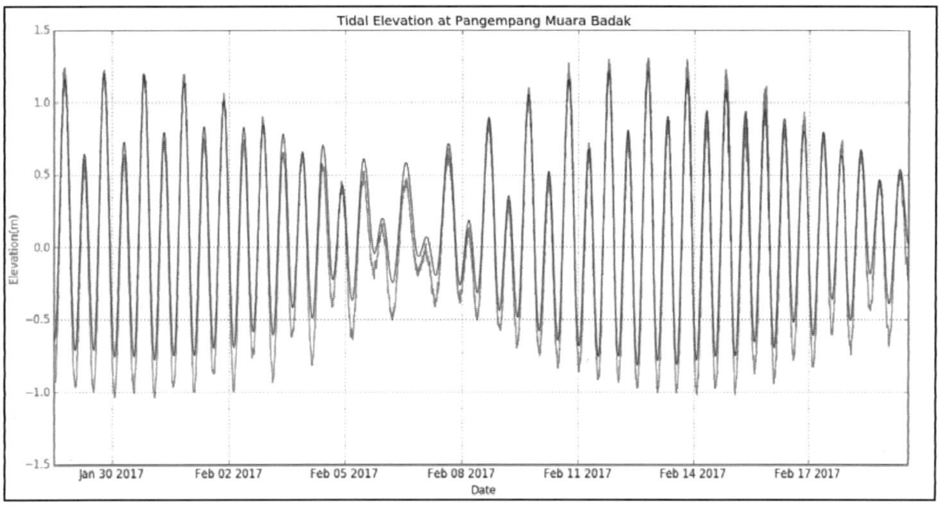

Figure 4. Model verification between tide observation data and ROMS results at Pangempang Muara Badak, west part of domain model. Observation data showed by red line and model result showed by blue line.

Model results showed that current always move along from north part to the south throughout the year. In July, August, and September with ONI is -0.7 at 2011 (weak La Niña condition) (figure 5 (c)), average current velocity is about 0.72 ms^{-1}. In April, May, June, where ONI value is -0.3 at normal condition, figure 6 (b), average current velocity is about 0.88 ms^{-1}. And when weak El Niño (January, February, March) with ONI value is 0.6, model showed ocean current velocity is about 0.64 ms^{-1}. It showed that velocity when La Niña is slower than El Niño condition but still less powerful than normal condition. When strong La Niña happens with ONI -1.1, depth-average value of velocity is about 1.36 ms^{-1}. And when normal condition at October, November, December with ONI -0.2, average velocity is about 0.89 ms^{-1}. And maximum velocity showed at strong El Niño condition with ONI -2.5 at ONJ with value 1.56 ms^{-1} (darker green in figure 7 (d)). Current velocity showed also that higher velocity is placed at deeper bathymetry thus far from north part of Makassar Strait to Labani Channel at southern part. Some turbulences induced at west-north part of the model because characteristic of coastal line, in this case at Sangkulirang Bay and shallow bathymetry. Water masses passing trough from west Pacific Ocean to Makassar Strait using narrow gap at Talok Cape and move to Sangkulirang Bay, East Borneo (figure 9). When west monsoon occurs, there are counter current that move from Java Sea to west part of Makassar Strait. This fresher water masses mixed with saltier water masses from Pacific Ocean therefore it became fresher and disappeared at southern part of Makassar Strait.

Table 2. Seasonal depth-averaged maximum current velocity between three scenarios

	Moderate La Nina (2011)		Very Strong El Nino (2015)		Normal (2013)	
	ONI	Velocity (ms^{-1})	ONI	Velocity (ms^{-1})	ONI	Velocity (ms^{-1})
JFM	-1.1	0.66	0.6	0.64	-0.3	0.83
AMJ	-0.5	0.73	1	0.8	-0.3	0.88
JAS	-0.7	0.72	1.8	0.81	-0.4	1.04
OND	-1.1	1.36	2.5	1.56	-0.2	0.89

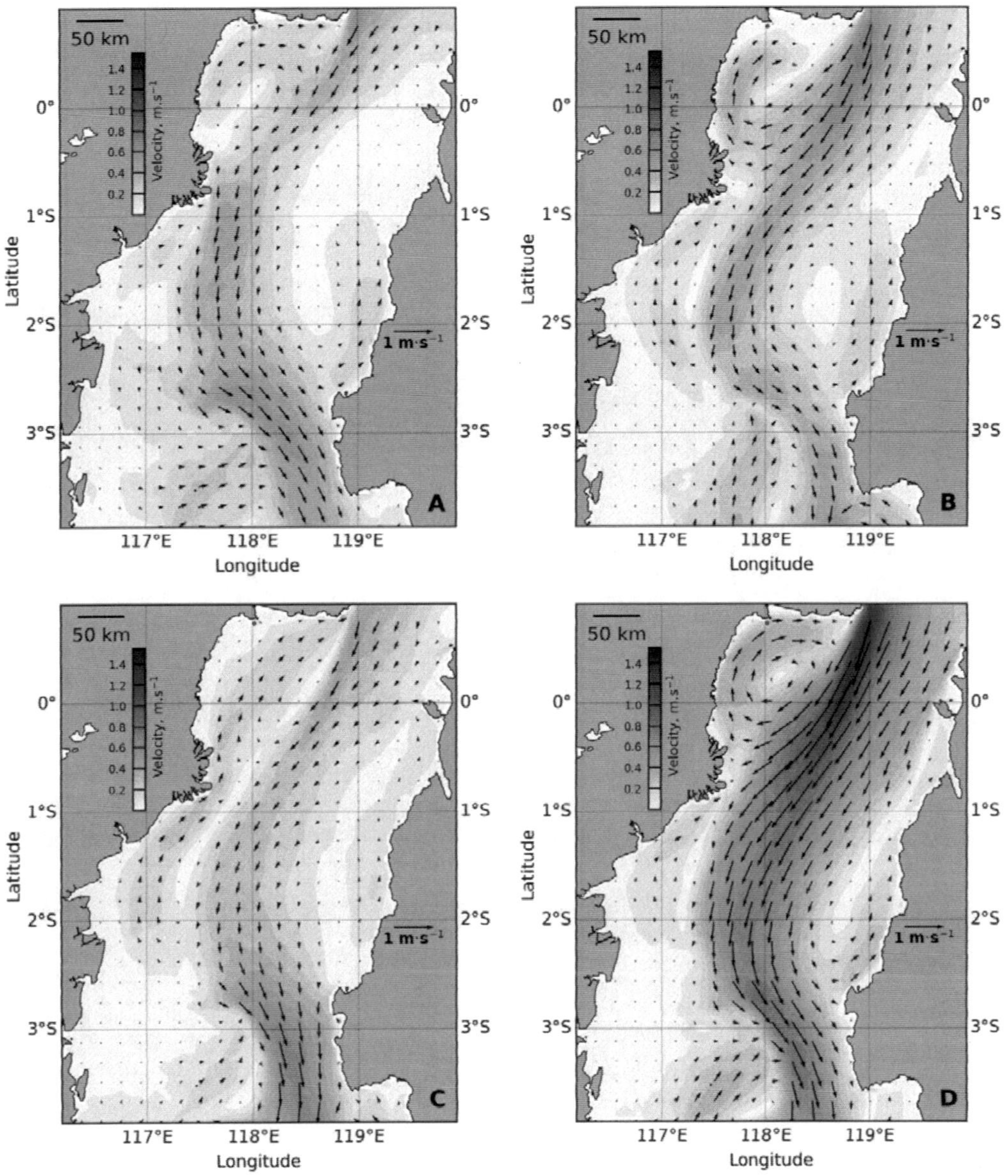

Figure 5. Seasonal depth-averaged ocean current pattern at Makassar Strait during La Niña 2011;(a) JFM, (b) AMJ, (c) JAS, (d) OND.

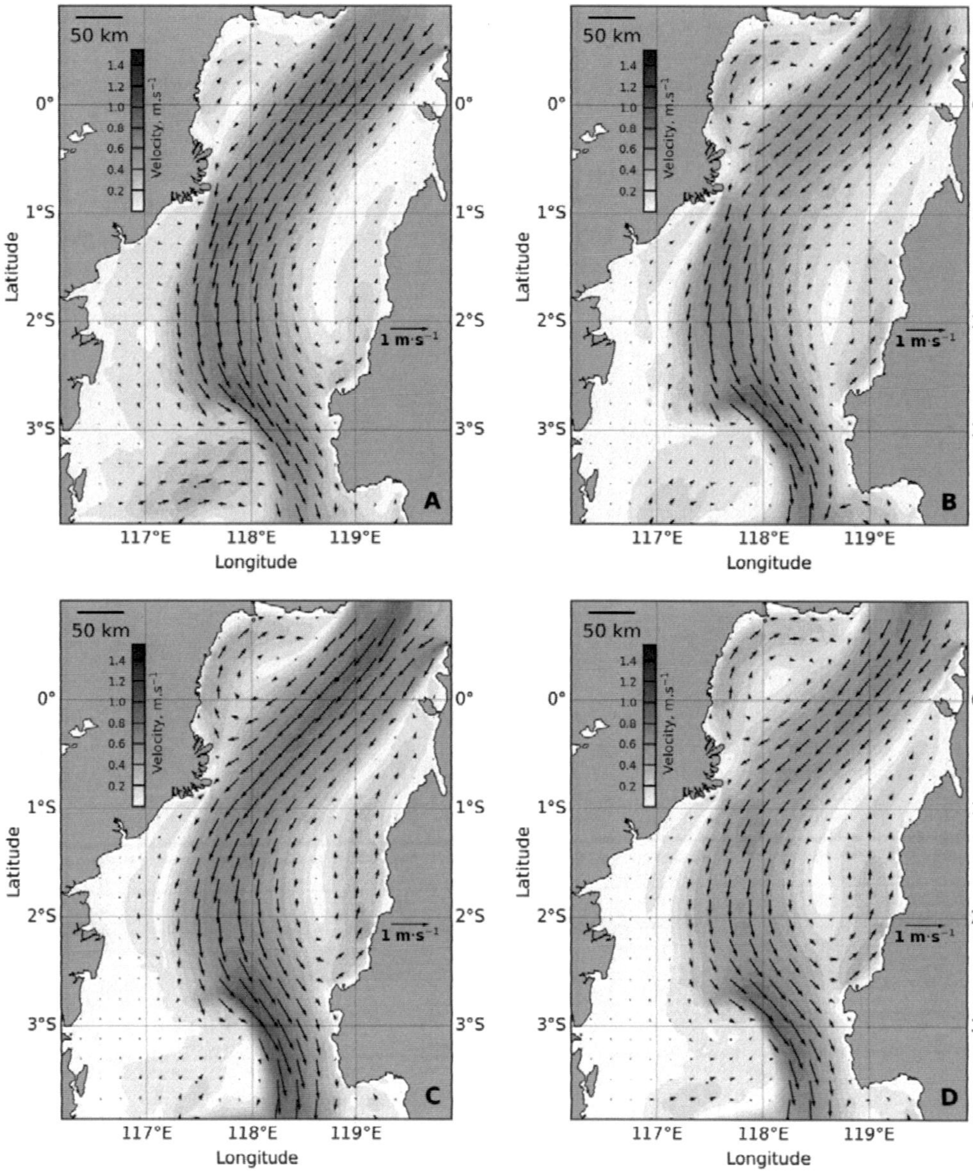

Figure 6. Seasonal depth-averaged ocean current pattern at Makassar Strait during normal condition 2013;(a) JFM, (b) AMJ, (c) JAS, (d) OND.

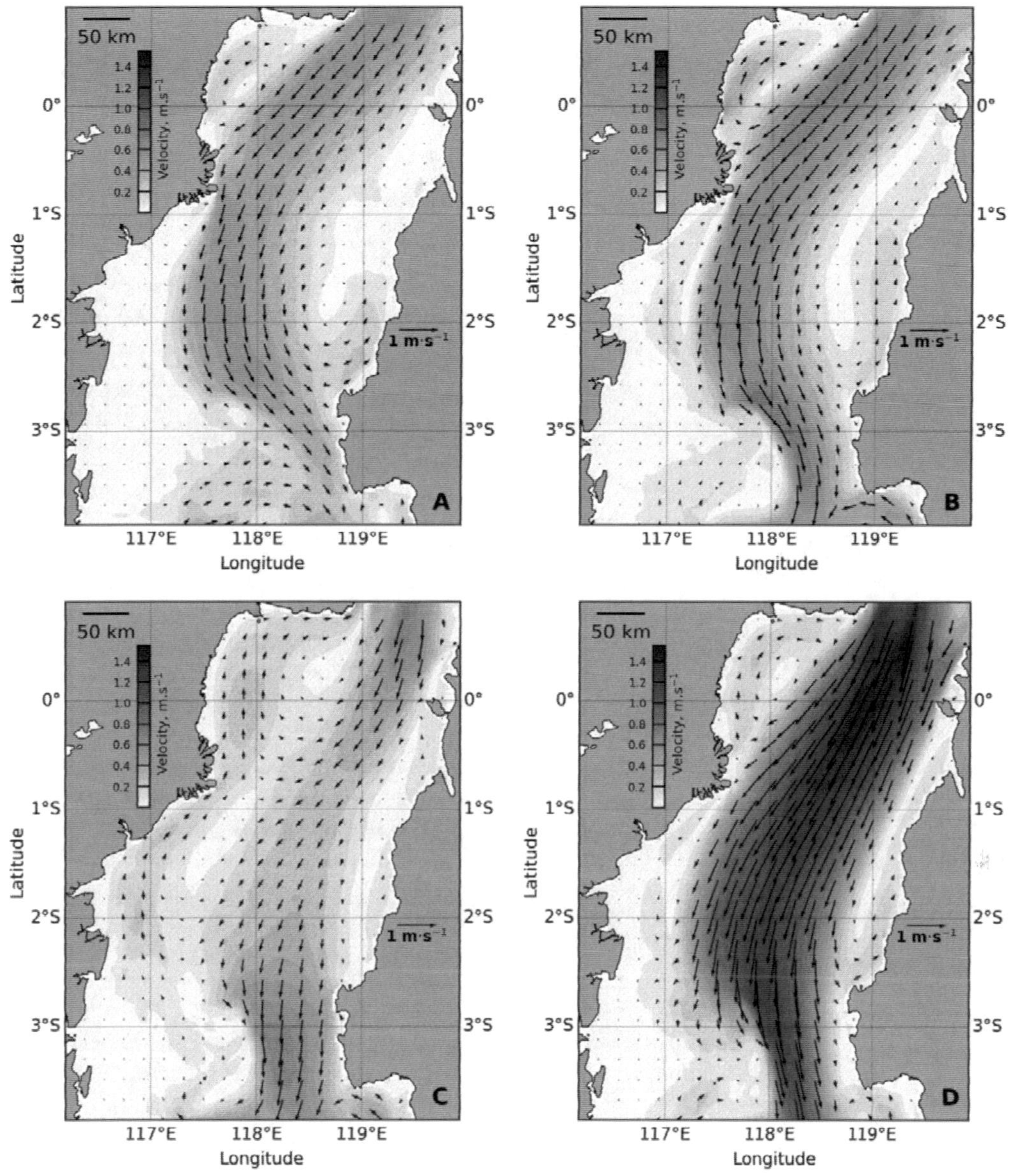

Figure 7. Seasonal depth-averaged ocean current pattern at Makassar Strait during El Niño 2015;(a) JFM, (b) AMJ, (c) JAS, (d) OND.

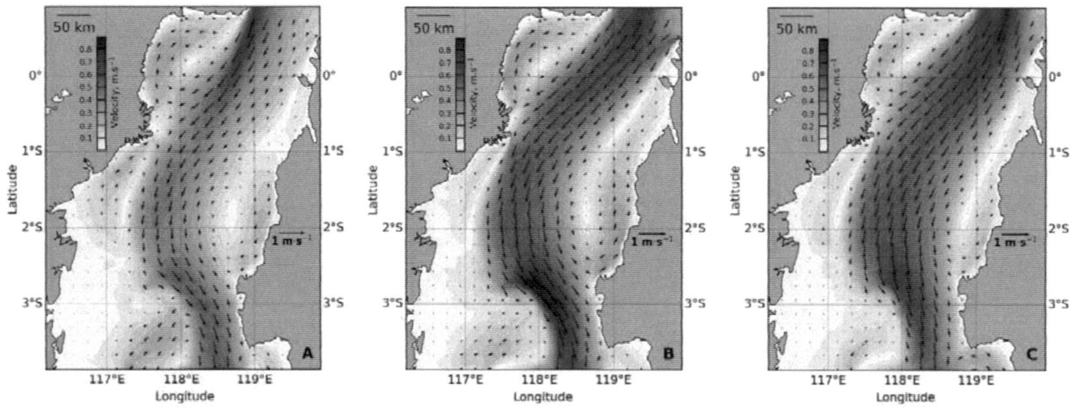

Figure 8. Yearly averaged of ocean current at Makassar Strait between (a) La Niña, (b) normal condition, (c) El Niño.

4.3 SST vs ENSO

Figure 9 (a), (b), (c), and (d) showed sea surface temperature seasonal averaged, January February March (JFM), April May June (AMJ), July August September (JAS), October November December (OND), respectively, for 2011 La Niña condition. Figure 10 (a), (b), (c), and (d) showed SST seasonal averaged, JFM, AMJ, JAS, OND, respectively, for 2013 Normal condition. Figure 11 (a), (b), (c), and (d) showed SST seasonal averaged, JFM, AMJ, JAS, OND, respectively, for 2015 El Niño condition. And figure 12 (a), (b), and (c) is SST and STT anomaly for yearly average for 2011, 2013, 2015, respectively.

Result from model showed that SST varying from 27.20 °C – 30.32 °C during La Niña condition 2011. During normal condition, SST is slightly warmer than La Niña with value varying from 27.5 °C – 31.15 °C. This result is not quite correct because SST supposes to be warmer during La Niña than normal condition with warm pool should move to west Pacific and effecting Makassar Strait water mass. This is caused by model input is not good enough to reproduce a warmer SST during La Niña condition. The atmospheric forcing from peak of La Niña period which is OND on previous year is neglected. Another explanation possible is during La Niña SST supposes to be warmer but peak location of La Niña is do not match in the model area in Makassar Strait possibly will get a delay in SST signal. So, SST during La Niña is colder than normal period which is model result do not explain the realistic condition. Reversely, during El Niño condition, SST is colder than La Niña and normal condition with 25.7 °C – 31.23 °C. There are some area that always colder than any other area where placed at south-east part of the model and believed that was caused by input of atmospheric parameter such as heat flux. A warmer area was found at west part of the model where is water mass from Java Sea passing trough along coastal area to Makassar Strait. Higher SST found when strong La Niña 2011 occurred with ONI = -1, January February March (JFM) period. Colder SST found when weak El Niño 2015 occurred with ONI = 2.2, October, November, December (OND) period.

To see distribution and variability of thermocline depth, we took 7 sample points and plot them in vertical profiles of temperature (figure 13). Seasonal vertical profiles of temperature showed by figure 13 (a) for La Niña, figure 13 (b) for normal condition, and figure 13 (c) for El Niño. We calculated upper threshold (UT) and lower threshold (LT) of thermocline layer using gradient criterion ≥ 0.05 °C/m [21]. This criterion is good to used at South China Sea and around seas with 200 m depth or more [21]. We also calculated thermocline thickness during 3 different conditions with simply minus upper threshold to lower threshold. Thermocline layer at Makassar Strait found between 1.5 m – 391 m with thickness 146.5 m – 373.5 m and temperature decreased from 29.92 °C to 8.44 °C [33]. From model results, thermocline layer found around

53.69 m to 145.34 m with average depth around 91.65 m during La Niña 2011 condition. During El Niño 2015, thermocline layer found around 30.3 m to 114.74 m with average depth around 87.29 m. Table 3 showed variability of thermocline thickness, upper and lower threshold during 3 different conditions. Model result showed that thermocline layer during La Niña condition in Makassar Strait is found deeper than during El Niño condition. During La Niña, the thermocline is stable and very stratified and sea level height is lower than normal over the eastern Pacific, resulting in an increase slope of the ocean surface across the basin. Water masses from west Pacific Ocean pushed down the water masses at Makassar Strait hence the thermocline layer affected by it and became deeper as in line with strong La Niña occurred.

Table 3. Upper threshold, lower threshold, and thickness layer of thermocline during El Niño, La Niña, and normal condition at Makassar Strait.

Condition	JFM	AMJ	JAS	OND
		Upper Threshold		
La Niña (2011)	62.47 m	75.51 m	26.37 m	50.40 m
Normal (2013)	41.47 m	9.52 m	31.78 m	38.44 m
El Niño (2015)	72.39 m	40.17 m	4.77 m	4.46 m
		Lower Threshold		
La Niña (2011)	161.01 m	151.01 m	114.71 m	144.63 m
Normal (2013)	149.44 m	143.09 m	149.45 m	149.44 m
El Niño (2015)	154.99 m	133.09 m	88.21 m	94.67 m
		Thickness		
La Niña (2011)	98.55 m	85.49 m	88.33 m	94.22 m
Normal (2013)	107.98 m	123.56 m	117.66 m	111.00 m
El Niño (2015)	82.61 m	92.92 m	83.44 m	90.20 m

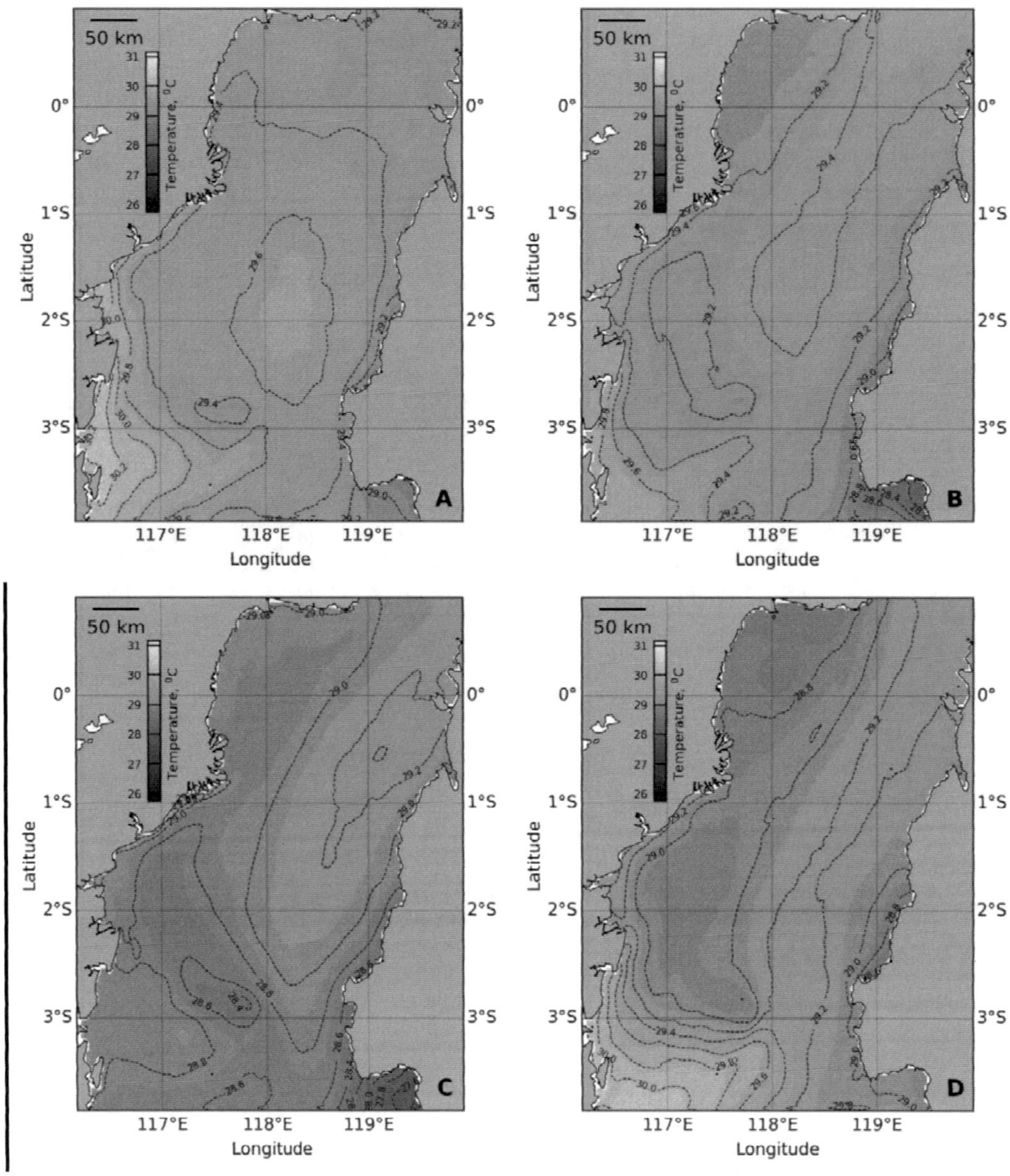

Figure 9. Seasonal averaged sea surface temperature at Makassar Strait during La Niña condition 2011;(a) JFM, (b) AMJ, (c) JAS, (d) OND.

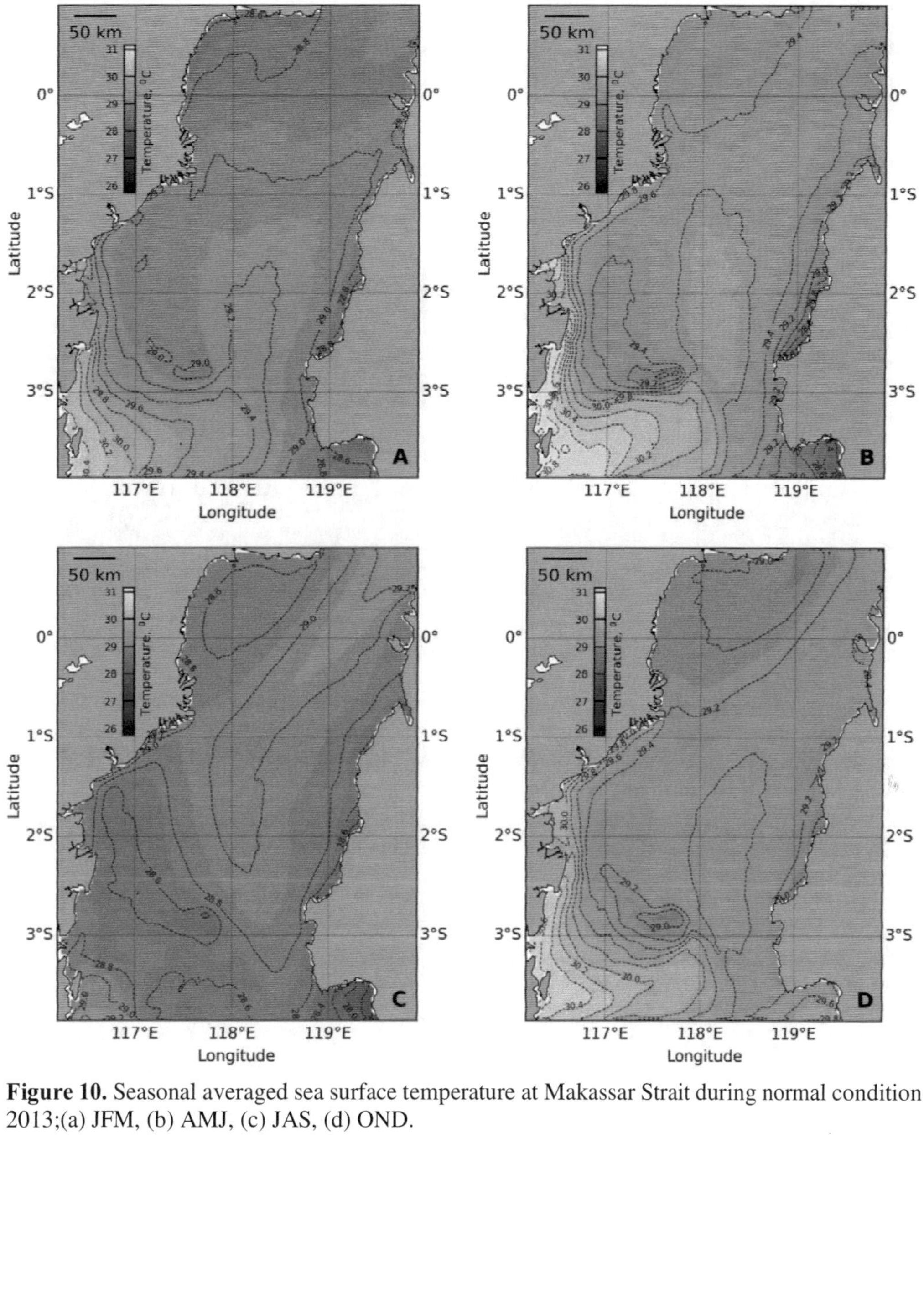

Figure 10. Seasonal averaged sea surface temperature at Makassar Strait during normal condition 2013;(a) JFM, (b) AMJ, (c) JAS, (d) OND.

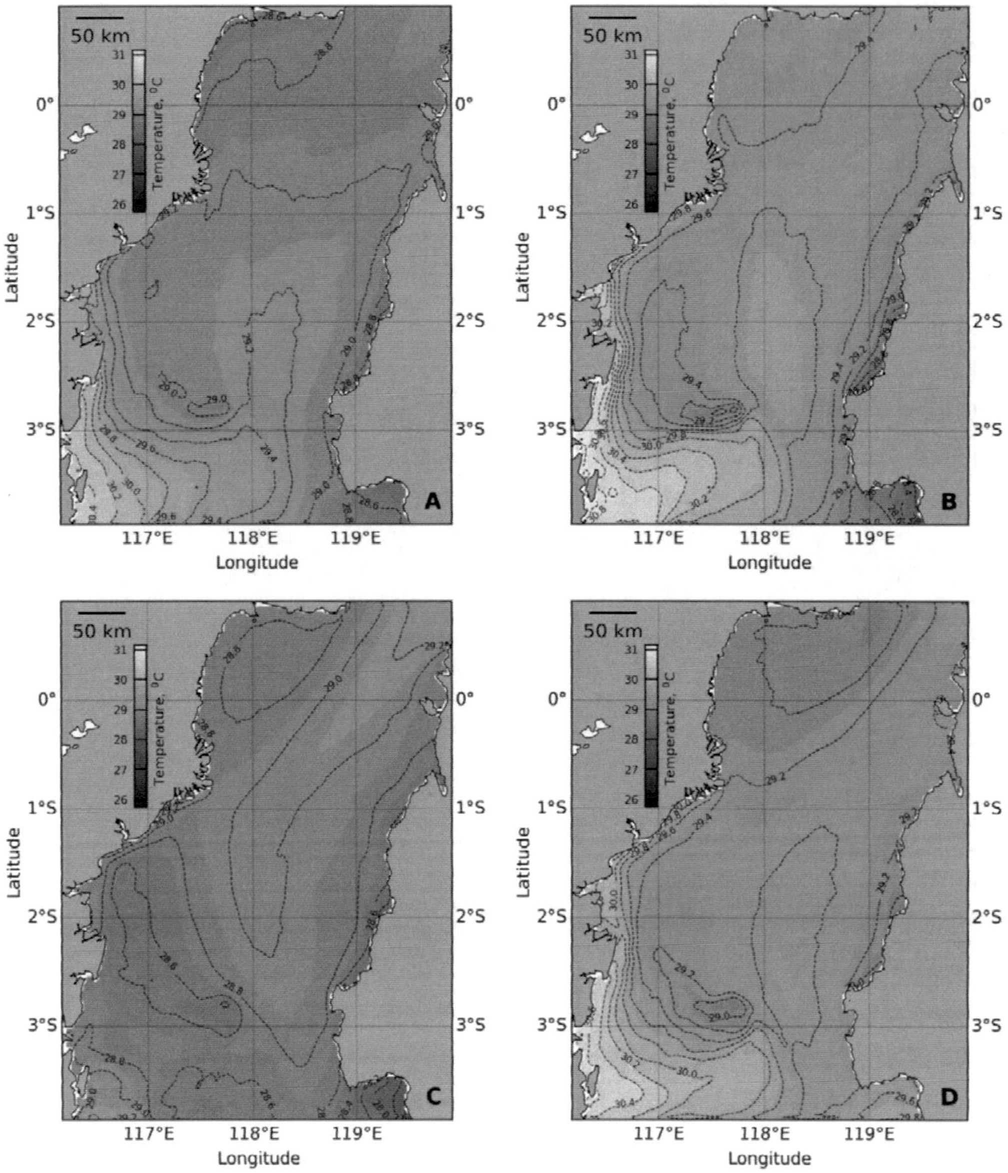

Figure 11. Seasonal averaged sea surface temperature at Makassar Strait during El Niño 2015;(a) JFM, (b) AMJ, (c) JAS, (d) OND.

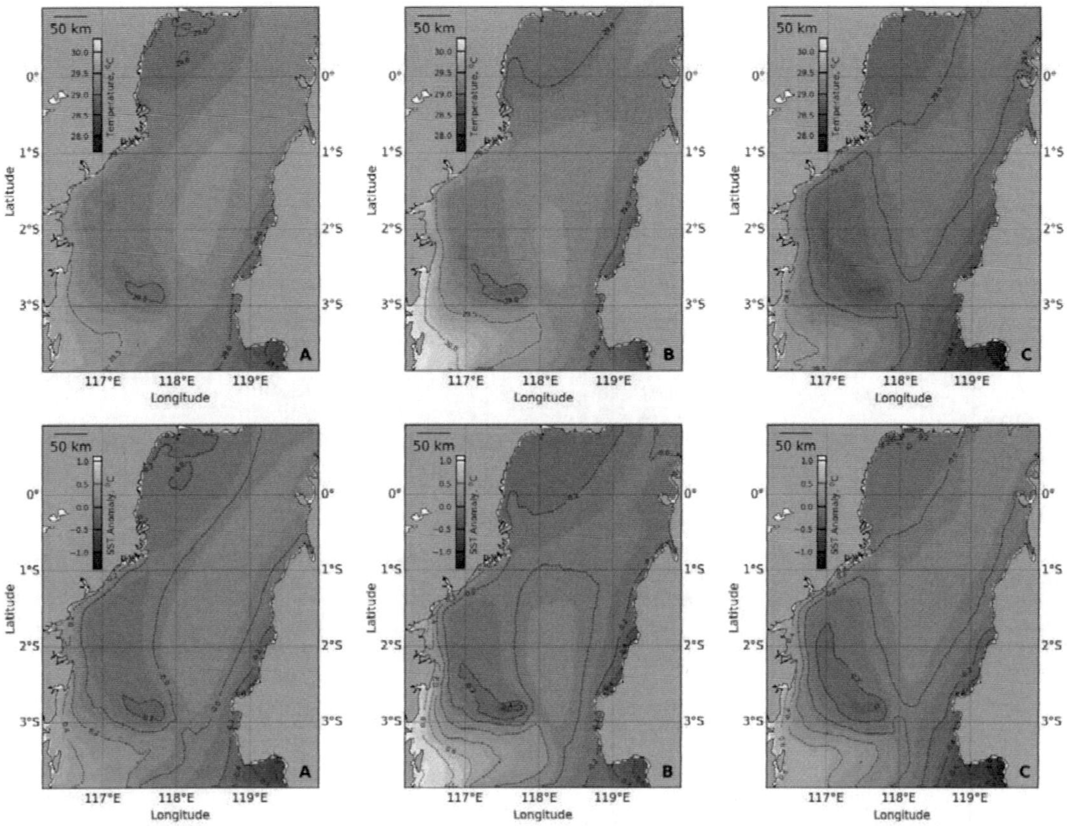

Figure 12. Yearly average of sea surface temperature (top) and SST anomaly (bottom) at Makassar Strait between (a) La Niña, (b) normal condition, (c) El Niño.

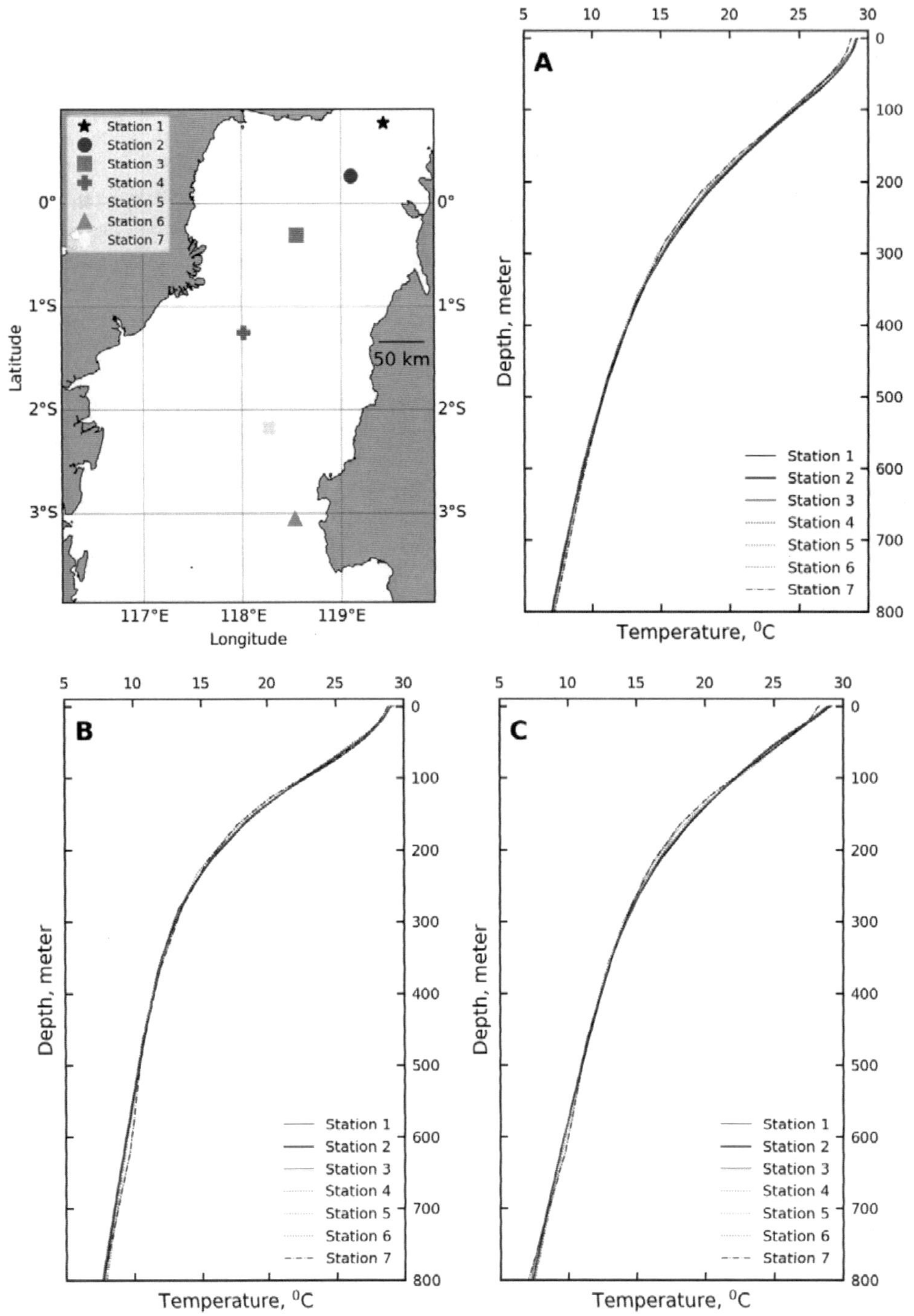

Figure 13. Seven sample points (top left) and annual averaged vertical profiles of temperature from surface to 800 m depth;(a) La Niña, (b) normal year, (c) El Niño

5. Conclusion

This study was an preliminary study of using model to described the effect of ENSO in changes of water mass characteristic and ocean circulation at Makassar Strait. Based on verification of tidal height, model showed a good result, but we cannot determine ocean current pattern and temperature value at Makassar Strait is good enough because lack of observation data of temperature and velocity. Ocean current flows from north to south part of the Makassar Strait. This flow was found stable throughout the year indicating the variability of ITF depends on ENSO phenomena in Pacific Ocean. Current velocity in El Niño is slightly faster than La Niña condition with different is about 0.1 ms^{-1}. Averaged SST showed that during La Niña is higher than El Niño condition but lower than normal condition. The model input is not good enough to reproduce warmer SST during La Niña compared to normal condition. Temperature vertical profiles showed that thermocline layer in La Niña found deeper than El Niño condition.

Acknowledgements

This research was partly supported by National University Operational Funding (BOPTN) 2017, Directorate General of Higher Education, Ministry of Education and Culture, Republic of Indonesia. This research also supported by Hydro-Atmosphere Environmental Research Group, Oceanography and Computational Laboratory, Faculty of Mathematics and Natural Sciences, Mulawarman University. Special thanks to : MSAT Committee for partially funding support. Funding for the development of TPXO OSU Tide has been provided by the National Science Foundation (NSF), the Office of Naval Research and the National Aeronautics and Space Administration (NASA). The output is publicly available at http://volkov.oce.orst.edu/tides/global.html. SRTM15_PLUS Global Bathymetry grid developed at Scripps Institution of Oceanography-University of California San Diego on its website and in its publications. The output is publicly available at http: //topex.ucsd.edu/WWW_html/srtm30_plus. Funding for the development of HYCOM has been provided by the National Ocean Partnership Program and the Office of Naval Research. Data assimilative products using HYCOM and the 1/12 deg global HYCOM+NCODA Ocean Reanalysis was funded by the U.S. Navy and the Modeling and Simulation Coordination Office. Computer time was made available by the DoD High Performance Computing Modernization Program. The output is publicly available at http://hycom.org.

References

[1] Gordon A L 2005 Oceanography of the Indonesian seas and their throughflow *Oceanography* **18** 14-27

[2] Susanto R D and Gordon A L 2005 Velocity and transport of the Makassar Strait throughflow *J. Geophys. Res.* **110** C01005 doi:10.1029/2004JC002425

[3] Ffield A and Gordon A L 1992 Vertical mixing in the Indonesian thermocline *J. Phys. Oceanogr.* **22** 184-195

[4] Sprintall J, Gordon A L, Murtugudde R and Susanto R D 2000 A semiannual Indian Ocean forced Kelvin wave observed in the Indonesian seas in May 1997 *J. Geophys. Res.* **105** (C7) 17 217-17 230 doi: 10.1029/2000JC900065

[5] Meyers G 1996 Variation of Indonesian Throughflow and the El Niño-Southern Oscillation. *J. Geophys. Res.* **101** 12 255-12 263

[6] Koch-Larrouy A, Lengaigne M, Terray P, Madec G and Masson S 2010 Tidal mixing in the Indonesian Seas and its effect on the tropical climate system *Climate Dynamics* **34** 6 891-904

[7] Hu S and Sprintall J 2016 Interannual variability of the Indonesian Throughflow: the salinity effect, *J. Geophys. Res. Oceans* **121** 2596–2615 doi:10.1002/2015JC011495

[8] Warner J C, Perlin N and Skyllingstad E 2008 Using the model coupling toolkit to couple earth system models *Environ. Model Softw.* **23** 1240–49

[9] Warner J C, Armstrong B, He R and Zambon J B 2010 Development of a coupled ocean atmosphere–wave–sediment transport (COAWST) modeling system *Ocean Model* **35** 230–44 doi:10.1016/j.oceanmod.2010.07.010
[10] Zambon J, He R and Warner J 2014 Investigation of hurricane Ivan using the coupled ocean atmosphere- wave-sediment transport (COAWST) model *Ocean Dynamics* **64** 1535–54
[11] Egbert G D, Bennett A F and Foreman M G G 1994 TOPEX/Poseidon tides estimated using a global inverse model *J. Geophys. Res.* **99** 24821–52
[12] Egbert G D and Erofeeva S Y 2002 Efficient inverse modeling of barotropic ocean tides *J. Atmos. Ocean Technol.* **19**(2) 183–204
[13] Liu T W, Katsaros K B, and Businger J A 1979 Bulk Parameterization of air-sea exchange of heat and water vapor including the molecular constraints at the interface, *J. Atmos. Sci.* **36**: 1722-1735
[14] Farr T G, Rosen P A, Caro E, Crippen R, Duren R, Hensley S, Kobrick M, Paller M, Rodrigues E, Roth L *et al* 2007 The shuttle radar topography mission *Rev. Geophys* **45** (RG2004)
[15] Becker J J, Sandwell D T, Smith W H F, Braud J, Binder B, Depner J, Fabre D, Factor J, Ingalls S, Kim S-H *et al* 2009 Global bathymetry and elevation data at 30 arc seconds resolution: SRTM30_PLUS *Marine Geodesy* **32**:4 355 — 71
[16] Sandwell D T, Müller R D, Smith W H F, Garcia E and Francis R 2014 New global marine gravity model from CryoSat-2 and Jason-1 reveals buried tectonic structure *Science* **346** (6205) 65-67 doi:10.1126/science.1258213
[17] Smith W H F and Sandwell D T 1997 Global seafloor topography from satellite altimetry and ship depth soundings *Science* **277** 1957-62
[18] Jakobsson M, Grantz A, Kristoffersen Y and Macnab R 2003 Physiographic provinces of the Arctic Ocean *GSA Bulletin* **115** (12) 1443–55
[19] Huang B, Banzon V F, Freeman E, Lawrimore J, Liu W, Peterson T C, Smith T M, Thorne P W, Woodruff S D and Zhang H M 2014 Extended reconstructed sea surface temperature version 4 (ERSST.v4): part I. upgrades and intercomparisons *Journal of Climate* **28** 911-30 doi:10.1175/JCLI-D-14-00006.1
[20] Liu W, Huang B, Thorne P W, Banzon V F, Zhang H M, Freeman E, Lawrimore J, Peterson T C, Smith T M and Woodruff S D 2014 Extended reconstructed sea surface temperature version 4 (ERSST.v4): part II. parametric and structural uncertainty estimations *Journal of Climate* **28** 931–51 doi:10.1175/JCLI-D-14-00007.1
[21] Huang B, Thorne P, Smith T, Liu W, Lawrimore J, Banzon V, Zhang H, Peterson T and Menne M 2015 Further exploring and quantifying uncertainties for extended reconstructed sea surface temperature (ERSST) Version 4 (v4) *Journal of Climate* **29** 3119–42 doi:10.1175/JCLI-D-15-0430.1
[22] Hao J, Chen Y, Wang F and Lin P 2012 Seasonal thermocline in the China Seas and northwestern Pacific *Ocean J. Geophys. Res.* **117** C02022 doi:10.1029/2011JC007246
[23] Abamowitz M and Stegun A 1965 *Handbook of Mathematical Function* Dover New York
[24] Haidvogel D B, Arango H G, Hedström K S, Beckmann A, Malanotte-Rizaoli P and Shchepetkin A F 2000 Model evaluation experiments in the North Atlantic Basin simulations in nonlinear terrain-following coordinates *Dynam. Atmos. Ocean.* **32** 239-281
[25] Shchepetkin A F and McWilliams J C 2005 The regional oceanic modeling system (ROMS) a split-explicit, free-surface, topography-following-coordinate oceanic model*Ocean Model.* **9**347-404
[26] Warner J C, Sherwood C R, Signell R P, Harris C K and Arango H G 2008 Development of a three-dimensional, regional, coupled wave, current, and sediment-transport model *Comput. Geosci.* **34** 1284-1306

[27] Large W G, McWilliams J C and Doney S C 1994 A review and model with a nonlocal boundary layer parameterization *Rev. Geophys.* **32** 363-403

[28] Durski S M, Glenn S M and Haidvogel D B 2004 Vertical mixing schemes in the coastal ocean: comparison of the level 2.5 Mellor-Yamada scheme with an enhanced version of the K profile parameterization *J. Geophys. Res.* **109** C01015

[29] Mellor G L and Yamada T 1982 Development of a turbulence closure model for geophysical fluid problems *Rev. Geophys. Space. Phys.* **20** 851-875

[30] Umlauf L and Burchard H 2003 A generic length-scale equation for geophysical turbulence models *J. Mar. Res.* **61** 235-265

[31] Warner J C, Sherwood C R, Arango H G and Signell R P 2005 Performance of four turbulance closure models implemented using a generic length scale method *Ocean Model.* **8** 81-113

[32] Courant R, Friedrichs K and Lewy H 1928 *Math. Ann.* **100**: 32. https://doi.org/10.1007/BF01448839

[33] Horhoruw S, Atmadipoera A, Purba M and Purwandana A 2015 Current structure and spatial variation of indonesian throughflow in makassar strait under ewin 2013 (struktur arus dan variasi spasial arlindo di selat makassar dari ewin 2013) *ILMU KELAUTAN: Indonesian Journal of Marine Sciences* **20** (2) 87-100 doi:http://dx.doi.org/10.14710/ik.ijms.20.2.87-100

The changes of water mass characteristics using 3-dimensional Regional Ocean Modeling System (ROMS) in Balikpapan bay, Indonesia

A A Nur, I Mandang, S Mubarrok and M Riza

Hydro-Atmosphere Environment Research Group, Physical Oceanography and Computational Modeling Laboratory, Program Study of Physics, Faculty of Mathematics and Natural Sciences, Mulawarman University, Samarinda, Indonesia

E-mail: ashadinur@fmipa.unmul.ac.id

Abstract. Balikpapan Bay is located in the East Borneo Province of Borneo, Indonesia and directly connected to the Makassar Strait and also has very important role in determining the dynamics of ocean current, heat and freshwater content in the Makassar Strait. Numerical model has been used to simulate the dynamic process such as the changes of water mass in the Balikpapan Bay. We have used the terrain following three-dimensional Regional Ocean Modeling System (ROMS). Tides, river discharges, and atmospheric forcing (surface fields) have been used as generating forces for the model. The model has been simulated for three months from 1 October 2012 to 1 January 2013. The model has been validated by computing the *RMSE, MAPE* and Willmott's index of agreement (d) using water level observation at the sampling station in Semayang port, Balikpapan. The verification showed a good agreement between model prediction and field observation data with *RMSE* = 7.8 cm, *MAPE* = 14.3% and d = 0.995 (with a value of perfect agreement equals to 1.0). The results analysed using vertical profile of salinity and temperature from 22 sampling stations with 11 stations located at the inner part of Balikpapan Bay. While the other stations located outside the Balikpapan Bay were used to analyse watermass. Water circulation is mostly dominated by the forcing from tides. The model results showed that the outer part of Balikpapan Bay is saltier than the inner part of Balikpapan bay. The highest average temperature occurred along coastal areas with value ~31 ^{0}C. The distribution of vertical salt transport showed that the water masses tend to be stratified during neap tides and mixed during spring tides condition.

1. Introduction

The Indonesian seas are the tropical regional sea that lies between two major oceans, Pacific and Indian oceans and also known as the Indonesian throughflow (ITF) because they provide a pathway for the Pacific and Indian inter-ocean exchange. They also the only tropical pathway global thermohaline circulation [1]. Makassar Strait is one of the Indonesian throughflow that has been known has major impact to the global ocean circulation. From the research [2], it reveals that the Pacific and Indian ocean heat and freshwater budgets have depended on the Indonesian sea throughflow. Heat and freshwater flux into the Indian ocean at the expense of the Pacific may affect tropical sea surface temperature (SST) pattern and air-sea coupling with potential impact on the *El Nino Southern Oscillation* (ENSO) monsoon phenomena [3]. Observation data also showed that the throughflow is composed mostly of North Pacific thermocline and intermediate water flowing through Makassar Strait [4].

In the middle part of Makassar Strait, there are two areas in the Borneo island that have become periodic sources of the heat and freshwater flux which always input these fluxes straight into the strait, they were Mahakam delta and Balikpapan bay. These two areas geologically have been formed in the different process and produced different coastal morphology respectively. Mahakam delta is an active delta system which has been formed by large tidal amplitude, low wave energy, and large fluvial input [5]. Meanwhile, Balikpapan Bay is semi-enclosed bay and strongly affected by waves distribution in the southern part of the bay. By its location, Mahakam Delta and Balikpapan Bay have their respective roles to the current and transport pattern in the Makassar Strait

Balikpapan Bay is located in the East Borneo Province of Borneo, Indonesia (figure 1) and is one of the bay that directly connected to Makassar Strait. That means, the hydrological conditions in the bay was strongly influenced by the Makassar Strait. Periodically, the freshwater inputs moving through the bay from the rivers in the north, west and east to the downstream area in the southern part of the bay, and at the same time the seawater from the strait comes into the bay and affecting the heat and salinity vertical structure of water masses in the Balikpapan bay. The width of the bay channels varying significantly and have ranges from ~ 6 km in the mouth of the bay to <1 km on upstream side. With a length of channel about >40 km, there are many possibilities of dynamical process that can be happened along the bay channel, like partial mixing, salt intrusion and sediment transport from the rivers.

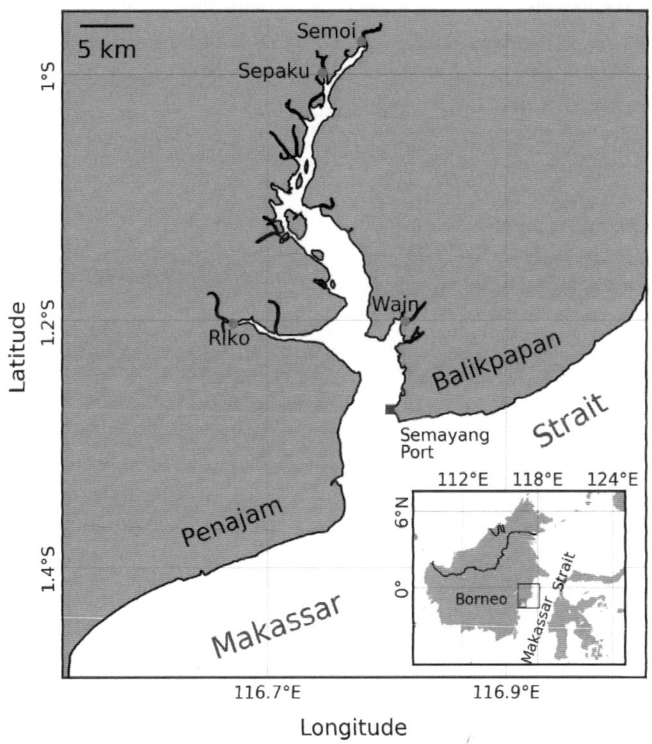

Figure 1. Map of Balikpapan Bay, East Borneo, Indonesia. The blue square denotes the water level station at Semayang port and red dot denotes the source of river discharges for the model. In the inset map, the blue line square indicates a parent domain (ECB) and red line square indicates a child domain (BPP).

Based on introduction above, we aim to know the effect of tidal mixing alternated water mixing column between freshwater from Balikpapan Bay and saltier water from Makassar Strait. We also like to know how temperature and salinity transform between two areas, inner part and outer part of Balikpapan Bay. Thus, in this research we have objective to studied the changes of water mass characteristics because of exchanges of heat and freshwater flux from Balikpapan bay into the Makassar Strait and vice versa.

To see its changes, we used the Regional Ocean Modeling System (ROMS) from COAWST modeling system distribution. COAWST is stands for Coupled Ocean Atmosphere Waves Sediment Transport and it is an agglomeration of open-source modeling components that has been tailored to investigate coupled processes of the atmosphere, ocean, and waves in the coastal ocean [6]. Further, we also planned to modified a high resolution ROMS model that could describe many physical phenomena at Makassar Strait especially Balikpapan Bay.

2. Data

2.1. *Field Measurement*

Hourly time-series of sea surface elevation data were conducted by *Geospatial Information Agency* (http://www.bakosurtanal.go.id) during October 2012 until December 2012. This data was measured using a tide gauge instrument to monitoring water level and deployed at the port of Semayang, Balikpapan that was located in the southernmost part of Balikpapan or south-eastern part of Balikpapan Bay. This data will be compared to the model prediction for the same variable to measure performance of the model. River discharges were obtained from *The Erosion and Sedimentation Working Group* (Kelompok Kerja Erosi dan Sedimentasi) which conducted the field observation in the watershed of Balikpapan Bay. Unfortunately, it has only a month-averaged data which made us use repetitive value for river discharges input. River discharges data are listed in detail at **Table 1**. We also could not find observation data for temperature and salinity at the model domain on this period. The closest observation point for those data is only Makassar mooring (INSTANT program), but it was located outside of the study area.

Table 1. River discharges data

River	Discharges (m^3s^{-1})
Wain	2.477
Semoi	83.496
Sepaku	42.189
Riko	16.852

2.2. *Model Input*

Usually, the data input demanded by ROMS consists of: bathymetry, temperature, salinity, velocity, sea surface elevation, atmospheric (surface) forcing and river discharge. In this research, bathymetry data was obtained from the high-resolution topography SRTM15 PLUS (http://www.topex.ucsd.edu) and has 15 arcsecond resolution or ~450 m. This topographic data was based on SRTM30_PLUS grid with some improvements. The Hybrid Coordinate Ocean Model (http://www.hycom.org) analysis system has been used for the two-dimensional and three-dimensional components (sea surface elevation, temperature, salinity and velocity). This system uses the Navy Coupled Ocean Data Assimilation (NCODA) and has 1/12 arcdegree grid resolution. The surface (atmospheric) forcing datasets obtained from the ECMWF (European Centre of Medium-Range Weather Forecast datasets (http://www.apps.ecmwf.int). The ECMWF dataset is chosen because the data is available on the study area (Balikpapan Bay). Tidal amplitude was obtained from global ocean tides model TPXO8-ATLAS (http://www.volkov.oce.orst.edu), and have a better resolution in the research domain (1/30 degree of resolution) compared to other solution of tide models. Unlike the previous model (TPXO7.2) solutions, the new tides model keeps resolution in the higher resolution rather than averaging them on coarser grid (TPXO grid). As a result, this provides much improvement in tidal predictions for coastal areas.

3. Method

The simulation of hydrodynamic process at Balikpapan Bay coastal domain were conducted by using the numerical model of Regional Ocean Modeling System (ROMS). Many literatures [7][8][9] have

IOP Conf. Series: Earth and Environmental Science **162** (2018) 012006 doi:10.1088/1755-1315/162/1/012006

described the model capability, especially in the regional ocean domain. ROMS itself is a three-dimensional, free surface, terrain-following numerical model that solves finite difference approximation of the Reynolds-averaged Navier-Stokes (RANS) equation using the hydrostatic and Boussinesq assumption with a split-explicit time stepping algorithm [10][11]. It uses a horizontal curvilinear Arakawa "C" grid and vertical stretched terrain-following coordinates [12]. This model also can be configured depending of user application which has several choices for advection schemes, pressure gradient algorithms, turbulent closure, wet-dry, and several types of boundary conditions.

3.1. Governing Equations

The governing equations used in ROMS were presented in flux form on the Cartesian horizontal coordinates and sigma vertical coordinates. For the momentum equations on the x- and y- axis directions are:

$$\frac{\partial(H_z u)}{\partial t}+\frac{\partial(uH_z u)}{\partial x}+\frac{\partial(\upsilon H_z u)}{\partial y}+\frac{\partial(\Omega H_z u)}{\partial s}-fH_z\upsilon=-\frac{H_z}{\rho_0}\frac{\partial p}{\partial x}-H_z g\frac{\partial \eta}{\partial x}$$
$$-\frac{\partial}{\partial s}\left(\overline{u'w'}-\frac{\nu}{H_z}\frac{\partial u}{\partial s}\right)-\frac{\partial(H_z S_{xx})}{\partial x}-\frac{\partial(H_z S_{xy})}{\partial y}+\frac{\partial S_{px}}{\partial s} \quad (1)$$

$$\frac{\partial(H_z \upsilon)}{\partial t}+\frac{\partial(uH_z \upsilon)}{\partial x}+\frac{\partial(\upsilon H_z \upsilon)}{\partial y}+\frac{\partial(\Omega H_z \upsilon)}{\partial s}+fH_z u=-\frac{H_z}{\rho_0}\frac{\partial p}{\partial y}-H_z g\frac{\partial \eta}{\partial y}$$
$$-\frac{\partial}{\partial s}\left(\overline{\upsilon'w'}-\frac{\nu}{H_z}\frac{\partial v}{\partial s}\right)-\frac{\partial(H_z S_{yx})}{\partial x}-\frac{\partial(H_z S_{yy})}{\partial y}+\frac{\partial S_{py}}{\partial s} \quad (2)$$

$$0=-\frac{1}{\rho_0}\frac{\partial p}{\partial s}-\frac{g}{\rho_0}H_z\rho \quad (3)$$

with the continuity equation:

$$\frac{\partial \eta}{\partial t}+\frac{\partial(H_z u)}{\partial x}+\frac{\partial(H_z \upsilon)}{\partial y}+\frac{\partial(H_z \Omega)}{\partial s}=0 \quad (4)$$

and scalar transport:

$$\frac{\partial(H_z C)}{\partial t}+\frac{\partial(uH_z C)}{\partial x}+\frac{\partial(\upsilon H_z C)}{\partial y}+\frac{\partial(\Omega H_z C)}{\partial s}=-\frac{\partial}{\partial s}\left(\overline{c'w'}-\frac{\nu_\theta}{H_z}\frac{\partial C}{\partial s}\right)+C_{source} \quad (5)$$

These equations are closed by parameterizing the Reynolds stresses and turbulent tracer fluxes as

$$\overline{u'w'}=-K_M\frac{\partial u}{\partial z},\quad \overline{v'w'}=-K_M\frac{\partial v}{\partial z},\quad \overline{\rho'w'}=-K_M\frac{\partial \rho}{\partial z} \quad (6)$$

where K_M is the eddy viscosity for momentum and K_H is the eddy diffusivity. Eddy viscosities and eddy diffusivities are calculated using one of five options for turbulence-closure models in ROMS: (i) Brunt-Väisälä frequency mixing in which mixing is based on the stability frequency; (ii) a user-provided analytical expression such as a constant or parabolic shape; (iii) the *K*-profile parameterization [13],

IOP Conf. Series: Earth and Environmental Science **162** (2018) 012006 doi:10.1088/1755-1315/162/1/012006

expanded to include both surface and bottom-boundary layers [14]; (iv) Mellor-Yamada level 2.5 (MY2.5) method [15]; and (v) the generic length-scale (GLS) method [16] as implemented in [17] that also includes the option for surface fluxes of turbulent kinetic energy due to wave breaking. For the details of each variables on the equations are listed in **Table 2**.

Table 2. List of variables

Variables	Description	Dimensions
u	Velocity x-direction	m s^{-1}
υ	Velocity y-direction	m s^{-1}
Ω	Velocity s-direction	m s^{-1}
s	Vertical sigma coordinate	–
z	Vertical elevation	m
η	Wave averaged free surface elevation	m
H_z	Grid cell thickness	m
f	Coriolis parameter	s^{-1}
u'	Turbulent velocity x-direction	m s^{-1}
v'	Turbulent velocity y-direction	m s^{-1}
w'	Turbulent velocity s-direction	m s^{-1}
c'	Turbulent concentration	°C, salinity, or kg m^{-3}
p	Pressure	N m^{-2}
ρ	Total density of seawater	kg m^{-3}
ρ_0	Reference density of seawater	kg m^{-3}
g	Gravity	m s^{-2}
v	Tracer kinematic viscosity	m^2 s^{-1}
v_θ	Tracer kinematic diffusivity	m^2 s^{-1}
C	Tracer (temperature, salt, or suspended-sediment concentration)	°C, salinity, or kg m^{-3}
C_{source}	Tracer source/sink term	C units m s^{-1}

3.2. Model scenario

Single grid refinement method has been used to increase resolution in Balikpapan bay coastal domain. It started from the coarser grid (referred as the "ECB" model) with horizontal resolution ~750 m and downscaling it to the area of interest (referred as the "BPP" model) with ~150 m of horizontal resolution. The ECB domain include more than 200 km eastern coastline of Borneo island extending from south of Balikpapan Bay to the north of Bontang city, whereas BPP domain only focusing at the Balikpapan bay area from south coastline of Penajam regency to the east of Balikpapan city. These model domains have different open boundaries respectively, the ECB model domain has open boundary at the north, south, and east section of the domain, whereas the BPP model domain only just has open boundary at the south and north section (figure 2).

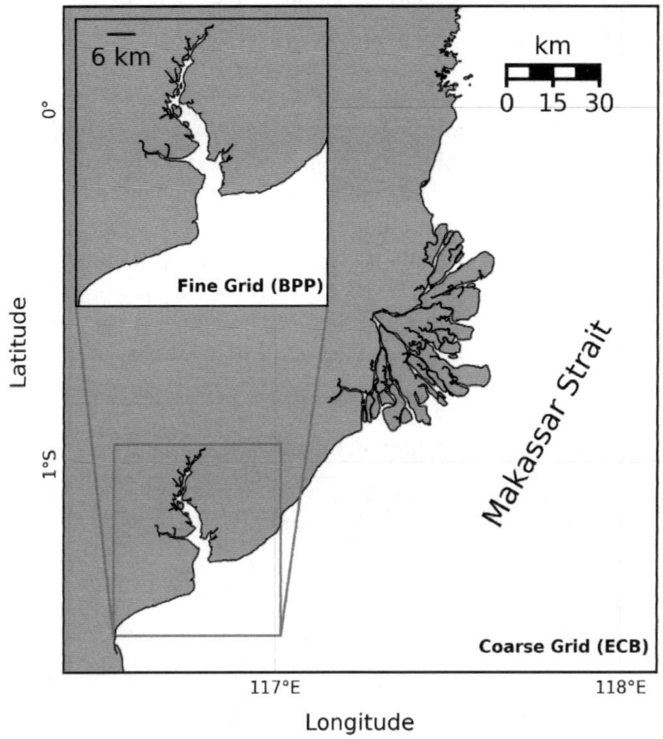

Figure 2. Model domain with nested grids (refinement ratio 1:5)

Bathymetric data for the BPP model grid was obtained using interpolation method by ratio 1:5 of its parent grid and then smoothed to limit the "Beckman and Haidvogel number" rx_0 (topographic stiffness ratio) less than 0.2 [18] and the "Haney number" rx_1 (hydrostatic instability number) under 10 [19]. To achieved the stability of computation, minimum depth (h_{min}) of ECB and BPP domain have been modified to be a positive value because the original bathymetry has negative values. In ROMS, the depth of sea floor (h) value has need to be a positive number because it represents the vertical water column thickness (meters) from ocean rest state ($\eta = 0$) and cannot have h value equal or less than zeros in any horizontal grid point even it masked by land. And also in the terrain-following vertical coordinates, h value is used to computed the vertical level thickness H_z and division by zero or other negative value is not allowed in the internal numerical kernel of ROMS model. As we know, in mathematics the division by zero is not defined and computers will lead to Inf or Nan. On the other hand, all the vertical levels in ROMS are located above h_{min} ($h_{\min} > 0$) to limit the vertical time-step of the model due to the Courant-Friedrichs-Levy (CFL) condition [20].

The hindcast simulation of parent grid were initialized beforehand to provide boundary condition for BPP model by using offline one-way nesting scheme. The hindcast of ECB model initialized from 28 September 2012 to 3 January 2013 meanwhile the BPP model running exactly for three months from 1 October 2012 to 1 January 2013. The offline nesting technique actually has a weakness compared to two-way nesting technique, because the interaction between both domain grid just occurred from the parent grid (ECB) to the child grid (BPP) causing no feedback of information from finer to coarser resolution model. However, we could not use two-way nesting scheme due to limited computational resource at the time and after several tests we found that one-way nesting scheme was good enough for this study. Without nesting, the only way to resolve physical phenomena in estuary region and coastal features at smaller scales is increasing the grid resolution on the entire ECB domain that would putting heavy task on computational process and also dealing with big output data afterwards.

Initial and lateral boundary condition of ECB for two-dimensional surface elevation fields and three-dimensional components such as velocity, temperature and salinity fields were extracted from HYCOM model analysis. The lateral boundary values for ECB have been imposed with the cycle period of once every five days by using mixed radiation-nudging boundary condition (RadNud) [21]. The RadNud boundary condition has been chosen because it has better solution compared to original radiation boundary condition (Rad) after doing several tests of the model. For the free-surface and two-dimensional momentum component like elevation and depth-integrated velocity, we have applied the Chapman-Shchepetkin boundary condition (Che-Shc) combination [22], which was the best option for this application. Shchepetkin boundary condition [23] is implemented a new solution, and more stable condition compared to Flather [24] boundary conditions because it has been founded that this boundary condition (Flather) has numerical instability under certain circumstance. Furthermore, the solutions from ECB model will be used for the initial and boundary values of the BPP model. The BPP model has the same boundary condition configuration just like its parent domain, except the nudging period was set to once every three days because of the smaller domain size it has. The detail of model parameters for ROMS configuration are showed in **Table 3**.

Table 3. Model variables of each domain model

Model parameter	Variable	Domain 1	Domain 2
Number of grids	*Lm, Mm, N*	254, 278, 30	360, 400, 20
Time step	*Dt*	30	30
Simulation steps	*Ntimes*	293760	285120
s-coordinate surface parameter	*theta_s*	7	4
s-coordinate bottom parameter	*theta_b*	2	0.2
Critical depth	*Tcline*	250	50
Factor between outflow and inflow open boundary condition	*Obcfac*	72 (every 5-days)	120 (every 3-days)

The momentum boundary condition for each model domain on the surface has been parameterized by activating ocean-atmosphere boundary layer, it means we tell ROMS to compute net heat flux and wind stress from atmospheric fields. Surface forcing was computed internally by ROMS using bulk-flux formulation, therefore the turbulent fluxes for wind, heat, and moisture are computed using Monin-Obukhov similarity theory [25]. The surface fields that should be prepared are wind component (usually 10 m above sea level), air temperature (2 m above sea level), air pressure, air relative humidity (2 m above sea level), cloud fraction, rain fall rate, and shortwave radiation flux. This surface fields have three hours cycles frequency, because the higher time-frequency of the surface forcing then it would be better to resolve the air-sea interaction.

ROMS has freely allowed user to obtain tidal elevation data from the tidal models. In this study, tidal forcing at the open boundary will be provided by output from ocean tides model TPXO8-ATLAS. In ROMS, this tidal forcing is super-imposed on the sea surface elevation if the user activated Flather or Shchepetkin boundary conditions. For the better barotropic solution of the models, we have been used 9 major tidal harmonic components: M2, S2, N2, K2, K1, O1, P1, Q1, and M4. The tidal forcing data for each tidal component consists of: angular period, elevation amplitude, elevation phase angle, current phase angle, current inclination angle (angle between ellipse major axis and u-axis), maximum tidal current (ellipse major axis), and minimum tidal current (ellipse minor axis) [20].

3.3. Model Validation

The hindcast model quality were quantitatively compared to the field observation by calculating of its root-mean-square-error (*RMSE*), mean absolute percentage error (*MAPE*), and index of agreement (*d*) as defined by [26]

$$skill = 1 - \frac{\sum \left| X_{model} - X_{obs} \right|^2}{\sum \left(\left| X_{model} - \overline{X_{obs}} \right| + \left| X_{obs} - \overline{X_{obs}} \right| \right)^2} \quad (7)$$

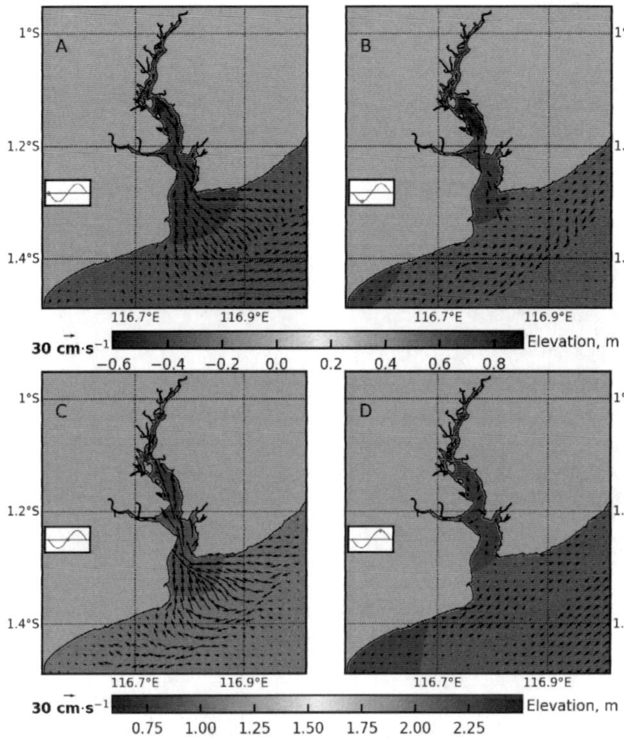

Figure 4. Spatial distribution of surface elevation and surface current during 4 different spring tide condition (a) flood, (b) low water, (c) ebb, and (d) high water.

where X is the variable being compared with $\overline{X}$ (time-averaged of X). Perfect agreement between the model prediction and field observation yields a skill of one whereas complete disagreement will yield a skill of zero. The variable being used to compute the *RMSE, MAPE*, and *d* is sea surface elevation data.

4. Results

4.1. Sea Surface Elevation and Current

The water level output from BPP model shows that tides at Balikpapan bay has semidiurnal type, marked by two high and two low tides pattern for one lunar day. Sea surface elevation has been analysed in the neap and spring tide condition because of these tidal conditions play an important role in the dynamic process in the model domain. At neap tide condition (figure 3), the ranges of sea level only changed about 10 cm from the lowest level at low tide to peak amplitude at high tide condition. Tidal current in the bay was dominantly moving outside and have magnitude <30 cm/s and then decreased near coastline area shortly after leaving the bay. The flow on the offshore of domain constantly directing from southwest to northeast with magnitude of >30 cm/s. At spring tide condition (figure 4), the sea level and current flow seems to have different attitude from the previous condition. During ebb tide condition, the tidal current flows strongly from inside to outside of the bay with the speed reaching >50 cm/s. When the low tide occurred, the direction of current has changes and the magnitude decreased to <30 cm/s. After that, the tidal current reaching its maximum speed by magnitude about >50 cm/s coinciding with the flood tide condition and mostly flowing from the northeast direction. At high tide condition, the tidal

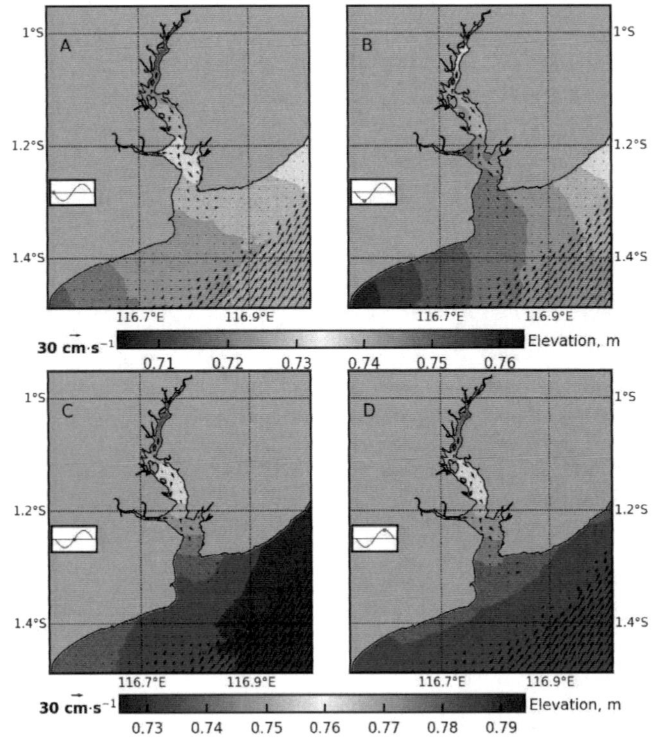

Figure 3. Spatial distribution of surface elevation and surface current during 4 different neap tide condition (a) flood, (b) low water, (c) ebb, and (d) high water.

current speed decreases to <30 cm/s and change direction from southwest to northeast. Oppositely from the neap tide, spring tide has a very large range about 3 m from low tide to high tide condition.

4.2. Spatial Temperature and Salinity

The time-averaged of temperature and salinity at surface level used to analyse the effect of air-sea flux exchanges to sea surface characteristics in the inner and outer part of the Balikpapan bay. Averaged Sea Surface Temperature (SST) at the model domain (figure 5a) has a range around 27.6 - 30.8 ^{0}C and the higher temperature occurred along the coastline of Penajam and Balikpapan where the temperature reaches the highest value compared to the offshore in the Makassar strait. Temperature increase gradually from inner part to the middle part and again decreased at the outer part of the bay, although the temperature also warmed up at the coastal area all over of model domain. The warmer area that located in the middle part of the bay and the coastal area are exactly occurred because of the net heat flux (surface forcing) penetration that occurred in the shallow-depth area. The wind stress also give effect for the occurrence of mixing at the surface layer which then leads to increase or decrease the surface temperature in the Balikpapan bay. This indicates that ocean-atmosphere interaction has a great contribution to changes surface temperature characteristic besides the advection-diffusion process from the ocean itself.

The three-month averaged Sea Surface Salinity (SSS) shows that the surface layer of Balikpapan Bay was strongly dominated by low salinity water masses (figure 5b). This could happen due to river discharges from the upstream which flows constantly into the bay, although values of river discharge should be varying with the time to produce more realistic solution. The visible surface velocity pattern also confirms the previous statement where mostly the current flow direction inside the bay is heading south and moving towards Makassar strait. Surface salinity from model output ranges from around 6 - 34 psu where inside the bay the salinity limited to ~26 psu. After leaving the mouth of the bay, the concentration of this low saline water spread at radius of ~6 km on the Balikpapan coastal area towards

Makassar strait. This means surface salinity along-channel of the bay was dominantly affected by air-sea flux interaction and freshwater input.

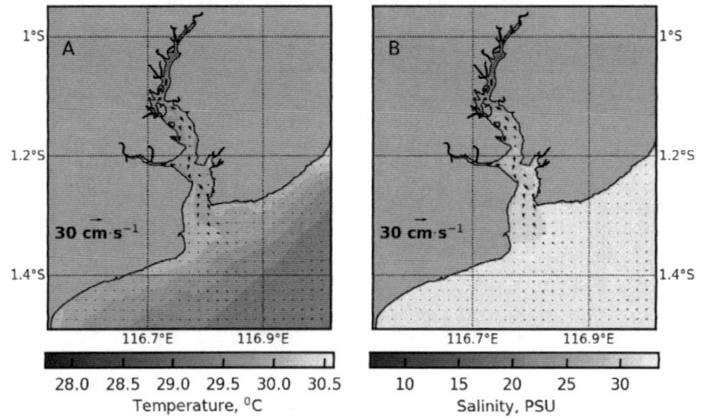

Figure 5. Three-month averaged of (a) sea surface temperature and (b) sea surface salinity overlay with three-month averaged of sea surface current.

4.3. Vertical profiles

Balikpapan bay coastal area have very complex bathymetry, making the maximum layer depth at every station sample also varying inside and outside with maximum depth about >40 m (figure 6a). The time-averaged of vertical temperature shows that the inner stations only have the temperature difference about 1 ^{0}C with minimum temperature about 29.7 ^{0}C – 30.8 ^{0}C (figure 7a) whereas at the outer part, temperature increases from 26.4 ^{0}C in the deepest layer to surface layer where one of the outer stations recorded maximum temperature about 30 ^{0}C (figure 7b). Similar with SST, vertical temperature inside the Balikpapan bay tending to be warmer than the outside because the influence of surface forcing like we mentioned before (figure 7c). Vertical salinity structure at every station samples tend to have the same trends, which have low salinity at the surface and gradually increases proportional to the depth. The surface fields influence the surface layer by freshwater flux and makes water mass in this layer to be more freshened. Generally, the outer part of the bay always much saltier than the inner part of the bay because of the existences of river inflow which can be reflected by the results of the model (figure 7d).

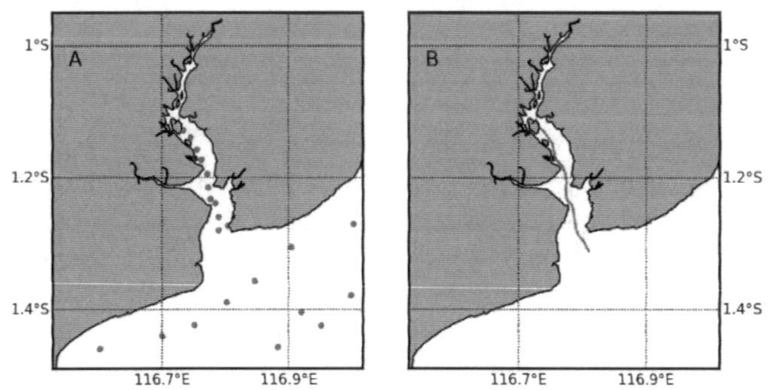

Figure 6. (a) Stations sample for vertical profiles and (b) the transect path of vertical distribution.

4.4. Vertical Distribution of Salinity

The vertical distribution of salinity shows variation values during neap and spring tidal condition. During neap tides the distribution pattern is changes slightly, and freshwater shown moving along surface layer from upstream to downstream of the bay. The higher salinity with value >34 psu is concentrated in the middle to deeper layer and extends almost vertical due to intrusion of freshwater toward outer of the bay (figure 8a). The mass of water which contains more freshwater from inner bay seems to holding this salt water to near km 9 and causes the salt distribution in water column is stratified uniform with depth (figure 8b). On the other hand, the salt distribution at spring tides act very different compare to previous tidal condition. At the lowest spring tide (figure 8c), the freshwater moving farther than neap tidal condition and the salinity decreased by >4 psu. At the highest spring period, two water masses from the opposite direction meet each other and causes the water column in this domain strongly mixed and as a consequence, the salt front tend to a vertical orientation near km 20 (figure 8d).

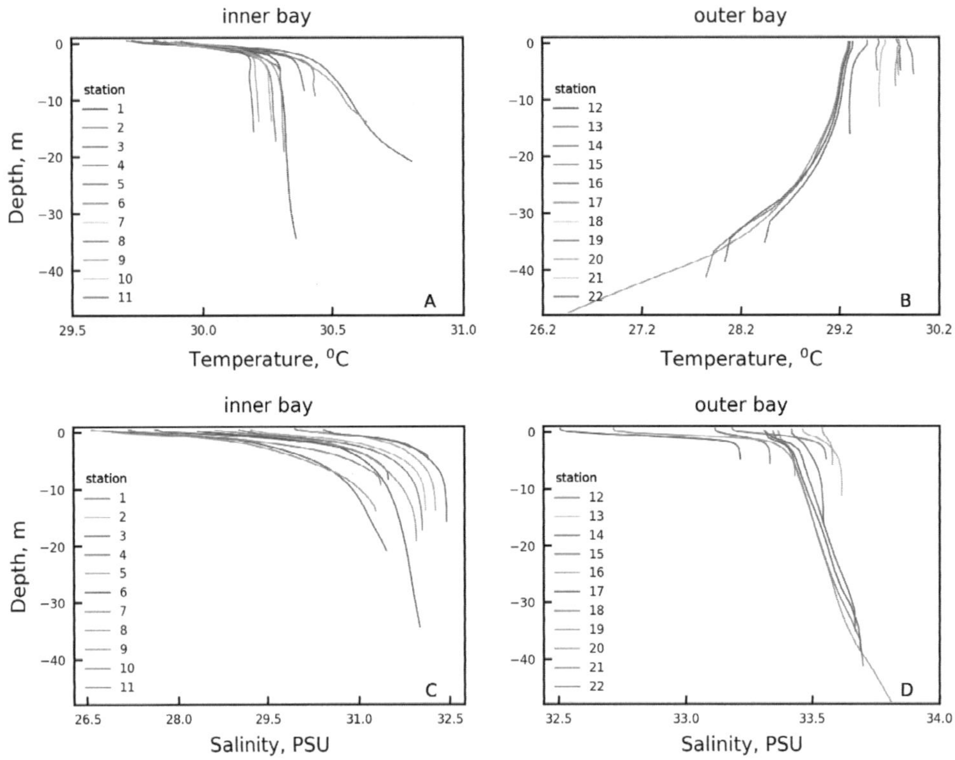

Figure 7. Vertical temperature and salinity profiles at Balikpapan Bay; (a) temperature profiles of inner stations sample, (b) temperature profiles of outer stations samples, (c) salinity profiles of inner stations sample, and (d) salinity profiles of outer stations sample.

4.5. Validation Result

The model predicts very well the observed water elevation at Semayang port (figure 9). The computed RMSE of water elevation between model solution and observation data is only about 7.8 cm. The MAPE is about 14.3 % and the model has very good agreement about 0.995 (perfect agreement = 1.0) indicating that the hindcast model has a great success to correctly simulates the observed sea level. However, an overestimation and underestimation also seen from data comparison due to unrealistic river inflow scenario (constant value of river discharges) which is used as input model.

5. Conclusions

A three-dimensional ROMS model has been successfully simulated the hydrodynamic process in the Balikpapan Bay on period 1 October 2012 – 1 January 2013. During the process, the one-way nesting method proven to offer the model a better solution when it was nested to coastal zone. In this study, the model is almost accurate to estimate the water level at Semayang port with good agreement between model solution and observed data.

IOP Conf. Series: Earth and Environmental Science **162** (2018) 012006 doi:10.1088/1755-1315/162/1/012006

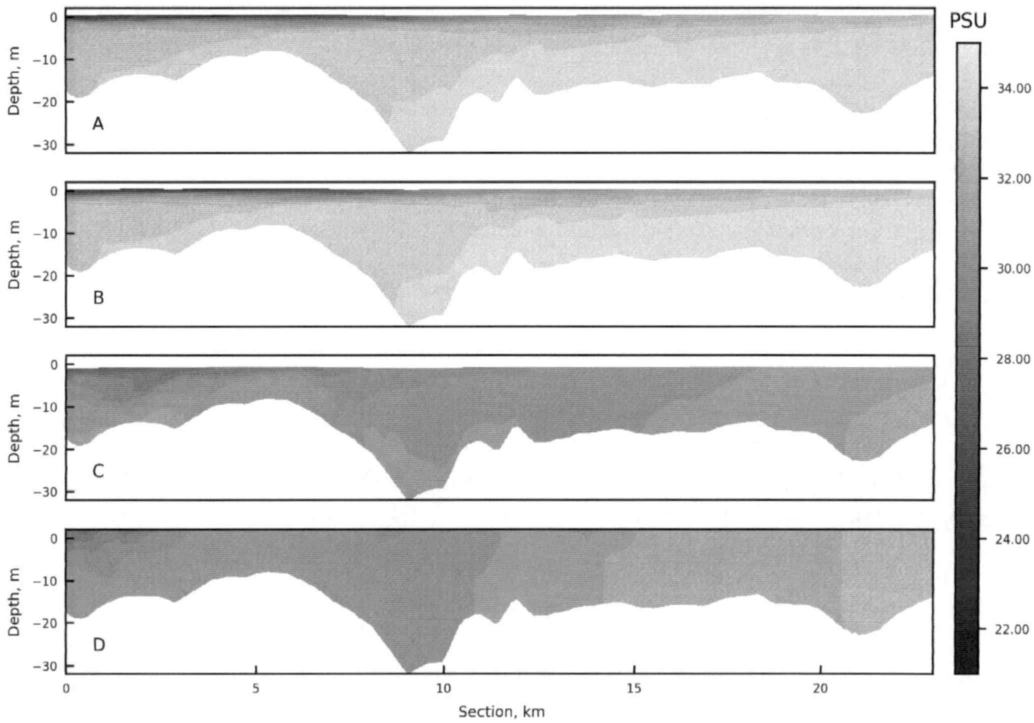

Figure 8. Cross-sectional distribution of salinity from inner to outer part of the Balikpapan bay during (a) low water and (b) high water at neap tide period and then (c) low water and (d) high water during spring tide condition.

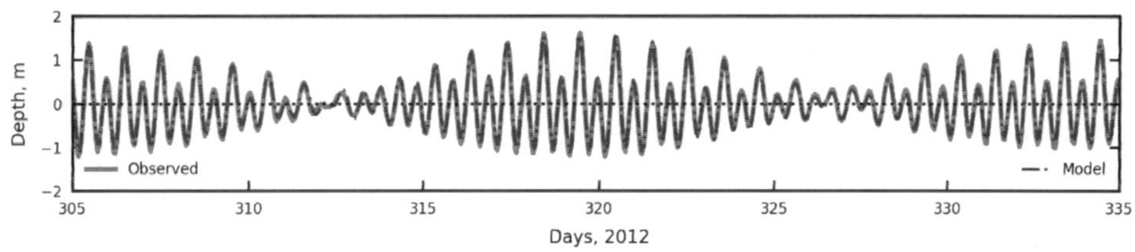

Figure 9. Comparison of observed and prediction time series of water level at the Semayang port, Balikpapan.

The model results showed that on the surface layer, the outer part of Balikpapan Bay is saltier than the inner part of the bay as expected with difference in salinity between inside and outside is about 8 psu. The same thing also happened for temperature though the difference is only about 1 ^{0}C where the warmer area occurred in the mid-section of the bay and along coastal areas. Freshwater from the rivers and salt water from the outer bay causes the water column near mouth of the bay stratified during neap tides and strongly mixed at spring tidal condition. Finally, we can conclude that the tidal mixing, atmospheric forcing and also input of river discharges have played important roles to dynamic process and water masses characteristic at Balikpapan bay.

References

[1] Sprintall J, Gordon A L, Koch-Larrouy A, Lee T, Potemra J T, Pujiana K and Wijffels S E 2014

The Indonesian Seas and their role in the coupled ocean-climate system *Nature Geosci.* **2188**
[2] Toole J and Raymer M 1985 Heat and fresh water budgets of the Indian Ocean *Deep-Sea Res.* **32** 917-28
[3] Schneider N 1998 The Indonesian throughflow and the global climate system *J. Climate* **11** 676-89
[4] Gordon A L and Fine R 1996 Pathways of water between the Pacific and Indian oceans in the Indonesian seas *Nature* **379** 146-9
[5] Mandang I and Yanagi T 2008 Tide and tidal current in the Mahakam Estuary, East Kalimantan, Indonesia *Coast. Mar. Sci.* **32** 1-8
[6] http://woodshole.er.usgs.gov/operations/modeling/COAWST/
[7] Warner J C, Geyer W R and Lerczak J A 2005 Numerical modeling of an estuary: A comprehensive skill assessment *J. Geophys. Res.* **110** C05001
[8] Penven P, Roy C, Brundrit G B, Colin de Vedière A, Frèon P, Johnson A S, Lutjeharms J R E and Shillington F A 2001 A regional hydrodynamic model of upwelling in the Southern Benguela *S. Afr. J. Sci.* **97**
[9] Zhou F, Huang D J and Su J L 2009 Numerical simulation of the dual-core structure of the Bohai sea cold bottom water in summer *Chinese Sci. Bull.* **54** 4520-4528
[10] Haidvogel D B, Arango H G, Hedström K S, Beckmann A, Malanotte-Rizaoli P and Shchepetkin A F 2000 Model evaluation experiments in the North Atlantic Basin simulations in nonlinear terrain-following coordinates *Dynam. Atmos. Ocean.* **32** 239-81
[11] Shchepetkin A F and McWilliams J C 2005 The regional oceanic modeling system (ROMS) a split-explicit, free-surface, topography-following-coordinate oceanic model *Ocean Model.* **9** 347-404
[12] Warner J C, Sherwood C R, Signell R P, Harris C K and Arango H G 2008 Development of a three-dimensional, regional, coupled wave, current, and sediment-transport model *Comput. Geosci.* **34** 1284-306
[13] Large W G, McWilliams J C and Doney S C 1994 A review and model with a nonlocal boundary layer parameterization *Rev. Geophys.* **32** 363-403
[14] Durski S M, Glenn S M and Haidvogel D B 2004 Vertical mixing schemes in the coastal ocean: comparison of the level 2.5 Mellor-Yamada scheme with an enhanced version of the K profile parameterization *J. Geophys. Res.* **109** C01015
[15] Mellor G L and Yamada T 1982 Development of a turbulence closure model for geophysical fluid problems *Rev. Geophys. Space. Phys.* **20** 851-75
[16] Umlauf L and Burchard H 2003 A generic length-scale equation for geophysical turbulence models *J. Mar. Res.* **61** 235-65
[17] Warner J C, Sherwood C R, Arango H G and Signell R P 2005 Performance of four turbulance closure models implemented using a generic length scale method *Ocean Model.* **8** 81-113
[18] Beckmann A and Haidvogel D B 1993 Numerical simulation of flow around a tall isolated seamount. Part 1: Problem formulation and model accuracy *J. Phys. Oceanogr.* **23** 1736-53
[19] Haney R L 1991 On the pressure gradient force over steep topography in sigma coordinate ocean models *J. Phys. Oceanogr.* **21** 610-19
[20] http://www.myroms.org
[21] Marchesiello P, McWilliams J C and Shchepetkin A F 2001 Open boundary conditions for long-term integration of regional ocean models *Ocean Model.* **3** 1-20
[22] Chapman D C 1985 Numerical treatment of cross-shelf open boundaries in a barotropic coastal ocean models *J. Phys. Oceanogr.* **15** 1060-75
[23] Mason E, Molemaker J, Shchepetkin A F, Colas F, McWilliams J C and Sangrà 2010 Procedures for offline grid nesting in regional ocean models *Ocean Model.* **35** 1-15
[24] Flather R A 1976 A tidal model of the north-west european continental shelf *Mèmoires Sociètè Royale des Sciences de Liège* 141-64
[25] Liu W T, Katsaros K B and Businger J A 1979 Bulk parameterization of the air-sea exchange of

IOP Conf. Series: Earth and Environmental Science **162** (2018) 012006 doi:10.1088/1755-1315/162/1/012006

heat and water vapor including the molecular constraints at the interface *J. Atmos. Sci.* **36** 1722-35

[26] Willmott C J 1981 On the validation of models *Phys. Geogr.* **2** 184-94

IOP Conf. Series: Earth and Environmental Science **162** (2018) 012007 doi:10.1088/1755-1315/162/1/012007

Numerical modeling of wave-current interaction in Merak Port, Indonesia

S Nurfitri, N S Ningsih, A N Sentanu and R Rachmayani

Department of Oceanography, Bandung Institute of Technology, Jalan Ganesha 10, Bandung 40132, Indonesia

E-mail: suliskania.n@oceanography.itb.ac.id

Abstract. This study aimed to investigate the effects of currents to wave height in the waters of Merak Port, Banten using wave spectrum model. The wave-current interaction simulation was performed by utilizing SWAN (Simulating Waves Nearshore) in stationary mode and using the unstructured triangular grid. Data of wind velocity and direction as model inputs were obtained from BMKG (Indonesian Agency for Meteorology, Climatology, and Geophysics), whereas tidal current velocity data was obtained from TMD (Tide Model Driver). The model was simulated in two different scenarios, i.e. without and with following (adverse) current which is same (opposite) with wave propagation. The model input of dominant west wind was used to represent the west monsoon, while the dominant wind from the north was used to represent the east monsoon. The simulation results show that the maximum wave height is 2.98 m in the west monsoon and 1.9 m in east monsoon. In the west (east) monsoon the following current decrease wave height to 9.26% (5.09%), while the adverse current increase wave height up to 9.07% (6.56%).

1. Introduction

Merak Port, Cilegon, Banten is the port which devoted ship transportation from Java Island to Sumatra and vice versa. The water transportation activity at this port is highly depends on the wave condition. Wave height is one of the major factors which affects the ship performance and safety [1]. Wave condition is strongly influenced by wind variability which is the main generating force (dominant) of sea wave [2, 3]. Ocean circulation of Indonesian sea, here in particular, Java Sea is affected by the monsoon cycle. From December to March the wind blows southeastward and eastward near the equator [4]. It is known as the northwest monsoon or in Java it is also called west monsoon. On the contrary, from June to September, the prevailing wind blows northwestward from Australia [4]. In Java Sea, the wind blows from the east, thus it is called east monsoon.

The previous study of ocean waves in Sunda Strait area using SWAN (Simulating Waves Nearshore) spectrum model has been conducted by Ramadhan (2014) and Adiprabowo (2015) [2, 5]. Adiprabowo (2015) conducted two stages of nested modeling using a regional model by Ramadhan (2014) as a large model with non-stationary and stationary mode [5]. The simulation was using the dominant wind field at Serang Station during 2003-2013. The result shows that the maximum wave height (*Hmax*) at southern Sunda Strait in west monsoon is 3.2 m and east monsoon is 1.6 m [5].

Study of wave-current interaction has been conducted by Baddour and Song (1990) and Wolf and Prandle (1999). Waves which propagate in the area of current circulation may experience the changes of wave height and wavelength [6, 7]. When the waves propagate in the same direction with current, significant wave height decreased and wavelength increased. Meanwhile, when the wave propagates in the opposite direction with current, wave height increased and wavelength decreased [6, 7, 8]. The previous study of wave-current interaction also conducted by Hein et al (2011) using SWAN coupled with HAMSOM in Ems-Dollard Estuary [9]. Generally, the results show that in shallow regions the difference between the simulation of waves, with and without currents is less than 10% [9]. On the other hand, inside tidal channels, the difference can exceed 25% [9].

This study aimed to investigate the effects of currents to wave height in the waters of Merak Port, Banten, Sunda Strait using wave spectrum model. The interaction of wave and current can produce extreme wave height which caused the disruption of ship transportation and also ship docking at the port. The research was conducted by utilizing SWAN. The scope of research is covering the Sunda Strait, here in particular Merak Port. Constant wind velocity (spatial and temporal) which used as generating force. In addition to that, the tidal current was taking into account.

2. Methodology

2.1. Wave equations

SWAN is based on the Eulerian formulation of the discrete spectral balance of action density that accounts for refractive propagation over arbitrary bathymetry and current fields [10]. SWAN solves the evolution of the wave spectrum by using the action density spectrum [8]. In the presence of ambient current, the action density is conserved while the energy density is not [8]. The action density N(σ; θ) is equal to the energy density E(σ; θ) divided by the relative angular frequency σ, i.e. $N(\sigma;\theta) = E(\sigma;\theta)=\sigma$. SWAN solves the evolution of the wave spectrum by using the action density spectrum [8]. The governing equation for Cartesian coordinates is [8, 11]:

$$N(\sigma,\theta) = \frac{E(\omega,\theta)}{\sigma} \tag{1}$$

$$\frac{\partial}{\partial t}N + \frac{\partial}{\partial x}c_x N + \frac{\partial}{\partial y}c_y N + \frac{\partial}{\partial \sigma}c_\sigma N + \frac{\partial}{\partial \theta}c_\theta N = \frac{S_{tot}}{\sigma} \tag{2}$$

where x and y are horizontal Cartesian coordinates, t is time, θ is the propagation direction of each wave component, c_x, c_y, c_σ, and c_θ are the propagation velocity in x-space, y-space, σ-space, and θ-space respectively [8]. S is the source term in terms of energy density, which include the effects of generation, dissipation, and nonlinear wave-wave interaction [8]. The first term on the left-hand side of equation (2) is the rate of change of action density in time, the second and third terms are the propagation of action in physical space [8]. The fourth and fifth terms show the shifting of the relative frequency and the refraction due to variations in depth and currents [8].

2.2. Model domain and parameters

Area of interest of the study is covering Java Sea, Sunda Strait, and Indian Ocean (figure 2) and focused on Merak Port (figure 2). Model inputs are dominant wind velocity, bathymetry, coastline and current velocity. Wind data obtained from BMKG (Indonesian Agency for Meteorology, Climatology, and Geophysics). Bathymetry data obtained from GEBCO (Global Bathymetric Chart of Oceans) resolution 30" and coastline from GSHHG (Global Self-consistent, Hierarchical, High-Resolution Geography Database) [12, 13]. Meanwhile, current velocity obtained from TMD (Tide Model Driver) [14].

The wave simulation utilized unstructured triangular grid as shown in figure 1 which generated by mesh generator MESH2D. Physical processes during the wave simulation are wave generation, quadruplet nonlinear interaction, dissipation and current-wave interaction. The boundary condition is the open boundary. There are 6 model scenarios which showed in Table 1. The scenarios are divided based on wind and current direction (table 1). The current is divided into two directions i.e. following (FC) and adverse current (AC) which has the same and opposite direction with wave propagation.

IOP Conf. Series: Earth and Environmental Science **162** (2018) 012007 doi:10.1088/1755-1315/162/1/012007

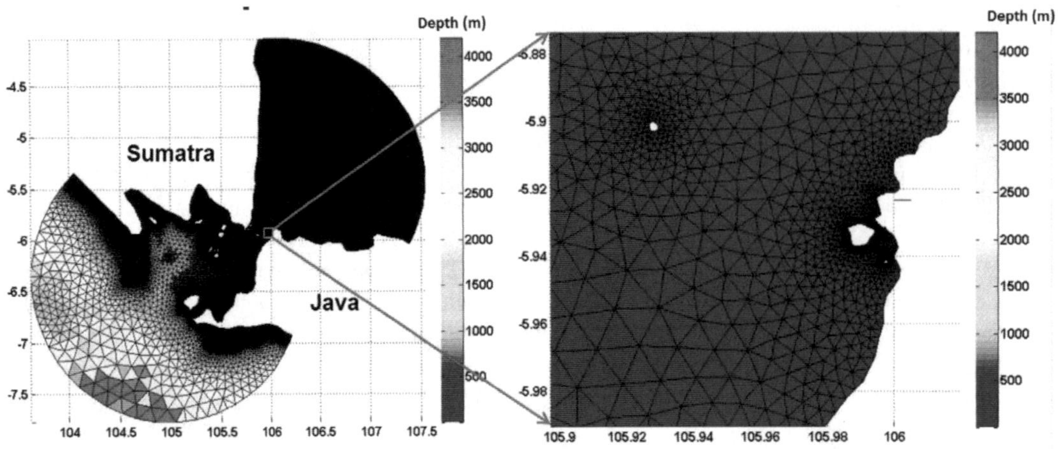

Figure 1. Model grid

Table 1. Model scenarios

Sc.	Wind velocity (from-)	Current
1	West (10.28 m/s)	No Current
2	North (7.72 m/s)	No Current
3	West (10.28 m/s)	Following Current (FC)
4	North (7.72 m/s)	Adverse Current (AC)
5	West (10.28 m/s)	Adverse Current (AC)
6	North (7.72 m/s)	Following Current (FC)

3. Results and discussion

3.1. Model verification

The model verification was done by compared the *Hs* (significant wave height) with reference model by Adiprabowo (2015) at 3 locations (figure 2). Wave simulation using west wind as generating force representation of west monsoon, meanwhile east monsoon represents the north wind. Generally, wave propagation is agreed with the wind direction. For example, the west wind generates eastward wave and north wind generates the southward wave.

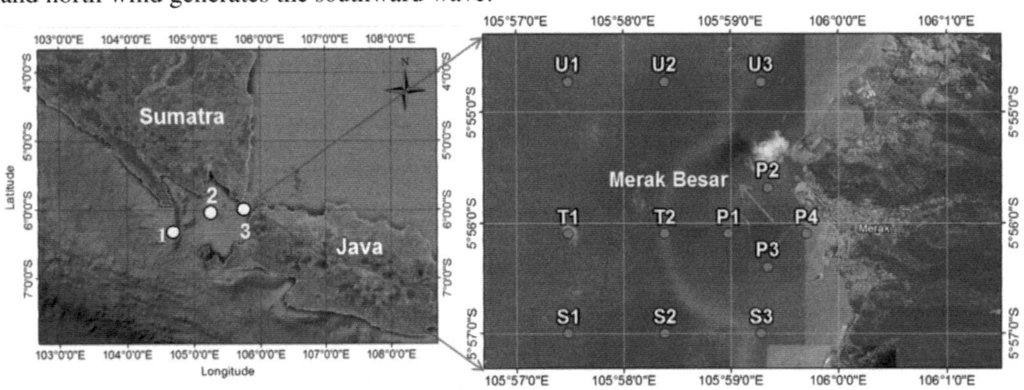

Figure 2. Area of interest (right). Location of model verification show by the yellow circles (left).

Table 2. Model verification results

	Adiprabowo (2015)				This study			
	West monsoon		East monsoon		West monsoon		East monsoon	
St.	H_s (m)	H_{max} (m)	H_s (m)	H_{max} (m)	H_s (m)	H_{max} (m)	H_s (m)	H_{max} (m)
1	1.6	3.2	0.8	1.6	1.5	3.0	0.8	1.6
2	1.5	3.0	0.6	1.2	1.5	3.0	0.7	1.4
3	1.2	2.4	0.5	1.0	1.4	2.8	0.7	1.4
			Bias (m)		0.03		0.01	
			RMSD (m)		0.13		0.13	
			CC		0.97		0.94	

Significant wave height in west monsoon is higher than east monsoon. It is caused by the dominant wind velocity is faster than east monsoon. The results agree with verification results conducted by Adiprabowo (2015) in Sunda Strait [5]. The model verification produce the bias 0.03 m in west monsoon and 0.01 m in east monsoon (table 2). These results show that the model overestimates compared to referenced model by Adiprabowo (2015). In addition to that, the RMSD (Root Mean Square Difference) is 0.13 both in west and east monsoon. Meanwhile, the coefficient correlation is relatively high, that is 0.97 and 0.94 in west and east monsoon.

3.2. Wave-current interaction

There are two different of current direction which used in this research i.e. following and adverse current. Following current (FC) was used when the current direction is same with wave propagation. While adverse current (AC) was used if the current direction is opposite with wave propagation. The vector of current and wind which generated wave is shown in figure 3. Scenario 3 represented the west monsoon when flood tide, hence the current is in the same direction with wave propagation (FC). Meanwhile, scenario 5 represent the ebb tide, so that the current direction is opposite with wave propagation (AC). Scenario 4 represented the east monsoon when flood tide, hence the current is opposite with wave propagation (AC). While scenario 6 represented the ebb tide so that the current is in the same direction with wave propagation (FC).

Figure 4 shows the *Hs* differences between three scenarios in west monsoon. The percentage of *Hs* difference between a simulation with and without current in the west monsoon is -9.26% (FC) and +9.07% (AC). Meanwhile in east monsoon, *Hs* alteration is -5.09% (FC) and +6.56% (AC). Negative (positive) sign show the decreasing (increasing) of *Hs*. These simulation results show that the current direction affects the significant wave height (*Hs*). When the wave propagates in the same direction with current, significant wave height decreased. Meanwhile, when the wave propagates in the opposite direction with current, *Hs* increased.

The condition which affects the port activities and ship transportation significantly is when the wave propagating in the opposite direction with the current. In this experiment, the adverse current occurs when ebb tide (flood tide) in west monsoon (east monsoon) as shown in scenario 5 (scenario 4). Both in west and east monsoon, *Hs* increment is below 10%. Despite the relatively small percentage, the *Hs* increment may affect the ship transportation, especially if the *Hs* with no current effects is equal or more than 2 meters. As stated on 'Guide the Marine Meteorological Services, Third Edition' that almost all the sailing ship will be disrupted by the wave which has $Hs \geq 2$ meters [15].

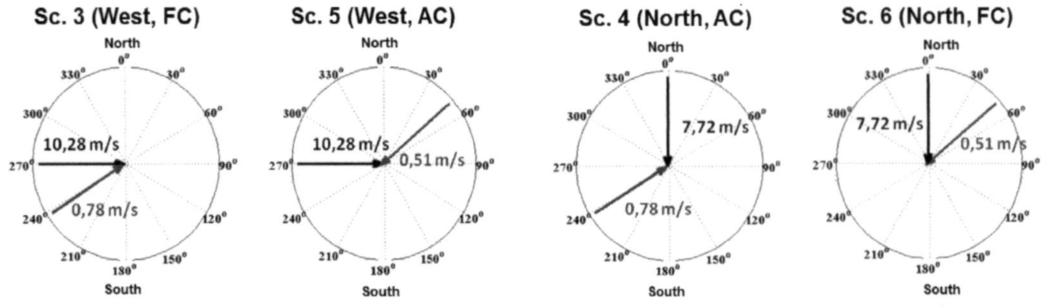

Figure 3. Wind (black) and current (blue) direction.

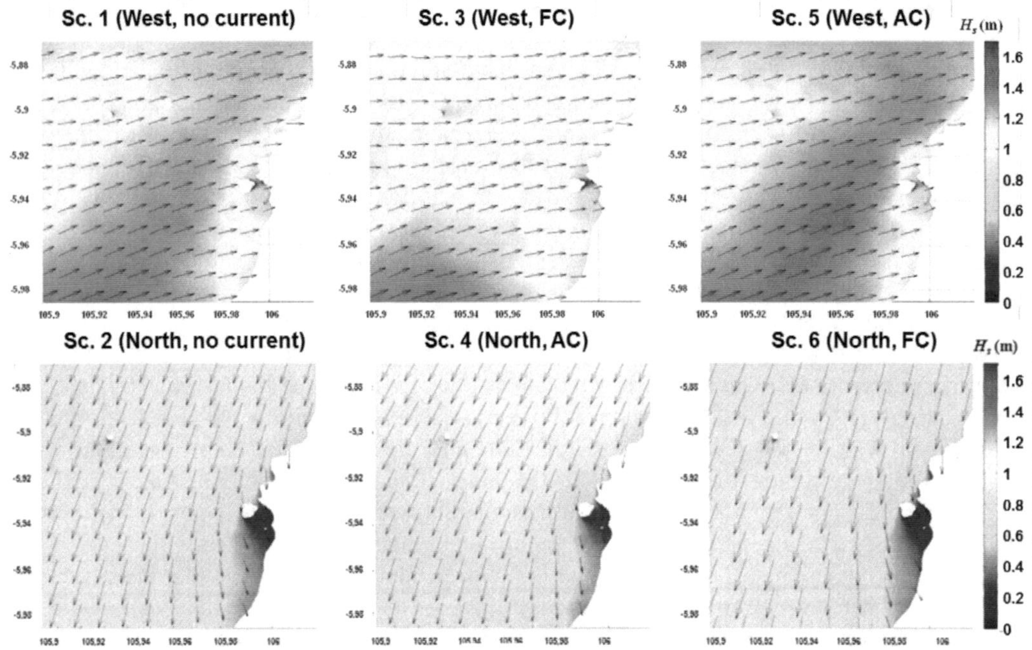

Figure 4. The significant wave height in the west monsoon (above) and the east monsoon (below). The vector arrows show the direction of wave propagation.

The graph in figure 5 and figure 6 show the comparison of energy spectrum in several points. The graph represents energy change during wave propagation from the open sea to shore. Generally, the energy density spectrum of the wave with adverse current is larger than without current, whereas the smallest is the following current. In the west monsoon, the energy spectrum from open sea (T1) to nearshore (P1) is decreasing (figure 5). The *Hs* reduction is up to 0.18 m. This happened because of energy dissipation due to shallower waters. Energy density is decreasing in P4 because this area is being covered by Merak Besar Island. The *Hs* reduction in P4 is 0.58 m which is higher than P1.

In east monsoon which the wave propagates from the north, the energy density is decreasing from U3 to P2 and P3, then increasing in S3 (figure 6). Point P3 (behind Merak Besar Island) is the 'covered' area, so it has lower energy than in P2 (in front of Merak Besar Island). Even S3 also located behind Merak Besar Island, but its location is further than P3. It causes S3 has longer fetch (wave generating area) and generates higher *Hs* than P3 (figure 6).

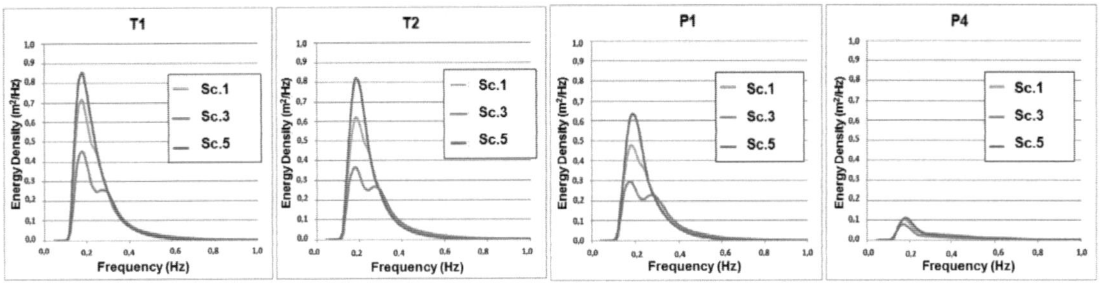

Figure 5. The energy density spectrum 1 dimension during the west monsoon.

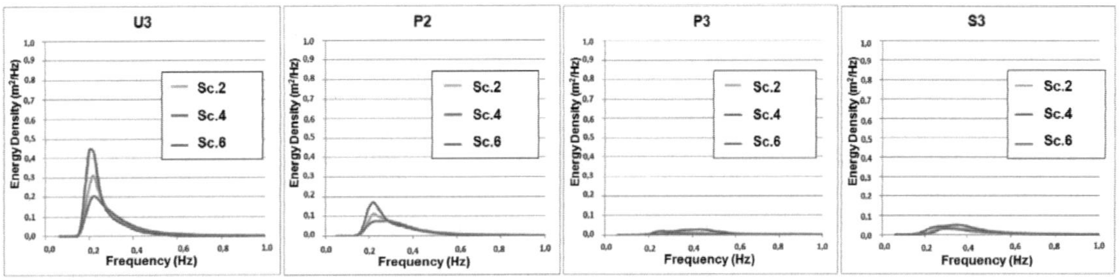

Figure 6. The energy density spectrum 1 dimension during the east monsoon.

The energy density of wave in west monsoon is higher than east monsoon. It caused by the wind velocity in west monsoon (10.28 m/s) is higher than east monsoon (7.72 m/s). The graph (figure 5 and figure 6) show that energy density of wave which propagates along with adverse current as scenario 4 and 5 has the highest value among other scenarios. In the contrary, the energy density of wave which propagates along with following current as scenario 3 and 6 has lower value compared to no current simulation as scenario 1 and 2.

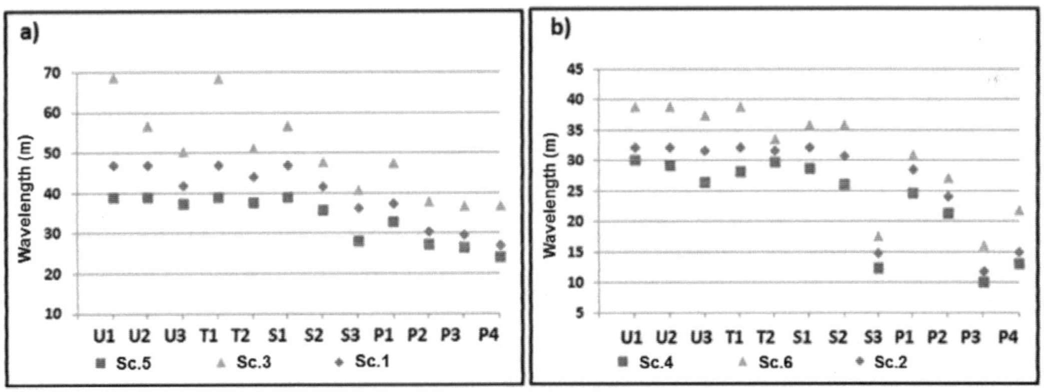

Figure 7. Wavelength calculated by Fenton equation in a) west; b) east monsoon.

The wavelength was calculated using the dispersive wave equation by Fenton (1988) [16, 17]. Based on that calculation, the wave which propagates along with adverse current has a shorter wavelength than following current (figure 7). The shorter wavelength correlate with higher wave height and vice versa. The correlation between wavelength and wave height is associated with the law of continuity or conservation of mass during the wave propagation. When the wave encounters the current in opposite direction, the water mass is accumulated and caused the wave height increase. As the result, the wavelength is decreasing.

4. Conclusions

Simulation of wave propagation with the following current produced *Hmax* (maximum wave height) in west and east monsoon by 2.27 m and 1.56 m. Furthermore, the simulation with adverse current produced the higher *Hmax* i.e. 2.98 m in the west and 1.90 m in the east monsoon. The following current decreased wave height by 9.26% in the west and 5.09% in the east monsoon. On the contrary, the adverse current increased *Hmax* by 9.07% in the west and 6.56% in the east monsoon.

References

[1] Cai Y, Wen Y and Wu L 2014 Ship route design for avoiding heavy weather and sea conditions *J. Mar. Navig. Saf. Sea Transport.* **8** 551-556
[2] Arinaga R A and Cheung K F 2012 Atlas of global wave energy from 10 years of reanalysis and hindcast *Renew. Energ.* **39** 49-64
[3] Ramadhan H 2014 *Characteristics and Analysis of Waves in West Indonesian Waters using Wave Spectrum Model* (Indonesia: ITB)
[4] Putri M R 2005 *Study of Ocean Climate Variability (1959-2002) in the Eastern Indian Ocean, Java Sea and Sunda Strait Using the HAMburg Shelf Ocean Model* (Germany: University of Hamburg)
[5] Adiprabowo S 2015 *Study of Wave Characteristics in Sunda Strait using Wave Spectrum Model* (Indonesia: ITB)
[6] Baddour R E and Song S 1990 On the Interaction between Waves and Currents *Ocean Eng.* **17** 1-21
[7] Wolf J and Prandle D 1999 Some Observation of Wave-Current Interaction *Coast. Eng.* **37** 471-485
[8] Qin W, Kirby J T and Badiey M 2005 Application of the spectral wave model SWAN in Delaware Bay *Res. Rep.* (Newark: Center for Applied Coastal Research)
[9] Hein H, Mai S and Barjenbruch U 2011 Interaction of wind-waves and currents in the Ems-Dollard Estuary *Int. J. Ocean Clim.* **2** 249-258
[10] Booij N, Ris R C and Holthuijsen L H 1999 A third-generation wave model for coastal regions 1. Model description and validation *J. Geophys. Res.* **104** 7649-7666
[11] Tolman H L 1988 Propagation of wind waves on tides *Coast. Eng.* 512-523
[12] OC, IHO, and BODC 2003 *Centenary Edition of the GEBCO Digital Atlas* (Liverpool: British Oceanographic Data Centre)
[13] Wessel P and Smith W H F 1996 A Global Self-consistent, Hierarchical, High-resolution Geography Database *J. Geophys. Res.* **101** 8741-8743
[14] Padman L and Erofeeva S 2005 *Tide Model Driver (TMD) Manual* (Seattle: Earth & Space Research)
[15] World Meteorological Organization (WMO) 2001 *Guide the marine meteorological services, Third Edition* (Geneva: Author)
[16] Holthuijsen L H 2007 *Wave in ocean and coastal waters* (New York: Cambridge)
[17] Fenton J D 1988 The numerical solution of steady water wave problems *Comput. Geosci.* **14** 357-368

Environmental assessment of Pari Island towards oil spill using Geographic Information System (GIS): a preliminary study

S Nurfitri[1], M Raihan[1], M R Putri[1], A Setiawan[2], M Gade[3] and T Pohlmann[3]

[1] Department of Oceanography, Bandung Institute of Technology, Jalan Ganesha 10, Bandung 40132, Indonesia
[2] Marine Research Centre, Ministry of Marine Affairs and Fisheries, Jalan Pasir Putih 1, Jakarta Utara 14430, Indonesia
[3] Institute of Oceanography, Universität Hamburg, Bundesstrasse 53, Hamburg 20146, Germany

E-mail: suliskania@gmail.com

Abstract. Major national and international ship traffic routes pass through the western part of Java Sea. Due to this reason, this area is vulnerable to oil spills from ships and offshore exploration. One of the most vulnerable areas to oil pollutant is the MPA (Marine Protected Area) of the Seribu Islands, particularly the region around Pari Island. The current study was part of IndoNACE (Indonesian seas Numerical Assessment of the Coastal Environment) project and aimed to assess the coastal vulnerability towards oil spill using a Geographic Information System (GIS). In this frame, a spatial analysis was conducted by overlaying coastline map, bathymetry map, and ecosystem distribution map. The overlay method was referred to the Environmental Sensitivity Index Guidelines created by NOAA (National Oceanic and Atmospheric Administration). The preliminary results showed that 38%, 24%, and 38% of Pari Island coastline were high, moderate, and low vulnerable towards oil spill pollution, respectively. The low and moderate vulnerable areas were located at the outer reef crest and inner lagoon. Meanwhile, the high vulnerable area was located at the vegetated shoreline which dominated by mangrove. Due to its biological value and its role in the environment, the vegetated area should receive priority protection against oil pollution.

1. Introduction

The western part of Java Sea is one of the several busiest ocean in the world. Several major international shipping routes pass through this region connecting the Pacific Ocean to the Indian Ocean [1]. Based on the research by Lu (2003), the areas of intense oil pollution correlate very well with the major shipping routes. Due to this reason, this area is considered to be vulnerable to oil spills from ships and offshore exploration. One of the most vulnerable areas to oil pollutant is the MPA (Marine Protected Area) of the Seribu Islands, and particularly the region around Pari Island. The previous study regarding the oil pollution in Indonesia by Bodennec and Joanny (1984) stated that Seribu Islands is two hundred times more polluted than Riau Islands, Makassar Strait, and South of Java [2]. Oil spills in Seribu Islands had been recorded since 2003, 2004, 2006, 2007, and 2008 [3]. In a year, there could be at least two oil spills which occurred on the transition monsoon between December-January and April-May [4, 5, 6]. During the IndoNACE observation in 6th November 2015, the oil slick was discovered at the southern shore of Pari Island which consists hydrocarbon concentration of 1.7 mg/l [6].

The fate and behaviour of spilled oils depend on the weathering process which reduces the concentration hydrocarbon in water and alters the chemical composition of spilled oils [7]. Naturally, the oil in the sea could also disperse by biodegradation process. Major environmental factors affecting oil biodegradation include weathering processes, temperature, availability and concentration of nutrients, availability and concentration of oxygen, and pH [8]. Meanwhile, the oil spill spreading is mostly affected by ocean physical processes such as currents, waves, and tides.

Contingency planning for potential oil spills is becoming a necessity for most of the coastal waterways [9]. One of the earlier attempts to estimate coastal oil spill vulnerability can be found in Gundlach and Hayes (1978) where a classification of coastal environments based on their physical and geological characteristics is presented [9, 10]. Many approaches were used and described to determine the shoreline sensitivity toward oil spills, including GIS (Geographic Information System) application and remote sensing techniques [11].

Seribu Islands, here in particular Pari Island is facing several environmental stresses such as oil pollution. Therefore, it is important to determine the shoreline vulnerability. The main question of the present research was to assess the level of coastal vulnerability of Pari Island toward oil spills. An integral part of this study was to determine which coastal area would be most seriously damaged by an oil spill so that they may receive priority protection [9]. Furthermore, the map of oil spill vulnerability is an important tool to develop the best-suited oil spill responses strategies [12].

2. Methodology

This research was utilized Geographical Information System (GIS) to determine the vulnerability index. The GIS allows analysis through the relationship linking the data together [13]. Then it becomes easy to build up any combination of some variables of different cartographic nature and spatial density [13]. There were two main maps which used in this research i.e. Base Map and Ecosystem Map. Base Map obtained from the satellite image of Lansat 8 (downloaded from Google Earth) which has the resolution by 15x15 m^2. This map was digitized using GIS to obtain the map of shorelines classification. Meanwhile, ecosystem map was obtained from Indonesian Institute of Sciences.

The scoring classes of vulnerability parameters are shown in table 1. There were four parameters which are taken into account to determine the vulnerability index, i.e. shoreline, ecosystem, depth area, and human activity (table 1). The choosing of vulnerability parameters was referred to NOAA's Environmental Sensitivity Index (ESI) with some modifications [14]. Each of parameters was divided into three classes and the score represent the level of vulnerability. The higher class score indicated the higher vulnerability towards oil spill and vice versa. Based on table 1, the shoreline was classified into three type i.e. vegetated sand, sandy beach, and human-made structure. Moreover, there were three main tropical ecosystems i.e. mangrove, seagrass, and coral [15]. Since the focus of the research was the coastal area, therefore the depth classification was limited to 5 meters. The last parameter is the human activities which classified into high, moderate, and low.

Table 1. Scoring classes of vulnerability parameters

No	Shorelines Classification	Ecosystem	Depth area	Human Activities	Class Score
1	Vegetated sand	Mangrove	Beach (0-1 m)	High (port)	3
2	Sandy beach	Seagrass	Shallow (1-5 m)	Moderate (tourism, beach, residence)	2
3	Human-made structure	Coral	Deep (> 5 m)	Low (uninhabited islands)	1

In addition to the parameters, the mapping of vulnerability area was also referred to NOAA's Environmental Sensitivity Index (ESI). The ESI is a spatial information system which is composed of three main components: a shoreline type, oil-sensitive biological resources, and human-use resources of commercial, recreational, or subsistence value [10, 14, 16]. Vulnerability index was calculated using the weighted overlay method. Every parameter on map layer was given a sensitivity score towards oil spill based on ESI Guidelines Version 3.0 with some modifications [16]. Vulnerability index was calculated using the equation (1) [16]:

$$X = \sum_{i}^{n} W_i \times X_i \qquad (1)$$

Where X denotes the vulnerability index, X_i is the class score of parameter-i, W_i is the weighted score of parameter-i, which are 0.3 for shoreline, 0.4 for habitat, 0.2 for bathymetry, and 0.1 for human activities. Symbol n shows the number of parameter which used to determine the vulnerability index.

3. Results

There were three main types of shoreline in Pari Island (figure 2) i.e. human-made structure (grey), sandy beach (brown), and vegetated shoreline (green). Human-made structure mostly made of concrete. There was also the sandy beach which mostly covered the eastern side of Pari Island. The sediment analysis in several beach around Pari Island showed that sand percentage varies from 80-96%, followed by silt from 2-14%, gravel from 1-6%, and clay which lower than 2.9% [17].

The sediment type of beach determines the rate of oil penetration [18]. The coarser grained sediment has a higher potential of oil penetration rather than compact and solid sediment [18]. At the coarse-grained sand beach, for example, the oil may sink and/or be buried rapidly making the clean-up process more difficult [9]. On the contrary, at the fine-grained sand beaches, oil does not penetrate into the sediment [9]. At the fine-grained sandy beach, oil was accumulated and formed an oil-slick-ribbon along the intertidal zone.

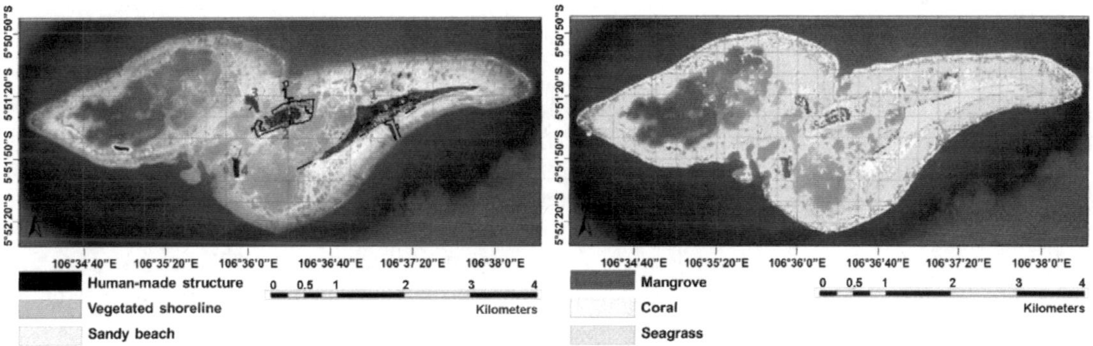

Figure 1. Shoreline (left) and ecosystem (right) classification map. Number 1=*Pari* Island, 2=*Tengah* Island, 3=*Kongsi* Island, 4=*Burung* Island, and 5=*Tikus* Island.

In addition to human-made and fine-grained sandy beach, the shoreline was also classified into the vegetated shoreline. The vegetated beach in Pari Island had a total length of 5.5 km which dominated by mangrove. The mangrove vegetation that lies on the shore was the most vulnerable to oil spill because of its biological value [19, 20]. Mangrove forest is an important ecosystem since its role for the carrying capacity of the environment such as a spawning or breeding and enlarging various types of nursery ground for aquatic biota (fish and shrimp) and terrestrial [12]. Based on the map of the ecosystem, mangrove which lies on the shorelines was dominantly located in the northern of Pari Island (figure 1).

The group of Pari Island was surrounded by extensive fringing reefs characterized by a sand flat, reef flat or lagoon, reef edge, and reef slope which drops to a depth of 20 m [21]. This coral reef was one of the main tropical ecosystem in Pari Island (figure 1) [15]. Oil spills in the ocean can disturb the coral reefs through oil slick on the sea surface and sink oil [3]. Oil slick on the sea surface will disrupt the living organisms on the surface, increasing the sea surface temperature, blocking the sunlight intensity and ocean-atmosphere gas exchange [3, 5].

In addition to the ecosystem, human activity also took into account to determine the vulnerability index. Human activities differ into three main type that is the high activity (port), medium (tourism, beach, residence area), and low (uninhabited islands). The highest of human activities located in the port. The port was one of the sources of the oil spill due to the high of ship activities.

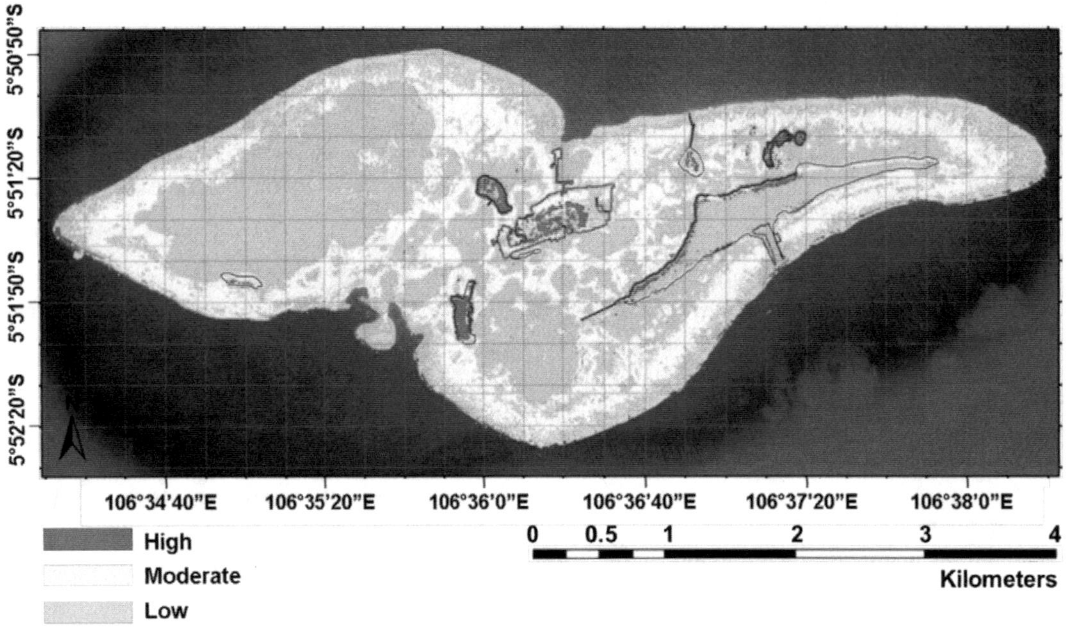

Figure 2. Map of vulnerability towards oil spill index in Pari Island.

The weighted overlay method of shorelines, ecosystem, and depth classification map produced the vulnerability index map (figure 2). These maps show the level of vulnerability toward oil spill in the group of Pari Islands. The index was classified into three classes i.e. high, moderate, and low. The low and moderate index were mostly located in the outer reef crest because they were facing the open sea. This area was exposed to the higher of water dynamics rather than the inner lagoon. The mixing process by the wave generated the faster flushing process. As stated by Utyanto *et al.*, 2003 that the coast which exposed to the strong wave and tides would recover faster from the oil spill, as the coast can be naturally cleaned by both wave and tides flux energy [18].

Besides the outer reef crest, the moderate index also located in seagrass area and sandy beach surround *Tikus* Island, also in northeast, southwest, and southeast area of *Pari* Island. Meanwhile, the high index located in north and west area of *Pari* Island. Dominantly the high index was located in vegetated shoreline (mangrove and seagrass). The high index cover at least 38% from 14.5 km of Pari Island shoreline. The high index was mainly caused by the existence of vegetation, here in particular mangrove which had the highest score among the other ecosystem. The oil spill impact on mangrove was determined by tidal condition. When flood tide, the oil could submerge to the shore and sticking to mangrove roots. In the event of the slight oil spill, the clean-up would occur naturally. However, in the event of the heavy oil spill, oil may be cleaned-up by oil dispersants.

Furthermore, the high index was also resulted from the low exposure of most types of beaches in Pari Island to natural clean-up process. The inner lagoon had relatively calm water than the outer reef slope because of its isolated morphology. It caused the longer of flushing process in the event of the oil spill. Based on this result, it can be considered that the vegetation area of the inner lagoon should receive priority protection against the oil pollution. This area was including *Burung* Island and northern coast of Pari Island. For future research, it is important to include the wave and tides as part of the vulnerability index parameters.

4. Conclusions

The vulnerability index showed that 38%, 24%, and 38% of Pari Island shoreline were considered to be high, moderate, and low vulnerable towards oil spill pollution, respectively. The low and moderate vulnerable area was located at the outer reef crest and inner lagoon. Meanwhile, the high vulnerable area located at the vegetated shoreline which dominated by mangrove. In the event of the oil spill, the vegetated area would take a longer time to clean-up especially by the natural processes. The further research is needed to consider the potential location of oil sources as well as the ocean dynamics such as tides and waves as the factor of the vulnerability index.

Acknowledgements

We express our thanks to IndoNACE (Indonesian seas Numerical Assessment of the Coastal Environment) for data observation and funding this research.

References

[1] Lu J 2003 Marine Oil Spill Detection, Statistics and Mapping with ERS SAR Imagery in South-East Asia *J. Remote Sens.* **24** 3013–32

[2] Bodennec G and Joanny M 1984 Occurrence of Tar Pollution along Shores in Indonesia *CNEXO-LEMIGAS Field Survey Preliminary Report May 1984* p 41

[3] Suhery N, Damar A and Effendi H 2017 Coral Reef Ecosystem Vulnerability Index to Oil Spill: Case of Pramuka Island and Belanda Island in Seribu Islands *J. Trop. Mar. Sci. Tech.* **9** 67–90

[4] Rosianna 2008 *Analysis of Oil Spill on Environmental Factors in Seribu Island* (Indonesia: ITB)

[5] Putri M R, Setiawan A, Sari T, Mayer B and Pohlmann T 2017 Trajectory Model for Identification of Oil Spill Around the Coast of Pari Island, Seribu Islands, North Jakarta *J. Trop. Mar. Sci. Tech.* **9** 657–64

[6] Putri M R, Setiawan A, Pohlmann T, Mayer B and Gade M 2016 The Assessment of Oil Pollution in Seribu Islands based on Remote Sensing and Numerical Models *Proc. Living Planet Symp. 2016, Prague, Czech Republic,3-16 May 2016*

[7] Wang Z, Fingas M, and Page D S 1999 Oil Spill Identification *J. Chromatogr. A.* **843** 369–411

[8] Zhu X, Venosa A D, Suidan M T and Lee K 2001 *Guidelines for The Bioremediation of Marine Shorelines and Freshwater Wetlands* (U. S. Environmental Protection Agency) p 163

[9] Gundlach E R and Hayes M O 1978 Vulnerability of Coastal Environments to Oil Spill Impacts *Mar. Technol. Soc. J.* **12** 18–27

[10] Castanedo S, Juanes J A, Medina R, Puente A, Fernandez F, Olabarrieta M and Pombo C 2009 Oil Spill Vulnerability Assessment Integrating Physical, Biological and Socio-Economical Aspects: Application to the Cantrabrian Coast (Bay of Biscay, Spain) *J. Environ. Manage.* **91** 149–159

[11] Adler E and Inbar M 2007 Shoreline Sensitivity to Oil Spills, the Mediterranean Coast of Israel: Assessment and Analysis *Ocean Coast. Manage.* **50** 24–34

[12] Hernawan U and Risdianto R K 2017 Coastal Protection of Southern Part of Bintuni Bay from Oil Spill: An Environmental Sensitivity Index Approach *B. Mar. Geol.* **32** 77–88

[13] Populus J, Moreau F, Coquelet D and Xavier J. P 1995 An Assessment of Environmental Sensitivity to Marine Pollutions: A Solutions with Remote Sensing and Geographical Information System (GIS) *Int. J. Remote Sens.* **16** 3–15

[14] NOAA 2002 *Environmental Sensitivity Index Guidelines Version 3.0* (Hazardous Materials Response Division Office of Response and Restoration NOAA Ocean Service) p 192

[15] Wouthuysen S, Abrar M, Mira S, Nikijuluw I, Suhardi A, Mansur and Reza 2013 Planning for Education of Tourism Zone and Development of Marine Protected Areas in Pari Island *Indonesian Institute of Sciences Research Report* p 31

[16] Jensen J R, Halls J N and Michel J 1998 A System Approach to Environmental Sensitivity Index (ESI) Mapping for Oil Spill Contigency Planning and Response *Photogramm. Eng. Remote Sens.* **64** 1003–14
[17] Putri M R, Setiawan A, Nurfitri S, Sari T, Tsani R N, Raihan M, Tatipatta W M and Pohlmann T 2016 Indonesia Numerical Assessment Coastal Environmental (INDONACE): Field Observation of Pari Island *Project Report* p 45
[18] Utyanto, Hartono and Sutikno 2003 GIS Application of Environmental Sensitivity Index Mapping Case Study in Cilacap Coastal Area and Segara Anakan *Hum. Environ.* **10** 131–40
[19] Muarif, Damar A, Hariyadi S, Sutrisno D and Boer M 2017 Development of Variable of Ecological Sensitivity Index (ESI) for Oil Spill Management in Mangrove Ecosystem *J. Kel. Nas.* **11** 21–34
[20] Prasetyo A, Santoso N and Prasetyo L B 2017 Environmental Sensitivity of Mangrove Ecosystem to Oil Spillage in Ujung Pangkah Sub-district, Gresik *J. Forest Nat. Conserv. Res.* **14** 91–98
[21] Brown B E and Suharsono 1990 Damage and Recovery of Coral Reefs Affected by El Nino Related Seawater Warming in the Thousand Island, Indonesia *Coral Reefs* **8** 163–70

IOP Conf. Series: Earth and Environmental Science **162** (2018) 012009 doi:10.1088/1755-1315/162/1/012009

Image side-scan sonar segmentation for seafloor detection in Lembeh strait, Bitung, North Sulawesi

[1]Rina Nurkhayati, [2]Henry M Manik, [2]Totok Hestirianoto, [3]Susilohadi, [3]Delyuzar Ilahude

[1] Marine Technology, Faculty of Fisheries and Marine Sciences, Bogor Agricultural University, Dramaga, Bogor, 16680, Indonesia
[2] Department of Marine Science and Technology, Faculty of Fisheries and Marine Sciences, Bogor Agricultural University, Dramaga, Bogor, 16680, Indonesia
[3] Marine Geological Research and Development Center, Bandung, Indonesia

E-mail: [1]rinanur53@gmail.com, [2]henrymanik@ipb.ac.id

Abstract. Segmentation is the process of dividing and grouping the pixel of an image into a segments or objects with the similar characteristics into homogenous areas. This method is expected to get better result with relatively faster result compared to manual (digitation) method. Side-scan Sonar image segmentation using Multi-resolution segmentation algorithm with considered many parameter and using pixel based threshold. The segment grouping using spectral reference by emphasizing the degree of brightness and texture where the two parameters are related to each. Segmentation with two parameters (brightness and texture) is effectively used for targets located near a towfish because of higher target spectral heterogeneity. TVG (Time Varried Gain) correction deliver more detailed result, image with TVG have highest accuracy of 0.987. This proves that the TVG correction that regulates the backscatter to affect the hue in the image has an important role in the seafloor classification process, in addition it can sharpen the appearance of the target which is far from the tow-fish range. Weakness of energy during measurement caused by angle and distance from instrument to target far from the towfish will affect the value of brightness and texture.

1. Introduction

Seafloor mapping with acoustic instruments have been done with many different method. One of the method using Side-scan Sonar instrument [1]. The images produced by acoustic instrument is difficult to interpret because its complexion of characteristic and processes during propagation and reflection in marine seafloor. Manual interpretation of Side-scan Sonar imagery usually take long processing due to data coverage and spatial resolution depending the instrument's frequency. The appearances of various target in different location affected by dynamic of seawater. Thus, more effective and efficient method are necessary to classify each target by taking into account the factors affecting characteristic of target. One method that can be used in this target classification is segmentation.

Segmentation is digital classification which divide data into homogenous area [2]. Generally, segmentation on Side-scan Sonar is performed on numerical data by applying processing signal using frequency variable, time, and other numerical data. In this research, image processing is tried to the raster data of Side-scan Sonar. The aim of this research to found the method that faster and efficient [3] to process Side-scan Sonar data with segmentation method.

2. Study Area & Data

The data uses in this research is a Side-scan Sonar measurement data, seafloor sampling with grab sampling method and underwater photograph/video. The instrument uses Side-scan Sonar Starfish 990F with 1 MHz frequency, with acquisition software is Scan-line Starfish version 2.1. This instrument mounted on ship and following the pre installment track using GPS. Measurement carried by four areas and each area contain five survey path. For preprocessing Side-scan Sonar data using Hypack 2015. The measurement conducted in April 2016 on Lembeh Strait, Bitung, North Sulawesi. The measurement is in collaboration with Marine Geological Research and Development Center, Ministry of Energy and Mineral Resources in the framework of research and development of marine potential energy in Lembeh Strait.

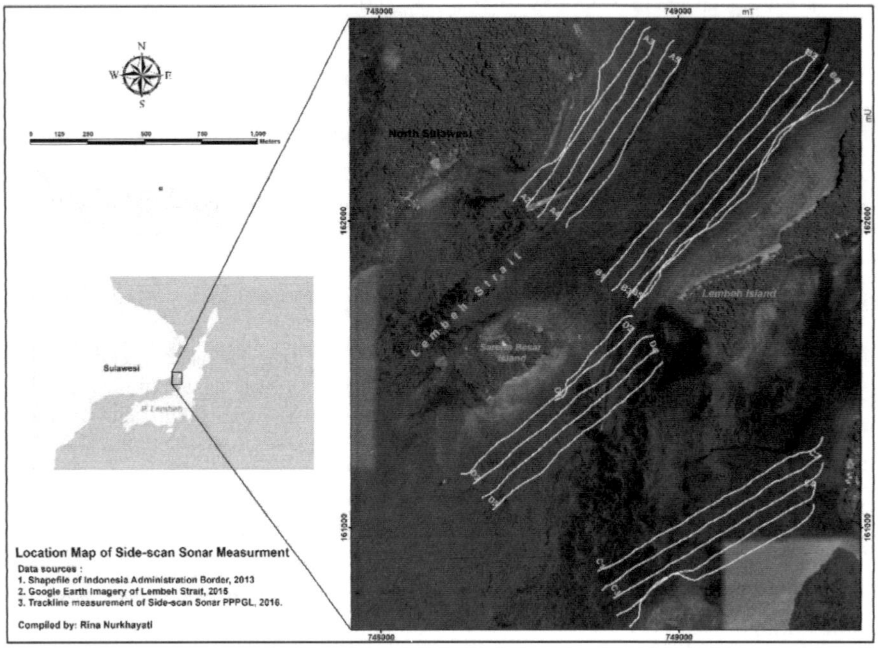

Figure 1. The location of research. Survey track line tagged by yellow line

Table 1. Specification of Starfish 990F

System Part Number	BP00181
Frequency	1 MHz CHIRP
Operating Range	35 m (115 ft) per channel
Horizontal Beam Width	0.3°
Vertical Beam Width	60°
Tranducer Angle	Tilted Down 30° from horizontal
Length	378 mm (14.88")
Width	110 mm (4.33")
Height	97 mm(3.81")
Depth Rating	50 m (164ft)

Figure 2. Starfish 990F instrument

3. Method

Side-scan Sonar image interpretation using digital processing by placing pixel into target class based on visual image. Digital classification can distinguish target by similarity of spectral characteristic and target geometry. Segmentation process are disintegrating and grouping the pixel with the same characteristics, and describes the segments into a particular class. Segmentation of Side-scan Sonar images using multi-resolution segmentation algorithm. This algorithm consider several parameter and using pixel-based threshold for grouping the segment [4]. The parameter in this research is brightness and texture. In the process of segmentation, there is a parameter scale determination in the formation of segments that group the targets based on the range of adjacent homogeneous pixels.

Segmentation with multi-resolution algorithm using pixel-based threshold to identified the direction of color change and pixel texture in the image to the segment [5]. For identification, the borders of color and texture change are influenced by the resolution affecting the pixel size. The higher the resolution will be gentle edge border between segments, for image Side-scan Sonar is used a high resolution image, each pixel size of 0.5 x 0.5 meters for frequency 1 MHz so that the results of segmentation for all three data has the degree of edge changes between segments are smooth or details.

In this research the test data conducted in Block A line 1 data. The data processing differ 3 categories are original image with water column, image with water column correction, and image with Time Varried Gain (TVG) correction. Processing data using E-Cognition software begin by design tree processing of class hierarchical with the analogy of decision tree [6]. The design tree processing shown below :

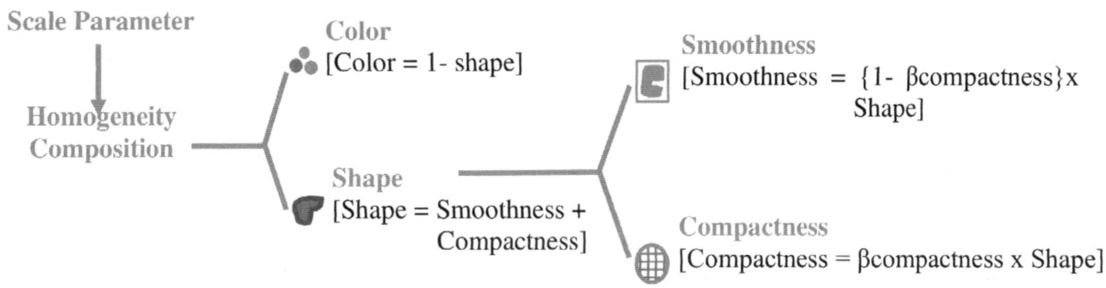

Figure 3. Multi-resolution method concept, E-cognition, Trimble [7]

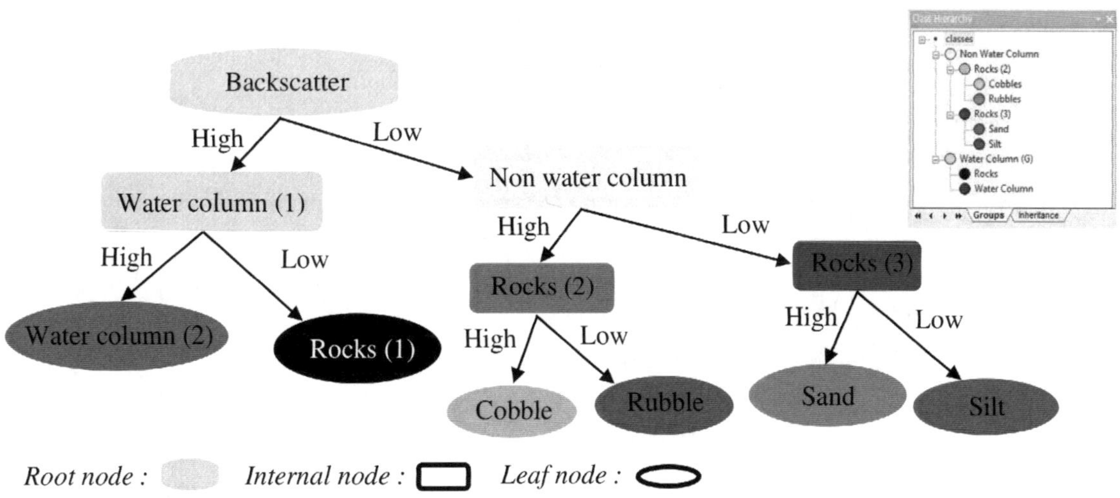

Figure 4. Decision tree analogy for seafloor classification

The decision tree analogy using backscatter force as the node root in decision-making. Seafloor backscattering phenomenon related to surface roughness and hardness [8]. Relatively strong backscattering associated with more roughness target and weak backscattering associated with low roughness target. The strong backscattering data is spelling into three level data classification: (1) water column and non-water column, (2) water column with the rock classes, and (3) rock classes classified into more specific as the seafloor classes.

Multi-resolution segmentation is tested on all three data using brightness and texture based on the pixel value where the pixel value is inversely proportional to the backscatter strength. The amount of pixel value between one data and the data may be different due to the correction of the water column and TVG. Figure 5 showing a technical processing of Side-scan Sonar which have two steps, first is segmentation and second is seafloor classification.

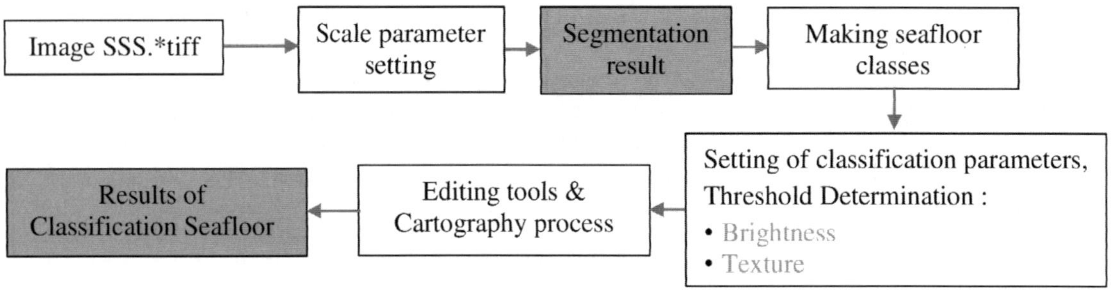

Figure 5. Technical processing of Side-scan Sonar which have two step, first is segmentation and second is seafloor classification

4. Result and Discussion

This is segmentation result (table 3) shown the difference Side-scan Sonar image and segmentation result each level of classification. Scale parameters for all three levels of data are determined by numbers that are differently ranging from 20 to 25, scale parameter numbers will further shrink or decrease the third iteration of segmentation more and more. In this research, iteration is done 3 times to produce 3 levels of data, iteration is done in order to detail the result of segmentation and classification. The segmentation result for three image shown that image with applied TVG have more segment than image with water column correction. Presence of water column can be determining factor of segmentation result. The table 2 is total the segment from all three images.

Table 2. Total segment from different methods

Image	Scale Parameter LV 1	Object Level 1	Scale Parameter LV 2	Object Level 2	Scale Parameter LV 3	Object Level 3
Original image with water column	25	15822	22	16153	20	16623
Image with water column correction	25	15305	22	15570	20	15820
Image with TVG correction	25	17362	22	17648	20	18014

IOP Conf. Series: Earth and Environmental Science **162** (2018) 012009 doi:10.1088/1755-1315/162/1/012009

Table 3. Segmentation result each level data classification:

Process	Side scan-Sonar with water column	Side scan-Sonar with water column correction	Side scan-Sonar with TVG correction
Before processing	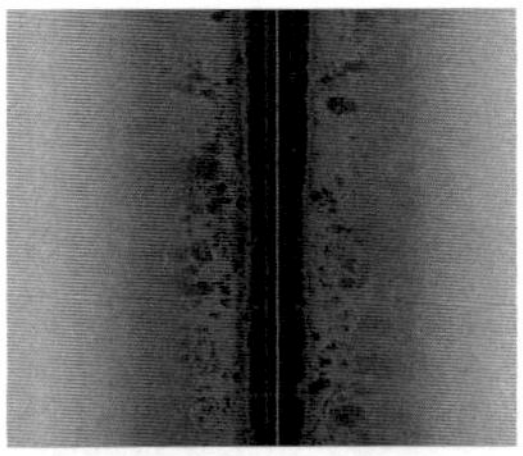	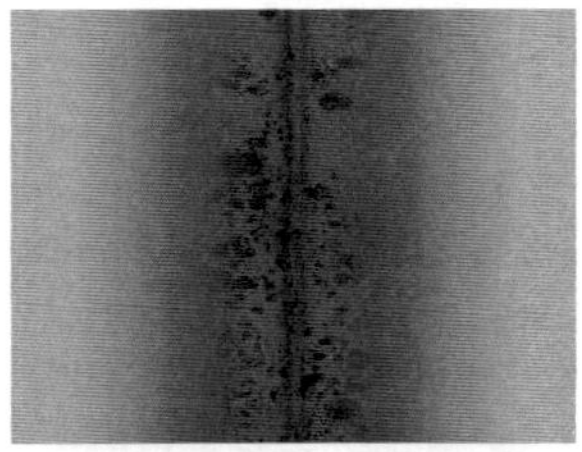	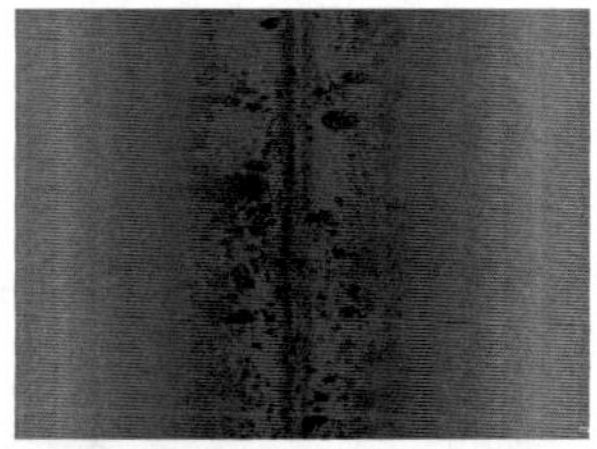
1st Segmentation	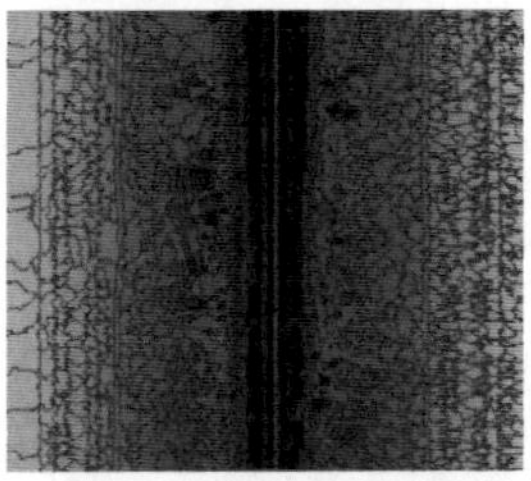	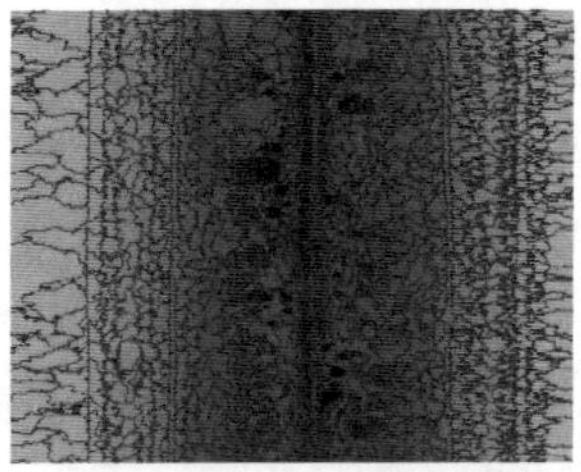	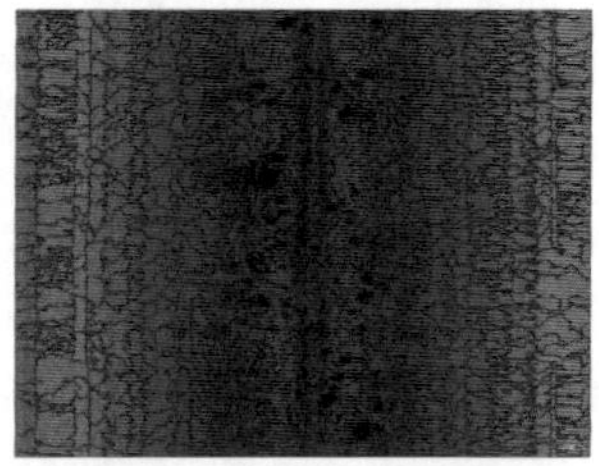
2nd Segmentation	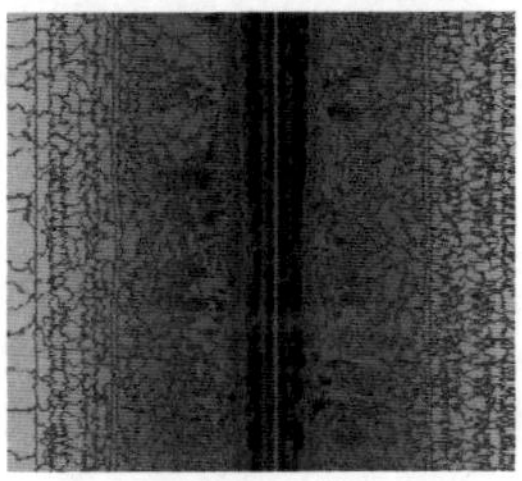	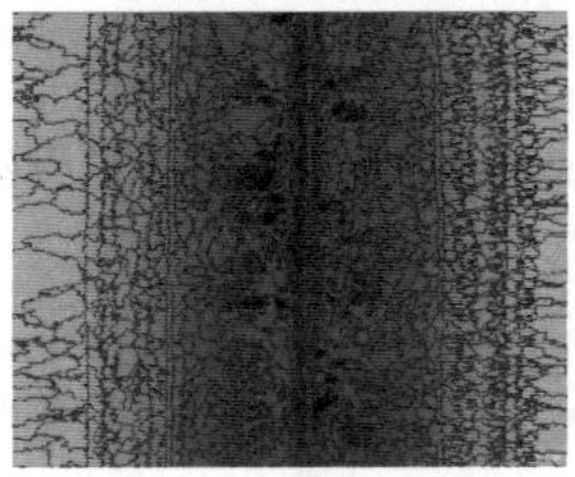	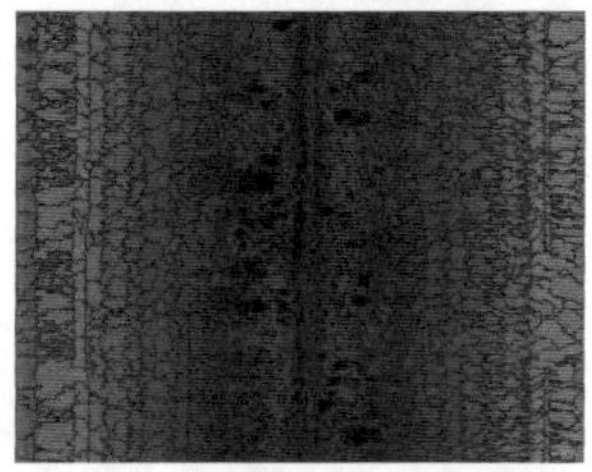
3rd Segmentation	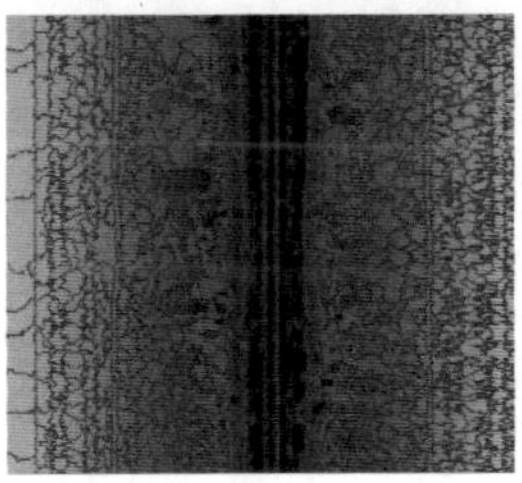	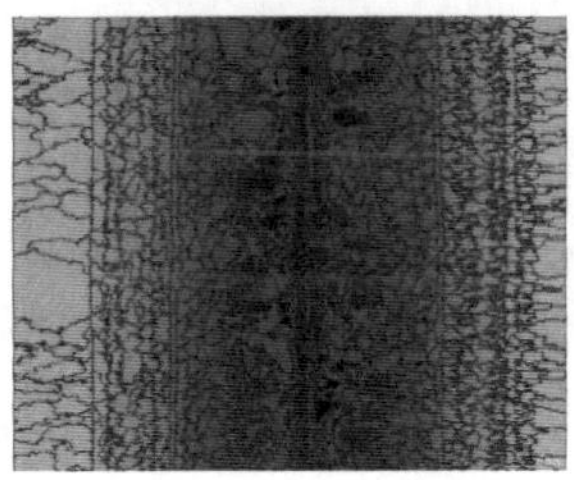	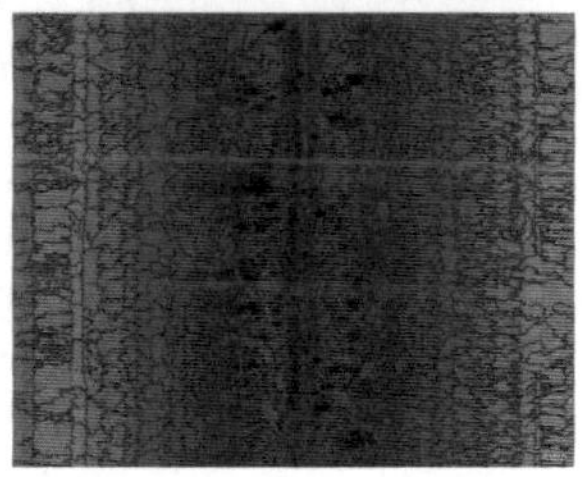

The segmentation result will affect the classification, the higher level of segmentation more detailed result, done by 3 time iteration to produce 3 classification level.. The three level data classification: (1) water column and non-water column, (2) water column with the rock classes, and (3) rock classes is classified into more specific as the seafloor classes; this classes is devided into water column, rock, cobble, rubble, sand, and silt. TVG correction deliver more detailed result compare to original image and image with water column correction. The accuracy of segmentation and classification for three classes, image with amplified (TVG) have highest accuracy of 0.987.

Table 4. Classification result each level of classification

Process	Side-scan Sonar with water column	Side-scan Sonar With water column correction	Side-scan Sonar with TVG correction
Before processing	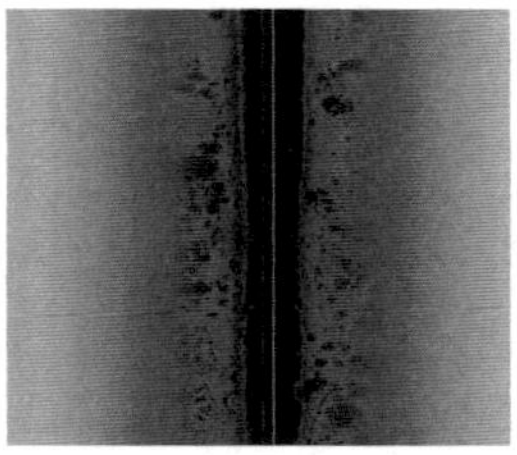	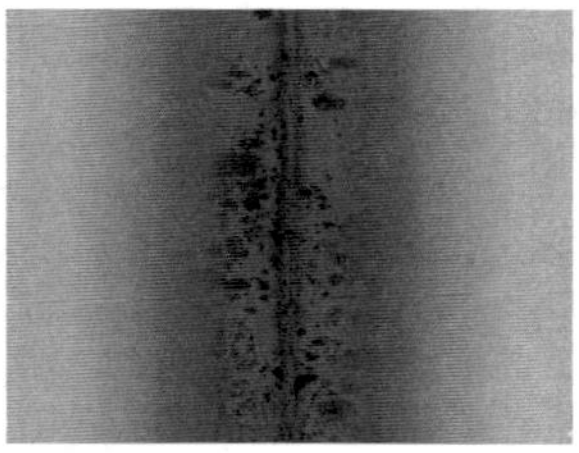	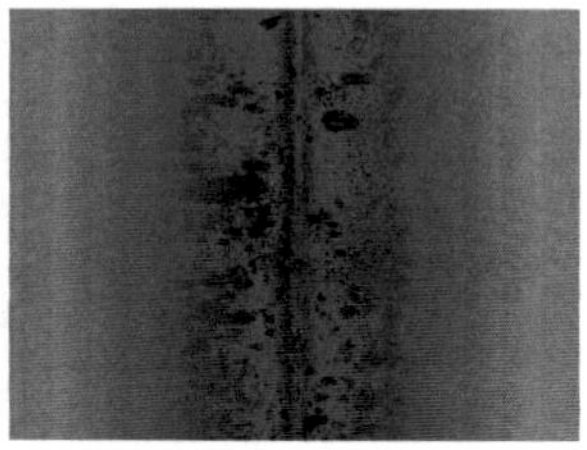
1st Classification	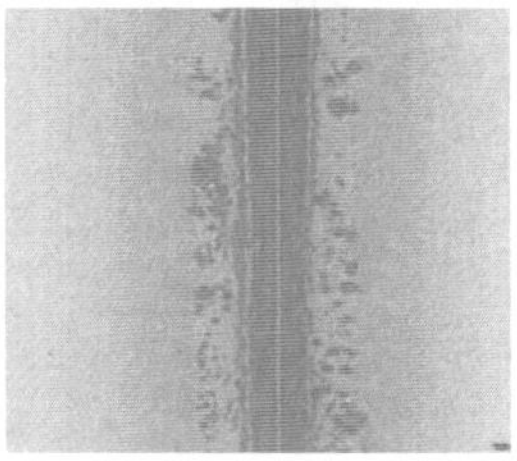	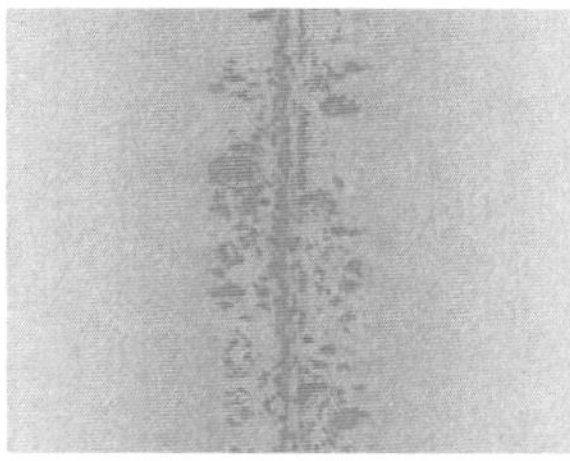	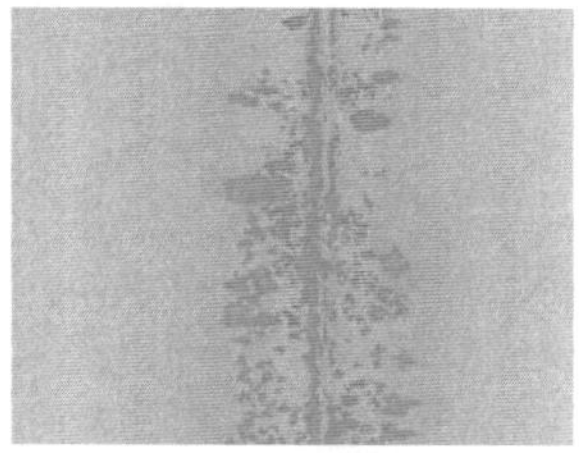
2nd Classification	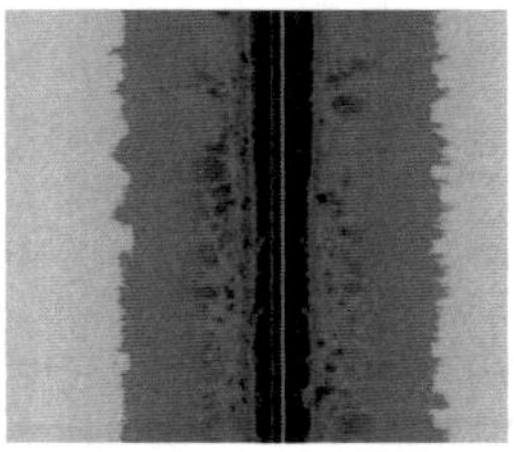	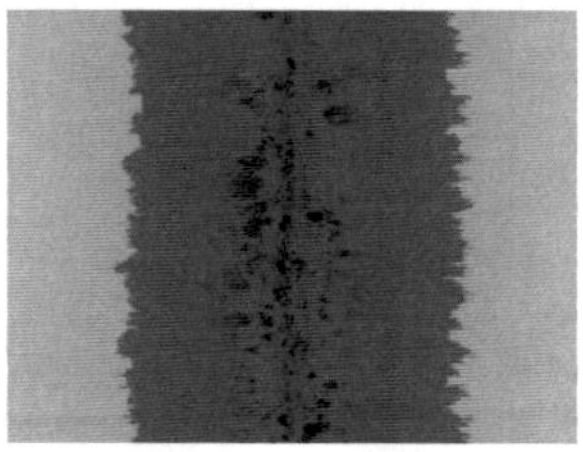	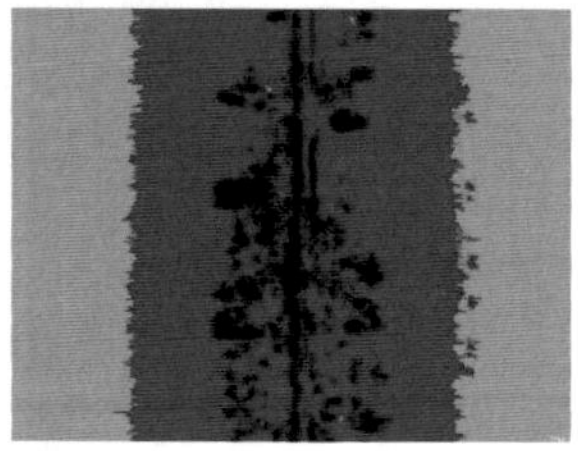
3rd Classification	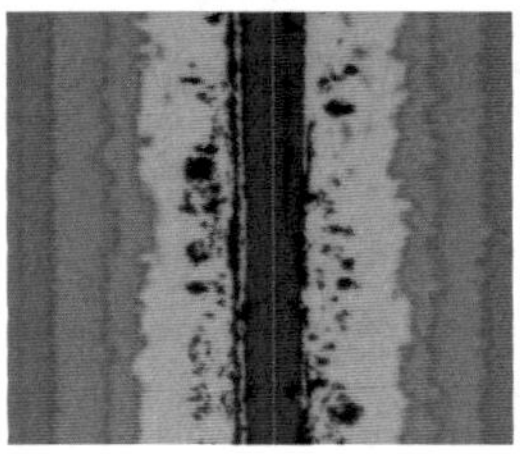	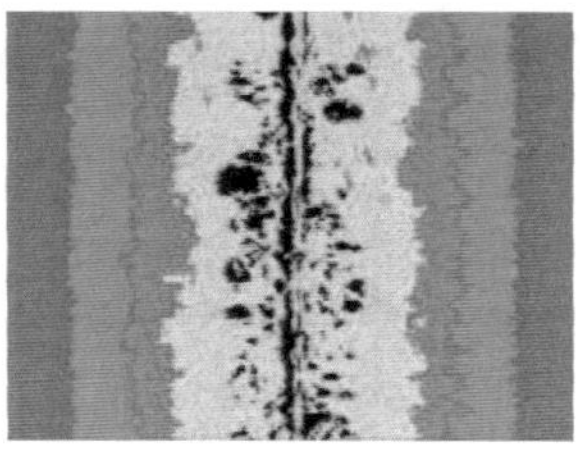	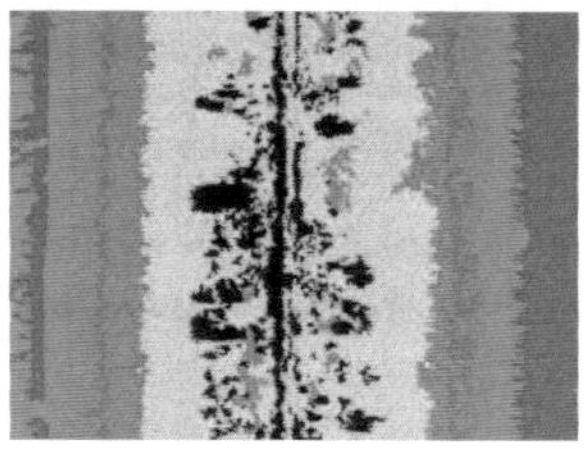

Water Column Rocks Cobble Rubble

Sand Silt

This proves that the TVG correction that regulates the backscatter to affect the hue in the image has an important role in the seafloor classification process, in addition it can sharpen the appearance of the target which is far from the tow-fish range. Water column presence become consideration in the process of segmentation and classification.

Table 5. Error Matrix based on TTA Mask (Image Side-scan Sonar with water column)

User\Reference Class	Silt	Sand	Rubble	Cobble	Rocks	Water column	Sum
Confusion Matrix							
Silt	137047	1973	0	0	0	0	139020
Sand	4256	6539	1742	0	0	0	16228
Rubble	527	2164	13185	5532	0	0	21410
Cobble	2174	0	6346	20552	508	0	29580
Rocks	0	0	0	18	4186	48	4494
Water column	0	0	0	0	573	3579	21728
Unclassified	0	0	0	0	0	0	0
Sum	144004	10676	21273	26104	5267	25136	
Accuracy							
Producer	0.951	0.612	0.619	0.787	0.794	0.841	
User	0.985	0.403	0.615	0.694	0.931	0.973	
Totals							
Overall Accuracy	0.871						
KIA	0.783						

Table 6. Error Matrix based on TTA Mask (Image Side-scan Sonar with water column correction)

User\Reference Class	Silt	Sand	Rubble	Cobble	Rocks	Sum
Confusion Matrix						
Silt	28286	4286	0	0	0	32572
Sand	1393	12430	0	0	0	13823
Rubble	0	3771	7234	1631	0	12636
Cobble	121	0	3221	16000	1346	20688
Rocks	289	0	0	133	6496	6918
Unclassified	0	501	0	0	0	501
Sum	30089	20988	10455	17764	7842	
Accuracy						
Producer	0.940	0.592	0.692	0.900	0.828	
User	0.868	0.899	0.572	0.773	0.939	
Totals						
Overall Accuracy	0.808					
KIA	0.747					

Table 7. Error Matrix based on TTA Mask (Image Side-scan Sonar with TVG correction)

User\Reference Class	Silt	Sand	Rubble	Cobble	Rocks	Water column	Sum
Confusion Matrix							
Silt	15507	0	0	0	0	0	15507
Sand	627	27796	500	0	0	0	28923
Rubble	0	0	16655	0	0	0	16655
Cobble	0	0	0	23865	0	0	23865
Rocks	0	0	0	0	6015	48	6063
Water column	0	0	0	0	0	3579	3579
Unclassified	0	0	0	0	0	0	0
Sum	16134	27796	17155	23865	6015	3627	
Accuracy							
Producer	0.961	1	0.970	1	1	0.986	
User	1	0.961	1	1	0.992	1	
Totals							
Overall Accuracy	0.987						
KIA	0.984						

Comparison of the accuracy of original image with accuracy of water column correction shown that the image with water column correction has a higher accuracy. Water column and the rock are a critical sample in the classification process. The pixel values of the two classes has no significant difference and equally high (dark), seen in the second level classification results for the original image (in the presence of the water column), the water column target known as rocks shown in black. Critical samples show the similarity of spectral characteristics of the target will affect the value of the target stability, critical samples show the target that will affect the results, showing which samples will be similar / critical and the number of samples.

The result of seafloor classification has been validated by integration of grab sampling distribution map and bathymetry of study area taken adjacent to Side-scan Sonar measurement, although the sediment distribution map is made in larger scale so it is more general. Map of sediment distribution shows sedimentation in the form of gravel material where the seafloor classification of segmentation result is in the range of sand class to cobble. Deep classes in bathymetry map indicated that large grain will be in the shallower depth class, the smoother and smaller grain sediment will be in the deeper depth class.

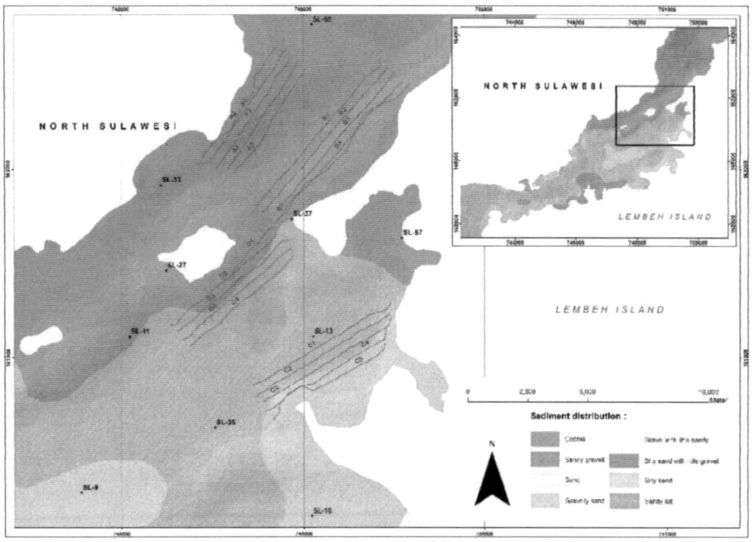

Figure 6. Sediment distribution map in Lembeh Strait, Bitung, North Sulawesi, grab sampling in April 2016

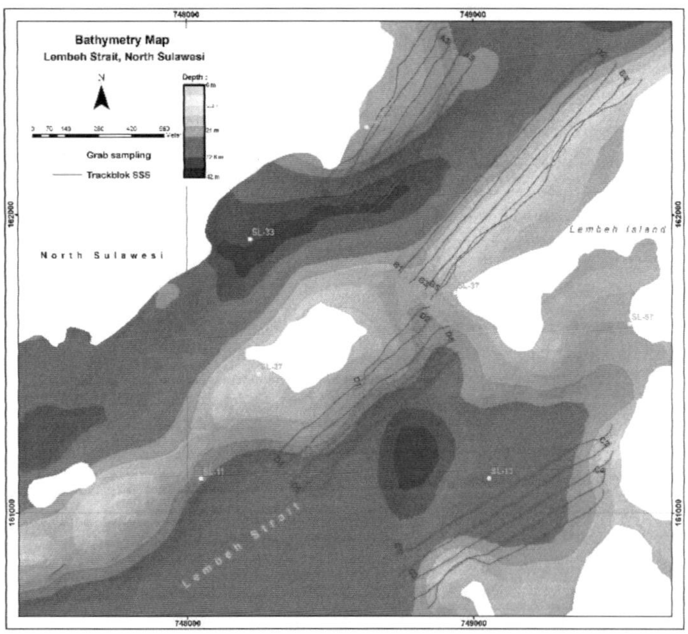

Figure 7. Bathymetry map in Lembeh Strait, Bitung, North Sulawesi, sounding in April 2016

IOP Conf. Series: Earth and Environmental Science **162** (2018) 012009 doi:10.1088/1755-1315/162/1/012009

5. Conclusion

Segmentation and classification of Side-scan Sonar can be done using integrating backscatter power and image processing with difference pixel value of target. Backscatter associated with the target material of hardness and roughness can be correlated in multi-resolution segmentation method using smoothness and compactness values. However, every pixel value each correction level varies for each type of data so that in the segmentation and classification process it is necessary to retrieve the training area to find out the critical target or target with a high probability of bias. The classification results of the segmentation process would be better if the data validation is done on the same scale.

Acknowledgements

This research is part of the thesis as one of the requirements to obtain a Master of Science degree in Marine Technology Studies major and first stage research of the Master Program of Education Leading to Doctoral Degree for Excellent Graduates. Acknowledments to the Ministry of Research Technology and Education and Research Center, also for Marine Geological Research and Development Center (PPPGL), Ministry of Energy and Mineral Resources for assistance in preparing this research.

References

[1] Blondel P 2009 *The Handbook of Side-scan Sonar* (New York: Springer)

[2] Jain A K 1989 *Fundamental of Digital Image Processing* (New Jersey: Prentice-Hall International Inc)

[3] Yogamangalam B R 2013 Segmentation techniques comparison in image processing *Intern. J. Eng.Technol* **5** 1

[4] S Klatt 2012 *Skid Trail Detection with Multiresolution Segmentation In Ecognition Developer* (https://www.scribd.com/document/255524648/wee-ecog)

[5] Anguilar M A, Vicente R, Anguilar F J, Fernandez A, and Saldana M M 2012 Optimizing object-based classificationin urban environment using very high resolution geoeye-1 imagery *Pho. Eng. Rem. Sen.* 1-7

[6] Lu K C and Yang D L 2009 Image processing and image mining using decision trees *J. Inf. Sci. Eng.* **25** 989-1003

[7] Ouyang Z 2015 *Object-Based Classification & eCognition* (Michigan: Department of Geography, Environment, and Spatial Sciences & Center of Global Change and Earth Observation)

[8] Gardner P D and James V 2004 Predicting seafloor facies from multibeam bathymetry and backscatter data *Pho. Eng. Rem. Sen.* **70** 1081–1091

Temperature-Salinity stratification in the Eastern Indian Ocean using argo float

N P Purba[1,2], W S Pranowo[3,4], I Faizal[2], H Adiwira[1,2]

[1] Marine Research Laboratory (MEAL), Padjadjaran University, Jatinangor, Indonesia
[2] Department of Marine Science, Faculty of Fisheries and Marine Science, Padjadjaran University, Jatinangor, Indonesia
[3] Marine and Coastal Data Laboratory, Marine Research center, Agency for Marine & Fisheries Research & Human Resources, Ministry of Marine Affairs & Fisheries, Indonesia
[4] Department of Hydrographic Engineering, Naval Postgraduate School (STTAL), Indonesia

E-mail: noirpurba@gmail.com

Abstract. The spatial and temporal of Temperature and Salinity in the Eastern Indian Ocean (EIO) are studied by using Argos data. EIO consisting of Pacific seawater mass which passing through Indonesian seas and mix with Indian Ocean itself that form complex formation. It reveals link among ITF, SEC, Leeuwin Current, and SJC and routes into global circulation. The purpose of this study is to describe the main physical properties variability based on Argo Floats data. The dataset is used from 80 Argo Floats during 1999-2016 within area south of Java to west of Australia with depth range 0-2000 meters. The result shows that vertical temperature and salinity profile quite homogeny near Sunda Island-West Australia. Monthly spatial distribution of temperature is founded 15-30^0 C, and varying in lesser Sunda. Vertical profile in 4 sections is shallow at near Java, but shows differences in 25 horizontal. Salinity profile is having range 25-34 psu. ITF is contributing to EIO mixing. Finally, salinity is a parameter that affects the EIO dynamics especially in near Australia basin. The SEC variability showed clear in between 10^{0}S to 15^{0}S.

1. Introduction

East of Indian Ocean (EIO), also known as South Indian Ocean (SIO) [1] is relatively complex and of the representation of unique characteristic in Indian Ocean. EIO covering south of Java Island, Timor Seas, and Western Australia. The presence of several inputs and its dynamics principally controlled by Pacific and Indian Ocean itself has been confirmed. Seawater mass via Indonesia i.e. ITF [2, 3], come from Indian Ocean itself (South Equatorial Currents; SEC), and Leeuwin currents flows southward [4, 5, 6, 7]; South Java Current (SJC) [8] and also monsoonal periodic [9]. EIO becomes area that water mass from Indian and Pacific Oceans meet [10] and flow water mass from east boundary to west. Both ITF and Leeuwin play important roles in global and regional circulation scales. The mixing and process lead to understand the pattern linkage into global circulation. A proper representation of EIO is particularly important. EIO consists of tropical and sub-tropical area [11] which temperature and salinity play important issues in term of physical, biological, and chemical processes in the ocean. A good representation of the mixed layer in the Southern Ocean is particularly important feature because various mechanisms add to force the circulation in EIO area.

From the previous research, subsurface data could not be collected over whole ocean basins simultaneously [12] because limited instrument and time. Argo Floats (AF) as dynamics in situ measurement brings a new era in oceanographic observation. The Argo program is a major component of the Global Ocean Observing System and strives to monitor the evolving temperature and salinity fields of the upper ocean [13, 14, 15]. Since AF launched in 1999, there are 3,814 AF worldwide operating (update July 2017). For wide range spectrum of research [16] such as for Mixed Layer Depth (MLD) stratification, [17] for mesoscale eddies, and near real time access data, AF becomes one of instrument to describe ocean phenomena. This method is used vary widely, and become important measurement for in situ observation. Argo observations in the Indian Ocean are creating new insights from many different authors on a variety of subjects especially in oceanography fields.

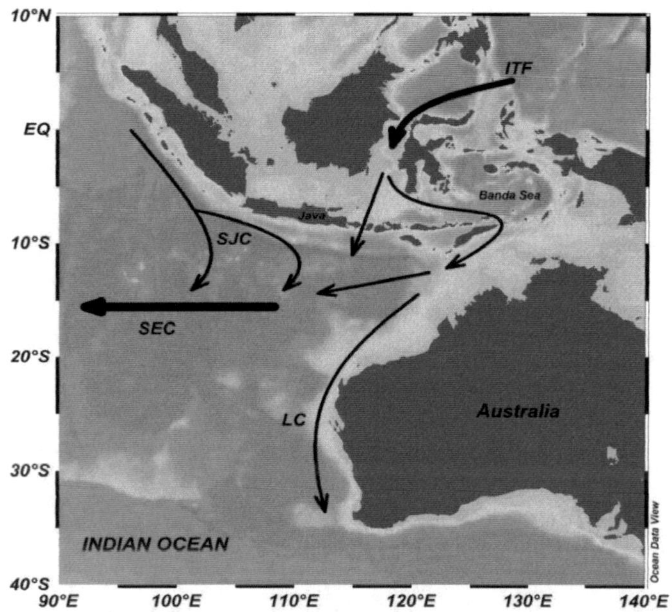

Figure 1. Schematic circulation in East of Indian Ocean (after [18]). SJC (South Java Currents) flows around Sumatera and Java, SEC (South Equatorial Currents), LC (Leeuwin Currents) from ITF flows southward, ITF (Indonesia Troughflow) is a part of Makassar flows.

Studying of Indian Ocean dynamics using AF is much more effective and cheaper than by using research vessels. ITF and SEC occurred along the year, but ITF is at its peak during Southeast Monsoon. A strong ITF current leads to LC current strengthen, meanwhile SJC at its peak on Northwest Monsoon season. The meeting of all the currents will result a changing pattern of monthly and yearly ocean variability in Indian Ocean. In this study we attempt to investigate the dynamics of EIO by investigating spatial and temporal temperature and salinity from Argos data. The objectives of this research are to enhance the water properties in EIO. Here we will concentrate on the different source and pattern near Indonesia-Australia basin. The EIO define between the northwest coast of Australia and the Indonesian islands of Java and Sumatra is termed the Indo-Australian Basin.

2. Method

Area of Study is laying in the Eastern Indian Ocean near basin Indonesia-Australia (90.0 E-130.0 E and 0.0-30.0 S) (figure 2). The northern boundary is limited to Sumatera, Java, and islands near Bali, while the eastern boundary is Australia. Vertical structure and spatial distribution investigated using Argos profile from August 1999 to June 2016. Each floats diving into a depth (typically 1,000 to 2,000 m) and drifts freely. In every 10 days, it swimming up to the surface and acquisitioning temperature, conductivity (salinity), and pressure. The accuracy of Argo for temperature is 0.005^{0}C and 0.01 PSU for salinity. Argo Files is downloaded from the Argos server www.argoweb.whoi.edu. These data were collected and made freely available by the International Argo Program and the national programs that contribute to it (http://www.argo.ucsd.edu, http://argo.jcommops.org). The Argo Program is part of the Global Ocean Observing System. We only use good quality coverage for analysis [19]. Complete documentation can be seen in http://www.argodatamgt.org.

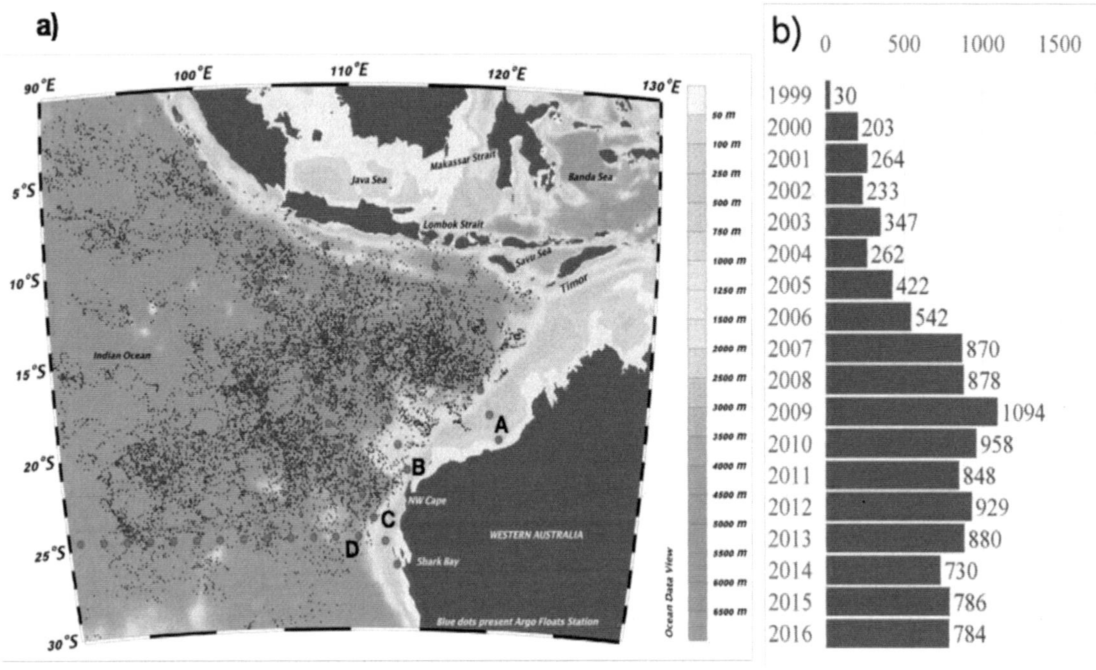

Figure 2. Area of EIO and Argos distribution, a) Distribution of Argos and four section in EIO area and b) Argo counts from 1999-2016 (bottom topography these water masses is dominated by the presence of Indonesia basin or south Java Basin, Perth Basin, and West Australia Basin with depth from 50-4000 meters

The number of profiles increases since launched in 1999. We also visualize with Ocean Data View [20]. We only use pressure (P), temperature (T), and salinity (S) in good quality data and probably good quality data to analyze [17]. Here we visualize vertical and spatial structure of temperature and salinity. We analyze 4 vertical sections to describe some phenomena in East of Indian Ocean. Three sections are over Indonesia-Australia, and 1 is over Australia. Section A (115^0 E) is located near Bali-NW Cape as ITF representative; Section B (110^0 E) is located in Lesser Sunda-NW Cape as ITF-SCJ-Leeuwin characteristics; Section C (100^0 E) is located in Sumatera-Shark Bay as SEC-Leeuwin; Section D is located at Shark Bay (25^0 S) as a Leeuwin boundary (22^0 S-34^{0}S). There are 11,332 data stations acquired from 80 AF. In this study, we only investigating 0 - 500 m depth and represent mixed layer, thermocline and halocline layers, and deep layer.

3. Results

3.1. T-S Profile

The location of water masses, and their relative position to each other can be identified using Temperature and Salinity (T-S) diagram. T-S profile shows the difference characteristic of density for each station. For Argo stations that close to each other like the ones around Savu and Java Seas has a similarity on its seawater characteristics. If it compared to station around West Australia Sea, shows a large range of characteristics differences. These because the presence of mixing forms a large range of characteristics differences in the EIO [21]. The water masses in EIO forms by Indonesian Upper Water (IUW), Indonesia Intermediate Water (IMW), South Indian Central Water (SICW), Antartic Intermediate Water (AAIW).

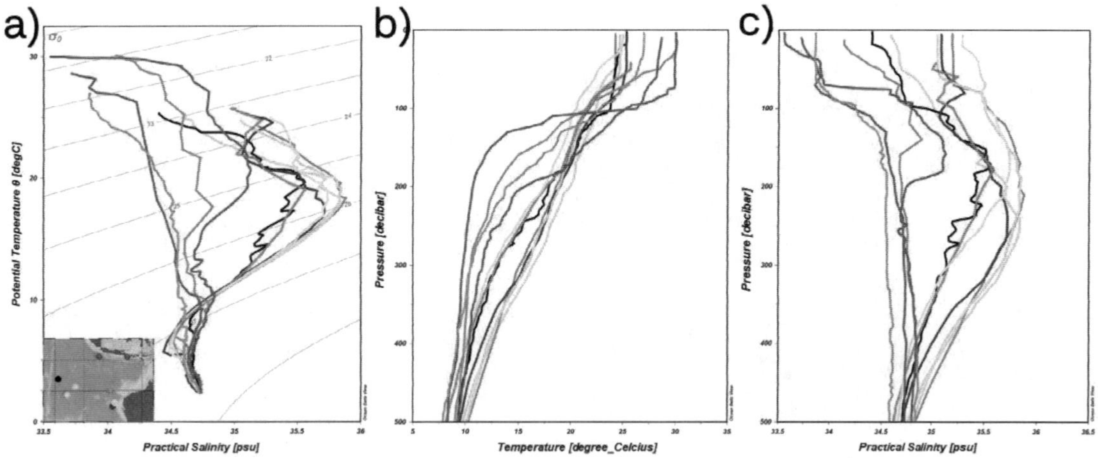

Figure 3. Temperature and Salinity profile from several path in EIO (inset) a) T-S diagram assumed some watermass properties. Four dots near Java island to describe South Java Currents (SJC) and ITF, three dots assumed near Australia (Leeuwin and ITF), and three dots in the centre of EIO to find out the SEC, b) and c) Temperature and Salinity vs depth profiles up to 500 meters.

See Figure 3 (inset picture). Seawater mass in the outflow area of ITF (red is near Lombok and green dot near Savu) directly affected by seawater mass from the North Pacific [2]. Seawater mass in this station has mean temperature about 28°C, with its mean salinity is 33.75 PSU. The same mean values have been found also from several research [8,22]. A drastic descending of temperature found in the ITF outflow area (15°C, in 200 m depth). Seawater mass characteristics in the EIO supplied from varies watermass. Two main source flows from Subtropical Indian Ocean (SIC) and Australasian Mediterranean Sea [23].

3.2. Temporal and vertical Stratification of Temperature and Salinity

EIO entrances and follows some different paths before link/joint to equatorial currents flows to Western Africa. From sections analysis in Figure 4, temperature and salinity throughout lesser Java and Bali having a very unique pattern profiles.

Temperature and salinity profile from 2000-2016 shows unique water mass characteristics. The thermocline layer in sections A, B, and C founded in 150-400 meter, in section D it found from 150-500 m depth. The temperature at the surface (mixed layer) ranged from 22.5°C-30°C with the lowest temperature in section D (Subtropical area). Vertical temperature profile tended to fluctuate in section C which is from South Sumatera Sea to Australian Sea with monthly interval. The value of temperature on the mixed layer changes monthly, with the mean value 27.5°C. The same pattern also found in section B and C.

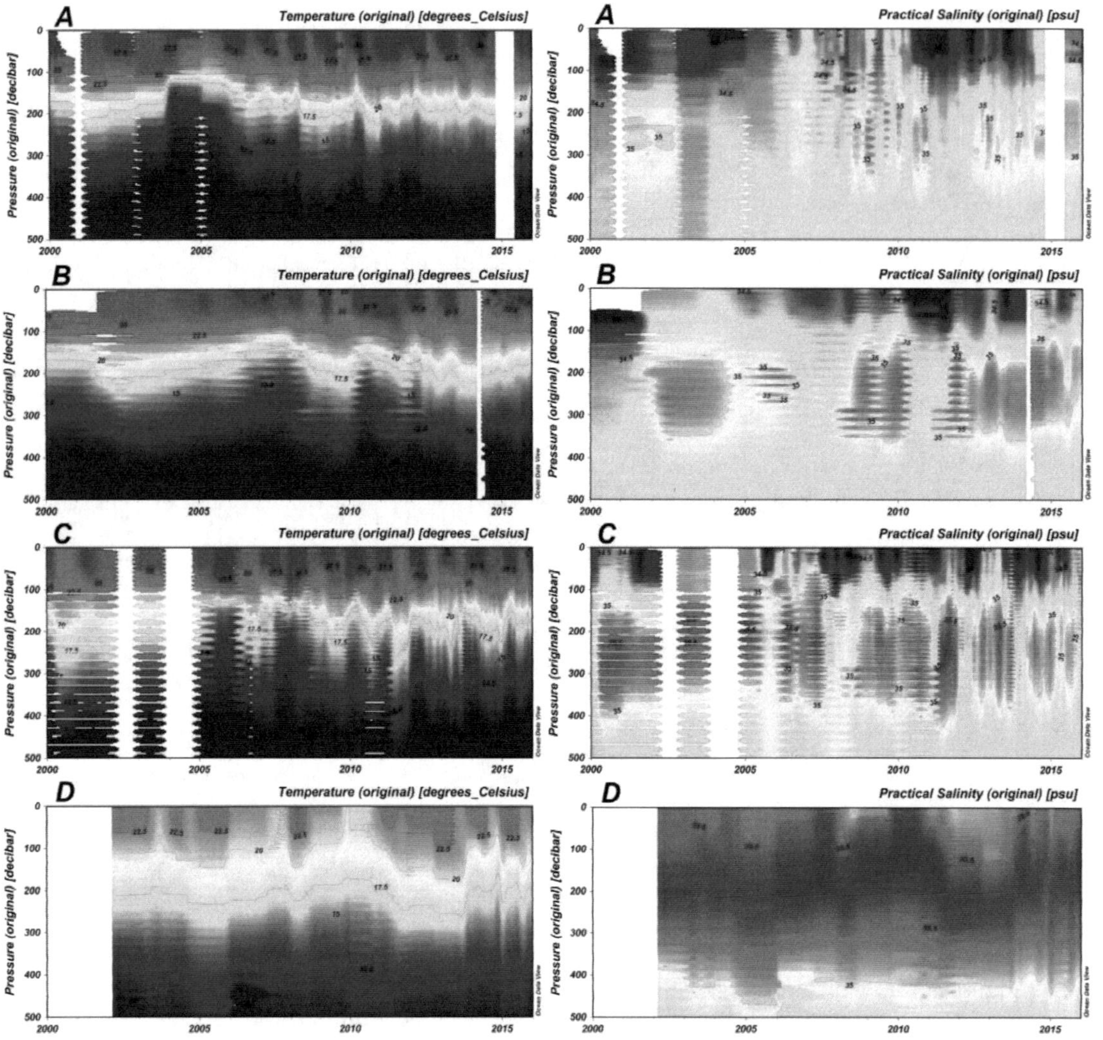

Figure 4. Vertical section over time (left) Temperature, (right) Salinity

The salinity profile (34 – 35.5 PSU) is more vary than temperature. Halocline layer found about 200 meter depth. The salinity profile doesn't have similar pattern with temperature profile. A fluctuated salinity distribution found in section C with halocline condition developed differently every month.

For section A, the profile of Bali Strait-NW Cape, with a deeper mixed layer (100-150 meter) and thermocline range from 50-75 meter depth. In the surface, water temperature is higher around Bali Strait shows that there is water mass added from Makassar and Banda Seas. Profiles A to C show insignificant differences, but profile D shows a very different water temperature characteristics. As regards the distribution of heat content, the distribution in the whole deep appears to be uniform, values ranging from 10-15°C.

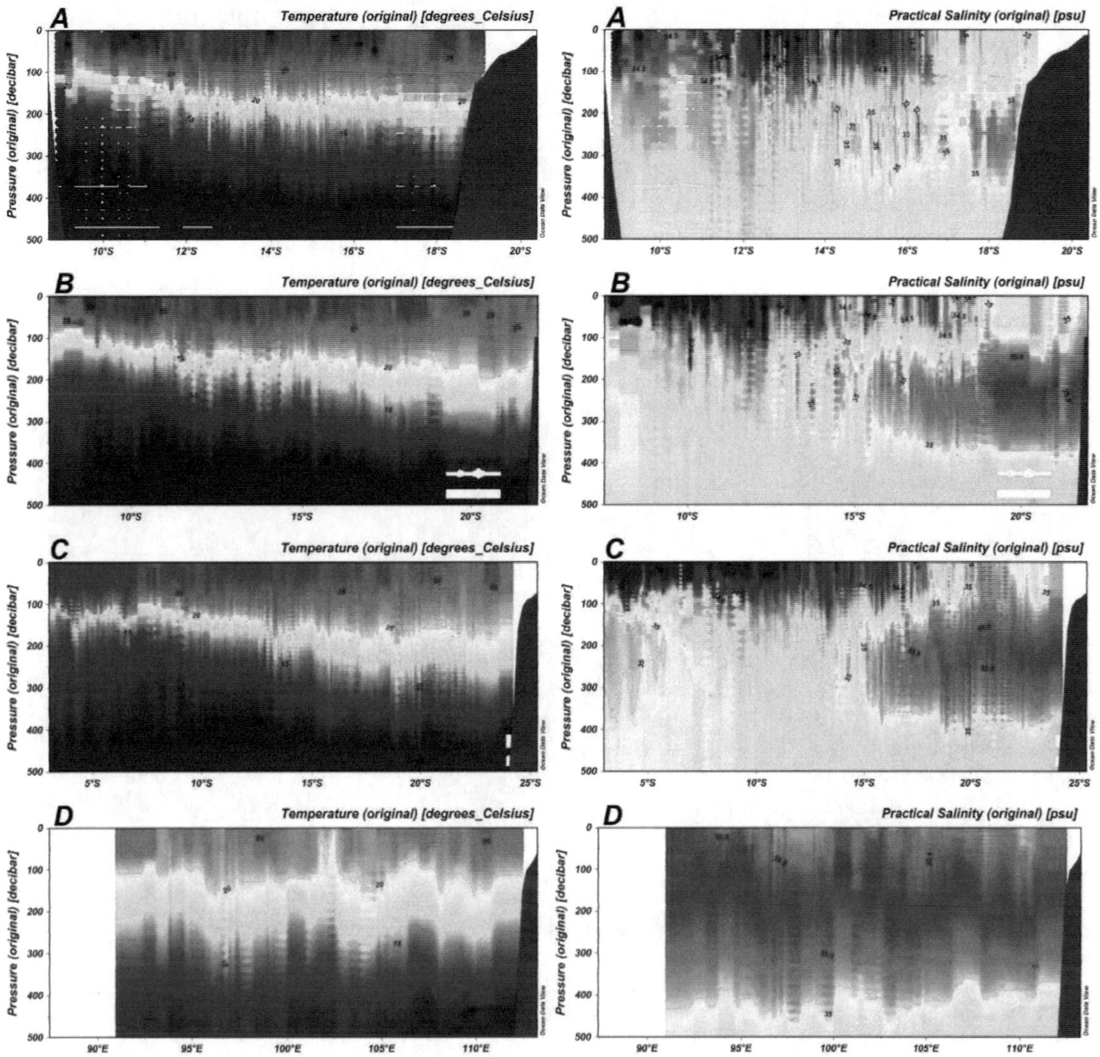

Figure 5. Stratification water mass (left) Temperature, (right) salinity

Salinity profile for section A to C displaying a higher salinity around Australia Seas (35-35.5 PSU), meanwhile salinity around Indonesia Seas is lower (34-34.5). The lower salinity around Indonesia Seas caused by the input of freshwater from SJC. The semi-annually reversing SJC plays an important role in distributing freshwater into and out of the southeast Indian Ocean [24]. For section D (25°S) shows a very different profile than the other section, it's because section D is approximately the boundary of subtropical and tropical zone. The vertical stratification of salinity shows the presence of a salinity maximum zone in the southern of the section. The minimum salinity value is observed near Java within the mixed layer. The vertical distribution of temperature and salinity content is very similar to that in section A and C, especially in the upper layers.

3.3. Spatial variability of Temperature

In the surface layer, SST shows a very distinctive difference in 15°S, 20°S, and 30°S. On January to April, ITF strengthen and causes the temperature arises to about 30°C. The changing temperature also shows in West Sumatra (5^0S), especially the rising temperature started from April. The rising temperature also altering the boundary at 17°S and then expands to about 20°S, but LC with maximum transport is around 22°S [1].

Spatially, the changing of temperature on the surface is very dynamic especially in the seas close to Indonesia region. The boundary of temperature is very clear in 15°, 20°, 25°. The heat carried by the (ITF), as it follows its circuit route through the basins and multiple narrow impacts of both the Pacific and IO [25] and moved water mass from Pacific to IO [26]. The total distance ITF in IO to southward path from 10^0S until 20^0S near the NW Cape 1110 km.

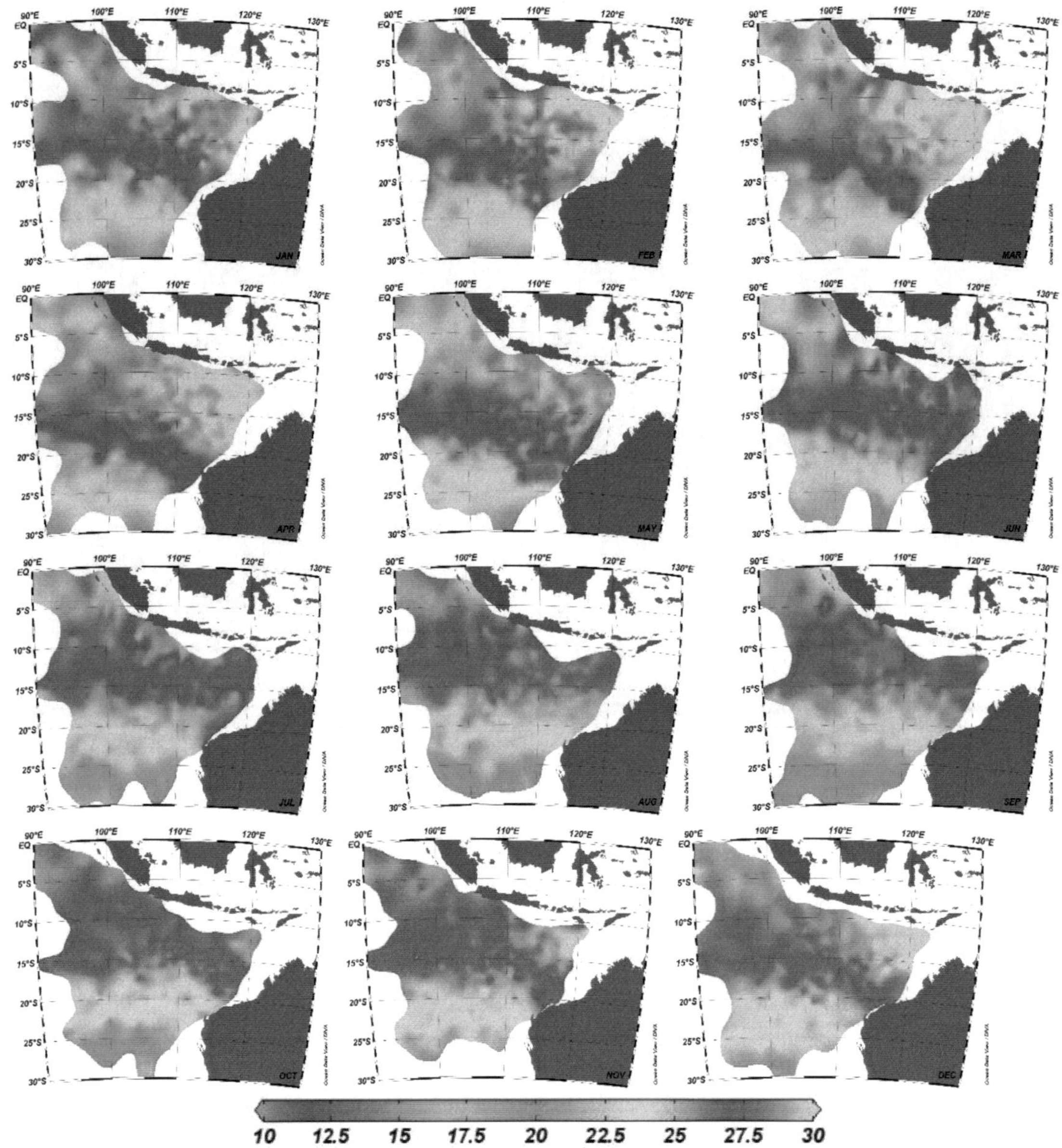

Figure 6. Monthly temperature profile in ~4 meter (4dbar)

The temperature of water mass coming from Pacific Ocean via Indonesia Troughflow (Bali Strait at 115^0 E and Savu at 121^0 E) as primary outflow expands to 20°S and promotes a stronger LC. In summer months, added warm from tropical and salt from Western Australia Current. Thereafter, the cold temperature water from subtropical south expands northward in May to about 20°S with minimum temperature 17.5°C. Range temperature near 30°S is 23-32°C [5]. The flow starts in February and peak in June. In August and September, water from Antarctic penetrates as far to 20°S [21]. Variability in the SEC in EIO region is weak during Jan-May, but is strong during Jun-Oct. The strong SEC liner with strong flows of ITF to Indian Ocean [2].

3.4. Spatial variability of Salinity

The SJC reverses to southeastward flow semi-annually around May and November, SJC has been found to consist of narrow cores of accelerated flow extending to 150-250 m depth [10, 27]. Near NW Cape, salinity rises. The lower salinity around Sumatera and Java occurred monthly, with range about 33-35 PSU. Low salinity water is carried out southward in some area at Java strait and near Bali.

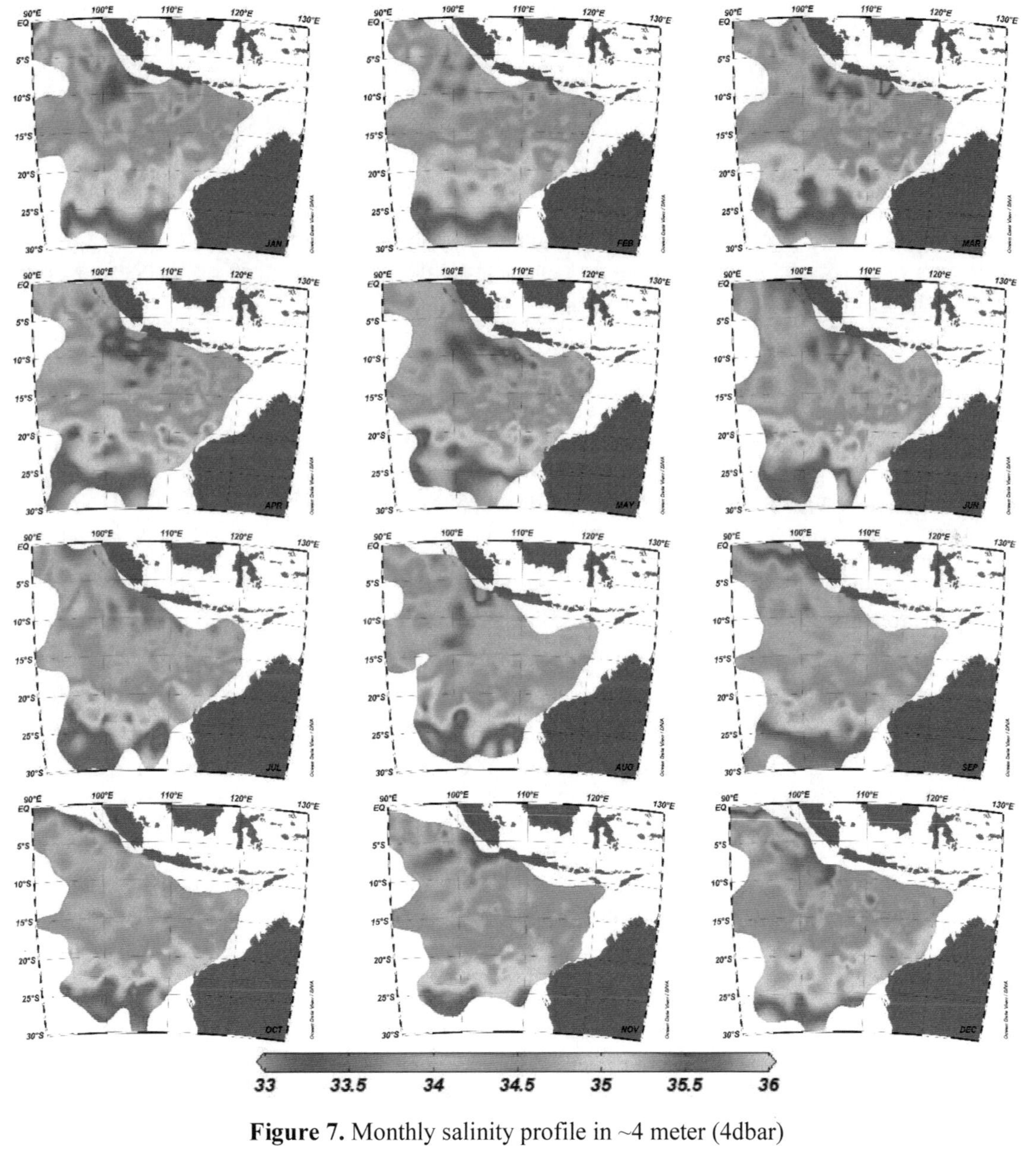

Figure 7. Monthly salinity profile in ~4 meter (4dbar)

The changing pattern near Indonesian Seas also followed by change pattern in the South Indian Ocean (20 °S -25°S). For the occasion of the decreasing salinity in the Indonesia Seas. It causes the high salinity moves southwardly. This probably caused by strengthen of SEC with salinity range 34.2-34.7 PSU and have been recognized [23]. Based on model [20] salinity in near 30°S with range 31.5-35.3 PSU. The eastern of Australia in May with strong salinity fronts [6].

4. Conclusion

We have described the characteristics and variability of temperature and salinity using Argo profiles in EIO between 1999-2016. This more precise description about using Argo's to assess the characteristics in Indonesia-Australia that linked to SEC. ITF contributes a huge impact on the changing of temperature and salinity in the EIO. The water mass profile on the seas close to Sumatera-Java-Bali tends to fluctuate highly, and there is a distinctive stratification in salinity and temperature profile. Salinity can play a strong role in the density stratification of EIO. Temperature and salinity varies in all seasons, and furthermore the temperature structure in the upper water column can vary depending on the season. SJC not able to be seen clearly from the temperature and salinity contour, this probably caused by the deficient resolution of Argo. But overall, AF defines temperature and salinity adequately in all sections. From all section, the description of SJC is not clear to define in the spatial contour. This is might be the position of Argos is not close to Java.

Acknowledgements

This work mostly conducted in the Computation Lab at Department of Marine Science, Padjadjaran University. In particularly, great thanks to KOMITMEN' Crew for data analysis. We also thank the anonymous reviewers for their insightful comments and for their suggestions regarding to improving the manuscript.

References

[1] Lambert E, Bars D L and Ruijter W P M 2016 The connection of the Indonesian Throughflow South Indian Ocean Countercurrent and the Leeuwin Current *J. Ocean Sci.* **12** 771–80 doi:10.5194/os-12-771-2016

[2] Atmadipoera A, Molcard R, Madec G, Wijffels S, Sprintall J, Larrouy A K, Jaya I and Supangat A 2009 Characteristics and variability of the Indonesian throughflow water at the outflow straits *J. Deep-Sea Res.* I **56**, 1942–54

[3] Lee T, Fukumori I, Menemenlis D, Xing Z and Fu L L 2002 Effects of the Indonesian Throughflow on the Pacific and Indian Oceans *J. of Phys. Oceanography* American Meteorological Society **32,** 1404-29

[4] Feng M, Meyers G, Pearce A and Wijffels S 2003 Annual and Interannual Variations of the Leeuwin Current at 32°S. *J. of Geophys Res.* **108**, C11 3355 doi:10.1029/2002JC001763

[5] Utamy R M, Purba N P, Pranowo W S and Suherman H 2015 The Pattern of South Equatorial Current and Primary Productivity in South Java Seas 5th International Conference on Environment Science and Biotechnology (ICESB) **81,** 1-5 doi: 10.7763/IPCBEE

[6] Thomson R O R Y 1984 Observation of Leeuwin Current off Western Australia *J. American Meteorology Society* **14,** 623-8 https://doi.org/10.1175/1520-0485

[7] Waite A M, Thompson P A, Pesant S, Feng M, Beckley L E, Domingues C M, Gaughan D, Hanson C E, Holl C M, Koslow T, Meuleners M, Montoya J P, Moore T, Muhling B A, Paterson H, Rennie S, Strzelecki J, and Twomey L 2007 The Leeuwin Current and Its Eddies: An introductory overview *J. Deep-Sea Res.* Pt. II **54** 789–96 https://doi.org/10.1016/j.dsr2.2006.12.008

[8] Sprintall J, Chong J, Syamsudin F, Morawitz M, Hautala S, and Bray N and Wijffels S 1999 Dynamics of The South Java Current in The Indo-Australian Basin *J. Geophys. Res. Letters* **26** (16): 2493-6

[9] Webster P J and Fasullo J 2003 Monsoon: Dynamical Theory in *Encyclopedia of Atmospheric*

Sciences. Eds. J. Holton and J. A. Curry Academic Press 1370-85

[10] Fieux M, Andrie C, Delecluse P, Ilahude A G, Kartavtseff A, Mantisi A, Molcard R, and Swallow J C 1994 Measurements within the Pacific-Indian Oceans throughflow region *J. Deep Sea Res.* Part 1 **41** 1091-130

[11] Pranowo W S, Philips H and Wijffels S 2005 Upwelling Event 2003 Along South Java Sea & The Sea of Lesser Sunda Islands *J. Segara* **1** (3): 116-123

[12] Roemmich D and Gilson J 2009 The 2004–2008 mean and annual cycle of temperature, salinity, and steric height in the global ocean from the Argo program *J. Prog. Oceanogr.* **82** 81– 100 doi:10.1016/j.pocean.2009.03.004

[13] Argo Science Team 1998 On the design and Implementation of Argo – An initial Plan for a global array of Profiling Floats International CLIVAR project Office ICPO Report No. 21. *GODAE Report* **5** (Melbourne: GODAE International Project office, c/o Bureau of Meteorology), 32pp

[14] Intergovernmental Oceanographic Commission (IOC) 1993 Toward a Global Ocean Observing System: The Approach to GOOS (Paris: UNESCO IOC-XVII/8) *Report,* 73 pp

[15] Riser S C, Freeland H J, Roemmich D, Wijffels S, Troisi A, Belbéoch M, ... and Jayne, S R 2016 Fifteen years of ocean observations with the global Argo array Nature Climate Change, 6(2), 145–153 https://doi.org/10.1038/nclimate2872

[16] Dong S, Sprintall J, Gille S T and Talley L 2008 Southern Ocean mixed-layer depth from Argo float profiles *J. Geophys. Res.* **113** C06013 doi:10.1029/2006JC004051

[17] Chaigneau A, Texier M L, Eldin G, Grados C and Pizarro O 2011 Vertical structure of Mesoscale Eddies in The Eastern South Pacific Ocean: A composite Analysis from Altimetry and Argo Profiling Floats *J. Geophys. Res.* **116** C11025 doi:10.1029/2011JC007134

[18] Talley L D 2013 Closure Of The Global Overturning Circulation Through The Indian, Pacific And Southern Oceans: Schematics and Transports. *J. Oceanography,* **26**(1), 80-97, http://dx.doi.org/10.5670/oceanog.2013.07

[19] Böhme L, and Send U 2005 Objective analyses of hydrographic data for referencing profiling float salinities in highly variable environments *J. Deep-Sea Res.* Part II **52**(3-4)**,** 651–64 doi:10.1016/j.dsr2.2004.12.014

[20] Schlitzer R 2016 Ocean Data View https://odv.awi.de

[21] Quadfasel D, Frische A and Cresswell G R 1996 The circulation in the source area of the South Equatorial Current in the eastern of Indian Ocean *J. of Geophysical Research,* **(1)** 12483-12488

[22] Sprintall J, Wijfels S, Chereskin T and Bray N 2002 The JADE and WOCE I10/IR6 Throughflow sections in the southeast Indian Ocean. *Part 2: J. Velocity and transports* **49**(7–8), 1363-1389, https://doi.org/10.1016/S0967-0645(01)00163-1

[23] Bray N A, Wijffels S E, Chong J C, Fieux M, Hautala S, Meyers G and Morawitz W M L 1997 Characteristics of the Indo-Pacific throughflow in the eastern Indian Ocean *J. Geophys. Res. Lett.,* **24,** 2569–2572

[24] Gordon A L, Ma S, Olson D B, Hacker P, Ffield A, Talley L D, Wilson D and Baringer M 1997 Advection and Diffusion of Indonesian throughflow water within the Indian Ocean South Equatorial Current *J. Geophysical Res. Letters,* **24(**11), 2573-2576

[25] Bryden H L and Imawaki S 2001 Ocean transport of heat In: Siedler G, Church J, Gould J. (Eds.), *Ocean Circulation and Climate* (New York: Academic Press) **77,** 736 pp

[26] Schneider N 1998 The Indonesian Throughflow and the global climate system. *J. of Climate* **11,** 676–89

[27] Meyers G, Bailey R J and Worby A P 1995 Geostrophic transport of Indonesian Throughflow *J. Deep Sea Res. Part 1* **42,** 1163-74

IOP Conf. Series: Earth and Environmental Science **162** (2018) 012011 doi:10.1088/1755-1315/162/1/012011

Monochromatic interfacial wave propagating over one and two bar(s)

L H Wiryanto

Department of Mathematics, Bandung Institute of Technology, Jalan Ganesha 10 Bandung, Indonesia

Email: leo@math.itb.ac.id

Abstract. Wave propagation is an interesting subject to be modelled and studied in oceanography. Interfacial wave is one of waves that can occur between two fluid-layers of different density. Since the complexity of the problem, simplification is made by considering a specific type of wave, i.e. monochromatic. In propagating, the transmission and reflection of the wave are considered as the effect of disturbing of the fluid depth. It can be made by putting a bar on the bottom of the channel. The dimension of the bar plays an important role in obtaining the minimum amplitude of the transmitted wave. This is then continued to determine the distance between two bars to get the minimum amplitude. The result was presented in the MSA-conference.

1. Introduction

Wave propagation in ocean is an interesting phenomena for entertainment or some attractions, but also carrying energy that can destroy places where the waves are passed by. As references, wave propagation can be read with various mathematical model, such as in Benjamin [1] (KdV equation), Joseph [2] and Kubota, et. al. [3] (Intermediate Long Wave (ILW) equation), Benjamin [4], Davis and Acrivos [5] and Ono [6] (Benjamin-Ono (BO) equation), Matsuno [7], Choi and Camassa [8, 9], and Camassa, et.al. [10] (combination between KdV and ILW equations).

Meanwhile, when the waves approach a beach, they need to be controlled so that they are not dangerous surrounding. Study on wave propagation indicates that energy is proportional to the amplitude squared, and amplitude of the wave can be reduced when the wave passes over a bar, since part of the wave will be reflected and some is transmitted. This can be seen for example in Mei [11]. Therefore, to control waves we can propose by putting some bars along the bottom and we simulate to calculate the percentage of reduction of the amplitude after passing those bars.

In this paper, we consider monochromatic wave, but involving the effect of the fluid above the ocean. The density of air is not neglected, so that we can consider the air is another fluid laying above the ocean, and the wave is then called as interfacial wave. Tuck and Wiryanto [12] modelled the interfacial wave into a single equation called Composite Long Wave equation, similar to BO equation, but involving Hilbert transform. They obtained a periodic interfacial wave as the solution of the model. In propagating, the wave satisfies a relation called dispersion relation, the relation between the frequency and the wave-number. Wiryanto [13] derived that relation based on potential function of fluid particle moving on flat bottom in one system fluid, but then Wiryanto and Jamhuri [14], also Lusia et. al. [15], extended the derivation of the dispersion relation of interfacial wave in two-fluid system. The dispersion relation, for one and two-fluid system, is mainly depends on the fluid depth, so that when the wave propagates passing over a bar, it can change wave-number. This is then used to observe the break up of the wave into reflection and transmission.

The dispersion relation in formulating the transmitted wave is presented in this paper. Since the wave propagates over one or two bars, the water depth changes and each area contains two waves in different direction. Therefore, at the changing depth, we use the continuation of surface and flux. When these conditions are applied we can calculate the amplitude of each wave. So that we can determine the reduction of the amplitude after the wave passing over the bars. Moreover, we can also determine the optimal dimension of the bar and the distance between two bars, giving the minimum amplitude of the transmitted wave. This paper summaries [13], [14] and [16].

2. Problem Description

2.1. Dispersion relation

We consider wave propagation between two fluids with different density in a canal. The lower fluid has density ρ_1 with depth h_1. Another fluid with density ρ_2 and having infinite depth lies down above the lower layer. A bar is placed at the bottom of the canal. Therefore, the wave propagating over the bar is partly transmitted and reflected. We illustrate this wave propagation and the medium in Figure 1.

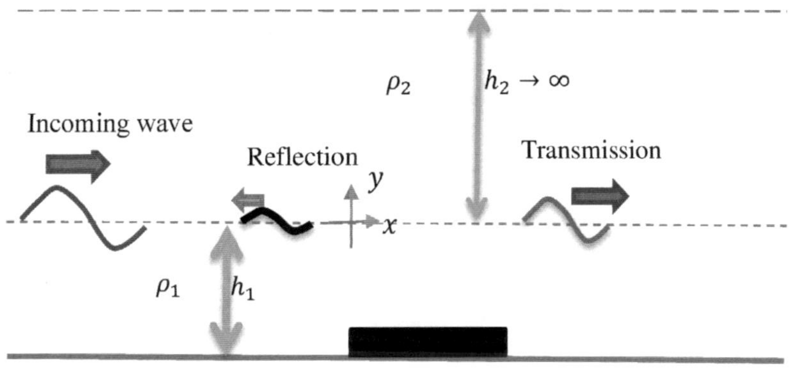

Figure 1. Sketch of the interfacial wave and coordinates

As the coordinates, we choose the x-axis is along the interface with $x = 0$ at the left side of the bar. The vertical y-axis is perpendicular to the horizontal one. So, the wave can be denoted by $y = \eta(x,t)$as the elevation of the still water interface.

In case the bottom is flat and the wave is monochromatic, i.e. $\eta(x,t) = Ae^{-i(kx-\omega t)}$ with constant amplitude A, the relation between the wave number k and the frequency ω is given by

$$\left(\frac{\omega^2}{k}\right)\left[-\rho_1\left(\frac{e^{-h_1 k}}{\sinh(h_1 k)}+1\right)+\rho_2\left(\frac{e^{-h_2 k}}{\sinh(h_2 k)}+1\right)\right] = (\rho_2-\rho_1)g \qquad (1)$$

where g is the acceleration of gravity. The derivation of (1) can be seen in [15]. When the upper fluid is not involved, we can take $\rho_2 \to 0$, and (1) agrees with the dispersion relation for one layer, as derived by Wiryanto [13]

$$\left(\frac{\omega^2}{k}\right) = g \tanh h_1 k$$

Meanwhile, for infinite depth of the upper layer, (1) becomes

$$\left(\frac{\omega^2}{k}\right)\left[-\rho_1\left(\frac{e^{-h_1 k}}{\sinh(h_1 k)}+1\right)+\rho_2\right]=(\rho_2-\rho_1)g \tag{2}$$

This dispersion relation is then used to calculate the amplitude of the transmitted and reflected waves.

2.2. Wave amplitude

We consider an incoming wave of amplitude A. In its propagation, the wave passes a bar of height d and length L. So, there are three regions:

1. in the left side of the bar $x < 0$,
2. above the bar $0 < x < L$,
3. in the right side of the bar $x > L$.

Since the fluid depth changes, the wave number at each region also change satisfying (2) for given ω. Also in region 1 and 2 there are two waves travel in different direction and different amplitude. Therefore, we can express the interface elevation

$$\eta(x,t)=\begin{cases} Ae^{-i(k_1x-\omega t)}+A_r e^{-i(k_1x+\omega t)}, & x<0 \\ B_t e^{-i(k_2x-\omega t)}+B_r e^{-i(k_2x+\omega t)}, & 0<x<L \\ C_t e^{-i(k_3x-\omega t)}, & L<x \end{cases} \tag{3}$$

A is the amplitude of the incoming wave, A_r is the amplitude of the reflected wave and k_1 is the wave number corresponding to fluid depth h_{11} of lower layer in the region 1. Similar to B_t, B_r and k_2 represent the amplitude of the transmitted wave, reflected wave and the wave number corresponding to fluid depth h_{12} of lower layer in the region 2. C_t is the amplitude of the transmitted wave, and k_3 is the wave number in region 3, the same fluid depth in region 1.

Our task is to determine A_r, B_t, B_r, and C_t, that can be calculated from physical situation should be satisfied, i.e. the elevation and the flux must be continue across $x = 0$ and $x = L$. Note that flux is the total horizontal velocity. Wiryanto [13] formulated these into

$$\left.\begin{aligned} A+A_r &= B_t+B_r \\ B_t e^{-ik_2L}+B_r e^{ik_2L} &= C_t e^{-ik_1L} \\ k_1(A-A_r) &= k_2(B_t-B_r) \\ k_1\left(B_t e^{-ik_2L}-B_r e^{ik_2L}\right) &= k_2\, C_t e^{-ik_1L} \end{aligned}\right\} \tag{4}$$

When we can solve that problem, we then observe C_t for various L. Wiryanto [13] showed that there is a relation between both quantities periodically, the same value C_t for many values L. The shortest L that gives smallest value of C_t is then called L optimal, denoted by L_{opt}.

Now we extend for two bars on the bottom of the channel, with distance S. So, there are five regions. Similar following the formulation above, each region contains two waves with different direction and different amplitude, except the last region containing only the transmitted wave, namely E_t. The interface elevation is then expressed by extended from (3), i.e.

$$\eta(x,t)=\begin{cases} Ae^{-i(k_1x-\omega t)}+A_r e^{-i(k_1x+\omega t)}, & x<0 \\ B_t e^{-i(k_2x-\omega t)}+B_r e^{-i(k_2x+\omega t)}, & 0<x<L \\ C_t e^{-i(k_1x-\omega t)}+C_r e^{-i(k_1x+\omega t)}, & L<x<L+S \\ D_t e^{-i(k_2x-\omega t)}+D_r e^{-i(k_2x+\omega t)}, & L+S<x<2L+S \\ E_t e^{-i(k_1x-\omega t)}, & 2L+S<x \end{cases} \tag{5}$$

Here we use the same length (L) and the same height ($h_{11}-h_{12}$) of the bars.

All type of amplitudes is formulated from elevation and flux continuation at each place where the fluid depth changes, We adopt from Wiryanto & Mungkasi [16]

$$\left.\begin{array}{c} A+A_r=B_t+B_r \\ B_te^{-ik_2L}+B_re^{ik_2L}=C_te^{-ik_1L}+C_re^{ik_1L} \\ C_te^{-ik_1(L+S)}+C_re^{ik_1(L+S)}=D_te^{-ik_2(L+S)}+D_re^{ik_2(L+S)} \\ D_te^{-ik_2(2L+S)}+D_re^{ik_2(2L+S)}=E_te^{-ik_1(2L+S)} \\ k_2(A-A_r)=k_1(B_t-B_r) \\ k_1\left(B_te^{-ik_2L}-B_re^{ik_2L}\right)=k_2(C_te^{-ik_1L}-C_re^{ik_1L}) \\ k_2\left(C_te^{-ik_1(L+S)}-C_re^{ik_1(L+S)}\right)=k_1(D_te^{-ik_2(L+S)}-D_re^{ik_2(L+S)}) \\ k_1(D_te^{-ik_2(2L+S)}-D_re^{ik_2(2L+S)})=k_2E_te^{-ik_1(2L+S)} \end{array}\right] \quad (6)$$

In this step, we are interested in determining S optimal, denoted by S_{opt}, i.e. the smallest distance between two bars that gives minimum value of transmitted amplitude E_t, by using L_{opt}. We present the result in the next section.

3. Results

The reflected and transmitted interfacial-waves are observed from an incoming wave of amplitude A in this section. We first discuss the result for one bar submerged beneath the water. In our calculation, we use the acceleration of gravity $g = 10$, wave frequency $\omega = 2$, the lower fluid depth $h_{11} = 3.5$, and the bar height $d = 1.5$ $(h_{12} = 2)$. The lower fluid density is $\rho_1 = 1$.

From those quantities, (2) with $\rho_2 = 0.1$ gives the wave number $k_1 = 0.5132$ corresponding to h_{11} and $k_2 = 0.6809$ corresponding to h_2 the region above the bar. We then calculate the amplitudes: A_r, B_t, B_r, and C_t satisfying (4), for giving the length L of the bar. But basically we are interested in determining C_t, proportional to the amplitude of the incoming wave A. In other word we choose $A = 1$. For $L = 2.3$, we obtain $A_r = 0.2754$ and $C_t = 0.9881$. For other L, we can repeat the calculation. Plot of C_t versus L, shown in Figure2, indicates that there is minimum value of C_t. This value can be obtained for some L, but for the smallest L we then call as the L optimal, denoted by L_{opt}. In this case, we obtain $L_{opt} = 3.0$ giving $C_t = 0.9856$. Similar curve in Fig.2 can be obtained for different density of the upper fluid, for example when we use $\rho_2 = 0.49$, the optimal length of the bar is $L_{opt} = 2.73$ giving $C_t = 0.9769$. We can see the effect of the upper fluid density to the reducing of the transmitted amplitude.

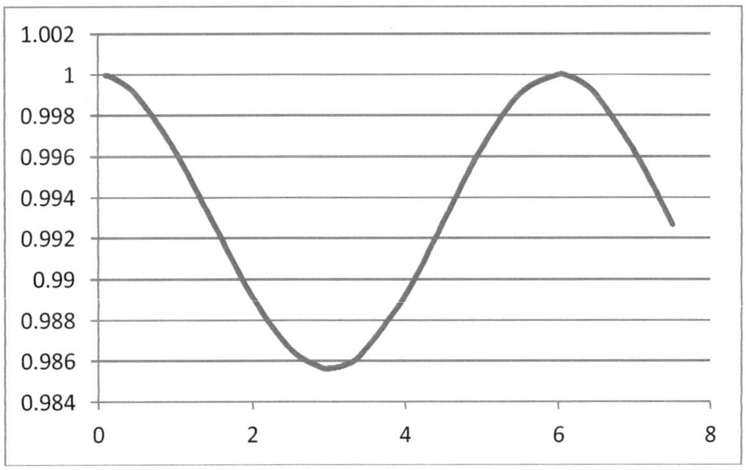

Figure 2. Plot of C_t (vertical) versus L (horizontal)

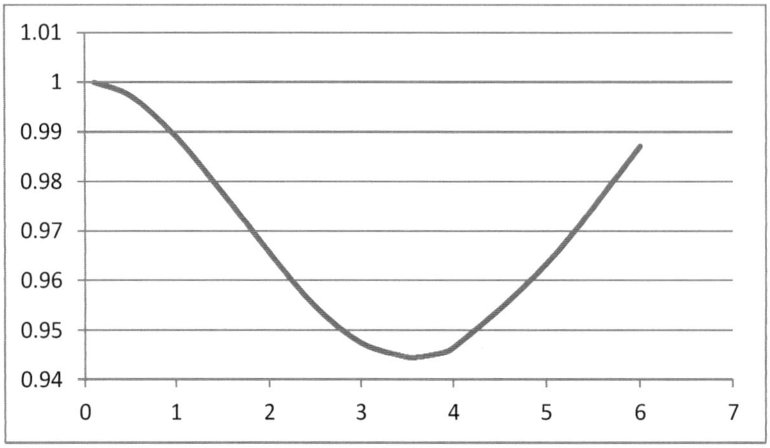

Figure 3. Plot of E_t (vertical) versus S (horizontal)

Now, we continue to observe the transmitted wave after passing two bars. From our previous calculation, for $\rho_2 = 0.1$, we put two bars of optimal length ($L_{opt} = 3.0$) side by side and with distance S between them. We then calculate the transmitted amplitude E_t after passing those two bars by solving (6). When the distance S is 1.5, (6) gives $E_t = 0.9777$ and the reflected amplitude is $A_r = 0.21000$. Meanwhile, for $S = 3.6$, we obtain the transmitted amplitude $E_t = 0.9445$. Compare to other value of S, for example $S = 4$, we have $E_t = 0.9465$. What is the optimal of S giving minimum E_t? To answer this question, we can calculate for various S and collects the data for E_t, The plot of E_t versus S is given in Figure 3. The optimal S is at 3.55 giving $E_t = 0.9445$.

We can compare our model of two bars by considering the bar separately. When the incoming wave passes the first bar, it reduces the amplitude becoming 0.9858. This is then used as a new incoming wave that passes to the second bar. The amplitude re-reduces again becoming quadrat of that number, i.e. 0.9718. Calculation of separate bar reduces smaller amplitude than using the optimal distance S_{opt}.

To see the effect of the upper-layer density, the above procedure can be repeated for various ρ_2.We present the result in following table

Table 1. The optimal length of bars and the optimal distance between two bars for various upper density

ρ_2	L_{opt}	S_{opt}
0.05	3.02	3.57
0.1	3.00	3.55
0.2	2.94	3.52
0.3	2.88	3.48
0.4	2.81	3.44
0.5	2.72	3.39

Meanwhile, by increasing ρ_2 the transmitted wave has smaller amplitude.

4. Conclusion

We have observed an interfacial wave of type monochromatic passing over one and two bars. The amplitude reduction is observed and the dimension of the bar plays an important role in obtaining the maximum reduction. The distance between two bars is also parameter that can be optimized. But all of those is also effected to the ratio of the fluid-layer density.

Acknowledgement
The research of the problem was supported by Research Gant from Indonesian Government Bandung, year 2017 and 2018. Therefore, the author acknowledged.

References

[1] T B Benjamin 1966 Internal waves of finite amplitude and permanent form *J. Fluid Mech.*, vol **25**(1966) p 241-270. http://dx.doi.org/10.1017/s0022112066001630.

[2] R I Joseph 1977 Solitary waves in finite depth fluid *J. Phys. A: Math. Gen.* vol **10**(1977) L225-L227. http://dx.doi.org/10.1088/0305-4470/10/12/002.

[3] T Kubota, D R S Ko and L D Dobbs 1978 Propagation of weakly nonlinear internal waves in a stratified fluid of finite depth AIAA *J. Hydrodyn.* vol **12**(1978) p 157-165.

[4] T B Benjamin 1967 Internal waves of permanent form in fluids of great depth *J. Fluid Mech.* vol **29**(1967) 559-592. http://dx.doi.org/10.1017/s002211206700103x.

[5] R E Davis and A Acrivos 1967 Solitary internal waves in deep water *J. Fluid Mech.* vol **29**(1967) p 593-607. http://dx.doi.org/10.1017/s0022112067001041.

[6] H Ono 1975 Algebraic solitary waves in stratified fluids *J. Phys. Soc. Japan*, vol **39**(1975) p 1082-1091. http://dx.doi.org/10.1143/jpsj.39.1082.

[7] Y Matsuno 1994 Phase shift of interacting algebraic solitary waves in a two-layer uid system, *Phys. Rev. Lett.* vol **73**(1994) p 1316-1319. http://dx.doi.org/10.1103/physrevlett.73.1316.

[8] W Choi and R Camassa 1996 Weakly nonlinear internal waves in a two fluid system *J. Fluid Mech.,* vol **313**(1996) p 83-103. http://dx.doi.org/10.1017/s0022112096002133.

[9] W Choi and R Camassa 1996 Fully nonlinear internal waves in a two-fluid system *J. Fluid Mech.,* vol **396**(1996) p 1-36. http://dx.doi.org/10.1017/s0022112099005820.

[10] R Camassa, W Choi and H Michallet 2006 On the realm of validity of strongly nonlinear asymptotic approximations for internal waves *J. Fluid. Mech.*, vol **54**9(2006) p 1-23. http://dx.doi.org/10.1017/s0022112005007226.

[11] C C Mei 1989 The applied dynamics of ocean surface waves World Scientific.

[12] E O Tuck and L H Wiryanto 1999 On steady periodic interfacial waves *J. Eng. Math.* vol **35** (1999) p 71-84. http://dx.doi.org/10.1023/a:1004395800854.

[13] L H Wiryanto 2010 Wave Propagation over a Submerged Bar, *ITB J. Sci.* Vol. **42** A, No. 2 p 81-90

[14] L H Wiryanto and M Jamhuri 2014 Monochromatic wave over one and two bars *Applied Math. Sciences,* **8**(61) p 3017-3025.

[15] Lusia K Budiasih, L H Wiryanto and Sudi Mungkasi 2015 Gelombang monochromatic antar muka pada dua lapisan fluida, *J. Indust. Appl. Math.* **2**(1) p 9-13

[16] L H Wiryanto and S Mungkasi 2015 Internal monochromatic wave propagating over two bars, *Applied Math. Sciences*, **9**(90), p 4479-4488.

IOP Conf. Series: Earth and Environmental Science **162** (2018) 012012 doi:10.1088/1755-1315/162/1/012012

Numerical simulation of internal wave propagation over a variable topography

R K Yasmin and S R Pudjaprasetya

Department of Mathematics, Faculty of Mathematics and Natural Sciences, Institut Teknologi Bandung, Jalan Ganesha 10, Bandung, 40132, Indonesia

E-mail: `rifadinaky@gmail.com`, `sr_pudjap@math.itb.ac.id`

Abstract. The Boussinesq-type of equation is considered here as a model for describing interfacial wave dynamics in a two-layer fluid system. The equation, which is derived under rigid-lid assumption, has first order nonlinear term and dispersion term, and it holds for interfacial wave with long wavelength and small amplitude, relative to depth. The second-order MacCormack scheme is implemented to solve this Boussinesq model. The numerical scheme is validated by simulating solitary wave as well as monotonic bore. Finally, the evolution of solitary wave propagating over a variable bathymetry with a shelf is examined.

1. Introduction

Internal wave are gravity wave that occur within ocean waters. Stratified ocean density support the existence of these internal wave. Typical ocean density is low near the surface and increases as depth increases. A layer of fluid with large density gradient is called a pycnocline. When pycnocline layer is relatively thin, ocean density can be well-described as a two-layer fluid system. In this case, internal wave propagate horizontally along the pycnocline and are called interfacial wave. Propagation of these interfacial wave is often described by the Boussinesq model. This model is a depth integrated two-layer model that incorporates weakly nonlinear and dispersive effects. In the model formulation, assumption of shallow water and rigid-lid approximation are applied.

Internal wave are ubiquitous. A comprehensive survey about this natural phenomena can be found for instance in Gerkema [1]. Several theoretical models have been derived to describe the dynamics of internal wave. To resemble the original situation in nature, Osborne and Burch [2], Liu [3], and Helfrich [4] observed the dissipation and shoaling effect of internal wave by using the Korteweg-de Vries (KdV) equation, whereas Holloway et al. [5] used the rotated-modified extended KdV equation. Duda et al. [6] used the extended KdV to study the behaviour of two kinds of internal wave: large amplitude solitary wave and internal tide. In the other hand, an experimental and numerical study of internal wave generation was conducted by Lin and Liu [7].

Tomasson and Melville [8], as well as Lynett and Liu [9], formulated a depth-integrated model to examine internal wave dynamics in a two-layer fluid. The Boussinesq-type equations were derived using the perturbation method of two small parameters representing the ratio of amplitude to depth and depth to wavelength. This high order Boussinesq model was then solved using the fourth-order predictor-corrector scheme. In this paper, the Boussinesq model of

Lynett and Liu [9] is adopted and restricted to one dimension model of first order term. Then, the finite difference Mac-Cormack method is implemented to solve these equations. Validation of the scheme is presented via two test cases: simulation of solitary wave propagation and monotonic bore. These test cases indicate that the proposed scheme can handle non-linearity and dispersion with good balance. Further, deformation of a solitary wave when it propagates over a shelf-topography is also presented.

2. Depth-averaged two-layer model

In this section, a direct derivation of one-dimensional Boussinesq equations will be discussed, starting from the full governing equation. Consider a fluid domain with a thin pycnocline layer, in such a way that it can be represented as a two-layer homogeneous fluid: the upper fluid layer with density ρ_1 and the lower fluid layer with density ρ_2, as illustrated in Figure 1.

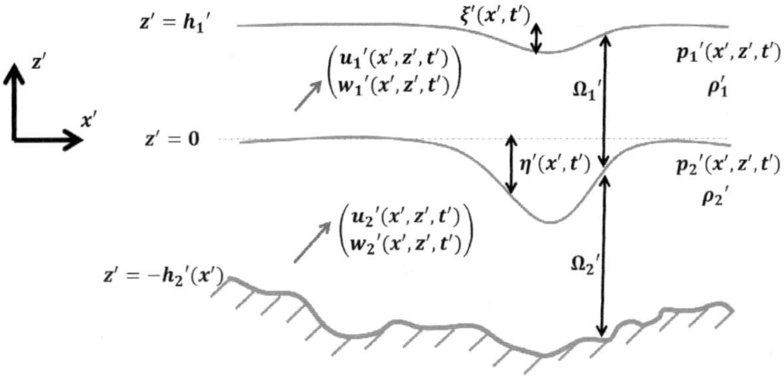

Figure 1. Sketch of a two-layer fluid system.

The undisturbed levels are: free surface $z = h_1$, interface $z = 0$, and bottom topography $z = -h_2(x)$. The interface displacement is denoted by η' and the free surface displacement is ξ', with a positive sign represents an upward displacement. The two-layer fluid domains are denoted as

$$\Omega'_1 = \{(x', z') \mid \eta' < z' < \xi' + h'_1, x' \in R\}, \text{ and}$$
$$\Omega'_2 = \{(x', z') \mid -h'_2 < z' < \eta', x' \in R\}.$$

Density, pressure, and particle velocities in the upper layer are respectively represented by ρ'_1, p'_1, and $\begin{pmatrix} u'_1 \\ w'_1 \end{pmatrix}$, whereas in the lower layer by ρ'_2, p'_2, and $\begin{pmatrix} u'_2 \\ w'_2 \end{pmatrix}$. Amplitude and wavelength of the internal wave are denoted by a_0 and l_0, respectively.

To formulate the depth averaged model for interfacial wave, the normalized variables are introduced first [9]. Let l_0 and h_0 be the horizontal and vertical length scales, respectively, $\dfrac{l_0}{\sqrt{g^* h_0}}$ be the time scale and a_0 be the scale of wave amplitude, therefore the non-dimensional variables will take form as follow

$$x = \frac{x'}{l_0}, \qquad z = \frac{z'}{h_0}, \qquad t = \frac{t'\sqrt{g^* h_0}}{l_0},$$

$$\eta = \frac{\eta'}{a_0}, \qquad \Delta\rho = \frac{\Delta\rho'}{\rho_0}, \qquad \xi = \frac{\xi'}{a_0 \Delta\rho},$$

$$(h_1, h_2) = \frac{(h_1', h_2')}{h_0}, \quad (p_1, p_2) = \frac{(p_1', p_2')}{\rho_0 g^* a_0}, \quad (\rho_1, \rho_2) = \frac{(\rho_1', \rho_2')}{\rho_0},$$

$$(u_1, u_2) = \frac{(u_1', u_2')}{\varepsilon\sqrt{g^* h_0}}, \quad (w_1, w_2) = \frac{(w_1', w_2')}{\frac{\varepsilon}{\mu}\sqrt{g^* h_0}},$$

where the parameters are

$$\rho_0 = \rho_2', \qquad g^* = g\frac{\Delta\rho'}{\rho_0}, \qquad h_0 = \frac{h_1' h_2'}{h_1' + h_2'}. \tag{1}$$

On those scalings, two small parameters were introduced as follow

$$\varepsilon = \frac{a_0}{h_0}, \quad \text{and} \quad \mu = \frac{h_0}{l_0}, \tag{2}$$

each represent nonlinearity and dispersion, respectively.

In dimensionless variables, the two-layer homogeneous fluid domains are Ω_1 and Ω_2, which represent the upper and lower layer, respectively. The Euler equations for each fluid domain are following

$$\mu^2 \frac{\partial u_j}{\partial x} + \frac{\partial w_j}{\partial z} = 0 \qquad \text{in } \Omega_j, \tag{3}$$

$$\frac{\partial u_j}{\partial t} + \varepsilon u_j \frac{\partial u_j}{\partial x} + \frac{\varepsilon}{\mu^2} w_j \frac{\partial u_j}{\partial z} = -\frac{1}{\rho_j}\frac{\partial p_j}{\partial x} \text{ in } \Omega_j, \tag{4}$$

$$\varepsilon\frac{\partial w_j}{\partial t} + \varepsilon^2 u_j \frac{\partial w_j}{\partial x} + \frac{\varepsilon^2}{\mu^2} w_j \frac{\partial w_j}{\partial z} = -\frac{\varepsilon}{\rho_j}\frac{\partial p_j}{\partial z} \text{ in } \Omega_j, \tag{5}$$

for $j = 1, 2$. In the above equations, indices $j = 1$ and $j = 2$ represent the upper and lower layer, respectively.

Furthermore, several boundary conditions are applied. Along the free surface ($z = h_1 + \varepsilon\xi\Delta\rho$) the total pressure vanishes

$$p_1 = \rho_1 \xi, \tag{6}$$

and the kinematic boundary condition along the surface reads

$$W = \mu^2 \Delta\rho(\xi_t + \varepsilon\xi_x U). \tag{7}$$

Meanwhile, along the interface ($z = \varepsilon\eta$), the pressure is continuous

$$p_2 = p_1 + \eta, \tag{8}$$

and the kinematic boundary condition read as

$$W = \mu^2(\eta_t + \varepsilon\eta_x U), \tag{9}$$

$$w = \mu^2(\eta_t + \varepsilon\eta_x u), \tag{10}$$

or it can be written

$$W - \mu^2 \varepsilon \eta_x U = w - \mu^2 \varepsilon \eta_x u.$$

Finally, along the bottom ($z = -h_2$), the impermeable boundary condition is applied

$$w = -\mu^2 h_{2_x} u. \tag{11}$$

To summarize, the governing equations for internal wave propagation in two-layer system fluids are equations (3)-(5), with boundary conditions (7)-(11).

Internal wave appears within the ocean body. Although there are internal wave with large amplitude within the ocean, the free surface above them is nearly undisturbed; only ripples may appears at the surface [3]. One common approach is assuming the free surface to be zero at all times, which is known as the rigid-lid assumption. Under the rigid-lid assumption, which is simply $\xi(x,t) = 0$, the upper fluid is $\varepsilon\eta(x,t) < z < h_1$, $x \in R$ and the lower layer fluid is $-h_2 < z < \varepsilon\eta$, $x \in R$.

Next, the depth averaged velocity for the upper layer and lower layer are defined as

$$\bar{u_1}(x,t) = \frac{1}{h_1 - \varepsilon\eta} \int_{\varepsilon\eta}^{h_1} u_1 dz, \tag{12}$$

$$\bar{u_2}(x,t) = \frac{1}{h_2 + \varepsilon\eta} \int_{-h_2}^{\varepsilon\eta} u_2 dz, \tag{13}$$

respectively. Later, the above depth averaged formulations together with boundary conditions will be used to derive the Boussinesq equations.

The following equation is obtained from the relation (13) and the lower layer boundary conditions

$$((h_2 + \varepsilon\eta)\bar{u_2})_x = \frac{\partial}{\partial x} \int_{-h_2}^{\varepsilon\eta} u_2 dz$$

$$\Leftrightarrow \quad (h_2 + \varepsilon\eta)_x \bar{u_2} + (h_2 + \varepsilon\eta)\bar{u_2}_x = \int_{-h_2}^{\varepsilon\eta} u_{2_x} dz + u_2|_{\varepsilon\eta} (\varepsilon\eta)_x + u_2|_{-h_2} h_{2_x}.$$

Under the shallow water assumption, the horizontal velocity is homogeneous over the fluid depth, so $u_2(-h_2, t) = \bar{u_2}(x,t) = u_2(\varepsilon\eta, t)$, therefore the first term on the left hand side and the second and third terms on the right hand side are cancelled out. Hence

$$(h_2 + \varepsilon\eta)\bar{u_2}_x = \int_{-h_2}^{\varepsilon\eta} u_{2_x} dz.$$

Integrating the continuity equation for the lower layer (equation (3) for $j = 2$) over the depth from $z = -h_2$ to $z = \varepsilon\eta$ yields

$$\int_{-h_2}^{\varepsilon\eta} \left(\mu^2 \frac{\partial u_2}{\partial x} + \frac{\partial w_2}{\partial z} \right) dz = 0$$

$$\Leftrightarrow \quad \mu^2 (h_2 + \varepsilon\eta)\bar{u_2}_x + w_2|_{\varepsilon\eta} - w_2|_{-h_2} = 0$$

$$\Leftrightarrow \quad \eta_t + ((h_2 + \varepsilon\eta)\bar{u_2})_x = 0.$$

The last relation is obtained after adopting boundary conditions (10) and (11).

Performing the analogue process for the upper layer, and using boundary conditions (7) and (9) yield

$$\eta_t - ((h_1 - \varepsilon\eta)\bar{u_1})_x = 0.$$

Let M denotes fluid discharge from the lower layer to the upper layer. Mass conservation for each layer requires that M should be equal to the reverse discharge: from the upper layer to the lower layer, which is expressed as follows

$$(h_2 + \varepsilon\eta)\bar{u_2} = M = -(h_1 - \varepsilon\eta)\bar{u_1}, \tag{14}$$

Then, the continuity equation expressed in depth averaged variables is

$$\eta_t + M_x = 0.$$

For the momentum equation, by assuming that density difference between layers is small, i.e. $\mathcal{O}(\mu^4)$, the equation can be obtained by means of perturbation method [9]. The resulting equation contains terms of successive order $\mathcal{O}(1)$, $\mathcal{O}(\varepsilon)$, $\mathcal{O}(\varepsilon^2)$, $\cdots$, as well as terms of $\mathcal{O}(\mu^2)$, $\mathcal{O}(\mu^4)$, $\cdots$. Keeping terms up to $\mathcal{O}(\varepsilon), \mathcal{O}(\mu^2)$, and neglecting higher order terms, the depth averaged model for interfacial wave are following

$$\eta_t + M_x = 0 \tag{15}$$

$$\begin{aligned} &M_t + a\eta_x + b\left(MM_x - (M\eta)_t\right) + cM^2 h_{2_x} \\ &+d\left((M\eta^2)_t - M(M\eta)_x - \eta M M_x\right) + eM_{xxt} \\ &+f\left(M_{xt}h_{2_x} - \frac{2}{h_2}M_t h_{2_x}^2 + (M_t h_{2_x})_x\right) = 0 \end{aligned} \tag{16}$$

where

$$a = \left(\frac{1}{h_1} + \frac{1}{h_2}\right)^{-1}, \qquad b = \varepsilon\left(\frac{1}{h_1} - \frac{1}{h_2}\right), \qquad c = -\varepsilon\frac{1}{h_1^3}a,$$

$$d = \varepsilon^2\left(\frac{1}{h_1^2} - \frac{1}{h_1 h_2} + \frac{1}{h_2^2}\right), \qquad e = \mu^2\frac{h_1 h_2}{3}, \qquad f = -\mu^2 a.$$

In further discussions, equations (15, 16) will be solved numerically. Then, various simulations will be presented under Boussinesq assumption, i.e. the non-linearity and the dispersion are in the same order $\mathcal{O}(\varepsilon) = \mathcal{O}(\mu^2)$.

3. Mac-Cormack method

In this section a numerical scheme to solve the equations (15, 16) is formulated. Here the Mac-Cormack scheme is implemented, which is a suitable scheme for solving hyperbolic problems [10]. The Mac-Cormack method is a finite difference second-order predictor-corrector scheme. Firstly, note that the equation (16) contains term with mixed derivative. To handle this situation, a new variable φ is introduced as follows

$$\varphi = M - bM\eta + dM\eta^2 + eM_{xx} + f\left(2h_{2_x}M_x - \frac{2}{h_2}h_{2_x}^2 M + h_{2_{xx}}M\right).$$

Written in the new variable φ, equation (16) becomes

$$\varphi_t + \left(a - dM^2\right)\eta_x + (b - 2d\eta)\, MM_x + cM^2 h_{2_x} = 0.$$

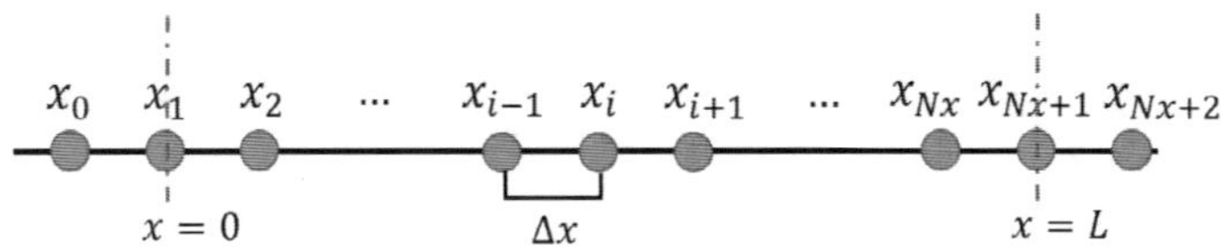

Figure 2. Space domain discretization.

Let Δt be the length of time step, and discrete time variable is denoted as $t_n = (n-1)\Delta t$, $n = 1, 2, \cdots, Nt+1$, whereas Δx is the length of space step, and discrete space variable is denoted as $x_i = (i-1)\Delta x$, $i = 1, 2, \cdots, Nx+1$, in the computational domain $[0, L]$, see Figure 2. Notation η_j^n represents the value of η at space grid j and time grid n, so do φ_j^n and M_j^n.

The predictor step is a forward-time forward-space approximation,

$$\overline{\eta_i^{n+1}} = \eta_i^n - \frac{\Delta t}{\Delta x}\left(M_{i+1}^n - M_i^n\right) \tag{17}$$

$$\begin{aligned}\overline{\varphi_i^{n+1}} = \varphi_i^n &- \frac{\left(a_i - d_i(M_i^n)^2\right)\Delta t}{\Delta x}\left(\eta_{i+1}^n - \eta_i^n\right) - \frac{c_i\Delta t}{\Delta x}(M_i^n)^2 h_{2_{x_i}} \\ &- \frac{\left(b_i - 2d_i\eta_i^n\right)\Delta t}{\Delta x}M_i^n\left(M_{i+1}^n - M_i^n\right).\end{aligned} \tag{18}$$

The overlines indicate predicted values. In the end of each time-step, the predicted values $\overline{M_i^{n+1}}$ for $i = 2, 3, \ldots, Nx$ can be obtained by solving $\overline{D}\overline{M^{n+1}} = \overline{\varphi^{n+1}}$, where

$$\overline{D} = \overline{g^*} + k^* + \frac{1}{\Delta x^2}e^*, \tag{19}$$

$$\overline{g^*} = \begin{pmatrix} \overline{g_2} & 0 & \cdots & 0 \\ 0 & \overline{g_3} & \cdots & 0 \\ \vdots & \vdots & \ddots & \vdots \\ 0 & \cdots & 0 & \overline{g_{Nx}} \end{pmatrix}, \qquad k^* = \begin{pmatrix} -k_2 & k_2 & 0 & \cdots & 0 \\ 0 & -k_3 & k_3 & \cdots & 0 \\ \vdots & \vdots & \ddots & \vdots & \vdots \\ 0 & \cdots & 0 & -k_{Nx-1} & k_{Nx-1} \\ 0 & \cdots & 0 & 0 & -k_{Nx} \end{pmatrix},$$

$$e^* = \begin{pmatrix} -2e_2 & e_2 & 0 & \cdots & 0 \\ e_3 & -2e_3 & e_3 & \cdots & 0 \\ \vdots & \vdots & \ddots & \vdots & \vdots \\ 0 & \cdots & e_{Nx-1} & -2e_{Nx-1} & e_{Nx-1} \\ 0 & \cdots & 0 & e_{Nx} & -2e_{Nx} \end{pmatrix},$$

$\overline{g_i} = 1 - b_i\overline{\eta_i^{n+1}} + d_i\overline{\eta_i^{n+1}}\ \overline{\eta_i^{n+1}} - \frac{2f_i}{h_2}h_{2_{x_i}}^2 + f_i h_{2_{xx_i}}$, and

$k_i = 2f_i h_{2_{x_i}}$.

Meanwhile, the corrector step is the forward-time scheme with the average between the backward-space of predicted-values at time $n+1$ and the forward-space of real-values at time

n,

$$\eta_i^{n+1} = \eta_i^n - \frac{\Delta t}{2\Delta x}\left[\left(M_{i+1}^n - M_i^n\right) + \left(\overline{M_i^{n+1}} - \overline{M_{i-1}^{n+1}}\right)\right] \tag{20}$$

$$\begin{aligned}\varphi_i^{n+1} = \varphi_i^n &- \frac{\left(a_i - d_i(M_i^n)^2\right)\Delta t}{2\Delta x}\left[\left(\eta_{i+1}^n - \eta_i^n\right) + \left(\overline{\eta_i^{n+1}} - \overline{\eta_{i-1}^{n+1}}\right)\right] \\ &- \frac{\left(b_i - 2d_i\eta_i^n\right)\Delta t}{2\Delta x} M_i^n\left[\left(M_{i+1}^n - M_i^n\right) + \left(\overline{M_i^{n+1}} - \overline{M_{i-1}^{n+1}}\right)\right] \\ &- \frac{c_i\Delta t}{\Delta x}(M_i^n)^2 h_{2_{x_i}}.\end{aligned} \tag{21}$$

The values of M_i^{n+1} for $i = 2, 3, \ldots, Nx$ can be obtained in the end of each time-step by solving $DM^{n+1} = \varphi^{n+1}$, where

$$D = g^* + k^* + \frac{1}{\Delta x^2}e^*, \tag{22}$$

$$g^* = \begin{pmatrix} g_2 & 0 & \cdots & 0 \\ 0 & g_3 & \cdots & 0 \\ \vdots & \vdots & \ddots & \vdots \\ 0 & \cdots & 0 & g_{Nx} \end{pmatrix}, \text{ and}$$

$$g_i = 1 - b_i\eta_i^{n+1} + d_i\eta_i^{n+1}\eta_i^{n+1} - \frac{2f_i}{h_2}h_{2_{x_i}}^2 + f_i h_{2_{xx_i}}.$$

Stencil of the numerical scheme above is illustrated in Figure 3. Note that both left and right boundary condition is needed to perform this numerical scheme.

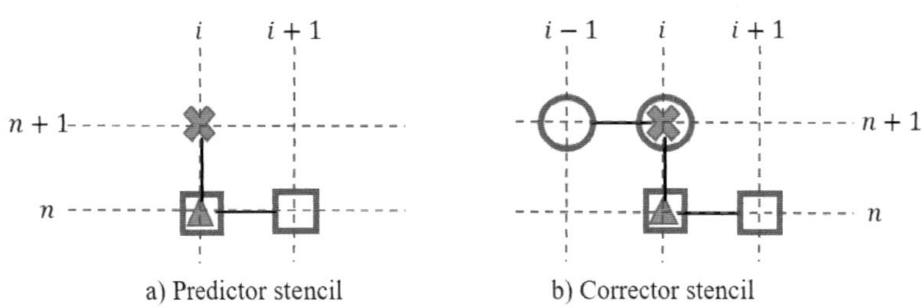

Figure 3. Stencil of Mac-Cormack Method.

4. Numerical simulation

To run the numerical simulation, as mentioned earlier, a left and right boundary condition is needed. In this paper, hard-wall boundary conditions are adopted for both ends, written as follow

$$\bar{u_1}(0,t) = 0 \quad \text{and} \quad \bar{u_1}(L,t) = 0,$$
$$\bar{u_2}(0,t) = 0 \quad \text{and} \quad \bar{u_2}(L,t) = 0,$$

for $t > 0$. Since $(h_2 + \varepsilon\eta)\bar{u_2} = M = -(h_1 - \varepsilon\eta)\bar{u_1}$, the hard-wall boundary conditions imply

$$M(0,t) = 0 \quad \text{and} \quad M(L,t) = 0, \quad \forall t > 0 \tag{23}$$

or in the discretized domain can be written as

$$M_1^n = 0 \qquad \text{and} \qquad M_{Nx+1}^n = 0, \qquad n = 1, 2, \cdots, Nt.$$

Afterwards, apply the Lax-Wendroff method to the continuity equation at both end of the space domain to get the boundary condition for η

$$\eta_1^{n+1} = \eta_1^n - \frac{\Delta t}{2\Delta x}(M_2^n - M_1^n) + \frac{1}{2}\left(\frac{\Delta t}{\Delta x}\right)^2 (M_2^n - 2M_1^n + M_0^n) \tag{24}$$

$$\begin{aligned}\eta_{Nx+1}^{n+1} &= \eta_{Nx+1}^n - \frac{\Delta t}{2\Delta x}\left(M_{Nx+2}^n - M_{Nx+1}^n\right) \\ &\quad + \frac{1}{2}\left(\frac{\Delta t}{\Delta x}\right)^2 \left(M_{Nx+2}^n - 2M_{Nx+1}^n + M_{Nx}^n\right).\end{aligned} \tag{25}$$

As illustrated in Figure 2, the point x_0 and x_{Nx+2} are outside the calculation domain. From (23), the value of M_0 can be obtained as an odd function interpolated from M_1 and M_2, or to be explicit $M_0 = -M_2$. And the same thing holds for M_{Nx+2}. With these results, equations (24, 25) can be written as

$$\begin{aligned}\eta_1^{n+1} &= \eta_1^n - \frac{\Delta t}{2\Delta x} M_2^n \\ \eta_{Nx+1}^{n+1} &= \eta_{Nx+1}^n + \frac{\Delta t}{2\Delta x} M_{Nx}^n,\end{aligned}$$

which are the boundary conditions for η as a result from the hard-wall boundary condition.

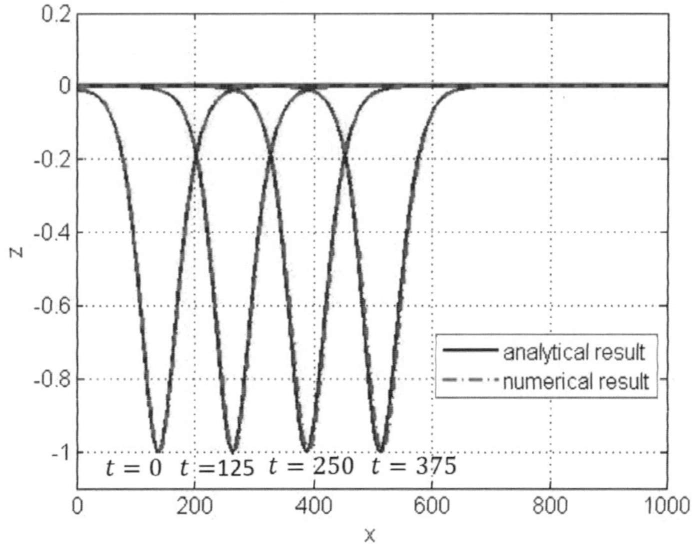

Figure 4. Solitary wave simulation using $\varepsilon = \mu^2 = 0.01$.

The numerical simulations will be conducted under Boussinesq assumption $O(\varepsilon) = O(\mu^2)$. Moreover, the higher terms such as $O(\varepsilon^2)$, $O(\varepsilon\mu)$, $O(\mu^3)$, etc. will be omitted. The first test case is performed to check whether the numerical solution is consistent to the known analytical solution at constant depth. Two known analytical solutions are used for this test, a solitary wave and a monotonic bore.

The solitary wave has a dimensionless form as follows

$$\eta(x,t) = \text{sech}^2\left(\frac{x-mt}{l}\right), \tag{26}$$

where $m = 1 + \frac{\varepsilon(h_2 - h_1)}{2h_1h_2}$ and $l = \frac{2h_1h_2}{\sqrt{3\varepsilon(h_2-h_1)}}$. A solitary wave with small nonlinearity, $\varepsilon = 0.01$, will be observed first. This simulation uses a constant depth $h_2 = 3$ and the ratio of the upper and lower layer depth $\frac{h_2}{h_1} = 2$. The domain is discretized by $\Delta x = 1$ and $\Delta t = 0.01$.

The analytic and numerical result are plotted simultaneously as it can be seen in Figure 4. The figure shows that the numerical calculation is stable and consistent to its analytical solution for a quite long time.

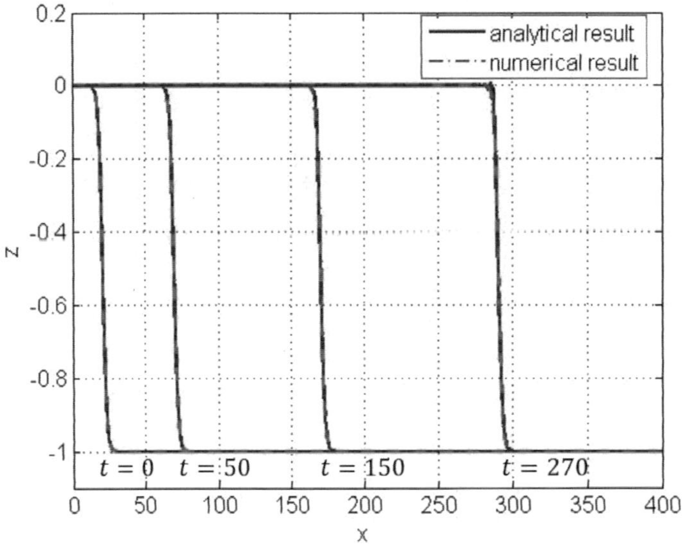

Figure 5. Monotonic bore simulation using $\varepsilon = \mu^2 = 0.01$.

A monotonic bore has the following dimensionless form

$$\eta(x,t) = \frac{1}{2}\left[1 + \tanh\left(\frac{x-mt}{l}\right)\right], \tag{27}$$

where $m = 1 + \frac{\left(h_2^2 - h_1^2\right)^2}{8\left(h_2^3 + h_1^3\right)}$ and $l = \sqrt{\frac{16\mu^2\left(h_2+h_1\right)\left(h_2^3+h_1^3\right)}{3\left(h_2^2-h_1^2\right)^2}}$. The simulation is conducted on a constant depth $h_2(x) = 2.1$ with the ratio between the upper and lower layers as $\frac{h_2}{h_1} = 1.1$. First, the simulation is performed for $\varepsilon = \mu^2 = 0.01$ using $\Delta x = 0.25$ and $\Delta t = 0.01$. The numerical result is stable and consistent to its analytical solution for eleven wavelengths of propagation as shown in Figure 5.

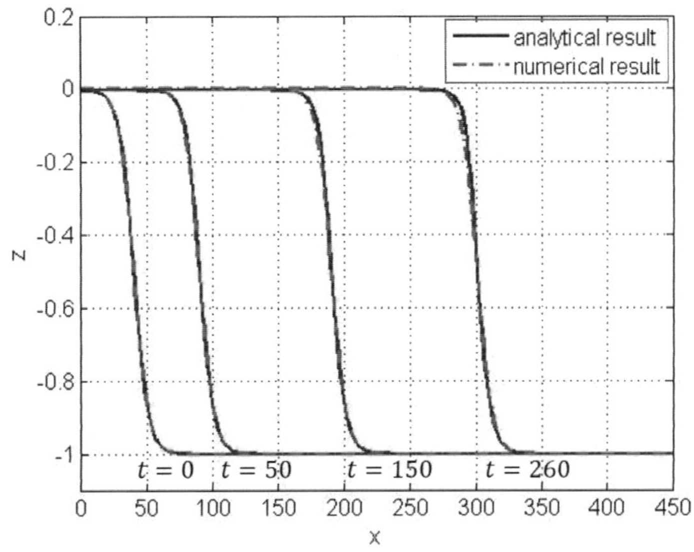

Figure 6. Monotonic bore using $\varepsilon = \mu^2 = 0.2$.

The second simulation is using larger parameter of nonlinearity, $\varepsilon = 0.2$. It was conducted using $\Delta x = 1.2$ and $\Delta t = 0.01$, and yields result in Figure 6. Again, it is shown that the simulated monotonic bore is in good agreement with the analytical bore.

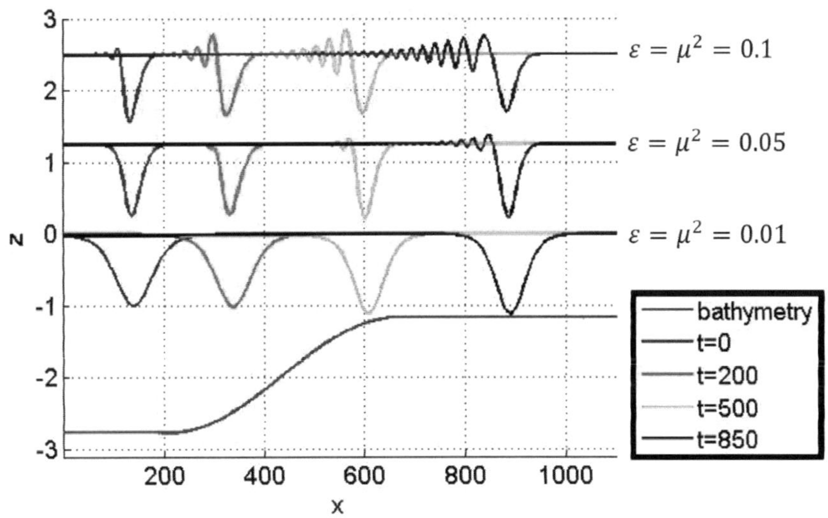

Figure 7. Simulation of a solitary wave evolving a shelf using $\varepsilon = \mu^2 = 0.1$, $\varepsilon = \mu^2 = 0.05$, and $\varepsilon = \mu^2 = 0.01$.

On the next test case, the behavior of a solitary wave that propagates over a shallower region will be examined, which is expected to give a shoaling effect. Here we take a topography which has a smooth transition that connects two constant depths (calculated from the interface level): 100 m and 40 m [11]. The interface is located on 60 m under the sea level. Dimensionless parameters correspond to the above physical setting are $h_1 = \frac{8}{5}$, with the shelf topography is

$$h_{2o} + \frac{1}{2}(h_{2o} - h_{2e})\left(\cos\left(\frac{\pi x}{L}\right) - 1\right),$$

where the upstream and downstream depths are $h_{2o} = \frac{8}{3}$ and $h_{2f} = \frac{16}{15}$, respectively. Simulations are conducted using several values of $\varepsilon = \mu^2$, i.e. 0.01, 0.05 and 0.1, and the results are plotted in Figure 7. For the case of $\mu = \varepsilon^2 = 0.01$, which is a small number, the initial solitary wave undergo the shoaling process which is almost linear. Simulation results of larger parameters $\mu = \varepsilon^2$ clearly demonstrate the effect of nonlinearity and dispersion; as the wave get steeper by nonlinearity, wave train appears due to dispersion, and hence the solitary wave develops into a negative leading wave followed by a wave train. Even though here the first order Boussinesq model is adopted, and solved using second-order Mac-Cormack method, the results are in good agreement with Lynett and Liu [9].

5. Conclusion

The implementation of Mac-Cormack method to the Boussinesq equations gave us a stable scheme. The scheme was validated by simulating the solitary wave solution that propagates undisturbed in shape with constant velocity. This solitary wave conform the analytical solitary wave. Another benchmark test was simulating the propagation of a monotonic bore. Simulations results demonstrated that the scheme can handle non-linearity and dispersion with good balance. Simulation of a solitary wave propagates over a topography with shelf displayed the shoaling phenomena. For a solitary wave with somewhat large non-linearity ε and dispersion μ^2 parameters, the wave undergo shoaling and dispersion effects, and the wave breaks down into a series of wave.

Acknowledgments

The authors acknowledge the financial support from Riset KK ITB 2017.

References

[1] Gerkema T and Zimmerman J T F 2008 *An Introduction to Internal Waves*(Texel : Royal NIOZ)
[2] Osborne A R and Burch T L 1980 Internal solitons in the andaman sea *Sci.* **208** 451-9
[3] Liu A K 1988 Analysis of nonlinear internal waves in New York Bight *J. Geophys. Res.* **93(C10)** 12317-29
[4] Helfrich K R 1992 Internal solitary wave breaking and run-up on a uniform slope *J. Fluid Mech.* **243** 133-54
[5] Holloway P E, Pelinovsky E and Talipova T 1999 A generalized Korteweg-de Vries model of internal tide transformation in the coastal zone *J. Geophys. Res.* **104(C8)** 18333-50
[6] Duda T F, Lynch J F, Irish J D, Beardsley R C, Ramp S R, Chiu C S, Tang T Y and Yang Y J 2004 Internal tide and nonlinear internal wave behavior at the continental slope in the northern south china sea *IEEE J. Oc. Engrg.* **29(4)** 1105-30
[7] Lin P and Liu P L F 1999 Internal wave-maker for Navier Stokes equations models *J. Wtrwy, Port, Coast., and Oc. Engrg.* **ASCE 125(4)** 207-16
[8] Tomasson G G and Melville W K 1992 Geostrophic adjustment in a channel : nonlinear and dispersive effects *J. Fluid Mech.* **241** 23-57
[9] Lynett P J and Liu P L F 2002 A two-dimensional, depth-integrated model for internal wave propagation over variable bathymatry *Wave Motion* **36** 221-40
[10] Garcia R and Kahawita R A 1986 Numerical solution of the St.Venant equations with the MacCormack finite-difference scheme *Int. J. Num. Methods in Fluids* **6** 259-74
[11] Liu A K, Chang Y S, Hsu M K and Liang N K 1998 Evolution of nonlinear internal waves in the East and South China Seas *J. Geophys. Res.* **103(C4)** 7995-8008

IOP Conf. Series: Earth and Environmental Science **162** (2018) 012013 doi:10.1088/1755-1315/162/1/012013

Study of coastal flooding in Kupang, East Nusa Tenggara due to Tropical Cyclone Frances (case study April 27th – 30th 2017)

I G A P P Astiduari[1], T H A Rasyiid[2], I R Nugraheni[1]

[1] School of Meteorology Climatology and Geophysics, Jl. Perhubungan 1 No.5, Pondok Aren, Tangerang Selatan 15221, Indonesia
[2] Fransiscus Xaverius Sikka Meteorological Station - Indonesian Agency for Meteorology Climatology and Geophysics, Jl.Angkasa,Maumere, Flores 86111, Indonesia

E-mail: astiduari@gmail.com

Abstract. Storm tide is one of the major natural hazard that can be endangered property and residents around the coastal region. A storm tide flooding has been occurred in Kupang, East Nusa Tenggara (NTT) on April 28th 2017 during the appearance of Tropical Cyclone Frances. The aim of this paper is to evaluate the coastal flooding events due to cyclone in order to develop the mitigation strategies in case of similar events in the future. The data from cyclone, such as the track and phase, were shown to see the cyclone stage and its impact to the changes of the waters on the coast of NTT. Three-hourly numerical hindcast of significant wave height (H_s) and primary swell from Indonesian Agency for Meteorology Climatology and Geophysics (BMKG) were used to analyze the waters condition of NTT area. The hindcast was modelled using GrADS with a resolution of 0.125° x 0.125° which has been operationalized by BMKG. Meanwhile, the sea level anomaly from AVISO Satellite Altimetry Data was used to examine the variation of regional sea level variation during the flooding occurrence. From the processing of several data and the analysis, we get that the significant wave and the swell have reaches 6 m high in on the peak of the cyclone event with the sea level anomaly is 10 – 15 cm. Thus, we conclude that this event was occurring because of the swell assisted with the strong steady wind, where the energy could transported the waters up into the mainland.

1. Introduction

Tropical cyclone is a form of extreme weather disturbance that begins with a low pressure center over the ocean which triggers the convection process and intensive cloud formation [1]. Tropical cyclone develops in convective areas and has at least 34 knots maximum wind speed of the region circling the center and at least lasts for six hours over warm waters. There are three conditions of cyclone formation according to Tjasyono.et.al (1983), namely: (1) there is convergence in the lower layers of the atmosphere that strong enough to raise the moist air bar, (2) there is divergence in certain layers to move air from the underlying layer and make the air fell. (3) the presence of sufficient energy to maintain the circulation [2]. Energy is obtained from the surface of the warm sea water because it contains a lot of water vapor as a fuel for cloud formation.

Cyclones are influenced by the Corioli force determined by the latitude position. Generally, the formation of tropical cyclones is effective in latitudes above 10° of north latitude and south latitude [3]. Therefore, tropical cyclone will be difficult formed in Indonesia because corioli force close to zero in equatorial region. Tropical cyclones can have an impact on the weather in the lane and surrounding areas. The territory of Indonesia which is geographically adjacent to the cyclone formation region will indirectly be affected. Cyclones can cause strong winds, heavy rains, rising average sea levels over a period of time, leading to flooding. Coastal flooding events may changing the coastline, harming the infrastructure, causing contamination of freshwater reserves, and in the worst cases can causing the death of human life [4].

Siklon Picu Gelombang Tinggi

Pusaran Angin Terbentuk di Barat Lampung

JAKARTA, KOMPAS — Siklon tropis Frances, yang terbentuk di Laut Arafuru pada Kamis (27/4) sore, mulai menguat pada Jumat (28/4) sore. Siklon ini juga melebarkan radius pusarannya. Dari pantauan Satelit Hinawari, sebagian pusaran itu berada di atas Pulau Timor, Nusa Tenggara Timur, dan memicu gelombang setinggi 6 meter di Laut Timor.

Air laut pun naik ke daratan dan menggenangi wilayah Desa Kolbano, Kotolin, dan Bokong di Kabupaten Timor Tengah Selatan. Berdasarkan informasi yang diterima *Kompas* dari Bupati Timor Tengah Selatan Paul Mella, air menggenangi tiga desa itu sekitar 20 menit.

Berdasarkan data Badan Meteorologi, Klimatologi, dan Geofisika (BMKG), siklon tersebut juga memicu gelombang tinggi berkisar 1,25-2,5 meter di perairan selatan Pulau Sawu, Samudra Hindia bagian barat Sumatera hingga selatan Nusa Tenggara Barat, perairan Kepulauan Tanimar, Aru, serta Arafuru bagian tengah dan timur. Gelombang dengan ketinggian 2,5-4 meter berpotensi terjadi di perairan selatan Kupang, Laut Arafuru bagian barat, dan Samudra Hindia sebelah selatan NTT.

"Pusat pusaran siklon berada di Laut Timor yang lebih dekat ke Australia, tetapi karena skala pusarannya yang besar, tepian pusarannya sampai di Pulau Timor," kata peneliti cuaca dan iklim ekstrem BMKG Siswanto, di Jakarta.

Karakter pergerakan siklon tropis, sebagaimana lazimnya, bergerak menjauhi wilayah tropis. Dengan penguatan intensitas siklon ini, kata Siswanto, sebagian besar Pulau Timor, terutama sisi selatan, berpotensi terdampak hujan dengan intensitas tinggi, selain dampak angin kencang dan gelombang tinggi.

Siswanto mengatakan, berdasarkan laporan dari Stasiun BMKG Eltari, Kupang, terjadi hujan ringan. "Dari pantauan satelit di sekitar pusat siklon, intensitas hujan mencapai 200 milliliter per hari. Namun, hujan lebat ini jatuh di sekitar laut," ujarnya.

Namun, embusan angin bisa mencapai lebih dari 40 knot. Meningkatnya kecepatan angin meningkatkan status topan tropis Frances naik dari siklon kategori 2 menjadi kategori 3.

Badan Otoritas Cuaca Australia (BoM) menyebutkan, siklon kategori 3 digolongkan sebagai jenis yang merusak. Kecepatan siklon jika terjadi di daratan bisa mencapai 165-224 kilometer per jam. "Beruntung siklon ini melintas di lautan," ucap Siswanto.

Deputi Bidang Meteorologi BMKG Yunus S Swarinoto mengatakan, siklon tropis ini memicu terjadi hujan dengan intensitas ringan hingga sedang di NTT bagian selatan dan Maluku bagian selatan. Hujan di NTT itu termasuk anomali karena sejak bulan ini daerah itu memasuki musim kemarau.

Pada Sabtu (29/4), siklon diperkirakan akan berada sekitar 265 kilometer sebelah selatan Kota Kupang, NTT, dan bergerak ke barat daya. "Analisis pada Jumat, posisi siklon tropis Frances di Laut Timor, sekitar 325 kilometer sebelah timur-tenggara Kupang, dan bergerak ke arah barat daya," kata Kepala Bidang Humas BMKG Hary Tirto Djatmiko, di Jakarta.

Pusaran angin

Selain keberadaan siklon tropis Frances ini, menurut Siswanto, saat ini juga ada pusaran angin di barat Lampung dan berpotensi menjadi badai tropis dalam dua hari ke depan. Dampaknya bisa memicu intensitas hujan dan gelombang tinggi di sekitar Sumatera dan Jawa bagian selatan.

Menurut kajian Siswanto, kondisi cuaca secara global saat ini makin dinamis seiring dengan terjadinya pemanasan global. Salah satu indikasinya adalah ada tren makin sering terjadi siklon tropis, terutama di wilayah Samudra Hindia.

Dengan menganalisis data kejadian pada 1960 hingga 2010, diketahui ada tren siklon tropis kategori 3 di Samudra Hindia bagian selatan meningkat secara signifikan, baik jumlah maupun durasi hidupnya. Rata-rata terjadi penambahan frekuensi satu siklon tropis per 10 tahun. Sementara durasi hidupnya bertambah dua hari per 10 tahun atau 5 jam per tahun.

Dalam April ini saja, sebelumnya, terbentuk siklon tropis Ernie di Samudra Hindia di selatan Cilacap pada Jumat (7/4). Adapun pada Maret terbentuk siklon tropis Caleb di Samudra Hindia, sebelah selatan Lampung pada Kamis (23/3). (AIK/KOR)

Figure 1. News from Kompas Newspaper about the cyclone and it's impact to some areas of Indonesia [5].

On April 28th, 2017, there was a coastal flooding on Kolbano beach due to Cyclone Frances. Residents from the three affected villages namely Kotolin, Kolbano and Boking villages must be evacuated (Figure 1) [5]. According to news of the Okezone dated April 28th, 2017, the waves are high enough that the water could flood the settlements [6].

On this research, we show the documentation and analysis of the coastal flooding that occurred in Kolbano Beach, Kupang. We use data of waves and swell modeling data from BMKG, and oceanographic observation data by satellite altimetry from AVISO. The purpose of this research is to know the effect of cyclone Frances that occur in western waters of Australia and moves to the waters in East Nusa Tenggara region. It is important to know the process that lead the flooding in order to cope the impact that will occur, especially for the island and maritime nation.

2. Methodology

2.1. *Study area*

The authors use a descriptive analysis for this coastal flooding case study in East Nusa Tenggara, April 28th, 2017. The incident location is Kolbano Beach, Timor Tengah Regency, Kupang (shown by the red dot). The eastern coastal area of East Nusa Tenggara is the study areas of this research (Figure 2). This beach area of East Nusa Tenggara is facing Timor Sea that separates Indonesia from Australia. Analysis will take place on April 26th 2017 until May 1st 2017.

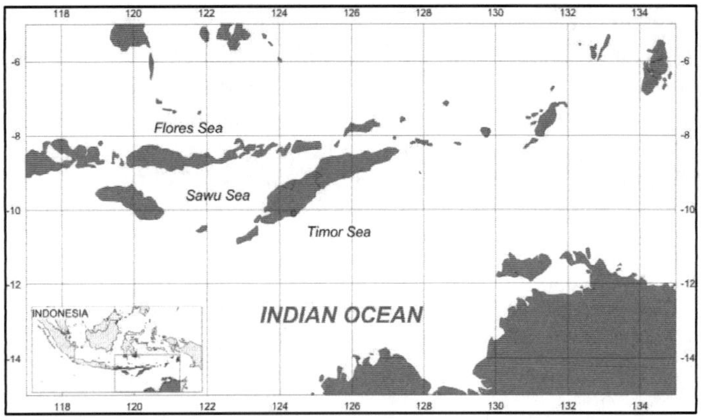

Figure 2. Map of study area, the red dot shows the location of coastal flooding during April 28th 2017.

2.2. Track of Cyclone

Track data of Cyclone Frances is obtained through the website of Unisys Weather (http://weather.unisys.com/hurricane/s_indian/2017/index.php) [7] released by Joint Typhoon Warning Center (JTWC). The data is presented every 12 hours symbolized by line path. Time usage is in Universal Time Coordinate (UTC). The category of cyclone based on Saffir-Simpson Hurricane Wind Scale is shown in (Table 1).

Table 1. Saffir-Simpson Hurricane Wind Scale [Unisys Weather. 2017].

Type	Category	Pressure (mb)	Winds (kt)	Winds (mph)	Line Color
Depression	td	-----	< 34	<39	Green
Tropical Storm	TS	-----	34 - 63	39 - 73	Yellow
Hurricane	1	> 980	64 - 82	74-95	Red
Hurricane	2	965 - 980	83 - 95	96 - 110	Light Red
Hurricane	3	945 - 965	96 - 112	111 - 130	Magenta
Hurricane	4	920 - 945	113 - 135	131 - 155	Light Magenta
Hurricane	5	< 920	> 135	>155	White

The Saffir-Simpson Hurricane Wind Scale (hereafter will be called as Saffir-Simpson Scale) is a scale for hurricane based the potential for property damage. Hurricanes reaching Category 3 and higher are considered major hurricanes because of their potential for significant loss of life and damage. Category 1 and 2 storms are still dangerous, and require preventative measures [8].

2.3. Sea Level Anomaly

Along-track multi-mission altimeter satellite product for sea level anomaly (hereafter referred to as SLA) will be used to examine the variation of regional sea level variation during the flooding occurrence [9]. The data was taken from AVISO Satellite Altimetry Data (with resolution 1/4° x 1/4°) [10]. This satellite product is gridded sea level anomalies data which computed with respect to a twenty-year mean. The SLA displayed by Hovmöller diagram to show the spatial and time distribution of the anomaly in period of April 26th – May 1st 2017.

2.4. Wave, swell, and wind
Significant sea level (H_s) and primary swell will be used to know how the condition of the waters during the event. Data were obtained from BMKG from April 26th - May 1st, 2017 [11]. This data is the result of calculation because the observational data is not acquired at that location. In addition we also used six-hourly wind speed and direction data to compare this to the wave direction. The numerical hindcast of H_s, primary swell and the wind will be displayed with GrADS.

3. Result and Discussion
3.1. Tracking Analysis
Tropical Storm begins to grow on Northern Australia (Timor Sea close to Tiwi Island) on April 27th, 2017 and spread to the Southwest throughout the Timor Sea until finally dissipated on April 30th, 2017 in Northeast Australia. The propagation of the cyclone can be seen in Figure 3.

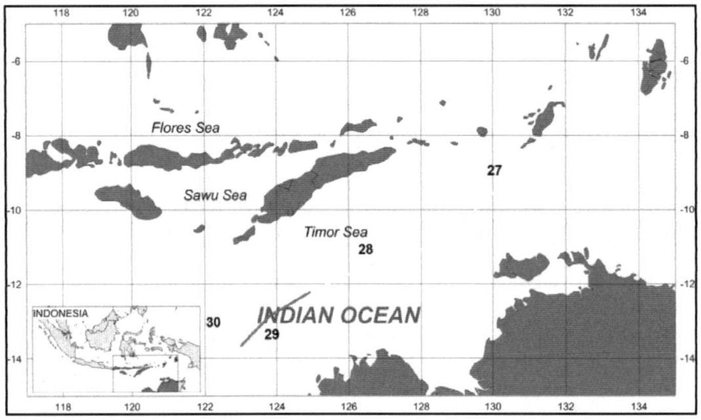

Figure 3. The track of Tropical Cyclone Frances.

Here is the life cycle stage of Cyclone Frances :

Table 2. Life stage of Cyclone Frances [Unisys Weather. 2017].

Time	Latitude	Longitude	Wind Speed (kt)	Status
04/27/00Z	-9.60	130.20	35	TROPICAL STORM
04/27/12Z	-10.90	128.40	45	TROPICAL STORM
04/28/00Z	-11.60	126.30	55	TROPICAL STORM
04/28/12Z	-12.10	125.00	70	CYCLONE-1
04/29/00Z	-12.80	123.80	70	CYCLONE-1
04/29/12Z	-13.60	122.90	60	TROPICAL STORM
04/30/00Z	-13.40	122.10	35	TROPICAL STORM

The maximum speed of Cyclone Frances is 70 kt, so according to the Saffir-Simpson Scale, Cyclone Frances is included as Category 1 (Table 2). At this stage of cyclone, citizens, stockbreeding, and pets that flying or falling debris probably get hurted or killed. Some housing that it being built badly may have major damage, involving loss of the roof cover and disposal of patio coverings. Swimming pool hedge with aluminum material could be breakaged. Some roof top of the apartment buildings and shopping malls can be partially removed. Failure to unprotected windows will be common. Large branches of trees will snap and shallow rooted trees can be detached [12].

3.2. *SLA Analysis*

The condition of SLA in Kolbano Beach can be seen with Hovmöller Diagram (Figure 4). SLA height at the location of flooding events (inside yellow box) ranged between 10 - 15 cm. Based on diagram, the peak of SLA in the Kolbano Beach occured on April 26th, and gradually decreased. During the event, the anomaly is around 14.0 - 14.5 cm. The height of SLA of the location was relatively similiar compared to the research that conducted in southern coast of Java, Bali, and West Nusa Tenggara that is <20 cm [13, 14].

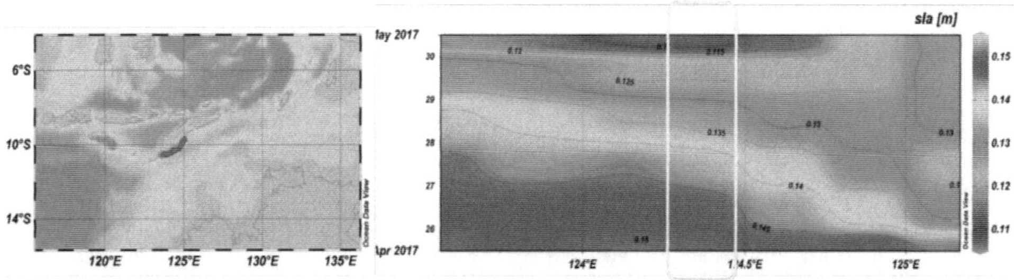

(a) (b)

Figure 4. (a) Domain for Hovmöller Diagram, the red area is the main section (with 30 km width) (b) Hovmöller Diagram that shows the SLA from April 26th – May 1st 2017 (x axis shows the longitude, y axis is days, and color contour is the height of SLA in meter).

3.3. *Wave and swell analysis*

On April 26th 2017, a spin of the water begins to appear between the Saumlaki island and the Aru island with the highest wave height of about 2 - 2.5 m near the center of the seedling of the cell. On the next day, the cell moves closer to Australia and has grown further with a wave height of 4.0 meters around its center. Then, the cell of cyclone moves to the southwest throughout Timor Sea. The peak of the cyclone is on April 28th 2017, which Hs reaches 6 m high on the center (Figure 5). The propagation of the swell rather similar with H_s, which the height of swell reaches 6 m high on the center (Figure 6

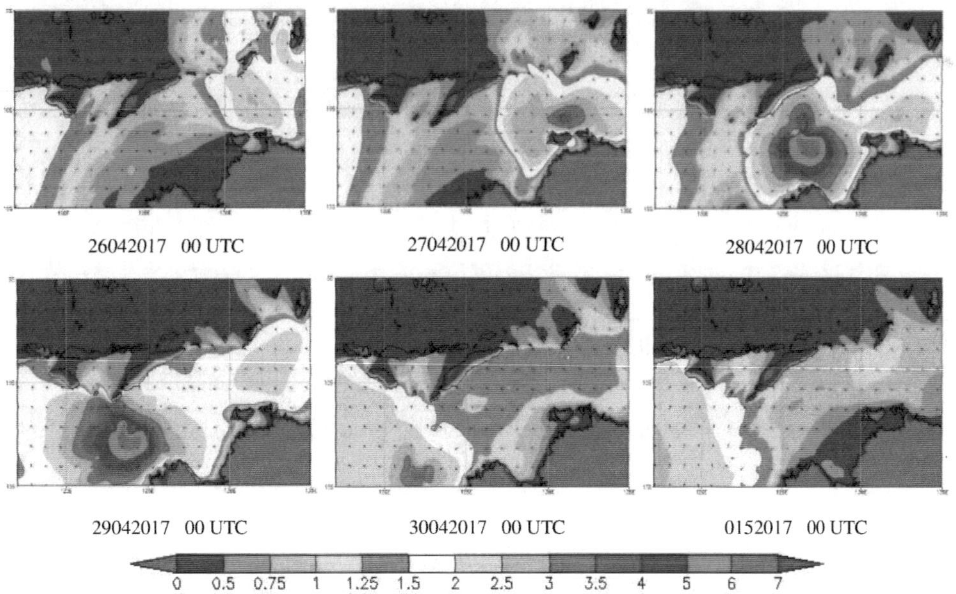

Figure 5. Spatial conditions of daily average of significant wave height (H_s) on April 26th – May 1st 2017. Waves are measured in meter.

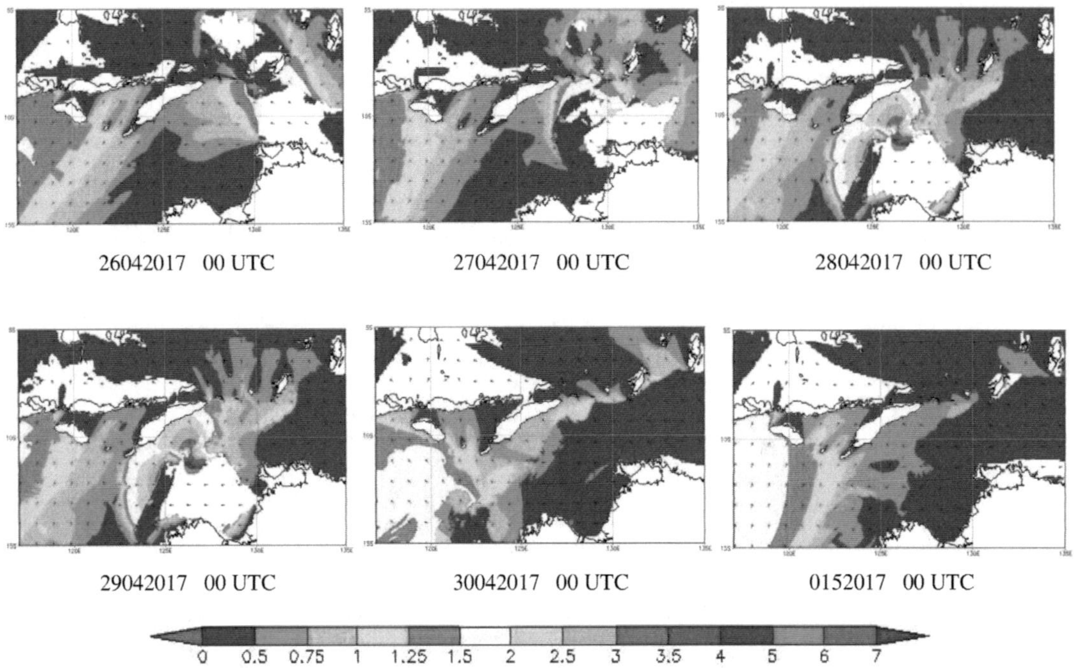

Figure 6. Spatial conditions of daily average of primary swell on April 26th – May 1st 2017. Swell measured in meter.

Based on these pictures below, the circulation is already formed with speed 40 – 50 m/s and develops until it reaches more than 60 m/s on April 26th. The cyclone began to dissipate on April 30th.

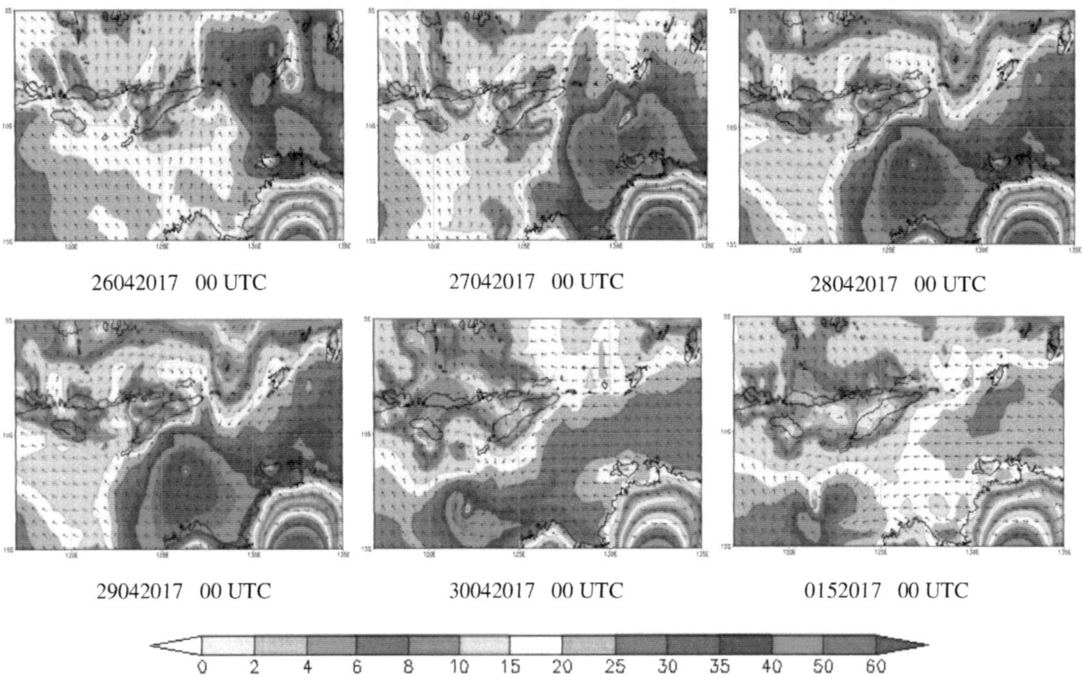

Figure 7. Spatial conditions of wind direction and speed on April 26th – May 1st 2017.

The peak of the cyclone is on April 28th 2017, where the peak of H_s is as high as 2.0 - 2.5 m and 1.5 - 2.5 m for swell in the incident site. The propagation of H_s and swell is relatively similar from April 26th to May 1st. The propagation of the cyclone moves from east to southwest. Reciprocally, the direction of the wind looks similar with the direction of H_s and swell, which can be seen on Figure 7.

4. Conclusion

From the simulation of several data in this study, it shows that although the cyclones Frances in April 2017 did not generate a high anomaly along the eastern coast of East Nusa Tenggara on Kolbano Beach, which its surge high is < 20 cm, the coastal area remains susceptible to the coastal flooding. Sea level anomaly on the incident site reaches 10 – 15 cm with the highest significant wave and swell about 6 m. The condition of significant waves is influenced by swell and the wind.

References

[1] Susilo E, Hadianti S 2015 Improvement of Arafura Sea fertility and surrounding after Tropical Cyclone Lam event *Journal of Science and Technology* **10** (1): 1-9

[2] Bayong T H K, Prawirowardoyo S and Saryono 1983 Hurricane effect tropical to the weather in Indonesia. Directorate Research Development and Community Service, Directorate General of Higher Education, Department Education and Culture, Jakarta

[3] Haryani N S and Zubaidah A 2012 Tropical Cyclone dynamics in Southeast Asia using remote sensing data. Widya, **29** (324):54-58

[4] Barnett J 2011 Dangerous Climate change in the Pacific island: food production and food security. *Journal of Regional Environmental Change* **11** (1): 229-237

[5] Kompas. 2017. Cyclone triggers the high waves. Kompas Newspaper on April 29th 2017, 13.

[6] Okezone. 2017. Impact of Cyclone Frances, residents of Kolbano NTT evacuated. Available at https://news.okezone.com/read/2017/04/28/340/1679149/imbas-badai-frances-warga-kolbano-ntt-diungsikan

[7] Unisys Weather. 2017. 2017 Hurricane/Tropical data for Southern Indian Ocean. Available at http://weather.unisys.com/hurricane/s_indian/2017/index.php

[8] NOAA. Saffir-Simpson Hurricane Wind Scale . Available at http://www.nhc.noaa.gov/aboutsshws.php

[9] Nugraheni I R, Wijayanti D P, Sugianto D N and Ramdhani A 2017 Study of inundation events along the southern coast of Java and Bali, Indonesia (case studies 4-9 June 2016). IOP Conf. Ser.: Earth Environ. Sci. **55** 012014

[10] AVISO. 2017. Aviso Global Sea Surface Height Products. Available at https://www.aviso.altimetry.fr/en/data/products/sea-surface-height-products/global/sla-h.html

[11] BMKG. 2017. Ina-Waves. Available at http://peta-maritim.bmkg.go.id/render

[12] Schott T, Landsea C, Hafele G, Lorens J, Taylor A, Thurm H, Ward B, Willis M and Zaleski W 2012 The Saffir-Simpson Hurricane Wind Scale. NOAA. Available at http://www.nhc.noaa.gov/aboutsshws.php

[13] Ningsih N S, Hadi S, Harto A B, Utami M D and Rudiawan A P 2010 Modelling of storm tide flooding along the southern coast of Java, Indonesia. Advances in Geosiences **24**: 87-103

[14] Syamsudin F and Kaneko A 2013 Ocean variability along the southern coast of Java and Lesser Suda Island. Journal Oceanography. Published online: July 16th 2013, The Oceanographic Society of Jpan and Springer Japan 2013

IOP Conf. Series: Earth and Environmental Science **162** (2018) 012014 doi:10.1088/1755-1315/162/1/012014

Characteristics of diabatic heating associated with mesoscale convective systems over Indian Ocean from radiosonde and GPS radio occultation

Khairunisa Febrina[1], Tri Wahyu Hadi[2], Noersomadi[3]

[1] Weather and Climate Prediction Laboratory, Faculty of Earth Science and Technology, Institut Teknologi Bandung, Indonesia
[2] Atmospheric Science Research Group, Faculty of Earth Science and Technology, Institut Teknologi Bandung, Indonesia
[3] Center of Atmospheric Sciences and Technology, National Institute of Aeronautics and Space, Indonesia

E-mail: kfebrina@meteo.itb.ac.id

Abstract. This study aims to investigate the characteristics of diabatic heating profiles associated with appearances of Mesoscale Convective Systems (MCSs) over Sumatra Island (SI) and Indian Ocean (IO) including the transitional offshore zone (TZ). We define the parameters of eccentricity, cloud lifetime, and cloud coverage from Multi-Functional Transport Satellite (MTSAT) infrared imageries to identify the MCSs. We derive the diabatic heating profiles over SI and IO using radiosonde data collected during Cooperative Indian Ocean Experiment on Intraseasonal Variability in the Year 2011 (CINDY2011) / Dynamics of the Madden Julian Oscillation (DYNAMO) campaign. We also estimate the diabatic heating profiles from the GPS Radio Occultation incorporated with the European Centre for Medium-Range Weather Forecasts (ECMWF) reanalysis data over TZ, where the radiosonde observations were limited. The results show that heating occurred throughout the entire troposphere during the mature stages of MCSs with more variations in the amplitude and maximum heating up to 20 K/day overland. Furthermore, the altitude of maximum heating over SI is lower than that of IO and TZ. Over SI and TZ, profiles of dissipations (postmature) stage are characterized by cooling in the lower altitude and an upward shift of maximum heating. By analyzing vertical velocity profiles, we confirm that cooling (heating) processes are associated with updraft (downdraft) in the cloud systems. In addition, the presence of cooling that occurs in layers close to the surface could indicate the formation of cold pool during the dissipation phase of the MCSs.

1. Introduction

Diabatic heating in the atmosphere appears from combination of latent heat released during the phase changes of water substance, convergence of radiative fluxes, and turbulence flux of sensible heat. This type of heating is a source of excitation of atmospheric waves that play important roles in generating and maintaining general-, as well as, local-circulations (e.g., Hartmann et al. [3], Mapes and Houze[14]). Knowledge of the diabatic heating profiles in the atmosphere is fundamental to understand the mechanism of wave generation and propagation, which are crucial information for the development and advancement of weather and climate models.

Mesoscale Convective System (MCS; Houze [7]), is an organized group of thunderstorms that produces a contiguous precipitation area measuring 100 km or more (Houze [7]). The MCS often evolves to form the convective towers at its mature stage, then transforms into convective-stratiform complexes, before further decays into stratiform-rain only. It is well known that convective systems produce significant diabatic heating due to a large amount of water vapor condensation in their development process. Houze [6] pointed out that convective and stratiform regions in MCSs strongly characterize tropospheric heating profiles. Additionally, Yuan and Houze [19] showed that there is high frequency of occurrences of MCSs in the Maritime Continent (MC) especially over the Eastern Indian Ocean (Trismidianto et al. [16]).

Diabatic heating profiles associated with MCSs have been previously studied in several works, as summarized in Table 1. It can be seen that researches on diabatic heating profiles have been mainly conducted over the oceanic environment and outside MC region. It should also be noted that the height of heating peaks vary from place to place and cooling peaks were only observed in some cases. Around MC, results from measurements conducted over the South China Sea (Johnson and Young [11]) and that of the Indian Ocean (Katsumata et al. [12]) show some differences in terms of the height of heating and cooling peaks. These results posed at least two questions. (a) Are heating profiles over MC distinctly different from those of other regions? And (b) what factors determine the height of heating/cooling peaks?

Table 1. Summary of previous research on diabatic heating profiles with information on the height of heating and cooling peaks as highlighted results.

Author(s)	Location	Heating Peak (hPa)	Cooling Peak (hPa)
Reed and Recker (1971)	West Pacific (Close to the island)	450	-
Yanai et al. (1973)	West Pacific including	450	-
	Marshall Island	500	-
Johnson (1976)	Florida (Land)	500	-
Houze (1982)	Atlantic (Sea)	300	600
Johnson and Young (1983)	South China Sea (Close to the Island)	400-300	700
Houze and Rapport (1984)	Atlantic (include Pohnpei Island)	400	800
Johnson and Ciesliski (2000)	Atlantic (Sea)	400	-
Katsumata et al. (2011)	Indian Ocean (Sea)	400	-

The aim of this study is to investigate the characteristics of tropospheric diabatic heating profiles associated with appearances of MCSs over Sumatra Island (SI) and eastern part of Indian Ocean (IO), including the transitional zone (TZ) between those two regions. These studied areas represent a gradual change of the environment, from land, to coastal zones, and open ocean, also from MC region to the area that is close to its perimeter. Thus, this study differs from previous similar research in terms of land-to-ocean coverage. In addition, a novel approach to utilize atmospheric profiles from both radiosonde and GPS Radio Occultation (RO) data for estimation of diabatic heating profiles is herein also highlighted. Detailed description about data and methods is presented and the next section.

2. Data and Methodology

In this research we use geostationary satellites data from the Multi-Functional Transport Satellite (MTSAT-2R), which is operated by the Japan Meteorological Agency (JMA) for identify the MCSs. For obtain the diabatic heating profiles, we use radiosonde data obtained during Cooperative Indian Ocean Experiment on Intraseasonal Variability in the Year 2011 (CINDY2011) / Dynamics of the Madden Julian Oscillation (DYNAMO) as reported in Yoneyama et al. [18]. In addition, we also use the Global Positioning System Radio Occultation (GPS-RO) technique, which is an active limb sounding of the atmosphere and ionosphere from the propagation of GPS signal to the low Earth orbit satellites (Anthes et al. [1]). The COSMIC Data Analysis and Archive Center (CDAAC) website produced several product data. There are several mission used in this study which is COSMIC wetPrf (Constellation Observing System for Meteorology Ionosphere & Climate), TerraSar-X (The German TerraSar-X mission), and GRACE (Gravity Recovery and Climate Experiment). The continuous GPS-RO observation may provide atmospheric profiles inside the MCSs that can resolve the problem of limited number of radiosonde data. The European Centre for Medium-Range Weather Forecasts (ECMWF, Dee et al [3]) reanalysis data were also incorporated with GPS-RO in TZ and SI where radiosonde observations were insufficient.

The area of this study can be seen in figure 1. We part the area into three sub-regions which is SI, TZ, and IO. The three sub-regions differ not only by location, but also by data usage. Specifically for TZ, we used one radiosonde data retrieved inland combined with GPS-RO; for IO we use sounding data from RV/Mirai combine with the nearest GPS-RO; and for SI we use ground-based sounding data from three stations in Medan, Padang, and Singapore.

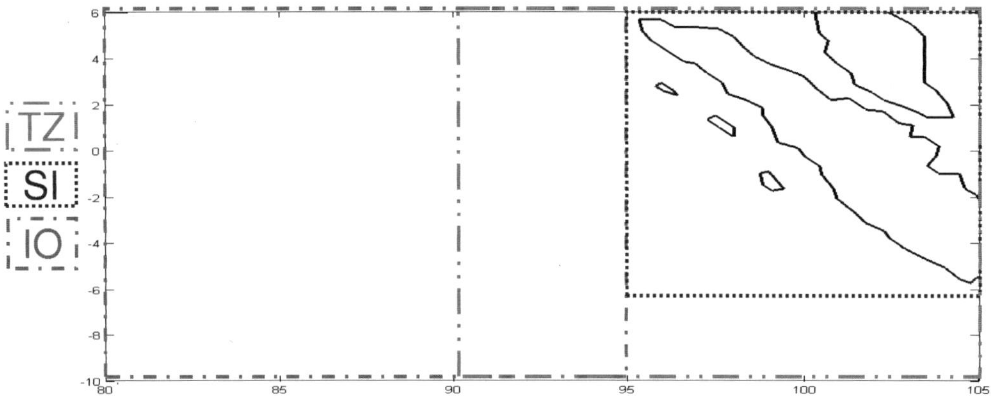

Figure 1. Area of study defined as Sumatra Island (SI; 95°E-105° E,6°N-6°S); Indian Ocean (IO; 78°E-95°E,6°N-10°S) that adjacent to the SI, including the Transitional Zone between those two regions (TZ; 90°E-105°E,6°N-10°S).

To identify the MCS in this study, we use the equivalent Temperature Black Body (hereafter T_{bb}) from the MTSAT-2 infrared imageries for data from 20 November 2011 to 11 December 2011. The satellite data has spatial resolutions 0.05° x 0.05° and at one hour time interval. This study adapts the method of Machado et al. [13] that automated-objectively classified MCSs in satellite images using the criteria as described in table 2 (defined in Jirak et al. [8]). Examples of the observed characteristics are shown in figure 2.

Table 2 MCS definitions based on the analysis of IR satellite imagery adapted from Jirak et al. [8].

MCS category	Size	Duration	Shape
MCC	Cold cloud region ≤ -52°C with an area ≥50 000 km^2	Size definition met for ≥6h	Eccentricity ≥ 0.7 at time of maximum extent
PECS			0.2 ≤ eccentricity < 0.7 at time maximum extent
MβCCS	Cold cloud region ≤ -52°C with an area ≥30 000 km^2 and maximum size must be ≥50 00 km^2	Size definition met for ≥3h	Eccentricity ≥ 0.7 at time of maximum extent
MβECS			0.2 ≤ eccentricity < 0.7 at time maximum extent

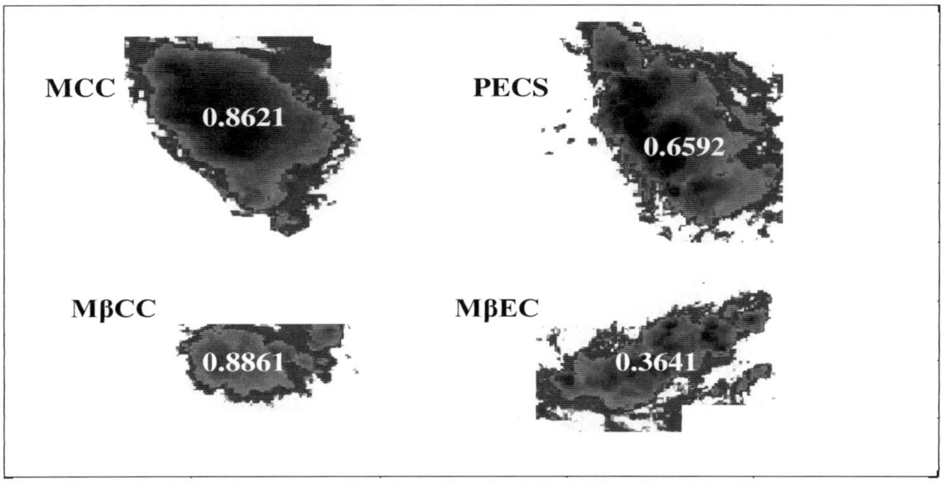

Figure 2. Examples of MCS identification results (with eccentricity determined using Machado method) for four cases of: MCC on November 22nd, 2011 at Sumatera Island, PECS on November 25th, 2011 in the region between Sumatera Island and the Indian Ocean, MβCC on November 20th, 2011 in the Indian Ocean, and MβEC on November 22nd , 2011 at Sumatra Island.

To determine the samples of MCSs for further analyses, we use the information of the appearances of MCS in the three sub-regions and examine coinciding sounding data. Using the radiosonde and GPS-RO temperature profiles, we can calculate the apparent heat source for each sample of MCS. The radiosonde data in SI retrieved from Radiosonde Sounding stations located in Medan (3.565°N, 98.75°E) [21], Padang (0.884°S, 100.35°E) [22], and Singapore (1.33°N , 103.8°E) [20]. Meanwhile, in region IO and TZ we used data from JAMSTEC RV/MIRAI (located within -8°N – 3°N, 78°E – 100°E) [23] and GPS-RO from several missions that match spatially and temporally with the cases of MCS. The spatial distribution of sounding data is illustrated in figure 3. Atmospheric parameters such as meridional wind and zonal wind produced by ECMWF ERA-INTERIM were used to complement GPS-RO data, which only contain temperature and pressure profiles.

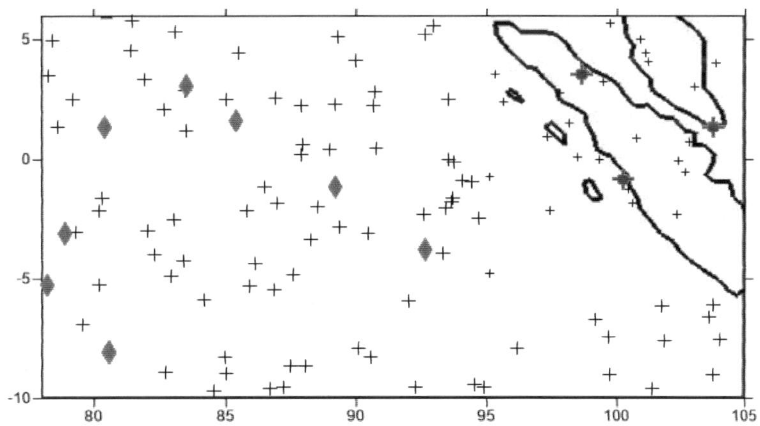

Figure 3. Distribution of GPS-RO data (+), the locations of JAMSTEC RV/MIRAI (♦), and the locations of stations inland (•).

To observationally estimate diabatic heating associated with MCS using upper-air soundings, we use the method of Yanai et al. [17]. Their method is to calculate the apparent heat source Q_1 by adding the horizontal and vertical advection components to the observed temporal variation of dry static energy (for Q_1). The apparent heating Q_1 of the large-scale motion system consists of the heating due to radiation, the release of latent heat by net condensation, and vertical convergence of the vertical eddy transport of sensible heat, which can be expressed as

$$Q1 \equiv \frac{\partial \bar{s}}{\partial t} + \bar{\boldsymbol{v}}\, \nabla \bar{s} + \frac{\partial \bar{s}\bar{\omega}}{\partial p} \tag{1}$$

In the above, most of the notations are conventional; Q_1 is the apparent heat source (K/day), $\partial \bar{s}(\partial t)^{-1}$ is the dry static energy lapse rate, $\bar{\boldsymbol{v}}\, \nabla \bar{s}$ is the horizontal divergence, $\partial \bar{s}\bar{\omega}(\partial p)^{-1}$ is vertical advection. The horizontal averages are denoted by $(\bar{\ })$. For calculating the horizontal averages, we used Delaunay Triangulation following that of other research (Reed and Recker [15], Johnson [9], Katsumata et al. [12]). This method need three point locations circumventing an isolated area. For region SI we used observations from stations in Medan, Padang, and Singapore; for region TZ, we used two point locations of GPS-RO mission that coincide with the observation station inland in either Medan or Padang; and for region IO, we used two point locations of GPS-RO mission that are nearest to the RV/MIRAI during its cruise.

The sensible heat budget of the area is taken into account in terms of the dry static energy s, defined as the sum of the sensible heat and potential energy per unit mass. That is,

$$s = c_p T + gz \tag{2}$$

where c_p is the specific heat of dry air at constant pressure, T temperature, g the gravitational acceleration and z geopotential height.

The area-averaged horizontal divergence over the triangular region were computed by,

$$\overline{\nabla \cdot V} = \frac{1}{A}\left(\oint u\mathrm{d}y - \oint v\mathrm{d}x\right) \tag{3}$$

where u and v are the eastward and the northward components of the winds along the periphery of the triangle whose area is denoted by A.

The average vertical p-velocity is obtained by,

$$\bar{\omega} = \int_{po}^{p} \overline{\nabla \cdot V}\, \mathrm{d}p \tag{4}$$

in which the original estimates of $\overline{\nabla \cdot V}$ were slightly corrected to make $\bar{\omega}$ vanish at 100 mb. All the computations were carried out for certain MCS case in each region and at intervals of 100 mb in pressure level. All the vertical differentiations were replaced by either of these finite differences:

$$f'(x) = (f(x+h) - f(x-h))(h)^{-1} \ \textit{(Central difference)} \tag{5}$$

$$f'(x) = (f(x+h) - f(x))(h)^{-1} \ \textit{(Forward difference)} \tag{6}$$

$$f'(x) = (f(x) - f(x-h))(h)^{-1} \ \textit{(Backward difference)} \tag{7}$$

depending on data availability and reliability.

The Q_1 measures the net heating effects of convective and radiative processes averaged over the area under investigation. Their vertical profiles have been determined in many regions of the tropics and subtropics for convective systems containing both convective-scale (cumulonimbus) and mesoscale components (Yanai et al. [17], Johnson [9]). In the tropics, vertical advection $\partial \bar{s}\bar{\omega}(\partial p)^{-1}$ is dominant in area with size relevant to this study. Therefore, since $\partial \bar{s}(\partial t)^{-1}$ is roughly constant in the tropics, the net heating may be interpreted as a direct indication of the mean vertical motion $\bar{\omega}$ in the area. In this study, we analyze diabatic heating profile associated with MCS in three sub-regions, without discussing further detail on each type of MCS.

3. Results and Discussion

3.1. Convective Activities in the Area of Study

In this section, we first discuss the results of MCS identification from satellite imageries acquired in the period of study, which corresponded to the active phase of Madden-Julian Oscillation (MJO) from 20th November 2011 to 11th December 2011. Basically, the MCS in the Indian Ocean has similar appearance as reported in previous studies (Trismidianto [16]; Yuan and Houze [19]) referring to the definition of MCS from Jirak et al [8]. Figure 4 shows the number of daily occurrences of MCSs that were successfully identified within the area as defined in figure 1. It can be seen from figure 4 that convective activity increased from 20th November 2011 and peaked around 28th November 2011, and then drastically reduced after 2nd December 2011. Results of MCS classification (data not shown) show that more circular type (eccentricity ≥ 0.7) of MCC and MβCC appearances dominated the sample with the total of eight to nine occurrences, while elongated PECS and MβEC occurred only three times. Due to such uneven distribution, we did not differentiate MCS types in the computation of apparent heating profiles.

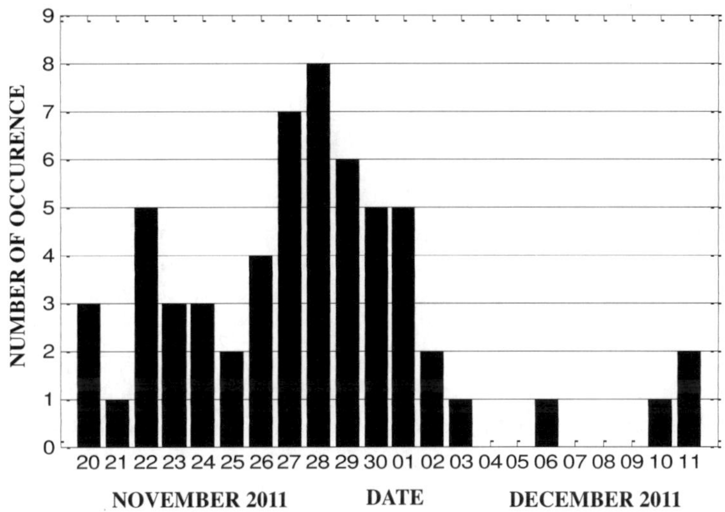

Figure 4. Number of identified MCSs in the area of study (6°N-6°S; 95°E - 105°E) during the period of 20th November to 11th December 2011.

Relatively frequent appearances of MCSs found in this study are consistent with results from previous works. As pointed out by Yuan and Houze [19] and Trismidianto [16], the Eastern Indian Ocean is convectively active region due to very contrast difference between SI and the warm seas of the IO. Such land-sea contrast could trigger strong diurnal cycle leading to the creation of large convective systems like MCS.

3.2. Diabatic Heating Profiles

We computed diabatic heating profiles associated with the appearances of MCS in three sub-regions by methods discussed in Section 2. As previously mentioned, we analyzed the results by grouping the heating profiles according to three sub-regions (SI, TZ, and IO) and two phases of MCSs (mature and dissipation) so that in total we have six groups of samples. Notwithstanding limitation in data availability, we managed to obtain four samples of heating profile for each group as shown in figure 5 and figure 6.

It is clear from figure 5 that, during the mature phase of MCS, warming occurs throughout the troposphere regardless of the sub-regions. This result is consistent with those of Yanai et al. [17],

IOP Conf. Series: Earth and Environmental Science **162** (2018) 012014 doi:10.1088/1755-1315/162/1/012014

Johnson [9], Johnson and Young [11], and Katsumata et al. [12]. However, heating profiles over land showed more variations in the amplitude with maximum heating up to 20 K/day, compared to other sub-regions where maximum values are below 20 K/day. Moreover, there are also marked differences in the height of layers where the maximum heating occurs.

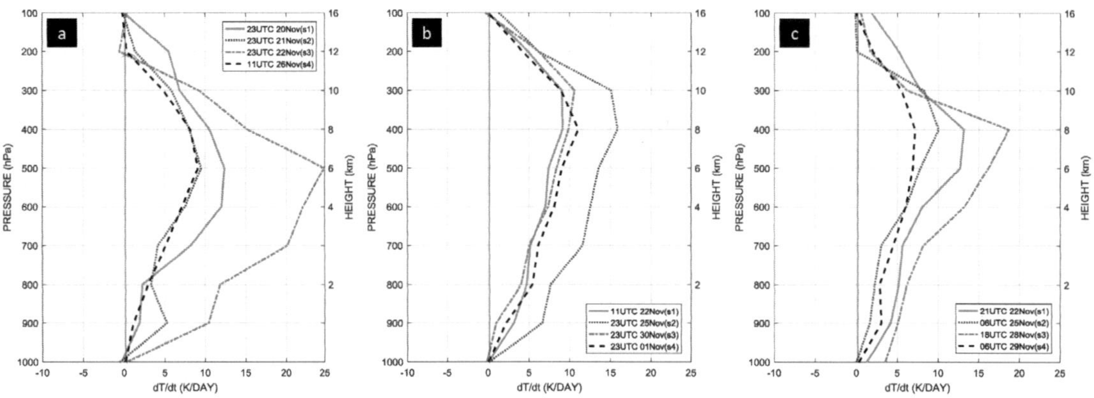

Figure 5. Profiles of diabatic heating computed as apparent heating Q_1 (see text) in three sub-regions of (a) Sumatra Island (SI), (b) Transition Zone (TZ), and (c) Indian Ocean (IO) corresponding to mature phase of MCSs at different times.

Diabatic heating profiles that correspond to dissipation phase of the MCSs (figure 6) show a general pattern, in which negative values (cooling) appear at the lower layers of troposphere between 1000 to 700 hPa. It should also be notable that maximum heating shifts upward to occupy middle and upper troposphere. The upper tropospheric maximum is known primarily to be a consequence of latent heating by condensation and freezing in the anvil, while the cooling in the lower troposphere is largely due to precipitation evaporation and melting (Houze [5]). Moreover, radiative heating and cooling become significant in the stratiform clouds that predominate during the transition phase from mature to dissipation.

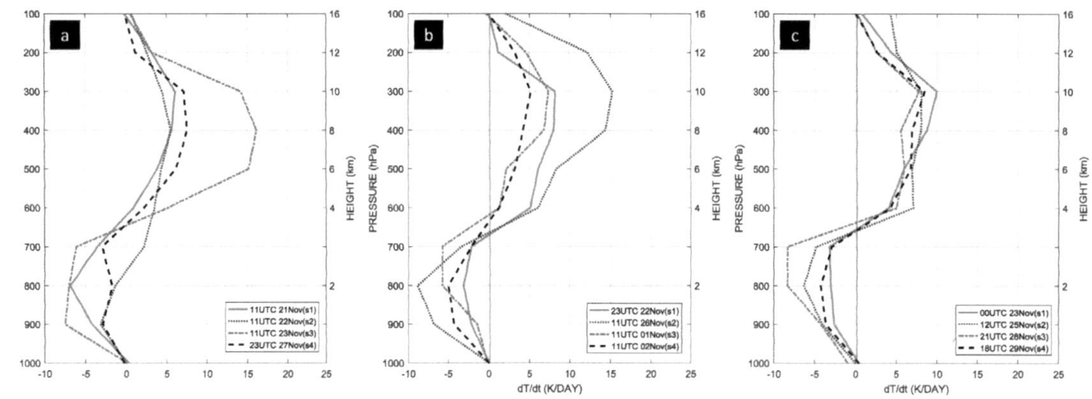

Figure 6. Same as figure 5 but for dissipation phases of MCSs.

Our results confirm that profiles of diabatic heating / cooling due to the presence of MCSs have similar general patterns throughout the MC region. Differences in the amplitude of the heating might be due to several factors such as differences in the atmospheric profile data source (radiosonde, GPS-RO, and reanalysis), the timing between mature and dissipation phases, and the classification of the MCSs.

Herein we restrain from further analysing those factors influencing the amplitude due to insufficient samples. However, differences in the height of heating/cooling peaks seem to be more related to land-sea contrasts.

3.3. Difference in the Height of Heating and Cooling Peaks

In order to see differences in the height of heating/cooling peaks, composite diabatic heating profiles, for the same data groups as in figure 5 and figure 6, are presented in figure 7 and figure 8. It should be clear that for the mature phase, warming peak is centered near 500 hPa for SI, 400-300 hPa for TZ, and 400 hPa for IO. It should be clear that there is a remarkable differences in the height of maximum heating over land and oceans corresponding to the mature phase of MCS. Figure 8 (also figure 5) shows that difference in heating peaks persist during the dissipation phase of MCS, while the height of cooling peaks is almost the same in all sub-regions.

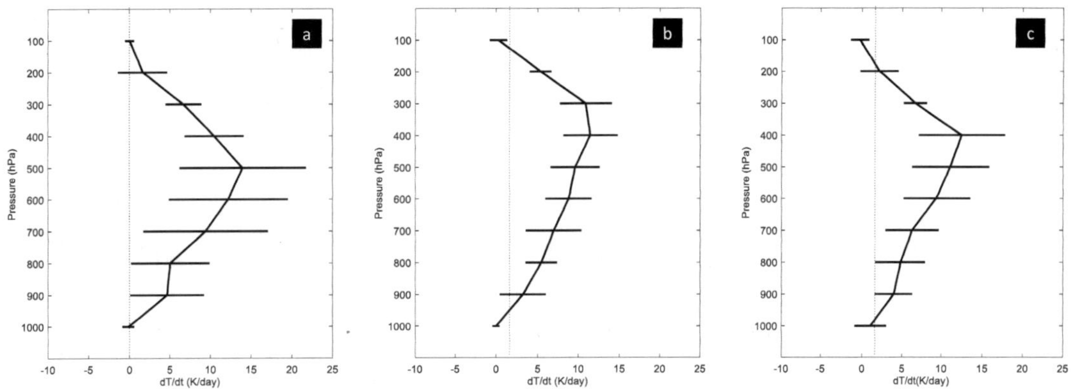

Figure 7. Composite result from the vertical structure of diabatic heating Q1 at the time of maximum / mature for the area a) SI, b) TZ, and c) IO. The horizontal line at each altitude level is the standard deviation calculated from each sample per region

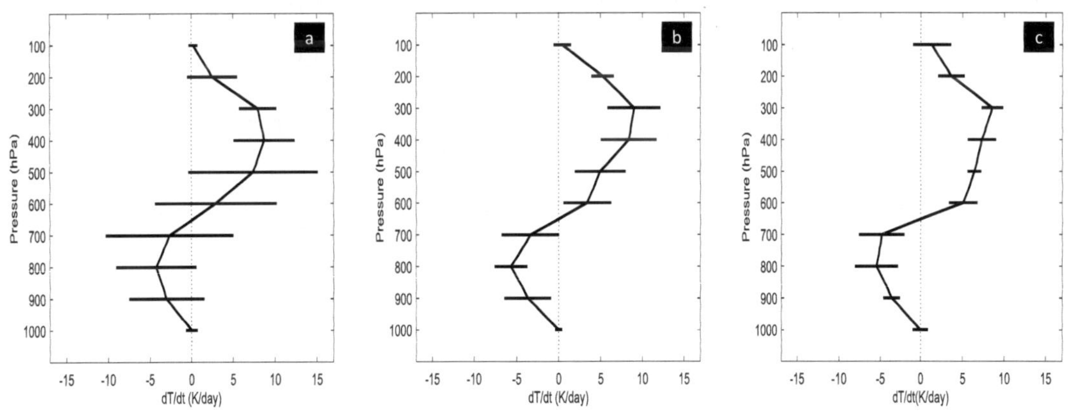

Figure 8. Same as figure 7 but for dissipation phase.

As it has been pointed out in Section 2, the apparent heating Q_1 in the tropics is dominated by vertical advection term. Figure 9 confirms that heating and cooling in the troposphere are in general explained by the occurrence of updrafts and downdrafts. Vertical movement over the ocean is mainly driven by large-scale convergence, whereas topography plays more significant role overland. Sumatra Island can cause strong and rapid development of MCS due to effects of orography and local circulations like sea/land breezes. The strong updrafts overland transport large amount of water vapor to lower condensation level, because of higher saturation, leading to the release of latent heat that support the development of deeper convention. On the contrary, updrafts over the ocean are normally weaker (see figure 9) causing massive condensation to occur at higher level due to slower rate of vertical moisture transport. This vertical motion pattern is consistent with the result by Cifelli and Rutledge [2].

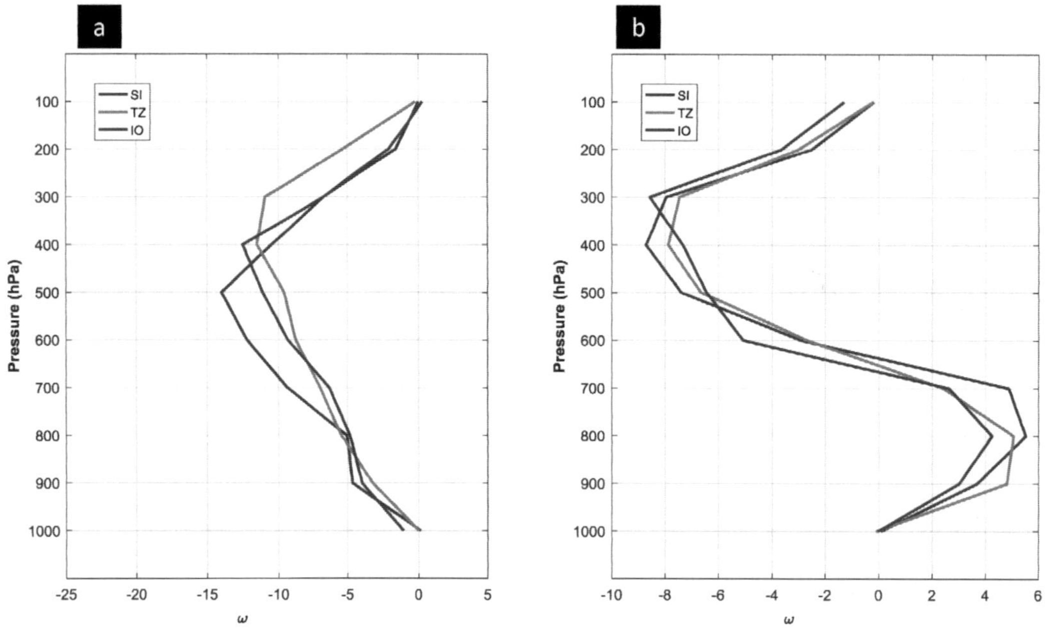

Figure 9. The result of composite vertical motion (ω) of SI, TZ, and IO (see text) in (a) mature phase and (b) dissipation phase.

4. Summary

We investigated the diabatic heating profiles on MCS using data of temperature, pressure, and horizontal wind components obtained from radiosonde observations around the Indian Ocean and the Maritime Continent. We also utilized the temperature profiles retrieved by GPS-RO and the wind data from ECMWF reanalysis to elaborate the characteristics of the diabatic heating over the three different regions of Sumatra Island (SI), Indian Ocean (IO), and the transition zone in between (TZ) during CINDY/DYNAMO campaign periods. In this study, samples of MCSs have been identified from satellite IR imageries using methods suggested by Jirak et al. [8] and diabatic heating profiles were computed following Yanai et al. [18]. A number of 11 MCSs appearances have been identified from the analyzed dataset.

The results show that in all areas (SI, IO, and TZ), during the mature stage of MCSs, heating occurred in the entire troposphere, consistent with the previous study by Yanai et al. [17] and Houze [5][6][7]. However, maximum heating over land is located at the lower altitude compared to that of the oceanic and transitional area. Heating profiles over land also have more variations in the amplitude with maximum value of up to 20 K/day. In the dissipation (postmature) stage, heating profiles of the dissipation stage were characterized by cooling in the lower troposphere and upward shift of maximum heating location for SI and TZ areas, and also sample over IO. The vertical velocity profiles showed that heating and cooling are mainly associated with updraft and downdraft in the cloud systems, this result is similar to the finding by Cifelli and Rutledge [2].

As additional remarks, the presence of cooling that occurs in layers close to the surface could indicate the formation of cold pool. The presence of cold pool may trigger atmospheric instability leading to the development of new convective cells. This result provides supporting evidence for the finding of

Trismidianto et al. [16] about cold pool formation in the dissipation stage of MCS as shown by the diabatic heating profiles. More detailed studies are needed to confirm whether numerical models are able to reproduce correct heating profiles, not only for MCS in general but also for each type MCSs that may appear in MC and Indian Ocean regions.

Acknowledgements
The radiosonde data from CINDY2011/DYNAMO are obtained from the Japan Agency for Marine-Earth Science and Technology (JAMSTEC) (http://www.jamstec.go.jp/iorgc/cindy) and Earth Observing Laboratory (EOL) of UCAR. GPS-RO data are provided by COSMIC Data Analysis and Archive Center (CDAAC) (http://cdaac-www.cosmic.ucar.edu/ cdaac/products.html). The author (KF) is grateful to the organizer of 1st International Conference on Maritime Sciences and Advanced Technology for providing the support.

References
[1] Anthes R A, Bernhardt P A, Chen Y, Cucurull L, Dymond K F, Ector D, Healy S B, Ho S, Hunt D C, Kuo Y, Liu H, Manning K, McCormick C, Meehan T K, Randel W J, Rocken C, Schreiner W S, Sokolovskiy S V, Syndergaard S, Thompson D C, Trenberth K E, Wee T, Yen N L and Zeng Z 2008 The COSMIC/FORMOSAT-3 Mission: Early Results *Bull. Amer. Meteor. Soc.,* **89**, 313–4.
[2] Cifelli R and Rutledge S A 1994 Vertical Motion Structure in Maritime continent mesoscale Convective Systems: Results from a 50-MHz Profiler. *J. Atmos. Sci.,* **51**, 2631–52
[3] Dee D P, Uppala S M, Simmons A J, Berrisford P, Poli P, Kobayashi S, Andrae U, Balmaseda M A, Balsamo G, Bauer P, Bechtold P, Beljaars A C M, van de Berg L, Bidlot J, Bormann N, Delsol C, Dragani R, Fuentes M, Geer A J, Haimberger L, Healy S B, Hersbach H, Hólm E V, Isaksen L, Kållberg P, Köhler M, Matricardi M, McNally A P, Monge-Sanz B M, Morcrette J -J, Park B -K, Peubey C, de Rosnay P, Tavolato C, Thépaut J -N and Vitart F 2011 The ERA-Interim reanalysis: configuration and performance of the data assimilation system *Quarterly Journal of the Royal Meteorological Society* **137** 553–597.
[4] Hartmann D L, Hendon H H and Houze R A Jr 1984 Some implications of the mesoscale circulations in tropical cloud clusters for large-scale dynamics and climate *J Atmos Sci* **41** 113–21.
[5] Houze R A 1982 Cloud clusters and large-scale vertical motion in the tropics *J Meteorology Society Japan* **60** 396–410.
[6] Houze R A 1989 Observed Structure of Mesoscale Convective Systems and implications for Large-Scale Heating *Quaterly Jurnal of The Royal Meteorological Society* **115** 425-61.
[7] Houze R A Jr 2004 Mesoscale Convective Systems *Rev. Geophys* **42**
[8] Jirak I L, Cotton W R, McAnnelly R L 2003 Satellite and radar survey of mesoscale convective system development *Monthly Weather Review* **131** 2428–49
[9] Johnson R H 1984 Partitioning tropical heat and moisture budgets into cumulus and mesoscale components: Implications for cumulus parameterization *Monthly Weather Review* **112** 1590–601
[10] Johnson R H and Ciesielski P E 2000 Rainfall and radiative heating rate estimates from TOGA COARE atmospheric budgets *J Atmos Sci* **57** 1497–514
[11] Johnson R H and Young G S 1983 Heat and moisture budgets of tropical mesoscale anvil clouds *Journal of Atmospheric Scientics* **40** 2138–47
[12] Katsumata M, Ciesielski P E, Johnson R H 2011 Evaluation of budget analyses during MISMO *Journal of Applied Meteorology and Climatology* **50** 241–54.
[13] Machado L A T, Rossow W B, Guedes R L and Walker A W 1998 Life cycle variations of mesoscale convective systems over the Americas. *Mon. Wea. Rev.,* **126** 1630–54.
[14] Mapes B E and Houze R A 1995 Diabatic divergence profiles in western Pacific Mesoscale convective systems *J Atmos Sci* **52** 1807–28.

IOP Conf. Series: Earth and Environmental Science **162** (2018) 012014 doi:10.1088/1755-1315/162/1/012014

[15] Reed R J and Recker E E 1971 Structure and properties of synoptic-scale wave disturbances in the equatorial western Pacific *J Atmos Sci* **28** 1117–33.
[16] Trismidianto, Yulihastin E, Satyawardhana H, Nugroho J T and Ishida S 2017 The Contribution of the Mesoscale Convective Complexes (MCCs) to total rainfall over Indonesian Maritime Continent *IOP Conf. Ser.: Earth Environ. Sci* **54** *012027*
[17] Yanai M, Esbensen S and Chu J 1973 Determination of bulk properties of tropical cloud clusters from large-scale heat and moisture budgets *Journal Atmospheric Science* **30** 611–27.
[18] Yoneyama K, Zhang C and Long C N 2013 Tracking pulses of the Madden-Julian Oscilation *Bull. Amer. Meteor. Soc.,* **94** 1971–91
[19] Yuan J and Houze R A Jr 2010 Global Variability of Mesoscale Convective System Anvil Structure From A-Train Satellite Data *Journal of Climate* **21** 5864–88.
[20] UCAR/NCAR - Earth Observing Laboratory. 2012. Singapore Radiosonde L3 Data. Version 1.0. UCAR/NCAR - Earth Observing Laboratory. https://data.eol.ucar.edu/dataset/347.030. Accessed 25 Aug 2017.
[21] Yoneyama, K., Japan Agency for Marine-Earth Science and Technology (JAMSTEC). 2012. Indonesia Site - Medan Radiosonde L3.0 Data. Version 1.0. UCAR/NCAR - Earth Observing Laboratory. https://data.eol.ucar.edu/dataset/347.053.
[22] Yoneyama, K., Japan Agency for Marine-Earth Science and Technology (JAMSTEC). 2012. Indonesia Site - Padang Radiosonde L3.0 Data. Version 1.0. UCAR/NCAR - Earth Observing Laboratory. https://data.eol.ucar.edu/dataset/347.097.
[23] UCAR/NCAR - Earth Observing Laboratory. 2012. R/V Mirai Radiosonde L3.0 Data (ESC Format). Version 1.0. UCAR/NCAR - Earth Observing Laboratory. https://data.eol.ucar.edu/dataset/347.014.

IOP Conf. Series: Earth and Environmental Science **162** (2018) 012015 doi:10.1088/1755-1315/162/1/012015

Semiannual variation of zonal current along the equatorial Indian Ocean

Faruq Khadami[1], Totok Suprijo[1] and Iskhaq Iskandar[2]

[1] Oceanography Study Program, Faculty of Earth Science and Technology, Institute Technology of Bandung, Indonesia
[2] Faculty of Science and Mathematics, Sriwijaya University, Indonesia

E-mail: fkhadami@oceanography.itb.ac.id

Abstract. Semiannual current variation in the equatorial Indian Ocean is examined using currents velocity records from Research Moored Array for African-Asian-Australian Monsoon Analysis and Prediction (RAMA) Program. Spectral analysis shows that semiannual variation is the strongest variation in equatorial Indian Ocean. Further analysis exhibits that this variation is associated with Wyrtki Jet phenomenon and forced by wind along the equator. Upward phase propagation (downward energy propagation) is clearly shown as indication of the propagation of equatorial Kelvin wave. Cross correlation of two mooring data captures westward propagation of reflected Rossby wave that interferes the propagation of Kelvin wave. During IOD positive event, stronger reflected Rossby wave occurs and causes weakening of upward phase propagation. This result reveals that interferences of equatorial wave have important role on forming vertical structure of currents along equatorial Indian Ocean.

1. Introduction

Semiannual current variation of tropical Indian Ocean was discovered early by an observation of the equatorial Wyrtki Jet in 1973 [1]. The jet is observed during the monsoonal transition, particularly in March-May and September-November [1, 8]. It is forced by westerly wind burst occurred due to the weakening of monsoon wind in the western Indian Ocean [1]. This jet is a narrow westward strong current in equator Indian Ocean which has an average velocity of 64 cm/s. The velocity decreased exponentially until ± 100 m in depth [1, 2].

The equatorial Wyrtki Jet is associated with the eastward propagation of the equatorial Kelvin wave [3]. Upon reaching the west coast of Sumatra, the Kelvin wave propagates to south and north of equator as coastal Kelvin wave [4]. Some of the Kelvin wave energy is reflected as the westward reflected Rossby wave [3, 4]. The westward propagation of reflected Rossby wave interferes the eastward propagation of forced Kelvin wave, causing a phase difference between wind and forced currents in the equator Indian Ocean [3, 4, 7].

On interannual scale, the Indian Ocean Dipole (IOD) along with the El Nino and Southern Oscillation (ENSO) play role in the dynamics of the equatorial Indian Ocean currents. IOD and ENSO led to the weakening of Wyrtki Jet in 1994, 1999, 2000 and 2006-2011 as well as the counter flow of Wyrtki Jet in 2008 [5]. The IOD and ENSO more affected the fall Wyrtki Jet than the spring Wyrtki Jet [6].

Vertical structure of the Indian Ocean water column is affected by the dynamics of downwelling Kelvin wave as a part of Wyrkti jet [1]. The propagation of downwelling Kelvin wave affects the shallowing of the thermocline in the west region of equator Indian Ocean, while in the east region, the thermocline is deeper [1]. On an intraseasonal scale, the Kelvin wave affects the vertical structure of equatorial zonal currents. The upward phase propagation of the vertical structure of the currents is an indication of Kelvin wave propagation's effect [4].

The aim of the study is to evaluate the variation of semiannual zonal currents in the equator Indian Ocean. In this paper we present the variation of currents using moorings observation data. The outline of the paper is present as follow: Section 2 describes data and method used in this study, Section 3 presents the variation of currents and the forcing dynamics and the summary is presented in last section.

2. Methodology

2.1. Data

The main data in this study are ocean currents velocity obtained from Acoustic Doppler Current Profiller (ADCP) moorings as a part of Research Moored Array for African-Asian-Australian Monsoon Analysis and Prediction (RAMA) program. The ADCP moorings are deployed in the equator at 80.5°E and 90°E (Figure 1). Ocean currents data at 80.5°E have a 5 m-interval vertical resolution with daily resolution from the near surface down to 175 m depth. ADCP moorings at 90°E provide ocean currents data velocity, which have a 10 m-interval vertical resolution with daily resolution from the near surface down to 400m depth. In order to get comparable result, only ocean currents data between 20 and 175 meters depth from both moorings, which were taken from November 2004 to September 2008were used.

Daily surface wind data were used to evaluate dynamics and forcing of ocean currents.. The surface wind data were derived from Quick Scatterometer (QSCAT), which has 0.25° x 0.25° in spatial resolution. Wind surface data along the equator between 50°E and 95°E from November 2004 to September 2008 (Figure 1) were used in this study.

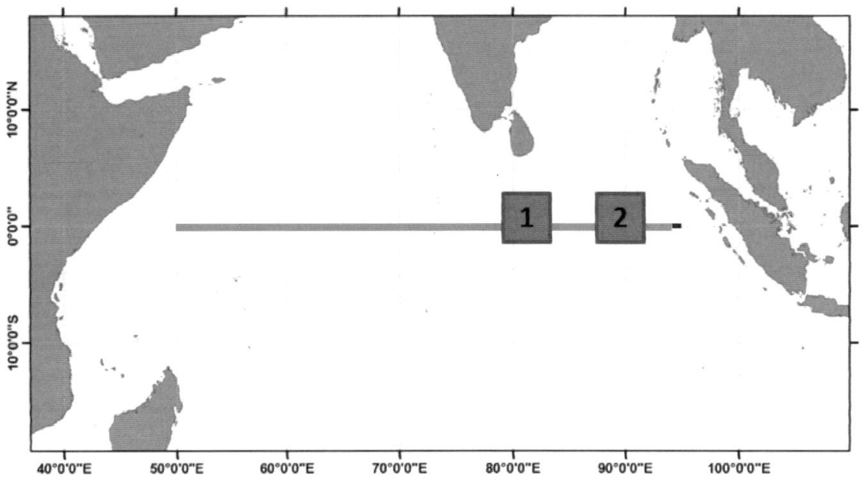

Figure 1. Locations of ADCP moorings used in this study are marked by number 1 (80.5°E) and 2 (90°E). The location of wind surface data are marked by the line.

2.2. Methods

In this study, various spectral analysis and statistical analysis were used. Fourier Transform analysis [10] was applied to evaluate the dominant frequency in both currents time series data and to describe the fluctuation of the data. Simple statistical analysis such as mean and standard deviation were applied to describe the data variation. We also applied bandpass filter using window from periods 150–200 days to separate semiannual variation from the raw data. To evaluate the dynamics of currents, cross correlation between currents data in 80.5°E and 90°E were calculated. To explore the forcing mechanism, calculation of cross correlation between currents data and wind were carried out.

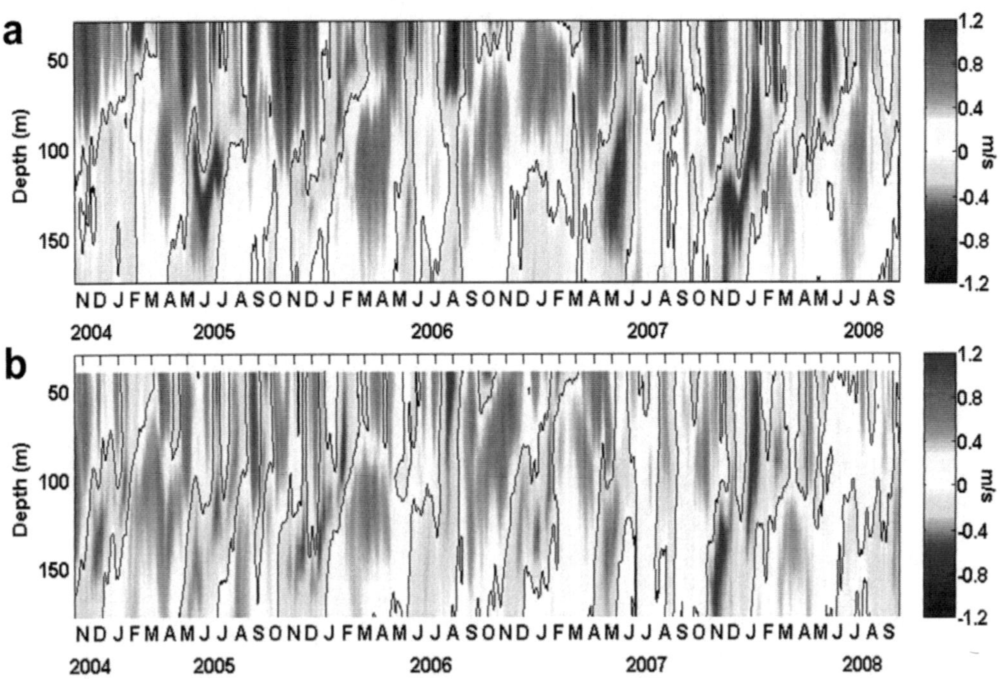

Figure 2. Time-depth diagrams of zonal currents at 0°S, 80.5°E (a) and at 0°S, 90°E (b). Positive (negative) value is eastward (westward). Black contour is mark of zero value.

3. Result and discussion

The depth-time diagrams of the observed currents (Figure 2) show that semiannual variations are clearly observed at 80.5°E and 90°E. Semiannual variations associated with the spring and fall equatorial jets. The equatorial jet occurred during strong westerly wind along the equator at monsoonal transition [1]. Spectral peak in semiannual period is clearly shown in both spectral diagrams (Figure 3a, b). The spectral energy of semiannual period at 80.5°E is higher compared to that at 90°E. Beside the semiannual period, intraseasonal period also has significant energy although it is not as high as semiannual period.

The change of standard deviation along the depth (Figure 3c) suggests currents variation along the depth. The variation of the equator zonal current before bandpass filtered at 80.5°E in the near surface is the highest, measured up to 0.5 m/s and decreases gradually to 0.3 m/s at 100 m depth. From 100-120 m depth, the variation is constant at 3 m/s and gradually declines after 120 m depth. At 90°E, the variation of zonal current measured at 40–120 m depth is 0.3 m/s. The gradual decreasing of standard deviation occurred beyond 120 m depth suggests that the variation of the currents is getting smaller by the depth.

After the bandpass filter applied on the currents data, the standard deviation along the depth is then changed (Figure 3d). The semiannual zonal currents at 80.5°E have high variation on near surface and weaken gradually until 100 m depth. The variation increases after 100 depth and has its peak at 120 m depth and subsequently decreases from then. Semiannual zonal currents at 90°E, on the other hands, slightly increase from near surface to at about 80 m depth. The increasing turns sharper up to 120 m depth and decreases gradually after 120 m depth. .

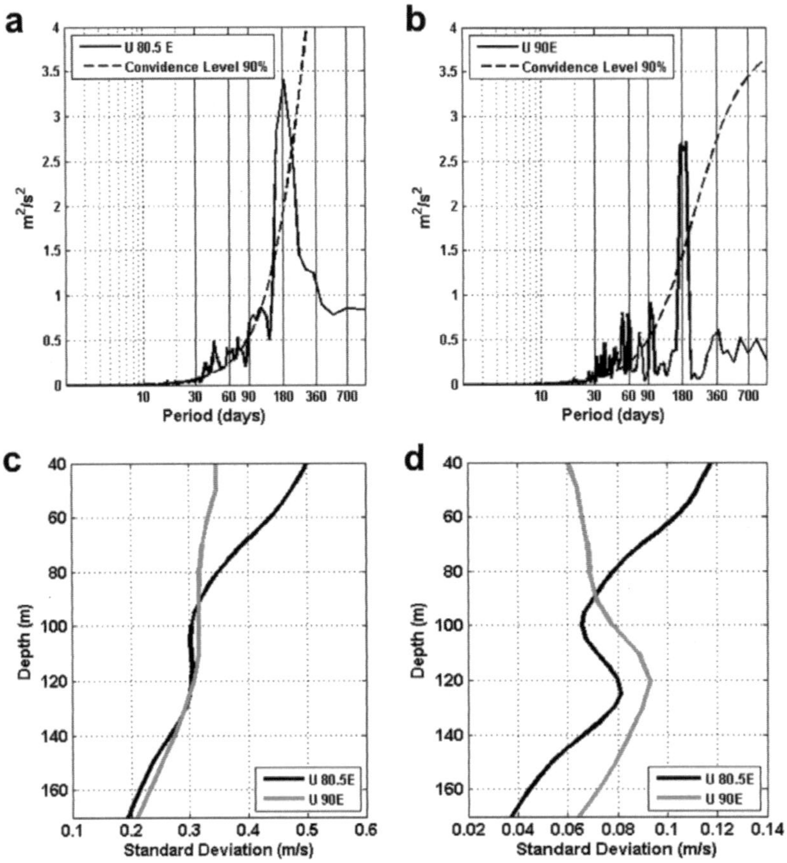

Figure 3. (a) Spectrum of zonal velocity averaged over 40-100 m depth in 0°S 80.5°E and (b) 0°S 90°E. Standard deviation of (c) the zonal currents in 0°S 80.5E and 0S 90°E and (d) semiannual zonal currents in 0S 80.5°E and 0S 90°E

Depth-time diagram of semiannual zonal currents velocity at 80.5°E and 90°E are presented in Figure 4. At both moorings, the eastward currents occur on the transition monsoon, which associated with the equatorial Wyrtki Jet. Depth-time diagrams show an upward phase propagation, where a gradual phase shift in depth with the deeper level leads the near surface level. Those upward phase propagations indicate the propagation of the equatorial Kelvin wave. At 80.5°E, near surface level has strongest current amplitude, while at 90°E, the strongest amplitude variability measured at 120 m depth. In relation with Indian Ocean Dipole (IOD), in normal IOD event November 2004–May 2006, semiannual zonal currents at 80.5°E had strong amplitudes at near surface. During the same interval, semiannual zonal currents at 90°E have maximum amplitude at 120 m depth. During IOD positive phase, in July 2006–February 2007, strong currents at 80.5°E were weakening and near surface currents at 90°E were strengthening.

To explain the dynamical process of the evidence, we calculated the cross correlation between semiannual currents in 80.5°E and 90°E (Figure 5). At the upper level (40-80 m depth), the signal of semiannual zonal currents in 90°E leads the 80.5°E signal. This result reveals the westward propagation of the signal and presumably indicates the propagation of reflected Rossby wave. In the deeper layer, the eastward propagation of the signal is clearly captured and it reveals the propagation of forced Kelvin wave. It can be explained that when the Kelvin Wave reaches the west coast of Sumatra, it propagates poleward both north and south of the equator and some energy is reflected and propagate as the reflected

Rossby wave. The reflected Rossby Wave then interferes the propagation of Kelvin wave [3]. This result shows that interferences of the reflected Rossby Waves on the propagation of forced Kelvin wave occurred in the near surface level. Interferences of the waves affect the vertical profile of semiannual currents. During the IOD positive events (2006-2011), as the westerly wind weakened and the forced Kelvin wave also became weaker, the strong currents at 80.5°E did not appear. On the contrary, during the same IOD positive event, the interference of reflected Rossby waves appeared clearer and the strong currents in the upper layer at 90°E was observed.

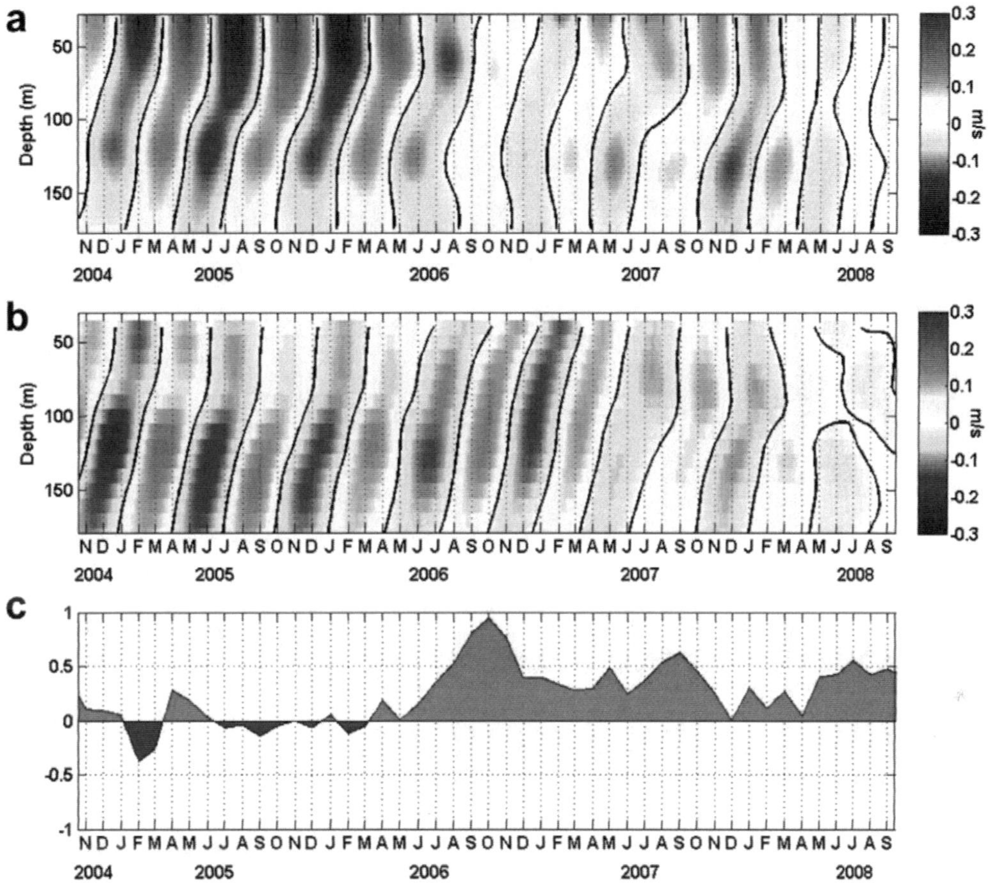

Figure 4. Time-depth diagrams of semiannual zonal currents at 0°S, 80.5°E (a) and at 0°S, 90°E (b). Positive (negative) value is eastward (westward). Black contour is mark of zero value. (c) Dipole Mode Index

In order to understand the forcing of the semiannual zonal currents, the relationship between semiannual zonal currents and wind in both data were examined. Correlation between semiannual zonal currents averaged 40–100 m depth and the semiannual wind along the equator in both data (Figure 6) indicate the forcing of currents. Currents at 80.5°E have significant correlation with wind at 70°E and have 0 days lag. This result indicates that semiannual currents at 80.5°E are forced directly by wind at 70°E. Currents at 90°E also have significant correlation with winds in 70°E but have 10 days lags, in which the wind led to the currents. This result illustrates that the semiannual currents are forced in the western Indian Ocean and propagated to the east.

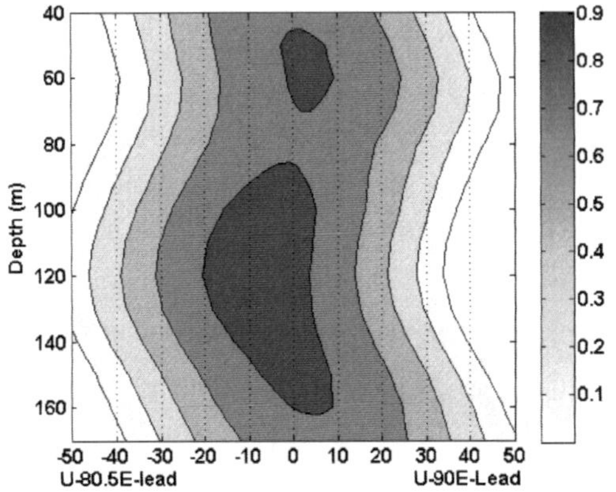

Figure 5. Correlation coefficient for semiannual currents between 80.5°E and 90°E.

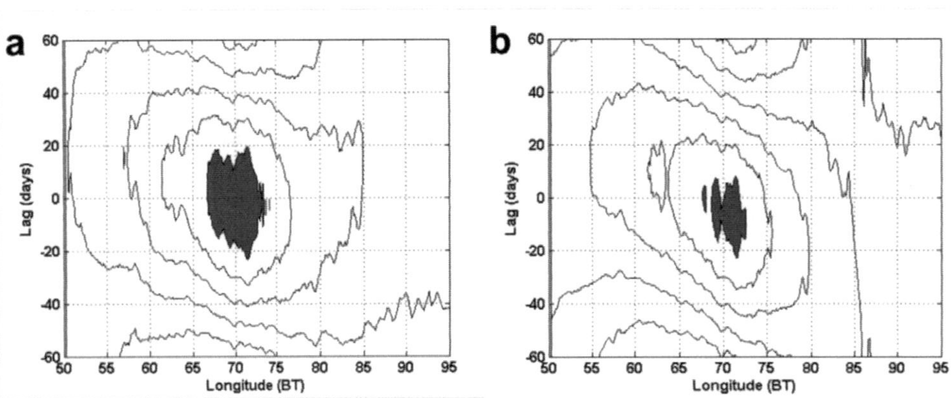

Figure 6. Correlation coefficients between wind along the equator and semiannual current (a) at 80.5°E, (b) at 90°E. Correlation above 95% confidence limit are shaded

4. Summary

Semiannual variation is the strongest variation in the equatorial zonal currents. The semiannual currents variability associated with the equatorial Wyrtki Jet and the propagation of Kelvin wave. The vertical structure of semiannual zonal currents is affected by the interferences of forced Kelvin wave and reflected Rossby wave. In addition, our preliminary result presents that during the IOD positive event, stronger reflected Rossby wave occurs and causes weakening of upward phase propagation. This study confirms that interferences of equatorial wave play important role on forming vertical structure of zonal currents along equatorial Indian Ocean.

References

[1] Wyrtki K 1973 An equatorial jet in the Indian Ocean *Science* **181** 262–264

[2] Reppin J, Schott F A, Fischer J, and Quadfasel D 1999 Equatorial currents and transports in the upper central Indian Ocean: Annual cycle and interannual variability *J. Geophys. Res.* **104(C7)** 15495–15514

[3] Nagura M and McPhaden M J 2010 Wyrtki Jet dynamics: Seasonal variability *J. Geophys. Res.* **115** C07009

[4] Iskandar I and McPhaden M J 2011 Dynamics of wind-forced intraseasonal zonal current variations in the equatorial Indian Ocean, *J. Geophys. Res.* **116** C06019

[5] Joseph S, Wallcraft A J, Jensen T G, Ravichandran M, Shenoi S S C and Nayak

S 2012 Weakening of spring Wyrtki jets in the Indian Ocean during 2006–2011 *J. Geophys. Res.* **117** C04012

[6] Gnanaseelan C, Deshpande A and McPhaden M J 2012 Impact of Indian Ocean Dipole and El Niño/Southern Oscillation wind-forcing on the Wyrtki jets *J. Geophys. Res.* **117** C08005

[7] Nagura M and McPhaden M J 2010 Dynamics of zonal current variations associated with the Indian Ocean dipole *J. Geophys. Res.* **115** C11026

[8] Qiu Y, Li L and W. Yu 2009 Behavior of the Wyrtki Jet observed with surface drifting buoys and satellite altimeter *Geophys. Res. Lett.* **36** L18607

[9] Schott F and McCreary J P 2001 The monsoon circulation othe Indian Ocean *Prog. Oceanogr.* **51**,1–123

[10] Emery, W J and Thomson R E 2004 *Data Analysis Methods in Physical Oceanography* (Amsterdam: Elseiver)

IOP Conf. Series: Earth and Environmental Science **162** (2018) 012016 doi:10.1088/1755-1315/162/1/012016

Interannual variability of sea surface height difference between western Pacific Ocean and eastern Indian Ocean and its effect to geostrophic current in Lombok Strait

H A Rejeki[1], Kunarso[2] and Munasik[2]

[1] Indonesian Agency for Meteorology Climatology and Geophysics, School of Meteorology Climatology and Geophysics, Tangerang Selatan, 15221 Indonesia
[2] Faculty of Fisheries and Marine Science, Diponegoro University, Semarang, 50275 Indonesia

E-mail: hastiamrihrejeki@gmail.com

Abstract. Interannual variability of sea surface height (SSH) in western Pacific Ocean and eastern Indian Ocean occurs as a result of the ENSO phenomena. This variability affects SSH difference between both of those oceans. In normal condition, SSH in western Pacific Ocean is higher than in eastern Indian Ocean which causes a current that passes through Indonesia which known as the Indonesian Throughflow (ITF). This study used SSH and geostrophic currents data from 1993 to 2015 which obtained from AVISO satellite altimetry data to determine the variability of SSH difference in ENSO condition which represented by the sea surface temperature (SST) anomaly of Niño 3.4 data from NOAA and its relation to the geostrophic currents in Lombok Strait which is one of the ITF crossing path. The result of the correlation calculation of ENSO condition to SSH difference between western Pacific Ocean and eastern Indian Ocean shows the negative value. It means that SSH difference at both oceans have opposite condition that when El Nino happened SSH in western Pacific Ocean is lower than the eastern Indian Ocean and when La Niña happened SSH in the western Pacific Ocean is higher than the eastern Indian Ocean. The SSH difference does not affect the direction of geostrophic currents but affects its speed.

1. Introduction

ENSO (El Niño Southern Oscillation) is a phenomenon of climate variability that affects Indonesia, equatorial Pacific and other parts of the world [1]. The ENSO phenomenon consists of two events, El Niño and La Niña. This phenomenon can give negative impacts to the community directly such as prolonged droughts and forest fires, and excessive rainfall which causes the flash flood. El Niño occurs when sea surface temperature (SST) in the eastern and central Pacific Ocean increase which associated to the reduction of rainfall in Indonesia whereas when La Niña occurs it shows the opposite condition [2, 3]. The phenomenon of El Niño and La Niña can be studied from several indicators such as sea surface temperature anomaly, southern oscillation index, sea level altitude, and Indonesian Troughflow or commonly called as ITF [1].

ITF is a system of marine circulation in Indonesian waters where a flow path takes the mass of water from the Pacific Ocean to the Indian Ocean [4]. The current occurs due to differences in sea surface height between the Pacific Ocean and Indian Ocean [5]. The occurrence of ITF is mainly caused by the tropical surface of the Western Pacific Ocean is higher than in the eastern Indian Ocean, it produces a pressure gradient which drives a flow of currents from the Pacific Ocean to the Indian Ocean as its result [6]. The important indicators to be observed for assessing the ITF are the sea surface height (SSH) in the Pacific and Indian Oceans and the variability of sea surface temperature (SST) in Indonesian waters [1].

SSH differences variability in the Pacific Ocean and Indian Ocean has been calculated by subtracting the SSH of Pacific Ocean with the SSH of Indian Ocean [7]. The result shows that the calculations during the El Niño years, the value of SSH difference is negative while at the time La Niña is positive. The results of the calculations in that study can be seen in table 1 and 2.

Table 1. Monthly sea surface height difference values to ENSO in January, February, March, April, May, and June 1993 - 2015[7]

Time	Jan	Feb	Mar	Apr	May	Jun
1993	-0.06	0.00	-0.02	-0.10	-0.08	-0.06
1994	-0.09	-0.05	0.00	-0.03	-0.05	-0.02
1995	-0.08	-0.08	-0.03	-0.03	-0.03	-0.01
1996	-0.03	-0.02	-0.02	0.02	-0.03	-0.04
1997	-0.10	-0.06	-0.09	-0.07	-0.10	-0.08
1998	-0.09	-0.07	-0.04	0.00	0.01	0.03
1999	0.01	0.05	0.07	0.04	0.00	0.00
2000	0.00	0.03	0.04	0.00	-0.03	-0.01
2001	-0.01	0.03	0.02	0.00	-0.02	-0.01
2002	-0.09	-0.03	-0.07	-0.06	-0.06	-0.04
2003	-0.12	-0.11	-0.03	-0.01	-0.01	-0.05
2004	-0.04	0.00	-0.02	-0.06	-0.08	-0.06
2005	-0.05	-0.05	-0.02	-0.03	-0.02	-0.01
2006	-0.02	0.01	0.05	0.03	0.01	0.02
2007	-0.04	-0.05	-0.02	-0.04	-0.05	-0.01
2008	0.07	0.10	0.08	0.02	-0.03	0.00
2009	0.02	0.03	0.02	0.01	-0.03	-0.01
2010	-0.12	-0.11	-0.06	-0.01	0.01	0.04
2011	0.01	0.01	0.01	-0.04	-0.04	-0.04
2012	0.02	0.00	0.03	-0.03	-0.02	0.02
2013	0.01	-0.01	0.02	0.05	0.02	0.00
2014	-0.02	-0.03	-0.08	-0.09	-0.10	-0.07
2015	-0.08	-0.13	-0.12	-0.11	-0.13	-0.14
Mean	**-0.04**	**-0.02**	**-0.01**	**-0.02**	**-0.04**	**-0.02**

Table 2. Monthly sea surface height difference values to ENSO in July, August, September, October, November, and December 1993 - 2015[7]

Time	Jul	Aug	Sep	Oct	Nov	Dec
1993	-0.06	-0.03	-0.02	0.01	-0.01	-0.07
1994	-0.01	0.04	0.01	0.05	0.03	-0.05
1995	0.03	0.06	0.06	0.05	0.05	0.01
1996	-0.01	0.01	0.04	0.03	0.03	-0.03
1997	-0.07	-0.04	-0.05	-0.06	-0.07	-0.10
1998	0.01	0.04	0.06	0.08	0.04	-0.01

Time	Jul	Aug	Sep	Oct	Nov	Dec
1999	0.04	0.05	0.07	0.10	0.04	0.00
2000	0.02	0.03	0.05	0.03	0.01	-0.03
2001	-0.02	0.01	0.04	0.05	0.01	-0.05
2002	-0.03	-0.04	-0.05	-0.07	-0.10	-0.14
2003	-0.02	0.00	0.02	0.06	-0.01	-0.04
2004	-0.05	-0.01	0.04	0.03	0.02	-0.03
2005	0.01	0.01	0.04	0.03	0.01	-0.03
2006	0.01	0.01	0.02	0.03	-0.01	-0.04
2007	0.03	0.05	0.08	0.11	0.09	0.06
2008	0.05	0.09	0.08	0.05	0.03	0.05
2009	-0.01	0.01	0.03	-0.01	-0.05	-0.11
2010	0.04	0.04	0.06	0.02	0.03	0.00
2011	-0.02	0.02	0.09	0.10	0.07	0.03
2012	0.04	0.08	0.07	0.06	0.00	0.00
2013	0.00	0.02	0.05	0.04	-0.01	0.01
2014	-0.04	-0.02	0.00	0.01	-0.01	-0.05
2015	-0.11	-0.11	-0.07	-0.07	-0.09	-0.13
Mean	**-0.01**	**0.01**	**0.03**	**0.03**	**0.00**	**-0.03**

Since the value of SSH difference yields a small value, so to compare the pattern between SSH difference with Nino 3.4 SST anomaly then SSH difference values were multiplied by 10 to see the opposite pattern in those comparison graph (figure 1). It shows that the Niño 3.4 SST anomaly depicted by a yellow line has an opposite graph pattern of the SSH difference of the western Pacific Ocean and the eastern Indian Ocean depicted by a blue line.

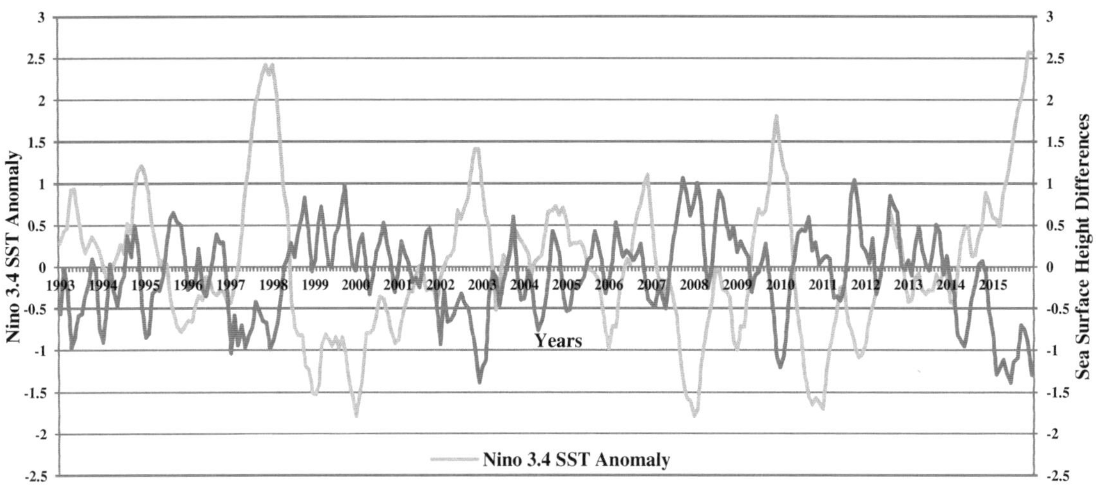

Figure 1. Comparative of Niño 3.4 SST Anomaly (yellow line) to SSH difference (grey line) between western Pacific Ocean and eastern Indian Ocean [7]. SSH difference is multiplied by 10 to better see the pattern.

This study extends the results of earlier work [7] and provides the correlation between interannual sea surface height in the western Pacific Ocean and eastern Indian Ocean to ENSO indices. ITF is a geostrophic transport at a maximum depth of 100 meters (upper layer) observed based on the difference

IOP Conf. Series: Earth and Environmental Science **162** (2018) 012016 doi:10.1088/1755-1315/162/1/012016

from the pressure gauge with absolute flow determined by the existing tidal and its speed section is corrected by the Acoustic Doppler Current Profiler [8]. Therefrom, we used the geostrophic velocity around Lombok strait to indicate the changing of SSH of both oceans due to ENSO events.

2. Data and method

The daily sea surface height (SSH) data from January 1993 to December 2015 with 0.25° x 0.25° resolution were provided by AVISO data center of CNES (Centre National d'Etudes Spatiales) based on TOPEX/Poseidon, Jason-1, and ERS-2 data is used in this study. The data have been corrected for all geophysical corrections includes ionospheric, dry and wet tropospheric delays, sea state bias, solid Earth, ocean and pole tides, ocean tide loading and inverted barometer correction. The data have also been corrected for instrumental drifts [9]. Those daily data were converted to monthly data by using monthly average calculation. The study areas were Western Pacific Ocean (125°-145° E; 10°-0° N) and Eastern Indian Ocean (110°-130° E; 10°-20° S) (figure 2) where those locations are based on the theory of the ITF attributed to differences in sea surface height between the Pacific Ocean and the Indian Ocean [5].

SSH in Western Pacific were compared to SSH in eastern Indian Ocean by computing the difference of SSH in both oceans. The assumptions used are based on the difference in sea level between the Pacific and Indian Oceans [5] where the Western Pacific Ocean is higher than that of the eastern Indian Ocean [6 in 4]. Therefrom the value of SSH in the western Pacific Ocean were subtracted by SSH in the eastern Indian Ocean [7]. A simple formulation to calculate the difference between SSH is as follows:

$$\Delta SSH = SSHpo - SSHio \qquad (1)$$

Where ΔSSH is the differences of sea surface height, $SSHpo$ is the sea surface height of the western Pacific Ocean, and $SSHio$ is the sea surface height of the eastern Indian Ocean.

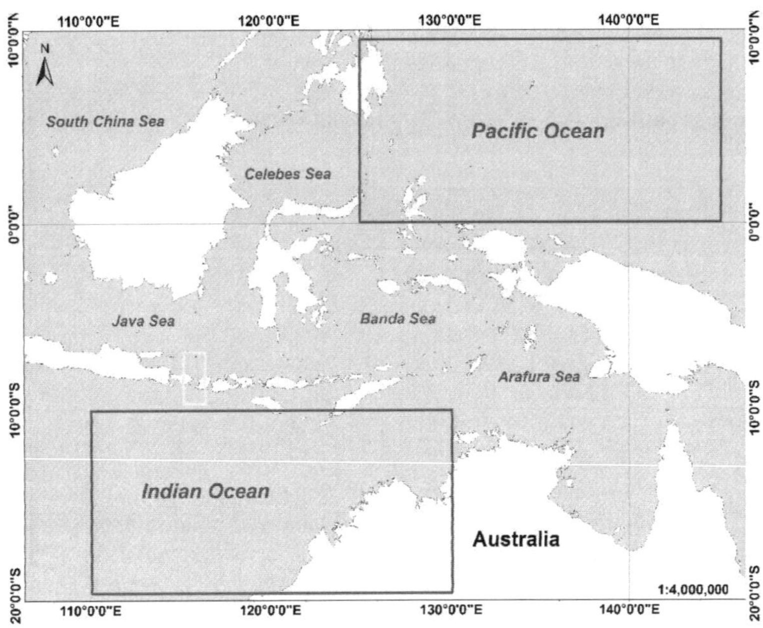

Figure 2. Research location. The red rectangles are the location of SSH data. The yellow box is the location of geostrophic current data which is Lombok Strait.

There are several indices used to monitor the tropical Pacific, all of which are based on sea surface temperature (SST) anomalies averaged across a given region. Usually the anomalies are computed relative to a base period of 30 years [10]. The Niño 3.4 index and the Oceanic Niño Index (ONI) are the most commonly used indices to define El Niño and La Niña events where the region of Niño 3.4 is located in 5N-5S, 170W-120W. Niño 3.4 index is the SST anomalies of Nino 3.4 region which may be thought of as representing the average equatorial SST across the Pacific from about the dateline to the South American coast. The Niño 3.4 index typically uses a 5-month running mean, and El Niño or La Niña events are defined as when the Niño 3.4 SST exceeds +/- 0.4°C for a period of six months or more. Whereas, ONI uses the same region as the Niño 3.4 index. It uses a 3-month running mean, and to be classified as a full-fledged El Niño or La Niña, the anomalies must exceed +0.5°C or -0.5°C for at least five months [10]. In this study, we used the monthly SST anomaly of Niño 3.4 from NOAA to determine the El Niño and La Niña episode, where the anomaly was taken toward the monthly average of SST Niño 3.4 during the period 1993 – 2015. When the monthly anomaly of SST Niño 3.4 is greater than 0.5 then is expressed as El Niño period and when monthly anomaly of SST Niño 3.4 value is lower than -0.5 then is expressed as La Niña period, respectively.

This study conducted a correlation analysis to find out how the relationship is between the parameters of sea surface height in the western Pacific Ocean and eastern Indian Ocean to the SST anomaly in Niño 3.4 region. In addition, correlation analysis was also performed on the SSH difference between the western Pacific Ocean and the eastern Indian Ocean to the occurrence of ENSO.

The variables which used in the correlation analysis are the anomaly of SST Niño 3.4 as the independent variable (x) which is considered to represent the incidence of El Niño and La Niña and the monthly average data of SSH, SSH difference in the western Pacific Ocean and the eastern Indian Ocean at the same event as the dependent variable (y). The used correlation equation is Pearson correlation (1749) which is formulated as follows [11]:

$$r_{xy} = \frac{n\sum XY - \sum X \sum Y}{\sqrt{[n\sum X^2 - (X)^2][n\sum Y^2 - (Y)^2]}} \qquad (2)$$

Where r is Pearson correlation, n is amount of sample, x is independent variable, and y is dependent variable. The level of relationship between the variables in the correlation analysis is based on the values in table 3 [12].

Table 3. Interpretation of Correlation Coefficients [12]

Interval of correlation	Level of relationship
0 – 0,199	Very low
0,20 – 0,399	Low
0,40 – 0,599	Moderate
0,60 – 0,799	Strong
0,80 – 1,00	Very strong

The Geostrophic current is caused by geostrophic balance occurring due to horizontal pressure gradient which is working on moving water masses, and balanced by the coriolis force [13]. The data of absolute geostrophic currents for 23 years from 1993 until 2015 with resolution 0.25° x 0.25° were also provided by AVISO. The difference of the sea water mass physical condition between one place and another produces a pressure gradient force which triggers the mass flow of seawater. The horizontal pressure force moves the current in a horizontal direction and in its movement will be influenced by the coriolis force which arising from the rotation of the earth. The geostrophic current at the sea surface consists of two main components such as u and v. The speed and direction of the geostrophic current is determined by calculating the resultant of the components u and v using the formula:

$$R = \sqrt{u^2 + v^2} \qquad (3)$$

IOP Conf. Series: Earth and Environmental Science **162** (2018) 012016 doi:10.1088/1755-1315/162/1/012016

Where R is the resultant of u vector and v vector (m/s), u is the speed of geostrophic current in the x direction (m/s), and v is the speed of geostrophic current in the y direction (m/s).

The measurement of the geostrophic current on the surface (z = 0) of the altimetry satellite is related to the slope of water surface, that is absolute dynamic or absolute dynamic topography. Dynamic topography is the difference in sea level above the geoid [14]. In AVISO, the mean dynamic topography (MDT) is obtained by averaging dynamic topography [15]. The calculation of absolute dynamic topography is as follows:

$$ADT = MDT + SLA \quad (4)$$

Where ADT is the absolute dynamic topography (m), MDT is the mean dynamic topography (m), and SLA is sea level anomaly (m).

The geostrophic current of Lombok Strait (7° - 9° S and 115° - 116° E) shown by red rectangle in figure 2 is chosen because that location is one of the exit pathway of ITF sea water mass to Indian Ocean [1]. The absolute geostrophic current velocity obtained from the calculation using absolute dynamic topography data [16] as the formula below:

$$V_s = \frac{1}{fg}\frac{\delta\zeta}{\delta x} \qquad U_s = -\frac{1}{fg}\frac{\delta\zeta}{\delta y} \quad (5)$$

Where Us is zonal surface geostrophic current velocity x (m/s), Vs is meridional surface geostrophic current velocity (m/s), ζ is absolute dynamic topography (m), g is gravity (m/s^2), and f is the coriolis parameters.

The selection of geostrophic currents as a medium to verify the SSH since geostrophic currents is the current occurring at sea surface due to the influence of horizontal pressure gradient forces and is balanced by the coriolis force [13, 17]. The sea level difference causes the horizontal gradient so that the geostrophic current can be used as a tool to verify the changing of SSH in western Pacific Ocean and eastern Indian Ocean when ENSO occurred.

3. Result and discussion

3.1. Correlation between sea surface height (SSH) and ENSO

The calculation results show that the correlation value between Niño 3.4 SST anomalies and SSH in the western Pacific Ocean with R value of -0.79 (figure 3). This suggests a strong relationship between the parameters of SSH in the western Pacific Ocean and Niño 3.4 SST anomalies.

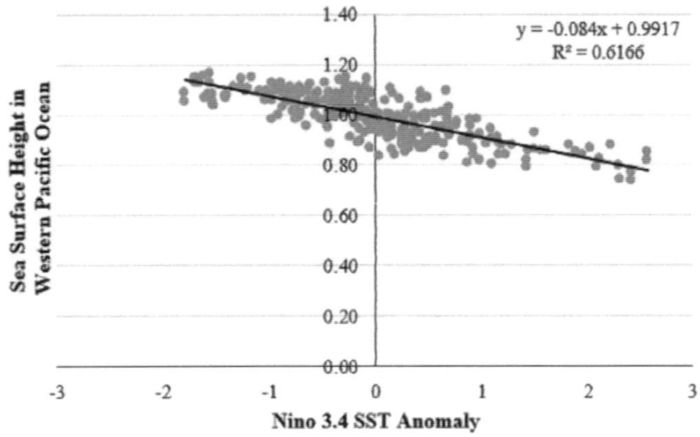

Figure 3. Scatter Plot of Niño 3.4 Sea Surface Temperature Anomalies and Sea Surface Height in Western Pacific Ocean.

IOP Conf. Series: Earth and Environmental Science **162** (2018) 012016 doi:10.1088/1755-1315/162/1/012016

The relationship between these two parameters indicates that the ENSO phenomenon affects SSH in the western Pacific Ocean region. The negative value on the correlation coefficient calculation results show the opposite relationship between the two parameters. The opposite relationship explains that when the SST anomalies in Niño 3.4 region increased (decreased), the SSH in the western Pacific Ocean decreased (increased) respectively.

The calculation of the correlation between SST Niño 3.4 anomalies with SSH in the eastern Indian Ocean showed a result of -0.62 (figure 4). This value indicates a close relationship between the two parameters studied.

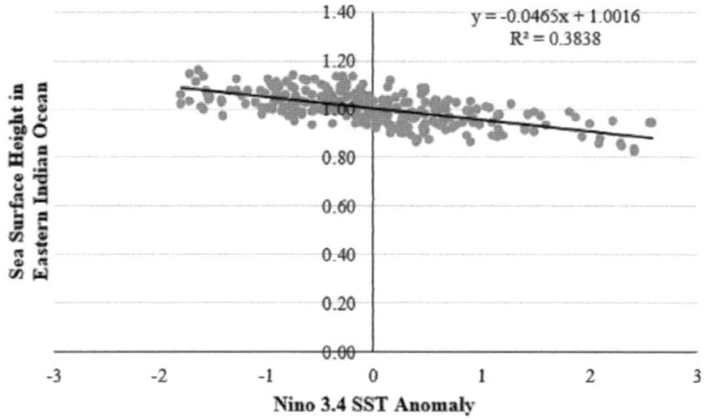

Figure 4. Scatter Plot of Niño 3.4 Sea Surface Temperature Anomaly and Sea Surface Height in Eastern Indian Ocean.

Negative values indicate an existing relationship has opposite meaning that when the Niño 3.4 SST anomaly is positive or when the El Niño occurs then the SSH in the eastern Indian Ocean becomes lower. When La Niña happened, the Niño 3.4 SST anomaly is negative then the SSH in the eastern Indian Ocean becomes higher.

3.2. Sea surface height condition in ENSO condition

The results of SSH data processing in the western Pacific Ocean and the eastern Indian Ocean indicate a variation in value throughout the year. The variation is influenced by the ENSO phenomena. Sea level can be used as a signal for the occurrence of ENSO and has a different response depending on the period of the ENSO itself [18]. This is similar to the results of this study. The influence of ENSO on SSH variations is clearly visible in both the western Pacific Ocean and the eastern Indian Ocean.

The higher Pacific Ocean SSH trend compared to the Indian Ocean is also affected by the ENSO phenomenon. Stated in the previous research [6], that the western Pacific Ocean is higher than the eastern Indian Ocean. The results of data processing conducted in this study is somewhat different with the previous references. The height of the SSH in the western Pacific Ocean can also be lower than the SSH elevation in the eastern Indian Ocean (figure 5).

IOP Conf. Series: Earth and Environmental Science **162** (2018) 012016 doi:10.1088/1755-1315/162/1/012016

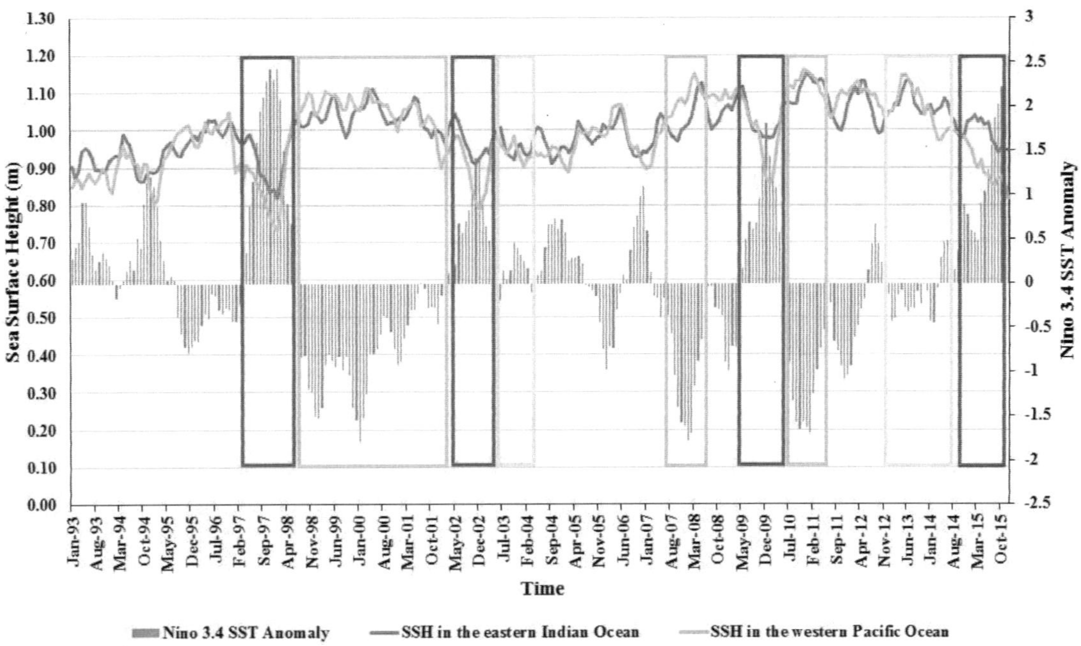

Figure 5. Comparative between SSH (green line: eastern Indian Ocean; orange line: western Pacific Ocean) and ENSO which showed by the blue bars as the value of Niño 3.4 SST anomaly (red rectangle: El Niño condition when the value of Niño 3.4 SST anomaly is greater than 0.5; green rectangle: La Niña condition when the value of Niño 3.4 SST anomaly is lower than -0.5; yellow rectangle: neutral condition when the value of Niño 3.4 SST anomaly is between 0.5 and -0.5) in 1993-2015.

Figure 5 shows that when the SST anomaly is greater than 0.5, the western Pacific Ocean SSH shown by the yellow line graph is lower than that of the East Indian Ocean SSH indicated by a green line graph. Anomalous values greater than 0.5 indicate the occurrence of El Niño phenomena. This is in accordance with the results of data processing which shows that the value of the SSH difference between the two oceans is negative when the El Niño occurs (table 1 and 2). An example of a comparison chart between SSHs in both oceans when the El Niño is marked with a red rectangle in figure 5.

In neutral condition, when the value of Niño 3.4 SST anomaly is between 0.5 and -0.5 the range of SSH in the western Pacific Ocean is mostly higher than the SSH in the eastern Indian Ocean except in the beginning of 2014 because it tends to be an El Niño condition. An example of a comparison chart between the SSH in both oceans when the neutral condition is indicated by a yellow rectangle in figure 5. This is in accordance with the previous reference which states that SSH in the western Pacific Ocean is higher than the eastern Indian Ocean normally [6].

As the normal condition of the western Pacific Ocean is higher than that of the eastern Indian Ocean, when the anomalous value of SST Niño 3.4 is smaller than -0.5 similar conditions occur. When La Nina onset applies, the western Pacific Ocean is higher than the Eastern Indian Ocean. This has an accordance with the results of data processing showing that the value of the SSH difference between those two oceans is positive (table 1). The comparison graph between the SSHs in both oceans when La Niña is marked with a green box in figure 5.

3.3. Correlation between SSH difference and ENSO

Correlation calculations were also performed on the difference between SSH in the western Pacific Ocean and the eastern Indian Ocean with an anomaly of SST Niño 3.4 to find out how strong the relationship between the SSH difference in both oceans to ENSO. The result shows that the correlation value is -0.63 (figure 6). It shows a strong opposite relationship.

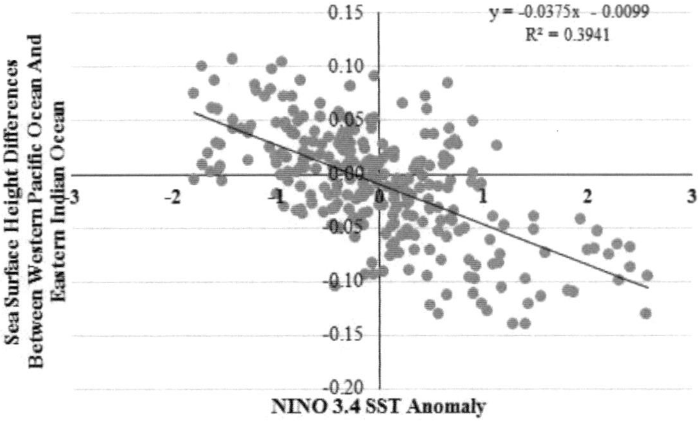

Figure 6. Scatter Plot of Niño 3.4 Sea Surface Temperature Anomaly and Sea Surface Height Difference of the Western Pacific Ocean - Eastern Indian Ocean.

When we compared between the mean values of the SSH difference during the years 1993 to 2015 against the average sea surface height difference at the time of El Niño and La Niña clearly shows that at the time of El Niño in the western Pacific Ocean will be lower than the eastern Indian Ocean And the opposite condition occurs at the time of La Niña (figure 6). Based on the graph in figure 7, it is shown that when El Niño occurs in April, the average value of the SSH difference between the western Pacific Ocean and the eastern Indian Ocean has a great negative value.

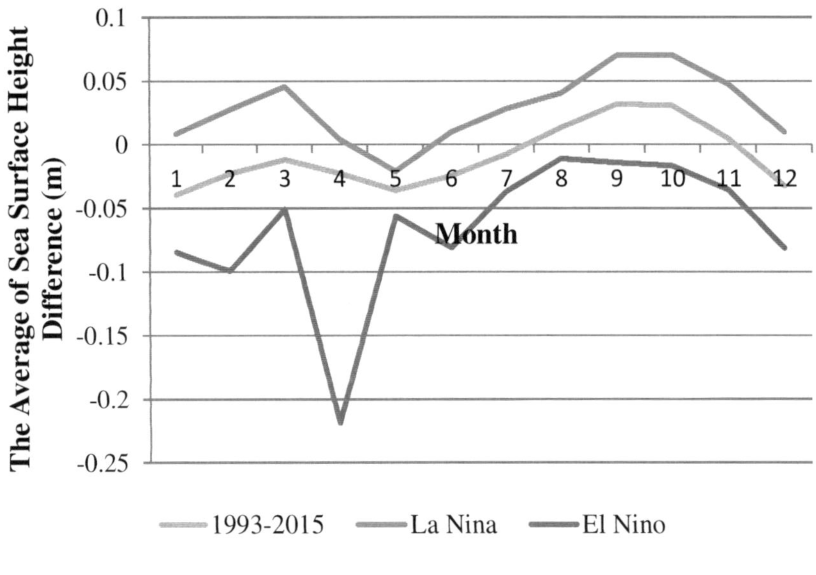

Figure 7. Comparative between the average of sea surface height difference in the Western Pacific Ocean and Eastern Indian Ocean. The green line is the average value along 1993 to 2015. The blue line is the average value when La Niña event along 1993 to 2015. The red line is the average value when El Niño event along 1993 to 2015).

The obtained result shows that the sea level in the western Pacific Ocean and the eastern Indian Ocean varies with the incidence of ENSO. Figure 7 shows that if El Niño occurs in April the decrease of SSH in the western Pacific Ocean can reach more than -0,2 m.

3.4. Verification of sea surface height difference value using geostrophic current velocity during ENSO

As our previous results [7], ENSO events applies increase/decrease in SSH difference. Taking the example of El Niño events in June and November 1997 and La Niña events in June and November 1999 (figure 8 and 9), shown the pattern of geostrophic currents direction in the Lombok Strait area (7° - 9° S and 115° - 116° E). The current moves from the higher SSH to the lower SSH however, based on the current pattern on both figures show no significant change in current direction at the time when El Niño occurred eventhough based on calculation in this research is SSH in western Pacific Ocean becomes lower than SSH in eastern Indian Ocean.

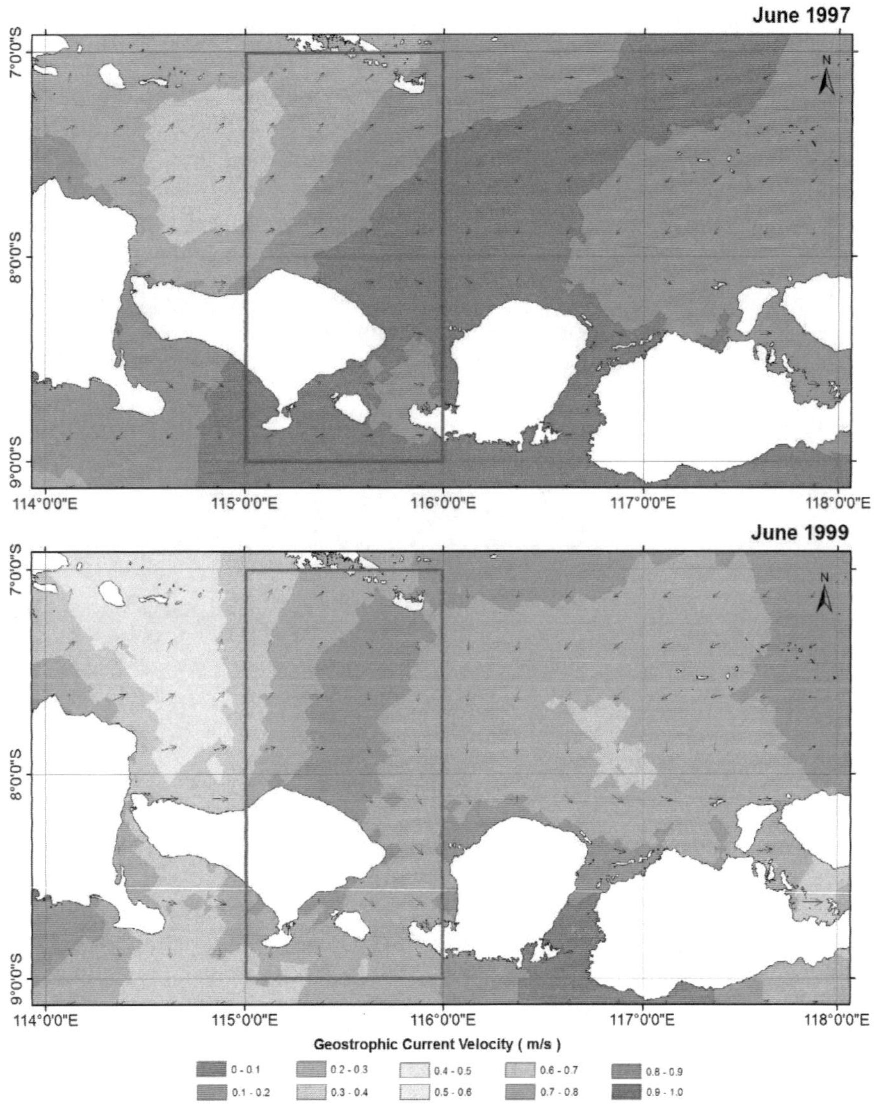

Figure 8. Geostrophic current pattern in surrounding Lombok Strait when El Niño in June 1997 (above) and La Niña in June 1999 (below).

Based on the example of ENSO events in June 1997 and 1999 in figure 8, there was no significant change in the direction of the geostrophic current. However, based on the color gradient display, the geostrophic current velocity in June 1997 tended to be lower than that in June 1999.

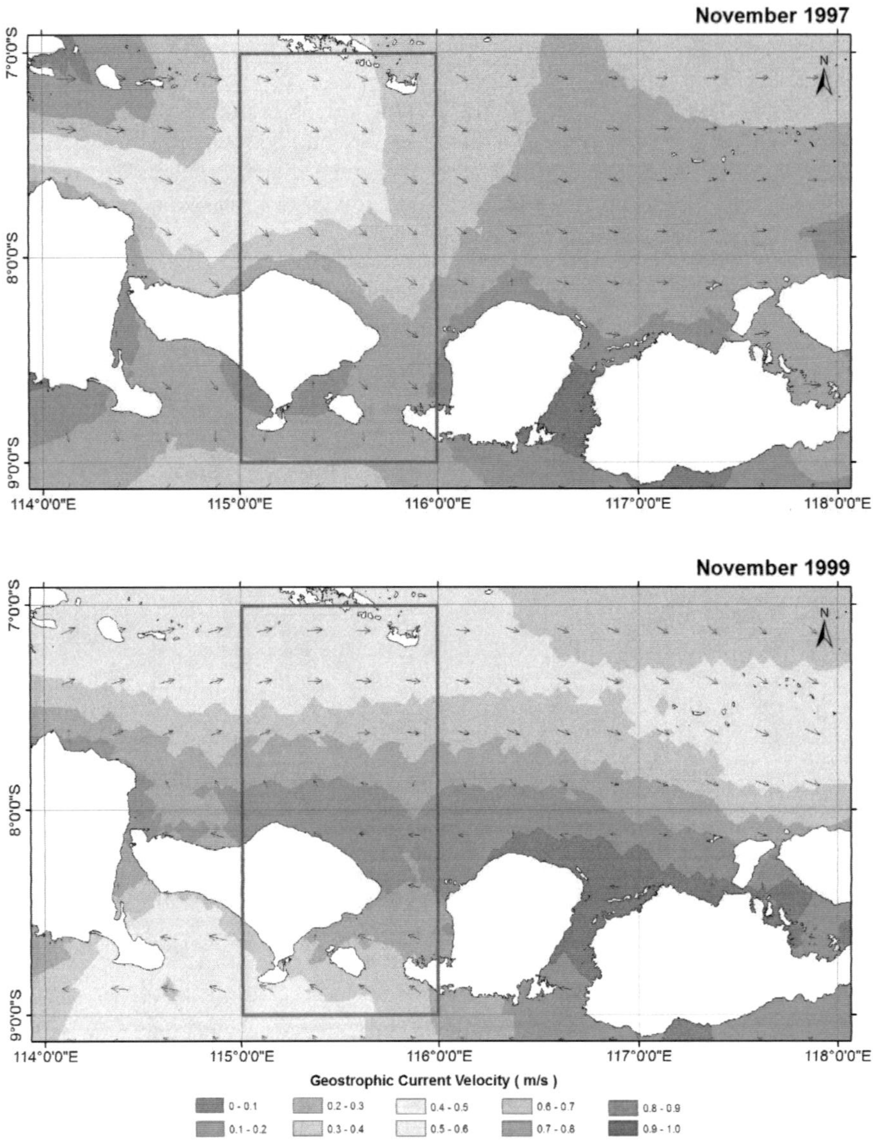

Figure 9. Geostrophic current pattern in surrounding Lombok Strait when El Niño in November 1997 (above) and La Niña in November 1999 (below).

Based on the example of ENSO events in November 1997 and 1999 in figure 9, there is change in the direction of currents that occur around the Lombok strait. It changes direction. The color gradient display shows the geostrophic current velocity in November 1997 is also tended to be lower than in November 1999.

Reviewing from the parameters of the geostrophic currents velocity, there is an influence on the phenomenon of the ENSO. Verification of SSH differences to the current velocity in Lombok Strait during ENSO can be seen in figure 10. The period of ENSO occurrence selected was from 1997 to 2001. The Elements of El Niño and La Niña were strong or weak.

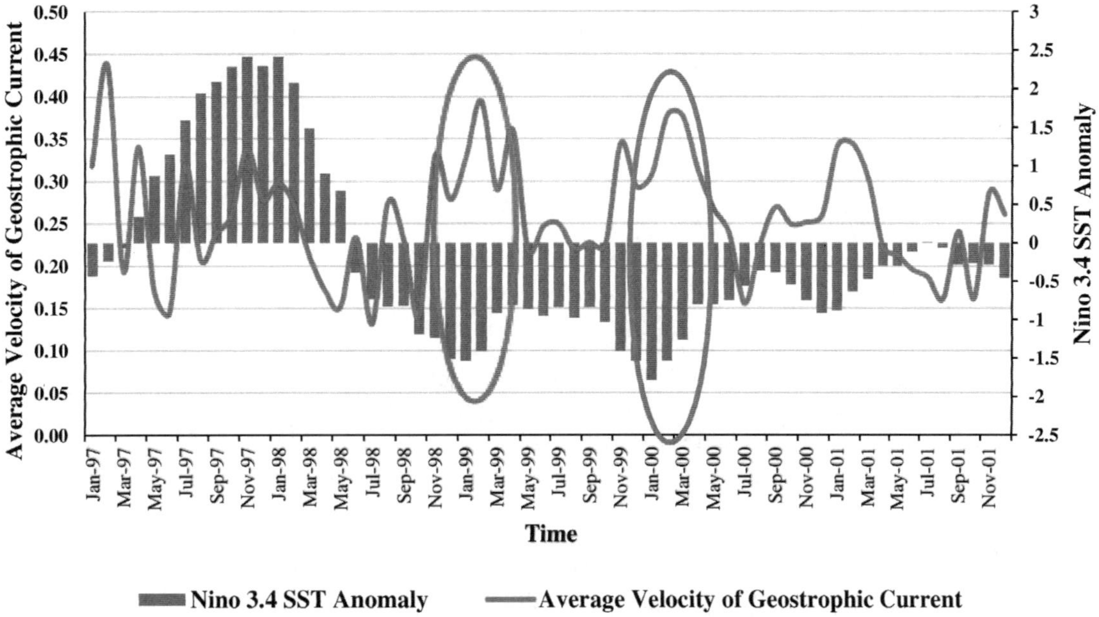

Figure 10. Geostrophic current velocity in Lombok Strait during ENSO in 1997 – 2001.

The average geostrophic current velocity in the Lombok Strait (7° - 9° S and 115° - 116° E) at the time of La Niña tends to be higher than at the time of the El Niño. The difference that occurred in the example of ENSO events in 1997 and 1998 was 0.05 m/s between the current velocity of El Niño and La Niña. The weaker current velocity conditions at the time of El Niño is due to lower SSH of western Pacific Ocean than the SSH in the eastern Indian Ocean.

Based on the graph of the average of geostrophic current velocity in figure 10 shows that there is a tendency of seasonal patterns in geostrophic currents around the Lombok strait. In the late and early months of the years the trend of geostrophic current velocity in the Lombok Strait was higher than in the mid-year period. While in November 1997, there was an increase in the velocity of geostrophic currents as in general, but the increase was smaller than when El Niño was absent. When La Niña occurs the SSH conditions of the western Pacific Ocean are higher than those in the eastern Indian Ocean. This is in accordance with normal conditions, but in the case of La Niña SSH in the western Pacific Ocean tends to be higher than normal conditions. This has an impact to the increasing of geostrophic current velocity in the Lombok Strait region.

The change of direction of geostrophic currents in the Lombok Strait in El Niño phase is as the nature of the geostrophic term itself, it will flow from the higher pressure to the lower one related to depth. The El Niño and La Niña modifies the mean sea level, hence the depth of the western Pacific, in term of sifting it to the east and west of the Pacific Ocean, respectively. Within El Niño phase the sea level in Indian Ocean is higher than in the Makassar Strait, therefrom we get reverse direction of the geostrophic flow.

4. Conclusion

ENSO (El Niño Southern Oscillation) affects the sea level in the western Pacific Ocean and the eastern Indian Ocean. When the El Niño occurs SSH in the western Pacific Ocean is lower than that in the eastern Indian Ocean whereas at the time of La Niña SSH in the western Pacific Ocean is higher than the western Pacific Ocean. The variability of SSH affects the geostrophic current velocity in Lombok Strait as the one of the exit pathway location of ITF water mass. The sea level difference in both oceans affects the geostrophic currents in the Lombok Strait. When sea level of the western Pacific Ocean is lower than the eastern Indian Ocean, the currents velocity in the Lombok Strait becomes lower than the normal condition. Meanwhile, when the sea level of the western Pacific Ocean is higher than the eastern Indian Ocean, the speed of geostrophic currents in the Lombok Strait becomes faster than the normal condition.

Acknowledgments
The authors would like to thank AVISO for providing the sea surface height (http://www.aviso.sea surface height.fr/en/data/products.html) and geostrophic current data (http://motu.aviso.altimetry.fr/aviso-gateway-servlet/Motu?action=productdownloadhome&service=AvisoDT&product=dataset-duacs-dt-global-allsat-madt-uv) and also to NOAA for providing the Niño 3.4 sea surface temperature anomaly (http://www.esrl.noaa.gov/psd/gcos_wgsp/Timeseries/Data/nino34.long.data).

References
[1] Pranowo W S, Kuswardhani A R T D, Kepel T L, Kadarwati U R, Makarim S and Husrin S 2005 *Revealing the Indonesia Throughflow* (Jakarta: Marine and Fisheries Resources Research Center-Marine and Fisheries Research Agency Department of Marine and Fisheries) pp 11-12
[2] Kane R P 1997 Relationship of El Niño-Southern Oscillation and Pacific sea surface temperature with rainfall in various region of the globe *Mon. Wea. Rev.* **125** 1792-1800
[3] Ropelweski C F and Halpert M S 1987 Global and regional scale precipitation pattern associated with the El Niño /Southern Oscillation *Mon. Wea. Rev.* **115** 1606-26
[4] Safitri M, Cahyarini S Y, Putri M R 2012. Variations of Indonesia Throughflow and oceanographic parameters in the Timor Sea as an indicator of ENSO events *Journal of Tropical Marine Science*
[5] Mayer B, Damm P E, Pohlmann T and Rizal S 2010 What is driving the ITF? An illumination of the Indonesian Throughflow with a numerical nested model system *Dynamics of Atmospheres and Oceans* **50**
[6] Hasanudin M 1998 Indonesia Throughflow *J. Oseana* **23:2** 1-9.
[7] Rejeki H A, Kunarso and Munasik 2017 The effect of ENSO to the variability of sea surface height in western Pacific Ocean and eastern Indian Ocean and its connectivity to the Indonesia Throughflow (ITF) *IOP Conf. Series: Earth and Environmental Science* **55** (2017)012066
[8] Hautala S L, Sprintall J, Potemra J C, Chong W, Pandoe N. Bray and Ilahude A G 2001 Velocity structure and transport of the Indonesian throughflow in the major straits restricting flow into the Indian Ocean. *J. of Geophysical Research* **106** 19527–546
[9] Tiwari, V M, Cabanes C, DoMinh K and Cazenave A 2004 Correlation of interannual sea level variation in the Indian Ocean from Topex/Poseidon altimetry, temperature data, and tide Gauges with ENSO *Global and Planetary Change* **43** 183-196
[10] Trenberth D 1997 The Definiton of El Niño *Bulletin of the American Meteorological Society* **78:12** 2771-77
[11] Sudjana 1996 *Statistical Method 6th Edition* (Bandung: Tarsito)
[12] Sugiyono 2009 *Quantitative, Qualitative and R & D Research Methods* (Bandung: Alfabeta)
[13] Bearman G and Brown J 1995 *Ocean Circulation* (United Kingdom: Pergamon Press)
[14] AVISO Archiving 2008 Validation and interpretation of satellite oceanographic *AVISO and PODAAC user Handbook IGDR and GDR Jason Product4th ed.* (France: AVISO)
[15] Ramadyani F and Radjawane I M 2013 Seasonal geostrophic flow in Arafura-Timor Waters *Journal of Tropical Marine Science and Technology* **5:2** 261 – 271
[16] Stewart R H 2008 *Introduction to Physical Oceanography* (Texas: Department of Oceanography Texas A & M University)
[17] Pond S and Pickard G L 1983 *Introductory Dynamical Oceanography* (United Kingdom: Pergamon Press)
[18] Luo W, Yi L, Yu Z, Sun H and Yuan L 2015 Barycenter reflected equatorial Pacific sea level

structure evolution and its indication of ENSO events. *Oceanologia* **57** 107—115

IOP Conf. Series: Earth and Environmental Science **162** (2018) 012017 doi:10.1088/1755-1315/162/1/012017

Water temperature variation of Segara Anakan, Sempu Island, South Malang, Indonesia in relation to the climate variation

A Sartimbul[1,3], H Nakata[2], E Rohadi[4], S H J Sari[1,3], M Najib[1], S Alisafira[1], S N Ikhsani[1] and D Listiyaningsih[1]

[1] Faculty of Fisheries and Marine Science, Brawijaya University, Jalan Veteran, Malang 65145, Indonesia

[2] Faculty of Fisheries, Nagasaki University, 1-14 Bunkyou-machi, Nagasaki 852-8521, Japan

[3] Marine Resources Exploration and Management (MEXMA) Research Group, Brawijaya University, Faculty of Fisheries and Marine Science, 5F, Jalan Veteran, Malang 65145, Indonesia

[4] Informatics Management, State Polytechnic of Malang, Jalan Sukarno Hatta 9, Malang 65141, Indonesia

E-mail: aida@ub.ac.id

Abstract. Water temperature variation and its correlation to meteorological factors and climate variation was studied at Segara Anakan Lagoon of Sempu Island during January 2012 to December 2013. Tidbit water temperature data, ECMWF meteorological data, climatic indices were analysed statistically. The result showed that the range of water temperature was between 23.7 to 30.4°C. Pearson correlation between water temperature and meteorological factors varied seasonally. The highest water temperature (30.4°C) occurred in Northwest Monsoon, was mainly affected by precipitation ($r = -0.999$). While, the lowest ones (23.7°C) in Southeast Monsoon, was affected by all meteorological factors. Furthermore, in transitional seasons (transitional-1: $r = 0.977$; transitional-2: $r = 0.744$), it was affected by solar radiation. This study also informed that there was no difference between water temperature and satellite SST ($t<t_{tab}$: $1.288<2.074$; sig: $0.211>0.05$) which showed that satellite SST were able to substitute the limited in situ data (Tidbit water temperature). PCA analysis indicated that climate factors (mainly SOI and DMI) affected water temperature of Segara Anakan. Present study suggested that good quality with direct and continuous measurement of local water temperature data and compiled with meteorological data and climate index were important for further prediction of water temperature variation as climate adaptation in South Malang.

1. Introduction

Temperature is a very important physical characteristic of seawater, as it can be used to identify seawater bodies. According to Patty [1], the distribution of sea water temperature is influenced by many factors, including surface current, upwelling, divergence and convergence especially in estuary area and along coastline. In semi-enclosed waters, the effect of sea water circulation such as upwelling occurs in minimum scale. However meteorological factors have more intense giving effect on water temperature variation as revealed by Takeshige [2].

Segara Anakan is one of semi-enclosed waters (lagoon) which located in Sempu Island, a small island in South Malang, East Java, Indonesia. Its water has a unique and different characteristic from other lagoons because of the existence of coral wall hole that separates inside (Segara Anakan) and outside (Indian Ocean) waters of Sempu Island. The discharge of sea water from Indian Ocean to Segara Anakan Lagoon through coral wall hole and tunnels beneath the surface water that occurred only in high tide while oppositely in low tide.

The previous study of water temperature of Segara Anakan has been done by analysing correlation between a year water temperature and some meteorological factors such as air temperature, air pressure, wind speed, and tidal current. The research showed that water temperature has very low correlation with tidal and strongly correlation with air pressure and wind speed [3]. This study only used limited meteorological parameters to determine the variation of water temperature in Segara Anakan Lagoon. As well known that meteorological parameters are not only limited as air pressure and wind speed. As a conservation area which important for sustainable marine life in Sempu Island, it is necessary to understand the variation of water temperature in detail with longer period of data, therefore this study is focused in understanding the effect of meteorological factors on temperature variations in Segara Anakan waters which based on the meteorological parameters such as air temperature, air pressure, wind speed, cloud cover, precipitation, evaporation, humidity, and solar radiation. In addition, climate data including Southern Oscillation Index (SOI), Dipole Mode Index (DMI), Niño 3.4 and Oceanic Niño Index (ONI) are also applied in this study.

The purpose of this research is to know the pattern of water temperature variation in Segara Anakan, Sempu Island, South Malang; understanding the effect of meteorological factors on water temperature variations in Segara Anakan; understanding the relation of each variable of meteorological factors with water temperature in Segara Anakan seasonally; and understanding the climate impact on water temperature variation in Segara Anakan.

2. Data and Methods

The research data includes water temperature of Segara Anakan, Sea Surface Temperature (SST) of Indian Ocean; meteorological data such as air temperature, air pressure, wind speed, cloud cover, humidity, precipitation, evaporation, and solar radiation; and climate data including Southern Oscillation Index (SOI), Dipole Mode Index (DMI), Niño 3.4 and Oceanic Niño Index (ONI). Water temperature of Segara Anakan data are collected by direct measurement using *TidBit* Temperature Data Logger (under collaboration research between Brawijaya University and Nagasaki University, Japan) which deployed at Segara Anakan Lagoon at 112°41'24"E and 08°27'19"S (Figure 1). Water temperature data collection was started from January, 2012 to December, 2013 with originally 30 minutes interval then analysed into daily and monthly mean data.

Sea Surface Temperature of Indian Ocean and meteorological data are provided by the European Centre for Medium-Range Weather Forecast (ECMWF) website. The originally 6 (six) hours interval of Indian Ocean SST and meteorological parameter (2012-2013) are analysed into daily and monthly mean data. Furthermore, climatic index data including Southern Oscillation Index (SOI), Dipole Mode Index (DMI), Niño 3.4 and Oceanic Niño Index (ONI) (2012-2013) are provided by National Oceanic and Atmospheric Administration (NOAA).

The research variables are divided into two variables, i.e. the dependent variable (water temperature) and the independent variables (meteorological and climate factors). All data were analysed using Microsoft Excel to find the mean, anomaly, and trends values based on daily mean data.

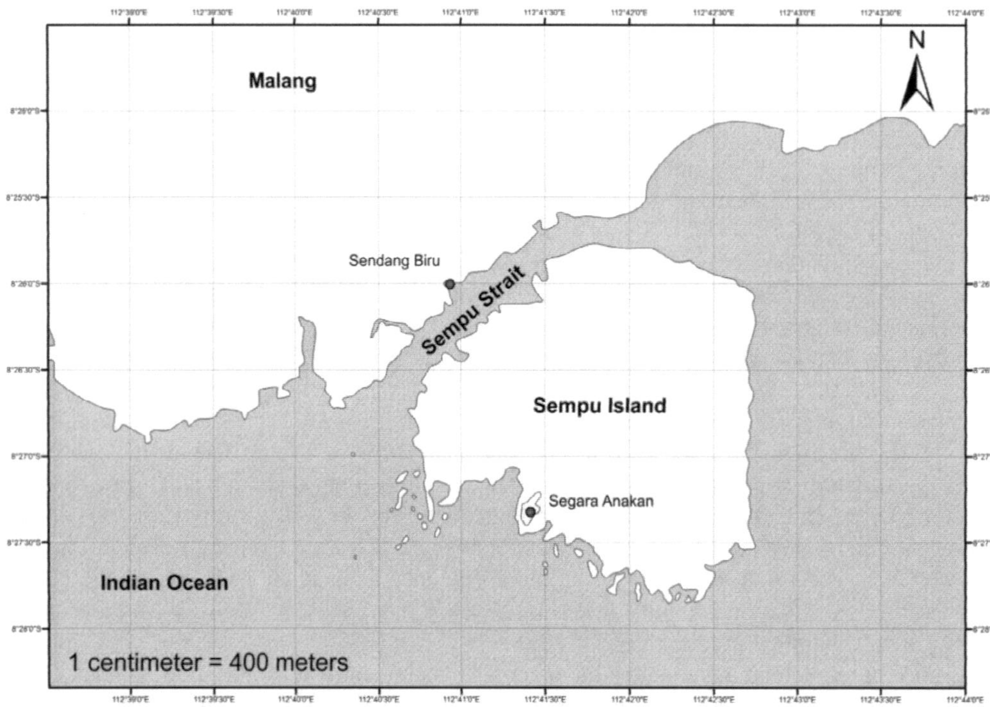

Figure 1. Map of Sempu Island, where Segara Anakan Lagoon exist. Segara Anakan Lagoon is located at the southern tip of Sempu Island and be potential affected by wider Indian Ocean.

All data are also analysed using T test of two unrelated samples of SPSS Statistics in order to know the difference between in situ water temperature of Segara Anakan Lagoon and Sea Surface temperature of Indian Ocean. Then Normality test has been done in order to know the normal distribution before done further analyses such as regression and correlation. Furthermore, Multiple linear regression analysis has been applied to determine the effect of meteorological variables on water temperature variables, while correlation analysis done to know the relationship between water temperature variables with each meteorological variable seasonally. Finally, the Principal Component Analysis (PCA) which run through XLSTAT software were applied for understanding the climate impact on water temperature of Segara Anakan Lagoon.

3. Result and Discussion

3.1 Daily Mean Variation of Water Temperature and Meteorological Factors

The two years (2012-2013) daily-mean data of water temperature and meteorological factors were analysed to obtain the mean, highest, lowest, anomaly, and trend values. The result showed that the mean value of water temperature of Segara Anakan Lagoon was about 27.84°C, with maximum value of 30.43 °C occurring on February 8, 2013 and minimum value of 23.7° C occurring on August 14, 2012. While the mean water temperature of Indian Ocean was detected as 27.08°C with maximum value on February 15, 2013 as 29.25°C and the minimum on August 14, 2012 as 23.4°C.

Then the water temperature data of Segara Anakan and Indian Ocean were analysed using T test of two unrelated samples on SPSS program. The result showed that there was no difference between mean of water temperature of Segara Anakan and SST of Indian Ocean, which shown by t value is smaller than t table (1.288 <2.074) and significance (0.211) more than 0.05. It indicates that the SST data obtained from ECMWF are able to substitute in situ water temperature data of Segara Anakan.

Beside water temperature and SST, here were the mean and minimum-maximum of meteorological data provided by ECMWF as follows: mean air temperature was 26.22°C with the maximum value occurring on April 3, 2013 at 28.73°C and the minimum value occurred on September 1, 2012 at 23.33°C. The mean air pressure value of 101.01 kPa with the maximum value occurred on October 2, 2012 was 101.42 kPa and the minimum value occurred on January 25, 2012 of 100,41 kPa (Table 1). Those values agree with Tjasyono [4] where Air pressure always decreases with increasing altitude and decreasing air temperature.

Table 1. Descriptive Analysis of Water Temperature and Meteorological Factor

Variables	Mean	Maximum	Minimum	Anomaly	Trend
Water Temperature (°C)	27.84	30.43	23.70	2.40	0.000536
SST (°C)	27.08	29.25	23.40	3.68	0.000030
Air Temperature (°C)	26.22	28.73	23.33	2.89	0.000100
Air Pressure (kPa)	101.01	101.42	100.41	0.61	0.000072
Wind Speed (m/s)	4.13	11.19	0.19	3.94	0.000878
Cloud Cover (%)	61.25	100.00	0.92	60.34	0.000511
Humidity (%)	83.85	96.88	70.01	13.84	0.001497
Precipitation (mm)	1.29	37.95	0.00	36.66	0.000163
Evaporation (mm)	0.97	2.35	0.31	1.38	0.000180
Solar Radiation (W/m^2)	44.71	72.08	12.35	27.3	0.006730

The wind velocity is obtained based on the calculation with the formula:

$$WS = \sqrt{}\,(U2 + V2) \tag{1}$$

where WS (wind speed) is the wind velocity value whereas U and V are the wind component value at 10m height. Based on the above equation, the result showed that maximum wind velocity (11.9 m/s) occurred on January 10, 2013, while the minimum wind velocity (0.19 m/s) occurred on December 14, 2012, with a mean wind velocity reached 4.13 m/s.

The mean cloud cover was found as 61.25%, with maximum value (100%) occurred on March 15, 2012, while the minimum value (0.92%) occurred on September 10, 2012. Furthermore, the mean humidity reached 83.85%, with the maximum value (96.88%) occurred on December 22, 2013 and the minimum value (70.1%) occurred on 13 August 2013. This high cloud cover may reduce penetration of sunlight entering the earth's surface and causes the increasing of water vapour content in the atmosphere and made it more humid [4]. High humidity may cause the declining air temperature (colder), as well as the water temperature.

As one of meteorological factors, the precipitation is important to be discussed. The mean precipitation at South Malang were recorded as 1.29 mm with maximum precipitation (37.95 mm) occurred on 23 December 2013. The minimum value (0 mm) means there was no precipitation on the day.

Furthermore, the average evaporation value reached 0.97 mm with maximum (2.35 mm) and minimum (0.31 mm) value occurred on March 15, 2012 and August 28, 2012, respectively. Precipitation that occurs on the sea usually can affect the decreasing of sea surface temperatures, while oppositely for evaporation can due to the flow of heat from the air to the surface layer of the water [4].

The value of solar radiation is obtained on the basis of the following formula:

$$QNET = QSW + QLW + QE + QH \tag{2}$$

Where QNET is the value of net heat flux, QSW is the shortwave radiation flux, QLW is the longwave radiation value radiation flux, whereas QE is the value of latent heat flux and QH is the value of sensible heat flux [2]. The average of solar radiation was 44.71 W/m^2, with the maximum (72.08 W/m^2) and minimum (-12.35 W/m^2) value occurred on 22 December 2013 (winter) and July 7, 2013 (summer), respectively. In addition, Ishii and Kondo [6] mentioned that solar radiation and horizontal heat transport played a primary role in the heating process in summer and latent and sensible heat fluxes become dominant in the cooling process in winter. In order to make it easier to interpret and understand the seasonal variations of water temperature and meteorological factors, the standardization analysis was applied. The standardization method is a simple statistical method which commonly used to interpret complex data with different scales or level in better and simple look [5]. The standardization analysis for water temperature and meteorology are shown in graph form which divided into four monsoon seasons: Northwest Monsoon (December-February), Transition Season I (March-May), Southeast Monsoon (June-August), and Transition Season II (September-November) (Figure 2).

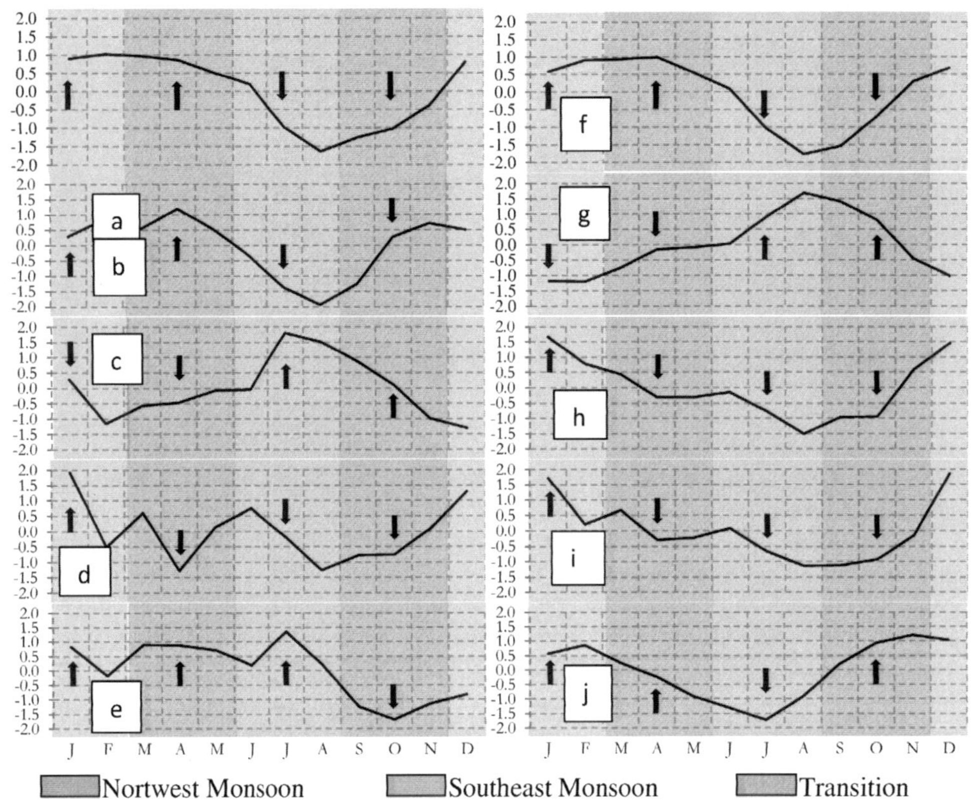

Figure 2. Standardization of Water Temperature of Segara Anakan Lagoon (a) and Meteorological Factors (air temperature (b); wind speed (c); humidity (d); evaporation (e); SST of Indian Ocean (f); air pressure (g); cloud cover (h); precipitation (i); solar radiation (j))

3.2 Annual Variations of Water Temperature and Meteorological Factors

The relationship between water temperature and meteorological factors was analysed using SPSS program with focused on multiple linear regression analysis. SPSS is the acronym of Statistical Package for the Social Science. SPSS is one of the most popular statistical packages which can perform highly complex data manipulation and analysis with simple instructions. It is designed for both interactive and non-interactive (batch) uses. Nowadays SPSS is commonly used for statistical analysis not only limited at social science. Current study we used multiple linear regression analysis using SPSS program and resulted regression equation as follows:

$$Y' = 55{,}362 + 0{,}436X1 - 0{,}361X2 - 0{,}603X3 + 0{,}025X4 - 0{,}071X5 + 0{,}096X6 + 4{,}5X7 - 0{,}003X8 \quad (3)$$

Where X = value of air temperature; X3 = value of wind speed; X4 = value of cloud cover; X5 = humidity value; X6 = Precipitation value; X7 = value evaporation; and X8 = the value of solar radiation.

The regression equation (Eq. 3) explains that if all meteorological variables are 0, the water temperature (Y ') is 84.434°C. The regression coefficient of variable air temperature (X1) of 0.436 means that if the temperature value increases 1°C, then the temperature value of the waters will increase by 0.436°C assuming other meteorological variables are fixed. The regression coefficient of air pressure variable (X2) of -0.361 means that if the value of air pressure increases 1 kPa, then the temperature value of the waters will decrease by 0.361°C. The wind velocity regression coefficient (X3) of -0.603 means that if the wind speed value increases 1 m/s, then the temperature value of the waters will decrease by 0.603°C.

Wind is caused by the difference in air pressure in two adjacent directions. The greater the air pressure the faster the wind will also be generated. The driving factor of air mass movement is the difference in air pressure between one place and another. This pressure difference is caused by the temperature of the air as a result of the heating of the earth's surface by the sun. The higher the value of wind speed, the water temperature will decrease, because the faster the wind that blows on the surface of waters will affect the sea surface is bumpier. This will reduce the penetration of heat into the sea water, and result in low surface temperatures [6]. Moreover the faster the wind will decrease sea temperature due to mixing process or upwelling [8].

Cloud cover variable coefficient of cloud cover (X4) of 0.025 means that if the cloud cover value increases 1%, then the temperature value of the waters will increase by 0.025°C.

The evaporation variable regression coefficient (X7) of 4.5 means that if the evaporation value increases 1 mm, the temperature value of the waters will increase by 4.5°C. However, in some studies, it was found that if sea surface temperature rises then the possibility of precipitation value will be higher and vice versa [5]. Rising sea-level temperatures show an increase in ocean energy that gives rise to higher levels of evaporation in the atmosphere. Therefore it is an indirect relationship between sea level rise in temperature and precipitation and evaporation. Changes in water temperature values caused by solar radiation throughout the year are generally not very significant.

The value of R is used to determine the relationship between independent variables (meteorological factors) to the dependent variable (water temperature) simultaneously. The value of R obtained is 0.986. Since R values are in the range of 0.800-1.000, it can be suggested that there is a very strong relationship between each of the meteorological factors to the water temperature. While, the value of R^2 used to know the percentage contribution of independent variable that influence (e.g. meteorological factors) simultaneously to the dependent variable (water temperature). The value of R^2 is 0.973. The Regression coefficient test is also conducted to determine the effect of meteorological factors on the water temperature whether the influence occurs partially or together. Based on T and F value, regression coefficient shows that all meteorological variables affect water temperature together, whereas partially, each variable has no effect on changes in water temperature value.

3.3 Seasonal Variation of Water Temperature and Meteorological Factors

The relationship among water temperature and meteorological factors was approached using simple correlation analysis run by SPSS program. The result of simple correlation analysis are presented in Table 2.

Table 2. Correlation Analysis of water Temperature and Meteorological Data

		Water Temp.	Air Press.	Wind Speed	Cloud Cover	Hum.	Prec.	Eva.	Solar Rad.
Northwest Monsoon									
Water Temp.	Pearson Corr.	0.878	-0.647	-0.267	-0.938	-0.938	-0.999	-0.026	0.040

	Sig. (2-tailed)	0.317	0.552	0.828	0.226	0.226	0.024	0.983	0.975
Transition I									
Water Temp.	Pearson Corr.	0.342	-0.732	-1.000	0.670	-0.065	0.628	0.993	0.977
	Sig. (2-tailed)	0.778	0.478	0.015	0.533	0.959	0.568	0.074	0.137
Southeast Monsoon									
Water Temp.	Pearson Corr.	0.996	-0.999	-0.831	0.993	0.990	0.999	-0.115	-0.418
	Sig. (2-tailed)	0.060	0.025	0.376	0.078	0.890	0.033	0.927	0.725
Transition II									
Water Temp.	Pearson Corr.	0.700	-0.958	0.933	1.000	1.000	0.989	0.633	0.744
	Sig. (2-tailed)	0.507	0.185	0.235	0.017	0.018	0.0.96	0.564	0.466

Based on the four monsoonal seasons, it is concluded that It has its own characteristics in each seasons. As revealed by Wyrtki [11] Indonesia well known has two seasons as called as west season (Northwest) and east season (Southeast) and other two intermediate (1 and 2) seasons in between, which strongly influenced by monsoon. Sartimbul [8] added that Northwest monsoon occur in December to February, while Southeast monsoon occur in June to August every year. Based on the correlation analysis, it is found that the meteorological factor has highly correlation with the water temperature in Segara Anakan and varies seasonally. During Northwest monsoon, water temperature in Segara Anakan has highly significant correlation with precipitation, with correlation coefficient of -0.999. The correlation coefficient shows a negative value which means the higher the value of precipitation, the lower the temperature value of the waters in Segara Anakan. According to Wyrtki [11], during December-February, the position of the sun is in the southern hemisphere so that the Asian Continent has a higher air pressure than on the Australian Continent. This leads to the occurrence of Northwest monsoon. This monsoon passes through vast sea areas bringing plenty of water vapour and causing rainy seasons on the islands of Borneo, Sumatra, Java and Sulawesi (the peak in December).

In contrast, during In June-August, the position of the sun is in the northern hemisphere so that the Asian Continent has lower air pressure than on the Australian Continent. This condition is called the Southeast monsoon, where the monsoon travels through the vast and dry desert, and causes a drought in Indonesia. The analysis shows that the water temperature of Segara Anakan does not correlate significantly to both air pressure and precipitation. However, during Southeast monsoon, all the meteorological factors contribute to the changes of the water temperature. There is no dominant factor.

Furthermore, during the transition season-1 (March-May) water temperature in Segara Anakan has strongly correlated with the variable of wind speed, evaporation, and solar radiation, while in the transition season-2 (September-November), water temperature in Segara Anakan has highly correlated with cloud cover and moisture variables. In the same time, during transition seasons (March and September), the position of the sun is around the equator. Since Segara Anakan Lagoon, Sempu Island just located at low latitudes, therefore the effects of solar radiation are more pronounced during the transition season. During this season, the sun is at low latitudes or around the equator. This condition is shown by the highly significant correlation between water temperature in Segara Anakan with solar radiation R=0.977 and R=0.744 in the transition season-1 and transitional season-2, respectively. The higher the value of solar radiation, the higher the temperature value of the water [8].

3.4 Variations of Water Temperature and Climate

Principal component analysis (PCA) was conducted to find out which the most principal variables influence to water temperature of Segara Anakan Lagoon. The PCA analysis shows that four existing climatic variables have been formed into 2 factors. The PCA analysis are expressed by Factor F1, F2, F3, to Fn. The F1 is defined as the most factors or component influence to the temperature variation of Segara Anakan Lagoon. Indeed, the Dipole Mode Index (DMI) and Southern Oscillation Index (SOI) are selected as the Factor 1, meanwhile factor 2 is the secondary factors were Niño 3.4 and Oceanic Niño Index (ONI). Sartimbul [8] mentioned that the Primary factor (F1) directly affect water temperature of Segara Anakan Lagoon, while secondary factor (F2) may modified the F1. Furthermore, negative DMI, Niño 3.4 and ONI affect the temperature of Segara Anakan waters as illustrated by loading factor as 0.677, 0.671 and 0.717, respectively. While positive SOI affects on Segara Anakan water temperature with loading factor as 0.677 (Table 3). The results showed that when DMI, Niño 3.4 and ONI are positive value then the temperature of Segara Anakan water become low. This mechanism was also explained by Sartimbul [8] that Niño 3.4 and a strong DMI indicated the occurrence of El Niño which lead to lower SST of South Indonesia.

Table 1. Factor Loading of Water Temperature, Meteorological Factors and Climate Indices. Principal Component or main factor in Principal Component Analysis (PCA) is expressed by Factor (F1, F2).

Variables	F1	F2
Water Temperature (WT)	0.856	0.037
Sea Surface Temperature (SST)	0.946	0.079
Air Temperature	0.776	-0.055
Air Pressure	-0.820	-0.016
Wind Speed	-0.738	0.541
Cloud Cover	0.892	0.131
Humidity	0.210	-0.565
Precipitation	0.772	0.047
Evaporation	0.094	0.904
Solar Radiation	0.153	-0.880
DMI	-0.677	0.134
SOI	0.674	-0.062
Nino 3.4	-0.313	-0.671
ONI	-0.363	-0.717

During Northwest monsoon, climatic index is found as the main factor that most influence the temperature of waters of Segara Anakan, i.e. SOI, Niño 3.4 and ONI with value 0,873; 0.576 and 0.762, respectively (Table 4). Furthermore, main climatic factors that influence water temperature of Segara Anakan during Southeast monsoon are DMI, SOI, and Nino 3.4 with a value of 0.983; 0.966 and 0.701, respectively (Table 5). In summary our study suggested that wider temperature variation modified by climate factor as shown by some climate indices which have remarkable effect on the smaller water column such as Segara Anakan Lagoon. This study consistence with previous study as mentioned by Sartimbul [5].

Table 2. Factor Loading of Water Temperature, Meteorological Factors and Climate during Nortwest Monsoon. The F1, F2 are principal factors.

Variables	F1	F2
Water Temperature (WT)	-0.816	0.283
Sea Surface Temperature (SST)	0.509	0.514
Air Temperature	-0.617	0.443
Air Pressure	-0.699	-0.344
Wind Speed	-0.332	0.237
Cloud Cover	0.927	0.358
Humidity	-0.348	-0.804
Precipitation	0.931	0.250
Evaporation	0.901	0.299
Solar Radiation	-0.914	-0.135
DMI	-0.103	0.901
SOI	0.873	-0.455
Nino 3.4	-0.576	0.569
ONI	-0.762	0.446

Table 3. Factor Loading of Water Temperature, Meteorological Factors and Climate Indices during Southeast Monsoon. The F1, F2 are principal factors.

Variables	F1	F2
Water Temperature (WT)	0.990	-0.084
Sea Surface Temperature (SST)	0.980	-0.150
Air Temperature	0.956	-0.263
Air Pressure	-0.969	-0.202
Wind Speed	-0.765	-0.634
Cloud Cover	0.985	0.145
Humidity	0.558	0.783
Precipitation	0.918	0.382
Evaporation	0.326	-0.910
Solar Radiation	-0.002	0.941
DMI	-0.983	0.075
SOI	0.966	0.110
Nino 3.4	-0.701	0.607
ONI	-0.637	0.688

4. Conclusion

The water temperature in Segara Anakan varies seasonally. The highest water temperature occurs during the Northwest monsoon (30.4°C), while the lowest water temperature in Southeast monsoon (23.7°C). Meteorological factors including air temperature, air presssure, wind speed, cloud cover, humidity, precipitation, evaporation, and solar radiation together influence water temperature of Segara Anakan. While partially, they are no significant effect on water temperature of Segara Anakan. The relation between water temperature and meteorological factors varies seasonally. During Northwest monsoon the water temperature in Segara Anakan is influenced mainly due to precipitation. While, in the Southeast monsoon, meteorological factors influence together. Furthermore, during transitional season, water temperature is primarily affected by solar radiation. The Principal Component Analysis shows that climate factors affect the water temperature in Segara Anakan (mainly SOI and DMI). Longer period of water temperature data of South Java Sea indicated the increasing trend due to climate variation. This study also suggested that good quality with direct and continuous measurement of local water temperature and compiled with indirect data of meteorological data and climate index are important for further prediction of water temperature variation for climate adaptation in South Malang.

Acknowledgement
The authors would like to thanks the Ministry of Research and Technology-Directorate General of Higher Education, Indonesia and Nakata Laboratory, Nagasaki University, Japan for research funding Thanks also be addressed to Fisheries and Marine Science Faculty, Brawijaya University, Indonesia and Marine Resources Exploration and Management (MEXMA) for laboratory and analysis support. Thanks to the European Centre for Medium-Range Weather Forecast for providing the SST and meteorological data; National Oceanic and Atmospheric Administration (NOAA) for providing of Southern Oscillation Index (SOI), Niño 3.4, Dipole Mode Index (DMI) and Oceanic Neritic Index (ONI).

References
[1] Patty S I 2013 Distribusi suhu, salinitas dan oksigen terlarut di Perairan Kema, Sulawesi Utara *Jurnal Ilmiah Platax* **1** 148-157
[2] Takeshige A, Takahashi T, Nakata H and Kimura S 2013 Long-term trends in sea surface temperature in coastal water in relation to large-scale climate change: A case study in Omura Bay, Japan *Cont Shelf Res* **66** 73-82
[3] Najib M A. 2015 *Hubungan variasi suhu dengan faktor meteorologi (tekanan atmosfer, kecepatan angin) dan pasang surut di Perairan Segara Anakan Pulau Sempu, Malang* (Malang: Fakultas Perikanan dan Ilmu Kelautan Universitas Brawijaya)
[4] Tjasyono B 2004 *Klimatologi Edisi Kedua* (Bandung: ITB) 348 h
[5] Fadholi A 2013 Persamaan regresi prediksi curah hujan bulanan menggunakan data suhu dan kelembapan udara di Ternate *Statistika* **13** 7-16
[6] Ishii T and Kondo J 1993 Seasonal variation of heat balance of the East China Sea *Tenki* **34** 517–526 (in Japanese)
[7] Sartimbul A, Nakata H dan Hayashi I 2007 Recent change in water temperature and its effect on fisheries catch of bottom gillnets in a coastal region of the Tsushima Warm Current *La Mer* **45** 1–13
[8] Rasyid A J 2013 Distribusi suhu permukaan pada musim peralihan barat-timur terkait dengan fishing ground ikan pelagis kecil di perairan spermonde **20** 1-7
[9] Talley L D, Pickard G L, Emery W J and Swift J H 2011 *Descriptive Physical Oceanography: an Introduction* (Elsevier) p. 555
[10] Aldrian D E 2008 *Meteorologi Laut Indonesia* (Jakarta: Badan Meteorologi dan Geofisika) h 252
[11] Wyrtki K 1961 Physical Oceanography of the Southeast Asean Water *Naga Report* **Vol II** In California: The University of California: Scrips Institution of Oceanography p 195
[12] Sartimbul A, Nakata H, Rohadi E, Yusuf B and Kadarisman H 2010 *Prog. Oceanogr* p 168–174.

Analysis of daily wind circulation toward sea level rise in Semarang

N Fitrianti[1], Y D Haryanto[2], E V S Simamora[3], H F A Bintari[4], A Hartoko[5], S Anggoro[6] and M Zainuri[7]

[1234] Indonesia Agency For Meteorology, Climatology, and Geophysics, Jakarta Indonesia
[257] Diponegoro University, Semarang, Indonesia

E-mail: fitriantinovibima@gmail.com

Abstract. Semarang is one of the coastal areas in Indonesia that frequently occur tidal flood. Tidal flood is a phenomenon where the sea water entering the land area. It can be influenced by astronomical factor that is gravity of moon and sun besides meteorological factor which also play a role in rising the sea level. Sea level rise can cause flood disaster in coastal areas if the height of coastal land is below sea level. At the time of minimum sea wave height, sea level rise is also low, so rising sea level is only played by high tidal fluctuations. Wind data used in this study. The data classified into land breeze, sea breeze and wind transition. From wind data, three classifications of wind speed are determined that is when the wind with minimum, average and maximum speeds. The method of calculating some variables of sea level rise (wave set up) as well as sea tidal observation and measurement. From this research it can be seen that sea level rise in Semarang besides caused by tidal factor is also caused by meteorological factors such as wind speed.

1. Introduction

The coastal area is an area that is highly vulnerable to environmental pressure from land or from the sea. The coastal area is the area between land and sea. One of the recent pressures that threaten the sustainability of coastal areas around the world is the increasing of sea levels. Towards the land of the coastal area includes part of the Mainland, either dry or submerged in water, which is still influenced by oceanic properties such as tides, sea breezes and saltwater permeation. While towards the sea, the coastal area includes part of the ocean which is still influenced by natural processes that occur on land such as sedimentation and freshwater flow or caused by human activities on the ground such as deforestation and pollution (Supriharyono, 2002).

Sea level rise events and the decline of land allegedly resulted in frequent floods in the coastal area of Semarang, during high tide in a period of ± 25 last year. The flood, known as the rob (tidal flood), inundated areas lower than sea level during the highest tide (Mean High Water Level). According Sarbidi (2002) water height due to the flood can reach 20-60 cm rob with the estimated inundation area of 32.6 km^2.

Rob or tidal flood is a phenomenon of overflowing sea water to the mainland[1]. The increase of sea water due to the tide is a natural phenomenon and can be predicted. Tidal events are due to the movement

[1] http://pusatkrisis.kemkes.go.id/what-is-rob-flood

of the sun, earth, moon and other celestial objects. Sea level rise caused by the tides, in addition it is also caused by other factors or external force such as water push, swell (waves caused from a distance), storms and tropical storms which is a phenomenon that frequently occurs at sea.

The movement of the wind caused the sea waves which makes the sea level rises. These are the possible causes for the occurrence of rob in Semarang. Wind conditions in the coastal area of Semarang is dominated by west monsoon winds.. year, the western monsoon blows at an average speed of 14.5 knots and an average maximum speed of 22.9 knots from the west, northwest and north (Prihatno, 2009). Therefore, it is necessary to analyse and calculate the wind velocity and fluctuation of tides to find out how high sea waves and sea level rise caused during the tidal period.

2. Data and method

This research is in the area of Semarang Maritime Meteorological Station , located at coordinate positions 6.95^0 S and 110.42^0 E. The data used in this study is the data direction and wind speed obtained from daily observation based on hourly surface observation. In addition, the data used are hourly tidal observation data with in period of 2012 - 2016. Several stages of data processing in this research, as follows:

- Classification of sea breeze, transition wind and ground wind based on penetration theory
- Determine the dominant wind direction to determine the fetch length.
- Calculation of effective fetch value at research point, fetch calculation is used to determine sea wave height (H) and wave period (T).
- After determining the height and period of wave then also determined the increase of sea level (wave set up) S_w because of waves or sea level rise reaching coastal areas.
- Counting the tides in the research area and then determined the highest tide.
- From the wave height data then accumulated with high during the tide period.
- Calculation of sea level high frequency at high wave and at the highest tide occurs simultaneously.

In the classification of land breeze and sea breeze obtained by using penetration theory that sea breeze can enter up to 100 km to the land area. therefore, the calculation of 100 km distance between Semarang maritime Meteorological Station with the coast and determine the angle of direction by using trigonometric theory, for transition wind is defined as 300 degree angle from sea breeze toward land and the rest is defined as land breeze.

After determining the wind classification, the dominant wind direction in the research area can be known. By calculating the dominant wind direction is used to measure the fetch length, fetch length is one of the factors that determine the wave characteristics generated by the wind (Davis 1991). Fetch is the distance travelled by the wind from the direction of wave generation . To calculate the fetch using the equation as follows:

$$F_{eff} = \frac{\Sigma X_i \text{ Cos a}}{\Sigma \text{ Cos a}} \quad (1)$$

F_{eff} = Effective Fetch (Km)
X_i = Fetch Length / Actual Distance (Km)
a = Angle of deviation on both sides of the dominant wind direction.

The calculation of the effective fetch used to convert wind into a wave, in this case the wave height (H_0) and period of a wave are converted by using the equation according to the SPM (shore protection manual), 1984 vol. 1 as follows:

$$H_0 = 5.112 \times 10^{-4} \times U_A \times F^{0.5} \quad (2)$$
$$T_0 = 6.238 \times 10^{-2} \times (U_A \times F)^{0.33} \quad (3)$$
$$R_L = \frac{U_W}{U_L} \quad (4)$$

IOP Conf. Series: Earth and Environmental Science **162** (2018) 012018 doi:10.1088/1755-1315/162/1/012018

$$U_A = 0.71 \times {U_w}^{1.23} \tag{5}$$

H_o = sea wave height (m)
T_o = period of sea wave (s)
F = Fetch/ Fetch Effective
U_A = voltage factor of wind
R_L = linkages of UL dan UW (velocity of sea breeze and velocity of land breeze)
U_W = wind velocity above sea surface (m)
U_L = wind velocity over land (m)
1 Knot = 0.5144 m/s

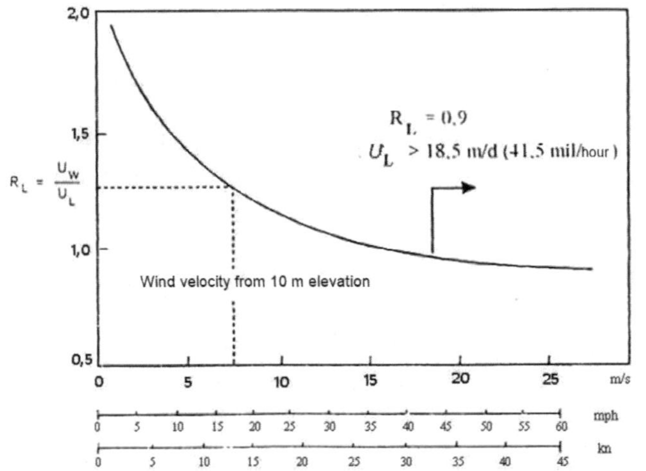

Figure 1. The relationship between wind velocity at sea and shore (Source: Triatmodjo, 1999)

Since the registering station is grounded to get the sea breeze velocity is approximated by the equation $R_L=U_w/U_L$ (Triatmodjo, 1999) as well as utilizing the graph of the relationship between wind speed at sea and land and based on Figure 1. Then the calculation of water level due to wave (S_w) is done by using several variables, like wavelength with equation Lo=1.56T2. Where the depth of the sea (d) is close to the coast of Semarang waters is 9 meters while the slope of the seafloor (m) with 1: 500 is 0.02. The depth of the sea (d) is based on research conducted by Satriadi (2012).

Table 1. Functions of d/L for even increments of d/L_0 (Source: Triatmodjo, 1999)

d/L_o	d/L	$2\pi d/L$	TANH $2\pi d/L$	SINH $2\pi d/L$	COSH $2\pi d/L$	H/H_o'	K	$4\pi d/L$	SINH $4\pi d/L$	COSH $4\pi d/L$	n	C_G/C_o	M
.006000	.03110	.1954	.1929	.1967	1.0192	1.620	.9812	.3908	.4008	1.077	.9875	.1905	133
.006100	.03136	.1970	.1945	.1983	1.0195	1.614	.9809	.3941	.4044	1.079	.9873	.1920	130
.006200	.03162	.1987	.1961	.2000	1.0198	1.607	.9806	.3973	.4079	1.080	.9871	.1935	128
.006300	.03188	.2003	.1976	.2016	1.0201	1.601	.9803	.4006	.4114	1.081	.9869	.1950	126
.006400	.03213	.2019	.1992	.2033	1.0205	1.595	.9799	.4038	.4148	1.083	.9867	.1965	124
.006500	.03238	.2035	.2007	.2049	1.0208	1.589	.9796	.4070	.4183	1.084	.9865	.1980	123
.006600	.03264	.2051	.2022	.2065	1.0211	1.583	.9793	.4101	.4217	1.085	.9863	.1994	121
.006700	.03289	.2066	.2037	.2081	1.0214	1.578	.9790	.4133	.4251	1.087	.9860	.2009	119
.006800	.03313	.2082	.2052	.2097	1.0217	1.572	.9787	.4164	.4285	1.088	.9858	.2023	117
.006900	.03338	.2097	.2067	.2113	1.0221	1.567	.9784	.4195	.4319	1.089	.9856	.2037	116
.007000	.03362	.2113	.2082	.2128	1.0224	1.561	.9781	.4225	.4352	1.091	.9854	.2051	114
.007100	.03387	.2128	.2096	.2144	1.0227	1.556	.9778	.4256	.4386	1.092	.9852	.2065	112
.007200	.03411	.2143	.2111	.2160	1.0231	1.551	.9774	.4286	.4419	1.093	.9850	.2079	111
.007300	.03435	.2158	.2125	.2175	1.0234	1.546	.9771	.4316	.4452	1.095	.9848	.2093	109
.007400	.03459	.2173	.2139	.2190	1.0237	1.541	.9768	.4346	.4484	1.096	.9846	.2106	108

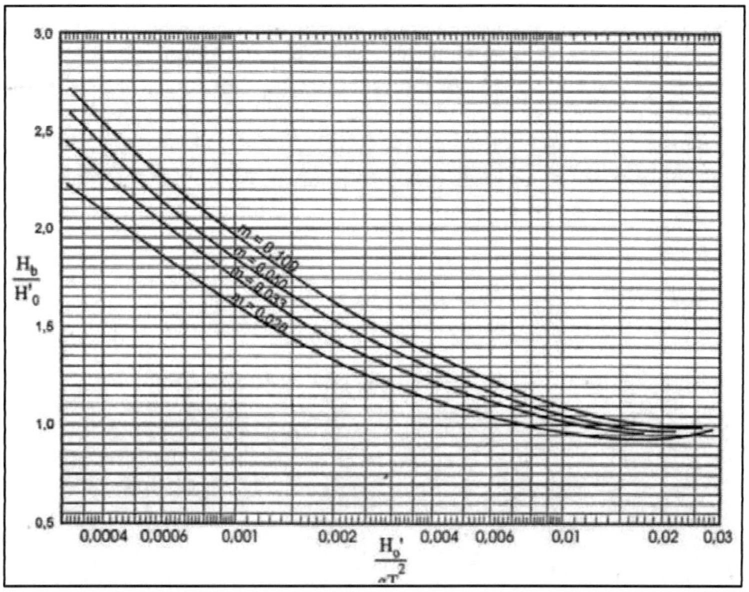

Figure 2. Graph of height of breaking wave (Triatmodjo,1999)

Table 1 is used to determine the value of L and n to determining the wavelength. after that rapid wave propagation has determinated by using the equation $C_0 = L_o / T$ (Keand $C = L / T$ the calculation of Rapid wave propagation is used to calculate the value of shoaling coefficient with the following equation:

$$K_S = \sqrt{\frac{1}{2} \times \frac{1}{n} \times \frac{1}{\frac{C}{C_O}}} \quad (6)$$

K_s = Shoaling Coefficient
n = the fraction of wave energy that travels forward with the wavefrom
C = Wave Speed (m/s)
C_0 = Depth water wave speed (m/s)
L = wavelenght(m)
L_0 = Depth water wavelength (m)

To determine the height of the breaking wave (H_b) in this case is done using the graph of the relationship between H'_0 / gT^2 and H_b / H_0 where to find the value of H'0 with the equation $H'_0 = H_0 / K_s$, after getting the breaking wave value then determined The value of wave set up in the coast because of the wave with the equation:

$$S_W = 0.19[\,1 - 2.82\sqrt{\frac{H_b}{gT^2}}]H_b \quad (7)$$

In general, the tides in various regions in Indonesia can be divided into 4 types (Triatmodjo, 1999), namely Semi diurnal tide, diurnal tide, mixed prevailing semidiurnal and mixed prevailing diurnal. In this study tidal data were obtained from observations and measurements in the period 2012 - 2016 observed and recorded hourly on a tidal observation form. The calculation is done by determining the value of Mean Sea Level (MSL), Mean High Water Level (MHWL) and Highest High Water Level (HHWL).

Frequency calculation is done by calculating of sea level in the coast at the high tidal wave along with the highest average tide. This is done to analyse sea water level in Semarang waters most often due to wind speed factor or high wind speed factor that occurs simultaneously at the highest tide.

3. Result and discussion

Based on the penetration theory that sea breeze can enter up to 100 km to the mainland. Thus, the point taken to classify the sea breeze in this study is the coastal point 100 km from the Semarang Maritime Meteorology station.

Measuring the distance of sea breeze into the mainland is calculated using Google Earth. After determining the direction of the sea breeze then classified land breeze and transition wind, where the transition wind is a wind that is calculated 30^0 from the direction of the sea breeze.

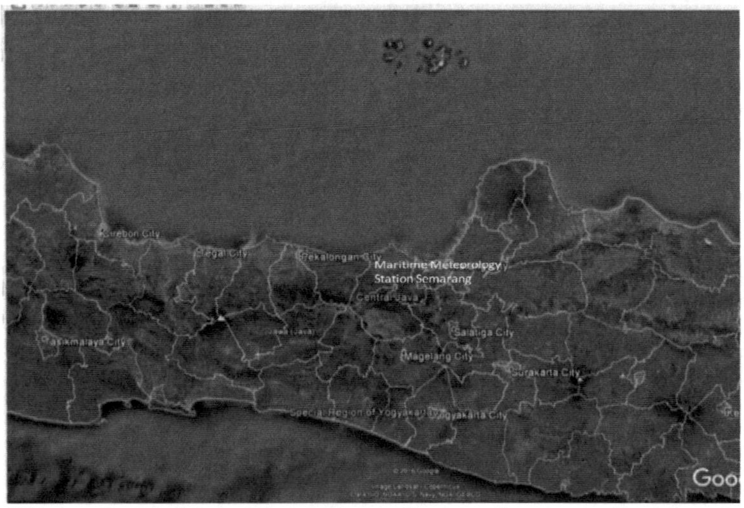

Figure 3. Maritime Meteorology Station Semarang Position (-6.95°S,110.41°E) In Google Earth.

Figure 4. Classification of Wind Direction by Penetration Theory

Based on Figure 4, the sea breeze is a wind that blows from 270^0 - 70^0 (West to East), for transition wind 1 of the wind direction from the direction of 70^0-100^0 (East), transition wind 2 comes from the direction of 240^0-270^0, while for the land breeze is the direction wind from 100^0 - 240^0(East to South

IOP Conf. Series: Earth and Environmental Science **162** (2018) 012018 doi:10.1088/1755-1315/162/1/012018

West). From the classification is obtained that the land breeze is the direction of wind blowing from the direction from west to northeast

Sea breeze is the wind that blows from the sea towards the mainland and generally occurs during the day, based on the wind penetration theory, the sea breeze in figure 5 occurs at 06.00 - 12.00 am. From the windrose diagram, we obtained that the dominant wind sea direction every 3 hours comes from the North-West wind direction.

Calculation of effective fetch that is by using the direction of the dominant wind that is northwest further the northwest direction we assume as 0^0 in the cartesiancoordinat. From the lines were made another line from the same point but with the angle from the previous line by 5^0 respectively. Then do as much as 4 times until you reach the corner of 20^0 as seen in Figure 6.

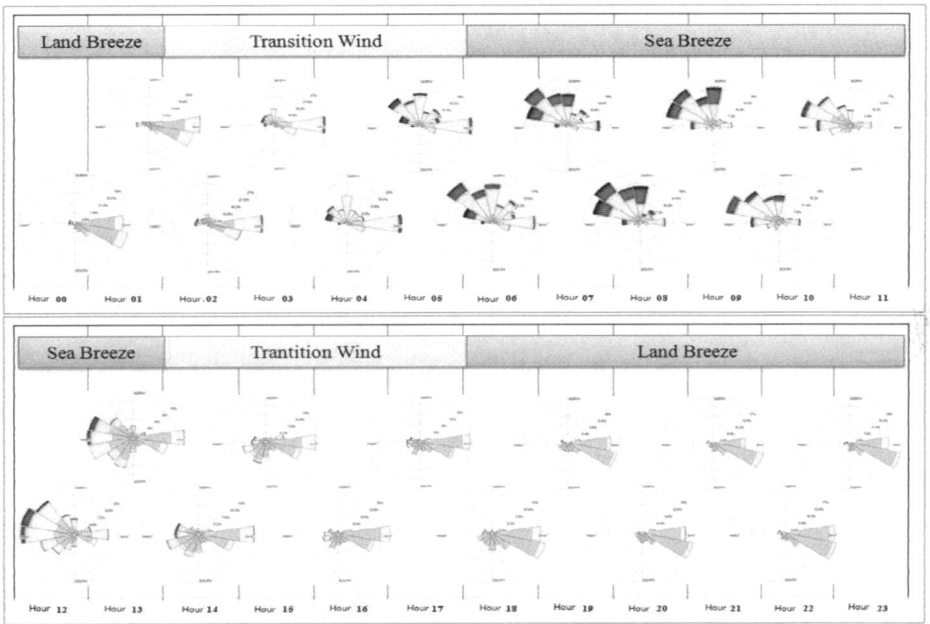

Figure 5. Windrose at 00 – 23 in June 2000 until June 2011

Figure 6. Map of fetch area of Semarang

Table 2. Fetch Distance Calculation

a (°)	Cos a	Xi (km)	Xi.Cos a
20	0.408	517	211
15	0.760	493	375
10	0.839	510	428
5	0.284	629	178
0	1.000	603	603
5	0.284	630	179
10	0.839	609	511
15	0.760	587	446
20	0.408	30	12
Sum	5.581		2942.7

$$F_{eff} = \frac{\sum Xi.cosa}{\sum cosa} = \frac{2942.736}{5.581} = 527.276\ km \quad (8)$$

Based on Table 2 the fetch effective of Semarang coast is 527.276 km.

Table 3. Ocean Waves Height Approach from observation wind data to Sea Surface Wind

Wind Speed	Land Breeze Speed (UL)	RL=Uw/UL	Sea Breeze Speed(UW)	Wave Height H (meter)	Wave Period (T)
1-10	0.5 - 5.1	1.9 - 1.4	0.98 - 7.20	0.26 - 2.99	4.27 - 9.60
11-20	5.1 - 10.3	1.4 -1.1	7.20 - 11.32	2.99 - 5.21	9.60 - 11.53
>20	>10.3	<1.1	>11.32	>5.21	>11.53

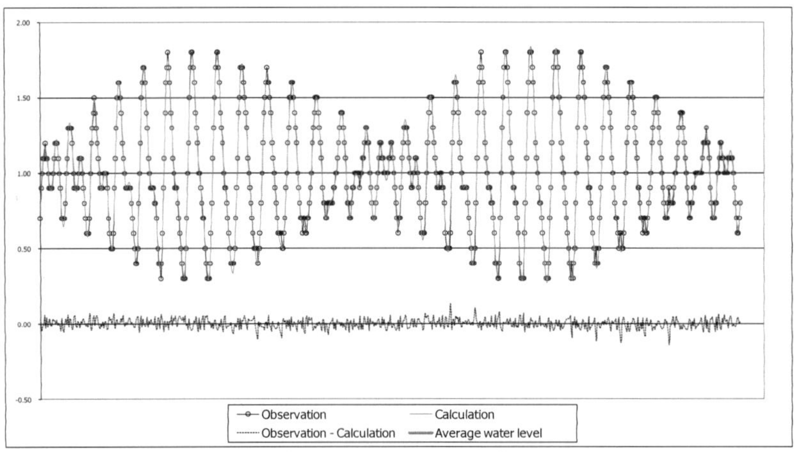

Figure 7. Tidal Waters Levels in Semarang

In Table 3 shows the value of wind speed of land breeze which is divided into 3 classes which are then converted into the sea surface wind approach, the wind speed in knots (kts) must be converted first into the unit m / s then obtained U_w to find the wave height and wave period. From the table obtained that if wind speeds range between 1 – 10 kts the result of wave height will be 0.26 – 2.99 m, wind speeds between 10 – 20 kts will result waves height of 2.99 – 5.21 m, meanwhile for a wind speed of 20 kts will result waves height over 5.21 m.

These values are used to determine the sea level rise (wave set up) that occurs on the coast as seenin Table 1 and Graph1. The result of the calculation of the wave set up 0.26 - 2.99 m after a silting process, sea level in coastal areas rising by 0.01 - 0.56 m, for sea wave height of 2.99 - 5.21 will causing sea level rise in coastal area of 0.56 - 0.87 m and for sea wave height> 5.21 will result in rising > 0.87 m in the coastal area.the result is obtained based on the calculation equation S_w and can be describe by Table 5.

Based on the data analysis of tidal wave in the period 2012 – 2016 which is done each hour, according to the calculations of Least Square then it can be noted that the type of tidal waters in Semarang is mixed, predominantly diurnal tide as indicated by the Figure 5 in addition to the average value of F for tidal in Semarang is 1.1. The results of this research also reinforced by research results from Adhitya (2003) and Darmono (2003) who also get the type of tide of semarang is a mixed, predominantly diurnal tide through the calculation of admiralty. This means that Semarang waters occur twice high and low tide receded during the day but different in high and time.

Based on table 4 obtained the value of HHWL, MHWL, E.G., MLWL and LLWL, but in this study used the value of the MHWL to see the condition of the average increase the sea level water at the time of the highest tide period on average. In the table 4 shows the results of the calculation of the mean sea level obtained 0.597 m while the value of the MHWL is 0.974 so the value of the tide will range from 0.38m above mean sea level.

Table 4. Calculations of Tide in Semarang's waters (2012-2016)

Symbol	Calculation	
HHWL	Z0+(M2+S2+K2+K1+O1+P1)	1.160
MHWL	Z0+(M2+K1+O1)	0.974
MSL	Z0	0.597
MLWL	Z0-(M2+K1+O1)	0.220
CDL	Z0-(M2+S2+K1+O1)	0.140
LLWL	Z0-(M2+S2+K2+K1+O1+P1)	0.034
LAT	Z0-(all constituents)	-0.024

Tides and sea level rise can occur concurrently when a tidal wave propagate to the shore. Therefore, doing the calculation of the total increase sea level rise which occur in conditions period of highest average tides. The calculation results show that when the sea level of 0.01 - 0.56 m occurs along with the highest tide, the sea level in the coast increases to 0.38 - 0.93 m. For sea level rise of 0.56 - 0.87 m and occurring at the highest tide will increasing in a sea level of 0.93 - 1.25 m in sea level while for sea water > 0.87 m and occur along with the highest tide will increase sea level in Coastline above 1.25 as presented in Table 5.

To know the influence of tidal wave events that caused by wind and conditions that occur simultaneously with the highest tide is done by calculating the frequency. On the frequency calculations carried out by using the sea breeze events with wind flow velocity based on 3 classes according to this study with wind speed and tidal conditions can be divided into 3 classes i.e. conditions at altitude <0.6m (MSL = 0.6) which means no tidal , Condition 0.6 - 1 m (MSL - MHWL) and condition> 1m (above MHWL).

Tabel 5. Sea level rise of coastal area at highest tide period

Wave Height H (meter)	Wave Set Up Sw	MHWL	Height Accumulation (Sw+MHWL) m
0.26 - 2.99	0.01 - 0.56	0.38	0.38 - 0.93
2.99 - 5.21	0.56 - 0.87	0.38	0.93 - 1.25
>5.21	>0.87	0.38	>1.25

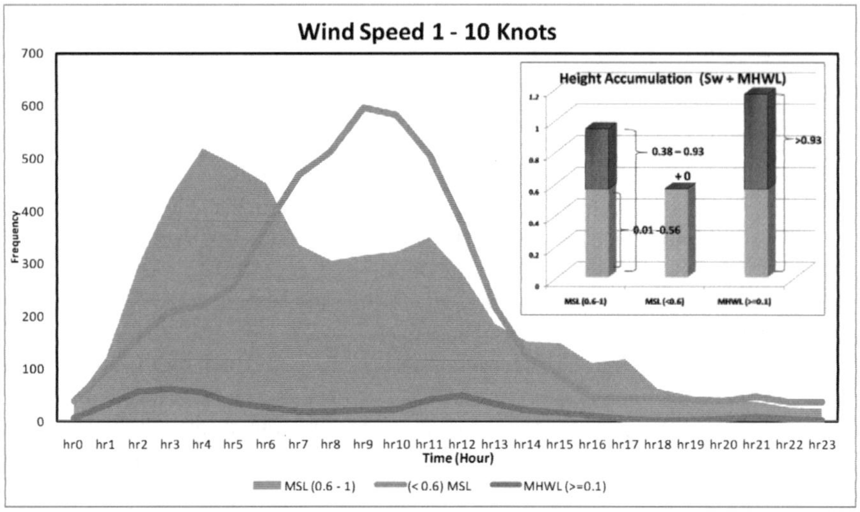

Figure 8. Frequency of Wind Speed (1-10), which occurred simultaneously with each tide height

Based on graph 3.it can be seen that the frequency of wind events from 270^0 - 70^0 (sea breeze) with wind speeds between from 1-10 knots occurring along with the highest average tide (0.6-1) is quite high at dawn (4 a.m – 6 a,m) which causes sea level in coastal area will ranging from 0.38 to 0.93 mthen the frequency of wind occurrence with same speed but not occur simultaneously with the highest tide (<0.6 MSL) are frequently occur in the morning (7 a.m – 12 a.m) with seal level in the coastal area still ranging from 0.01 m to 0.56 m (there is no addition of high tidal)) whereas the wind events with the same velocity that occurs along with the highest tide (> = 0.1 MHWL) has a fairly low incidence when compared to the previous two criteria, but if it happens will raise the sea level in the coastal area as much as> 0.93 m.

It can be seen in graph 4. that the frequency of wind occurrence from direction 270^0 - 70^0(sea breeze) with wind speeds between 10-20 knots that coincides with the highest average tide (0.6-1) is quite high in the morning (5 a.m – 9 a.m) causes sea level in the coastal area ranging from 0.93 to 1.25 m then the frequency of wind occurrence with same speed but not occur simultaneously with the highest tide (<0.6 MSL) are frequently occur in the morning (5 a.m – 11 a.m) with seal level in the coastal area still ranging from 0.56 m to 0.87 m (there is no addition of tidal) whereas the wind events with the same velocity that occurs along with the highest tide (> = 0.1 MHWL) has a fairly low incidence even almost no events but if it happens will raise the sea level in the coastal area as much as> 0.93 m.

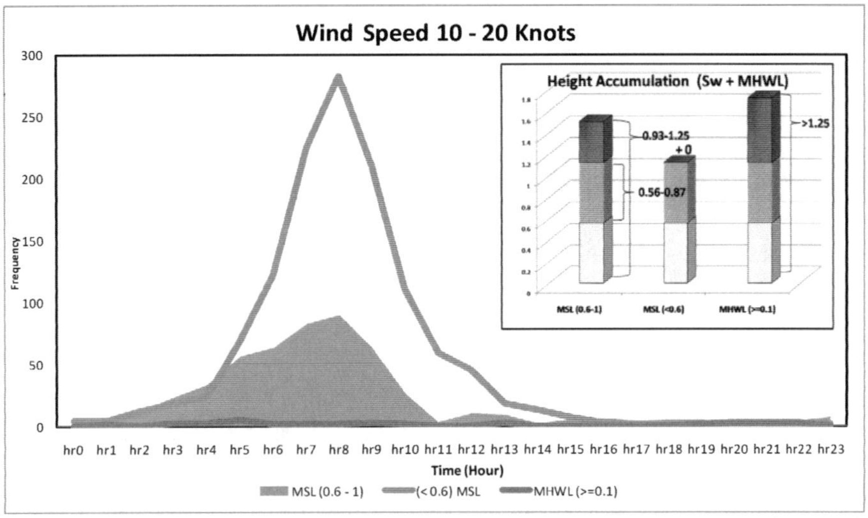

Figure 9. Frequency of Wind Speed (10-20), which occurred simultaneously with each tide height

Based on graph 5. it can be seen that the frequency of wind occurrence from direction 270^0 - 70^0(sea breeze) with wind speeds > 20 knots that coincides with the highest average tide (0.6-1) is very rare if it happens will cause sea level in the coastal area >1.25 m then the frequency of wind occurrence with same speed but not occur simultaneously with the highest tide (<0.6 MSL) are frequently occur compared by other criteria, the wind velocity >20knots will cause seal level in the coastal area >0.87m, whereas the wind events with the same velocity that occurs along with the highest tide (> = 0.1 MHWL) has never happened yet if it happens will raise the sea level in the coastal area much higher than 1.25m

.

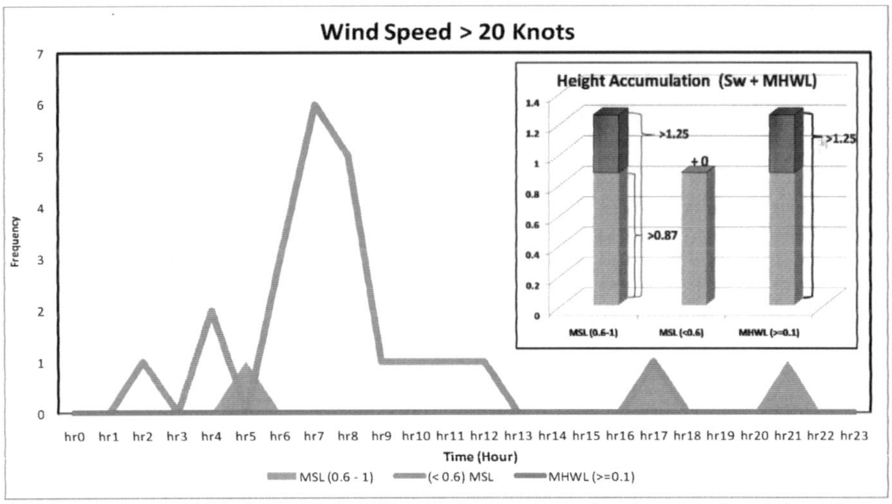

Figure 10. Frequency of Wind Speed (> 20), which occurred simultaneously with each tide height

Sea level rise can cause flood disaster in coastal areas if the height of coastal area below sea level. Ocean waves parameters caused by wind and tidal are dominant factors that influence sea level rise, in this study it is found that the highest sea level rise not only occurs when the wind blows tight that occurs along with the highest tide but the high waves that occur due to strong winds (not coinciding with tides) will also affect sea level rise that will impact on coastal flooding in Semarang.

4. Conclusion

From the research it is found that sea level rise which occurred in Semarang area is the local dynamics with each dominant factor influenced it (Ocean waves and tidal). In the above results obtained that the frequency of wind events 1-10 kts very frequent in Semarang region, this is causing tides on the coastal areas, especially at the highest tide occurs. For 10 - 20 kts wind speed is quite common but this can be a significant when it occurs at the highest tide. while for the frequency of high wind events> 20 kts is very rare in the area of Semarang but if it occurs at the time of maximum tide will cause a significant flood of rob. The results of this study is expected to be a representation related to high sea wave information and become a reference on the influence of meteorological factors to the rising water level.

References

[1] Adhitya FW 2003 analisis banjir rob di kecamatan semarang utara dan kecamatan semarang timur pada saat pasang tertinggi *Skripsi Jurusan llmu Kelautan* FPIK UNDIP (Semarang)

[2] Darmono 2003 aplikasi metode admiralty untuk analisa pasang di perairan semarang periode desember 2002 *Laporan Praktek Kerja Lapangan Jurusan Ilmu Kelautan FPIK Universitas Diponegoro* (Semarang)

[3] Davis R A 1991. oceanography: an introduction to the marine environment *New Jersey: Wcb Publisher International Pub A reference*

[4] Prihatno H 2009 riset kerentanan pesisir akibat perubahan paras muka laut, studi kasus semarang *Laporan Riset. Pusat Riset Wilayah Laut Dan Sumberdaya Non Hayati* BRKP-DKP2009. ISBN 978-979-3768-23-6.

[5] Sapriharyono 2002 Pelestarian dan pengelolaan sumber daya alam di wilayah pesisir tropis Cetakan ke-2. Pt Gramedia Pustaka Utama (Jakarta)

[6] Sarbidi 2002 Pengaruh rob pada pemukiman pantai (kasus Semarang) Prosiding Kerugian pada bangunan dan kawasan akibat kenaikan muka air laut pada kota-kota pantai di Indonesia (Jakarta)

[7] Satridi A 2012 Studi batimetri dan jenis sedimen dasar laut di perairan marina Semarang Jawa Tengah *Buletin Oseanografi Marina Oktober* Vol. 1 p 53 -62

[8] Shore Protection Manual 1984 Vol I-II. USAGE (US)

[9] Triatmodjo B 1999 Teknik Pantai Vol 2 Beta Offset (Yogyakarta–Indonesia)

Spatial multi criteria approach for tsunami risk assessment

Abu Bakar Sambah[12], Miura F [3], Guntur[1] and Fuad[1]

[1] Faculty of Fisheries and Marine Science, Brawijaya University, Indonesia;
[2] Marine Resources Exploration and Management Research Group, Brawijaya University, Indonesia;
[3] Faculty of Engineering, Yamaguchi University, Japan

E-mail: absambah@yahoo.com

Abstract. Tsunami risk is defined as mathematical product between vulnerability and hazard. Tsunami risk also refers to the estimated loss from a given hazard to a given element at risk, and it is defined as a combination of tsunami hazard, tsunami vulnerability, exposure, and event probability. Besides tsunami vulnerability, tsunami risk is assessed by a number of criteria. This study applies the combination of element at risk to assess tsunami risk area along coastal area of East Java Indonesia. The study introduces the spatial multi-criteria analysis into tsunami risk assessment along with the existing perspective evolving the role of GIS. In this study, the existing vulnerability parameter was analysed and evaluated. The analysis was done using GIS-based spatial multi criteria approach. All parameter were analysed through weighted overlay using analytical hierarchy process and geospatial analysis. The results are provided as thematic maps of tsunami vulnerability map and tsunami risk map. Tsunami risk maps as a result from this study illustrate five levels of risk from very low to very high based on geospatial analysis. It describes that coastal area with low elevation was at high risk to tsunami, similar to that indicated in tsunami vulnerability map. Coastal area with high density of vegetation describes a low level tsunami risk. The existence of river and other water stream along coastal area are also important parameters in generating tsunami risk map. Risk map highlights the coastal areas with a strong need for evacuation capacities, including evacuation route and evacuation building.

1. Introduction

The application of satellite remote sensing and Geographical Information System (GIS) has become an integrated, well-developed, and successful tool in disaster research for effectiveness of risk management and disaster mitigation. An overview is given of the use of spatial data with emphasis on satellite remote sensing data, and of the approaches used for hazard assessment [1][2]. The development of satellite remote sensing and its applications enables the use of satellite imagery for identifying the distribution of areas damaged by tsunami disaster. Satellite images have the advantage of being able to deliver simultaneous images of large areas [3][4][5].

In earthquake related research, post-event hazard-related applications deal mainly with the quantification and measurement of earthquake-induced changes of land surface. Several aspects of pre-event earthquake hazard analysis are tackled by means of satellite remotely sensed data, especially by applying high spatial resolution of satellite imagery. In pre-event geological observations, remote sensing addresses the need for quantitative observational parameters on landforms or land cover

changes. In addition, with the aid of GIS, spatial multi-criteria analysis helps prioritize the decision-making process using geo-reference data. Spatial multi-criteria analysis is vastly different from conventional multi-criteria decision making techniques, due to the inclusion of an explicit geographic element. In contrast to conventional multi-criteria decision making techniques, spatial multi-criteria analysis applies information on both the criterion values and the geographical positions of alternatives, in addition to the decision-maker's preferences with respect to a set of evaluation parameters [6][7].
A variety of methods exist to assess the risk of different hazards [8]. However, applications on tsunami risk are very limited. There is still no generally adopted method to assess tsunami risk. The assessment of tsunami hazard and risk is therefore an important issue that must be addressed to identify the risk to populated areas and the surrounding land use.

Based on the geological aspect, the south coast of Java Island is in the confluence of two major plates meet each other, Eurasian and Indo-Australian, where the movement of tectonic plates in this area will cause an earthquake as a trigger of tsunami event. An earthquake event that is followed by a potential destructive tsunami in the period of 1991 to 2006, recorded a tectonic earthquake in the Indian Ocean which triggered the tsunami on the southern coast of East Java, namely on June 3, 1994. A magnitude of 7.8 Mw, earthquake triggered a tsunami that affect southern coastal areas of Banyuwangi, East Java with estimated death toll reached 215 people [9]. It can be stated that the coastal area of East Java is vulnerable to tsunami disaster, and based on the historical event mentioned above this event have high possibility of occurrence in near future. It is expected that future tsunamis can have a higher impact due to the land use changes, increasing population and built areas and decreasing vegetation areas. To mitigate future tsunami events, it is necessary to understand and analyze this phenomenon in more detail using integrated approaches.

The study aim is to assess tsunami risk in the coastal area of East Java using spatial multi-criteria analysis together with the analysis of satellite imagery.

2. Methods

2.1. Dataset

Tsunami vulnerability area in the coastal of East Java was created from the Digital Elevation Model (DEM) collected form SRTM data. DEM data also applied 10 meters resolution from NextMap World 10. Seismic data of the study area from 1992 to 2014 collected from The United States Geological Survey (USGS), and downloaded from http://earthquake.usgs.gov/earthquakes/search/ was used as a supporting parameter for tsunami vulnerability which will be used to generate seismic map. In order to map the land use of study area, Landsat 8 OLI satellite imagery with the spatial resolution of 30 m was analyzed. Moreover, vector base map of East Java Indonesia was applied to prepare vector data of coastal morphology, coastal line, and river.

Table 1. Dataset

	Location	Product	Resolution		Source
Digital Elevation Model	Coastal area of East Java	SRTM V-4.1	3 arc-second	90m	CGIAR-CSI
	Coastal area of Malang District	NextMap World 10		10m	NextMap
Land use	Coastal area of Malang District	Landsat 8 OLI		30m	
Vector map	East Java	Vector map ; point, polyline, polygon		UTM Zona 49S	BIG Indonesia

2.2. Study Area

The study was applied at the south coastal area of East Java, focusing on the coastal area of Malang District, Indonesia (Figure. 1). The coastal area of Malang district was known as one of important fishing port in East Java. This area was also affected by 1994 tsunami event along coastal area of East

Java. The tsunamigenic earthquake occurred on June 3, 1994 in the Indian Ocean about 200 km south of Java. The earthquake, which had a surface-wave magnitude of 7.2 and a moment magnitude of 7.8 at 10.51°S and 112.87°E, generated a devastating tsunami that took the lives of more than 200 East Java coastal residents; [10][11]. Moreover, the general method of the study as described in Figure 2.

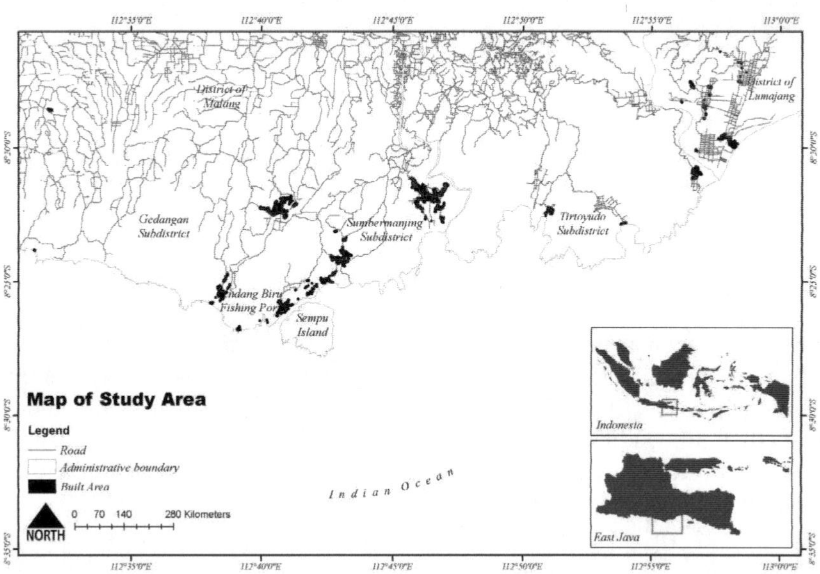

Figure 1. Study area.

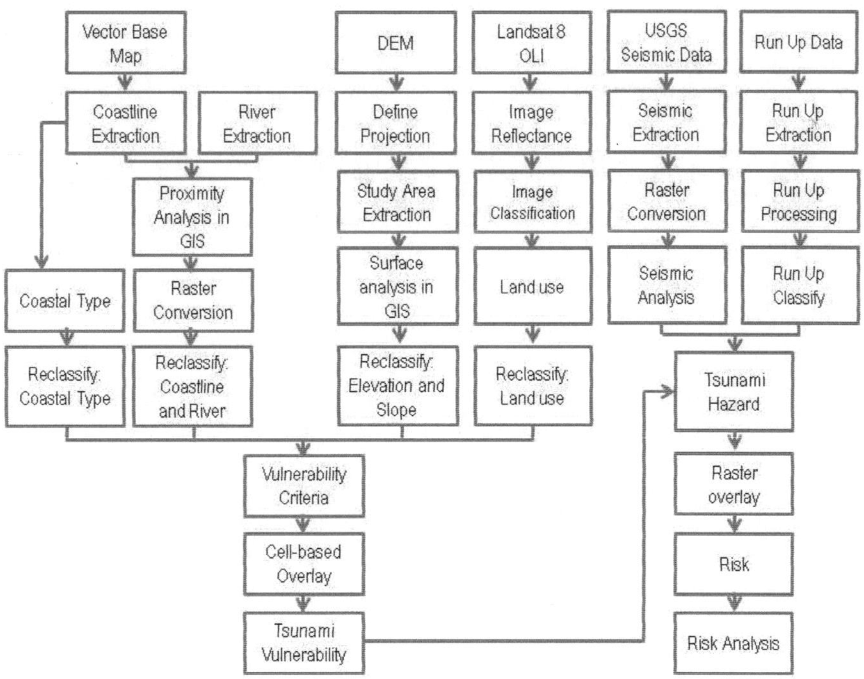

Figure 2. General method

In general, the analysis consists of two main goals, assessing tsunami vulnerability areas and assessing tsunami risk. To create tsunami vulnerability map, the parameter of elevation, slope, coastal proximity, river proximity, and land use was analyzed and overlaid based on cell-base modelling in GIS. In order to assess tsunami risk area, seismic data and historical run up data was calculated. Raster overlay approach in GIS was applied in assessing tsunami risk map in which both tsunami vulnerability map and tsunami hazard data were analyzed.

3. Result and Discussion

3.1. GIS Approach for Vulnerability Mapping

The vulnerability to a tsunami is a function of a number of physical as well as social parameters that include amongst others: distance from the shore, depth of flood water, construction standards of buildings, preparedness activities, socio-economic status and means, level of understanding and hazard perception and amount of warning and ability to move away from the flood zone. Thus, a tsunami vulnerability analysis should be developed that includes as many of these factors as possible in order to gain a more realistic picture of spatial and temporal patterns of physical and social vulnerability [12]. In order to derive a number of parameters describing physical vulnerability a digital surface model and a Digital Elevation Model (DEM) were generated and consequently the building height derived. The information layers have been integrated in a GIS together with other geographical information layers (building outlines, coastline, etc.) and an integrative physical vulnerability analysis was performed. The vulnerability analysis in this study is composed of two parts, field data collection and satellite data processing which are described in Sections 2.

Vulnerability mapping has been generated using the parameters elevation, slope, coastal proximity, river proximity, and land use. Elevation and slope were created from SRTM elevation and also Nextmap world 10 elevation data (Figure 3). Together with hazard or capacity, tsunami vulnerability is one of the parameters in assessing tsunami risk. Cell-based analysis was applied in combining all parameter through analysis of geo-spatial.

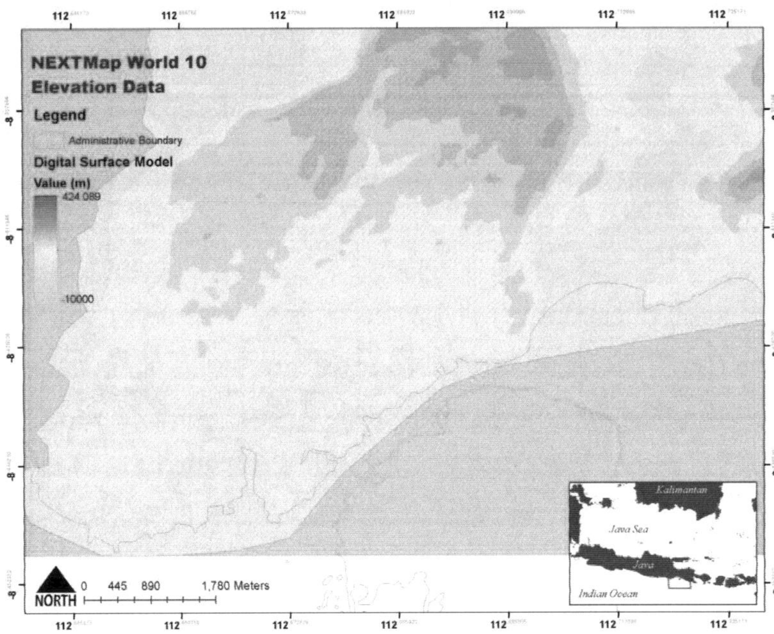

Figure 3. Nextmap world 10 elevation data of Malang district coastal area

Moreover, land use map was generated from supervised classification process of Landsat 8 OLI satellite image. Reflectance value of satellite image applied in maximum likelihood classification and generated five classes of land use in the study area. Each parameter constructing tsunami vulnerability was applied to raster reclassification based on tsunami vulnerability classes using the criteria as described in Table 2.

Table 2. Vulnerability classes for each parameter [2][13][14][15] (from 1 for least vulnerable to 5 for most vulnerable)

Parameters	Vulnerability Classes				
	5	4	3	2	1
Elevation (m)	<5	5 – 10	10 – 15	15 – 20	>20
Slope (percent)	0 – 2	2 - 6	6 - 13	13 - 20	>20
Coastal Proximity (m)	<200	200 - 500	500 - 1000	1000 - 1500	>1500
River Proximity (m)	0 -100	100 - 200	200 - 300	300 -500	>500
Land use	Urban	Agriculture	Bare soil	Water	Forest

Raster overlay for tsunami vulnerability mapping was done based on the weight of parameters. Weights were calculated using Analytical Hierarchy Process (AHP). Weighted overlay is a technique for applying a common measurement scale of values to diverse and dissimilar inputs to create an integrated analysis. Weighted overlay also one of the suitability analysis based on spatial multi-criteria processing [2][16]. Vulnerability map was generated by applying a weight value to each parameter in raster data format. Figure 4 shows tsunami vulnerability map based on different parameters.

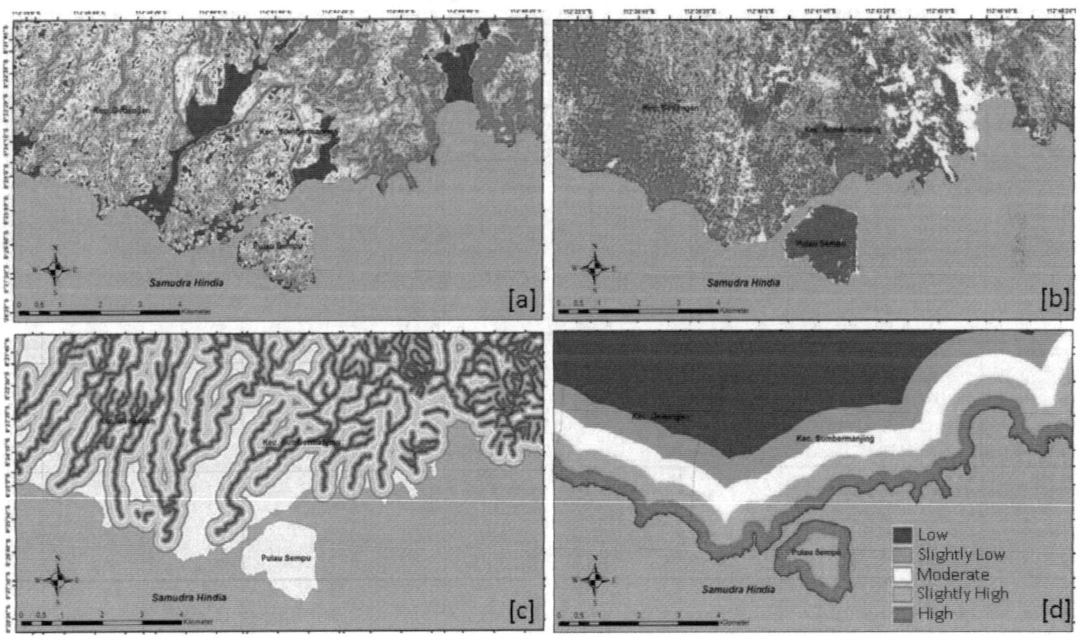

Figure 4. Tsunami vulnerability map based on; [a]slope; [b] land use; [c] river proximity; [d] coastal proximity.

The use of AHP analysis in generating weight of parameters in which to be applied in weighted raster overlay describes five classes of vulnerability due to tsunami. It describes that the western part of study area mostly falls in the class of slightly high to high class of tsunami vulnerability. This area was identified as flat area with land use class of urban area, bare soil, and agriculture. Within the coastal area that is exposed to tsunami inundation, the buildings and infrastructure are not uniformly at risk within the flood zone [17][18]. The probability of damage is related to both vulnerability and the tsunami wave energy. Damage level to buildings depends on building type and on inundation depth [19] or could depend on the density of vegetation around the coast that is assumed will reduce tsunami impact. Assessing tsunami vulnerability in the urban area has considered that individual buildings will interact differently with a tsunami depending on a number of parameters [17][18]. Various parameters affecting the resistance of the building interact to generate a real class of building vulnerability [20]. Although land boundary reflection brought an effect on the position of the maximum wave height to a certain extent, as the limits of the incident waveform and distances between the observation points and shore, it was not the dominant influence factor of the special waveform [21].

3.2. Seismic And Run Up Analysis

Seismic data are physical observations or measurements of seismic sources, seismic waves, and their propagating media. The purpose of processing seismic data is to learn about the Earth's interior. It needs to figure out some specific relations between the intended targets and measurable parameters in order understand certain aspects of the Earth [22].

All initial tsunami warnings are based on rapid detection and characterization of seismic activity. Because of the fundamental differences in nature between the solid earth in which an earthquake takes place and the fluid ocean where tsunami gravity waves propagate, the vast majority of earthquakes occurring on a daily basis do not trigger appreciable or even measurable tsunamis. It takes a large event (magnitude >7.0) to generate a damaging tsunami in the near-field and a great earthquake (magnitude >8.0) to generate a tsunami in the far-field [23].

The study area was identified with a high number of seismic points with the depth average of 1 to 5 km and the range of magnitude 1 to greater than 6. As a result, coastal area of East Java can be classified as highly vulnerable to tsunamis. Based on historical data, it was a magnitude of 7.8 Mw in the depth of 18 km and latitude of -10.477° and longitude of 112.835° that caused big tsunami and affected the coastal area of East Java (see seismic map on Figure 5).

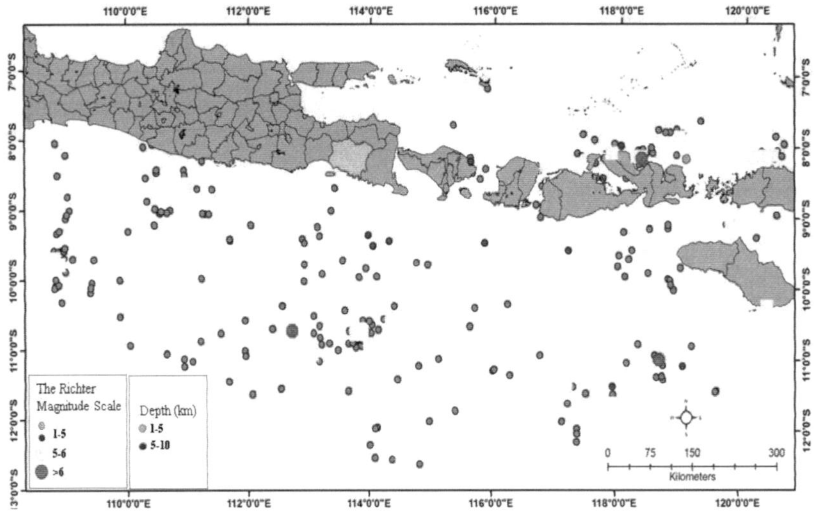

Figure 5. Seismic data in Indian Ocean close to coastal area of East Java from 1992 to 2004 from U.S. Geological Survey

Figure 5 illustrates that based on the U.S. Geological Survey data of seismic intensity around Indonesia, it was recorded that the range of magnitude was 1 to 6.3 Mw. It describes the typical effects of earthquakes of various magnitudes near the epicenter. Intensity and thus ground effects depend not only on the magnitude, but also on the distance to the epicenter, the depth of the earthquake's focus beneath the epicenter, the location of the epicenter and geological conditions. Magnitude of 1 – 6.3 was classified as micro to strong effect or I to VII in Mercalli intensity. It was categorized based on the damage to a moderate number of well-built structures in populated areas. Earthquake-resistant structures survive with slight to moderate damage. Poorly designed structures receive moderate to severe damage.

The historical tsunami event of 1994 was used as a basic data for run up analysis. Maximum run up was recorded at Tempurejo subdistrict, Jember district (an area in the eastern part of study area) (11.2 m), and minimum run up was 3.1 m in the area of Puger subdistrict. Tsunami run up parameter is one of the important parameters in determining tsunami risk due to this parameter being the main parameter in hazard criteria. In general, rivers can play an important role in expanding the impact of the damage during tsunami event. The run-up of the tsunami reaches the hinterland not only through the low elevation of the area, but also through rivers. Rivers also act as flooding strips transporting inundation [1][2].

3.3. *Tsunami Risk*

The term risk can have a different meaning in different applications or languages therefore it is necessary to clarify some basic definitions related to risk. In our understanding, the risk is the product of the likelihood that an undesired event such as a tsunami will happen within a specific period of time, and its consequences. In order to measure risk, it is necessary to define its characteristics, which are described in the process of risk analysis. Tsunami risk analysis is a very complex process which requires knowledge of the tsunami In mathematical form, the following general expression can be used $R = H \times V$, in which R = risk, H = probability of tsunami hazard occurrence, and V = vulnerability of the elements at risk (as described on Figure 6).

Tsunami risk assessment is a relatively new and growing discipline that is being developed from "generic" risk approaches, which are usually applied to the general field of technological risk management [8]. Tsunami risk map in the study area (Figure 7) describe that high and slightly high risk of tsunami was found in the area of low to flat elevation. These areas were identified as bare soil and urban area. Sempu island can act as a barrier island in reducing the tsunami impact to the main land as long as the high density of vegetation including mangrove also can act as the green sea belt to reduce the impact of tsunami wave when it hits the coastal area.

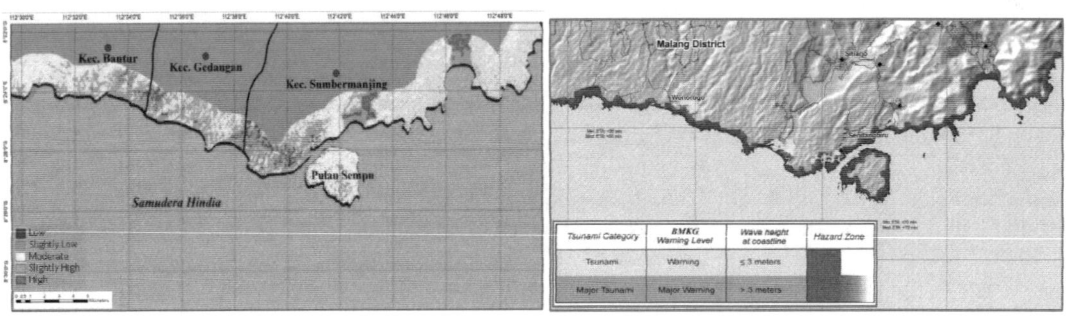

Figure 6. Tsunami vulnerability/V (left) and hazard map/H [24] (right)

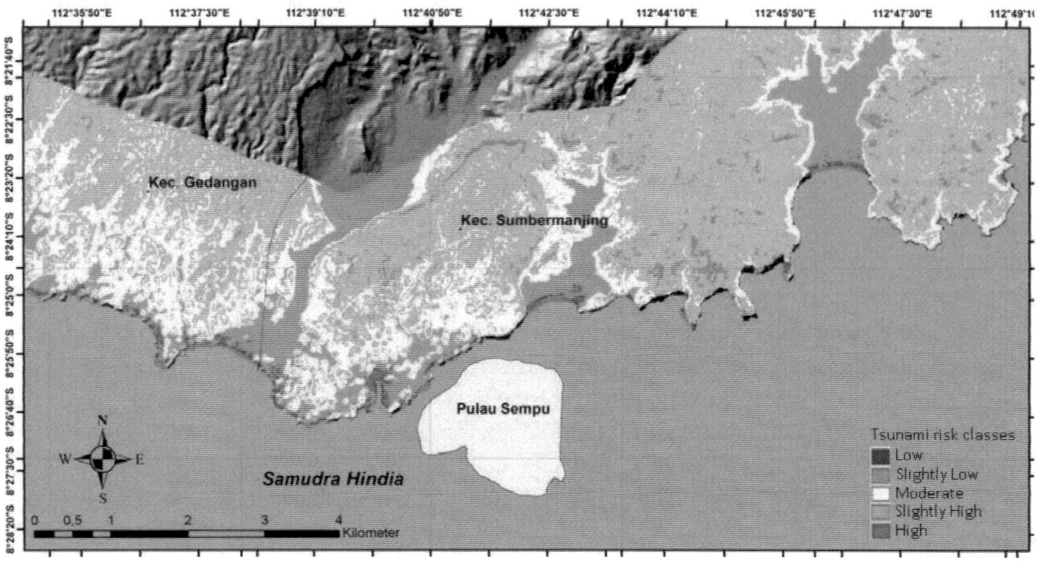

Figure 7. Tsunami risk map

4. Conclusion

Tsunami risk can be assessed using the application of multi criteria analysis in GIS in which DEM data was applied. The result obtained here can be used for the evacuation route and evacuation building planning. The result is also important as basic information in effectiveness of tsunami mitigation. The combination of raster weighted overlay in the geo-spatial data analysis indicated the vulnerability and risk area due to tsunami and described the possibility area that could be affected by tsunami. The weight of each parameter was calculated by pair-wise comparison matrix of expert judgment, in which every parameter was weighted not equally. By applying overlay process for tsunami risk map and existing land use, it will describe which area that first needs to be evacuated when tsunami comes. The result of weighted overlay illustrated that high class of tsunami vulnerability and tsunami risk occurs mostly in the class of urban area, while forest area was indicated in the low class. The more parameters that are applied, the more detailed is the assessment that can be analysed and displayed.

Acknowledgements

Authors are thankful to Ministry of Research, Technology and Higher Education Indonesia for financial support, USGS for SRTM DEM data and Landat satellite data, BIG Indonesia for providing the basic map of the study area, and U.S. Geological Survey for seismic data. We also thank to Laboratory of Disaster Prevention System, Yamaguchi University, Japan and Laboratory of Marine Resources Exploration, Brawijaya University Indonesia.

References

[1] Sambah AB and Miura F, "Spatial Data Analysis and Remote Sensing for Observing Tsunami-Inundated Areas", *International Journal of Remote Sensing*, 2016, **37**, Issue 9, pp. 2047-2065.

[2] Guntur, Sambah A B, Miura F, Fuad, and Arisandi DM, "Assessing Tsunami Vulnerability Areas Using Satellite Imagery and Weighted Cell-Based Analysis", *International Journal of GEOMATE*, **12**, Issue 34, 2017, pp. 115-122.

[3] Karen E J, Wright K C, Samsonov S V and Ambrosia V G (2009), "Remote sensing and the disaster management cycle", in Gary, J. (Ed.), *Advances in Geoscience and Remote Sensing*,

ISBN:978-953-307-005-6, available at: www.intechopen.com/books/advances-ingeoscience-andremote-sensing/remote-sensing-and-the-disaster-management-cycle

[4] Yamazaki F, Kouchi K and Matsuoka M (2006), "Tsunami damage detection using moderateresolutuion satellite imagery", *Proceeding of the 8th US National Conference on Earthquake Engineering*, Paper No. 465, April 18-22, San Fransisco, California, CA.

[5] Yamazaki, F. and Matsuoka, M. (2007), "Remote sensing technology in post-disaster damage assessment", *Journal of Earthquakes and Tsunamis, World Scientific Publishing Company*, **1** No. 3, pp. 193-210.

[6] Carver S J (1991), "Integrating multi-criteria evaluation with geographical information systems", *International Journal of Geographical Information Systems*, **5** No. 3, pp. 321-339.

[7] Jankowski P(1995), "Integrating geographic information systems and multiple criteria decision making methods", *International Journal of Geographic Information System*, **9** No. 3, pp. 251-273.

[8] Jelínek R and Krausmann E, Approaches to Tsunami Risk Assessment, 2008, European Commission Joint Research Centre Institute for the Protection and Security of the Citizen. Luxembourg: Office for Official Publications of the European Communities. European Communitie.

[9] Ramasamy S M, Kumanan C J, Sivakumar R, and BhoopSingh, "Geomatics in Tsunami", 2006, New India Publishing Agency.

[10] Maramai A, and Tinti S, "The 3 June 1994 Java Tsunami: A Post-Event Survey of the Coastal Effects", *Natural Hazards*, 1997, **15**, pp.31–49.

[11] Storchak D A, Giacomo D Di, Bondár I, Engdahl E R, Harris J, Lee W H K, Villaseñor A and Bormann P, "Public Release of the ISC-GEM Global Instrumental Earthquake Catalogue (1900-2009)". *Seism. Res. Lett.*, 2013, **84**, 5, pp. 810-815.

[12] Papathoma M, Dominey-Howes D, Zong Y and Smith D (2003), "Assessing tsunami vulnerability, an example from Herakleio, Crete", *Nat. Hazards Earth Syst. Sci.*, **3** No. 5, pp. 377-389.

[13] Bretschneider C L and Wybro P G, "Tsunami Inundation Prediction", *Proceedings of The 15th ASCE Conference on Coastal Engineering*, Honolulu, July 11-17, 1976, pp. 1006-1024.

[14] Iida K, "Magnitude, energy and generation mechanisms of tsunamis and a catalogue of earthquakes associated with tsunamis", *Proceeding of Tsunami Meeting at the 10th Pacific Science Congress*, 1963, pp. 7-18.

[15] Van Zuidam R A, "Guide to Geomorphologic - Aerial Photographic Interpretation and Mapping", International Institute for Geo-Information Science and Earth Observation, Enschede, 1983, The Netherlands.

[16] ESRI, "Environmental Systems Research Institute, Inc. Analyse Site Conditions Using Weighted Overlay", 2015, ESRI.

[17] Papathoma M and Dominey-Howes D, "Tsunami Vulnerability Assessment and Its Implications for Coastal Hazard Analysis and Disaster Management Planning, Gulf of Corinth, Greece", Natural Hazards and Earth System Sciences, 2003, **3** No. 6, pp733-747.

[18] Papathoma M, Dominey-Howes D, Zong Y and Smith D, "Assessing Tsunami Vulnerability, an Example from Herakleio, Crete", *Nat. Hazards Earth Syst. Sci.*, 2003, **3** No. 5, pp. 377-389.

[19] Tinti S, Tonini R, Bressan L, Armigliato A, Gardi A, and Guillande G : Handbook on Tsunami Hazard and Damage Scenarios, Schema (Scenarios for Hazard-Induced Emergen-Cies Management), Project No. 030963, Specific Targeted Research Project, Space Priority, Irc Scientific and Technical RePort (www.schemaproject.org), 2010.

[20] Alberico I, V. Fiore D, Iavarone R, Petrosino P, Piemontese, L, Tarallo D, Punzo Michele, and Marsella E, "The Tsunami Vulnerability Assessment of Urban Environments Through Freely Available Datasets: The Case Study Of Napoli City (Southern Italy)", J. MAR. SCI. ENG. 2015, 3, pp 981-1005

[21] YUAN Chun-guang,WANG Yi-gang,HUANG Hui-ming,CHEN Cheng and CHEN Da-ke. Propagation Mechanisms of Incident Tsunami Wave in Jiangsu Coastal Area, Caused by Eastern Japan Earthquake on March 11, 2011. China Ocean Engineering, An International Journal, 2016,(1):123~136

[22] Hua-Wei Zhou, "Introduction to seismic data and processing", Practical Seismic Data Analysis, Cambridge University Press, 1991, http://assets.cambridge.org/97805211/99100/excerpt/9780521199100_excerpt.pdf

[23] Committee on the Review of the Tsunami Warning and Forecast System and Overview of the Nation's Tsunami Preparedness; Ocean Studies Board, "Tsunami Warning and Preparedness: an Assessment of the U.S. Tsunami Program and the Nation's Preparedness Efforts", Division on Earth and Life Studies; National Research Council, 2011, (https://www.nap.edu/read/12628/chapter/7#111)

Critical-phase sea dike construction of NCICD program in Jakarta as national capital city

H Suprayogi[1], A Rudyanto[1], H Bachtiar[2] and L M Limantara[3]

[1] Directorate General of Water Resources, Ministry of Public Works and Housing, Jakarta Selatan, Daerah Khusus Ibukota Jakarta 12110, Indonesia
[2] Research Centre for Water Resources, Ministry of Public Works and Housing, Bandung, Jawa Barat 40135, Indonesia
[3] Department of Water Resources Engineering, Brawijaya University, Malang, Jawa Timur 65145, Indonesia

E-mail: huda.bachtiar@gmail.com

Abstract. National Capital Integrated Coastal Development Program (here in after called NCICD Program) in Jakarta as National Capital City has main function for flood protection and to countermeasure the land subsidence purposes. According to Bappenas (National Planning Agency), the Updated-Masterplan NCICD program has three phases; 1) Critical-phase 2) Mid-term phase 3) Option-phase. The critical-phase sea dike construction is projected to protect Jakarta until 2030. Land subsidence is the main factor that the critical-phase of sea dike must be constructed. Jakarta sinks down with the range of 2-20 cm/year. Therefore, the existing dike level also becomes lower along the time. Three focus area of sea dike construction under Ministry of Public Works and Housing (here in after called PUPR) are Muara Kamal, Kalibaru, and Muara Baru. At the moment, PUPR concentrates in Kalibaru and Muara Baru with the physical construction progress around 48% from the total length 4-8 km and it will be finished in 2019. These are essential due to many people live there. The construction of typical sea dike is spoon-pile with shifting construction to the sea between 50 m-100 m from mainland with the crest level +4.8 m from Lowest Water Spring. This type is chosen due to its efficiency on spatial and preventing social conflicts. In addition, the space between the main land and the spoon pile will be nourished by the sand to use it for public infrastructure, such as park. The material behind the dike and the toe protection in front the dike is used for stability purposes, avoid scouring, and reducing wave energy interact the dike itself.

1. Introduction

Jakarta is the national capital city of Indonesia, where the activity center of economy from regional, national, and international scale is located in this city. Jakarta has population about 9.5 million people [1]. According to Presidential Decree no. 54, 2008 Jakarta is one of the strategically area. Jakarta has coastline about 32 km and geographically has borderline with Bekasi City in Eastern part and in the Western part with Tangerang City. Nowadays, more than half of population in Jakarta lives in the coastal area.

The coastal zone area in the Northern Part has potential to be developed. In this area, port, warehouse, and trade center are located in the Northern Part of Jakarta. However, Jakarta in the Northern Part is also vulnerable of flooding from the river side and flooding from the sea side. In 1990 only 12% or about

1600 Ha land in Northern Jakarta is below sea level, the projection in 2030 almost 90% or 12500 Ha land in Northern Jakarta will be below sea level. In addition, the rate of land subsidence about 2-20 cm/year will cause river cannot flow to the sea gravitationally and hence it will lead to flood event from river side [1].

The current flood defenses in the coastal zone of Jakarta do not provide sufficient protection against flood. The coastal zone has subsided in the last decades and it is expected to continue doing so at an average rate of about 7.5 cm per year.. At the same time, the crest level of sea dike will also decrease and it will lead to the flooding from sea side.

At the moment three focus areas of sea dike construction under Ministry of Public Works and Housing are Muara Kamal, Kalibaru, and Muara Baru. The selection of these are essential due to many people live there. This paper discusses the critical-phase of physical sea dike construction; parameter for boundary condition, design consideration, and progress construction. NCICD program has two main program; (1) No-regret Measure (2) Conditional Measure. No-regret Measure is obligatory program, where the program has to be conducted. Conditionl Measure is the program, where the program should be conducted if land subsidence can not be controlled. Figure 1 shows the development plan of NCICD program with conditional measure included.

Figure 1. NCICD masterplan program.

2. Literature Review

2.1. Updated-masterplan of NCICD

According to National Planning Agency of Indonesia, the Updated-Masterplan NCICD program has three phases; 1) Critical-phase 2) Mid-term phase 3) Optimum-phase [2]. The Critical-phase of sea dike is a compulsory phase, where the sea dikes have to be constructed, elevated, and strengthened the existing sea dike. The length of critical sea dike is about 20.1 km and it should be finished by 2019. The critical sea dike is constructed by the government (DKI Regional Government and Ministry of Public Works and Housing) and the private company.

The main functional requirement of NCICD is that to provide protection against flooding in the long-term period. The master plan has to be integrated solution from the upstream to the downstream area; beside engineering aspect, the environmental, regulation, socio-economic conditions are also considered. Figure 1 shows the design concept of how the NCICD will be developed in the long-term period plan.

2.2. *Boundary condition*

The crest elevation of the dike is based on the boundary condition of the coastal area [3]. The dike design is based on the tidal condition in this case we use Highest Water Spring (HWS), Extreme Wave Condition (EWC), Sea Level Rise (SLR), Land Subsidence (LS), Free Board (FB), and Wave Run-up (WR). HWS is the tidal analysis results, where we pick the highest value of the tidal during spring tide period. EWC is wave design with the return period 1:10000. Rate of SLR, the dike design is based on the IPCC projection. Rate of land subsidence is based on the land subsidence modeling.

$$\text{Crest Elevation} = HWS + EWC + SLR + LS + FB \quad (1)$$

2.3. *Land subsidence*

Coastal regions are increasingly exposed to flood damages due to growing population assets, rising sea levels and possibly more frequent and intense storms. Flooding in the context of future storm variability, sea-level rise and shoreline change is one of the most important issues facing coastal populations today. Flooding in Indonesia comes from the upstream due to heavy rainfall or from the sea side due to the high tides event as well as significant sinking of the land down below the sea level. There is a correlation between population growth and land subsidence in coastal areas (see Figure 2) [4]. They argued that the subsidence is primarily caused by excessive ground water extraction for industrial, water use, and agricultural use. According to population data density in 2000 consensus, population density in Java Island (about 250 to 999 person/km2) is higher than SumateraIsland (about 25 to 249 person/km2). Therefore, it indicates that the higher dense of population, the higher rate of land subsidence due to groundwater depletion.

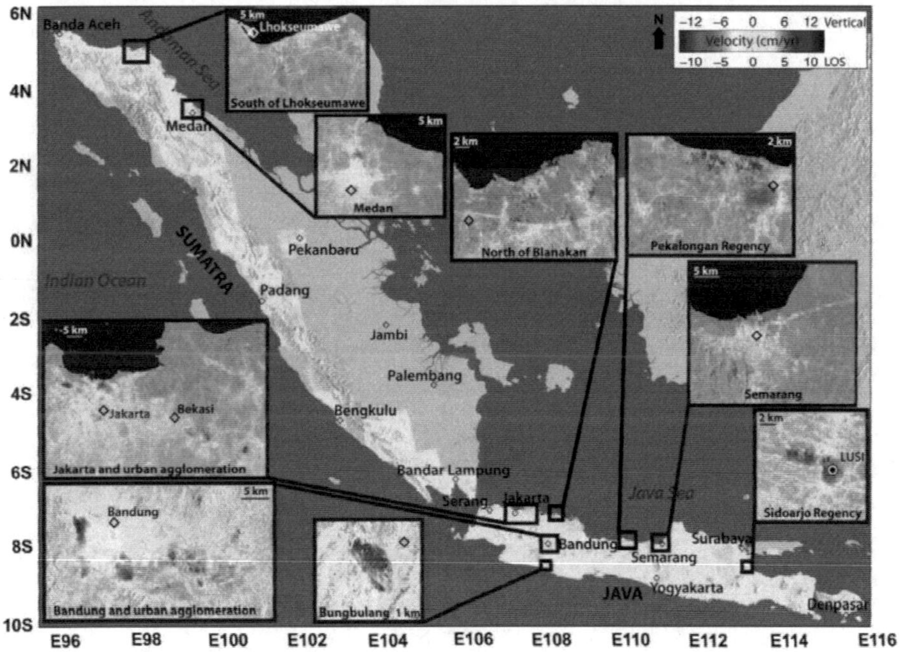

Figure 2. Land subsidence over Java and Sumatera Area (source: Chaussard, et al., 2012).

3. Results and Discussion

3.1. *Field situation and existing sea dike condition in 2013*

Figure 3 shows the first survey was conducted in March-April 2013 and covered public areas within the area of DKI. The second survey was conducted in September 2013 covered privately owned areas,

Tangerang, and Bekasi. Figures 4 and 5 show field situation of Jakarta coastline in Muara Baru and Kalibaru respectively. These both figures show the area condition where the critical dike before constructed. In addition, the areas are also dense populated. The existing dike showed that the inland area is below the sea level. Therefore, the critical situation has to be solved by constructing the dike.Red circle and green circle show the area where the coastal dike wil be constructed but by the private sector.

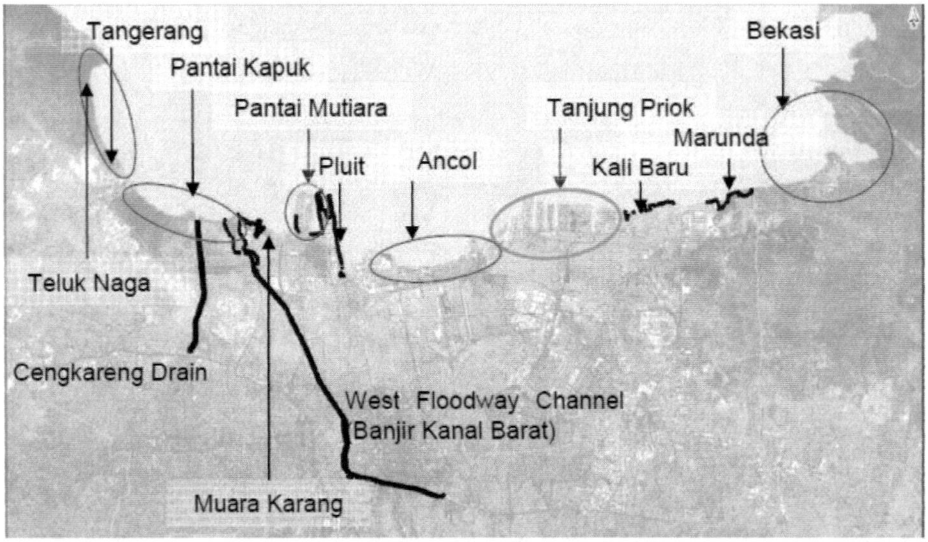

Figure 3. Field survey map location in 2013.

Figure 4. Field situation of dike in Muara Baru (Pluit Area).

Figure 5. Field situation in Kalibaru.

Table 1. Average height of coastal protections (from west to east)

Section	Average height	Type of protection	Remarks
Teluk Naga (Survey II)	1.0 m +LWS	Mangrove	Fish ponds
Tangerang (Survey II)	1.4 m +LWS	Mangrove	Fish ponds
Pantai Indah Kapuk (Survey II)	1.1 m +LWS	Mangrove	Mangrove reserve, dike further inland
Pantai Indah Kapuk (Survey I)	3.7 - 4.0 m +LWS	Dike	High end housing
Muara Karang	1.4 - 1.7 m +LWS	Seawall	Housing
Muara Karang (Survey I)	3.0 m +LWS	Seawall	PLTU
Pantai Mutiara (Survey II)	1.2 m +LWS	Seawall	High end housing
Pluit (Survey I)	1.6 - 1.9 m +LWS	Seawall	Fishing port
Ancol (Survey II)	1.5 m +LWS	Waterfront/Beach	Recreational area
Tanjung Priok	No data	No data	No data
Kalibaru (Survey I)	2.5 m +LWS	Sea dike/land reclamation	Land reclamation from crab-materials
Marunda (Survey II)	2.7 m +LWS	Quay walls	Port area, PLTU
Bekasi (Survey II)	0.7 m +LWS	Mangrove	Fish ponds

At some locations (Teluk Naga, Tangerang, Pantai Indah Kapuk, Bekasi) the actual coastal protection is located further inland. At these locations the approximate topographical level at the coastline is provided in Table 1. The current design water level is approximately 1.7 m +LWS2012. Crest levels should be higher to accommodate for wave run-up. Crest levels are close to the current design water level at most locations. This means that a combination of high water and strong waves may lead to overtopping and inundation of the areas. To the east (Bekasi) and west (Tangerang and Kapuk) large mangrove areas and fishponds are present. These rely on a frequent inundation of seawater. A coastal flood defence might be found further inland. However, topographical elevations may be high enough so inundation will not occur further inland at present.

At Muara Karang, Pantai Mutiara, Pluit and Ancol existing protection levels are dangerously low. At present they are already below the design water level. During combined high water and wind/wave setup or seasonal water level increase, the protections may overflow and inundate the populated hinterland.

At Pantai Indah Kapuk the coastal protection level is above the current design water level. Here the risk of overtopping is small. Also, the new port area of Marunda is constructed well above current (still water) the current design water level and overtopping is not likely on the short-term.

The dike protection level at Kalibaru is approximately 2.5 m +LWS2012. It is above the current design water level. However, this land reclamation is developed using coarse material. The permeability of the material is high and therefore the construction may not be able to protect the (low-laying) hinterland from inundation.

3.2. Crest elevation

The crest elevation is based on the boundary condition of the coastal area. The critical-phase sea dike is projected to protect the inland until 2030. The parameter and the value for the dike height calculation is showed in the Table 2 below. For the wave calculation, the wave model with 10,000 years return period yields significant wave height 2.27 m around location and at the shore around 1,75 m. According to IPCC projection the SLR projection is about 8mm per year and until 2030 it will predict linearly around 0.12 m [7]. The land subsidence around location is about 100 mm per year and projection until 2030 linearly is 1.4 m.

Table 2. Crest elevation.

Parameter	Value (meter)
Highest water spring	1,19
Extreme wave elevation	1,59
Sea level rise	0,12
Land subsidence in Muara Baru and Kalibaru	1,40
Free board	0,50
Dike height	+4,80 LWS

3.3. Sea dike design

The dike design at Kalibaru and Muara Baru use vertical dike at the sea side and slope dike at the land side (see Figure 6). The vertical dike at the sea side is chosen because it will reduce the space, whereas at the land side the dike has slope due to public area development later on. The dike is made from series of spun pile made from concrete with the strength K500. Between the concrete pile there is steel pile and joint the series of the concrete pile. It is constructed to provide additional strength among the series of pile form the dike and avoid seepage. The capping beam also to unite the concrete pile series and to splash the wave impact. Toe protection and armor are located in front of the dike. This is to reduce the wave impact and to avoid local scouring due to return flow of the undertow current. Geotextile is used as the mattress below the armour. In order to minimize the seepage as well, the pile is constructed with the depth around -20 m from LWS. In addition, inside the dike is filled by clay and covered by sand and grass.

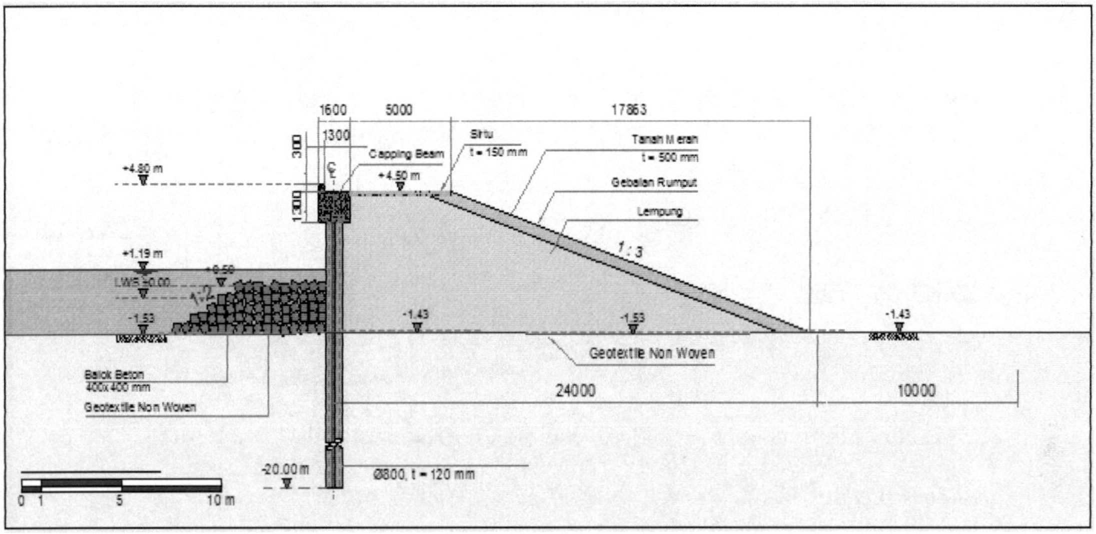

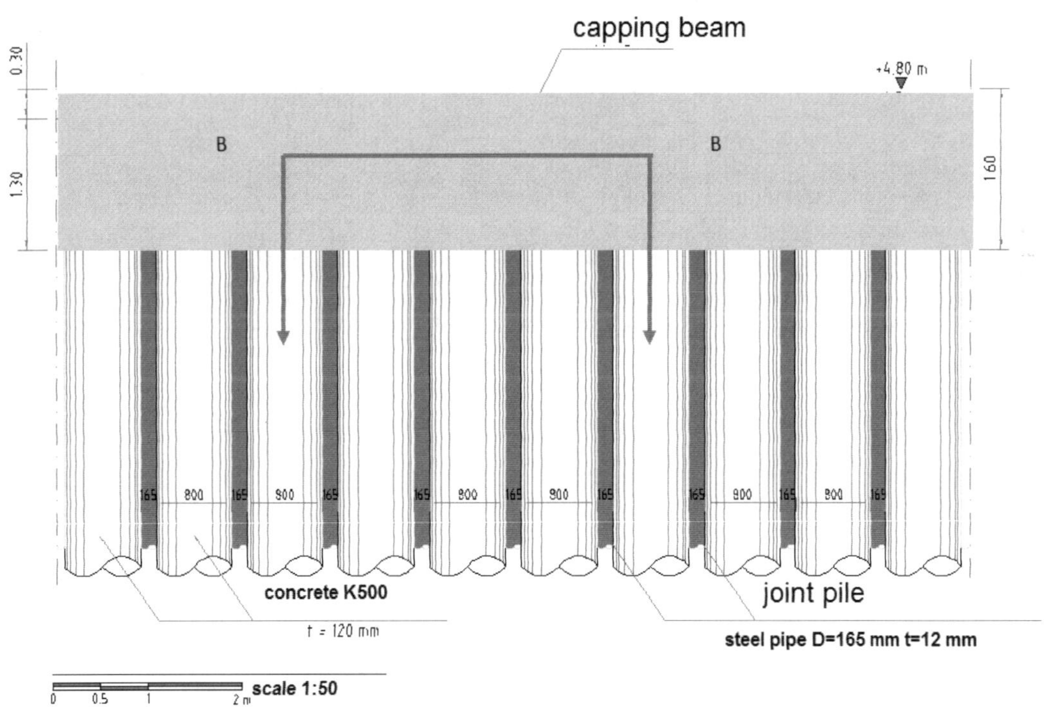

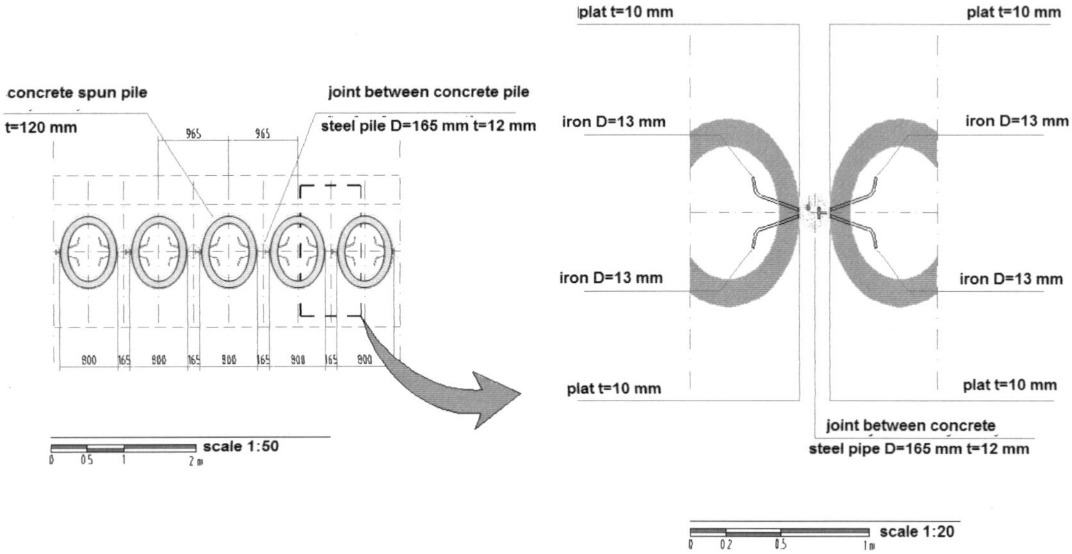

Figure 6. Critical-phase of sea dike design for Kalibaru and Muara Baru.

3.4. Physical construction

At the moment, PUPR concentrates the physical construction of sea dike at Kalibaru and Muara Baru. Figure 6 shows dike construction at Muara Baru and the green line is under PUPR responsibility and red line shows the construction progress. The construction shift to the sea side around 80-200 m from the current coastline due to the relocation issue. The determination of the distance is based on the field condition where the dikes are possible to be constructed. The physical construction dike is started by piling the concrete one by one until the depth -20 m from LWS. The distance between the concrete pile is around 165 mm.

Figure 7 shows physical dike construction in Muara Baru around Nizam Zachman Port. The pile is shipped by boat and one by one the pile is drilled to the ground until -20 m depth reached. The design of the dike also will accommodate the port design. The length of the dike in the inner site is longer than in the East side with the length difference is around 2,2 km. The dike design is the same along the coastal area using vertical dike design at the sea side and the slope dike at the land side.

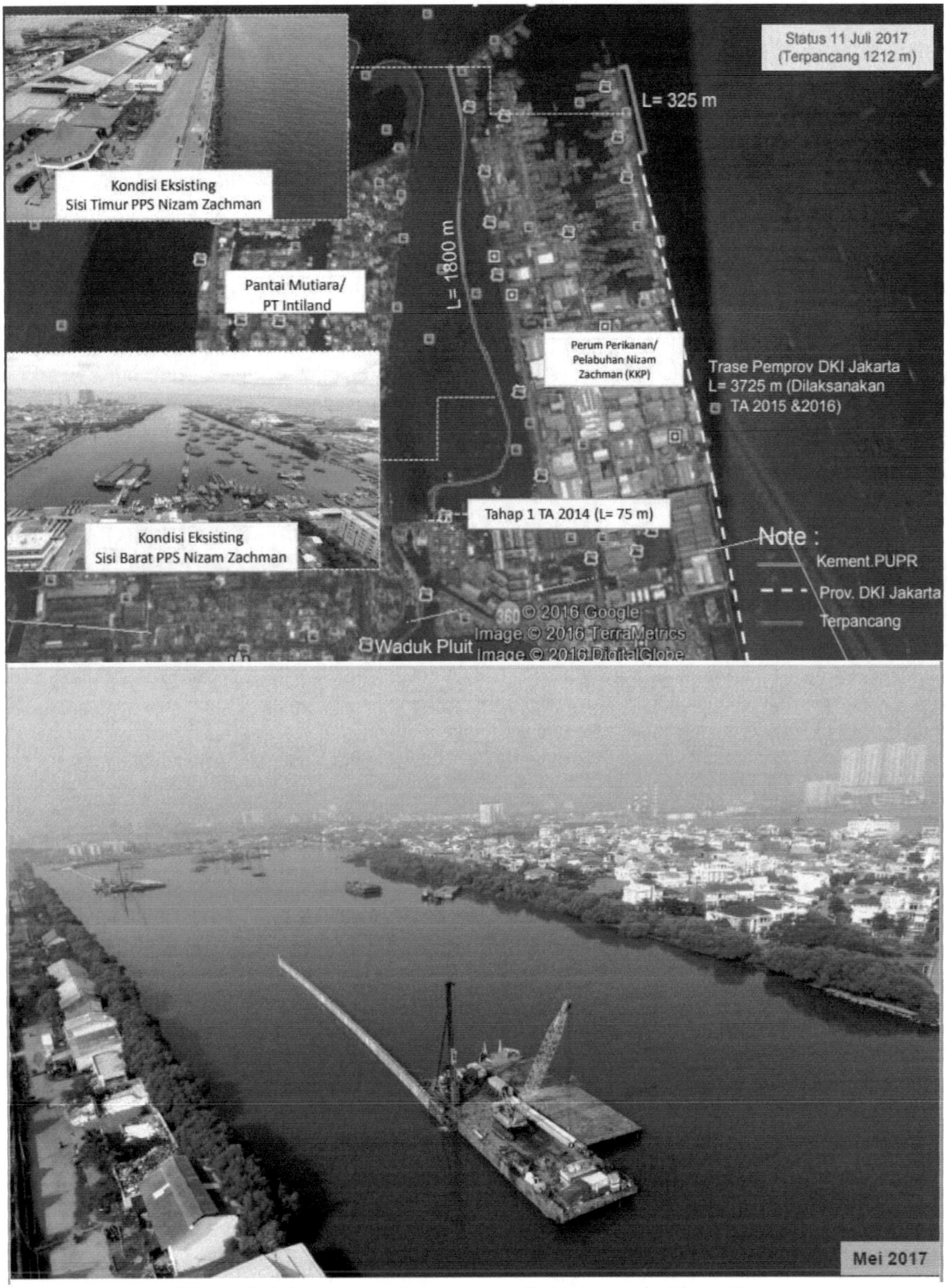

Figure 7. Dike construction area and their process in Muara Baru.

Figure 8 shows the physical dike construction progress until July 2017. The pile is already constructed then the joint steel pile will fill up between the concrete pile and it will be tied by the ring at the same time. Everyday we are able to pile 6 to 8 piles of concrete pile, depend on the field situation as well. Until July 2017, the physical dike construction progress in Muarabaru is about 1.2 km from 1.8 km the total length of the dike. It is both located in the western and eastern part. In particular area, if the soil is hard less than -20 m then the concrete pile will be excavated continuously until the depth reached. After the series pile are constructed then the cap beam is installed and the armour unit in front the dike will be deployed.

According to Presidential direction, the physical dike construction has to consider social, environment, economical, and regulation aspects. There are social issue at the site, where the fisherman cannot park their boat. During the dike construction they are facilitated by temporary port.. Therefore, all the development design must be integrated with their program. For example, if the reclamation area behind the dike is already reclaimed by the sand then the land will be used by the Ministry for fisherman facility or another processing industry for fish packaging. However, the regulation to legalize the program, especially in spatial planning, has to be discussed with regional government DKI, PUPR, and Ministry of Marine and Fisheries. Figure 9 shows spatial planning to utilize the space behind the dike, where some areas are used for relocation, fish industry, and public space.

Figure 8. Physical dike construction progress in Muara Baru.

Figure 9. Spatial planning after dike constructed.

Figure 10 shows the physical dike construction progress at Kalibaru. The dike design and their construction process are nearly the same with Muara Baru since the typical dike design is also the same, where there is concrete pile series and steel pile in between the concrete with the same dimension. The total length of dike in Kalibaru is 4.1 km, until July 2017 the progress construction is 1.3 km. In Kalibaru behind the dike will be used for public area, for example the jogging track with the sea and apartments building as the landscape around the area, inspection road, and park. In addition, there will be prototype green facility, such as garden, around 100 m length in the certain area (see Figure 11).

Social and environmental aspects are still the big challenges at the field. We put an extra effort to convince people about the urgency of the program. Communication public is the best way to communicate before the dike constructed. In addition, the social media is also used to publish about NCICD program. Sometimes social conflict is occurred at the field, where people think that the dike construction is directly related with reclamation program. Therefore, before physical construction of the dike socialisation stage of the program is compulsory process to make people understands about the critical-phase program and to have one perspective about NCICD program. In addition, the regulation and the procedural process.

Figure 10. Dike construction progress in Kalibaru.

Figure 11. Mock-up dike construction integrated with green area in Kalibaru.

4. Conclusion

Critical-phase of NCICD program is a no regret policy, where the program has to be conducted in a current situation. Critical-phase of sea dike construction is one of the program. PUPR concentrates in Muara Baru and Kalibaru for the critical dike construction at the moment with the physical construction progress around 48% from the total length 4.8 km and it will be finish in the beginning of 2019. The critical-phase of sea dike is projected enable the flood defense from sea side until 2030.

The two areas selection because these are essential and many people live there. The shifting of the dike construction to the sea side is to avoid relocation issue. The sea dike design has vertical profile at the sea side and slope at the land side. The unit armors are placed to reduce wave impact and local scouring.

The additional space behind the dike, where the reclamation is conducted, will be used for public space or residential area. However, the decision for utilization of the space has not decided yet and will wait for the agreement of another ministry.

References

[1] National Capital Integrated Coastal Development (NCICD) 2014 Technical report of NCICD

[2] NCICD-Updated Masterplan Report 2016 Technical report of NCICD

[3] Shore Protection Manual (SPM) 1984 Boundary condition of sea dike design guideline

[4] Chaussard E, Amelung F, Abidin H and Hong S 2012 Sinking cities in Indonesia: ALOS PALSAR detects rapid subsidence due to groundwater and gas extraction. *Remote Sensing of Environment* **128** 150-161

[5] Wöppelmann G and Marcos M 2016 Vertical land motion as a key to understanding sea level change and variability. *Reviews of Geophysics* **54** 64-92

[6] National Research Council (NRC) 1987 *Responding to Changes in Sea Level: Engineering Implications* (Washington DC: National Academy Press)

[7] IPCC Report 2013 Projection of sea level rise.

IOP Conf. Series: Earth and Environmental Science **162** (2018) 012021 doi:10.1088/1755-1315/162/1/012021

Nitrate and phosphate from rivers as mitigation of eutrophication in Benoa bay, Bali-Indonesia

Y Suteja[1], A I S Purwiyanto[2]

[1] Marine Science Department, Faculty of Marine and Fisheries, Udayana University, Indonesia
[2] Marine Science Department, Faculty of Math and Science, Sriwijaya University, Indonesia

E-mail: yuliantosuteja@unud.ac.id

Abstract. Benoa Bay is estuary of six rivers in Bali. In addition to carrying water, river flows also carry nutrients (nitrate and phosphate) derived from activities along the watershed. High nutrient concentrations have potential for eutrophication that leads to the Harm Alga Booms. The aims of this study were to determine the concentration, load, and tropic level of nutrients (nitrate and phosphate) in river which ends at Benoa Bay. Measurement of nutrients in water used brucine method for nitrates and ascorbic acid method for phosphate. The results were then compared with ministerial decree of the environment No 51 in 2004 for marine biota. The result show concentration of nitrate and phosphate is fluctuated along years. Nitrate concentration fluctuated form undetected to 16.950 mg/l, meanwhile phosphate concentration from undetected to 3.275 mg/l. The highest concentration of nitrate and phosphate found in dry season and the lowest concentration found in wet season. Nitrate and phosphate load to benoa bay also fluctuated in each season and river. The highest nitrate load found in Mati River during second transition season (4.387 tons/day) and also in Buaji River during dry season (4.366 tons/day). The highest phosphate load to Benoa Bay also comes from Mati River in dry season (6.980 tons/day). The Eutrophic condition for nitrate generally found in Second Transition Season, meanwhile for phosphate generally found in all river and season. The concentration of nitrate and phosphate found in all rivers and season which ends at Benoa Bay generally exceeded the quality standard base on Ministerial decree of environment No 51 in 2004 for marine biota. These conditions are potential for trigger eutrophication process in Benoa Bay.

1. Introduction

Eutrophication is a process of nutrients enrichment in aquatic environment caused by the human activities [1]. The danger effect of eutrophication is Harm alga blooms (HABs) which led to the creation of anoxic and toxic conditions in the aquatic environmental [2]. Eutrophication dominantly occurs in a closed or semi closed area (such as bay) that has high nutrient inputs and lack of water circulation [3, 4].

Benoa Bay is a semi-enclosed bay in Bali Island which has a very important function in terms of ecological and economical. Ecologically, the Benoa Bay is surrounding by mangrove (Taman Hutan Raya Ngurah Rai) which is the largest mangrove forest ecosystem in Bali. Economically, Benoa Bay is used as a port, cultivation area, catching area and for tourism activities. The condition of Benoa Bay waters is very fluctuated and strongly influenced by the activities in land or sea. The results from model show that Benoa Bay during flood got water from the Indian Ocean, while during ebb, water dominantly come from the land, especially through the river [5]. Nutrient concentration is increasing in the river estuary in Benoa Bay [6].

The river is an area that consistently drains water from upstream to downstream areas, both during the rainy and dry seasons. In addition to carrying water, the river also carry pollutants that originating from activities along watershed [7]. Population activities either directly or indirectly along the watershed such as household waste disposal, agriculture, and industrial activities will greatly affect the pollution level in river water [7, 8, 9, 10]. The downstream of river are generally embouchure in the sea, so the pollution in the river area will be transfer to the sea also [11].

Benoa Bay is estuary of six major rivers: Bualu, Sama, Mati, Badung, Buaji and Loloan Rivers (Tukad in Balinese language). The pollution conditions in these six rivers will greatly affect the condition of waters in Benoa Bay. Previous research obtained the nitrate and phosphate in Mati River estuary generally exceed quality standards based on Ministerial decree of environment No 51 in 2004 for tourism, marine biota and port especially either at during ebb and flood [12]. This indicates that the concentration of pollutants, especially nutrients that enter the Benoa Bay is quite high.

Monitoring of nutrient input into the Benoa Bay is needed to prevent nutrient enrichment that high potential for HABs. Surely this HABs event will greatly disrupt all the activity at Benoa Bay. If a preventive action is taken, the possibility for the occurrence of this disaster can be minimized. The aims of this study were to determine the concentration, load, and tropic level of nitrate and phosphate in the river which ends at Benoa Bay.

2. Methods

This research was conducted for 1 year with 4 times sampling that cover all season. Samples were collect once in every season: wet season on January 2017, first transition season I on April 2017, dry season on July 2016, and second transition season on October 2016. The sampling location was conducted in Bualu, Sama, Mati, Badung, Buaji and Loloan River (figure 1). Water Discharge from Badung River was measure in outlet of Badung Dam.

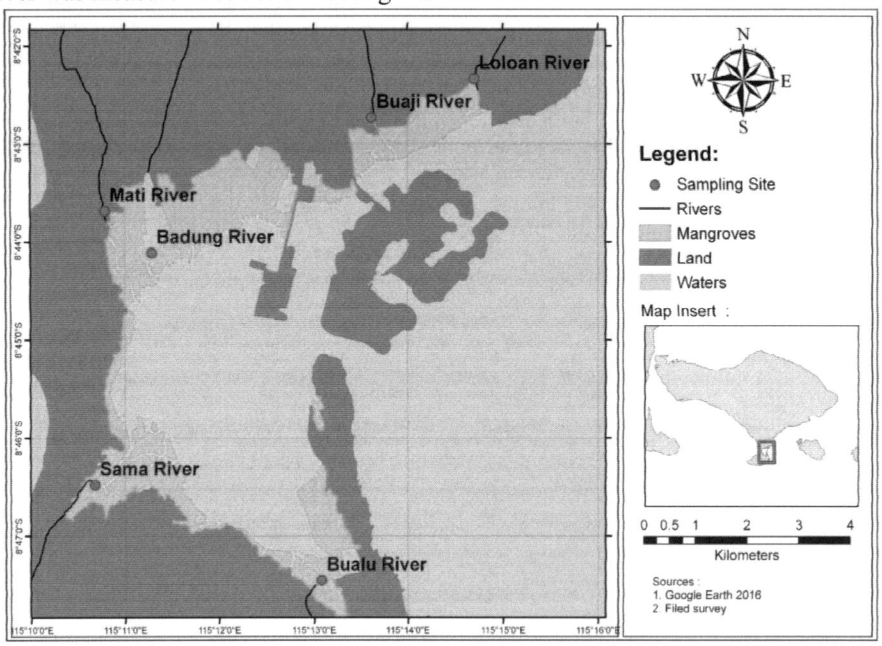

Figure 1. Map of research location, the red dot are sampling site for each river

Hydrological investigation was conducted before collecting water samples in each river. Salinity measurement using conductivity, temperature and depth (CTD) instrument is first step in hydrological investigation to ensure the water samples are all from the river. Accuracy of Castway CTD is 0.01 for salinity. If the water sample has 0 psu salinity, then measured the river water discharge using flow meter for current (v) in m/s and area measurement using ruler for width (m) and depth (m) of the river. To have uniform form for water discharge, the measurement was dividing in several areas and flows in one river. The water discharge (Q) calculated using [13]:

$$Q = v\ x\ depth\ x\ widht$$

After this step, we were collecting 600 ml water sample and stored in the cool box for later brought to the laboratory. We keep the temperature in cool box around 4°C using ice. The sample in the laboratory was filtered first using Whatman filter paper no 42, then the nutrient concentration was measured using Spectrophotometer. Nitrate was measure using brucine method and phosphate using ascorbic acid method [14]

The load of nutrient (F) from all rivers which ends at Benoa Bay was calculated using [15]:

$$F = C\ x\ Q\ x\ 10^6\ x\ 86400$$

where C is nutrient concentration (Nitrate or Phosphate) (mg/l), Q is water discharge (m^3/s), 10^6 is conversion factor from mg/l to ton/m^3, and 86400 is conversion factor from second to day).

Tropic level status of river which ends at Benoa Bay was determining using table 1 [16, 17]. Total N and P were only calculated from nitrate and phosphate only. These approaches were done with the consideration that phosphate and nitrate are the main nutrients that cause eutrophication [18]. Nitrate and phosphate concentration were also compared to the Ministerial decree of the environment No 51 in 2004 for marine biota standard. Potential eutrophication based on Ministerial decree of the environment No 51 in 2004 obtained when the nitrate and phosphate concentration has exceeded the standard of marine quality for marine biota.

Tabel 1. Trophic level in river [16, 17]

	Oligotrophic	Mesotrophic	Eutrophic
Total N (mg/L)	<0.700	0.700 – 1.500	>1.500
Total P (mg/L)	<0.025	0.025 - 0.075	> 0.075

3. Result and Discussion

3.1. Water Discharge

The water discharge form all rivers during the research are fluctuated in each river and season (figure 2). The lowest discharge found in Bualu River (0 m^3/s) in all season because Bualu River is a temporal river that only flows during the rainy season. During research, the water in Bualu River never flowed and the water samples were only taken from the stagnant water. The highest water discharge was found in Mati River (10.3 - 46.2 m^3/s) and Badung River (4.6 to 2.5 m^3/s). The high discharge in both river allegedly because these river is the longest river with the largest catchment area. Mati River is 12 km length and Badung River is 17 km length [18]. Although Badung River are longer than Mati River, the discharge form Badung River are smaller than Mati River, it is because the downstream of Badung River dammed by the Regional Water Company (PDAM) as a water source. The water discharge from Buaji and Loloan River is smaller than Mati and Badung River, this is allegedly because both rivers have short watershed. Buaji River is 4 km long and Loloan River is 5.2 km long [19].

Seasonally, the highest river floods were found during the wet season (average 16.2 $\pm$ 17.93 m^3/s) and the lowest water discharge during the first transition season (3.82 $\pm$ 4.51 m^3/s) (figure 3). The pattern of water discharge in rivers is strongly associated with the amount of precipitation, the higher precipitation the higher water discharge [20]. The precipitation pattern in Bali region is affected by the monsoon system, with the highest rainy (wet) season on December to February related to north-west monsoon characterized by moist warm air blow and the lowest precipitation in dry season on June to

July related to south-east monsoon characterized by dry wind blow [21, 22, 23]. The lowest water discharge generally found in dry season, but in this research the lowest water discharge distorted and found in first transition season. The distorted assumes do to the El Nino even in 2015-2016 that occurred earlier in the central Pacific [24]. El Nino southern oscillation (ENSO) even in Pacific Ocean also close related to the rainfall in Indonesia region [25] and driving rainfall anomalies [26].

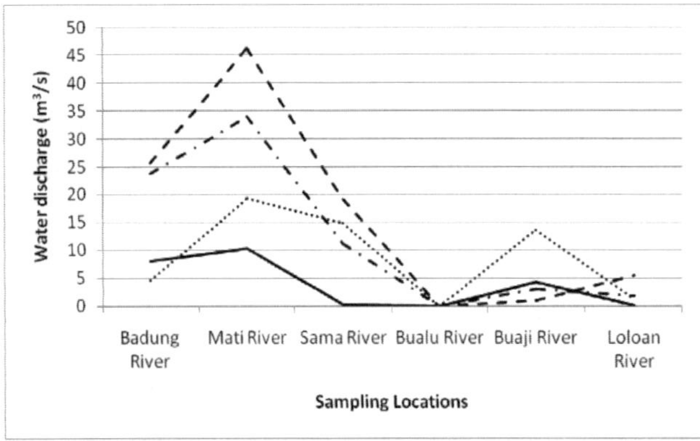

Figure 2. Water discharge in all rivers during wet season (dash line), first transition season (solid line), dry season (dash dot line), and second transition season (dot line).

3.2. Nitrate and Phosphate Concentration

Nitrate and phosphate concentration was measure in all rivers was fluctuated along year. Nitrate concentration fluctuated form undetected to 16.950 mg/l, meanwhile phosphate concentration from undetected to 3.275 mg/l (figure 4). The highest concentration of nitrate found in Buaji River in dry season (figure 4 left panel) and phosphate in Loloan River during second transition season (figure 4 right panel). The lowest nitrate (undetected) concentration found in Buaji and Loloan River in first transition season, meanwhile undetected phosphate concentration found in Bualu, Buaji and Loloan Rivers in second transition season. The fluctuation of nitrate and phosphate in Buaji River need further research and monitoring because in dry season we found the highest concentration but in the other season nitrate and phosphate concentration was undetected. During dry season measurement, the water form Buaji River is black and gave strong odor, but in other season the water was clean and less smelling. Fluctuated concentration of nitrate and phosphate in Buaji and Loloan Rivers, it's supposed there is no fix source of domestic waste in both rivers. Domestic waste affects the concentration of nutrients in waters [27].

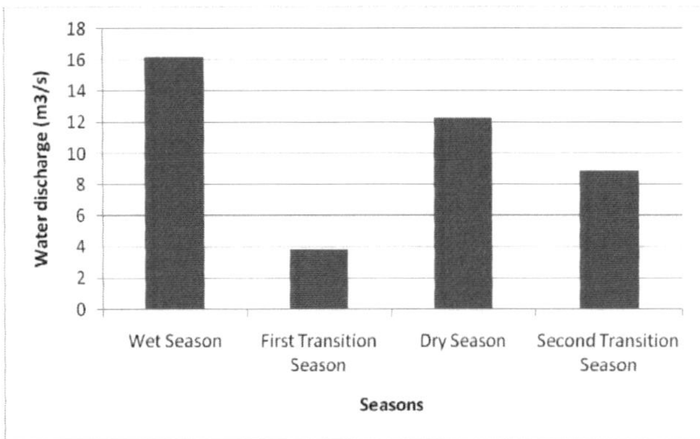

Figure 3. Average water discharge form all season

The highest concentration of nitrate in Buaji River in dry season supposed of low discharge water (3,0 m^3/s) and accumulation of nitrate from first transition season. Low discharge causes low dilution of material along river [28]. Otherwise the concentration of nitrate and phosphate are lower and relative constant in Badung and Mati Rivers which have high discharge water (figure 2). Similar result also found in Mississippi River as high river discharge has low nutrient concentration [29]. Relative constant of nitrate and phosphate concentration in both river presume because the upstream of these river is agricultural area [30] and through Denpasar City. The main source of water pollution recognize from agriculture [9, 10] that come from residues of fertilizer [31]. In organic fertilizer contain nitrogen and phosphate that use in agricultural system reach water body through leaching, flow and drainage [32].

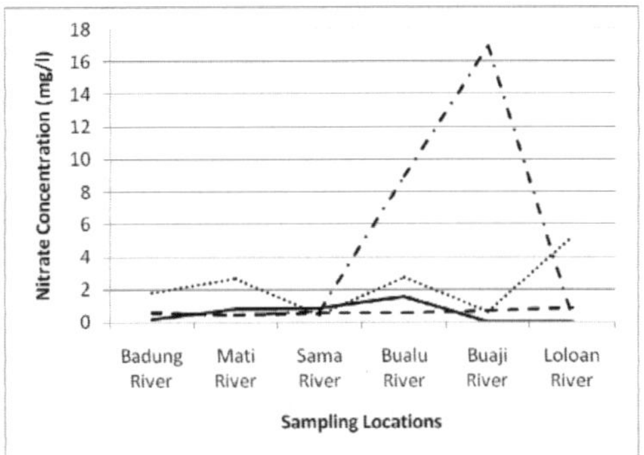

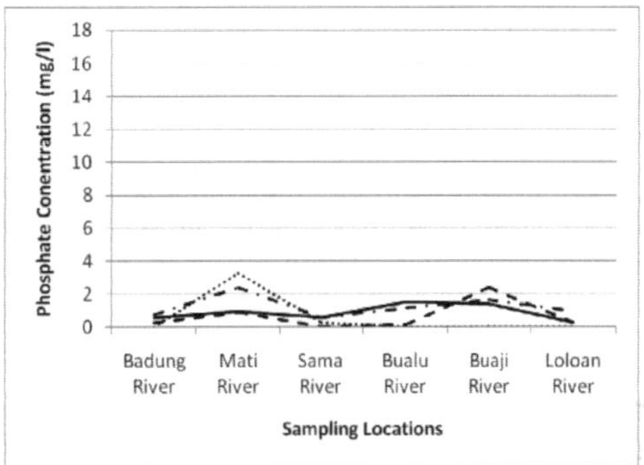

Figure 4. Nitrate (left) and phosphate (right) concentration in all rivers during wet season (dash line), first transition season (solid line), dry season (dash dot line), and second transition season (dot line).

The highest average concentration of nitrate (4.45 mg/l) and phosphate (1.26 mg/l) was found in dry season and the lowest nitrate (0.60 mg/l) dan phosfat (0.68 mg/l) was found in wet season (figure 5). The lowest concentration in wet season and first transition season assume due to high dilution by water in the river. Similar result also found in Lena River that dilution process during flood season decrease dissolved salt concentration [28]. The examples of opposite result in Yangtze River, the highest concentration found in wet season that link with dissolution of fertilizer and mother rock through runoff [33]. The average concentration of nitrate (2.00 $\pm$ 1.91 mg/l) is higher than phosphate (0.85 $\pm$ 0.30 mg/l) in all river and season. The high concentration of nitrate is thought due to characterize of nitrate in some type of soil is easier leached than other contaminant [34].

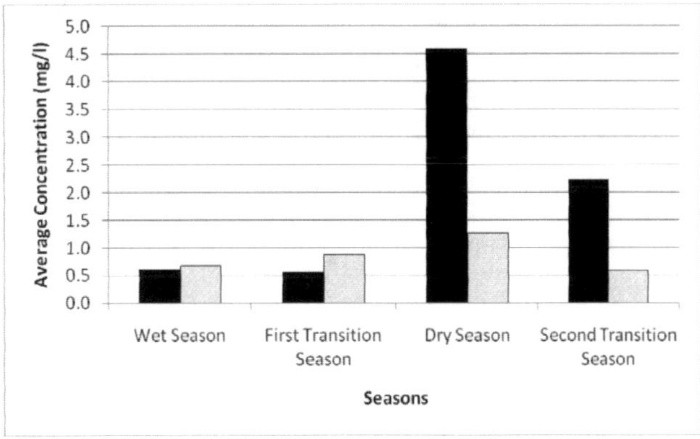

Figure 5. Average concentration of nitrate (black) and phosphate (grey) in all seasons

3.3. Nitrate and Phosphate Load

Nitrate and phosphate load along year is fluctuated in each season and river (figure 6). The lowest nitrate and phosphate load (0 tons/day) found in Bualu River in all season, it's because zero water discharge from this river. The highest nitrate load found in Mati River during second transition season (4.387 tons/day) and also in Buaji River during dry season (4.366 tons/day). The highest phosphate load to Benoa Bay also comes from Mati River in dry season (6.980 tons/day). Badung and Mati River need more special attention because it gives a great load throughout the season

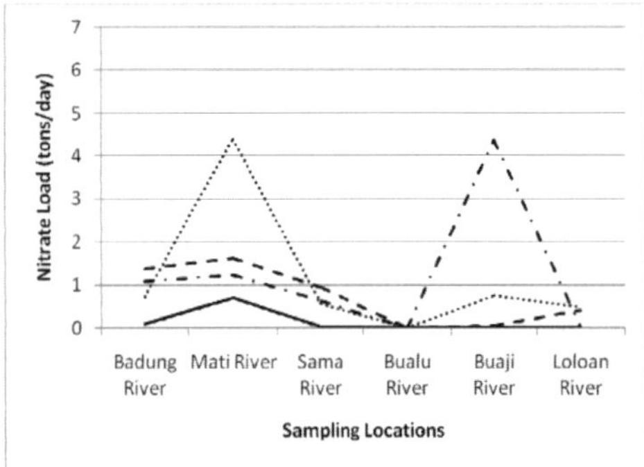

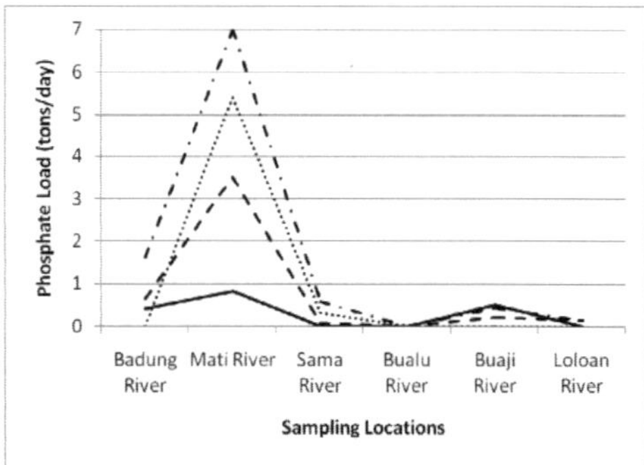

Figure 6. Nitrate (left) and phosphate (right) loads in all rivers during wet season (dash line), first transition season (solid line), dry season (dash dot line), and second transition season (dot line).

The average load of nitrate and phosphate in all season is fluctuated along year (figure 7). The lowest nitrate (0.82 tons/day) and phosphate (1.76 tons/day) loads found in first transition season, meanwhile the highest of nitrate (7.38 tons/day) and phosphate (9.77 tons/day) loads found in dry season. Opposite result found in Jakarta bay, the nutrient load is higher two times in rainy season than dry season because of high water discharge from river [35]. The nutrient load fluctuation to Benoa Bay not only affected by the concentration of nutrient, but also by water discharge from each river.

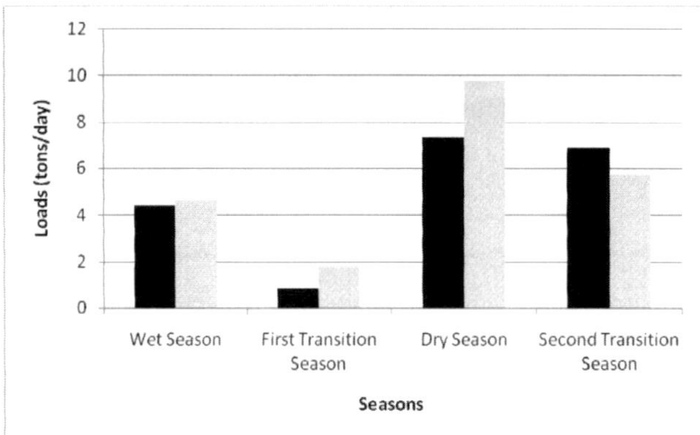

Figure 7. Average loads of nitrate (black) and phosphate (grey) in all seasons

3.4. Eutrophication Status

Euthropication status for nitrate and phosphate is fluctuated in all river and season (table 2). Nitrate concentrations during the wet season almost all in oligotrophic status, opposite result in second transition season show almost all of the rivers in eutrophic condition. In dry season, eutrophic status found in Buaji dan Bualu River, while the other river in oligotrophic condition. Bualu River almost eutrophic throughout the season except in wet season, but there is no load to Benoa Bay. Eutrophic status in Second Transition Season in both Badung and Mati River need more attention because this river has constant water discharge along year. Trophic level status based on Total N in all rivers and season can be increased considering that the data in this research calculated only from nitrate only without nitrite, ammonia and organic bound nitrogen data. The status level of phosphate indicates that phosphate is in eutrophic condition almost in all rivers and season. Oligotrophic status of phosphate only found in second transitional season. High phosphate concentrations in Badung and Mati River are thought derived from agricultural waste in the upstream [36].

The concentration of nitrate and phosphate found in all rivers and season which ends at Benoa Bay generally exceeded the quality standard base on Ministerial decree of environment No 51 in 2004 for marine biota. The Indonesian environment ministry only allows maximum concentration of nitrate is 0.008 mg/l and phosphate is 0.015 mg/l for marine biota. The nitrate and phosphate concentration from all rivers are potential for trigger eutrophication process in Benoa Bay.

Table 2. Nitrate and phosphate trophic level status base on Total N and P during Wet Season (WS), First Transitional Season (FTS), Dry Season (DS) and Second Transitional Season (STS)

	Nitrate				Phosphate			
	WS	FTS	DS	STS	WS	FTS	DS	STS
Badung River	*	*	*	***	***	***	***	*
Mati River	*	**	*	***	***	***	***	***
Sama River	*	**	*	*	**	***	***	***
Bualu River	*	***	***	***	***	***	***	*
Buaji River	*	*	***	*	***	***	***	*
Loloan River	**	*	*	***	***	***	***	*

Oligotrophic (*), Mesotrophic (**), Eutrophic (***)

4. Conclusion

The concentration of nitrate and phosphate is fluctuated along years. Nitrate concentration fluctuated form undetected to 16.950 mg/l, meanwhile phosphate concentration from undetected to 3.275 mg/l. The highest concentration of nitrate and phosphate found in dry season and the lowest concentration found in wet season. Nitrate and phosphate load to Benoa Bay along year also fluctuated in each season and river. The highest nitrate load found in Mati River during second transition season (4.387 tons/day) and also in Buaji River during dry season (4.366 tons/day). The highest phosphate load to Benoa Bay also comes from Mati River in dry season (6.980 tons/day). The concentration is strongly affected by concentration and water discharge. The Eutrophic condition for nitrate generally found in Second Transition Season, meanwhile for phosphate generally found in all river and season. The concentration of nitrate and phosphate found in all rivers and season which ends at Benoa Bay generally exceeded the quality standard base on Ministerial decree of environment No 51 in 2004 for marine biota. These conditions are potential for trigger eutrophication process in Benoa Bay.

Acknowledgments

Author gratefully appreciated to all Laboratorium Analitik technician and student who assist in sampling and measuring process.

References

[1] Nixon S W 1995 Coastal marine eutrophication: A definition, social causes, and future concerns *Ophelia* **41** 199–219

[2] Anderson D M, Glibert P M, Burkholder J M 2002 Harmful Algal Blooms and Eutrophication: Nutrient Sources, Composition, and Consequences *Estuaries* **25(4b)** 704–726

[3] Yang X-e, Wu X, Hao H-l, He Z-l 2008 Mechanisms and assessment of water eutrophication *J. Zhejiang Univ. Sci.* B **9(3)** 197-209
[4] Karydis M 2005 Eutrophication Assessment Of Coastal Waters Based On Indicators: A Literature Review *Global NEST J.* 7(2) 228-235
[5] Hendrawan I G, Asai K 2014 Numerical study on tidal currents and seawater exchange in the Benoa Bay, Bali, Indonesia *Acta Oceanol. Sin.* **33(3)** 90-100
[6] Rahayu N W S T, Hendrawan I G, Suteja Y 2018 Distribusi Nitrate dan Fospat secara Spasial dan Temporal saat Musim Barat di Permukaan Perairan Teluk Benoa, Bali *J. Mar. Aquat. Sci.* **4(1)** 1-13
[7] Bhuiyan A B, Mokhtar M B, Toriman M E, Gasim M B, Ta G C, Elfithri R, Razman M R 2013 The Environmental Risk And Water Pollution: A Review From The River Basins Around The World *Am-Eurasian J. of Sustai. Agr.* **7(2)** 126-136
[8] Bogen J 2009 The impact of environmental changes on the sediment loads of Norwegian rivers *CATENA* **79(3)** 251-256
[9] Billen G, Garnier J, Lassaletta L 2013 *Philos. T. R. Soc. B.* **368** 1–13
[10] Fowler D, et al. 2013 The nitrogen cascade from agricultural soils to the sea: modeling nitrogen transfers at regional watershed and global scales *Philos. T. R. Soc. B.* **368** 20130164.
[11] Owa FD 2013 Water Pollution: Source, Effects, Control and Management *Mediterr. J. Soc. Sci.* **4(8)** 65-68
[12] Suteja Y and Dirgayusa I G N P 2015 Analisis Beban Pencemar dan Kapasitas Asimilasi Muara Tukad Mati – Bali *Proc. Seminar Nasional Sains dan Teknologi 29 – 30 Oct* (Bali)
[13] Michaud J P and Wierenga M 2005 *Estimating Discharge and Stream Flows. A Guide for Sand and Gravel Operators*. Ecology Publication Number 05-10-070
[14] APHA 2005 *Standard Method for the Examination of Water and Wastewater. 21st edition* Eaton A D, Clesceri L C, Rice E W, Greenberg A E, editor. United State of America: American Public Health Association.
[15] XiuLi Y, WeiDong Z, HuaSheng H, Yan L, WeiDong G, Xiao H 2012 Distribution, fluxes and decadal changes of nutrients in the Jiulong River Estuary, Southwest Taiwan Strait *Oceanology* **57(8)** 2307 – 2318
[16] Dodds W K, Jones J R, Welch E B 1998 Suggested classification of stream trophic state: distributions of temperate stream types by chlorophyll, total nitrogen, and phosphorus W*ater Res.* **32** 1455–1462
[17] Dodds W K, Smith V H 2016 Nitrogen, phosphorus, and eutrophication in streams *Inland Waters* **6** 155-1164
[18] Khan F A, Ansari A A 2005 Eutrophication: An ecological vision *The Botanical Rev.* **71(4)** 449–482
[19] Central Bureau of Statistics (BPS) Bali Province 2015 *Bali In Figures 2015* Integrated Processing and Statistic Dissemination Divison (Bali) Statistic of Bali Province p 616
[20] Pandzic K, Cesarec K, Grgic B 1997 An Analysis of the Relationship Between Precipitation and Discharge Fields Over a Karstic River Basin *Int. J. Climatol.* **17** 891-901
[21] Aldrian E and Djamil Y S 2008 Spatio-temporal climatic change of rainfall in East Java, Indonesia. International Journal of Climatology *Int. J. Climatol* **28** 435–448.
[22] As-syakur A R, Tanaka T, Prasetia R, Swardika I K, Kasa I W 2011 Comparison of TRMM multisatellite precipitation analysis (TMPA) products and daily-monthly gauge data over Bali *Int. J. Remote Sens* **32(24)** 8969–8982
[23] Prasetia R, As-syakur A R, Osawa T 2013 Validation of TRMM Precipitation Radar satellite data over Indonesian region *Theor Appl Climatol* **112** 575–587
[24] Jacox M G, Hazen E L, Zaba K D, Rudnick D L, Edwards C A, Moore A M, Bograd SJ 2016 Impacts of the 2015–2016 El Niño on the California Current System: Early assessment and comparison to past events *Geophy ResLett* **43(13)** 7072–7080
[25] Aldrian E, and Susanto R D 2003 Identification of three dominant rainfall regions within

Indonesia and their relationship to sea surface temperature *Int. J. Climatol* **23** 1435–1452.

[26] Hendon H H 2003 Indonesian rainfall variability: impacts of ENSO and local air–sea interaction *J. Climate* **16** 1775–1790

[27] Ololade I A, Adewunmi A, Ologundudu A, Adeleye A 2009 Effects of household wastes on surface and underground waters *Int. J. Phys Sci.* **4(1)** 022-029

[28] Gordeev V V and Sidorov L S 1993 Concentrations of major elements and their outflow into the Laptev Sea by the Lena River *Mar Chem* **43** 33-45

[29] Dinnel S P and Bratkovich A 1993 Water discharge, nitrate concentration and nitrate flux in the lower Mississippi River *J. Mar. Syst.* **4** 315-326

[30] As-syakur A R, Suarna I W , Adnyana I W S, Rusna I W, Ulksmiwati IAA, Diara IW 2010 Identifikasi Hubungan Fluktuasi Nilai SOI Terhadap Curah Hujan Bulanan Di Kawasan Batukaru-Bedugul, Bali *Jurnal Bumi Lesatri* **10(2)** 200-207

[31] Lawniczak A E, Zbierska J, Nowak B, Achtenberg K, Grześkowiak A, Kanas K 2016 Impact of agriculture and land use on nitrate contamination in groundwater and running waters in central-west Poland *Environ. Monit. Assess.* **188** 172

[32] Savci S 2012 An Agricultural Pollutant: Chemical Fertilizer *Int. J. Env. Sci. Dev.* **3(1)** 77-80

[33] Li M, Xu K, Watanebe M, Chen Z 2007 Long-term variations in dissolved silicate, nitrogen, and phosphorus flux from the Yangtze River into the East China Sea and impacts on estuarine ecosystem *Estuar. Coast. Shelf S.* **71** 3-12

[34] Devito K J, Fitzgerald D, Hill A R, Aravena R 2000 Nitrate Dynamics in Relation to Lithology and Hydrologic Flow Path in a River Riparian Zone *J. Environ. Qual.* **29** 1075-1084

[35] Nugrahadi M S, Yanagi T, Tejakusuma I G, Adi S, Darmawan R A 2010 Seasonal variations of nutrient budgets in Jakarta Bay, Indonesia *Mar. Res. Indones.* **35(1)** 9-20

[36] Wijana I M S, Dewi I G A K S P, As-Syakur A R 2012 Studi Dan Pemantauan Kualitas Air Di Daerah Aliran Sungai (DAS) Badung *Ecotrophic* **7(1)** 54-58.

IOP Conf. Series: Earth and Environmental Science **162** (2018) 012022 doi:10.1088/1755-1315/162/1/012022

Identification of *Decapterus sp.* potential fishing grounds in Java and Western Kalimantan Seas

Avrionesti and M R Putri

Department of Earth Science, Faculty of Earth Science and Technology, Bandung Institute of Technology, Bandung, Indonesia

E-mail: avrionesti.ramadhan@gmail.com

Abstract. *Decapterus sp.* is one of the main targeted species for small pelagic fish in Java and western Kalimantan Seas (Karimata Strait and South China Sea). *Decapterus* sp. mainly catch at around 40-70 meters depth. They live in warm waters (more than 26°C) with salinity greater than 30 ppt, pH greater than 7.75 , and concentration of dissolved oxygen (DO) greater than 4.2 ml/l. Datas were obtained from World Ocean Atlas 2013, NCEP, and result of numerical model. The aim of the research was to investigate the seasonal variation of decapterus sp. potential fishing ground. The results show that variation of seasonal oceanographic conditions has an impact to the potential fishing ground pattern of decapterus sp. The catching prediction area was at 20 m depth, except when second transition season which was at 40 meters depth. Fishing period should be done during the rainy season according to the widest area of potential fishing ground.

1. Introduction

Decapterus sp. (scad fish) is the main catch fish in Karimata Strait and Java Sea for small pelagic fishes. Based on fishery statistics of Ministry of Marine Affairs and Fisheries Indonesia, over ten years (2004-2014), scad fish has accounted for almost 23% (346,645.9 tons / year) from total capture of small pelagic fish or about 7.75% of the total Indonesian catch [1]. Growth of scad capture itself reached 1.81% per year within that period. In terms of production value, this species contributed almost Rp 2,356,000,000,000 per year or 5.1% of the total value of fish production in 2004-2014 [1]. This suggests that scad fish is one of the national major catch and has a play important role in the coastal economy enhancement.

There are several species of scad fishes in Indonesia, such as Decapterus russelli, D. macrosoma, D. macarellus, and D. kurroides. Decapterus russelli and Decapterus macrosoma are commonly found in the exposed waters of Sunda Sea (South China Sea, Java Sea) and Arafura Sea, as well as in coastal waters [2]. Based on Pastoral et al, Decapterus sp. lives in warm waters (greater than 26°C) with salinity greater than 30 ppt, pH value greater than 7.75, and concentration of dissolved oxygen (DO) greater than 4.2 ml/l [3].

Oceanography characteristic plays an important role in Indonesia Sea, which is influenced by the seasons. Therefore, the characteristic of potential fishing ground will also affected by seasonal changes. The objective of the research was to investigate the seasonal variation of Decapterus sp. potential fishing ground.

2. Methods

The study area was located at Fisheries Management Areas (FMA) 711 and 712 (Figure 1) which was covered western Kalimantan Seas and Java Sea. Several sources of data were used in this study including oceanography and climatology data. Dissolve oxygen (ml/l), temperature (°C), and salinity (ppt) data were obtained from World Ocean Atlas 2013 with 1°x1° resolution. Those datas is average condition from 2005-2012 whose calculated by World Ocean Atlas 2013 (WOA 13) [4].

Results of numerical model was used to extract pH value [5]. We assumed the value of pH is considered equal from the surface to the seabed, adjusting to the numerical model we referred. The month grouping refers to the seasonal groupings used by WOA 13 where rainy season (boreal winter) consist of January-February-March, first transition season (boreal spring) consist April-May-June, dry season (boreal summer) consist July-August-September, and second transition season (boreal autumn) consist October-November-December. The rainfall data from NCEP was also used in this study as supporting data analysis [6]. The average rainfall calculated from 2005-2012 in accordance with average year used by WOA 13.

Fishing ground map was depicted using Ocean Data View at 20 m, 40 m, and 70 m depth. This approach was based on Pastoral et al. [3]; whereas the highest catch of Decapterus sp. in Sulu Sea predicted about 40-70 m depth, while in the South China Sea found in shallow water about 20 m depth.

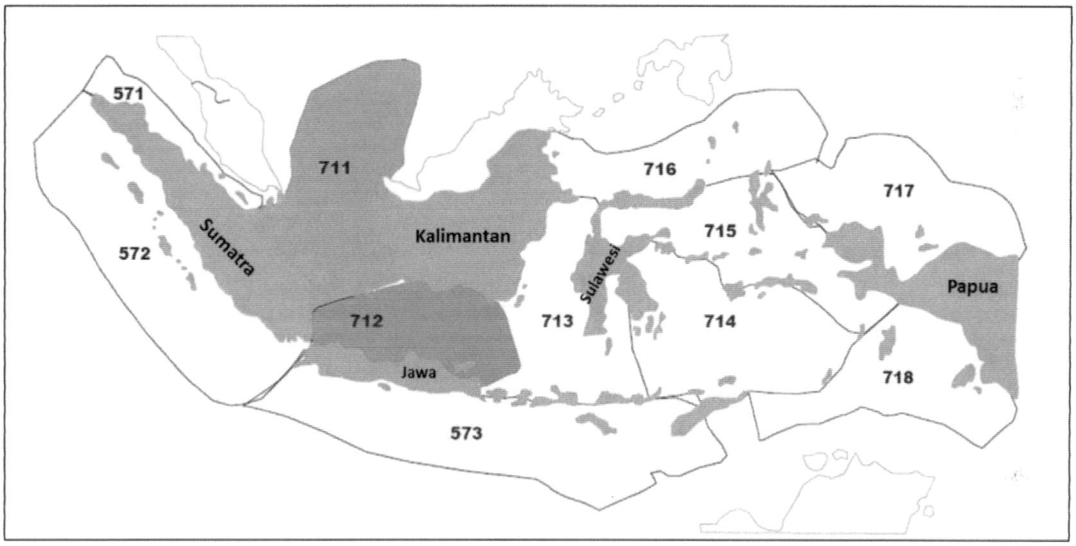

Figure 1. Zoning of Fisheries Management Area in Indonesia [7]

3. Results and Discussion

3.1. Characteristic of study area

Java Sea and western Kalimantan seas have shallow bathimetry with varying depths. The depth in the eastern waters of Kalimantan reaches 70 m around the bay of Thailand and is shallower to the south. While in the Java sea has a depth of 20 m off the coast of south Sumatra to more than 60 m in its eastern part [8]. This shallow bathymetry causes the sea to tend homogeneous. The study area also affected by monsoon so the oceanography paramaeters will have seasonal variation. During rainy season, the wind blows northwest and the current will move southward. This current will bring freshwater from study area penetrated eastern part of Indonesian waters and affected another parameter, such as salinity and chlorophill [8][9].

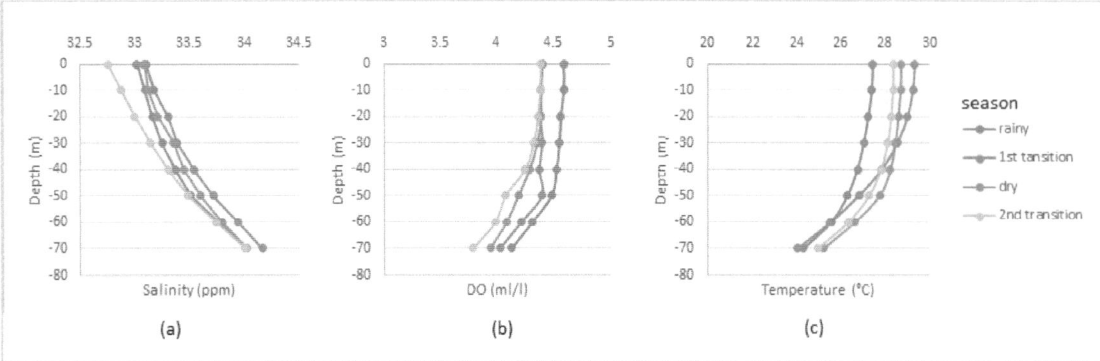

Figure 2. Vertical distribution of (a) salinity, (b) dissolve oxygen (DO), and (c) temperature in study area

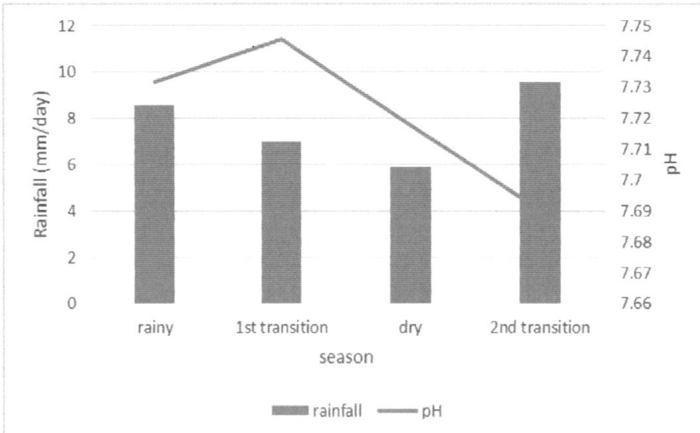

Figure 3. Seasonal Average Rainfall in Study Area (2005-2012)

Vertical distribution of several parameters we used can be seen in Figure 2 along with its seasonal variation. The data were obtained from WOA 13 which is the average condition in 2005-2012. During rainy season, the temperature reach its lowest value and tend to high value of DO. High precipitation during the rainy season (figure 3) will also decrease the salinity. But, as shown in figure 2 and 3, the highest precipitation occur in second transition season. It makes the salinity in second transition season has lowest value among all season. The pH value also has seasonal variation. The high precipitation will decrease pH value because the addition of freshwater with lower pH value (figure 3).

3.2. Potential fishing ground

Based on definition from Indonesia Ministry of Marine Affairs and Fisheries, fishing is an economic activities to catch or collect fishes/other aquatic animals/aquatic plants that grow naturally in inland openwater/marine areas and no belong to the property of any person [1]. Whereas, based on definition from Philippine Statistic Authority, fishing ground refers to areas in any body of water where fish and other aquatic resources congregate and become target of capture [10]. So, we can conclude that potential fishing ground is areas in any body of water where potentially profitable for economic activities to catch or collect fishes/other aquatic animals/aquatic plants. In this paper, we depict the potential fishing ground for decapterus sp.

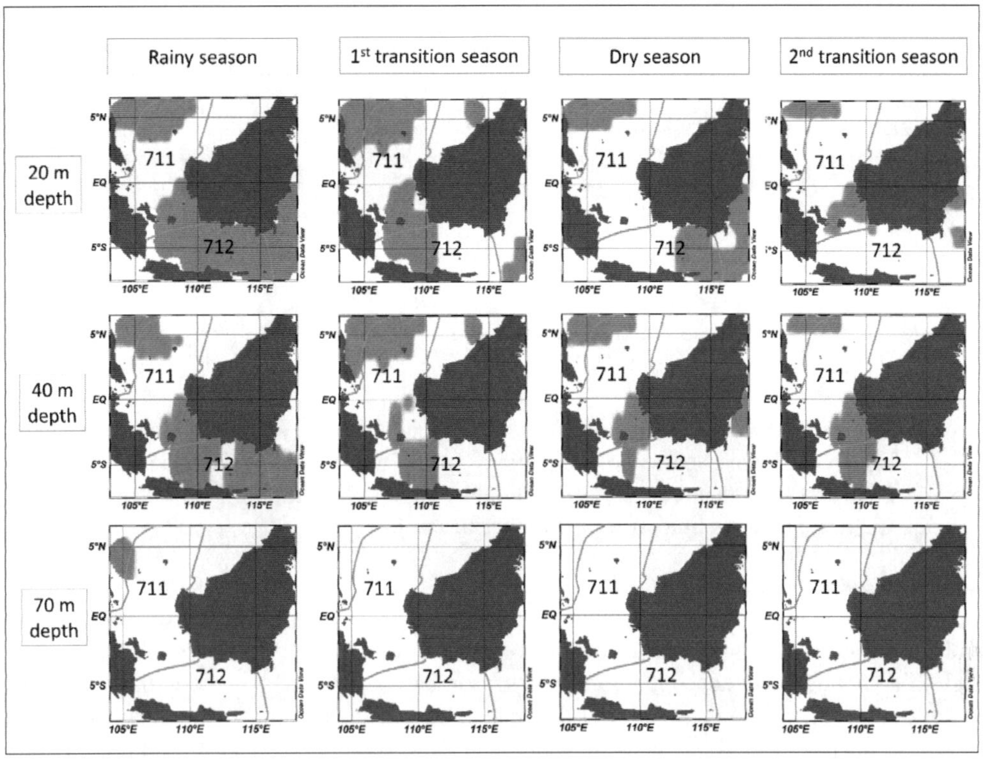

Figure 4. Seasonal Potential Fishing Ground In Study Area At 20 m, 40, and 70 m depth

The green area in figure 4 is the potential fishing ground for decapterus sp based on average condition during 2005 – 2012. Based on Figure. 4 each season had a different pattern of potential fishing ground. In general potential fishing ground in FMA 712 was distributed wider than FMA 711. Consequently the amount of capture fisheries production of FMA 712 was showed higher based on the fishery statistics of Indonesia Ministry of Marine Affairs and Fisheries. As shown in figure 5, the capture in FMA 711 reached 2,564 tons in 2010 and 5.694 tons in 2014. This total capture was far below the capture of FMA 712 which reached 81,690 tons in 2010 and 84,443 in 2014 [1][11]. In general, Java sea has higher pH value than at western Kalimantan. In dry season and autumn, the area with pH value less than 7.75 spreads in the western waters of Kalimantan. This indicated that the Java sea to be a better habitat for the scad fish. During the rainy season, lowest SST has correspond to highest DO among all season (figure 2). It makes the highest potential fishing ground occurs in rainy season season. Not also wider, the potensial fishing ground also reach 70 m depth in small part of west Karimata Strait (figure 4). While the narrower potential fishing ground occurred at second transition season correspond to the low pH value due to high rainfall (figure 3).

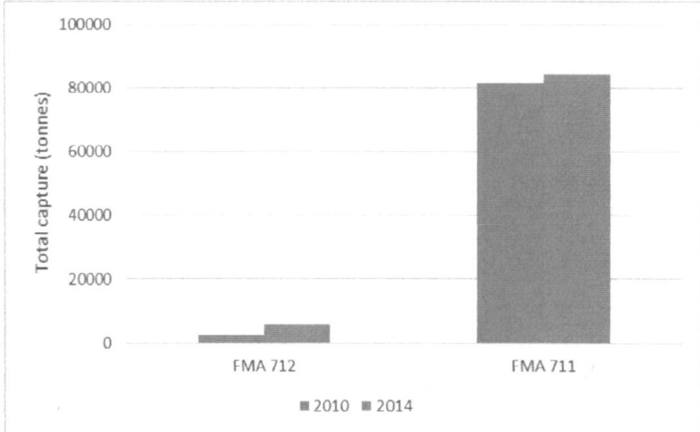

Figure 5. Total captured of decapterus sp in FMA 711 and FMA 712 at 2010 and 2014 (source: Capture Fisheries Statistics Of Indonesia, 2010 and Capture Fisheries Statistics Of Indonesia, 2014)

4. Conclusions and Further Work

Based on the results the fishing period should be done during the rainy season. This is based on the area of potential fishing ground that almost covers the entire study area. We also conclude that the least potential fishing period occur at second transition season based on the narrow potential fishing ground.

This study still has many deficiency. For the next research, the authors will include the chlorophyll factor as the scad's food and also ocean currents to get a more accurate map. In addition, the total fish catch referred to is calculated based on the landing port of those fish. So, it is not quite accurate to describe the abundance of fish in certain waters.

Acknowledgement

This research was funded by Research and Innovation, Institut Teknologi Bandung – 2017 entitled "The Ocean Current System of Indonesian Waters and its effects on Marine Fisheries Production". We also thank to Master Program of Earth Science, Faculty of Earth Science and Technology, Institut Teknologi Bandung.

References

[1] Directorate General of Capture Fisheries 2011 Capture Fisheries Statistics Of Indonesia, 2010 ISSN :1858-0505

[2] Suwarso and Zamroni, A 2014 Population Structure Analyses of Three Species of Layang (Decapterus Spp.) In The Java Sea and Makassar Strait: Sustainable Management Options Of Small Pelagic Fishes And Evaluation Of FMA

[3] Pastoral P C, Escobar S L and Lamarca N J 2000 Round Scad Exploration by Purse Seine in the South China Sea, Area III: Western Philippines. *Proceedings of the SEAFDEC Seminar on Fishery Resources in the South China Sea, Area III: Western Philippines*

[4] Boyer T P, J I Antonov, O K Baranova, C Coleman, H E Garcia, A Grodsky, D R Johnson, R A Locarnini, A V Mishonov, T D O'Brien, C R Paver, J R Reagan, D Seidov, I V Smolyar and M M Zweng 2013: World Ocean Database 2013, NOAA Atlas NESDIS 72, S Levitus, Ed., A Mishonov, Technical Ed.; Silver Spring, MD, 209 pp., http://doi.org/10.7289/V5NZ85MT

[5] Putri M R, Setiawan A and Safitri M 2015 Variation of Ocean pH in Indonesia Waters *AIP Conference Proceedings 1677, 060021 (2015); doi: 10.1063/1.4930701*

[6] Kalnay et al., The NCEP/NCAR 40-year reanalysis project, Bull. Amer. Meteor. Soc., 77, 437-470, 1996

[7] Suman A, Wudianto, Sumiono B, Badrudin, Nugroho D, Merta G S, Suwarso, Taufik M, Amri K, Kembaren D, Priyatna A, Setiaji E, Prihantara S, Prihatiningsih, Chodrijah U, Fauzi M, Ernawati T and Rahmat E 2014 Potensi Dan Tingkat Pemanfaatan Sumberdaya Ikan Di Wilayah Pengelolaan Perikanan Republik Indonesia (WPP RI). ISBN : 978-602-17996-3-5

[8] Wyrtki K 1961 Physical oceanography of the Southeast Asian waters, Naga Report 2. Scripps Institution of Oceanography, 195 pp.

[9] Gordon A L 2005 Oceanography of the Indonesian Seas and Their Throughflow. *Oceanography* Vol. **18**, No. 4.

[10] www.countrystat.psa.gov.ph, accessed at December 7, 2017

[11] Directorate General of Capture Fisheries (2015). Capture Fisheries Statistics Of Indonesia, 2014. ISSN :1858-0505

IOP Conf. Series: Earth and Environmental Science **162** (2018) 012023 doi:10.1088/1755-1315/162/1/012023

Microplastics in Sumba waters, East Nusa Tenggara

M R Cordova and U E Hernawan

Research Center for Oceanography, Indonesian Institute of Science (LIPI). JL. Pasir Putih 1, Ancol, Jakarta Utara, Jakarta, 14430, Indonesia

E-mail: muhammad.reza.cordova@lipi.go.id

Abstract. The accumulation of plastic debris in the oceans has been widely recognized as a threat to marine environment. A recent study estimated that Indonesia is one of the biggest sources of plastic wastes in the ocean, but directly-measured abundance data from the seawater in Indonesia is lacking. We documented the abundance and distribution of microplastics (size <5mm) in sub-surface seawaters of Sumba, a pristine region in Indonesia. Water samples were collected from 5 m, 50 m, 100 m, 300 m depth and near the sea bottom. Samples were examined for microplastics using flotation and filtration methods. We found microplastic in all sampling locations, consisting of fibers (45.45%), granules (36.36%) and other plastic form (18.18%). Most of microplastic particles were found at water depths less than 100 m (81.82%), which was the thermocline area. Our finding corroborates the believe that plastics has widely invaded marine environment in different parts of the seas and oceans, including pristine, remote, and unknown areas.

1. Introduction

Plastics are a successful story with regard to their characteristics that make them suitable for a wide variety of products [1]. By 2014 the total plastic production reached 311 million tons [2]. High consumption rate but low recovery rates has driven plastics to be a potential threat to the environment. [3] reported that most of plastic packaging is not recovered, of which 40% goes landfilled and 32% leaks to the environment including marine ecosystem. Plastic pieces that ended up in the environment remain still with very slow degradation (up to 100 years). Recent studies estimated that plastic debris in the ocean are between 7,000 – 250,000 metric tons [4,5]; land based input to the oceans are between 4.8 - 12.7 metric tons; and Indonesia was listed as one of the top sources of land based input [6]. This estimation, however, lacks real abundance data in the ocean since it used mathematical model based on solid waste production, population density and economic status of the countries.

Plastic pollution was initially seen as an aesthetic problem [7,8], however recent studies have shown that marine animals can be negatively affected by the presence of plastics. Threats posed by plastics with large size are obvious, clear environmental risks by physical impairment after swallowing, entanglement [9]. [10] stated that from the 1960s to 1990s, physical impairment caused by plastics had increased almost three times (267 to 693 species). Furthermore, [9] reported that all seabird species (395 species) contained plastics in their digestive systems.

Plastic waste can be degraded by UV thermal oxidation or mechanic processes up to the microscopic size [11,12]. Microplastics (<5mm) are formed by the physical, chemical and biological fragmentation of larger items of plastics. Besides commonly-used plastics, microplastic may originate from cosmetic or fabric, in the form of microbeads [13,14]. The impact of microplastics in the marine environment are not clearly known. Studies in the last decade have indicated potential environmental risks from microplastics, but real consequences are mostly unknown. Microplastics accumulation in gastrointestinal on marine organism has been reported [15–19], but the impact of the accumulation to the organisms are not known. Microplastics could enter the food web by organism ingestion [20], could interference digestive tract function [17], and may act as a carrier for organic material and heavy metals [21–23].

Sumba, located in East Nusa Tenggara, Indonesia, is a pristine region in the outlet of Indonesian Through Flow (ITF) connecting the Pacific Ocean and the Indian Ocean. It is in the southern part of the Wallacea transition zone, where Indo-Malaya and Australasian biogeographic characteristics meet. It has unique marine ecosystems and high biodiversity. This area is an important migration corridor to many marine megafaunas and pelagic fishes. Here, we documented the abundance and distribution of microplastics (size <5mm) in sub-surface seawaters of Sumba.

2. Introduction

2.1. Study area

Sumba is located in East Nusa Tenggara, eastern Indonesia (Figure 1). Sumba waters area is believed to have high marine life, but minimal scientific information. From the perspective of oceanography, this area is also believed to be unique because of the interaction between Indonesian Through Flow (ITF/Arlindo). The presence of four water masses on the western side, north and south of Sumba Island (Northern Pacific Subtropical Water-NPSW, North Pacific Intermediate Water-NPIW, Northern Indian Subtropical Water-NISW and Northern Indian Intermediate Water-NIIW), proves that this area is an outlet of ITF [24–26]. A front near the western tip of Sumba Island potentially trigger an eddy and might be influenced by the South Java Current (SJC) [24–26]. This current is believed to trigger upwelling which is important for fishery resources. In addition, oceanographic process also believed to be important in Sumba region is mixing, that is the mixing of water masses that can occur due to tidal currents, bathymetry, and internal waves.

2.2. Sample collection

Sampling was conducted during the 9th Ekspedisi Widya Nusantara (EWIN IX) in August 2016 using RV Baruna Jaya VIII. Sampling sites are shown in Figure 1. Water samples (10 liters per depth per site) were collected using Rosette Water Sampler at 5 m, 50 m, 100 m, 300 m and near the sea bottom, then were filtered using a sterile Whatman® cellulose nitrate filter papers (diameter 47 mm; pore size 0.45μm). To accelerate filtration process, we used Gast vacuum DOA-P504-BN. All samples in the filter papers were stored at 4±2 °C before analysis in sterile Petri dish and covered with ParaFilm® sealing film. To prevent samples from being contaminated, we wore laboratory latex gloves and eyeglasses for filtering, sorting and counting. All sterile glassware were used and filters were stored in Petri dishes, sealed with Para Film.

2.3. Sample analysis

Microscope Leica M205C was used to examine microplastics particles on the filters. We were using the following characteristics [17,27,28] to identify microplastics, namely (1) particle size ≤ 5mm, (2) homogeneous colour, not shiny or sparkling and no cellular or organic structure, (3) fiber particles are unbranched and not segmented. The microplastics identified were counted and measured. The types of microplastic then were classified as fibers, granule, fragment, and foam; and were categorized to the size of <300μm; 300-500μm; 500-1000μm; >1000μm. We analyzed microplastics with sizes more than 250μm using Fourier Transform Infrared (FT-IR). FT-IR can be used to analyze microplastics samples directly [29]. Polymer analysis was done using Nicolet™ iS5 FT-IR Spectrometer, equipped with a laminated diamond crystal Thermo Scientific™ iD5 attenuated total reflectance (ATR) accessory, and corrected using Omnic^{tm} Software. The instrument was operated based on [30] at a range of 600 and 3800 cm-1, a resolution 8 cm and at a rate of 16 scans per analysis, in single reflection mode. Before we analyzed using FT-IR, all potentially plastic particles were rinsed with ethanol 96%. To prevent samples from airborne contamination, we analyzed all particles based on a report by [31]. Afterwards, we did not quantify the fiber particle of the same polymer type as the lab clothes we have worn. Microplastics concentration are presented as n/m^3 unit (Table 1) to compare with other research result, simple statistical tests were performed on the data collected using Microsoft Excel are presented mean values ± standard deviations (SDs).

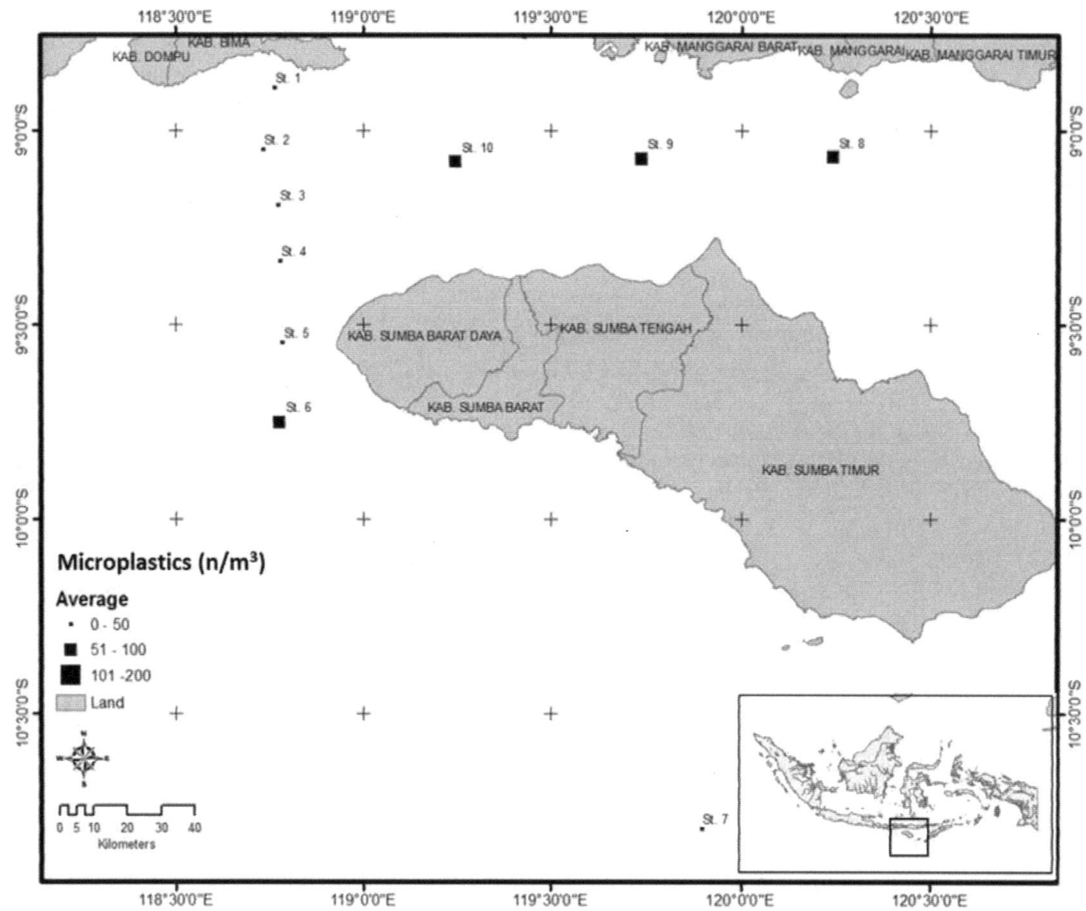

Figure 1. Sampling sites and average abundance of microplastics particles in five different depth (5 m, 50 m, 100 m, 300 m depth and near the sea bottom)

3. Results and discussion

3.1. Study area

Plastic particles were found in all sampling location (10 sampling locations). Average microplastics concentration per station in five different depth were 44 ± 24.59 n/m^3. The highest concentration of microplastics was observed in Sumba Strait (St-8 and St-10). At 5m depth, microplastics particles were found in all sampling locations (Figure 2). However, microplastics was not observed at 100m depth. This might be related to the presence of thermocline zone, ranging from 53m to 144m depth.

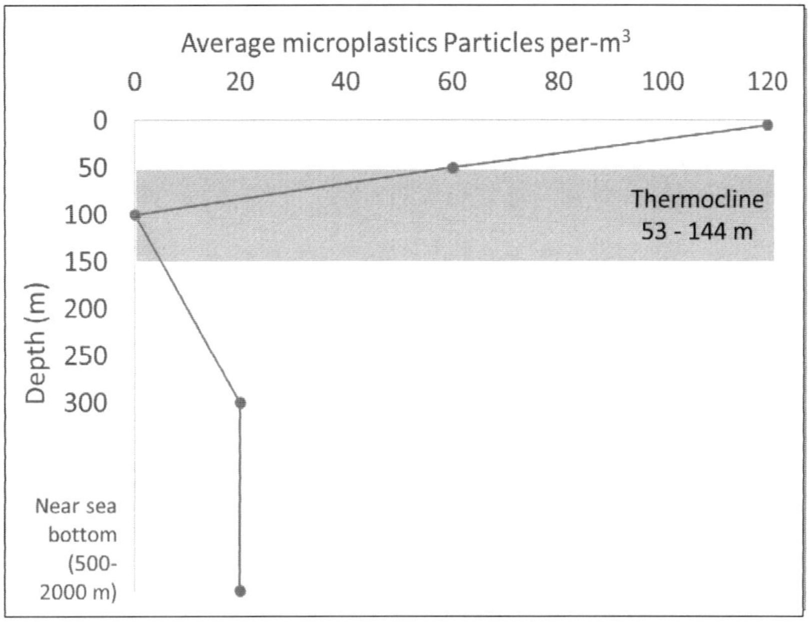

Figure 2. Average of microplastics abundance (n particles/m^3) in all station at five different depth in Sumba Sea

Microplastics abundance at 5m depth was nearly similar with that of the coastal area in Yangtze estuary China [32,33]. The abundance in this area was higher than that of other open oceans, for example North East Atlantic Ocean (2.46 n/m^3; [34]), East China Sea (0-1.44 n/m^3; [33]), and North West Mediterranean (0.116 n/m^3; [35]. The wide variety of microplastic abundance in different oceans might be related with the fact that microplastics tend to be heterogeneously distributed in a water mass [36]. Because Sumba region is one of the ITF outlets [24–26], characterized by four water masses (NPSW, NPIW, NISW, NIIW), we believe that microplastics observed in Sumba might not only come from anthropogenic activities around [28] Sumba, but also from other parts of the oceans in the Pacific.

Table 1. Microplastics occurrence, characteristic and polymer composition

Size	Form			Percentage	Polymer				Percentage
	Fibers	Granule	Other	(%)	PE	PS	PA	PP	(%)
<300µm	1	1	0	9.09	1	0	0	0	4.55
300-500µm	3	2	4	40.91	4	2	1	1	36.36
500-1000µm	4	5	0	40.91	5	0	0	4	40.91
>1000µm	2	0	0	9.09	4	0	0	0	18.18
Percentage (%)	45.45	36.36	18.18		63.64	9.09	4.55	22.73	

PE- Polyethylene; PS- Polystyrene; PA- Polyamide; PP- Polypropylene

3.2. Microplastics occurrence, characteristic and polymer composition

We classified microplastics from Sumba into four size categories: <300µm, 300-500µm, 500-1000µm, and >1000µm; and into three forms: fibers, granule, and other type (fragment, foam) (Table 1). In all sampling stations, microplastic size ranges from 280µm to 1120µm. Most of identified microplastics (81.82%) were 300-1000µm in size. Fibers were the most abundant form (45.45%), followed by granule (36.36%) and other type (18.18%). We identified four dominated categories of plastic polymers:

Polyethylene (PE), Polystyrene (PS), Polyamide (PA) and Polypropylene (PP). PE was the most common dominated polymer type, followed by PP, PS, and PA (Table 2).

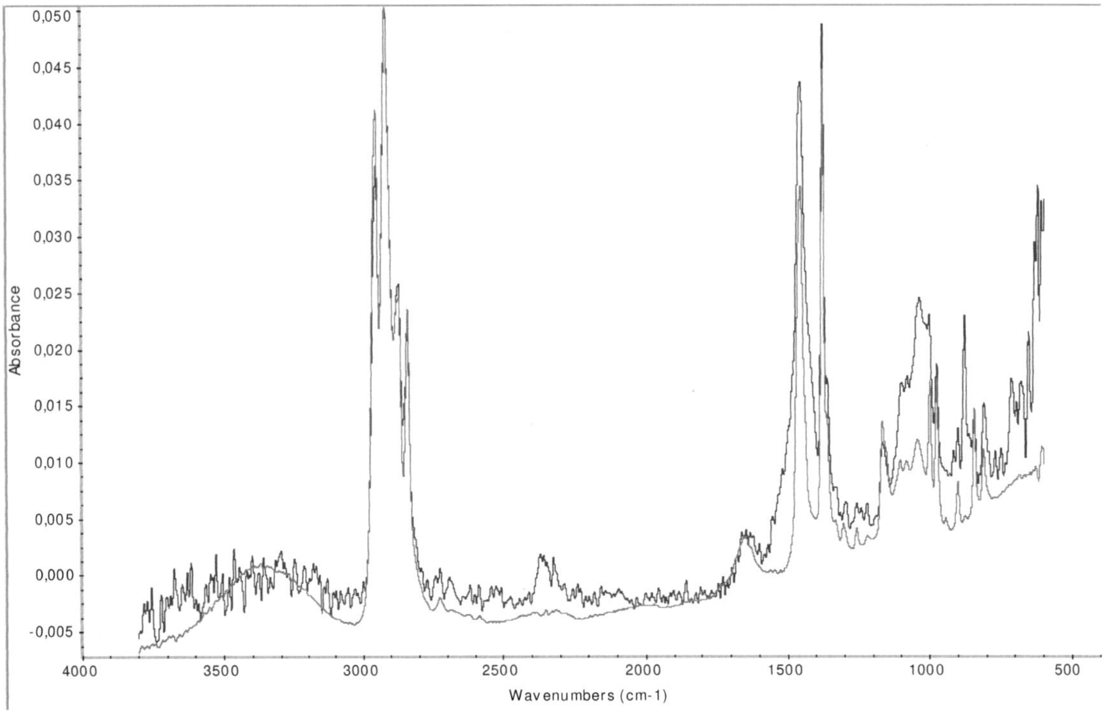

Figure 3. Example of comparison of the spectrum of polypropylene polymer standard from Thermo-scientific (red) and a spectrum obtained from the measurement of a dominated polypropylene fragment particle found in Sumba waters (blue) by ATR-based FTIR spectroscopy

Figure 3 shows examples of the polymers found, which dominated with Polypropylene (PP). From fragment particle sample, prominent presence peak at wavenumber 2950 cm^{-1}, 2916 cm^{-1}, 2837 cm^{-1} and 2868 cm^{-1}. Similar observation by [37] for PP samples indicated peak at 2951 cm^{-1}, 2911 cm^{-1} and 2844 cm^{-1} indicating an asymmetrical type of vibration (v_a-CH_2) [38]. Those were similar displayed vibration bands of polymer standard at 2950 cm^{-1}, 2917 cm^{-1}, 2869 cm^{-1} and 2837 cm^{-1}. In this study, fragment particle found in Sumba waters and polypropylene polymer standard (Figure 3) showed an absorption band at 1375 cm^{-1} and at 1454 cm^{-1}. Spectrum of PP showed symmetrical deformation vibrations (δs–CH_3) at 1372 cm^{-1} and asymmetrical deformation vibrations (δa-CH_3) at 1458 cm−1[37,38].

Size of microplastics could determine the potential impact to the organisms [28]. Smaller size increase the possibility of microplastics ingested by the organisms [33,39]. This situation may pose potential negative impacts to various marine megafauna and important pelagic fishes. Microplastics with size <1000 μm frequently founded on marine organism digestive tract [16,18,19,40], suggesting that marine organisms likely mistake microplastics as their food [41]. Microplastics could disrupt digestive tract function [17] and could also act as a vector for organic and heavy metal pollutants [21–23,42,43]. The most common form was fiber microplastics, similar to the findings by [28,33,44,45]. Fibrous plastic particles in Sumba might derive from fishing nets and rope materials; granule comes from hard plastics and granulated cleaner [45]. Based on FTIR analysis, most polymers identified were dominated polyethylene, followed by dominated polypropylene and dominated polystyrene. The primary use of polyethylene is packaging materials, e.g. food and beverage containers, geomembrane plastic bags and film. Polypropylene is widely used in food and beverage containers, clothing industry, ropes, and re-usable containers [46–48]. Surprisingly, polyamide fiber was also found in Sumba. This fiber might derived from fishing nets and rope material [49,50]. Despite the fact that plastics are newly developed materials (early nineteen centuries made), our finding, that microplastics was found in seawater of Sumba at all surveyed depth, indicated that plastic has invaded marine areas, including pristine areas. It confirms the common believe that plastic waste has spread widely to different parts of the seas and oceans, including remote and unknown areas [51].

4. Conclusion

Our study reports that microplastics were found in all sampling locations, in the form of fibers (45.45%), granules (36.36%) and other form (18.18%). Most of the microplastic particles (81.82%) were found at water depths less than 100m, which were thermocline area. However, no microplastics was observed at 100m depth. Most of identified microplastics (81.82%) were in 300-1000μm in size. Dominated plastics polymers identified were polyethylene, followed by polypropylene, polystyrene, and polyamide. Further studies are planned to examine microplastic distribution in the region where ITF presence and to investigate the potential impact of microplastics in that area.

Acknowledgement

This study was part of Ekspedisi Widya Nusantara IX (EWIN-IX) funded by DIPA LIPI from the Government of Indonesia. We would like to thank the support from all the crews of RV. Baruna Jaya VIII during the EWIN-IX cruise. We also thank to Mr. Sumijo Hadi Riyono for the sample collection.

5. References

[1] Derraik J G B 2002 The pollution of the marine environment by plastic debris: A review *Mar. Pollut. Bull.* **44** 842–52
[2] PlasticsEurope 2015 Plastics - the facts 2014/2015: An analysis of European plastics production, demand and waste data *PlasticsEurope* 1–34
[3] Ellen MacArthur Foundation 2016 The New Plastics Economy: Rethinking the future of plastics *Ellen MacArthur Found.* 120
[4] Cozar A, Echevarria F, Gonzalez-Gordillo J I, Irigoien X, Ubeda B, Hernandez-Leon S, Palma .A T, Navarro S, Garcia-de-Lomas J, Ruiz A, Fernandez-de-Puelles M L and Duarte C M .2014 Plastic debris in the open ocean *Proc. Natl. Acad. Sci.* **111** 10239–44
[5] Eriksen M, Lebreton L C M, Carson H S, Thiel M, Moore C J, Borerro J C, Galgani F, Ryan P G and Reisser J 2014 Plastic Pollution in the World's Oceans: More than 5 Trillion Plastic. Pieces Weighing over 250,000 Tons Afloat at Sea *PLoS One* **9**
[6] Jambeck J R, Geyer R, Wilcox C, Siegler T R, Perryman M, Andrady A, Narayan R and Law K L 2015 Plastic waste inputs from land into the ocean *Science* **347** 768–71
[7] Galgani F, Hanke G, Werner S and De Vrees L 2013 Marine litter within the European Marine Strategy Framework Directive *ICES J. Mar. Sci.* **70** 1055–64
[8] Gregory M R 2009 Environmental implications of plastic debris in marine settings entanglement, ingestion, smothering, hangers-on, hitch-hiking and alien invasions. *Philos. Trans. R. Soc. Lond. B. Biol. Sci.* **364** 2013–25
[9] Gall S C and Thompson R C 2015 The impact of debris on marine life *Mar. Pollut. Bull.* **92** 170–9
[10] Laist D 1997 *Impacts of marine debris: entanglement of marine life in marine debris including a comprehensive list of species with entanglement and ingestion records.* (New York, Springer) chapter 3 pp 99-139
[11] Andrady A L 2011 Microplastics in the marine environment *Mar. Pollut. Bull.* **62** 1596–605
[12] Wagner M, Scherer C, Alvarez-Muñoz D, Brennholt N, Bourrain X, Buchinger S, Fries E, Grosbois C, Klasmeier J, Marti T, Rodriguez-Mozaz S, Urbatzka R, Vethaak a, Winther Nielsen M and Reifferscheid G 2014 Microplastics in freshwater ecosystems: what we know and what we need to know *Environ. Sci. Eur.* **26** 12
[13] Fendall L S and Sewell M A 2009 Contributing to marine pollution by washing your face:

Microplastics in facial cleansers *Mar. Pollut. Bull.* **58** 1225–8
[14] Browne M A, Crump P, Niven S J, Teuten E, Tonkin A, Galloway T and Thompson R 2011 Accumulation of microplastic on shorelines woldwide: Sources and sinks *Environ. Sci. Technol.* **45** 9175–9
[15] Betts K 2008 Why small plastic particles may pose a big problem in the oceans *Environ. Sci. Technol.* **42** 8996
[16] Farrell P and Nelson K 2013 Trophic level transfer of microplastic: Mytilus edulis (L.) to Carcinus maenas (L.). *Environ. Pollut.* **177** 1–3
[17] Cole M, Lindeque P, Fileman E, Halsband C, Goodhead R, Moger J and Galloway T S 2013 Microplastic ingestion by zooplankton *Environ. Sci. Technol.* **47** 6646–55
[18] Browne M A, Dissanayake A, Galloway T S, Lowe D M and Thompson R C 2008 Ingested microscopic plastic translocates to the circulatory system of the mussel, Mytilus edulis (L.) *Environ. Sci. Technol.* **42** 5026–31
[19] Van Cauwenberghe L and Janssen C R 2014 Microplastics in bivalves cultured for human consumption *Environ. Pollut.* **193** 65–70
[20] Peng J, Wang J and Cai L 2017 Current understanding of microplastics in the environment: Occurrence, fate, risks, and what we should do *Integr. Environ. Assess. Manag.* **13** 476–82
[21] Teuten E L, Saquing J M, Knappe D R U, Barlaz M A, Jonsson S, Björn A, Rowland S J, Thompson R C, Galloway T S, Yamashita R, Ochi D, Watanuki Y, Moore C, Viet P H, Tana T S, Prudente M, Boonyatumanond R, Zakaria M P, Akkhavong K, Ogata Y, Hirai H, Iwasa S, Mizukawa K, Hagino Y, Imamura A, Saha M and Takada H 2009 Transport and release of chemicals from plastics to the environment and to wildlife. *Philos. Trans. R. Soc. Lond. B. Biol. Sci.* **364** 2027–45
[22] Rochman C M, Hentschel B T and The S J 2014 Long-term sorption of metals is similar among plastic types: Implications for plastic debris in aquatic environments *PLoS One* **9**
[23] Brennecke D, Duarte B, Paiva F, Caçador I and Canning-Clode J 2016 Microplastics as vector for heavy metal contamination from the marine environment *Estuar. Coast. Shelf Sci.* **178** 189–95
[24] Gordon A L 2005 Oceanography of the Indonesian Seas and Their Throughflow *Oceanography* **18** 14–27
[25] Pranowo W S, Kuswardhani A R T D, Kepel T L, Kadarwati U R, Makarim S and Husri S 2005 *Ekspedisi INSTANT 2003-2005 Menguak Arus Lintas Indonesia* ed A Supangat, I S Brodjonegoro, A G Ilahude, I Jaya and T R Adi (Jakarta, Indonesia: Pusat Riset Wilayah Laut & Sumberdaya Non-hayati, Badan Riset Kelautan dan Perikanan, Departemen Kelautan & Perikanan)
[26] McCreary J P, Miyama T, Furue R, Jensen T, Kang H W, Bang B and Qu T 2007 Interactions between the Indonesian Throughflow and circulations in the Indian and Pacific Oceans *Prog. Oceanogr.* **75** 70–114
[27] Hidalgo-Ruz V, Gutow L, Thompson R C and Thiel M 2012 Microplastics in the Marine Environment: A Review of the Methods Used for Identification and Quantification *Environ. Sci. Technol.* **46** 3060–75
[28] Mohamed Nor N H and Obbard J P 2014 Microplastics in Singapore's coastal mangrove ecosystems *Mar. Pollut. Bull.* **79** 278–83
[29] Käppler A, Windrich F, Löder M G J, Malanin M, Fischer D, Labrenz M, Eichhorn K J and Voit B 2015 Identification of microplastics by FTIR and Raman microscopy: a novel silicon filter substrate opens the important spectral range below 1300 cm(-1) for FTIR transmission measurements *Anal. Bioanal. Chem.* **407** 6791–801
[30] Löder M G J and Gerdts G 2015 Methodology used for the detection and identification of microplastics-a critical appraisal *Marine Anthropogenic Litter* pp 201–27
[31] Nuelle M-T, Dekiff J H, Remy D and Fries E 2014 A new analytical approach for monitoring microplastics in marine sediments *Environ. Pollut.* **184** 161–9

[32] Zhang K, Gong W, Lv J, Xiong X and Wu C 2015 Accumulation of floating microplastics behind the Three Gorges Dam *Environ. Pollut.* **204** 117–23
[33] Zhao S, Zhu L, Wang T and Li D 2014 Suspended microplastics in the surface water of the Yangtze Estuary System, China: First observations on occurrence, distribution *Mar. Pollut. Bull.* **86** 562–8
[34] Lusher A L, Burke A, O'Connor I and Officer R 2014 Microplastic pollution in the Northeast Atlantic Ocean: Validated and opportunistic sampling *Mar. Pollut. Bull.* **88** 325–33
[35] Collignon A, Hecq J H, Glagani F, Voisin P, Collard F and Goffart A 2012 Neustonic microplastic and zooplankton in the North Western Mediterranean Sea *Mar. Pollut. Bull.* **64** 861–4
[36] Dubaish F and Liebezeit G 2013 Suspended microplastics and black carbon particles in the Jade system, southern North Sea *Water. Air. Soil Pollut.* **224**
[37] Syakti A D, Bouhroum R, Hidayati N V, Koenawan C J, Boulkamh A, Sulistyo I, Lebarillier S, Akhlus S, Doumenq P and Wong-Wah-Chung P 2017 Beach macro-litter monitoring and floating microplastic in a coastal area of Indonesia *Mar. Pollut. Bull.* **122** 217–25
[38] Fotopoulou K N and Karapanagioti H K 2015 Surface properties of beached plastics *Environ. Sci. Pollut. Res.* **22** 11022–32
[39] Moore C J, Moore S L, Leecaster M K and Weisberg S B 2001 A comparison of plastic and plankton in the North Pacific Central Gyre *Mar. Pollut. Bull.* **42** 1297–300
[40] Foekema E M, De Gruijter C, Mergia M T, van Franeker J A, Murk A J and Koelmans A a 2013 Plastic in North Sea Fish *Environ. Sci. Technol.* **47** 8818–24
[41] Van Cauwenberghe L, Claessens M, Vandegehuchte M and Janssen C 2012 *Occurrence of microplastics in Mytilus edulis and Arenicola marina collected along the French-Belgian-Dutch coast* (Ghent)
[42] Hirai H, Takada H, Ogata Y, Yamashita R, Mizukawa K, Saha M, Kwan C, Moore C, Gray H, Laursen D, Zettler E R, Farrington J W, Reddy C M, Peacock E E and Ward M W 2011 Organic micropollutants in marine plastics debris from the open ocean and remote and urban beaches *Mar. Pollut. Bull.* **62** 1683–92
[43] Rios L M, Moore C and Jones P R 2007 Persistent organic pollutants carried by synthetic polymers in the ocean environment *Mar. Pollut. Bull.* **54** 1230–7
[44] Wright S L, Thompson R C and Galloway T S 2013 The physical impacts of microplastics on marine organisms: A review *Environ. Pollut.*
[45] Claessens M, Meester S De, Landuyt L Van, Clerck K De and Janssen C R 2011 Occurrence and distribution of microplastics in marine sediments along the Belgian coast *Mar. Pollut. Bull.* **62** 2199–204
[46] Miller R C 1990 Polypropylene *Mod. Plast.* **67** 84,86
[47] Arutchelvi J, Sudhakar M, Arkatkar A, Doble M, Bhaduri S and Uppara P V 2008 Biodegradation of polyethylene and polypropylene *Indian J. Biotechnol.* **7** 9–22
[48] Allahvaisi S 2012 Polypropylene in the industry of Food Packaging *Polypropylene* pp 1–20
[49] Thompson R C, Olsen Y, Mitchell R P, Davis A, Rowland S J, John A W G, McGonigle D and Russell A E 2004 Lost at sea: where is all the plastic? *Science* **304** 838
[50] Dekiff J H, Remy D, Klasmeier J and Fries E 2014 Occurrence and spatial distribution of microplastics in sediments from Norderney *Environ. Pollut.* **186** 248–56
[51] Van Cauwenberghe L, Vanreusel A, Mees J and Janssen C R 2013 Microplastic pollution in deep-sea sediments *Environ. Pollut.*

IOP Conf. Series: Earth and Environmental Science **162** (2018) 012024 doi:10.1088/1755-1315/162/1/012024

Measurement and analysis of acoustic backscattering strength for characteristics of seafloor sediment in Indian Ocean WPP 572-573

Fauziyah[1], Priatna A[2], Prakoso W F[1], Hidayat T[2], Surbakti H[1] and Nurjuliasti E[1]

[1] Marine Science Study Programme, Sriwijaya University, Indonesia

[2] Marine Fisheries Research Agency (MFRA), Ministry of Marine Affairs and Fisheries - Republic of Indonesia

E-mail: siti_fauziyah@yahoo.com

Abstract. Understanding and predicting the acoustic backscattering from seafloor substrates was essential to support the management and development of coastal area as well as offshore. The aims of this study were to quantify the acoustic backscattering strength from the seafloor substrates,classify the types of seafloor substrates, as well as correlation between the seafloor substrates and the value of surface backscattering strength (SS). The data collection was done in April to June 2016 in the Indian Ocean WPP 572-573 using hydroacoustic instrument SIMRAD EK60 with operating frequency of 38 kHz. Substrate sampling was used for data validation using Van Veen Grab. The results showed that the SS value for the type of sand substrate (-8.05 dB to -5.25 dB), silty sand (-8.15 dB to -7.03 dB), sandy silt (-11.30 dB to -11.35 dB), silt clayey (-14.71 dB to -12.17 dB) and clay substrate -23.55 dB. Between the SS value and the sand percentage had a very strong positive correlation, and a very strong negative correlation between the surface backscattering strength and the clay percentage. In contrast, a poor negative correlation was also found between the SS value and the silt percentage.

1. Introduction

Indian Ocean is a part of Fisheries Management Area WPP-572 and 573.These waters have a lot of potential for marine resource development. In order to support the management and development is required a comprehensive data and information both related to fish resources and its environment. This is important because Indonesian waters have multispecies fish resources and each region has different environmental characteristics [8]

One of important environmental factors is the data and information about condition and distribution of substrate type on the seabed. The information of seabed substrate is very important because in addition to the seabed substrate useful as a habitat, feeding ground, and nursery ground to the most aquatic organism, seabed substrate also has a complex composition ranging from a small substrate to a rock. In relation to coastal management, data and information about the substrate type is also useful for coastal infrastructure building planning, even did not close the possibility to be used in high seas.

Mapping of bottom sediments in marine is commonly based on sonar data and backscatter analysis [3] (e.g., multibeam, splitbeam and side scan), calibrated using sediment samples.

In the hydroacoustic method, the characteristics of the bottom substrate of the waters can be determined by assessing the characteristics of seabed backscattering because seabed have the characteristics of reflecting and scattering sound waves. According to [16], the hydroacoustic method is a technology that utilizes the echo principle that can detect targets in the aquatic columns such as plankton, larvae, fish, and bottom waters.

Research on substrate type of seabed by hydroacoustic method has been studied in various countries, [19] analyzed the relationship between *backscattering strength* measurement on seabed and geotechnical parameters, [20] classified the seabed using the data of backscatter from multibeam echosounder, [9] examined the effect of the relationship of backscattering values on the composition of sediment particles in New Jersey, while [11] focused on sound propagation at low frequencies for sediment characteristics in Kinneret Lake.

Research on substrate type of seabed using hydroacoustic technology has been done in Indonesia were [17] assessed the effect of grain size, roughness, and hardness of seabed on backscattering value of hydroacoustic detection, [5] mapped and classified the sediment with side scan sonar instruments, [2] reviewed the relationships of backscattering on seabed substrate (E1 and E2) to the distribution of demersal fish (Sv fishes), [1] classified the bottom sediment used mosaic data of backscatter, and [13] reviewed the measurement and analysis of acoustic backscattering to classify the seabed and its relation to makrozoobentos.

This study aims to determine the value of acoustic backscattering strength on seabed, classify the substrate of seabed, and the correlation between substrate type on seabed and the value of surface backscattering strength. The value of acoustic backscattering strength consists the value of volume backscattering strength (VS) and surface backscattering strength (SS).The value of VS consists of the value E1 and E2. Roughness was measured from the integral of the portion of the first echo that occurs after the returns from the initial incident angles (the initial returns contained strong, undesirable amplitude contributions from bottom reverberations). Hardness was defined as the integral of the second echo. The ratio of roughness (E1 for first echo) to hardness (E2 for second echo), hereafter referred to as E1:E2, was also used for bottom-sediment typing as an alternative to using plots of E1 versus E2 to identify clusters related to sediment type [6]. While SS is the backscattering strength on seabed surface [13].

2. Materials and methods

2.1 Study site

All data reported in this article were collected by the Marine Fisheries Research Agency (MFRA), Ministry of Marine Affairs and Fisheries - Republic of Indonesia, Its concerning in characteristics of fisheries biology, resource habitat, and potentialof production in WPP 572 and 573 on year 2016. The acoustic sounding and substrate sampling in the Western Waters of Sumatra (WPP 572) and Southern waters of Java (WPP 573). The sounding data has been processed in Fisheries Acoustic Laboratory, Marine Fisheries Research Agency (MFRA). The following is the location of the seabed sediment sampling as a research station (Fig 1) as well as the coordinates of the position of the research station (Table 1).

Table 1.The position of seabed sediment sampling

Station	Name of Area	Position	
		Latitude	Longitude
1	Waters of West Aceh	5^{0}11'59.7552" N	95^{0}09'22.9861" E
2	Waters of West Aceh	4^{0}27'07.4663" N	95^{0}38'25.4590" E
3	West Waters of North Sumatera	1^{0}25'10.6940" N	98^{0}34'59.4182" E
4	Waters of West Padang	0^{0}02'39.3903" N	99^{0}10'54.0540" E
5	Waters of West Bengkulu	3^{0}36'16.3079" S	101^{0}57'04.7124" E
6	Waters of West Bengkulu	4^{0}15'15.4038" S	102^{0}30'26.6433" E
7	South Waters of West Java	7^{0}50'43.1578" S	108^{0}20'54.2493" E
8	South Waters of West Java	7^{0}48'39.3109" S	108^{0}51'56.8118" E
9	South Waters of Cilacap	7^{0}50'52.6898" S	109^{0}40'50.2393" E

Station	Name of Area	Position	
		Latitude	Longitude
10	South Waters of Cilacap	8⁰13'32.2574" S	110⁰52'51.0958" E
11	South Waters of Cilacap	8⁰03'59.3028" S	110⁰16'56.5956" E
12	South Waters of Cilacap	7⁰51'54.7429" S	109⁰16'05.2106" E
13	South Waters of West Java	7⁰46'29.4480" S	108⁰38'03.4670" E

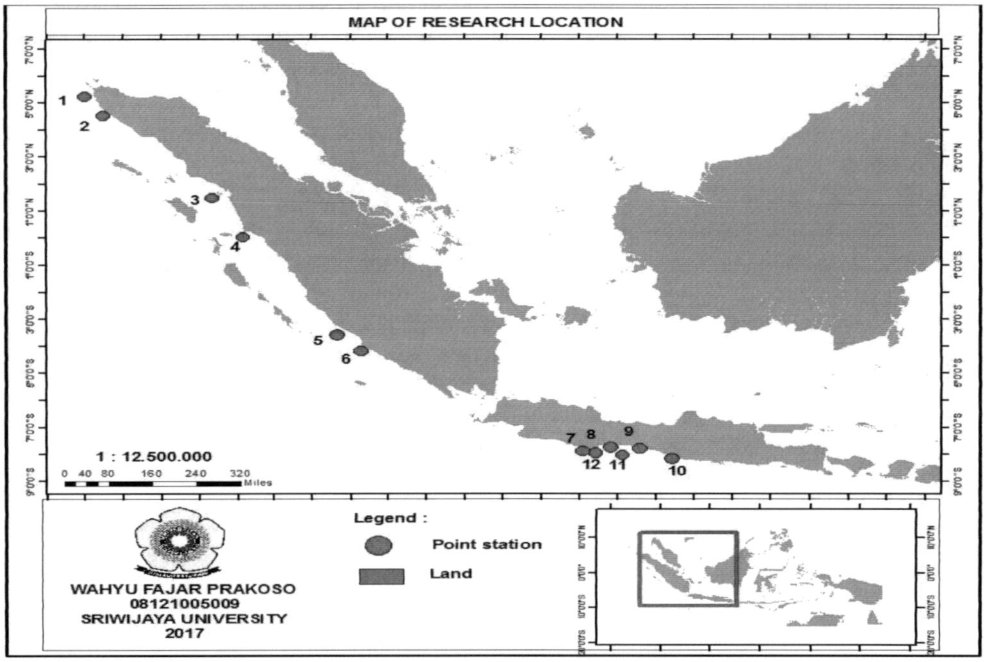

Figure 1. map of research location sediment sampling station

2.2. Acoustic data

Acoustics data is collected using a scientific split beam echosounder system SIMRAD EK 60 frequency 38 kHz, research vessel Baruna Jaya IV. The acoustic data was recorded using an ER60 software that will record and integrate all detected targets in a single file. Recording is performed about 10-20 minutes before sampling the substrate at each observation station.

Sediment samples were collected at 13 locations using a Van veen grab (dimensions 60 cm x 30 cm). Samples of substrate then analyzed by its physical properties, that is substrate texture in Soil Physics Laboratory, Land Research Institute, Ministry of Agriculture, Bogor-Indonesia. Texture of the substrate used as in situ data as well as comparable data from the results of acoustic measurement data. Substrate texture analysis was done by wet screen method of stratified. Substrate texture analysis was divided into 3 fractions i.e. sand fraction (0.05 - 2.00 mm), silt (0.05- 0.002 mm), and clay (> 0.002 mm). Classification of substrate types using Shepard triangle classification. The diagram is constructed with corners consisting of sand-silt-clay.

2.3. Acoustic data processing

The acoustic data were post-processed using Echoview 4.0 software to obtain water depths and backscatter measurements of bottom roughness (tail of first echo) and hardness (second echo). The first reflected data (E1) was processed using a minimum threshold of -50.00 dB (decibel) and a maximum of 0 dB. The second reflected data (E2) was processed using a minimum threshold of -70.00

dB and maximum threshold of0 dB (Pujiyati, 2008). Elementary Sampling Distance Unit(ESDU) was 100 ping. The integration thickness of E1 and E2 was 0.30 m.

According to [11]), the value of acoustic surface backscattering strength (SS) obtained using ring surface scattering model (RSS Model). The RSS Model show a relationship between the raw (or not averaged) SV value of the bottom echo (S_{VB}), and the raw bottom backscattering strength (S_S), as;

$$S_s = \frac{S_{VB}\,\psi(c\,\tau/2)}{\phi} \tag{1}$$

where ϕ and ψ are the equivalent beam angle for surface and volume scattering, respectively, c is the sound speed, and τ is the pulse width. At the peak of the bottom echo, we have:

$$\phi = \phi_0 \cong \psi \tag{2}$$

where ϕ_0 is the asymptotic value of the equivalent beam angle for surface scattering. Introducing Eq. (2) into Eq. (3) gives

$$S_s = (c\,\tau/2)\,S_{VB} \tag{3}$$

An application of this formula for all S_{VB} yields a convenient measure called "instantaneous" SS. The SS value was obtained by using a logarithmic equation, as:

$$SS = 10log(c\,\tau/2) + S_{VB} \tag{4}$$

2.4. Calibration Data

The data calibration is collected to establish relationships between acoustic data and bottom sediment type. The sample sites that encompassed a range of roughness and hardness combinations.

The relation of SS parameter with sediment type was annalized by best fit model approach to linear regression model, logarithmic regression and polynomial regression. Best fit modelwas chosen based on the highest coefficient of determination (R2). The coefficient of determination shows how much percentage of independent variable (sediment type) can explain the non-free (surface backscattering strength).

Scatter and trendline chart types in Microsoft Excel 2007 software were used to determine the best fit relationship model between each type of sediment with SS value. Based on the regression model graph can be known the direction and magnitude of the correlation between the independent variables (percentage of substrate type) and the dependent variable (the value of surface backscattering strength).

3. Result

3.1. Seabed substrate

The result of physical survey analysis to the seabed substrate on 13 stations shows that the average of sand fraction has percentage of 40.07%, the silt fraction has a percentage of 36.23% while the clay fraction has a percentage of 23.69% of the total sediment percentage (Table 2 and Fig 2). The highest sand fraction composition is found at stations 4 and 6 respectively of 90% with a depth of 37.82 m and 35.25 m. The highest silt fraction was found in station 13 with a composition of 65% at 48.38 m depth while the highest clay fraction composition was found at station 1 of 93% at 50.62 m depth.

Table 2. Percentage of substrate fraction at research location based on physical survey

Station	Depth (m)	Fraction (%)			Substrate Type
		Sand (%)	Silt (%)	Clay (%)	
1	50.62	2	5	**93**	Clay
2	33.74	22	62	16	Sandy Silt
3	50.72	46	37	17	Silty Sand
4	37.82	**90**	8	2	Sand
5	18.65	16	63	21	Clayey-Silt
6	35.25	**90**	3	7	Sand
7	44.20	58	32	10	Silty-Sand
8	61.28	11	56	33	Clayey-Silt
9	49.93	39	42	19	Sandy Silt
10	65.86	52	30	18	Silty Sand
11	58.44	85	6	9	Sand
12	54.38	6	62	32	Clayey-Silt
13	48.38	4	**65**	31	Clayey-Silt
Average		40.07	36.23	23.69	

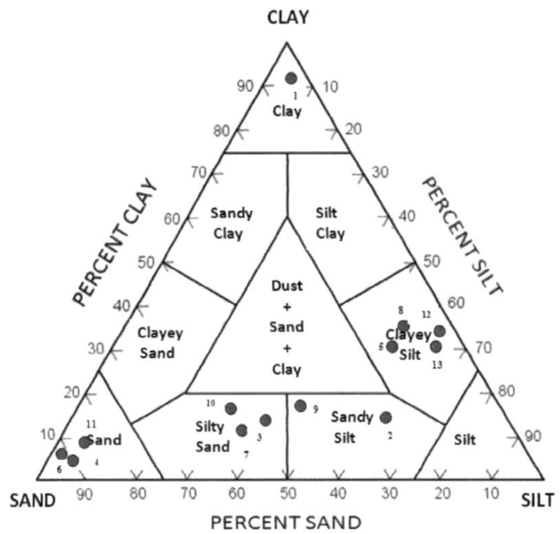

Figure 2. Classification of substrate fraction using Shepard triangle system

3.2. Bottom surface backscattering strength (SS)

The value of backscattering volume (SV) obtained as in Table 3 show the values of E1 (roughness) and E2 (hardness). The type of sand substrate has an E1 value ranging from -16.16 dB to -12.57 dB and E2 ranges from -56.89 dB to -44.05 dB. The type of silty sand has an E1 value ranging from -16.55dB to -15.29 dB and E2 value ranges from -60.72 dB to - 56.79 dB, sandy silt substrate type has an E1 value ranging from -19.40 dB to -18.49 dB and E2 -61.72 dB to -59.76 dB, Type silt -22.82 dB to -18.97 dB and E2 -67.89 dB to -46.21 dB and Clay substrate ranges between less than or less than -36.24 dB and E2 -58.94 dB.The obtained E1 and E2 values show that the sand fraction has a greater roughness and hardness value than the clay fraction and silt fraction. As mentioned on [17] and [15].

Table 3. The value of bottom backscattering strength.

Station	SV		SS (dB)	Substrate Type	WPP RI
	E1 (dB)	E2 (dB)			
1	-36.24	-58.94	-23.55	Clay	572
2	-18.49	-61.72	-11.35	Sandy silt	572
3	-15.88	-57.33	-7.93	Silty sand	572
4	-16.16	-44.05	-8.05	Sand	572
5	-18.97	-46.21	-12.53	Clayey-Silt	572
6	-13.58	-56.61	-6.13	Sand	572
7	-16.55	-56.79	-8.15	Silty sand	573
8	-21.46	-67.89	-14.48	Clayey-Silt	573
9	-19.40	-59.76	-11.30	Sandy silt	573
10	-15.29	-60.72	-7.03	Silty sand	573
11	-12.57	-56.89	-5.25	Sand	573
12	-22.82	-65.35	-14.71	Clayey-Silt	573
13	-19.52	-65.78	-12.17	Clayey-Silt	573
Ratio	1	3.3			

The bottom surface backscattering strength (SS) is a development model to figure the value of surface backscattering strength on the bottom. The SS value was obtained by using a logarithmic equation that connects the backscattering volume (SV) of the bottom, the sound speed (c), and the pulse width (τ). The value of SS is not directly proportional to the SV (E1 and E2) values. This is because the place of observation station has a long distance and difference characteristics among environmental conditions such as oceanographic factors such as temperature, salinity, the sound speed and also the bottom material characteristic will affect the difference in backscattering strength values [14]. Table 3 shows SS value for sand substrate ranging from -8.05 dB to -5.25 dB. Substrate type of silty sand ranging from -8.15 dB to -7.03 dB. Substrate type of sandy silt ranging from -11.30 dB to -11.35 dB. Substrate type of Clayey-Silt ranging from -14.71 dB to -12.17 dB as well as SS Value for clay substrate -23.55 dB.

SS influenced by the particle size of the substrate fraction (Fig 3). Graph of SS value will increase when the composition of clay fraction is very dominant while the graph of SS value will decrease if the composition of sediment at the station has the dominant sand fraction. The substrate of sandy silt has a higher SS value compared with Clayey-Silt substate. This is due to the influence of sand fraction which contributes a stronger reflection value than clay fraction.Based on the SS value pattern obtained in the study illustrates that the smaller and smoother the grain size of the sediment the acoustic backscattering strength value will be weaker and if the larger and harder grain size of the sediment the acoustic backscattering strength value will be stronger. As mentioned on [7], [17] and [1] that the value of backscattering strength is strongly influenced by the size of the grains of sediment.

3.3. The type of bottom substrate based on echogram display

The bottom substrates detected can be distinguished qualitatively based on the color spectrum of the echogram layer (Fig 4). Echogram display taken on each type of substrate obtained in this study as an example.The echogram on the substrate of sand, silty sand, sandy silt and Clayey-Silt has a pulse length 512 μs while clay substrate are1024 μs.

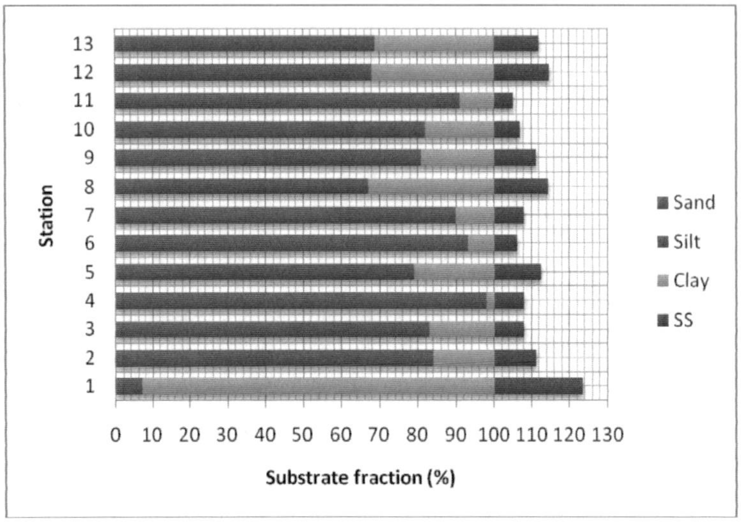

Figure 3. Graph of substrate fraction ratio with SS value.

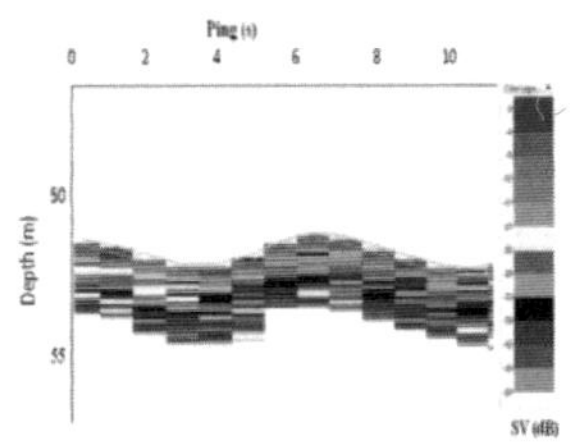

(a). Sand

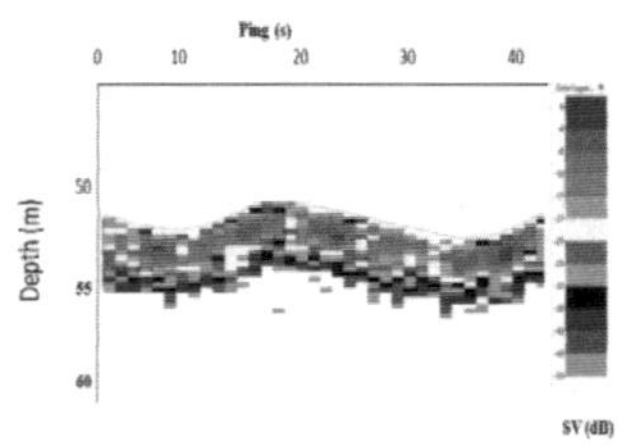

(b). Silty sand

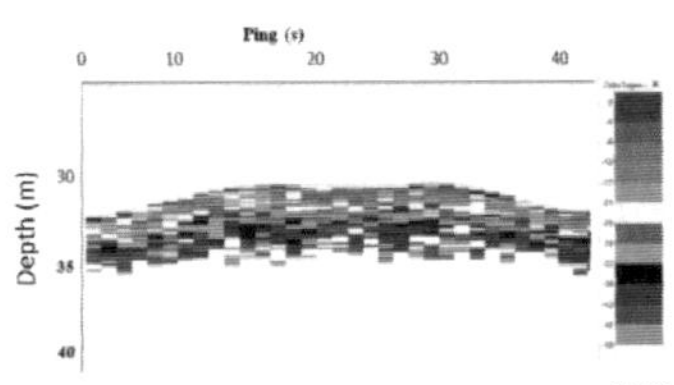

(c). Sandy silt

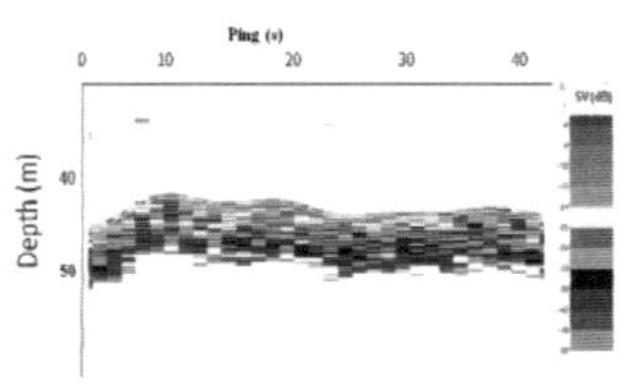

(d). Clayey-Silt

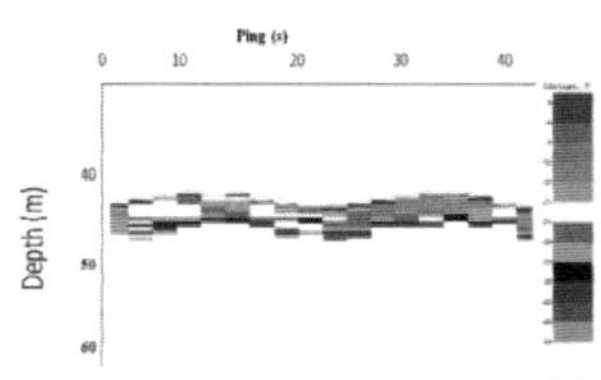

(e). Clay

Figure 4. Echogram display on the type of bottom substrate

4. Discussion

The composition of the grain size of the sediment will be smaller with the increasing depth of the water (Table 2). This is in accordance with the value of backscattering volume (SV) obtained as in Table 3 show that the values of E1 (roughness) and E2 (hardness) are influenced by substrate type and substrate particles. The obtained E1 and E2 values show that the sand fraction has a greater roughness and hardness value than the clay fraction and silt fraction.

The E1:E2 ratio provided the clearest information on sediment type (Table 3), showing distinct hydroacoustic sediment types. The sediments containing coarser fractions (sand, silt, clay) exceed an E1:E2 ratio of 2.00 [21]. Hence, there was information on where coarse and hard sediments occur, but there was a zone of overlap where all sediment classes were represented. The overlap was the main challenge to this case study and demonstrates that acoustic-based hardness and roughness measurements may have limited function.

Meanwhile, SS represents the return value of the bottom surface of the waters which is the derived equation of SV. Sand has a grain size larger than silt and clay [15] so the sand is able to provide the

IOP Conf. Series: Earth and Environmental Science **162** (2018) 012024 doi:10.1088/1755-1315/162/1/012024

strongest reflection according to the SS equation using SVmax. The roughness, hardness and substrate size of the substrate greatly affect the baseline SS value of the water [17], [1].

Figure 5 indicates the existence of two types of best fit models used, namely 1) logarithmic model for regression between sand and SS, and 2) polynomial model for regression between silt and SS as well as regression between clay and SS. Logarithmic model between sand fraction with SS value has R square value equal to 0,812. This shows that the sand fraction influences the SS value by 81.2% (significant) and another 18.8% is influenced by other factors outside this study. Similarly, between clay fraction and SS value having R square value of 0.890, it means that the clay fraction has an effect of 89% (significant) to SS value and the rest (11%) is influenced by other factors outside this research. Conversely, between the silt fraction and the SS value has R square value of 0.137, meaning that the silt fraction only affects 13.7% (not significant) to the SS value and 86.3% is influenced by other factors outside this study. Based on R square value, it can be concluded that the fraction of sand and clay significantly influence the value of surface backscattering strength and vice versa silt effect is not significant.

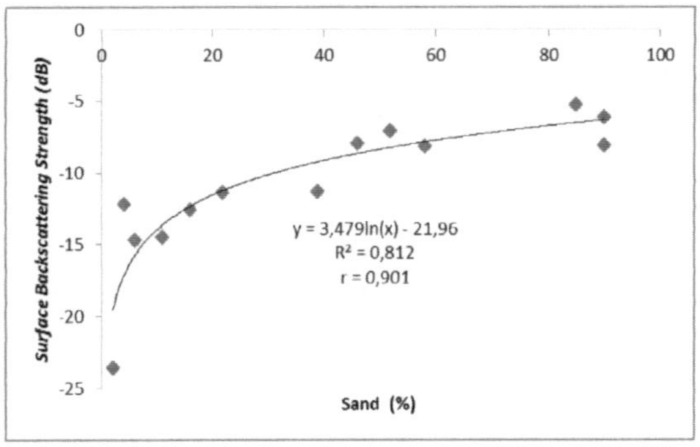

(a). Logarithmic Equation of Sand-SS

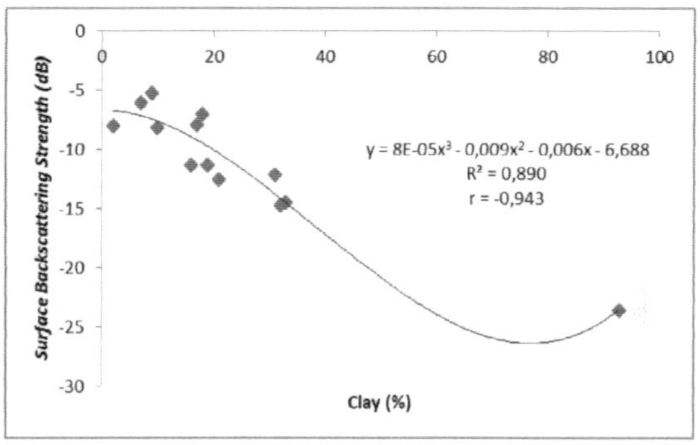

(b). Polynomial Equation of Clay-SS

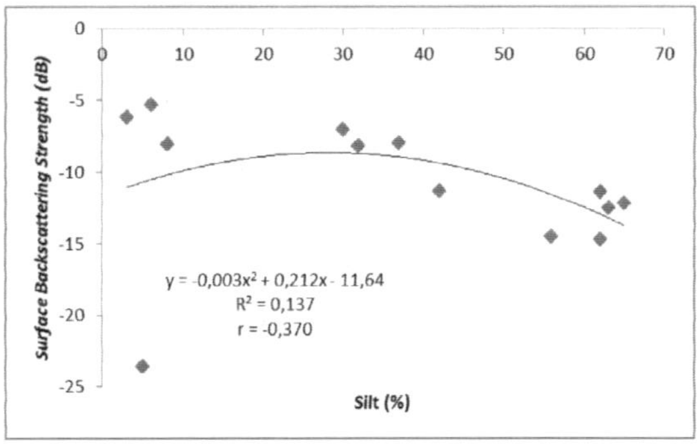

(c). Polynomial Equation of Silt-SS

Figure 5. Regression graph relation of bottom surface backscattering strength (SS) to sediment fraction

Based on the value of the correlation coefficient (r) obtained as shown in Figure 5, it can be classified how strong the relationship (correlation) between each type of sediment (substrate) with the value of SS. Refers to the criteria of [18], the correlation of sand and clay faction to SS value is included in very strong category ($0,8 \leq r \leq 1$). Conversely, the correlation of the silt fraction to the SS value is included in the low category ($0,20 \leq r \leq 0,39$). Based on the tendency of regression graph it can be seen that the sand fraction has a positive correlation (the greater % of sand fraction the greater the value of SS), otherwise fraction fraction of clay and silt have negative correlation. These results are in line with the results of the study of [4] which states that statistically, the sediment of sand and calcium carbonate reservoir has a significant correlation to 210 kHz backscatter signal. This shows that SS values have a strong correlation with subtratic type of bottom waters [10], [12].

Clay fraction has the smallest particle size compared to the sand and silt fraction so it has a high porosity and has little air [15]. It affects the energy of the reflected sound waves so that sound waves will be more absorbed into the bottom substrate of the water and lose much of the energy that is re-emitted to the tranducer.The silt fraction value has a low correlation of all sediment fractions with r value of 0.37. The silt fraction obtained in this study is suspected to spread and mix with sand and clay fractions. It can be seen that from13 substrate sampling stations there is no dominant silt fraction as in stations 4 and 6 for sand and station 1 for clay fractions.

5. Conclusion

The results showed backscattering strength (SS) values for the type of sand substrate -8.05 dB to -5.25 dB, Silty sand -8.15 dB to -7.03 dB, Sandy silt -11.30 dB to -11.35 dB, Clayey-Silt -14.71 dB to -12.17 DB and clay substrate -23.55 dB. The sand fraction has a strong positive correlation with the surface backscattering value, the clay fraction has a very strong negative correlation with the surface backscattering value, whereas the silt fraction has a weak negative correlation with the surface backscattering strength value.

Acknowledgements

This research is fully funded from the Marine Fisheries Research Agency (MFRA), Ministry of Marine Affairs and Fisheries - Republic of Indonesia. The authors would like to thank the Head of the Marine Fisheries Research Agency (MFRA), Head of Research Project in the Indian Ocean. This paper is a contribution of research activities of assessment of biological characteristics of fisheries, resource habitat, and potential production in WPP 572 and 573 in 2016, at the Center for Marine Fisheries Research, Jakarta Indonesia.

References

[1] Adi A P, H M Manik, S Pujiati 2016. Integrasi data multibeam batimetri dan mosaik backscatter untuk klasifikasi tipe sedimen *Jurnal Teknologi Perikanan dan Kelautan* **7**(1):77-84.

[2] Akbar H, S Pujiati, M Natsir 2013 Hubungan tipe dasar perairan dengan distribusi ikan demersal di Perairan Pangkajene Sulawesi Selatan *Jurnal Teknologi Perikanan dan Kelautan* **4**(1):31-39.

[3] Brown C J, S J Smith, P Lawton and J T Anderson 2011 Benthic habitat mapping: A review of progress towards improved understanding of the spatial ecology of the seafloor using acoustic technologies *Estuar, Coast, and Shelf Sci* **92**: 52-520.

[4] Chakraborty B, V Mahale, G Navelkar, B R Rao, R G Prabhudesai, B Ingole and G Janakirinam 2007 Acoustic characterization of seafloor habitats on the western continental shelf of India *ICES Journal of Marine Science* **64**(3): 551-558.

[5] Charnila D, H M Manik 2010 Pemetaan dan klasifikasi sedemen dengan instrumen side scane sonat di Perairan Balongan, Indramayu-Jawa Barat *Jurnal Teknologi Perikanan dan Kelautan* **1**(1):105-112.

[6] Chivers R C, N Emerson and D R Burns 1990 New acoustic processing for underway surveying *Hydrograph J* **56**: 8-17.

[7] Falco G A, R Tonielli, G D Martino, S Simeone and J M Parnum 2010 Relationships between multibeam backscatter, sediment grain size and Posidonia oceanica seagrass distribution. *Continental Shelf Research* **30**(18):1941-1950.

[8] Fauziyah, Hartoni, A Agussalim 2010 Karakteristik shoaling ikan pelagis menggunakan data akustik split beam di Perairan Selat Bangka pada Musim Timur *Jurnal Ilmu Kelautan* **15**(1):17-22 ISSN 0853-7291.

[9] Goof J A, B J Kraft, L A Mayer, S G Schock, C K Sommerfield, H C Olson, S P S Gulick and S Nordfjord 2009 Seabed characterization on the New Jersey middle and oter shelf; correlatability and spatial variability of seafloor sediment properties *Marine Geology* **209**:147-172.

[10] Haris K B, Chakraborty, B Ingole, A Menezes and R Srivastava 2012 Seabed habitat mapping employing single and multi-beam backscatter data: A case study from thewestern continental shelf of India *Continental Shelf Research* **48**:40-49.

[11] Katsnelson B, R Katsman, A Lunkov, and I Ostrovsky 2015 Low-frequency sound propagation and sediment characterization in Lake Kinnerat *Proceedings of the Institute of Acoustics* **37**(1):102-109.

[12] Kloser R J, J D Penrose, and J Butler 2010 Multi-beam backscatter measurements used to infer seabed habitats *Continental Shelf Research* **30**:1772-1782.

[13] Manik H M 2011 *Underwater acoustic detection and signal processing near the seabed* Croatia: Intechweb.

[14] -------------. 2012. *Seabed identification and characterization using sonar*. Hindawi Publishing Corporation: Advances in Acoustics and Vibration. doi: 10.1155/2012/532458

[15] Ningsih E N, F Supriyadi, dan S Nurdawati 2013 Pengukuran dan analisis nilai hambur balik akustik untuk klasifikasi dasar perairan dan hubungannya dengan makrozobentos di Delta Mahakam *Jurnal Penelitian Perikanan Indonesia* **19**(3):139-146.

[16] Priatna A, B Sadhotomo 2011 Sebaran plankton dan larva ikan di Perairan Kepulauan RajaAmpat: kajian metode hidroakustik dan survei konvensional. *BAWAL* **3**(5):345-350.

[17] Pujiyati S, S Hartati, W Priyono 2010 Efek ukuran butiran kekasaran, kekerasan dasar perairan terhadap nilai hambur balik hasil deteksi hidroakustik *Jurnal Ilmu dan Teknologi Kelautan Tropis* **2**(1):59-67.

[18] Sugiono 2010 *Statistika untuk penelitian* Bandung: Alfabet.

[19] Simons D G, M Snellen and M A Ainslie 2007 A Multivariate correlation analysis of high-frequency bottom backscattering strength measurements with geotechnical parameters *IEEE Journal of Oceanic Engineering* **32**(3):640-650. doi: 10.1109/JOE.2007.891890.

[20] Simons D G and M Snellen 2009 A Bayesian approach to seafloor classification using multi-beam echo-sounder backscatter data *Applied Acoustics* **70**(10):1258-1268.

[21] Venteris E R and C J May 2014 Cost-effective mapping of benthic habitats in inland reservoirs through split-beam sonar, indicator kriging, and historical geologic data *PLoS ONE* **9**(4): e95940. doi:10.1371/journal.pone.0095940.

IOP Conf. Series: Earth and Environmental Science **162** (2018) 012025 doi:10.1088/1755-1315/162/1/012025

Primary productivity of coastal ecosystems in the Seribu Islands (case study on Kelapa Dua Island, Pramuka Island and Pari Island)

Nurrahman Y A[1] and Nurdjaman S[2]

[1] Department of Marine Science, Faculty of Mathematics and Natural Sciences, Universitas Tanjungpura, Indonesia
[2] Department of Oceanography, Faculty of Earth Sciences and Technology, Institut Teknologi Bandung, Indonesia

E-mail: yusuf.budhysatrya@gmail.com

Abstract. A determination of the primary productivity in coastal ecosystems, namely mangrove, seagrass and coral reef was done in the Seribu Islands (Kelapa Dua Island, Pramuka Island and Pari Island). Data collection was performed by oxygen methods with a light-dark bottle. Data were collected on 11 - 16 February 2016. The data were Mangrove density, seagrass density, seagrass cover, and, coral reefs cover. Physical-chemical parameters of the water were water temperature, salinity, dissolved oxygen, and nutrient content. Mangrove density in Pramuka Island is the highest among the others with density 19,500 trees/ha. The highest seagrass density is found in Kelapa Dua Island of 1,099 stand/m^2 . Life corals in Kelapa Dua Island has the highest coverage by percentage of 53.2 %. The calculation results of primary productivity values show the highest value in Pulau Kelapa Dua up to 3.0062 gC / m^2 / day. The lowest value obtained in Pari Island that was equal to 0.6 gC / m^2 / day. Based on research, mangrove density, seagrass density and life corals coverage have a positive correlation to primary productivity.

1. Introduction

Indonesian waters cover 17% of the world's marine areas with high primary productivity. This is because Indonesia has a high level of biodiversity, especially in their coastal ecosystems. Coastal areas and oceans, in terms of various kinds of designation, is a very productive area. Primary productivity in coastal areas such as estuary, mangrove, seagrass, and coral reef ecosystems can reach more than 10,000 gC / m^2 / yr, which is about 100-200 times greater than the primary productivity in the open ocean. The high primary productivity of coastal ecosystems enables high levels of secondary productivity (fish and other marine animals). The Seribu Islands consists of 110 islands, and 11 of them are populated. The other islands are used for recreation, nature reserves, cultural heritage and other designations. The Seribu Islands area is approximately 108,000 ha, located off the north coast of Jakarta with an elongated position from north to south marked by small white sandy islands and coral reefs. Seribu Islands have a high marine biodiversity including a variety of coastal ecosystems such as mangroves, seagrass and coral reefs. This has led to high primary productivity in the Seribu Islands. Measurement of primary productivity of phytoplankton is a basic requirement to study the structure and function of aquatic ecosystems [1]. This study aims to determine primary productivity in some coastal ecosystems and the relationship between primary productivity with coastal ecosystem conditions such as mangrove, sea grass and coral reefs in Seribu Islands.

In the sea, especially open water, phytoplankton is the main autotrophic organism that determines the primary productivity of waters. The primary productivity of waters is the amount of organic matter produced by autotrophic organisms, i.e. organisms capabled of producing organic materials from inorganic materials with the help of solar energy. Primary productivity is often estimated as the amount of carbon present in living material and is generally expressed as the number of grams of carbon produced in one square meter of water per day (g C / m^3 / day)[1].

In a humid tropical climate, high primary productivity due to the high intensity of sunlight and evenly throughout the year[1]. The high intensity of light causes the increase in the speed of photosynthesis. The existence of the effect of light intensity on the speed of photosynthesis causes primary producers in the deepwater environment is getting lower.

2. Research location

The research location is located in the waters of Pari Island, Pramuka Island and Kelapa Dua Island, which is administratively located in the Seribu Islands Administrative District, DKI Jakarta Province (Fig 2.1). On each island are placed 3 research stations for each ecosystem located on site.

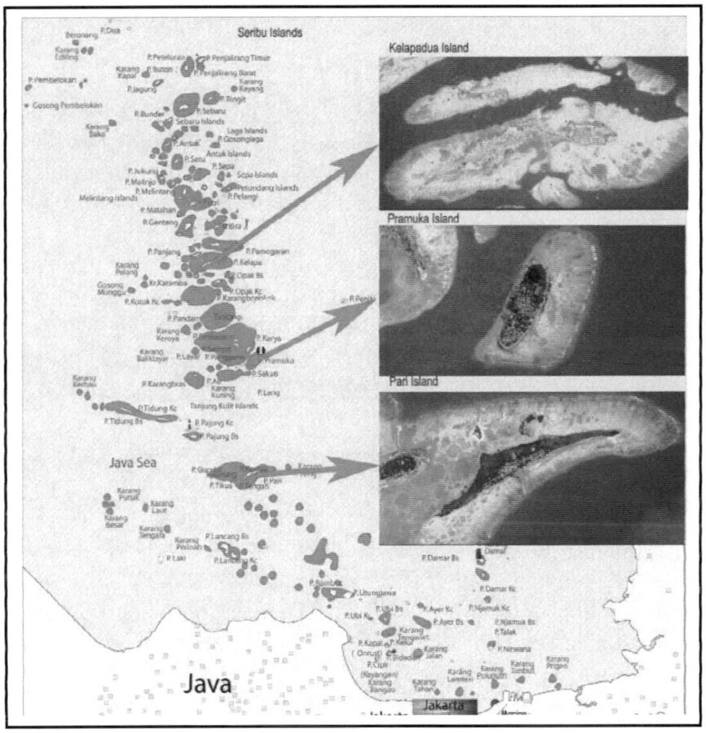

Figure 1. Seribu Islands map

Kelapa Dua Island is a fairly small island and nearby are Kelapa Island and Harapan Island: There are offices and mess of Kepulauan Seribu National Park, Hawksbill turtles captive breeding (*Eretmochelys imbricata*), and mangrove arboretum. The mangrove ecosystem is found in the east of the island along with the seagrass ecosystem. Meanwhile, coral reef ecosystems is found in the west, south, and the east of the islands.

Pramuka Island is the capital of Kepulauan Seribu Administrative District and is also known as a tourist island. Mangroves and seagrasses are found to the east of the Island, adjacent to the offices of the Kepulauan Seribu National Park Authority. Mangrove contained in Pramuka Island is the result of planting. Coral reefs are found around the island and 3 locations in the south, east and north are taken as observation stations.

Pari Island is the most crowded tourist island visited by tourists every weekend with the existence of marine tourism object and Pasir Perawan Beach. The seagrass observation station 1 is located in the south of the island, with a relatively small growth area, seagrass 2 and 3 observation stations are located around the Pasir Perawan Beach. Mangroves are found in the north of the island, starting from Pasir Perawan Beach to the west. Mangrove that was found in Pari Island is a natural mangrove. Meanwhile,coral reef data are taken in the north, east, and south of the islands.

3. Research methods

Research stations are determined based on satellite imagery observations. After obtaining the image data, the location of the station will be determinated based on the type of ecosystem and its conditions in the field.

3.1 Primary productivity measurement

Primary productivity measurements was conducted using oxygen, water sampling in each ecosystem contained in the research area. Equipment needed is water sampler, light and dark bottles size 300 ml and Oxygen meter. Incubation was carried out at the best time interval from 9.00 to 14.00 [12].

3.2 Coral reefs measurement

The method used to describe the condition of coral reefs is by Line Intercept Transect (LIT) method [9]. The clustering of coral community structures was carried out using lifeform categories by looking at the morphology of live coral cover, dead coral, substrate form (sand, mud), algae and other biota existence. Coral specs are expected to be recorded as a form of coral growth (lifeform).

3.3 Seagrass measurement

Data collection was conducted to determine the condition of seagrass (type, density and percentage of seagrass cover). Observations of seagrass beds will be conducted at low tide, with a depth of less than 1 meter. Method for observation of seagrass ecosystem used quadratic transect method. The squares to be used are 1 x 1 m and the transects will be horizontally aligned with the coastline starting from the first region of Seagrass up to the closest to the reef [2].

3.4 Mangroves measurement

Mangrove vegetation data was collected using survey technique and mangrove vegetation analysis. Analysis of mangrove forest vegetation in this research activity is done by a combination method of transect method (transect method) and quadrat method (quadrat method) [3][8].

Water quality data were taken in each ecosystem at each station using refractometer to measure aquatic salinity, luxmeter to measure light intensity, and thermometer to measure water temperature and DO meter to measure dissolved oxygen.

4. Data Processing

4.1 Primary productivity data processing

The primary productivity value of the waters was measured using the following formula [1]:

$$\frac{(O_2BT - O_2BA) \times 1000 \times 0{,}375}{PQ(t)} \tag{1}$$

Where NPP is the net primary productivity (mg C / m^3 / hr), O_2BT is oxygen in the bright bottle (BT) after incubation (mg / l), O_2BA is oxygen at the initial bottle (BA) (mg / l), PQ is the coefficient Photosynthesis, t is the incubation time (hr), 1000 is the conversion of liter to m^3, and 0.375 is the oxygen conversion coefficient to carbon.

IOP Conf. Series: Earth and Environmental Science **162** (2018) 012025 doi:10.1088/1755-1315/162/1/012025

4.2 Mangrove data processing

The calculation of the quantitative value of the mangrove vegetation parameters, especially in the determination of the density is done by using the following formula: [3]

$$cr = \frac{ci}{c} \times 100\% \tag{2}$$

Where:
cr : Relative density
ci : Species density
c : All species density

4.3 Seagrass data processing

To determine the extent of certain seagrass cover areas compared to the total area of cover for all seagrass species, a method was adapted from the Saito and Atobe Methods [2].

$$c = \frac{\sum seagrass}{\sum A} \tag{3}$$

Where:
c = Seagrass density
A : Total area

4.4 Coral reefs data processing

Coral reef condition is obtained from direct measurement result in field by Line Intercept Transect method. Furthermore, data processing is done by calculating Percentage of Coverage [9], i.e. :

$$n_i = \frac{l_i}{L} \times 100\% \tag{4}$$

Where:
n_i = life hard corals coverage percentage
l_i = life form transect length
L = total transect length

5. Results and discussion

5.1 Mangrove ecosystem condition

Mangrove ecosystems in the Seribu Islands particularly among the sites are generally a result of mangrove planting in the region of Kepulauan Seribu National Park on the island of Kelapa Dua and Pramuka. The species found was *Rhizopora Stylosssa*. While for the natural Mangrove found in Pulau Pari. The highest mangrove density was found in Pramuka Island precisely at Station 1 with the value of density reach 19,500 stands / hectares and the lowest is in the island of Pari precisely at Station 1 with a value of 1267 stands / hectare. This significant difference is caused by differences in Mangrove condition in Kelapa Dua island and Pramuka island where the a mangrove planting is very dense in distance, unlike the mangrove in Pari island that grows naturally with it condition which has big stem diameter.

Table 1. Mangrove density

Station	Density (stands/ha)	Category [7]
Kelapa 2-1	8,600	Good
Kelapa 2-2	12,060	Good

IOP Conf. Series: Earth and Environmental Science **162** (2018) 012025 doi:10.1088/1755-1315/162/1/012025

Kelapa 2-3	10,220	Good
Pramuka-1	19,500	Good
Pramuka-2	15,500	Good
Pramuka-3	14,233	Good
Pari-1	1,267	Good
Pari-2	1,567	Good
Pari-3	3,133	Good

5.2 Seagrass ecosystem condition

Seagrass species found in the study sites are *Enhalus acoroides, Thalasia hemprici, Halophila spinulosa,* and *Halodule unninervis*. The condition of Seagrass ecosystem at the highest research location is found in Kelapa Island precisely at Station 3 which is 1,099 stands / m^2. Meanwhile, for the lowest density is found in Pari Island precisely at station 2, however, the Seagrass plants in Pari Island have larger leaves than those found on other islands.

Table 2. Seagrass density and coverage

Station	Density (stands/m^2)	cover (%)	Category/status[6]
Kelapa 2-2	469	34	Damaged/less
Kelapa 2-3	1099.33	34	Damaged/less
Pramuka-1	690.33	40	Damaged/less
Pramuka-2	671	33,33	Damaged/less
Pramuka-3	502.67	24	Damaged/poor
Pari-1	413.33	15	Damaged/poor
Pari-2	30	4	Damaged/poor
Pari-3	71	5	Damaged/poor

5.3 Coral reefs ecosystem condition

The condition of coral reef ecosystem at the best research sites is found in Harapan Island with live coral cover reach up to 52.3% and the lowest is in Pramuka Island station 2.

Table 3. Coral reefs coverage

Station	Coverage (%)	Category [4]
Kelapa	26.76	Average damaged
Kelapa Dua	36.78	Average damaged
Harapan	53.20	Good
Pramuka-1	42.40	Average damaged
Pramuka-2	21.70	Heavily damaged
Pramuka-3	50.90	Good
Pari-1	45.46	Average damaged
Pari-2	29.32	Average damaged
Pari-3	41.3	Average damaged

5.4 Primary productivity

Data retrieval of primary productivity in the waters of the Seribu Islands carried out at each coastal ecosystems in each island. In general, the largest primary productivity found in mangrove ecosystem in Kelapa Dua Island, with an average of 3.0062 gC / m^3 / day or 1097.3 gC / m^3 / year. Meanwhile, the lowest primary productivity value was found on the seagrass ecosystem in Pari Island with an average primary productivity of 0.6 gC / m^3 / day or 225.26 gC / m^3 / year.

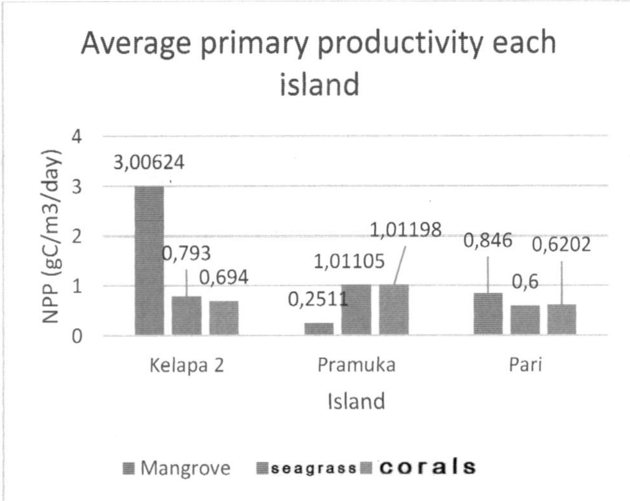

Figure 2. Primary productivity in Seribu Islands

5.5 Relationship of Primary Productivity with Aquatic Chemical-Physical Parameters

The intensity of light greatly affects the productivity of the slicker primer. At the maximum light intensity of 77,000 lux, the primary productivity also reached the highest value of 3.354 gC / m^3 / day. Salinity in the study sites ranged from 28.5 to 33 psu, where the highest salinity was found in the waters of Kelapa island and surrounding areas precisely in the mangrove community. And the lowest is in the waters of the Pramuka island, precisely in the community of mangrove and seagrass. In the process of data collection it is clear that at the time of data collection in the Pramuka island, the weather conditions were rainy while at the time of measurement in the Kelapa island , the weather conditions were bright.

Water temperatures in the study sites ranged from 28.65 - 33.3 ^oC. This surface temperature is strongly influenced by the high solar irradiance. This water temperatures are closely related to the primary productivity of waters.

Nitrate concentration in the three islands is very high, exceeds the quality standard that is 0.008 mg / l and exceeds the tolerance limit of 0.2 mg /l except in Kelapa Dua Island. This excess of nitrate triggers the growth of algae in large quantities. With the growth of these algae, there can be a shortage of oxygen in the water. In the field observations there are many seagrass covered by algae and it is found DCA (Dead Coral Algae) or dead corals that have been overgrown with algae on the coral reef ecosystem. Phosphate content in the waters of the three islands exceeds the standard of quality and the tolerance limit except in some stations that exhibit very small phosphate levels of less than 0.01 mg /l. Phosphate concentrations below 0.03 mg / l can stimulate the growth of phytoplankton, but in larger concentrations can cause algae growth that interfere with the life of phytoplankton in the sea. The content of nitrate and phosphate is opposed to its primary productivity value, but in fact nutrients are one of the main factors affecting the primary productivity of waters. Nitrate and phosphate is a nutrients and used by phytoplankton for its growth process, so with little nutrient content, it can be said that the optimum growth of phytoplankton in the area, so the value of primary productivity is high and vice versa if excess nutrients can trigger the growth of algae interfere with the growth of phytoplankton.

5.6 Relationship between Primary Productivity and Mangrove

Mangrove density is related to primary productivity, as shown in Figure 5.2. It exhibits that the mangrove ecosystem in Kelapa Dua Island and Pari Island in line with its primary productivity value. However, there is a difference on Pramuka Island compared to the other islands, it because primary productivity is closely related to sunlight intensity and at the time of data collection in Pramuka Island, the weather conditions were cloudy and rain, it causing primary productivity decrease due to reduced light intensity and mangrove tree canopy meeting is blocking the light.

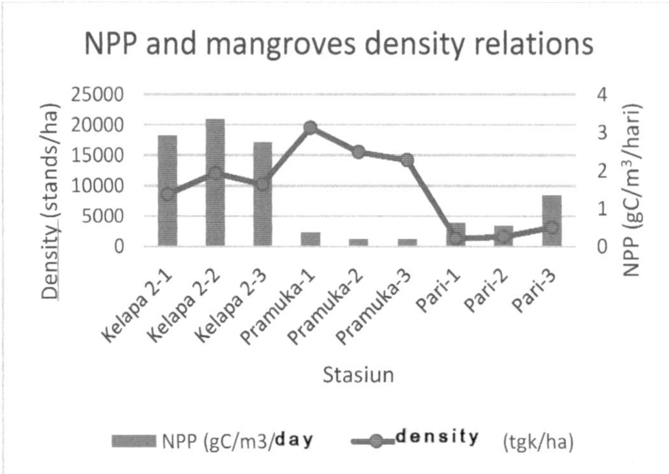

Figure 3. NPP and mangroves density relations

5.7 Relationship between Primary Productivity and Seagrass

According to the obtained data, the higher seagrass density, the higher primary productivity value of the waters. The relationship between primary productivity value and seagrass density is presented in Figure 5.3. Seagrass is a plant that can perform photosynthesis so that the productivity of waters is produced by phytoplankton and is also produced by seagrass. However, the primary productivity value obtained only came from photosynthesis of phytoplankton because the water that incubated in bright and dark bottles did not including seagrasses.

Based on the obtained data, the density of seagrass in Kelapa Dua Island is higher than that of seagrass in Pramuka Island, but its primary productivity is higher in Pramuka Island, this is because the type of seagrass found in Pramuka Island is only one type of *Enhalus acoroides* while in Kelapa Dua Island found more than one type. *Enhalus acoroides* is known to have the highest primary productivity among other seagrass species [12]. This causes the primary productivity of seagrass in Pramuka Island is higher than the productivity of seagrass in Kelapa Dua Island.

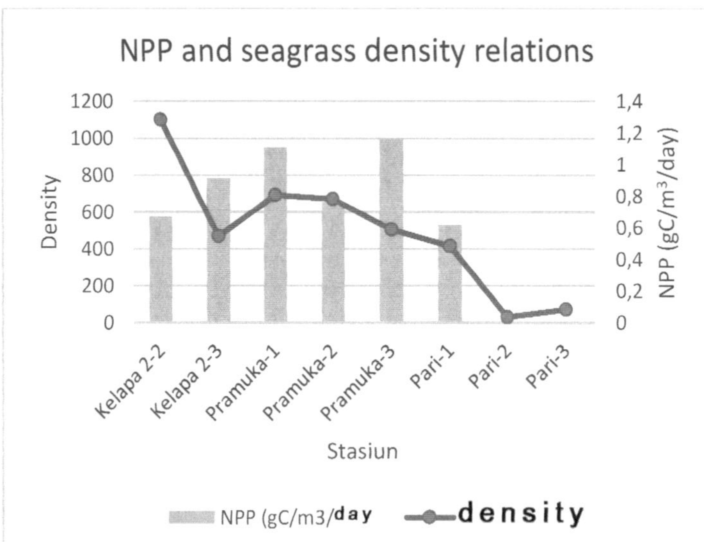

Figure 4. NPP and seagrass density relations

5.8 Relationship between Primary Productivity and coral reefs

According to the obtained data, primary productivity value is closely related to the value of live coral cover in the study sites, as shown in the figure, increase in primary productivity value followed by increase in live coral cover value and also decrease in primary productivity value followed by decrease in live coral cover value. This proves that the high value of primary productivity is influenced by the height of life coral cover and its species diversity [10].

The highest primary productivity value of the coral reef ecosystems is found in Pramuka island at station 3, but the highest coral cover is found in Kelapa Dua island at station 3. This is due to the diversity of species diversity, based on visual observation in the field, the diversity of coral lifeform in Pramuka island at station 3 is larger than Kelapa Dua island at station 3.

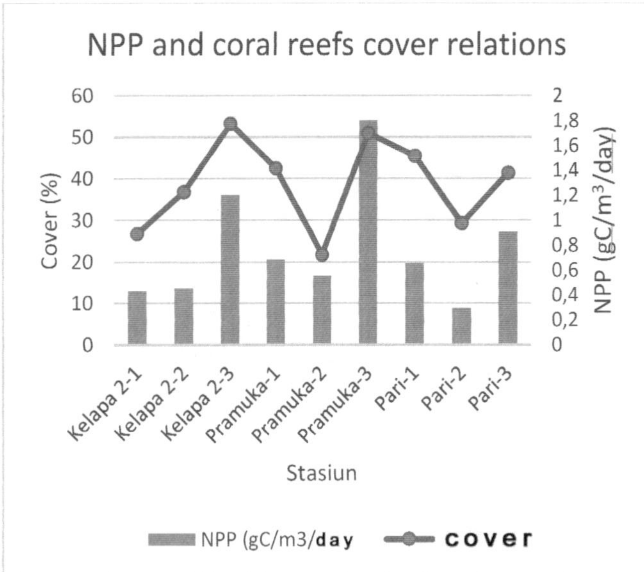

Figure 5. NPP and coral reefs cover relations

6. Conclusion

Based on the results,the research can be concluded as follows: (1) Generally, the condition of coastal ecosystems in Seribu Islands have various conditions, from severely damaged to still in good category. The position of the island from the mainland (Java) is getting much higher primary productivity with the highest in Kelapa Dua island in the range of 3.0062 gC / m^3 / day, followed Pramuka Island with a range of 1.01 gC / m^3 / day, while The lowest was found on Pari Island with primary productivity values in the range of 0.6 gC / m^3 / day. (2) Mangroves at the research sites (Pulau Kelapa Dua, Pulau Pramuka and Pulau Pari) are categorized as good. The best mangrove condition is found in Pramuka Island with the density of 19,500 stands / hectare and the worst condition is found in Pari Island with 1,267 stands / hectare. (3) Seagrass at the research site is categorized as damaged. The best seagrass condition is found in Kelapa Dua Island with a density of 1,099 stands / m^2, whereas the worst condition is found in Pari Island with a density of 30 stands / m^2. (4) The best coral reefs found in Kelapa Dua Island with live coral cover reach 53.2% and the heavily damaged category is found in Pramuka Island with live coral cover only 21.7%. (5) Primary productivity in the highest mangrove ecosystem is found in Kelapa Dua Island 2.74 - 3.35 gC / m^3 / day and the lowest is in Pramuka Island with primary productivity 0.18 - 0.37 gC / m^3 / day. (6) The primary productivity of the highest seagrass ecosystems was found in Pramuka Island with a value of 0.7-1.11 gC / m^3 / day and the lowest was in Pari Island with a primary productivity of 0.6 gC / m^3 / day. (7) The primary productivity of the highest coral reef ecosystems was found on Pramuka Island in the range of 0.55 - 1.8 gC / m^3 / day and the lowest was found on Pari Island with primary productivity of 0.29 to 0.91 gC / m^3 / day.

Acknowledgement
This research is funded by a magister scholarship from the Ministry of Research Technology and Higher Education - Republic of Indonesia. The authors would like to thank the Minister of Research Technology and Higher Education.

References

[1] Alianto Adiwilaga E M and Damar A 2008 *Produktivitas Primer Fitoplankton dan Keterkaitannya dengan Cahaya dan Unsur Hara di Perairan Teluk Banten.* Jurnal Ilmu-ilmu Perairan dan Perikanan Indonesia, Juni 2008, Jilid 15, Nomor 1: 21-26
[2] Ansyori A I 2014 *Kemampuan Komunitas Padang Lamun dalam Menyerap Karbon yang Telah Terdegradasi di Teluk Banten.* Thesis. PSMIL Universitas Padjadjaran. Bandung.
[3] Fachrul M F 2008 *Metode Sampling Bioekologi* Bumi Aksara. Jakarta. 198 hlm.
[4] Keputusan Menteri Negara Lingkungan Hidup No. 04 Tahun 2001 tentang Kriteria Baku Kerusakan Terumbu Karang
[5] Keputusan Menteri Negara Lingkungan Hidup No. 51 Tahun 2004 tentang Baku Mutu Air Laut
[6] Keputusan Menteri Negara Lingkungan Hidup No. 200 Tahun 2004 tentang Kriteria Baku Kerusakan dan Penentuan Status Padang Lamun
[7] Keputusan Menteri Negara Lingkungan Hidup No. 201 Tahun 2004 tentang Kriteria Baku Kerusakan Mangrove
[8] Nurrahman Y A 2012 Struktur dan Komposisi Vegetasi Mangrove di Kecamatan Sungai Raya Kepulauan Provinsi Kalimantan Barat *Jurnal Akuatika* **II** 2012
[9] Sudiono G 2008 *Analisis Pengelolaan Terumbu Karang Pada Kawasan Konservasi Laut Daerah (KKLD) Pulau Randayan dan Sekitarnya Kabupaten Bengkayang Provinsi Kalimantan Barat.* Thesis. ePrints Undip. Semarang.
[10] Supriharyono 2002 " *Pelestarian dan Pengelolaan Sumber Daya Alam di Wilayah Pesisir Tropis.* PT. Gramedia Pustaka Utama. Jakarta.
[11] Tambaru dkk 2008 *"Penentuan Selang Waktu Inkubasi Yang Terbaik Dalam Pengukuran Produktivitas Primer Di Perairan Kepulauan Spermonde.*" Jurnal Torani 18.3. 2008.
[12] Tomascik T, Mah A J, Nontji A and Moosa M K 1997 *The ecology of the Indonesian seas.* Oxford University Press, Oxford, 829 – 898

IOP Conf. Series: Earth and Environmental Science **162** (2018) 012026 doi:10.1088/1755-1315/162/1/012026

Study of biophysical status and resources support marine tourism area of mangrove in Indramayu Karangsong

Donny Juliandri Prihadi, Indah Riyantini and Mochamad Rudyansyah Ismail

Marine Science Department, Faculty of Fisheries and Marine Science, Universitas Padjadjaran, Jalan Raya Sumedang-Bandung KM.21 Jatinangor, UBR 40600, Sumedang, Indonesia

E-mail: donny.juliandri.prihadi@unpad.ac.id

Abstract. Karangsong beach is located in the north of Indramayu district at Karangsong village. It has an area of approximately 25 Ha for mangrove forests conservation. This research had been conducted from April to June 2017 with a purpose of surveying mangrove's diversity and density, evaluating the status of mangrove ecosystems, and assessing the management of mangrove marine tourism area in Karangsong, Indramayu. We found out that throughout the whole ecotourism area, there are three types of mangrove: *Rhizophora mucronata, Rhizophora apiculata* and *Aviciena marina*. The biophysical status of mangrove ecosystem proves to be suitable for ecotourism, operations are well managed and could be accounted for its sustainability.

1. Introduction

One of the potentials possessed by coastal areas is marine tourism. Mangrove ecosystems have high ecological and economical values. Their roles as protectors of coastlines, can prevent sea water intrusions. They contain a wide range of biodiversity (habitats) where they function as feeding, nursery and spawning grounds for various aquatic biota. They also have economical functions that allow the production of household and industrial goods, and could be used as recreational areas for tourist attractions.

In 2017, mangrove forests in Indramayu Regency have sustained the worst damage in West Java province. Based on the analysis of satellite imagery data, in 2008, mangrove forests in Indramayu has an area of 1,103 Ha [2]. In the past 9 years, from 2008 until 2017, there should be an increase in its land area.

Mangrove management should consider methods that are not destructive neither directly nor indirectly, and their usage can be used as an additional source of income for people living around the area. Ecologically, mangrove ecosystems can protect the physical environment from anchoring waves, wind and are the habitat of a variety of wildlife such as monkeys, snakes, monitor lizards, birds, fish, and shellfish.

Karangsong beach is located in the north of Indramayu, a city where its village lies. Its coastal regions have approximately 25 Ha of mangrove forests' conservation areas. From 2008 till 2014, mangrove trees were being planted by State officials, Governor of Indramayu and Pertamina. These areas are then given access to public for tours.

IOP Conf. Series: Earth and Environmental Science **162** (2018) 012026 doi:10.1088/1755-1315/162/1/012026

Mangroves are usually damaged due to the lack of planning in formulating the management procedures and also due to the pressure resulting from economic necessities. Some of the main reasons that lead to the damages are: (1) a high number of populations, therefore the demand to convert mangroves to human goods increase; (2) planning and managing coastal resources in the past were very expensive; (3) low level of public awareness towards conservation and functions of mangrove forests; and (4) poverty of coastal communities.

With proper planning in conservation, mangrove ecosystems will be healthier, better and can affect the well-being of local communities. One of the plans that is currently being made is tourist sites. Marine tourism is of specific interest to which maritime-related activities, above sea level (marine) and under the sea surface (submarine), are conducted. This potential of coastal resources can be used for tours, hence it's referred to as marine tourism. Nowadays marine tourism is specially activity on tourism.

Ecotourism is a form of marine resource management from the development of sea and coastal conservation [4]. Its concepts are not emphasized on economic growth factor, but maintaining a balance between resource preservation and utilization [13]. Therefore, marine tourism is a concept which connects tourism with conservation. Ecotourism is a form of activity that involves tourists enjoying themselves and at the same time, being responsible towards the environment.

2. Materials and methods

The research was carried out in mangrove areas which are meant for tourists in Karangsong, Indramayu Regency. It started from February-July 2017. The source of data used are primary and secondary data. Data was obtained by objective random sampling.

Random sampling is done to obtain samples from each population. Based on the total area of mangrove tourism, this method is used to allow the same opportunity for each existing populations to be taken as samples. For density measurements and mangrove types, the data that were obtained are physical and chemical parameters of water, and biodiversity along with its associations. Purposive sampling was used to interview visitors and determine observation points based on specific objectives and the researchers' considerations, such as data collection of visitors (tourists), length measurement of the river, its depth and substrate. Secondary data are data from related services and research reports in advance.

2.1. Sampling Location

Sampling was done in 8 (eight) stations which represent all the characteristics of mangrove forests in Karangsong (Figure 1).

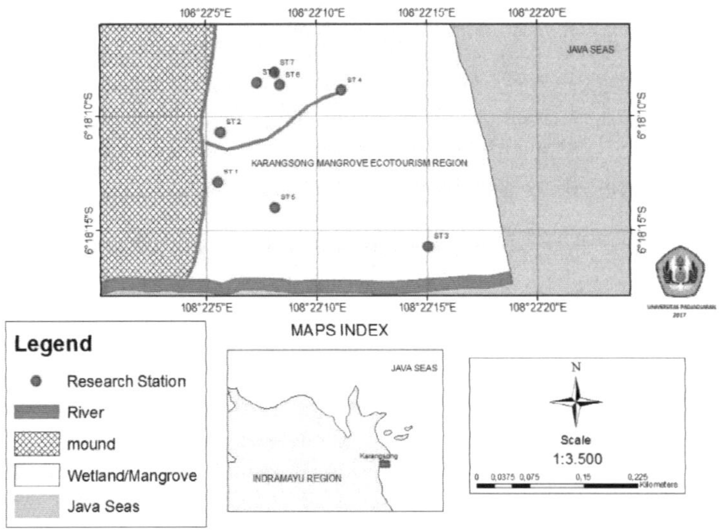

Figure 1. Map of Mangrove Research Station in Karangsong

2.2. Transect Sampling

To obtain transect sampling data, a line is drawn from a point located in the outermost area of mangrove trees, in the direction perpendicular to the coastline. The transect line is pulled as long as 50 m. Each line was made so that the observation plots are in accordance with its growth rate (Figure 2). The size of tree trunks and stakes are known by measuring the circumference of chest-high stalks (GBH), whereas the rate is only to log the type and number of stands.

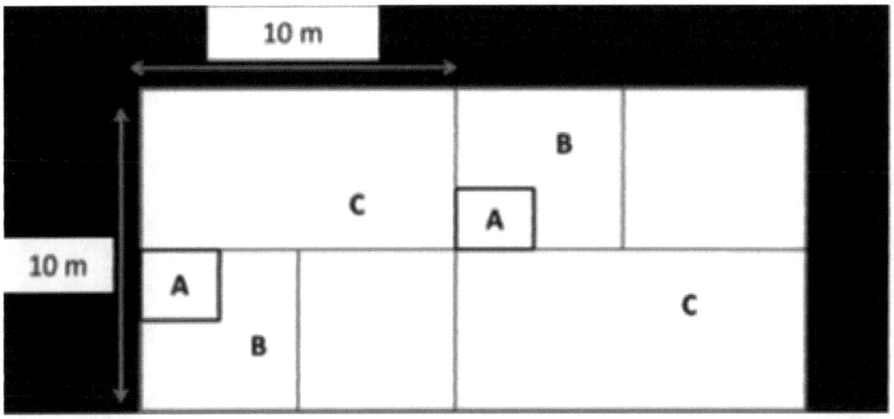

Figure 2. The observation design of vegetation in the field

Description:

A: Plots for observation (1 m x 1 m)
B: Plots for stake observation (5 m x 5 m)
C: Hiding plots for trees observation (10 m x 10 m)

2.3. Water Quality

The data required to study water quality has been obtained. It is used to study the physical and chemical parameters of the water, such as: temperature, salinity, dissolved oxygen (DO) and acidity (pH).

2.4. Biota Association

Biota data would be calculated and collected from direct observation and interviews with society/fishing in the surrounding area to get information they may not found or seen directly at the time. Observations were done to see objects that were not predetermined on biota criteria assessment which are based on object biota in the table according to the ecotourism suitability matrix [13]. Biota-object are observed in every research station. They found on photo and identification based on the macrozoobenthos book on mangroves.

2.5. Mangrove Substrate

Substrate and sediment samples were taken from stations that represent mangrove ecosystem areas in every location that has been defined and transected. The methods used for substrate analysis is sieve (Granulometry), using large scale grain Wentworth.

2.6. Suitability of Ecotourism Area

Based on the suitability calculations, the data obtained are as follows (table 1):
Maximum value = 88
S1 = Very fit, with a value of 80 - 100%
S2 = Fit, with a value of 60 - < 80%
S3 = Corresponding conditional, with a value of 35 - < 60%

N = no match, with a value of < 35%

IKW = Σ (Ni/Nmax) x 100%
IKW = index of suitability for mangrove ecosystems

Great fit/ really suitable : 80 - 100%,
Suitably fits : 60 - 80%,
Corresponding Conditional : 35 - < 60% ,
Not suitable : < 35 %).
NI = the value of the parameter to-i (Weight x Score).
Nmax = the maximum value of category of mangrove tour

Table 1. Matrix of Suitability For Mangrove Ecosystem Areas

No.	Criteria	Density	Score 4	Score 3	Score 2	Score 1
1	Density (trees/100 m^2)	3	>15-25/100 m^2	>10-15/100 m^2	5-10/100 m^2	<5/100 m^2
2	Mangrove type	5	>5	3-5	1-2	0
3	Thickness (m)	3	>500 m	>200-500 m	50-200 m	<50 m
4	Biota Association	5	4	3	2	1
5	River's Length (km)	1	>3 km	3 km	2 km	1 km
6	River's Width (km)	1	>500 m	201-500 m	4-200 m	<4 m
7	River's depth (m)	3	>3-5m	>2-3m	1-2m	<1m
8	Substrate	1	Rough, a little rough and soft sand	Rough and a little rough sand	Soft sand	Sand

3. Results

This research identified 3 types of mangroves in Karangsong, namely *Avicennia marina*, *Rhizophora mucronata* and *Rhizophora stylosa*.

3.1. Water Quality and Substrate

Water quality in mangrove ecosystems are divided into 4 physical and chemical parameters, including temperature, salinity, dissolved oxygen (DO) and power of hydrogen (pH). The testing was done in 8 stations in Karangsong and the results are represented by mangroves (table 2).

Table 2. The results of water quality and substrate measurements

Parameter	**Station**							
	I	**II**	**III**	**IV**	**V**	**VI**	**VII**	**VIII**
Temperature (°C)	30,33	29,5	30,5	30	30	29	30	30
Salinity (%)	28	27	29	28	29	28	29	29
DO (mg/L)	6,4	5,2	5.8	5,3	5,3	5,4	5,3	5,3
Water pH	7,5	7,4	7,33	7,5	7,4	7,5	7,4	7,5
Substrate type	Soft sand (mud)		Sand			Sand	Sand	

3.1.1. Temperature

The results of the measurements done on each station show that the temperature range in mangroves of Karangsong are : Station I 30.33 ° C, Station II 29.5 ° C, Station III 30.5 ° C, Station IV 30 °C, Station V 30 °C, Station VII 30 °C and Station VIII 30 °C. Generally, the temperature in each station is not much different from the temperature range of observation stations. This condition supports life biodiversity like mangrove crabs (*Scylla* sp) and violin crabs (*Uca* sp.). These temperatures are also suitable for mangroves to survive. The difference in time when measurements were taken in each station is related to the intensity of sunlight received by water bodies, weather conditions, and shade (closure) by plants.

3.1.2. Salinity

The results of the measurements done on each station show that the salinity range in water bodies of Karangsong are 28 ppt in Station I, IV & VI, 27 ppt in Station II, and 29 ppt in Station III, V, VII & VIII. Generally, the salinity on each station is not much different from the salinity range of observation stations. This condition supports the survival of violin crabs (*Uca* sp.). The results are in accordance with Kepmen LH No. 51 Year 2004 which states that the salinity suitable to support life for mangrove crabs (*Scylla* sp) and violin crabs (*Uca* sp) are up to 34 ppt. The salinity range in observation sites still shows corresponding values for mangroves' lives. The optimum salinity for the growth of mangroves ranges between 10-30 ppt [10].

3.1.3. Dissolved Oxygen (DO)

Based on the results of dissolved oxygen (DO) measured from each station, the data obtained are : Station I 6.4 ppm, Station II 5.2 ppm, Station III 5.8 ppm, Station VI 5.4 ppm and 5.3 ppm for Station IV, V, VIII & VIII. These are good conditions for the life of mangroves. Mangroves can live in waters with oxygen levels > 4 mg/l [10]. The amount of dissolved oxygen in water is affected by the process of aeration, photosynthesis, respiration, and oxidation of organic materials [9]. There is a relationship between oxygen levels with temperature which is : an increase in temperature reduces the solubility of oxygen. DO is indispensable for the survival of aquatic organisms, including mangrove [9].

3.1.4. pH

Based on the results of pH measured from each station, the data obtained are : 7.55 for Station I, 7.4 for Station II, III & IV, 7.33 for Station 3, and 7.5 for Station IV, VI & VIII. In general, there is a slight difference in pH. The pH of the water in mangrove observation sites were assessed according to mangrove plants. Generally, mangroves are still able to grow at a pH range of 5 – 8.5 which is in compliance with Quality Raw on Kepmen LH No. 51 Year 2004 which stated that the ideal pH water for mangroves is 7 – 8.5.

3.1.5. Substrate

Based on the identification results of substrate samples conducted in the laboratories of marine science and technology, the data obtained can be seen in table 3.

Table 3. Results of Mangrove Forest's Substrate in Karangsong

Station	Gravel (%)	Sand (%)	Mud Texture (%)	Texture class
1	0,1	74,3	25,6	Muddy sand
2	0	82,9	17,1	sand
3	0	96	4	Sand
4	0	93,6	6,4	Sand
5	0	96	4	Sand
6	0	93,5	6,5	Sand

7	0	95	5	Sand
8	0	95	5	Sand

The substrate observation results from each station are : Station I consists of 0.1% gravel, 25.6% muddy sand, and 74.3% sand, Station II consists of 0% gravel, 17.1% muddy sand and 82.9% sand, Station III consists of 0% gravel, 4% muddy sand and 96% sand, Station IV consists of 0% gravel, 6.4% muddy sand and 93.5% sand, Station V consists of 0% gravel, 4% muddy sand and 96% sand, Station VI consists of 0% gravel, 6.5% muddy sand, 93.6% sand, Station VII consists of 0% gravel, 5% muddy sand and 95% sand, and Station VIII consists of 0% gravel, 5% muddy sand and 95 % sand. The station has a pad sand. Sand is classified as little stones/ pebbles. The presence of different types of sediment in each station is probably due to the source of sediments. Coarse-sized particles generally are deposited in a location not far from their source, whereas finer particles will be further transported by currents and waves, and deposited even further from the source.

Mangroves in Karangsong are the result of rehabilitation by the villagers which started in 1998. Data observations from 8 stations showed that mangroves at criteria are rarely included in Karangsong (broken) with density (trees/ha) presented at < 1000 and only station 1, 5, and 6 are the ones that have medium damage; see table 4.

Table 4. The Measurement Results of Mangrove Density

Station	Density (trees/ha)
1	1040
2	380
3	560
4	620
5	1160
6	1060
7	860
8	920

Based on the decision of Kepmen LH number 201 year 2004 about raw damage, mangroves in station I-VIII have damaged ecosystem conditions with sparse density. The data are presented in table 5.

Tabel 5. Raw Damage Criteria for Mangroves

Criteria	Density (trees/ha)
Solid	>1500
Medium	>1000 - <1500
Rare (damage)	<1000

Source: Decree of the State Minister of Environment No. 201 Year 2004

Station I and II are located in the mangrove forest, near Jetty boats which ferry tourists. According to Kepmen LH No. 21 Year 2004, a density of 1040 trees/Ha are dominated by *Avicennia* sp mangroves. Station II has 380 trees/Ha and are dominated by *Rhizopora* types, while Station III station has 560 trees/Ha which is located close to the beach and is dominated by *Avicennia* species. Station IV has 620 trees/Ha and is conveniently located near a pond which is dominated by *Rhizopora* sp. Station V to VIII has 860 – 1160 trees/Ha and are dominated by *Rhizophora* sp.

Station V contains the highest, most valuable density of mangroves (1160 ind/ha) and Station II has the lowest density (380 ind/ha). The mangroves in Karangsong are still largely in the stake phase. The low density vegetation and the amount of mangrove species in the territorial waters may diminish even further due to the threat caused by the plans of converting the area into fishponds, logging and settlements. The criteria for raw damage of mangroves have changed from medium to rare.

Mangrove forests have been largely influenced by human activities, therefore high sedimentation and habitat changes can affect mangrove zoning. Availability of propagule are allegedly more influential in the reproduction process [8]. If the environmental conditions are suitable or appropriate, the mangroves will reproduce. This is related to the adaptation of mangrove resources against extreme conditions, for example, shoals of mud will be dominated by the propagule up at the venue.

3.2. Biota Association

There are 5 objects of biota which are associated with mangrove forests in Karangsong, Indramayu. In these five objects, there are 14 types of biota (table 6).

Table 6. The Observation Results Of Biota Association

No.	Object Biota	Type of Biota
1	Fish	Gelodok fish (*Periopthalamus sp*)
		Mullet fish (*Mugil dosumieri*)
2	*Crustacea*	Mangrove crab (*Scylla serrata*)
		Violin crab (Uca sp)
3	Mollusc	Conus shell (Conus sp)
		Cassidula aurisfelis
		Cerithidea cingulate
		Cerithidea Pomacea
		Telescopium sp
		Mangrove shell (*Polymesoda bengalensis*)
4	Reptile	Monitor lizard (*Varanus salvator*)
		Mangrove snake (*Chryosopolea sp.*)
		Lizard (*Emoia crotostata*)
5	Bird	White Heron (*Bubulcus ibis kuntul*)

The mangrove biota's diversity such as fish, crabs, molluscs, reptiles (monitor lizards and snakes), monkeys and birds add appealing values in mangrove habitats. Terrestrial fauna does not have any special adaptations to live in mangrove forests, like insects, snakes, birds and primates. Aquatic fauna consists of two types : (a) fauna living in water column, particularly various types of fish and shrimps; and (b) the ones that occupy substrates especially crabs, shells and various types of invertebrate [7]. Mangrove communities have an important role for growth, reproduction, feeding and spawning, source of nutrients for fauna which is generally a commodity economy for human communities around the beach. Fish biota include: fish, shrimp, crab, mussels, other fisheries biodiversity and non-commercial use [1].

Biota living in mangrove areas need to be preserved. The most appropriate management for biota to be able to live is to involve local communities and district managers to be more peaceful and helpful - they can live and breed. As for the benefits for public and tourists, they receive a lot of information which are derived from the existence of biodiversity in mangrove areas, like how the plants and animals live, their way of living and breeding, how their habitats are, their beauty and forms. All of this can create a feeling of happiness.

Mangrove is an organism that has numerous functions from various aspects of biology, physics and chemistry. One of them is ecotourism. To support the biophysical components for marine ecotourism, land needs to be provided [11]. The biophysical relation with mangrove ecotourism can be described using the suitability matrix with these parameters: density, type, thickness, width and length of the associated river, and substrate [13]. These parameters will produce values based on the biophysical parameters of suitability. Mangrove ecotourism itself can be used as an alternative way to use mangroves in a positive manner besides using them just to produce wood [10]. This activity can also be used as a means of surveillance against the utilization of mangrove woods so as to preserve and maintain the natural vegetation (See Table 7).

Table 7. The Result of Suitability Matrix For Mangrove Ecosystem

No	Parameters	Density	Result	Score	Density X Score
1	Density (trees/100m^2)	3	8.25	3	9
2	Mangrove type	5	3 types	3	15
3	Thickness of the mangrove (m)	3	>500m	4	12
4	Biota Association	5	5 types of biota (fish, *Crustacea*, Mollusc, Reptile, Bird)	4	20
5	Substrate	1	Rough, medium and soft sand	4	4
6	River's Width (km)	1	4-200m	2	2
7	River's Length (km)	1	2 km	2	2
8	River's Depth (m)	3	>2-3m	3	9
Total		=			73
IKW		= Σ (73/88) x 100%			
IKW=Σ(Ni/Nmax) x 100%		82.95%			
Karangsong mangrove areas are really suitable for Tourism					

IKW = index of suitability for mangrove ecosystems

Mangrove tourism is a very suitable activity to gather tourists. The activities include walking, seeing birds, going to the beach, biota sightseeing and boat riding. Tourism activities that are developed should be tailored to the potential resources and allocation. Each activity requires resources and appropriate environmental attractions which should be developed [12]. Ecotourism is a form of marine resource management from the development of sea and coastal conservation [3]. A form of tourism that is based on environmental sustainability and socio-culture can also be referred to as marine ecotourism. In term on this research marine ecotourism is mangrove tourism.

To increase the suitability index for mangrove ecotourism, reforestation needs to be done to increase the mangrove's density and thickness. To use mangrove forests in a friendly and environmentally manner, ecotourisms need community involvement to protect, maintain and manage mangrove ecosystems; that is in itself a concept of ecotourism. In addition, these activities are expected to support the livelihood of communities.

4. Conclusions

There are 3 types of mangroves found, namely *Avicennia marina, Rhizophora apiculata* and *Rhizophora stylosa*. The parameters and quality of the existing mangrove environment are still within the appropriate raw quality of Kepmen LH. We found 14 biota which are associated with mangroves in Karangsong, Indramayu.

The associated biota found in mangrove forests are Gelodok fish (Periopthalamus sp), mullet fish (*Mugil dosumieri*), mangrove crabs (*Scylla serrata*), mangrove oysters (*Polymesoda bengalensis*), mangrove snake (*Chrysopelea* sp), monitor lizard (*Varanus salvator*), lizards (*Emoia acrotostata*) and White Heron (*Bubulcus ibis*, herons). The substrate composition on Princess Island is fine sand, medium sand and silt.

The suitability index for Mangrove Tourism is 82.95%, which proves that it really is suitable for ecotourism. This shows that the mangrove ecosystem's biophysical components are worthy and have the potential for mangrove ecotourism. This fact is based on biophysical analysis.

Further research needs to be done to learn more about tourists' carrying capacity and create new tourism activities.

References

[1] Direktorat Jendral Rehabilitasi Lahan dan Perhutanan Sosial 2008 Inventarisasi dan Identifiksi Pemanfaatan Sumberdaya Hutan Mangrove
[2] BPLHD Prov. Jawa Barat 2008 *Laporan Kajian Status Mutu Laut* Bandung
[3] Ketjulan R 2011 Daya Dukung Perairan Pulau Hari sebagai Obyek Ekowisata Bahari *Jurnal Aqua Hayati* **7** (3): 183-188
[4] Ketjulan Romy, Arip Bayu and Ahmad Mustafa 2013 Kajian Potensi dan Kesesuaian Ekosistem Terumbu Karang di Pulau Lura Untuk Pengembangan Ekowisata Bahari *Jurnal Mina Laut Indonesia* Vol. **01** No. 01: 49-60
[5] Keputusan Menteri Negara Lingkungan Hidup No. 51 Tahun 2004 tentang Baku Mutu Air Laut
[6] Keputusan Menteri Negara Lingkungan Hidup No. 201 Tahun 2004 Tentang Kriteria Baku Dan Pedoman Penentuan Kerusakan Mangrove
[7] Mulya M B 2000 *Kelimpahan dan Distribusi Kepiting Bakau (Scylla sp.) serta Keterkaitannya dengan Karakteristik Biofisik Hutan Mangrove di Suaka Margasatwa Karang Gading di Langkat Timur laut, Provinsi Sumatra Utara* Thesis Institut Pertanian Bogor Bogor
[8] Onrizal and C Kusmana 2005 *Ekologi dan manajemen mangrove Indonesia* Buku Ajar Departemen Kehutanan FP USU Medan
[9] RSNI 2001 *Survey dan Pemetaan Mangrove* RSNI
[10] Saparinto C 2007 *Pendayagunaan Ekosistem Mangrove* PT. Dahara Prize Semarang
[11] Safri, Chandra Joei and Andi Zulfikar 2015 Analisis Biofisik Ekosistem Mangrove Untuk Kawasan Ekowisata Mangrove Sungai Kawal *Jurnal FIKP UMRAH*
[12] Shiddieqy Ash, Falmi Yandri and Arief Pratomo 2014 Kelayakan Ekowisata Mangrove Arungan Sungai Di Sungai Carang Berdasarkan Pada Biofisik Mangrove *Jurnal FIKP UMRAH*
[13] Yulianda F 2007 *Ekowisata Bahari Sebagai Alternatif Pemanfaatan Sumberdaya Pesisir Berbasis Konservasi.* Bogor: Institut Pertanian Bogor

Quantitative assessment of vulnerability in aquaculture: climate change impacts on whiteleg shrimp (*Litopenaeus vannamei*) farming in East Java Province

A D Puspa[1], T Osawa[2] and I W Arthana[3]

[1] Department of Environmental Remote Sensing, Udayana University, Bali, Indonesia
[2] CReSOS, Center for Remote Sensing and Ocean Sciences, Udayana University, Bali, Indonesia
[3] Faculty of Marine Science and Fisheries, Udayana University, Bali, Indonesia

E-mail: orindwipuspa@gmail.com

Abstract. Whiteleg shrimp (*Litopenaeus vannmaei*) commodities has a high economic value and a huge market demand. Changes in temperature and precipitation seems to be the most significant factors climatic that effect shrimp aquaculture production. The objectives are to understand of climatic factors influence on the whiteleg shrimp farming sector in Indonesia based on IPSL(Institut Pierre Simon Laplace)-CM5A-MR(Medium Resolution) model and to understand adaptability of whiteleg shrimp to climate change on vulnerable area. The projection of the model includes four Representative Concentration Pathway (RCP) scenarios, encompassing RCP 2.6, RCP 4.5, RCP 6.0, and RCP 8.5. These vulnerable map were created by overlaying the exposure maps (precipitation, maximum temperature and minimum temperature) and sensitivity map (production maps) in various time. Banyuwangi District was the most vulnerable area in East Java Province. The questionary has analysed descriptively and found that 92% of the respondents are aware of climate change. The respondents reveal that increased on rainfall (34%) and decreased of temperature (29%) has negative impacts on shrimp production. Decrease in production influenced by disease and climate factors. Either future threats or current threats related to climate change and disease. More deeply there are insignificant positive correlation between temperature and precipitation with production in Banyuwangi.

1. Introduction

Indonesia is an archipelagic country, traversed by the equator which makes Indonesia has rainy and dry season, remain throughout every year. Obviously it is very profitable for shrimp farming activities. Shrimp farming in Indonesia is a major export commodities help in increasing state revenues, compared to other fishery products such as tuna and blue swimmer crabs. Climate change that occurs as a result of excessive use of fossil fuels has a devastating effect on Indonesia, including for shrimp farming activities. In 2010 – 2012 whiteleg shrimp in East Java increased, but not in the following year, possibly due to reduced water quality and ecosystem environment. In this case, the uncertain climate change will affect the water quality in ponds [1].

The impacts of climate change could occur directly or indirectly on shrimp farming and cannot be attributed as one single factor, in the case of many aquaculture practices. Changes in average of temperature and average precipitation drives of climate change on shrimp aquaculture. Increased temperatures will affect pond of evaporation rates and the resultant temperatures for aquaculture on water quality in source water bodies. It could also increase the intensity and frequency of disease outbreaks. Change in amount of rainfall in this case decrease in the salinity, depletion of dissolved oxygen, algae blooms, particularly in summer months. It becomes difficult in inland and coastal areas of brackish water shrimp ponds can significantly impact to farm production [2].

The research predicts about the level of vulnerability in shrimp farming activities. The climate change scenarios (temperature and precipitation) for period 25 years are RCP 2.6, RCP 4.5, RCP 6.0, RCP 8.5. The goal of working with scenarios is not only to predict the future but also to provide a better understand about uncertainties and alternative futures. Geographic Information System (GIS) is employed in this study to conduct spatial analysis of the parameters associated with the process of identifying the impact of climate change that occurs in shrimp farming. The objective of this research are: 1) Determine vulnerable area of whiteleg shrim in response to climate change in East Java Province and concern to undestand perceptions of the shrimp farmer in selected vulnerable area, 2) Determine a relationship between climate factors with production in the most vulnerable district.

2. Metodology

2.1. Framework of Research

In climate change adaptation, there are two types of vulnerabilities: outcome vulnerabilities and contextual vulnerabilities [3,4]. This research uses outcome vulnerability to determine impact assessments. Sensitivity and adaptive capacity came as contextual vulnerability.

2.2. Conceptual Research

Vulnerability index as approached in this study has conducted by using quantitative assessment. In the case of vulnerability index, all the indicators were chosen by researcher that has potential impact on vulnerability of the region to climate change. The vulnerability map is calculated by summing the level of exposure and sensitivity. In the research, the exposure is indicated with precipitation and temperature (projection). Otherwise, the sensitivity indicated with production of shrimp farming. Adaptive capacity as additional information explains anything about shrimp farming condition related to climate change in a region.

2.3. Research Procedures

This research was divided into three main parts: data collection, data processing, and analysis process. 1) Data collection is belong to primary data such as; production of shrimp in East Java Province (2011-2015), production of shrimp in Banyuwangi District (2015-2016), projection climate data (2018-2042) consist of precipitation, maximum temperature, and minimum temperature in East Java Province, 2) Data processing: collected data of primary data and secondary data, arrangement in the form of rectangular matrix with columns representing indicators and rows representing regions, interpolate processing of climate data as indicators (i.e., maximum temperature, minimum temperature, and precipitation) using Geographic Information System (GIS) technical. Geostatistical analysis carried out by using kriging method, calculated normalization of indicators is using normalized formula (1), and it is important to identify the functional relationship between vulnerability and indicators previously (more vulnerable or less vulnerable).

$$X_{ij} = \frac{X_{ij}-Min\{X_{ij}\}}{Max\{X_{ij}\}-Min\{X_{ij}\}} \quad (1)$$

$$VI = \frac{\sum_{z=1}^{z=0} X_{ij}\ (z)}{K} \quad (2)$$

Where; X_{ij} = $|\uparrow|$ more vulnerable (Functional relationship between vulnerability and indicators), K = number of indicators.

Calculated vulnerability index was used equal weighting in method (2), to know impact of climate change in each region. Overlapping climate data and productivity data was able to make vulnerability map. Interview with stakeholders using questionnaire to know adaptive capacity. 3) Analysis process: integration between vulnerability map and questionary data were employed to know about climate change impact on shrimp farming.

2.4. Research Sites

This research is located in East Java where the largest of shrimp farming founded (Figure 1.). East Java Province consists of 38 district, 22 districts are belong to coastal area. They are Banyuwangi, Situbondo, Probolinggo, Probolinggo city, Jember, Lumajang, Malang, Lamongan, Pacitan, Tulungagung, Tuban, Bangkalan, Pamekasan, Surabaya city, Blitar, Pasuruan, Pasuruan city, Sumenep, Trenggalek, Gresik, and Sidoarjo.

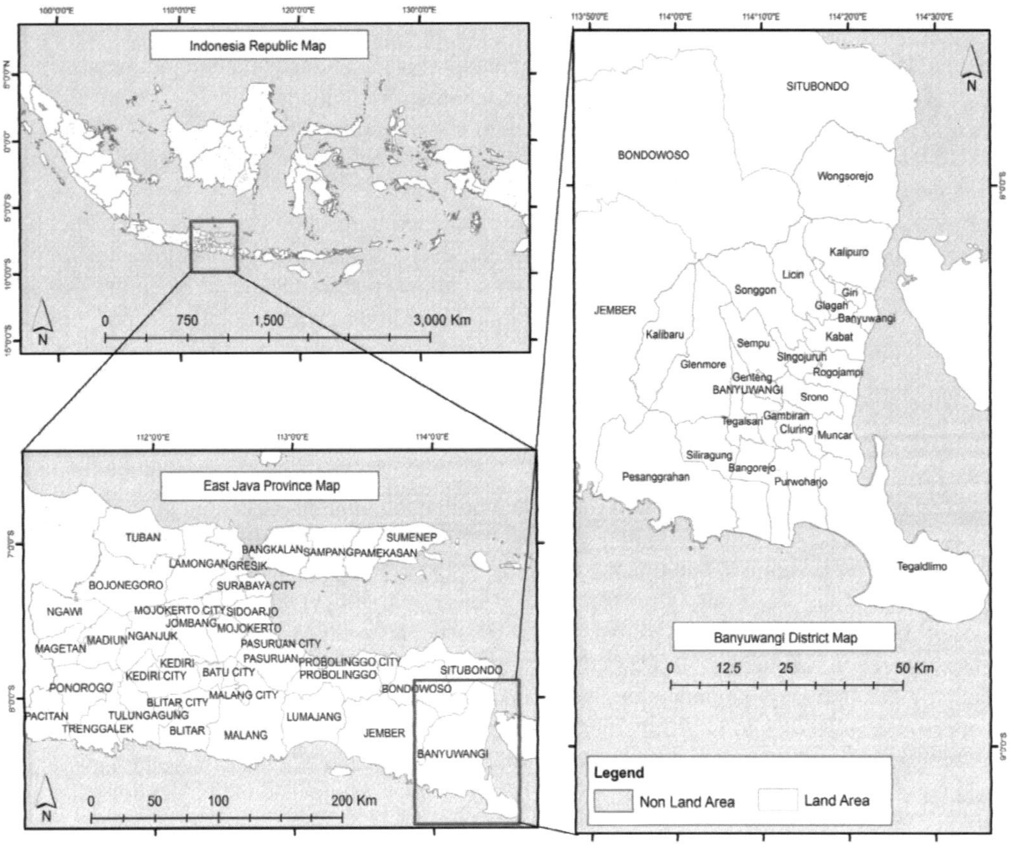

Figure 1. Research Location

2.5. Data Analysis

Analysis of this reseach is comprised into three parts: 1) Projection analysis of projection climate change (period 2018-2042) to determine characteristics of climate hazards for the future, 2) analysis of shrimp production in East Java Province and Banyuwangi District, 3) analysis of vulnerability maps with perception of farmers in East Java to know adaptive capacity on climate change in the region, and analysis of correlation using monthly climate data and monthly production data.

3. Result and discussion

3.1. Climate Variables Map Exposure Paramaters

One of key elements of vulnerability is exposure. The future climate factor used a climate models (i.e. IPSL-CM5A-MR models). In this case, exposure indicates with precipitation, minimum temperature and maximum temperature. The values of the figures are described as a color bar, where red color identify as a high value, and green color identify as a low value.

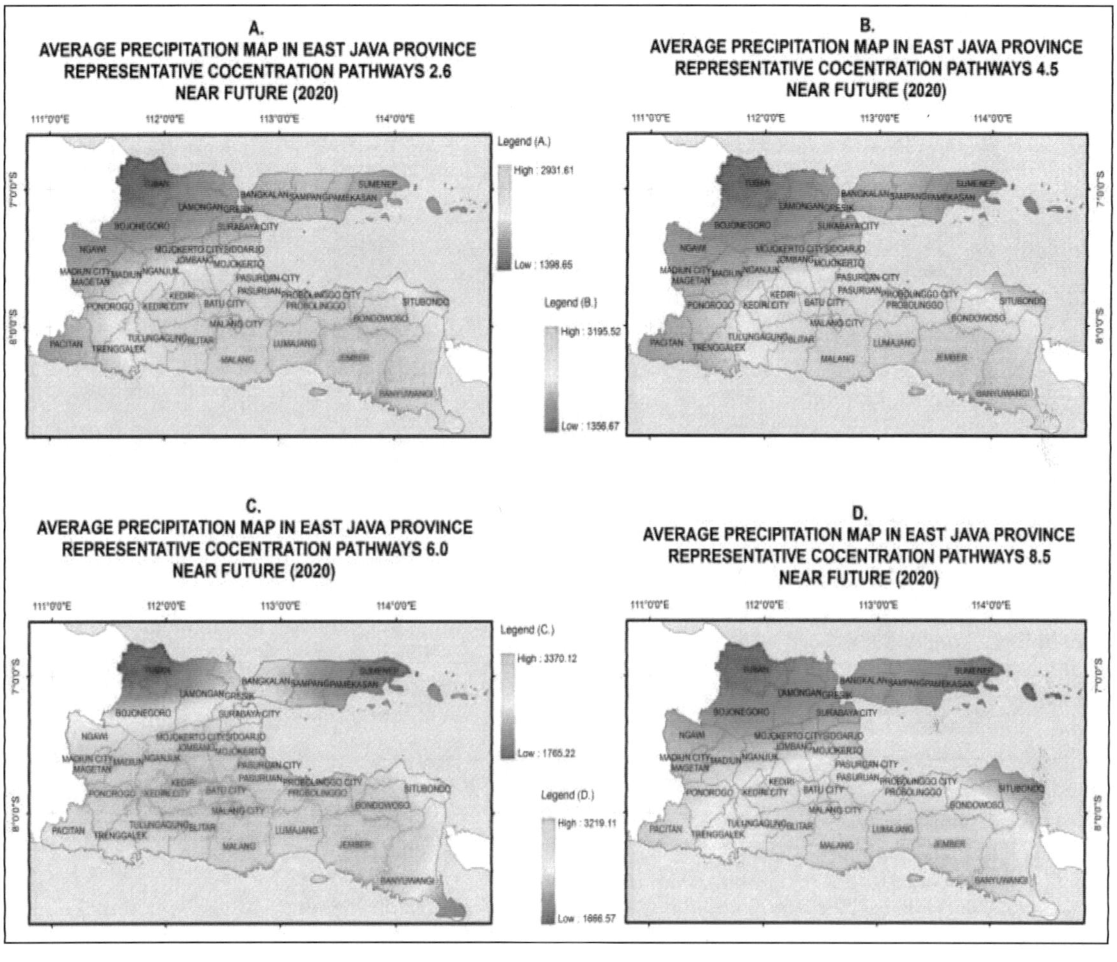

Figure 2. Exposure maps: average precipitation maps in 2020

Figures 2 give some information that the highest value of precipitation from period 2020 is 3370.12 mm/day under RCP 6.0 and the lowest value of precipitation is 1356.67 mm/day under RCP 4.5.

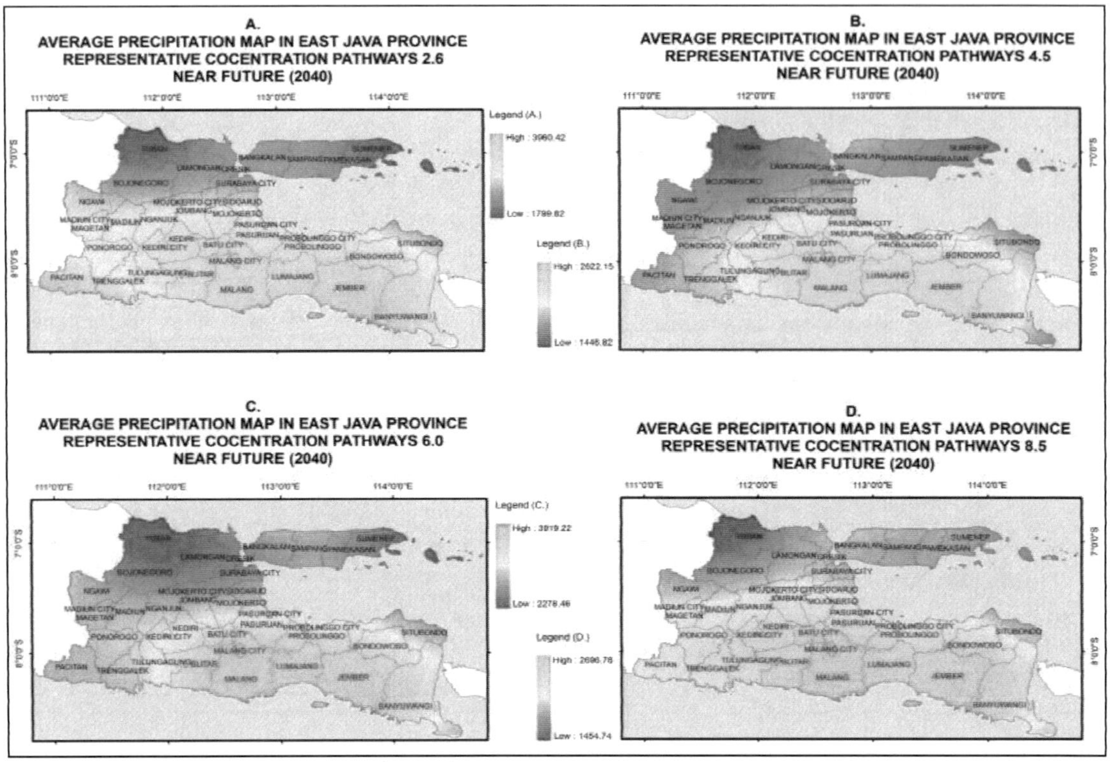

Figure 3. Exposure maps: average precipitation maps in 2020

In the period 2040, the highest values is 3960.42 mm/day under RCP 4.5 (Figure 3). Average precipitation for 25 years (2018-2042) explained that the highest values is 3022.73 mm/day under RCP 6.0 and the lowest values is 1371.08 mm/day under RCP 4.5 (Figure 4).

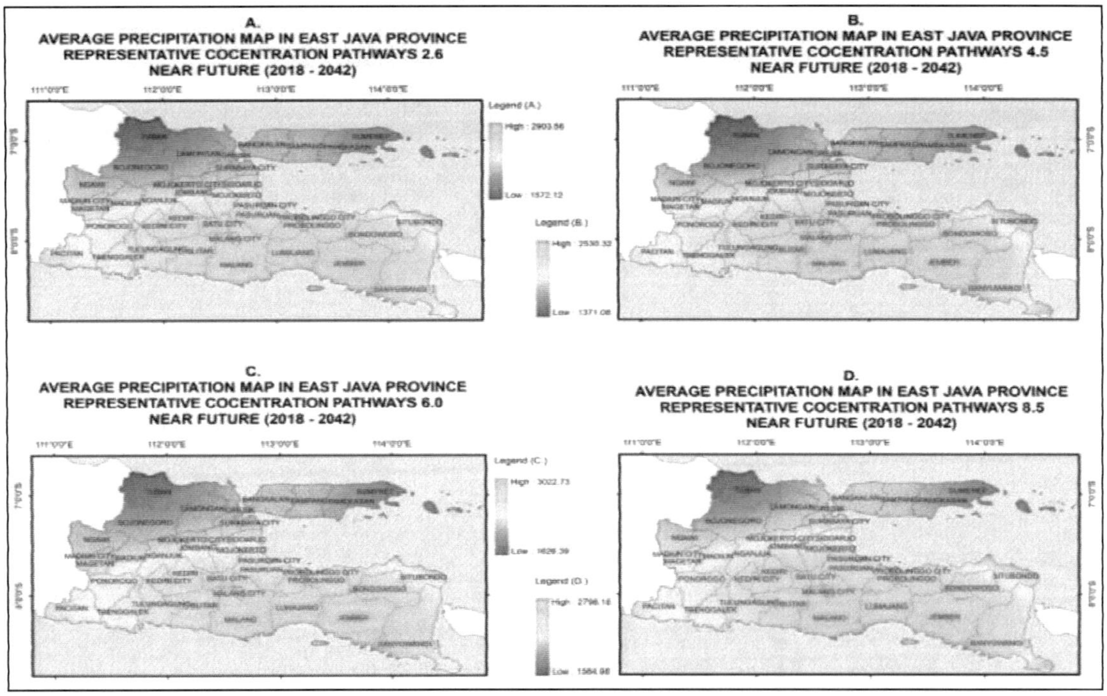

Figure 4. Exposure maps: precipitation maps for 25 years (2018 – 2042)

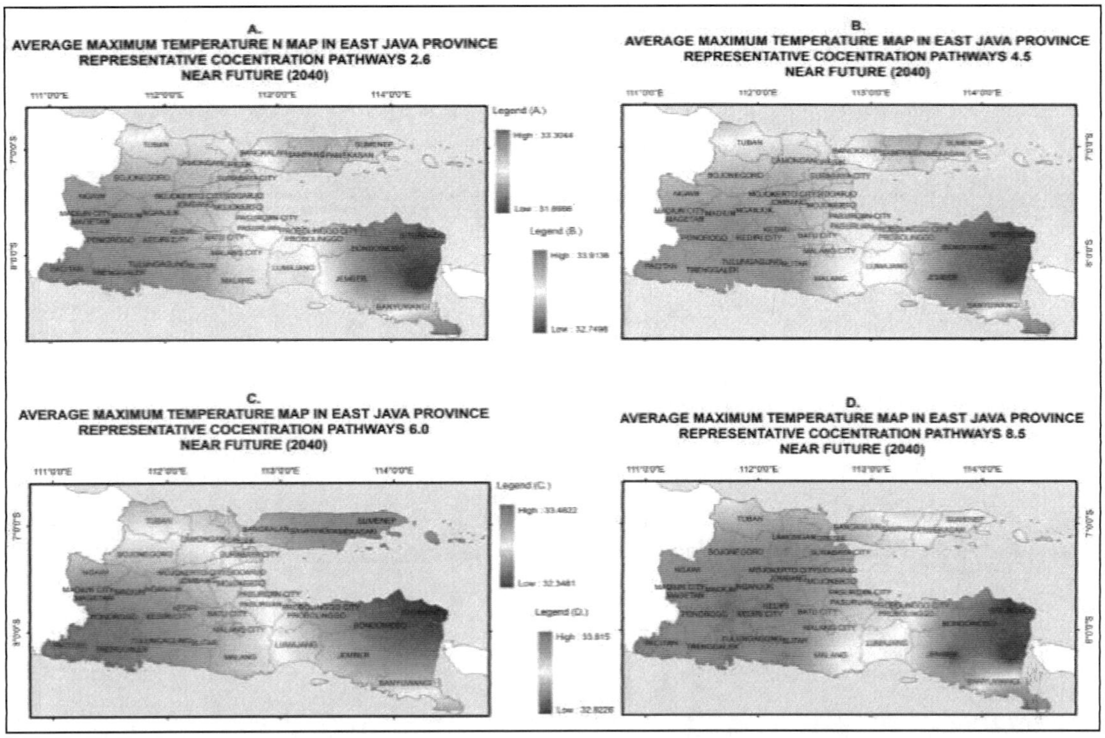

Figure 5. Exposure maps: average maximum temperature maps in 2020

The result of average maximum temperature map in East Java Province uses color bar to describing the level of average maximum temperature. The dark green color indicated low value, and the red color indicates as high value. The largest values of maximum temperature in 2020 is 33.27°C under RCP 8.5 and the lowest values is 31.74°C under RCP 2.6 (Figure 5).

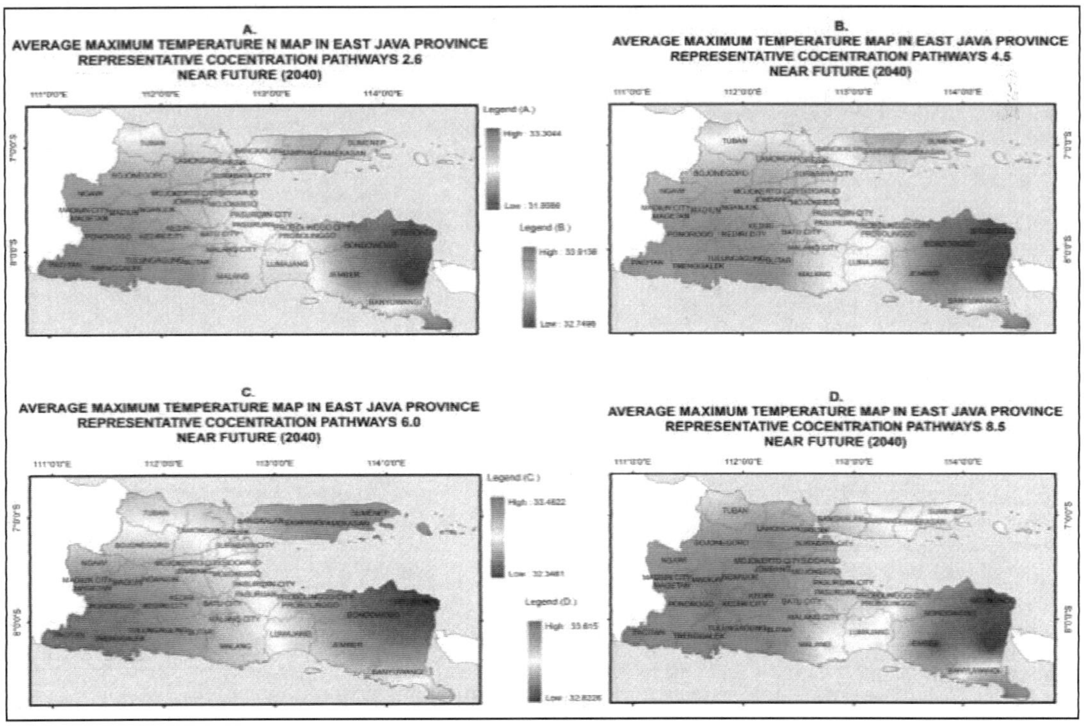

Figure 6. Exposure maps: average maximum temperature maps in 2040

In 2040, the largest values of maximum temperature is 33.91° under RCP 4.5 and the smallest values of maximum temperature is 31.89°C under RCP 2.6 (Figure 6). Map of averages for 25 years described that the highest values is 33.35°C under RCP 4.5 and the lowest values is 32.08°C under RCP 2.6 (Figure 7).

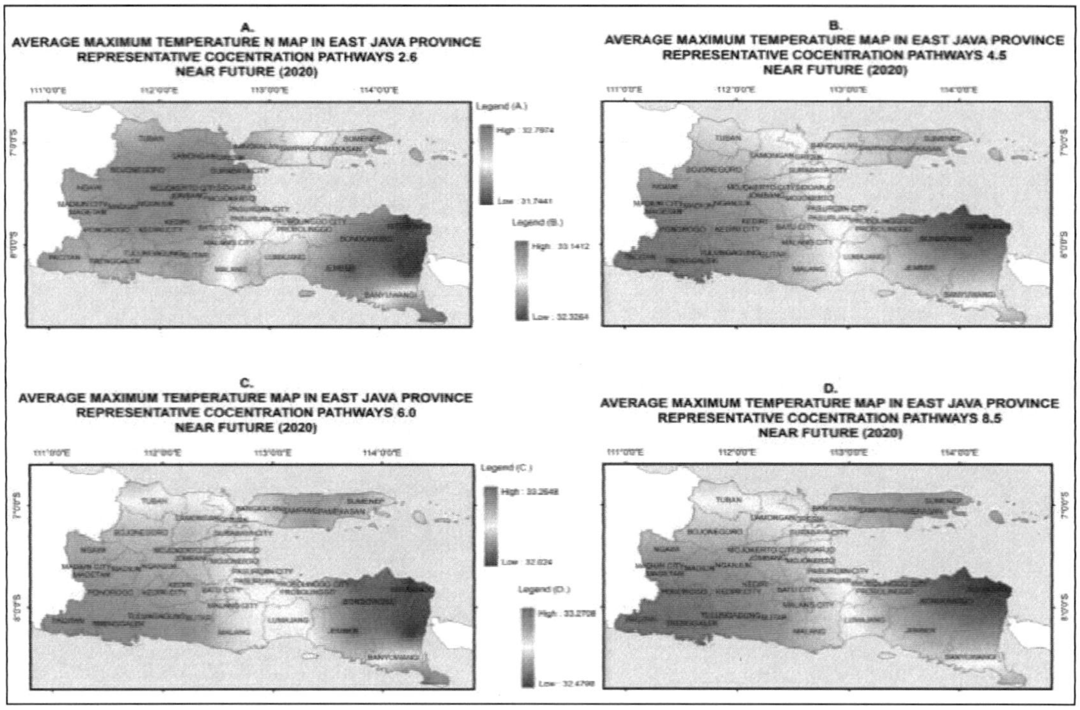

Figure 7. Exposure maps: average maximum temperature maps for 25 years (2018 – 2042)

The minimum temperature map in figure 8 for period 2020 shows that the highest values is 25.15°C under RCP 4.5 and the lowest value is 21.89°C under RCP 2.6.

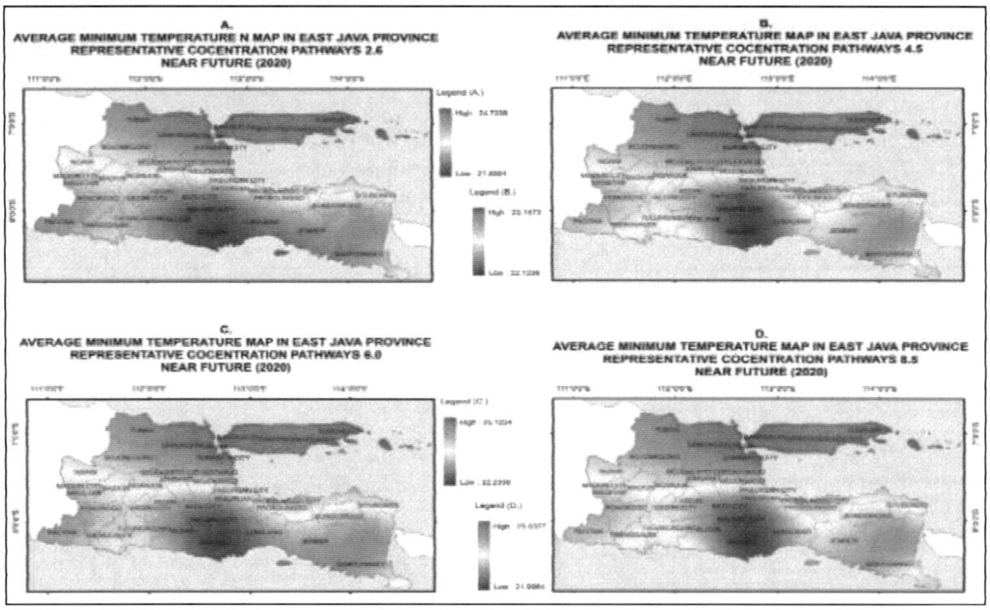

Figure 8. Exposure maps: average minimum temperature maps in 2020

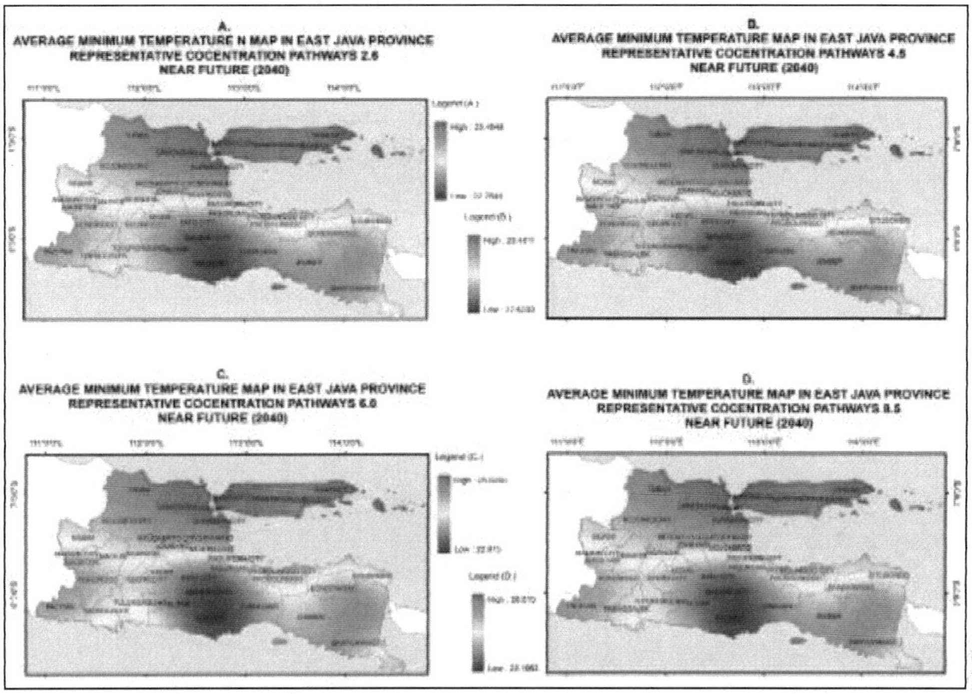

Figure 9. Exposure maps: average minimum temperature maps in 2020

The minimim temperature map in 2040 shows that the highest value is 26.01°C under RCP 8.5, while the lowest value is 22.62°C under RCP 4.5 (Figure 8). The average period for 25 years shows that the largest value is 25.37°C under RCP 8.5., and the lowest value is 22.12°C under RCP 4.5 (Figure 9)

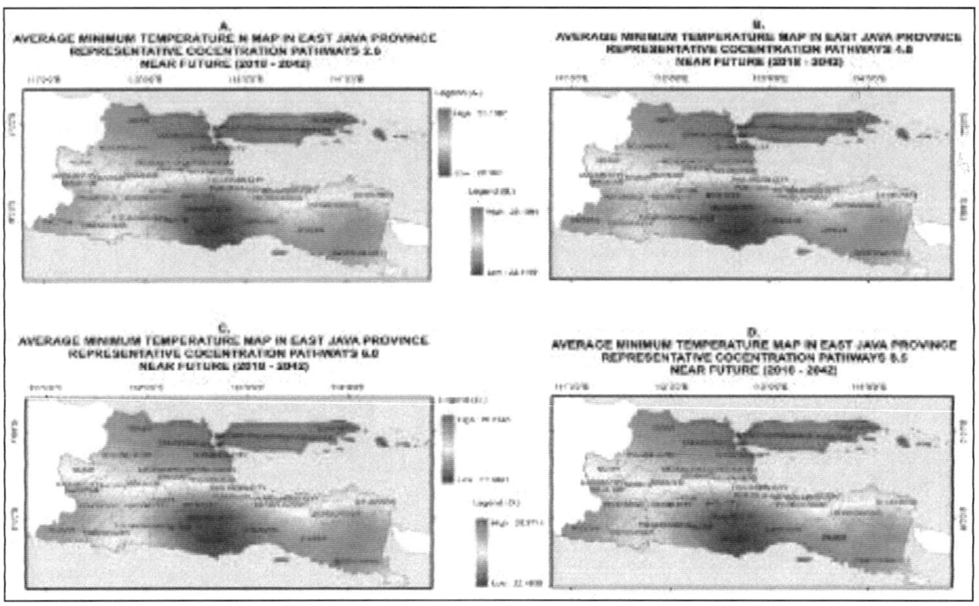

Figure 10. Exposure maps: average minimum temperature maps for 25 years (2018-2042)

3.2. Production Map as Sensitivity Paramaters

There are two kinds of production data. First is production data of whiteleg shrimp in East Java Province, which taken from Ministry of Marine and Fisheries (Figure 11a). The data shows production in each district of coastal area in East Java Province. Second is production data of whiteleg shrimp in Banyuwangi District, which taken from Fisheries and Food Security Department (Figure 11b).

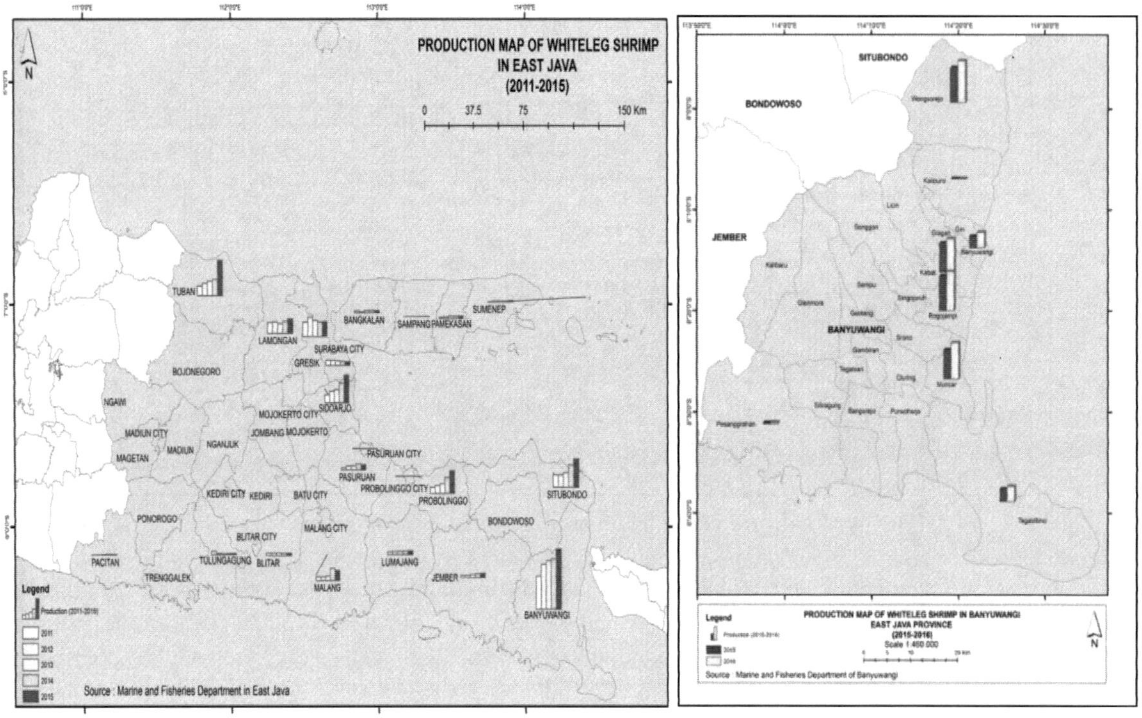

a. b

Figure 11. a) Production map of whiteleg shrimp in East Java Province during 2011-2015, b) Production map of whiteleg shrimp in Banyuwangi District during 2015-2016

Figure 11a shows that the biggest value from the average of production data during 2011 – 2015 in East Java Province is Banyuwangi District (27.72%), follows by Situbondo (10.35%), Tuban (10.24%), Gresik (9.33%), Sidoarjo (9.28%), Probolinggo (7.32%), Lamongan (6.80%), Malang (4.14%), Surabaya (2.66%), Pasuruan (2.41%), Jember (2.04%), Lumajang (1.86%), Blitar (1.37%), Bangkalan (1.17%), Pamekasan (1.14%), Tulungagung (0.98%), Sumenep (0.71%), Pacitan (0.31%), Probolinggo (0.11%), Sampang (0.07%), Pasuruan (0.01%), and Trenggalek (0%).

Figure 11b shows that Wongsorejo sub-district is the highest value of average production data during 2015 – 2016 in Banyuwangi District, i.e., Wongsorejo (22.60%), Rogojampi (21.81%), Muncar (19.34%), Kabat (18.71%), Tegaldlimo (8.35%), Banyuwangi (8.09%), Pasanggrahan (0.88%), Kalipuro (0.23%).

3.3. *Vulnerability Area in East Java Province*

From production data in East Java only 21 District had production data. In Trenggalek District was not found production of whiteleg shrimp. Trenggalek District is captured as region with mountains area (48.31%) which is used as cultivation. Vulnerability map of East Java shows that in Banyuwangi District is the most vulnerable area under all scenarios data (Figure 12).

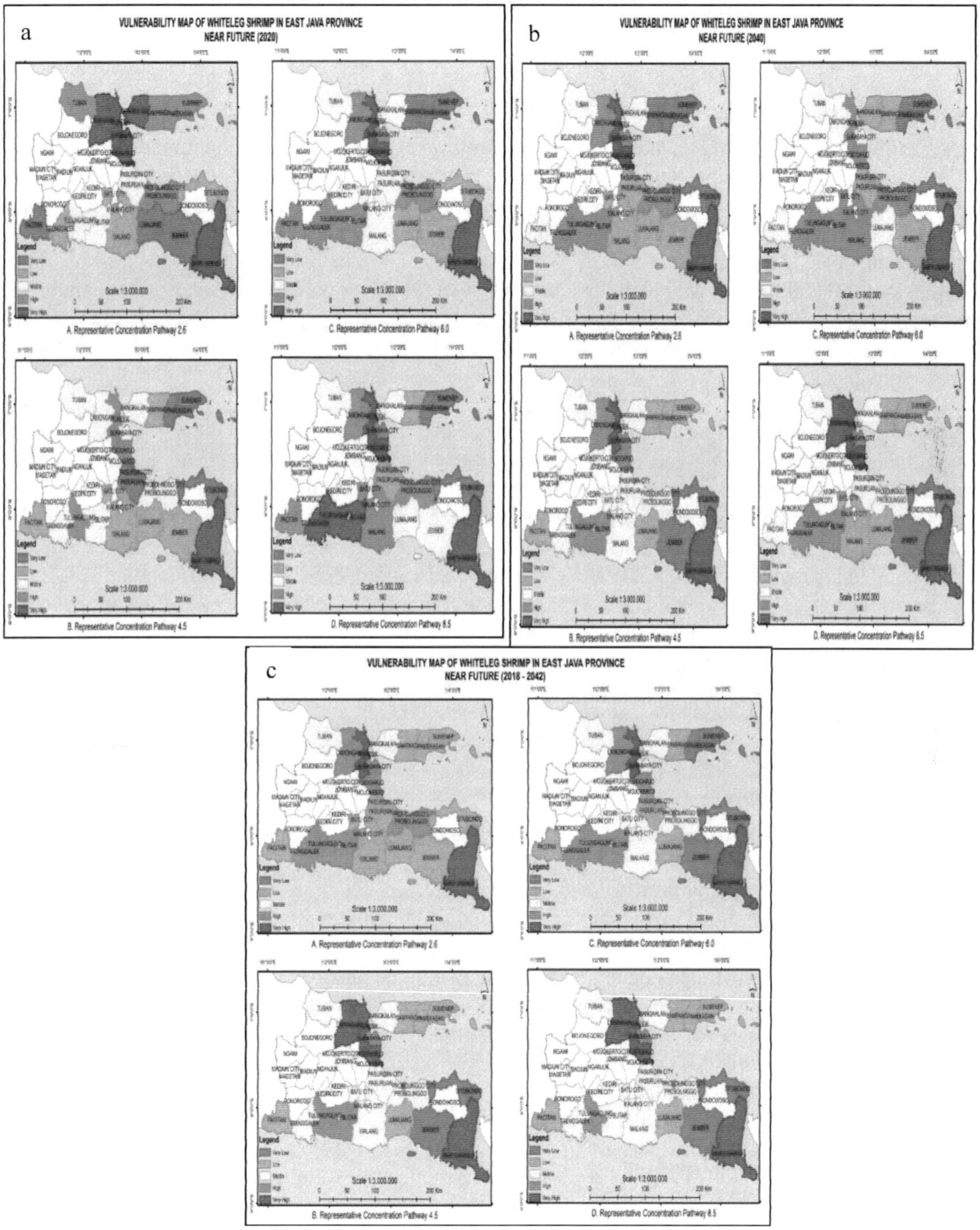

Figure 12. Vulnerability map in coastal area in East Java Province : a) 2020, b) 2040, c) 2018 – 2042 under RCP scenarios

3.4. Perception of the Whiteleg Shrimp Farmers in Banyuwangi District

3.4.1. General Information of Respondent.

Total of the respondents were 25 respondents. All the repondents were men and already married with had different background. The main job of the respondents were shrimp farmer (55.88%), fisherman (5.88%), others (38.24%) (as a farmer, technical man, distibutor, trader, and operator in the pond . It implies that married men were able to support the family with the main job as a shrimp farmer. It mentioned that 80% of the respondents (men) were married and implied that for income and livelihood to their family could be supported from aquaculture [5].

3.4.2. Trend of Productivity.

Trends shrimp production has fluctuated. The respondents mentioned that the production decrease more than 30% since 2014. All the respondents said there is influence between disease and climate factors. Some disease such as IMNV (Infectious Myo Necrosis Virus), WFS (White Feses Syndrome), WSD (White Spot Disease), and TSV (Taura Syndrome Virus) cause decrease in production. Temperature plays a crucial role in the occurrence of IMNV disease, and it can be attacked in all seasons. The possibility of the condition where the disease more virulent than ever occurred as a result of changes in temperature extremely or continuously [6]. 52% respondents mentioned that in rainy season disease more often arise. The whiteleg shrimp had ability to live against fluctuations in water quality, especially in dry season [7].

3.4.3. Awareness of Climate Change.

Some of the respondents (40%) had been informed about the weather and the rising sea levels. The unpredictable climate and absence of information relevant to the climate's issue create a problem for sustainability of shrimp production. The respondents acknowledged that the climate has changed. Most of the respondents (92%) were aware of climate change. In 2016, they feel climate was having an impact on production, increase precipitation (34.09%), and decrease temperature (2.55%) were the most factors influence of decreasing shrimp production. The heavy rain causes water salinity and temperature level decrease, affected to shrimp growth and productivity which was low water salinity with fluctuations water temperature (3°C - 4°C) causes the outbreak of WSSV in shrimp. This climatic condition affect to all pond systems [8]. For development skills, some of the respondents have received training related to the way of cultivation.. the respondents education and training expected to consider global climate change to be a more able to adapt in serious risk than the respondents with typical nonscientists [9].

3.4.4. Adaptive Ability.

Most of the respondents aware that the biggest challenges that threaten the sustainability of aquaculture today is related to the weather, such as rain, wind, and temperature. There is relationship between aquaculture production with environmental conditions. It agreed which was mentioned that changes in the environment may have direct or indirect effects on the abundance of pathogens and parasites [10]. Also, it mentioned that the fundamentally success in aquaculture depend on aquatic environment, which is a major concern considering the wide rangin impacts on climate change [11]. In 2016 the farmers said that flood happen frequently. Some of farmers (12%) said that the flood happen no related to climate change but it was happen caused the improper of infrastructure. While treat in the future mentioned that is not different from the exist today, which is related to weather and disease. Some respondents also explained that land, capital, and water quality as another treat in the future. Respondents were hoping for increasing technical, financial and skill for help them to adapt. Others mentioned need for supervision and mentoring by an expert in any farming activities. This is because climate change as an attention so hard to detect if only based on personal experience, they thought that learning through an expert can be better to make decision [9].

3.4.5. Mitigation of Climate Change.

The mitigation action to the climate change is important for sustainability in aquaculture. One of the respondents had been taking appropriate action to reducing the impact of climate change, by using biodiesel as a substitute for diesel fuel. Respondents are ready to make changes to reduce the impact of climate change that occurred. Although they also says, that the reduction of fuel and electricity can affect to production. The chosen land for pond need to consider about mangrove as a buffer zone. 36% respondents said that their ponds were not in area of buffer zone, and 64% were in area of buffer zone. The buffer zone used to protect ponds from pollutant or spreading of parasites and disease [12].

3.5. Correlation Between Production Data and Climate Data in Banyuwangi District

The collected data from Fisheries and Food Security Deparment in Banyuwangi District shows that the number of farm during 2015-2016 are not changed (i.e., 498). The area of ponds also no changes during 2015 – 2016, which are 1381.76 ha. The study finds that total of production in 2015 – 2016 is increasing although the area and number of farms experienced no changes. This study finds that the monthly maximum temperature data during 2015 – 2016 and the monthly whiteleg shrimp production is positively insignificant correlated ($R^2 = 0.2137$) (Figure 13a).

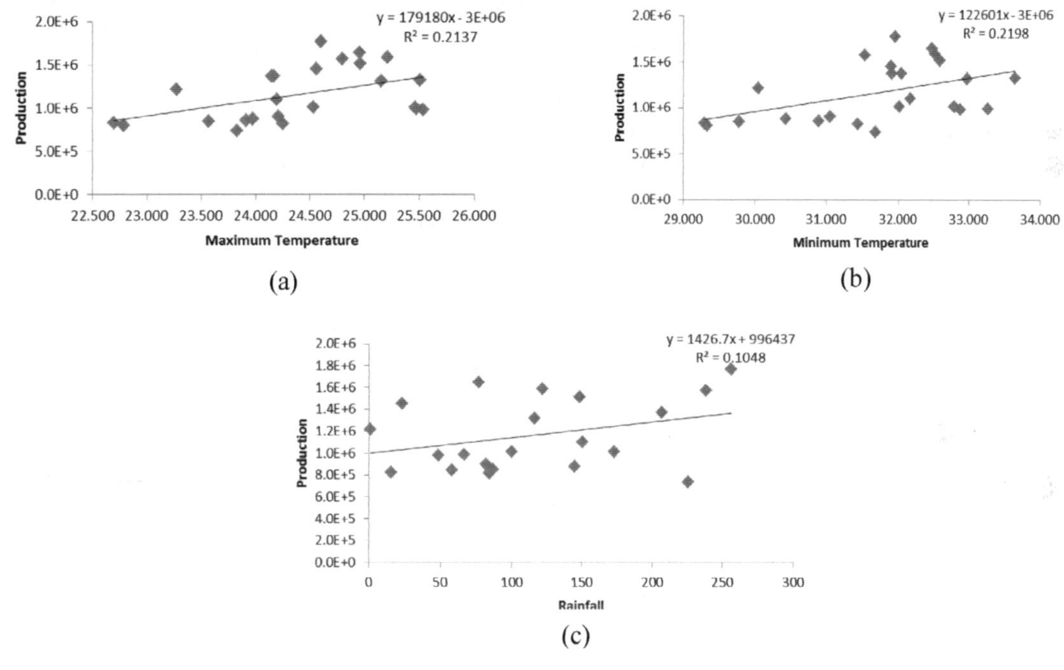

Figure 13. Correlation pattern between exposure parameters (temperature and rainfall) and sensitivity parameters (production data): a) maximum temperature, b) minimum temperature, c) rainfall

Figure 13b. explains that there are correlation between minimum temperature and production in Banyuwangi District during 2015 – 2016. The correlation is insignificantly positive increasing ($R^2 = 0.2198$). The study also finds that the monthly rainfall data during 2015 – 2016 and the monthly whiteleg shrimp production is positively correlated, with $R^2 = 0.1048$ (Figure 13c).

The other researcher explained about correlation of shrimp production with climatic parameters [13]. Within the research, the researcher also found that the annual shrimp production and temperature was insignificant positively correlated. It was also found for correlation between the annual shrimp production and rainfall, which was showed positively correlated and insignificant [13].

4. Conclusion

This study concludes several things, as follows : the most vulnerability area in East Java Province is Banyuwangi District, under projection climate data (encompassing RCP 2.6, RCP 4.5, RCP 6.0, RCP 8.5). Most of the respondents (92%) aware of climate change. The respondents reveal that increased rainfall (34%) and decreased temperature (29%) are the most contribute to cause a negative impacts on shrimp production. Decrease in production influenced by disease and climate factors, in rainy seasons the disease more frequently. Treat in the future, it is mentioned not much different from the treat that exist today, which is related to climate and disease. There are positively correlation between climate parameters (precipitation, maximum temperature, and minimum temperature) with production data (in Banyuwangi District) but it was insignificant.

References

[1] Ahmed N and Diana J 2015 *J. Ocean and Coast.* 42 - 52.
[2] De Silva S S and S Doris 2009 *Climate Changeand Aquaculture : Potential Impacts, Adaptation and Mitigation. In : Climate Change Implication for Fisheries and Aquaculture* ed. By Cochrane K de Young C Soto D and Bahri T pp 151-212
[3] Adger W N Brooks N Bentham G Agnew M and S Eriksen 2004 *New Indicators of Vulnerability and Adaptive Capacity* (UK : Tyndall Centre for Climate Change Research Norwic)
[4] O'brien K S Eriksen L P Nygaard and A Schjolden 2007 *Why Different Interpretations of Vulnerability Matter in Climate Change Discources* pp 73 – 88
[5] Adeleke M L and Omoboyeje V O 2016 *J. Fish and Aqua* 50 -58
[6] Prasad K P Shyam K U Banu H Jeena K Krishnan R 2016 *J. Aquaculture* 632460
[7] Budiardi T Batara T and Wahjuningrum D 2005 *J. Aqua in Indonesia* 89 – 96
[8] Tendencia E A and Verreth J A J 2011 *J. of Aquaculture* 63 : 548
[9] Weber E U 2010 *What Shapes Perceptions of Climate Changes? Wiley Interdiciplinary Reviews (WIREs): Climate Change* 3 pp 332 – 342
[10] Karvonen A Rintamaki P Jokela J Valtonen E T 2010 *International Journal for Parasitology* 40 : 1483 – 1488
[11] IPCC 2007 *Climate Change 2007 : The Physical Science Basis. Summary for Policymakers* Fourth Assessment Report of Intergovenmental Panel on Climate Change
[12] Primavera and Jurgenne H 2005 *Mangroves and Aquaculture in Southeast Asia*. Report of Regional Technical Consultant of Code of Practice for Responsible Aquaculture in Mangrove Ecosystems pp 25 - 37
[13] Rimi R H Farzana S Sheikh Md S Abedin Md Z and Bhownmick A C 2013 *J. Word Environment* 116 - 125

Developing one map of national marine resources of Indonesia

Nurwadjedi, Sri Hartini, Lien Rosalina

Geospatial Information Agency
Jl. Raya Jakarta - Bogor Km. 46, Cibinong, Indonesia

E-mail: nurwajedi@gmail.com, sri.hartini@big.go.id, liensis2@gmail.com

Abstract. Indonesia is the largest archipelagic country in the world, with two third of waters containing large marine resources. However, the coastal and marine resources have not been optimally utilized due to the limitation of Spatial Zoning Plan for Coastal and Small Islands (RZWP3K). In addition, there were many institutions which produce thematic maps of the coastal and marine resources (MGI) using different basemap, standard and technical specifications. Hence, same thematic maps provide different information. For these reasons, One Map Policy (OMP) is required to overcome the problems. The objective of this research was to develop datasets of one map of the coastal and marine resources needed for developing RZWP3K by using thematic layer maps produced by their respected Ministries/Agencies as the custodianships. The current custodianships of thematic maps in Indonesia is defined based on the Head of Geospatial Information Agency Decision Number 54 Year 2015. The method used the OMP approach was based on one reference, one standard, one database, and one geoportal. The research resulted the dataset of MGI integrated to the national basemap in a geodatabase format. From this research, it was concluded that the thematic maps of the coastal and marine resources which were produced in many institutions can be integrated to a national basemap and standard specification. The integrated thematic maps are then useful for developing the RZWP3K.

1. Introduction

1.1. Physiography of Indonesia and its Marine Resources

Indonesia is the largest archipelagic country worldwide, extending between the Australian and Asian Continents and the Indian Ocean and the Pacific Ocean. Geographically, Indonesian archipelago is located between the 94 and 141 degrees East Longitude and between the 6 degrees North and the 11 degrees South Latitude. Indonesia's claim as an archipelagic nation has documented as part of important national history. Historically, Indonesia has always regarded the seas within its archipelago as internal waters, hence the internal water become territorial sovereignty water. Indonesia's word for 'fatherland' - 'tanah air', means 'the land and the water', so that the seas between island bounding the islands together to form the territory of Indonesia. As an archipelagic nation, Indonesia also has right on claiming the Exclusive Economic Zone (EEZ) as regulated on the United Nations Convention on the Law of the Sea (UNCLOS) 1982.

That has been commonly stated that Indonesia has 17,508 islands. Recent update, Indonesia submitted 16,056 islands during the 9th meeting of United Nations Conference on the Standardization of Geographical Names (UNCSGN) in New York, 21-30 August 2017. This submission was an update of the previous submission of the number of islands to UNCSGN which reported that Indonesia had submitted 13,466 islands. Those islands submitted were islands that had been verified with names and coordinates. Since the geospatial survey of islands continues, the number of the islands submitted to UNCSGN will possibly change in the future.

Furthermore, total area of the Indonesian archipelago is 4,986,325 square kilometers, with land area is 1,904,569 square kilometers and water area is 3,081,756 square kilometers [1]. Meanwhile, the Geospatial Information Agency (BIG) stated that the water area of Indonesia including the EEZ was 6,315,222 square kilometers with 99,093 kilometers of coastline [2]. Thus, the seas and coastal area dominated the physiography of Indonesian archipelago.

Since the Indonesian archipelago is in a strategic location across major sea lanes from the Indian Ocean to the Pacific Ocean, Indonesia's sea-lanes have long been a key highway for naval vessels sailing between the Asia-Pacific region and the Indian Ocean. Almost all maritime traffic transiting South East Asia must enter Indonesian waters at some points. There are three designated sea lanes, called as Archipelagic Sea Lane I, II, and III that running along a north-south axis the archipelago. The Indonesian submission of the sea lanes came under consideration at the 67th Meeting of the Maritime Safety Commission of the International Maritime Organization (IMO) in 1998, where the IMO accepted the partial designation of Indonesian archipelagic sea lane passages [3]. The designation of the sea lanes is part of an obligation of the recognition of Indonesia as an archipelagic country according to UNCLOS 1982, which is an archipelago is "a group of islands, including parts of islands, interconnecting waters and other natural features which are so closely interrelated that such islands, waters and other natural features form an intrinsic geographical, economic and political entity, or which historically have been regarded as such."

With such geographical area, Indonesia has abundant coastal and marine natural resources. Mangroves forests are found in most of the coastlines of Indonesia. Papua, Kalimantan and Sumatra are three big islands of important mangroves ecosystems [4]. The stand of mangroves varies between type of species and region, some can reach up to 50 meters in height. Mangroves forest found in intertidal area, at height between mean sea level and high tide. Therefore, mangroves area is flooded during the high tide. In general, the ground is layer of dense mud that containing thick levels of organic materials that exposed during the low tide. However, some mangroves can grow on sand substrate as that found in the eastern islands of Indonesia such as in Sumba and Kei islands. The area of mangroves Indonesia is about 3 million hectares nationwide. This is about 23 percent of all mangroves ecosystems in the world [5]. Other important ecosystem in shallow water area is coral reef and seagrass. Scientists have noted that Indonesia is the center of coral diversity. It was reported that 82 genera and about 590 species of scleractinian corals have been recorded in the Indonesian and its surrounding waters [6][7][8]. The seagrass coverage in Indonesia was estimated for 30,000 kilometers square, is the largest among other Asian countries [9].

In terms of fisheries, Indonesia is one of the main fish producers in Southeast Asia. Capture fisheries and aquaculture occur in marine, brackish and freshwater. Marine capture at the second largest position with 6.01 million tons in 2014 [10]. Besides that, there is also considerable potential of offshore oil and gas mining and other minerals also contribute to national economics. The economic potential of the marine resources worth as much as US$ 1.2 trillion/annually or about 10 times the state budget [11]. With those potential marine resources, the president of Indonesia, Joko Widodo, declare that the future of the nation is in the sea.

However, there are also several issues on the coastal and marine conditions. The coastal and marine resources deterioration has occurred in many places due to development practices and unfavorable exploitation of the natural resources. Some issues like tidal flood, mangroves degradation, coral bleaching, coastline abrasion, and illegal fishing occur in many places in Indonesia. In the long run, those issues could result in reducing the fish production.

1.2. Policy on provision of Geospatial Information in Indonesia

The management of natural resources in Indonesia is carried out by many ministries/agencies and local governments, with reference to their duties and functions or their respective interests. Since it is part of natural resources management activities, provision of Geospatial Information (GI) especially thematic geospatial information is also carried out by many ministries/agencies and local government. This condition has resulted in the implementation of GI in institutional manner. This institutional GI implementation then has led to overlapping of land use or spatial use conflicts, uncertainty on spatial information, spatial allocation in restricted areas, social conflicts, and others. The impact of institutional GI implementation is certainly also can hinder the development of the region or infrastructure.

Law Number 4 Year 2011 on Geospatial Information (GI Law) has been enacted on April 21, 2011. According to the GI Law, the implementation of GI is grouped into 2 main outputs, namely the implementation of Basic Geospatial Information (BGI) and implementation of Thematic Geospatial Information (TGI). The implementation of BGI is conducted by the Geospatial Information Agency (BIG) (Article 22 paragraph 2 of the GI Law), while the implementation of TGI is carried out by government agencies (ministries/agencies), local government, and/or person (Article 23 Paragraph 1 of the GI Law). One Map Policy (OMP) is a further arrangement on the implementation of TGI by ministries/agencies and local government so that TGI between ministries/agencies and local government can be integrated with reference to the same baseline map. The benefits of TGI from OMP products are to realize transparent and efficient governance, accelerate regional-based development (spatial and infrastructure), avoid spatial conflicts, guarantee investment location issuance, and so on.

BIG's empirical experience in assessing the quality of the Regional Spatial Planning (RTRW) maps concludes that most of existing district/municipal RTRW maps are not compiled using the thematic maps that have been specified in the Public Works Minister's Decree Number 20 Year 2007 on Technical Guidelines of Physical and Environmental Aspect Analysis, Economics, and Socio-Culture in the preparation of Spatial Planning. The unavailability of thematic maps of in scale of 1:50,000 both ministries/agencies and local government is the cause of this situation. This has resulted in many in problematic District/City RTRW maps, resulting in many spatial use conflicts in the regions.

Considering those problems stated above, the Government of Indonesia issued OMP which is part of the Economic Policy Package VIII to address the spatial utilization conflicts. The purpose of this OMP Regulation is to accelerate the provision of 1:50,000 scale thematic maps, so that the District/City RTRW map can be accurately compiled. The District/City RTRW maps along with their accurate supporting thematic maps are fundamental TGI data that can ensure a sustainable land use allocation, including for infrastructure development.

1.3. Problems

Unfortunately, the thematic geospatial data about those environmental issues as mentioned above at *Section 1.1* and *Section 1.2* are limited both in terms of their availability and quality. Likewise, data about coastal and marine natural resources and habitats considered limited in terms of the availability. Some data provided based on a specific purpose and hard to be accessed, so that is not useful for others. Some types of data might be provided by several institutions but using different standards and technical specifications. But, the most important is that the data are already dated and need to be updated. On the other hand, demand for the data, especially geospatial data is increasing following the awareness on management the natural resources to ensure sustainable development by ensuring continuous functions of the coastal and marine environment.

Meanwhile, spatial planning is considered as a suitable tool for managing marine and coastal area. The need of spatial planning for coastal and marine has been expressed, among others by the enactment of Law number 27/2007 about Coastal Area and Small Islands Management. The idea is to organize coastal and marine area into several zones by considering its physical conditions and potential uses. By doing this, hopefully there is a balance on the spatial usage of the marine and coastal area, between exploitation and conservation of natural resources and environment. This spatial planning development obviously requires good geospatial data to provide accountable planning.

IOP Conf. Series: Earth and Environmental Science **162** (2018) 012028 doi:10.1088/1755-1315/162/1/012028

1.4. Objectives

The objective of this research is to develop the datasets of one map of the coastal and marine resources needed for developing Spatial Zoning Plan for Coastal and Small Islands (that translated from "Rencana Zonasi Wilayah Pesisir dan Pulau-Pulau Kecil" and abbreviated as RZWP3K) on the basis on OMP requirement each thematic layer which is produced by its custodian. It is hoped that the related institutions can start developing RZWP3K so that the conflict of spatial use in the coastal region can be avoided.

2. Principle of One Map Policy

OMP is a strategic direction for the fulfillment of a single map referring to one geospatial reference, one standard, one database, and one geoportal (Presidential Regulation on OMP, 2016). Those four components of OMP are the requirement to produce an accurate, accessible, and accountable TGIs. The requirement of using single geospatial reference is a fundamental factor for producing integrated TGIs. Article 19 of GI Law states that TGI made by ministries/agencies and local governments must refer to BGD (Geodetic Control Network and basemap). A national single reference system used in Indonesia is SRGI 2013.This is a national coordinate system that consistent and compatible to global coordinate systems [12]. BIG has already distributed the 1:50,000 scale topographic map of Indonesia (RBI) covering the whole Indonesia to the 18 ministries/agencies to be used as the basemap for the implementation of thematic mapping.

'One Standard' is a component in OMP that plays important roles in standardizing the mapping method so that all geospatial information produced are accurate and accountable. In addition, this 'one standard' component is useful for accelerating the mapping process in Indonesia considering the vast territory of Indonesia to be mapped. To face these challenge, both ministries/agencies and local governments are encouraged to work together within Working Group, even though they have role as the custodian of thematic map. In this case, the mapping standard can be used by other parties as a quality control for the output maps as well.

A single database requirement in the other OMP component that is required to ensure that geospatial information is easily accessible and can be shared using national geoportal. Geospatial Information System (GIS) technology is required in the implementation of OMP to provide accurate geospatial information. The utilization of GIS technology is essential particularly for data operability and sharing since geospatial information dataset is unlikely to be produced using a single geospatial information provider or software. Database management system (DBMS) is one of the major components of GIS software DBMS will facilitate data integration and exchange [13].

The final goal of the OMP is that the output TGIs prepared by ministries/agencies as the custodianship of the TGIs can be shared through the Geospatial Information Network (JIGN) or the national geoportal. The output of TGIs must be "clean and clear" so that it can be used together and integrated each other using GIS technology in the JIGN platform. As stipulated in Presidential Regulation Number 27/2014, the JIGN consists of geospatial information nodes in the central government (ministries/agencies) and in local government. All the network nodes are integrated by connecting to the existing network nodes in BIG. The JIGN's operation for TGI data sharing is useful to support the realization of transparent and efficient governance. Moreover, the sharing data on JIGN will facilitate the synchronization process of TGIs to facilitate problem solving on spatial conflicts [14].

2.1. Scope of One Map Policy

OMP aims to accelerate the provision of TGI which fulfills one reference, one standard, one database, and one geoportal, for the development of spatial planning in district level (at the scale of 1: 50,000). Benefits that can be gathered from OMP is an improvement of thematic data on each sector according to regulatory standard/regulation, and control planning and development policies, such as by implementing integrated spatial planning. Besides that, OMP will increase accuracy of spatial data, hence avoid overlapping of licensing among investment sectors controlling the utilization of natural resources toward sustainable development. Based on Presidential Regulation Number 9 Year 2016, the target of OMP implementation (2016-2019) is 85 TGIs by involving 19 ministries/agencies in 34 provinces. It is acknowledged that the number of thematic maps required is far beyond the scope of the OMP implementation targeted within 2016-2019, both in terms of scale and the themes. However, the principle of OMP shall be applied to the production of all thematic maps so that the maps are useful for supporting governments' decision making.

2.2. Implementation of One Map Policy

The implementation of OMP based on Presidential Regulation Number 9 Year 2016 aims to support the government agenda on national development, such as infrastructure development. The geospatial information (GI) development activities carried out in this OMP consists of 3 phases, namely compilation, integration and synchronization. The achievement target of each phases is set out on the action plan. The 85 TGI targeted can be divided into TGI Status, TGI Spatial Planning and TGI Potential. TGI Status is TGI that contain legal aspects related to land tenure. TGI Spatial Planning is TGI which contains aspects of spatial utilization and planning. Whereas TGI Potential is TGI which contain information on Transportation and Utilities, Environment and Region.

The TGI compilation and integration activities are intended to solve map geometry problems, while the TGI synchronization phase is intended for resolving the overlapping or conflict between land usages. BIG is responsible for handling mapping problem related to map geometry in cooperation with related ministries/agencies which is incorporated in TGI Working Group (Satgas-1). Meanwhile, the Ministry of Coordinating Economy is responsible for resolving spatial conflicts by involving related ministries/agencies that act as the TGI custodian (Satgas-2). Solution on spatial use conflicts will be formulated based on agreement among ministries/agencies. The TGI synchronizing process will be hierarchically based on the type and degree of the conflicts to be addressed from the policy makers from echelon 2, echelon 1, minister, to the president. The ministries and agencies also urge to allocate sufficient budget necessary to fulfil their responsibility to the implementation of OMP in their own institutions.

3. Methods

3.1. Approach to Develop MGI One Map

The approach on development of one map of the coastal and marine thematic maps (MGI) was summarized in Figure 1 below. The process of one map development started with coordination among stakeholders. The stakeholders included ministries/ agencies which were responsible for particular TGI(s) or refer as the custodians and other ministries/agencies, NGO and other parties that also required the TGI(s). Those stakeholders then form a working group to accommodate discussion on technical and administration substance subjected to the development of the TGI. The technical substance included standard of integrated one map as well as classification and database standard of the one maps.

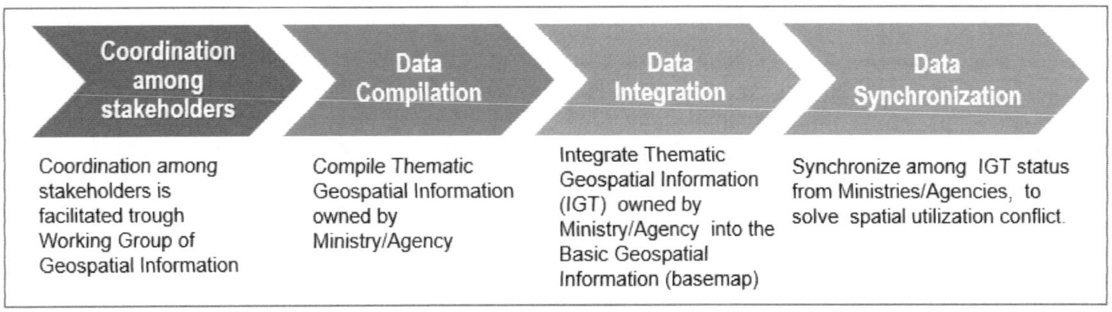

Figure 1. Approach of one map development process.

The next step was data compilation process. In this step, the thematic maps were collected from the ministries/agencies to be compiled and verified. After that, all the thematic maps in a shapefile or

IOP Conf. Series: Earth and Environmental Science **162** (2018) 012028 doi:10.1088/1755-1315/162/1/012028

geodatabase file format were integrated to the national basemap produced by BIG which uses Indonesia Geospatial Reference System (SRGI). SRGI is an Indonesian coordinate reference system which was consistent and compatible with the global coordinate. Meanwhile, the national basemap consisted of Indonesia Topography map (RBI), National Marine Environment map (LLN), and Indonesian Coastal Environment map (LPI). Those basemaps were produced by BIG. In the data integration process, all of the topological errors of the thematic maps were corrected and the file formats were standardized into the format of ArcGIS geodatabase. In case topological errors of the thematic maps found, the topological errors were evaluated and necessary corrections were performed until the topological errors were cleaned. These data integration processes resulted one map for each thematic map as the main product of OMP, which was ready to be synchronized. The data synchronization was to harmonize among the status thematic maps. This synchronization process was the toughest process because this process was heavily in policy negotiation rather than technical mapping. In this paper, the data synchronization process was not elaborated.

4. Results and Discussion

4.1. Data availability

Most of the coastal and marine thematic maps (MGI) are not yet included in the targeted one map in the action plan of the Presidential Regulation Number 9/2016 about the Acceleration of One Map Policy. However, the provision of the MGI must follow the principle and process of the one map development. The availability of MGI resulted from several projects that had been done in the past. Among the projects dedicated for providing the MGI are Digital Marine Resource Mapping (DMRM) (1996 - 1999), Marine Resource and Evaluation and Planning (MREP) (1993-1998) and Marine and Coastal Resource Management (MCRM) (2001-2006) projects. Those projects implemented by the National Coordination Agency for Surveys and Mapping (BAKOSURTANAL) – now BIG. The implementation of the MCRM project then was conducted in collaboration with the Ministry of Marine Affairs and Fisheries (MoMAF), following the establishment of this ministry in 2000. In addition, there is also Coral Reef Rehabilitation and Management Program (COREMAP) for monitoring condition of coral reef and related benthic habitat. Besides that, BAKOSURTANAL and now BIG has continuously work on the provision of basemap and thematic map of coastal and marine geospatial data.

The data produced from those projects mentioned above varied in scale, theme and conditions. The available MGI in the form of basemap are Maritime Boundary, LLN, and LPI. The Maritime Boundary map form the national boundary in adjacent to other countries form the territory of Indonesia. Most of the maritime boundaries are already settled down but some are still in negotiations. Indonesia border at the sea with 10 neighboring countries, namely Malaysia, Thailand, Vietnam, Singapore, Australia, the Philippines, India, Republic of Palau, Democratic Republic of Timor-Leste and Papua New Guinea. Some of the boundary delimitations are already definite but some still in progress of negotiation.

The LPI and LLN maps provide information specifically for coastal and marine areas, especially bathymetry, coastal types (sandy, muddy, or rocky), as well as other basic information related to navigation and administration in the marine areas. The difference between LLN and LPI maps is at the portion of the land and water area to be mapped in each map sheets. In general, the LLN map cover mostly the water area, while the LPI map cover the coastal area with land portion up to 40% of the map sheet area. The LLN map is available at the scale of 1: 500,000, 1: 250,000 and 1: 50,000, with coverage of 100%, 6.67% and 4.84% respectively. Meanwhile, the LPI map is available at the scale of 1: 250,000, 1: 50,000, 1: 25,000 and 1: 10,000, with coverage of 74.06%, 49.52%, 2.37%, and 0.33% respectively. The map index of LLN and LPI can be seen in Figure 2 and Figure 3.

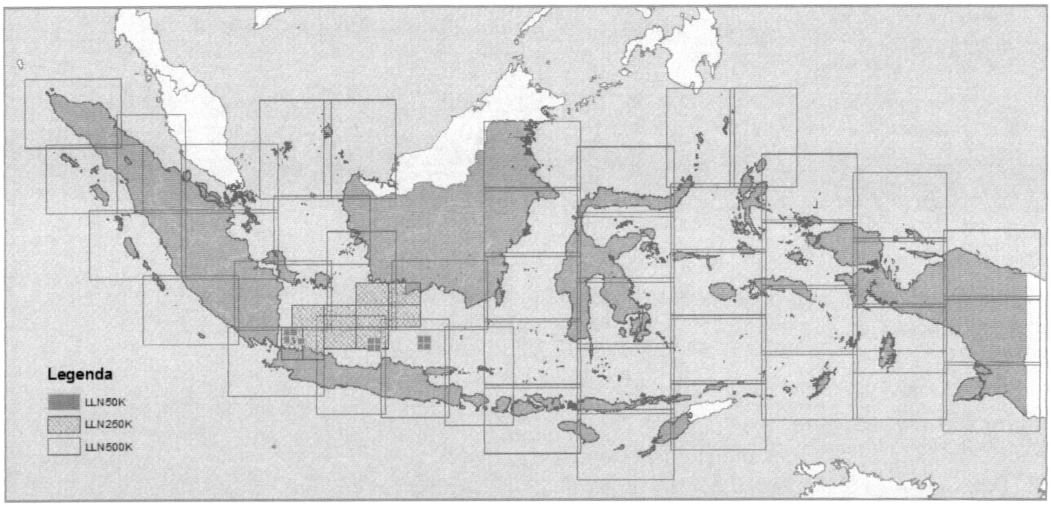

Figure 2. Index of National Marine Environment (LLN) map produced by BIG (*source: accessed from BIG's product services, 2017*).

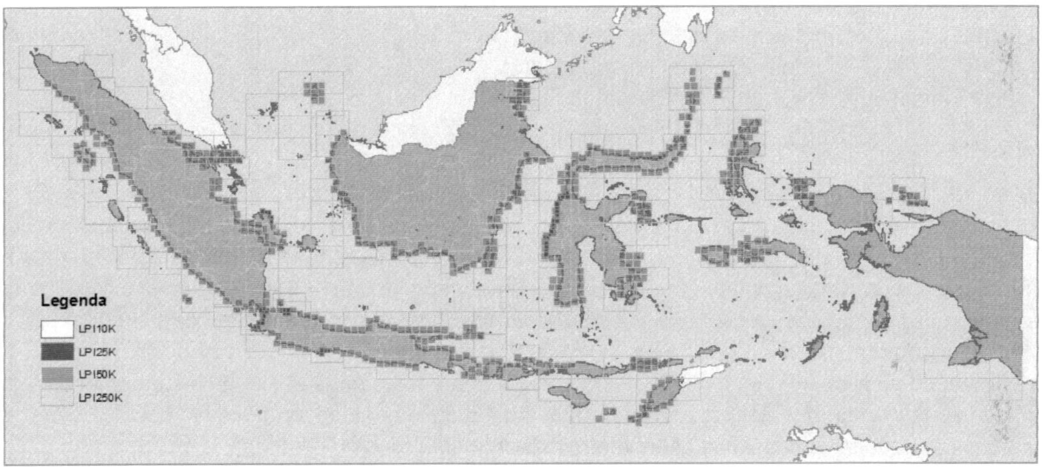

Figure 3. Index of Indonesian Coastal Environment LPI map produced by BIG (*source: accessed from BIG's product services, 2017*).

In contrast to the basemap that provide only by BIG, the thematic maps are provided by many institutions. There are several working groups of thematic one map that work together to provide the national coastal and marine resources geospatial data. The thematic maps and the contributor institutions are listed in Table 1.

4.2. Data integration

Data integration is another important part of the one map development including the one map of coastal and marine resources. Data integration ensures thematic map has single reference which is the basemap produced by BIG. Therefore, all thematic maps compiled from Ministries/Agencies must be verified by BIG. The data integration is the responsible of the respected Ministries/Agencies (the Custodianship), even though the members of Working Group of the thematic map shall participate during the integration process. The process of data integration also includes verification and integration of the database so the thematic maps follow the standard database rule. This is important for data sharing and data utilization.

IOP Conf. Series: Earth and Environmental Science **162** (2018) 012028 doi:10.1088/1755-1315/162/1/012028

Table 1. List of thematic maps and the contributor institutions.

	Thematic Map	Source of Data
1.	Mangroves	KLHK, BIG, KKP, LAPAN, Wetland International
2.	Coral reef	LIPI, BIG, KLHK, LAPAN, KKP, TNC
3.	Seagrass	LIPI, BIG, KLHK, KKP, LAPAN
4.	Oceanography	ESDM, BIG, KKP, BPPT, LIPI
5.	Seabed cover	ESDM
6.	Salt pond	BIG, KKP
7.	Sea lane	Kemenhub, Pushidrosal
8.	Seismic	ESDM and BIG
9.	Tides	BIG, Pushidrosal, BMKG
10.	Marine Ecoregion	KKP, KLHK, BIG, LAPAN, Wetland International

Abbreviation:

BIG	: Geospatial Information Agency
BMKG	: Bureau of Meteorology and Geophysics
KKP	: Ministry of Marine Affairs and Fisheries
KLHK	: Ministry of Environment and Forestry
ESDM	: Ministry of Energy and Mineral Resource
Kemenhub	: Ministry of Transportation
LIPI	: Indonesian Science Institute
LAPAN	: Indonesian Aerospace Agency
Pushidrosal	: The Navy Hydrographic Survey
TNC	: The Nature Conservancy

4.3. One map product development

The availability of coastal and marine thematic maps in Indonesia is already stated previously in Section 3.1. However, some of the thematic maps do not pass the one map criteria since those spatial data has yet fulfill the standard of one map. Among the MGI that already satisfy the one map standard are one map mangroves, coral reef, and seagrass. The mangroves mapping in Indonesia have been conducted by several institutions and NGO (Wetland International Indonesia). The mangroves mapping that has been conducted nationally started by BAKOSURTANAL using Landsat satellite imageries in 2009. Based on this mapping, the mangroves was estimated for 3,244,018.46 hectares [15]. This mangroves mapping is continuously updated by using remote sensing approach on the mapping process. Field work has been done supplementary to validate the mangroves mapping. Following the advancement of remote sensing technology and the availability of satellite imageries, the mangroves mapping continuously updated using better resolution of satellite imageries. Some areas have been mapped using high resolution satellite imageries (HRSI) such as Quickbird, GeoEye, WorldView, and Pleiades for Java and Bali islands. The mangroves mapping carried out by BAKOSURTANAL/BIG in coordination and cooperation with respected ministries and institutions such as the Ministry of Environment and Forestry, MoMAF, LAPAN, LIPI, and the Wetland International Indonesia. Those institution are among the institution that form the National Working Group of Mangroves. The custodian of one map mangroves is the Ministry of Environment and Forestry. Figure 4 shows the development of one map of Mangroves Indonesia, the scale and main source of satellite data used for the mapping.

One map of coral reef and one map seagrass are the other MGI product. The custodian of one map coral reef in Indonesia is LIPI. The coral reef map has been carried out by BAKOSURTANAL/BIG and MoMAF through MREP and MCRM projects, and by LIPI though COREMAP project. The mapping also conducted by NGO such as TNC. All of the coral reef and seagrass data have been compiled and integrated by BIG. Currently, BIG is conducting coral reef and seagrass mapping in the province of Maluku and Maluku Utara, as well as at some areas in the islands of Sumatra region. Meanwhile, LIPI

IOP Conf. Series: Earth and Environmental Science **162** (2018) 012028 doi:10.1088/1755-1315/162/1/012028

designated several stations for coral reef monitoring. Figure 5 and Figure 6 shows the one map coral reef and seagrass Indonesia.

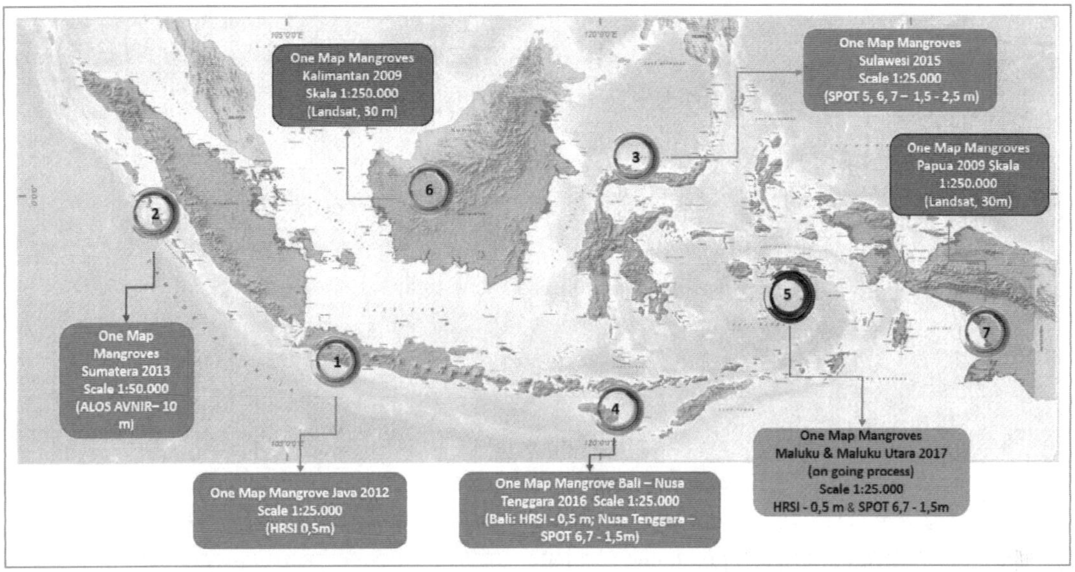

Figure 4. One Map Mangroves Indonesia and the development (*source: compiled by authors from BIG thematic map database, 2017*).

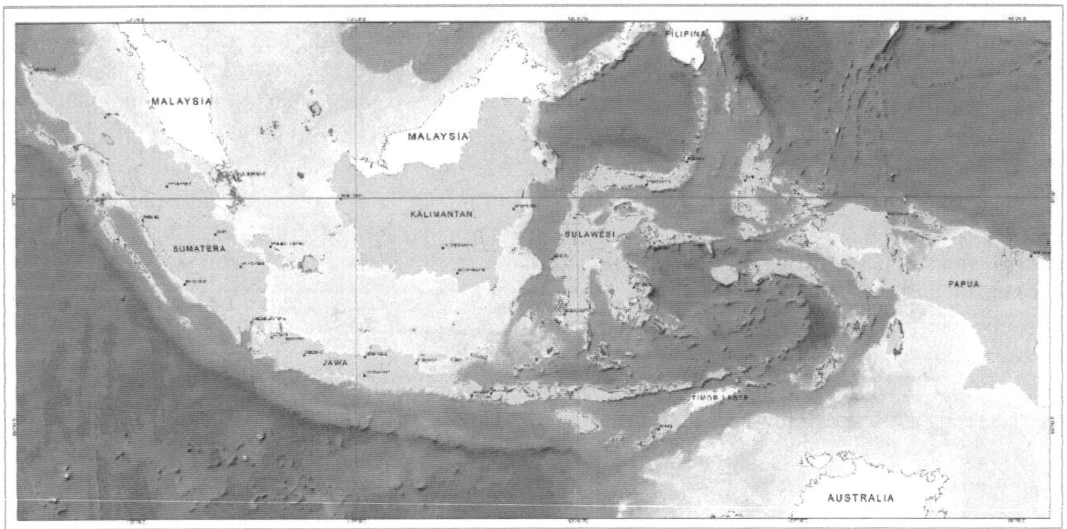

Figure 5. One Map Coral Reef Indonesia (*source: compiled by authors from BIG thematic map database, 2017*).

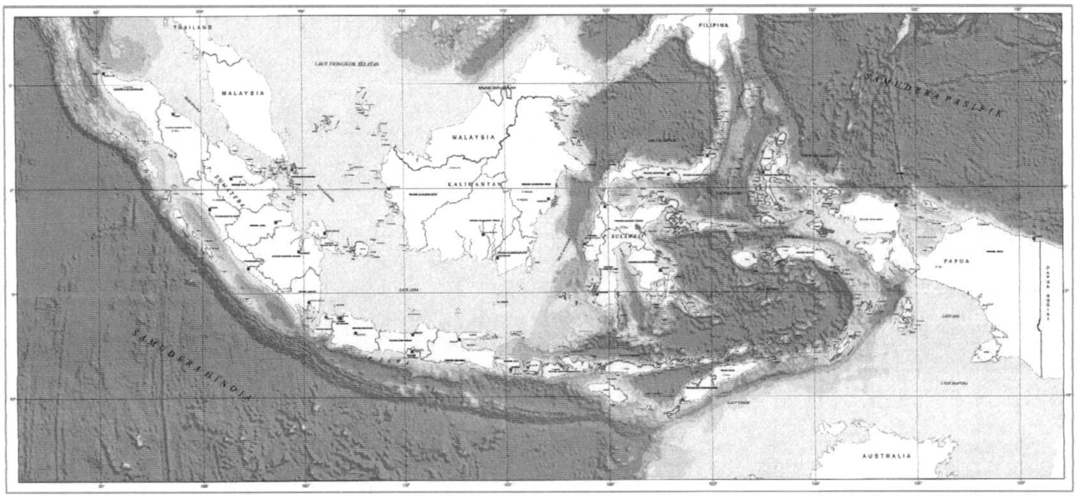

Figure 6. One map Seagrass Indonesia (*source: compiled by authors from BIG thematic map database, 2017*).

The other coastal and marine geospatial data that already compiled and integrated to the basemap are oceanography (water quality), salt pond, and sea-lane, and seabed-cover maps that are used as input for the development of Coastal and Small Island Zonation Plan (RZWP3K). The oceanographic water quality data collected consisted of brightness, depth, temperature, conductivity, turbidity, salinity, Dissolve Oxygen (O), and potential of hydrogen (pH). The oceanographic data represented in point, with attribute of time and date of the data measurement. Meanwhile, the mapping of Salt Pond was carried out by BAKOSURTANAL/BIG between 2010 and 2014 to support one of the MoMAF program which was the "salt-farmer empowerment program" or known as PUGAR to achieve self-sufficiently of domestic salt consumption. Besides that, this map also used to calculate potential production of salt, so that the government can predict the amount of imported salt to meet the domestic consumption needs. High resolution spatial imageries were used for the mapping. Some areas in Java were mapped in a parcel base completed with "by name by address" attributes. The Sea-lane map provided by the Ministry of Transportation, whereas the Seabed-cover map provided by the Centre for Marine Geological Research, the Ministry of Energy and Mineral Resources. The Seabed-cover map shows information about marine sediment exist at the seafloor. There are four basic types of marine sediments, all of which are grouped and ordered by the origin of their particles, the grain sizes, and where they are deposited. The area has been map covering 228 sheet maps in scale of 1:250.000.

All of the one map products mentioned above were developed by considering scientific aspects of geospatial information development which are useful for developing RZWP3K.

5. Conclusion

There are several points that can be underlined regarding the development of one map of Marine Geospatial Information (MGI) in Indonesia. First, the implementation of One Map Policy (OMP) is important to standardize the marine resources data to provide accurate, accessible, and accountable geospatial data. Second, the products of one map marine resources that have been developed are useful for integrated spatial planning. Therefore, stakeholders related to the coastal and marine spatial planning development could start to develop RZWP3K map on the basis of one map policy. It is also understood that the implementation of One Map Policy requires strong coordination and collaboration among stakeholders. Advancement on spatial information technology would facilitate improvement on data quality and data sharing. Development of One map marine resources require improvement of Geospatial Information human resources.

References

[1] United States Department of State 2014 *Limit in the Seas No. 141. Indonesia: Archipelagic and other Maritime Claims and Boundaries* (Office of Ocean and Polar Affairs Bureau of Oceans and International Environmental and Scientific Affairs U.S. Department of State)

[2] BIG 2015 *Pentingnya Informasi Geospasial untuk Menata Laut Indonesia* (Badan Informasi Geospasial) [cited 2017 Aug 31] Available: www.bakosurtanal.go.id/berita-surta/pentingnya%09informasi-geospasial-untuk-menata-laut-indonesia.

[3] R Cribb and M Ford ed 2009 *Indonesia Beyond the Water's Edge: Managing an Archipelagic State* (Singapore: Institute of Southeast Asian Studies) p 248

[4] Republik Indonesia 2016 *Peraturan Presiden Republik Indonesia Nomor 9 Tahun 2016 tentang Percepatan Pelaksanaan Kebijakan Satu Peta Pada Tingkat Ketelitian Peta Skala 1 : 50.000*

[5] Giri C, E Ochieng, L L Tieszen, Z Zhu, A Singh, T Loveland, J Masek and N Duke 2011 Status and Distribution of Mangrove Forests of the World Using Earth Observation Satellite Data *Global Ecology and Biogeography* vol **20** no 1 pp 154-59 https://doi.org/10.1111/j.1466-8238.2010.00584.x.

[6] M B Best, B W Hoeksema, W Moka, H Moll, Suharsono and I N Sutarna 1989 Recent Scleractinian coral species collected during the Snellius-II expedition in Eastern Indonesia *Netherlands Journal of Sea Research* vol **23** no 2 pp 107–115

[7] A L Green and P J Mous 2008 Delineating the Coral Triangle, its Ecoregions and Functional Seascapes Version 5.0. (TNC Coral Triangle Program Report 1/08) p 44

[8] T Tomascik, A J Mah, A Nontji and M K Moosa 1997 *The ecology of the Indonesian seas* (Oxford University Press)

[9] E P Green and F T Short 2003 *World atlas of seagrasses* (The UNEP World Conservation Monitoring Centre, Berkeley USA: University of California Press)

[10] FAO 2016 *The State of World Fisheries and Aquaculture 2016 Contributing to food security and nutrition for all* (Rome) p 200

[11] A Eko 2014 Wow, Kekayaan Laut Indonesia Capai 1,2 Triliun USD/Tahun *TEROPONGSENAYAN.com* [cited 2017 Aug 31] Available from: http://www.teropongsenayan.com/4556-%09wow-kekayaan-laut-indonesia-capai-12-triliun-usdtahun

[12] BIG 2013 *Peluncuran Referensi Tunggal Informasi Geospasial. Mewujudkan Informasi Geospasial Nasional yang Terintegrasi. Menuju Indonesia yang Lebih Baik* (Cibinong)

[13] P A Burrough, R A McDonnell and C D Lloyd 1986 *Principles of Geographical Information System for Land Resource Assessment* (Oxford, New York: Clarendon Press)

[14] Nurwadjedi 2016 Kebijakan Satu Peta Dalam Mendukung Pembangunan Infrastruktur Nasiona *Buletin Tata Ruang dan Pertanahan* (Jakarta: Direktorat Tata Ruang dan Pertanahan Kementerian PPN/Bappenas)

[15] S Hartini, G B Saputro, M Yulianto and Suprajaka 2010 Assessing the Used of Remotely Sensed Data for Mapping Mangroves Indonesia *SELECTED TOPICS in POWER SYSTEMS and REMOTE SENSING in 6th WSEAS International Conference on REMOTE SENSING (REMOTE '10)* (Japan: Iwate Prefectural University) October 4-6) pp 210-215

Aesthetic quality assessment in Santolo Beach, West Java Province, Indonesia

Ankiq Taofiqurohman[1], Ivonne M Radjawane[2], Yayat Dhahiyat[3]

[1] Environmental Science Doctoral Program, Padjadjaran University
Jl.Dipatiukur 35, Bandung,West Java, Indonesia
[2] Departement of Oceanography, Bandung Institute of Technology
Jl.Ganesha 10, Bandung,West Java, Indonesia
[3] Fisheries and Marine Science Faculty, Padjadjaran University
Jl.Raya Bandung Sumedang Km.21,Sumedang, West Java, Indonesia

E-mail: ankiq109@gmail.com

Abstract. The Santolo beach is one of the tourism beaches in Regency of Garut, West Java, Indonesia. The position of Santolo beach which facing towards the Indian Ocean, caused limited water activities that makes beautiful scenery is the main attraction for tourists. This research aims to determine the value of aesthetic quality Santolo beach scenery based on the viewshed. Three analysis approaches are used to solve the research problem, that are the Kernel Density, the Viewshed Analysis and the Euclidean Classification.The results show there are 5 viewshed areas with an average area of 0.14 km^2. Each viewshed has a varied aesthetic quality value between moderate to good. In generally the Santolo Beach can be developed become the coastal tourism.

1. Introduction

Santolo Beach is the coastal part of Garut Regency in West Java Province, Indonesia (figure 1). The Indian Ocean is located in the south of Santolo Beach, while in the northern part, Santolo Beach is surrounded by mountains (figure 2), which makes Santolo Beach has unique scenery. Santolo Beach is a tourism beach has been visited by 205.595 tourists for the last decade. The total number of tourist visits Santolo Beach less than the number of tourist visits to Cipanas which is the tourist area in the mountains of Garut Regency [1]. The Number of tourist visits to Santolo Beach, prove that Santolo Beach has not developed into tourism beach. One of the solutions for Santolo Beach condition is developing coastal tourism based on aesthetic quality. The aesthetic quality is chosen because recreation in Santolo Beach are limited by oceanographic condition such as high wave, Rip current and coastal depth. On the other hand, Santolo Beach has the beautiful and natural seascape that can attract tourist to visit.

The aesthetic quality parameter of a coastal tourism development must be measured since it is one of the main reason for tourists to visit an area. Seascape aesthetic quality forms a unity and harmonization between sea and land, which attract tourist to enjoy beach situation. Seascape is the important part in sustainable of social, economy, and environment development [2]. The value of an area can depend on its attributes and one of the attributes can increase value an area is the aesthetic quality. The high value of aesthetic quality will create beautiful scenery. The aesthetic quality is

IOP Conf. Series: Earth and Environmental Science **162** (2018) 012029 doi:10.1088/1755-1315/162/1/012029

relating to the view, and good views are often associated with easy access to nature. Good view should not be limited to natural features such as the ocean or a forest, but should also include with attractive landscaping in the area [3].

Figure 1. Santolo Beach Position

Figure 2. Views of the mountains in santolo beach

The aesthetic quality of seascape is an attribute that is generally considered important for those visiting coastal areas to enjoy the ocean and coastal environments. The aesthetic quality assessment assumes the objects viewed have a negative and positive impact on views [4]. The Concise Oxford Dictionary defines Seascape as a picture or view to the sea. In general the concept of seascape is assumed as views from land to sea,views from sea to land,views along coastline, the effect on landscape of the conjunction of sea and land [5]. The aesthetic quality is used as an alternative solution for developing coastal tourism in Santolo Beach. The aim of this research is to assess aesthetic quality of Santolo Beach seascape and then aesthetic quality values are used to determine the tourist zones. This research also seeks the reason behind the attraction of Santolo Beach to tourism and another factors that can impact on aesthetic quality value in Santolo Beach.

2. Data and Methods

2.1. Data

Location of study site along the Santolo Beach of Garut Regency,West Java Province,Indonesia. This research uses Panoramio data photograph that connected to Google Earth (figure 3). Panoramio photograph data positions are assumed as representative of tourist positions. The photo used as the data must capture the figure of sea or land scenery.

Figure 3.a. Santolo Beach at the morning [6]

Figure 3.b. A part of Santolo Beach scenery [7]

Figure 3.c. Sunrise on Sayang Heulang beach [8]

Figure 3. The spread of Panoramio photograph data position (left) and the examples of Panoramio photograph data that capture Santolo Beach scenery (right)

2.2.Methods

The research uses a field survey and three analytical methods to solve the research problem. The analysis approaches are the Kernel Density, the Viewshed Analysis and the Euclidean Classification.

2.2.1.The Kernel Density

This research uses The Heatmap Qgis Plugin to get the recreation spot zone in the study area and the Heatmap Plugin uses Kernel Density estimation to create a density raster of Panoramio data photograph. The density is calculated based on the number of points in a location, with larger numbers of clustered points resulting in larger values [9].

2.2.2.The Viewshed Analysis

The viewshed is the coverage that can be seen from the specified location. The viewshed analysis compares the height of an object on earth surface against the height of the earth's surface along the line of sight (figure 4). If the height of the object in earth surface is less than that earth's surface, then the line of sight will be blocked and has no visibility. If it is higher, the object will be seen from the current position [10].The Viewshed is used to create the boundary of the tourist area. This research uses The Viewshed Analysis Qgis Plugin.

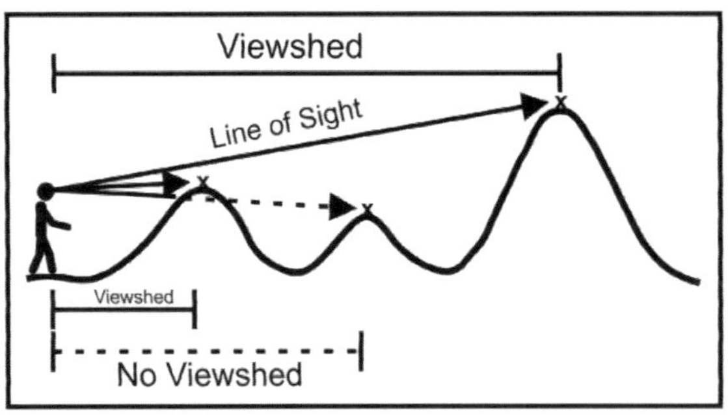

Figure 4. The viewshed illustration

IOP Conf. Series: Earth and Environmental Science **162** (2018) 012029 doi:10.1088/1755-1315/162/1/012029

2.2.3.The Euclidean Classification

Views from land to sea or sea to land have equal weight to classify the seascape quality value. This research uses that statement to classify seascape aesthetic quality based on Euclidean Algorithm. Euclidean Algorithm is adopted from Arkema research [11] about habitat risk assessment. According to Arkema, Euclidean Algorithm can be used to classification if the two factor of classification has the equal effect and interrelated. The two factors in this research are sea view and land view factor. Each factor is consisted by beach physical parameter (table 1). The parameter in this research is adopted and modified from Ergin [12] [13] [14] and Botero [15].

Table 1. The views parameter scores

Parameter		**Score**				
Sea Views		**1**	**2**	**3**	**4**	**5**
				Criteria		
Wave	Type	Plunging and collapsing		Spilling		Surging
Tides	Type	Semi diurnal and mixed tide prevailing semi diurnal		Mixed tide prevailing diurnal		Diurnal tide
	Height	Macro (>4m)		Meso (2-4m)		Micro (<2m)
Water Colour		Muddy brown or grey	Dark green	Clear green or dark blue	Clear blue	Very clear
Sea objects		Building	None		Reefs	Group of islands
Biotic elements		None	Coastal vegetation	Reefs only	Biotics reef	Sea animals (such as : dolphin, whale, sea turtle)
Water activities		Motor vessel	Water sport vehicles	Sailing	Canoeing or kayaking	No activities
Land Views						
Cliff	Height (m)	None	5-30	30-60	60-90	>90
	Slope (^{0})	None	45^0	60^0	75^0	90^0
Beach face	Type	Estuary	Mud	Cobble/ Boulder	Pebble/ Gravel	Sand
	Width (m)	None	< 5 ; > 100	5-25	25-50	50-100
	Colour	None	Dark	Dark tan	Light tan/ bleached	White
Dune		None	Remnants	Fore-dune	Secondary ridge	Several
Coastal vegetation		None	Bush	Meadow	Short lived trees	Long lived trees
Litter		Continuous accumulations	Full strand line	Single accumulations	Few scattered items	Virtually absent
Beach activities		Heavy industry and/or heavy port	Traditional port	Light tourism	Sensitive tourism	Preserve and/or none

the Euclidean algorithm :

$$E=\sqrt{(M-1)^2+(T-1)^2}$$

$$M=\frac{\sum_{i-1}^{N}\frac{m_i}{d_i}}{\sum_{i-1}^{N}\frac{1}{d_i}} \qquad T=\frac{\sum_{i-1}^{N}\frac{t_i}{d_i}}{\sum_{i-1}^{N}\frac{1}{d_i}}$$

where :

E : Seascape aesthetic quality index
M : Views from land to sea value
T : Views from sea to land value
m_i : Sea view parameter scores
t_i : Land view parameter scores
d_i : Data quality

The result of the Euclidean algorithm is classified using the aesthetic quality index (table 2) or using the Euclidean chart (figure 5). Classification uses the Euclidean chart is obtained by post the 'M' and 'T' value on the Euclidean chart axis. Euclidean chart classification is useful to show the distribution of value.

Table 2. The aesthetic quality index classification

Aesthetic quality criteria	Aesthetic quality index (E)
Very high	5,65 – 7,07
High	4,24 - < 5,65
Medium	2,82 - < 4,24
Low	1,41 - < 2,82
Very low	0 – < 1,41

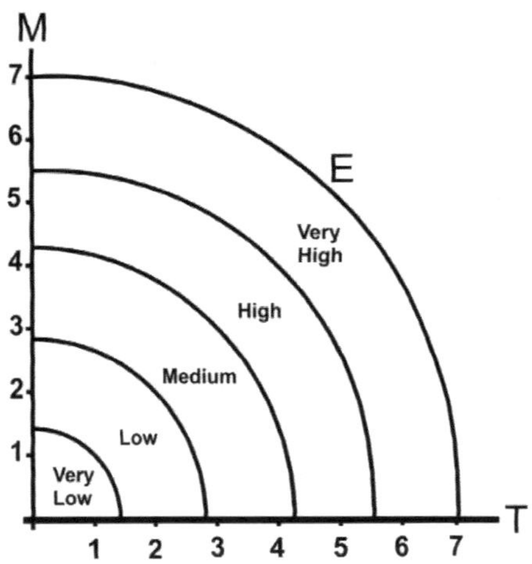

Figure 5. The Euclidean chart

3. Results

According to outcome of The Heatmap Qgis Plugin, tourists spreading density is divided into 4 classes, there are High density, Medium density, Less density and No density (figure 6). Furthermore high and medium density are selected to determine the tourist position and generate five recreation spot zone, i.e red spots for position A, orange spots for position B, green spots for position C, pink spots for position D, and turquoise spots for position E (figure 7).

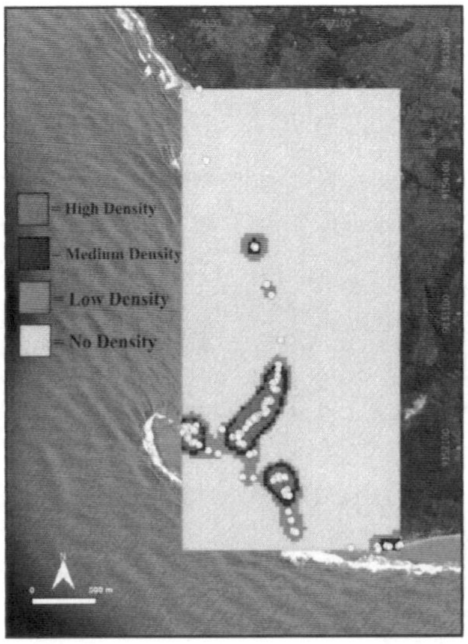

Figure 6. Tourist spreading density

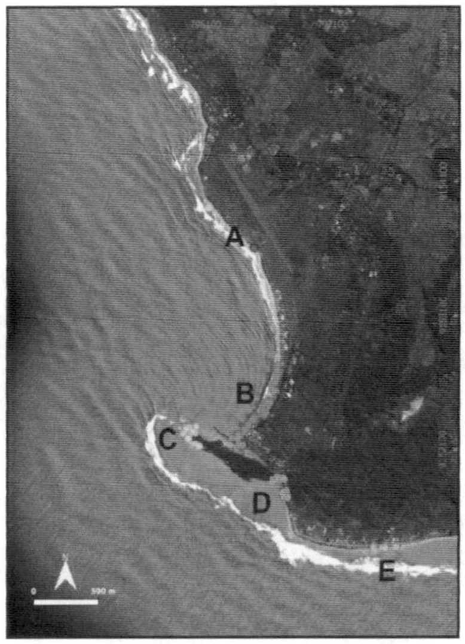

Figure 7. The recreation spot zones

The recreation spots zones are used as input data for The Viewshed Analysis Qgis Plugin, and the result shows the viewshed boundaries of the tourist zone (figure 8). Each zone has the coverage area as follows, zone A area is 0.2 km^2, zone B area is 0.23 km^2, zone C area is 0.07 km^2, zone D area is 0.14 km^2, zone E area is 0.04 km^2. Besides as tourism zone boundary, in this research, the viewshed is used as the boundary of aesthetic parameter filed survey.

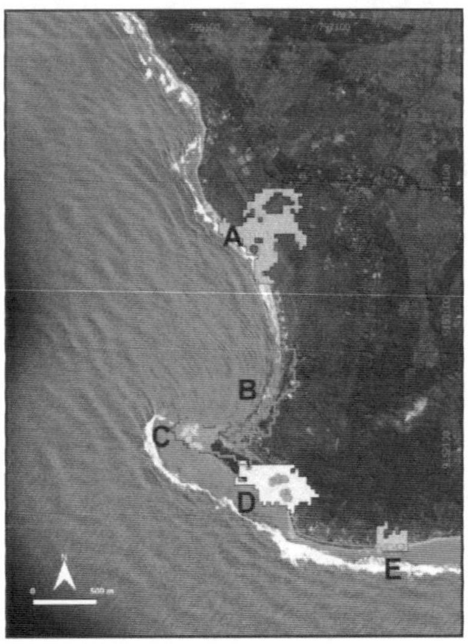

Figure 8. The viewshed zones which are shown by polygon

IOP Conf. Series: Earth and Environmental Science **162** (2018) 012029 doi:10.1088/1755-1315/162/1/012029

Field survey result shows the area of study is formed by white sand and consists of spilling wave type and also mixed tide prevailing diurnal type with micro height tide type. In general, each parameter criteria of aesthetic quality on every zone almost equal. Only water activity parameter that shows an extreme difference. A, B and C zone are fisherman vessel ships activity (figure 9) so that makes low criteria score . While D and E zone are the area without activity which has high criteria score for water activity (figure 10). Score of each parameter criteria of aesthetic quality on every zone is explained on table 3.

Figure 9. Fisherman vessel ships port

Figure 10. No water activity at E zone

Table 3. Aesthetic quality parameter scores

Parameter		**Zone**				
Sea Views		**A**	**B**	**C**	**D**	**E**
				Score		
Wave	Type	3	3	3	3	3
Tides	Type	3	3	3	3	3
	Height	5	5	5	5	5
Water Colour		3	3	3	5	5
Sea objects		2	2	4	4	4
Biotic elements		3	3	3	4	4
Water activities		1	1	1	5	5
Land Views						
Cliff	Height (m)	1	1	1	2	1
	Slope (^{0})	1	1	1	5	1
Beach face	Type	5	5	5	5	5
	Width (m)	4	4	4	3	3
	Colour	5	5	5	5	5
Dune		3	3	1	3	3
Coastal vegetation		4	2	2	5	2
Litter		4	3	4	4	3
Beach activities		3	3	3	4	3

IOP Conf. Series: Earth and Environmental Science **162** (2018) 012029 doi:10.1088/1755-1315/162/1/012029

The Euclidean Algorithm classification shows the tourist zone in the Santolo Beach are rated in medium and high aesthetic quality (Figure 11). The high aesthetic zone is located in southern of the Santolo Beach, on D and E zone. The main reason that makes D and E zone have high aesthetic quality is the existence of rocks on the sandy beach and also no water activity on its area. When the tidal flood is coming on D and E zone, the rocks became clear water natural pools with biotic elements like fish, molluscs and seagrass (Figure 12). In the medium aesthetic zone (A,B,C), fisherman boat activity and the litters are decreasing the aesthetic value, especially in B zone (Figure 13).

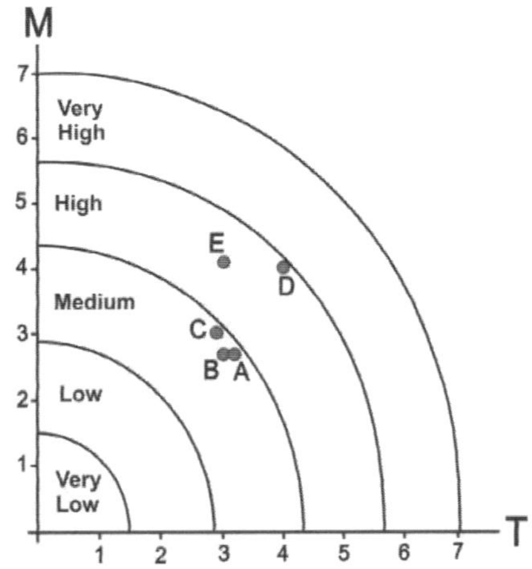

Figure 11. Santolo Beach aesthetic quality classification

Figure 12. Clear water natural pool at D zone [16]

Figure 13. Litters at B zone

4. Discussion

The Santolo beach has the relatively homogeneous diversity of views parameter, therefore aesthetic quality value is almost similar for each zone. Because of the remote coastal situation, the seascape modification is minimum so the Santolo Beach can be categorized as natural beach condition. The natural coastal condition will increase aesthetic quality value, and aesthetic quality ca n be used as a tourism identity for the coastal which the undersea conditions cannot be utilized because it is dangerous. Coastal aesthetic quality is affected by land and sea views factors [17] [18]. Ergin [12][13][14] have done some research using fuzzy logic and produce detailed parameters to assess coastal aesthetic quality. The aesthetic quality Ergin method requires many respondents so it can not be used for desolate beaches such as Santolo Beach area, therefore in this research using Euclidean classification method to assess the aesthetic quality. The Euclidean method is selected because this method is simple to assess a condition which is composed of two equal criteria [19]. The criteria in this research are views of land and sea from coastal. In order to apply aesthetic quality assessment using the Euclidean method as a general method, further research is required on some beaches in Indonesia.

5. Conclutions

The Santolo Beach has five zones that often visited by tourist. Three zones are categorized as medium quality and located in the northern part of Santolo Beach. On the other hand, the two other location has high quality aesthetic and located in the south of the Santolo Beach. Overall, the white sandy beach is the main reason that makes the Santolo Beach is attractive for tourism and the best time to visit the Santolo Beach is during the tidal flood in the afternoon. However, be required limits the number of activities in Santolo Beach, because too many activities might decrease the aesthetic quality value and increasing litters from tourism activities is another thing to consider. In general, Santolo Beach has good aesthetic quality, so the beach is potential to be developed as coastal tourism.

References

[1] Rinaldi, R N and Soewardikoen, D W 2015 *Identitas Visual dan Media Promosi Pantai Wisata Garut Selatan* (Bandung : Telkom University) p 117

[2] Tudor C 2012 *An approach to seascape character assessment* (Bristol : Natural England Commissioned Report) p 47

[3] Bourassa S, Hoesli M, Sun J 2004 What's in a view? *Environment and Planning* **8** 1427-1450

[4] Verutes G 2012 *InVEST 2.4.5 User's Guide:Integrated Valuation of Environmental Services and Tradeoffs* (Standford, CA: The Natural Capital Project) p 205

[5] Hill M, Briggs J, Minto P, Bagnall D, Foley K, Williams A 2001 *Guide to best practice in seascape assesment* (Dublin : The Marine Institute) p 57

[6] Yayu 2014 *Pantai santolo* http://www.panoramio.com/photo/108182075

[7] Maulany R 2014 *Pantai santolo* http://www.panoramio.com/photo/106167676

[8] Victory E 2013 *Sunrise on sayang heulang beach* http://www.panoramio.com/photo/89394447

[9] Thiede R, Sutton T, Duster H, Sutton M 2014 *Qgis Training Manual: Release 2.14* https://docs.qgis.org/2.14/en/docs/user_manual/index.html

[10] Bruy A, Svidzinska D 2015 *QGIS by Example* (Brimingham : Packt Publishing Ltd) p 412

[11] Arkema K K, Verutes G, Bernhardt J R, Clarke C, Rosado S, Canto M, Wood S A, Ruckelshaus M, Rosenthal A, McField M, de Zegher J 2014 Assessing Habitat Risk From Human Activities to Inform Coastal and Marine Spatial Planning : A Demonstration in Belize *Environmental Research Letters* **9** 1-11

[12] Ergin A, Karaesmen E, Micallef A, Williams A T 2004 A New Methodology for Evaluating Coastal Scenery : Fuzzy Logic Systems *Area* **36** 367-386.

[13] Ergin A, Karaesmen E, Ucar B 2011 A Quantitative Study for Evaluation of Coastal Scenery *Journal of Coastal Research* **27** 1065 – 1075.

[14] Ergin A, Williams A T, Micallef A 2006 Coastal Scenery : Appreciation and Evaluation *Journal of Coastal Research* **22** 958-964.

[15] Botero C, Anfuso G, Williams A T, Palacios A 2013 Preception of Coastal Scenery along the Caribbean Littoral of Colombia *Journal of Coastal Research* **65** 1733-1738

[16] Arirhd 2009 *Santolo beach* http://www.panoramio.com/photo/22126976

[17] DeWan T, DeWan T J 2008 *Scenic Assessment Handbook* (Augusta : Maine Coastal Program) p 80

[18] Morgan R, Williams A T 1999 Video Panorama Assessment of Beach Landscape Aesthetics on the Coast of Wales. *Journal of Coastal Conservation* **5** 13-22.

[19] Marco V R, Young D M 1987 The Euclidean Distance Classifier : an Alternative to the Linear Discriminant Function *Communications in Statistics- Simulation and Computation* **16** 485-505

IOP Conf. Series: Earth and Environmental Science **162** (2018) 012030 doi:10.1088/1755-1315/162/1/012030

Potential of methanol extract from the stem bark of mangrove *Rhizophora mucronata* against bacteria *Escherichia coli* and *Aeromonas hydrophylla*

Mahmiah[1], Giman[1], Nanik Siti Aminah[2], Mulyadi Tanjung[2]

[1] University of Hang Tuah Surabaya, Indonesia
[2] Faculty of Science and Technology, University of Airlangga, Indonesia

E-mail: mahmiah@hangtuah.ac.id

Abstract. *Escherichia coli* and *Aeromonas hydrophylla* are well known pathogenic bacteria causing diseases in both human and animals. Since the popular antibacterial drugs in the market lead to resistance, other alternatives antibacterial need to be searched from the active ingredient found in marine and terrestrial vegetation. The potential use of secondary metabolites from marine vegetation is currently being developed. One of marine vegetation that has been expected to have antibacterial activity is mangrove. This study aims to provide information related to the phytochemical and antibacterial activity of methanol extract of the bark of mangrove *Rhizophora mucronata*. The phytochemical screening and disc testing methods were used in this study.The results showed that the methanol extract of *R. mucronata* containedphenolics, alkaloids, and terpenoids. Furthermore, the positive outcome of active antibacterial potential of the methanol extract against *E. coli* and *A. hydrophilla* has been indicated by inhibition of discs in a clear zone at 7,56 and 7,00 mm.

1. Introduction

Mangrove is an estuary plant living in a high salinity habitat, influenced by temperature, wind, wave, and ocean current. This extreme environmental condition leads the plant to produce various secondary metabolite compounds that function as a protective effect to surrounding environment. The secondary metabolite compounds contained in these mangroves can potentially be developed as antibacterial, antiviral, and antioxidant agents. [1] explains that mangrove plants in Indonesia are the largest in the world, both in terms of territory (± 42,550 km^2) and number of species (± 45 species). On the basis of this mangrove, Indonesia certainly has a high species diversity and possibly potential as a medicinal material.

Escherichia coli and *Aeromonas hydrophylla* are pathogenic bacteria causing various diseases in human, fish and other animals. The use of antibiotic drugs such as chloramphenicol has been proven to cause resistance to these bacteria. Therefore, it is necessary to develop alternative antibiotics as antibacterial agents derived from terrestrial and aquatic plants.

One of the alternative aquatic plants that have antibacterial potentials is mangrove plant. Research on antibacterial activity of mangrove plants has been shown on the leaves extract of *Avicennia marina* and *R. stylosa* living in Sudanese waters. These plants demonstrated the potential reactions to inhibit the activity of *E. coli* and *Candida albicans*. The antibacterial activity of the plantswas suspected because of the secondary metabolite compounds; tannins, alkaloids, flavonoids, steroids, and glycosides contained in extract of *A. marina* and *R. stylosa* [2].

One of the mangroves living in East-coast Shore of Surabaya (Pamurbaya), East Java, Indonesia is *R. mucronata*. This mangrove species is a true mangrove commonly found and ethno-botanically utilized some parts of the plant as a pain reliever and natural wood dyeby the surrounding community [3]. This study focused on finding alternative utilization of methanol extract of plant stem bark of *R. mucronata* in inhibiting the activity of *E. coli* and *A. hydrophylla* bacteria.

Figure 1. *Rhizophora mucronata*

2. Method

2.1. *Materials Research*

Dry samples of mangrove rod bark of *R. mucronata* were collected from Pamurbaya area, Surabaya, East Java. The taxonomic test of *R. mucronata* sampling was conducted at Purwodadi Botanical Garden. The chemicals used in the extraction process include pro-analyzed (p.a) and technical organic solvents which has been distillated such as methanol, ethyl acetate, ethanol and n-hexane. The materials for antimicrobial assays include agar media, *E. coli* and *A. hydrophylla* bacteria.

2.2. *Extraction*

Methanol solvent was used in extraction of *R. mucronata*'s stem bark. This solvent was chosen because it is a universal solvent which can dissolve all the active ingredients of secondary metabolite compounds.

2.3. *Antibacterial Bioactivity Test*

The method used in the antimicrobial test is the disc diffusion method (agar diffusion method) with pour order.

2.4. *GC-MS Analysis of Methanol Extract R. mucronata*

Methanol extract of*R. mucronata*'sstem bark was analyzed using GC-MS HP 6890. 1 μL sample was injected into GC-MS operated using capillary model number agilent column 19091S-433 HP-5MS 5% Phenyl Methyl Siloxane, 30 m long, diameter 250 μm, and 0.25 μm thickness. The oven temperature was programmed between 100-220 ° C with a temperature rise rate of 15 ° C / min, a 10.5 psi pressure helium carrying gas, a total rate of 140 mL / min and a split ratio of 1:50.

3. Results

R. mucronata is one of the many mangroves living in the East-coast of Surabaya. The utilization of mangrove stem part of *R. mucronata* in this research was based on chemotaxonomic study of secondary metabolite compounds from the mangrove. The extraction process of secondary metabolite compound from *R. mucronata*'s bark was performed by maceration method using methanol solvent at room temperature.

A total of 1.005 kg *Rhizophoramucronata*'srod bark powder was proceededin maceration method by immersing the powder with methanol solvent for 3 (three) days. The extraction obtained as much as 235.0543 g brown paste-shaped. The antibacterial activity of methanol extract of *R. mucronata* in this research was conducted to measure the ability of the extract in controlling microorganism / bacterium of *A. hidrophylla* and *E. coli*. Chloramphenicol 1% was used as comparator. The 1% dosage of chloramphenicol is the usual dosage in antibacterial (antibiotic) drugs production.

3.1. Antibacterial activity of extract on A. hydrophylla

A. hydrophila is one of the bacteria causing disease in fish. The existence of these bacteria cause the fish quickly suffered to the protein damage so that it quickly experience decay. The magnitude of the methanol extract inhibition zone against *A. hydrophila* test bacteria was presented in Table 1.

Table 1. Inhibition zone diameter of methanol extract of *R. mucronata* against *A.hydrophila*

No	*Extract content (ppm)*	*Inhibitory rate (mm)*			*Average (mm)*
		1	2	3	
1	100	6.65	6.55	7.8	7.00
2	200	4.45	4.15	4.75	4.45
3	400	7.65	5.2	3.75	5.53
4	600	4.05	6.8	6.6	5.81
5	800	5.95	6.85	5.75	6.18
6	1000	7.2	5.95	6.05	6.4
7	Chloramphenicol	6.25	5.05	7.3	6.28

Based on Table 1, each sample test clearly has an inhibitory activity against *A. hydrophylla*. The largest inhibitory rate was at 100 ppm of methanol extract with average inhibitory of 7.00 mm. Compared with chloramphenicol,100 ppm of methanol extract has greater inhibitory rate. The comparison of inhibitory rate of methanol extract at various concentrations waspresented in Fig. 2.

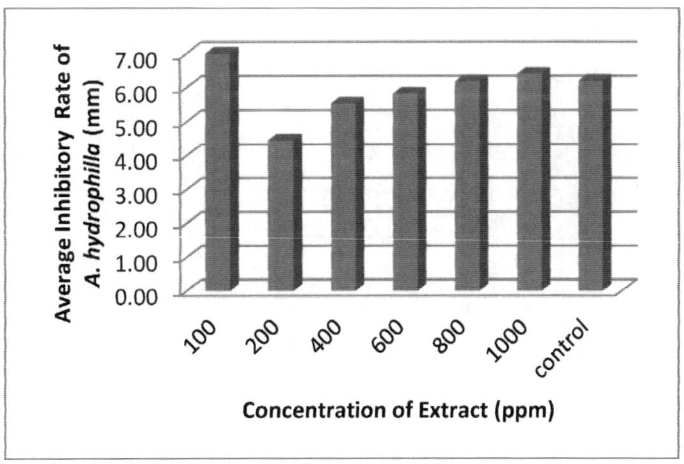

Figure 2. Comparison of concentration test against *A. hydrophilla* inhibition zone

According to [4], the formation of inhibition zone is a measurement of the strength of an antimicrobial agent against the examined bacteria. The resistance around the disc depends on the absorption capacity of the used active ingredient. If the antibacterial agent positively inhibit, the bacterial growth will stop. This condition was indicated by the clear circle around the disc that is not overgrown with bacteria after incubated for 18-24 hours.

3.2. Antibacterial activity of extract on E. coli

E. coli is one of the many types of bacteria found in fresh water. Consuming the bacterial-contaminated food or beverages can cause gastrointestinal diseases. The magnitude of methanol extract inhibition zone against *E. coli*was shown in Table 2.

Table 2. Inhibition zone diameterof extract methanol of *R. mucronata* bark against *E. coli*

No	*Concentration*	*Inhibitory rate (mm)*			*Average (mm)*
		1	**2**	**3**	
1	100	8.65	6.15	6.95	7,23
2	200	5.05	7.8	8.45	7,1
3	400	8.15	6.85	6.5	7.16
4	600	6.3	7.7	5.45	6.48
5	800	8.25	6.5	5.75	6.8
6	1000	8.65	6.85	7.2	7.56
7	Chloramphenicol	8.15	7.95	6.4	7.56

Based on table 2, it was recorded that the extract has ainhibition zone diameter of > 6 mm which means that the test material show positive activity in inhibiting the growth of *E. coli*. The highest concentration was seen at 1000 ppm with 7.56 ppm inhibition as well as chloramphenicol positive control was. Inhibitory rate comparison of methanol extract at various concentrations is shown in Fig. 3.

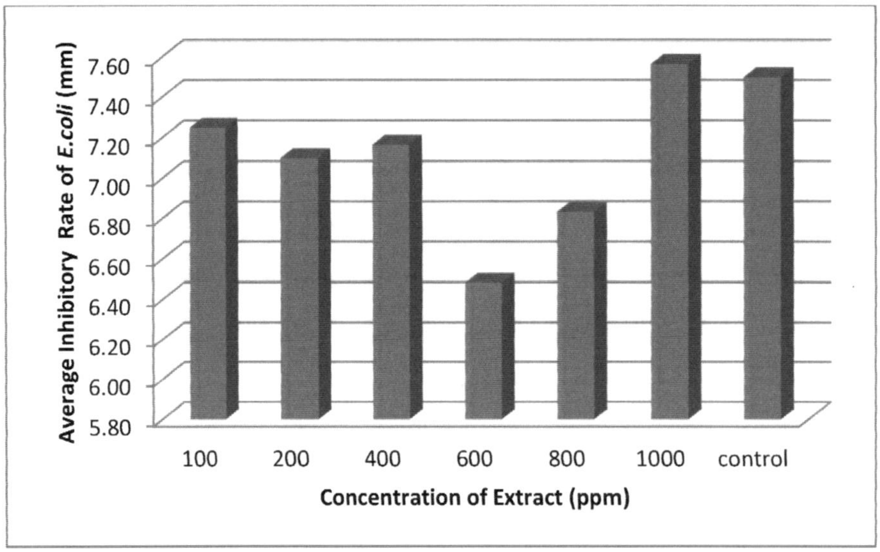

Figure 3. Comparison of concentration test against *E. coli* inhibition zone

The size of the inhibition areas produced by *R. mucronata* extract might be affected by the sensitivity of the organism, culture medium, incubation conditions and agar diffusion rate. This was in accordance with a statement that suggests factors affecting the speed of diffusion, they are; the concentration of microorganisms, the composition of the media, the incubation temperature and the incubation time [5]

The inhibitory zone differentiations produced by each extract can be affected by differences of the active ingredients content in each extract. The active compound will diffuse into the agar and will result in a concentration gradient. The highest concentration occurs in the area near the paper disc and decreases with increasing distance from the paper disc [5].

3.3. The content of secondary metabolite compounds of methanol extract of R. mucronata

Based on GC-MS chromatogram data (Fig. 4) it is known that there are 25 secondary metabolite compounds contained in methanol extract of *R. mucronata* stem bark.

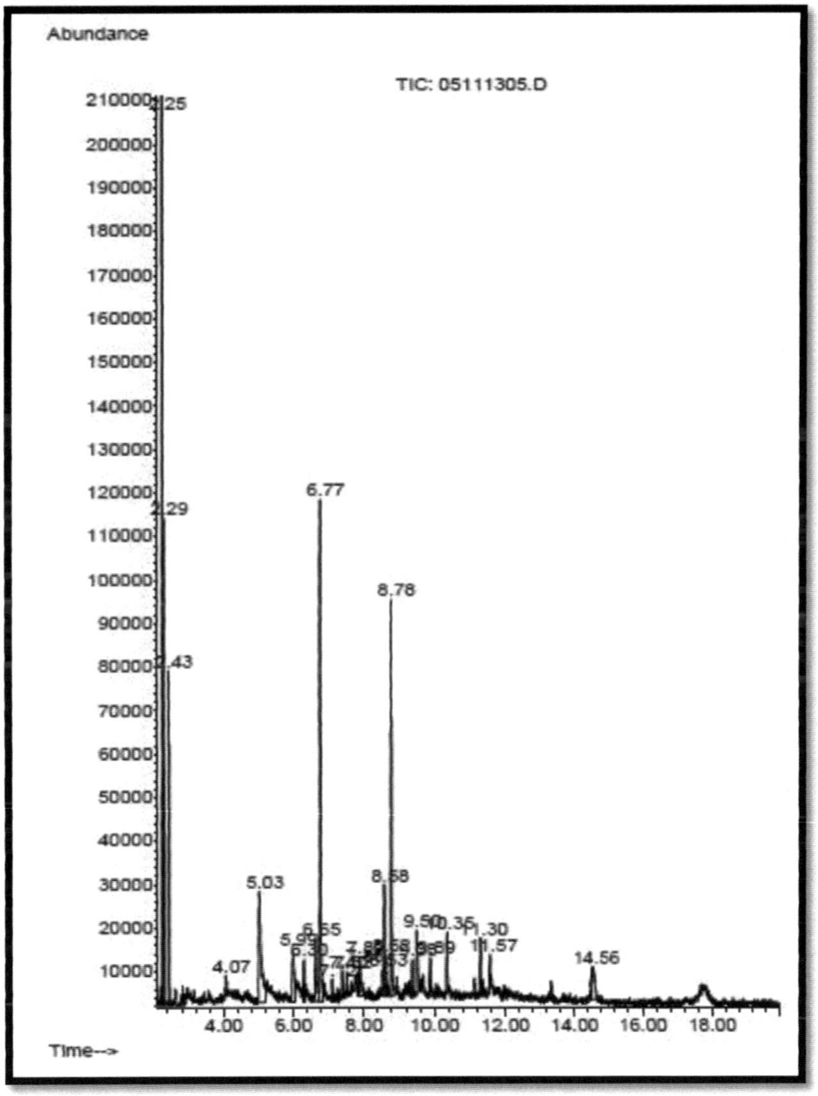

Figure 4. GC chromatogram of *R. mucronata* stem bark extract

The 25 secondary metabolite compounds contained in *R. mucronata* stem bark extract in accordance with the GC-MS tool literature are presented in Table 3.

Table 3. GC-MS literature ofsecondary metabolite compounds contained in *R. mucronata* bark extract

No	*Compound name*	*RT*	*% Area*
1	Benzene, ethyl- (CAS) $$ EB $$ ...	2,25	16,00
2	Benzene, 1,3-dimethyl- (CAS) $$...	2,29	8,63
3	XYLENE	2,43	6,86
4	3-Cyclohexene-1-carboxaldehyde,...	4,06	0,78
5	Phenol, 2-ethoxy- (CAS) $$ Guet...	5,03	9,05
6	1,2-Benzenediol, 4-methyl-	5,99	3,53
7	Phenol, 3-methyl-5-(1-methyleth...	6,30	2,01
8	Benzaldehyde, 4-hydroxy-	6,65	2,96
9	TRANS-2-(1-PENTENYL)FURAN	6,78	14,29
10	Vanilin	7,11	0,65
11	1,4-Benzenediamine, N,N-dimethyl-	7,40	0,87
12	Benzoic acid, 3-hydroxy-, methy...	7,53	0,82
13	Benzenamine, 3-azido-N,N-dimeth...	7,86	0,99
14	Benzoic acid, 3-hydroxy-	7,89	1,70
15	4-Methyl-2,5-dimethoxybenzaldehyde	8,53	0,73
16	Phloroglucinol $$ 1,3,5-Benzene..	8,58	4,06
17	Phenol, 4-(3-hydroxy-1-propenyl...	8,62	1,34
18	2-propyl-5-vinylthiophene	8,78	11,03
19	Phenol, 2-methoxy-4-(methoxymet...	9,38	1,06
20	2-Trimethylsilyl-1,3-dithiane	9,50	3,23
21	Phenol, 4-(3-hydroxy-1-propenyl...	9,89	1,45
22	2-Hydroxy-5-methylbenzaldehyde	10,35	1,87
23	Hexadecanoic acid, methyl ester...	11,30	1,75
24	2-Methoxy-5-amino-4,6-diphenylp...	11,57	0,63
25	Di-n-octyl phthalate	14,56	3,70

The above GC-MS analysis results provide information that in the stem bark of *R. mucronata* mangrove plant obtained from the East Coast coastline Surabaya is rich of secondary metabolite compounds of phenolic, alkaloid, and terpenoid groups. These compounds are a widely used chemical in the health field as antioxidant, antibacterial, and antibiotic agents.

The decrease of bacterial population used in the testing process at all concentrations given by the test compounds showed that *R. mucronata* extract was able to suppress the growth of *A. hydrophila* and *E. coli*indicated by the increasing of inhibition zone diameter. The methanol extract inhibition zone against *A. hydrophila* and *E. coli* was higher than that of chloramphenicol. This means that the extract is active to kill bacteria and can be used as an alternative to natural chloramphenicol.

Internal factors decreasing the density or population of bacteria is caused by *R.mucronata* extract that produce secondary metabolite compounds contained alkaloid group, or its derivatives that become inhibitors for the growth of *A.hydrophila* and *E. coli.* [6] explained that the steroid group has antibacterial activity. Alkaloids and their derivatives are also widely used as antibiotics that are able to inhibit bacterial growth [7].

4. Conclusions

The results of the research on the extract of methanol bark of *R. mucronata* obtained from the East Coast of Surabaya showed that the methanol extract was active as an antibacterial indicated by the inhibitory rate against *A. hydrophila* and *E. coli* of 7.00 and 7.56 mm. The activity of antibacterial extract is higher than that of the positive control of chloramphenicol, therefore it can be used as an alternative natural antibacterial agent. The 25 secondary metabolite compounds of phenolic, alkaloid, and terpenoid groups contained in *R. mucronata* stem bark extract. The active compounds that act as antibacterial agents are thought to belong to the alkaloids.

Acknowledgement

Special thanks to RISTEKDIKTI for the PEKERTI Grant Funding 2016.

References

[1] Spalding M D, Ravilious C and Green E P 2001 World Atlas of Coral Reefs University of California Press Berkeley USA

[2] Mouafi F E, Abdel-aziz S M, Bashir A A and Fyiad A A 2014 Phytochemical Analysis and Antimicrobiaal Activity of Mangrove Leaves (*Avicennia marina* and *Rhizophora stylosa*) Against Some Pathogens, *World Appied Science Journal* **29** (4): 547-554

[3] Mahmiah, Giman, Aminah N S, Tanjung and Mulyadi 2016 Antioxidant Activity of Methanol Extracts From The Stem Bark Of Mangrove Plants Rhizophoramucronata Surabaya: Proceeding ICMHS

[4] Jawetz E, Melnick J L and Adelberg E A 2005 *Mikrobiologi Kedokteran* diterjemahkan oleh Mudihardi, E., Kuntaman, Wasito, E.B., Mertaniasih, N., M., Harsono, S., Alimsardjono, L., Edisi XXII, Penerbit Salemba Medika, Jakarta

[5] Salosso Y and Jasmanindar Y 2014 Potential of Patikan Kerbau (Euphorbhia hirta) as Antibacterial on Aeromonas hydrophilla and Vibrio algynoliticus in Fish Culture *Aquatic Science and Technology* **2**(1)

[6] Kusmiyatidan Ni Wayan S A 2006 Uji Aktivitas Senyawa Antibakteri dari Mikroalga *Porphyridium cruentum* Indonesia: BIODIVERSITAS

[7] Ndwigah F I, Boga H, Wamyoike W and Kachiuri R 2010 Enzymatic Activity, and Secondary Metabolites of Fungal Isolates From Lake Sonachi in Kenya Kenya: University of Agriculture and Technology

Effects of artificial and natural materials ropes as substrates for settling and metamorphosis of pediveliger pearl oyster, *Pinctada maxima* (Jameson, 1901)

Medy Ompi[1], Fontje Kaligis[1], Stephanus Mandagi[1], and Kathe Jensen[2]

[1]Department of Natural Water Resources Management, Faculty of Fisheries and Marine Sciences University of Sam Ratulangi, Indonesia
[2]Zoologisk Museum, Copenhagen, Denmark

E-mail: ompimedy@unsrat.ac.id

Abstract. A study aimed at investigating the effects of four different materials ropes as substrates for settling and metamorphosis of tropical oyster, *Pinctada maxima*, pediveligers was conducted at laboratory scale. The number of pediveligers attached and metamorphosed on each substratum were counted at the termination of the experiment. Increment in shell length during settling and metamorphosis was also measured. Settling and metamorphosis were analyzed using One-Way ANOVA with substratum as the main effect. Two-Way ANOVA with time and substratum as the main effect on growth was also tested. The results showed that the settling and metamorphosis of pediveliger were affected by the substratum. The number of settled and metamorphosed pediveligers on polypropylene rope, natural ropes of *Arenga piñata* and *Cocos nucifera* was significantly higher than on the bottom of Pyrex glasses. No significant difference between numbers of larvae settled on polypropylene and natural ropes of *Arenga piñata* and *Cocos nucifera* appeared. A significant effect of time, but not of substratum, on growth occurred. Overall, settling and metamorphosis of *P. maxima* pediveliger larvae were strongly affected by artificial and natural ropes, while metamorphosis occurred haphazardly on the Pyrex glass. Growth occurred after settling and metamorphosis.

1. Introduction
Pearl oysters, *Pinctada maxima* (Jameson, 1901), have been actively cultured and developed to be the main pearl oyster industry in North Sulawesi. Spat of this pearl oyster is mainly supplied from hatcheries or collected from the oyster bed using spat collectors. Nowadays the use of polypropylene and monofilament as substrate for spat collectors has been intensified. Yet the unused materials are being dumped into the sea and become wastes, affecting benthic organisms as it may be ingested, accumulated and possibly transferred to higher throphic levels. According to Chae & Joo An (2017),

plastic debris, including micro and nanoplastics, can affect metabolism, fertility, and mortality of aquatic organisms.

To minimize the negative effects of plastic materials to the benthic ecosystem, it will be necessary to find alternative materials as spat collectors. Two types of natural fibers as organic materials obtained from sugar palm three, *Arenga pinnata* (Wurmb) Merr, and coconut tree, *Cocos nucifera* L. (Ferita et al 2015) can be considered to be such alternative materials. These materials are abundantly found in tropical areas and have been used for different purposes (Mogea et al 1991). Being organic materials, they are easily decomposed, and may be dumped to the sea bottom with no ill effects (Karlianzah et al 2010). The question is whether pearl oyster pediveliger larvae of *P. maxima* can be induced to settle and metamorphose on these materials?

Many researchers have shown that oyster larvae trigger on environmental cues, particularly on specific substrata when reaching the stage competent for settling and metamorphosis. For example, a significantly higher number of *P. margaritifera* attached on shade meshes than other available substrates, while larvae of *P. maculata* were found more abundant on plastic sheeting than other substrates (Friedman et al 1998). Another example, the larvae of *P. maxima* settled in significantly higher numbers on the combined PVC slat and rope collectors than on either nylon or PVC slat collectors (Taylor et al 1998). A higher number of larvae of *P. margaritifera* settled on garden shade spirals than on any other materials provided (Libini et al 2013). Interestingly up to now, information on using natural organic ropes from fibers of sugar palm and coconut as pearl oyster spat collectors is not available.

The purposes of this study were (1) to test whether the late larval stage, which is called pediveliger, of *P. maxima* settled and metamorphosed when exposed to various artificial and natural substrates such as polypropylene rope, ropes from natural fibers of *Arenga pinnata* and *Cocos nucifera*, and Pyrex glass (=no added substrate), (2) to test if there is any preferred substrate on which settling and metamorphosis would take place, and (3) to describe growth of pediveliger larvae after settling and metamorphosis had taken place.

This study provides valuable information for understanding the mechanism of the larval settling and metamorphosis of this species, particularly when natural substrates were provided, and it might be useful information of using more environmentally friendly materials as a spat collectors in the future.

2. Materials and methods

2.1. Settling and metamorphosis

These experiments used larvae of *Pinctada maxima* which had been spawned and reared until 18 days at the hatchery of Pt Samudra in Lembe Strait, Bitung Town, North Sulawesi. Four types of substrate, namely (1) polypropylene rope, (2) natural fiber rope from sugar palm tree, *Arenga piñata*, (3) coconut fiber rope from *Cocos nucifera* tree and (4) Pyrex glass, were soaked with seawater for one week before use. Three replicates for each type of substrate were performed. Each type of rope was cut into sections of 9.5 cm length and fit into an aluminum tube of 13 cm length. One end of the rope was fitted at the same level as one end of the aluminum tube. No rope was attached to the rest of 3.5 cm length. However, this end was bent into a curve shape. The curved end of the tube was wrapped around an aluminum cable lying horizontally across the top of a 500 ml Pyrex container. The substrates were hung vertically, and the other end was attached to the Pyrex glass bottom. The Pyrex glass container was filled with filtered sea water before adding the larvae. About 1 larva per ml was added into each of the Pyrex containers. All containers were placed into a box chamber system culture with circulating sea water system at 28 – 29°C. For control, three Pyrex glass containers were filled with filtered sea water and larvae but no substrates and placed into the same culture chamber. Larvae were fed every day with cultured diatom, *Chaetorus* sp.

The attached and metamorphosed larvae were observed after 96 hours at the termination of the experiment. Each rope was removed from the containers and observed under the microscope connected to a monitor and digital camera. The larvae on the glass bottom were also siphoned out from the container and placed in a counting chamber and observed under the microscope to find any metamorphosed larvae. Larval metamorphosis can be identified when eyes and umbo have appeared, and at the same time, velum and cilia are lost and the gill begins to develop. A larva capable of undergoing this metamorphosis change successfully was identified as competent. Larvae with velum and cilia swimming close to the bottom were not metamorphosed. Here, the larvae were not counted.

2.2. Growth
Shell length of attached and metamorphosed larvae was also measured. Substratum was removed quickly and placed under the microscope, connected with a monitor to measure shell length. Three to five larvae were measured from each substratum. Substratum with attached larvae was brought back as soon as possible to the culture system after the measurement was finished.

2.3. Data analysis
Settling and metamorphosis of larvae were analyzed using One-Way ANOVA with types of substrate as the main factor. Shell length of attached and metamorphosed larvae was tested using Two-Way ANOVA with time and substrates as the main factors. In order to fulfill the assumptions of analysis variance, the data were tested for homogeneity of variance using the F_{max} test (Fowler et al 1990). In cases where the ANOVA-test showed significant treatment effects, the means were compared using the SNK-test (Sokal & Rohlf 1981).

3. Results
3.1. Settling and metamorphosis
Larvae of *P. maxima* settled and metamorphosed on substrata of polypropylene rope, natural fiber of *Arenga pinnata* and *Cocos nucifera* rather than on Pyrex glass. Here, presence of eyes, foot, and umbo was observed when settlement took place. Pediveligers crawled on the substrata by using their foot. Sometimes pediveligers swam back again into water column, where velum and cilia extended from the shell valves. Metamorphosis occurred when velum and cilia were lost, gills developed, hard shell valves produced, and juveniles were attached on the substrata as shown at Fig. 1.

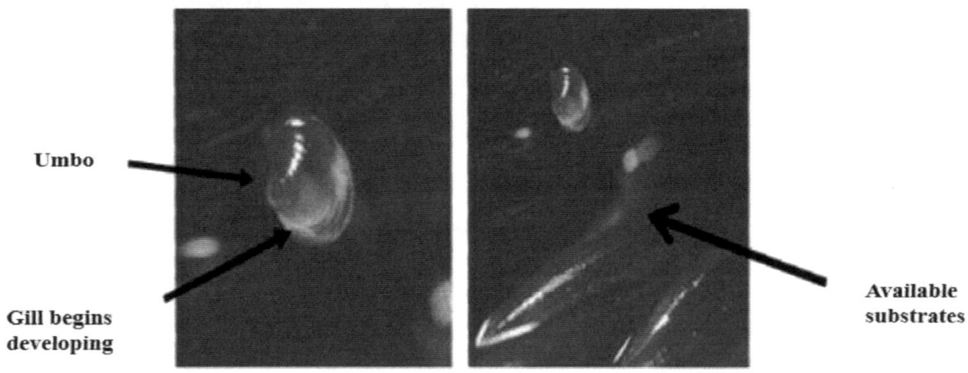

Figure 1. Pediveliger *P. maxima* attached on substrata available

The number of settled and metamorphosed pediveligers were analyzed using One-Way ANOVA with substratum as the main effect (Table 1). The results showed a significant effect of substratum on settling and metamorphosis (One-Way ANOVA, P<0.01).

Table 1. One-Way ANOVA, number of settled and metamorphosed pediveligers, with substratum as the main effect (***: P<0.01).

Sources	DF	SS	MS	F	P
Substrates	3	12.564	4.188	112.272	0.000***
Error	8	0.298	0.037		

A significantly higher number of settled and metamorphosed pediveligers occurred on each of the added substrates than on Pyrex glass (PG) (SNK-test, P<0.05) (Fig. 2). There was no significant difference in number of settled and metamorphosed pediveligers between polypropylene rope (PE), sugar palm rope (IP), and coconut rope (IK) (SNK-test, P>0.05) (Fig. 2).

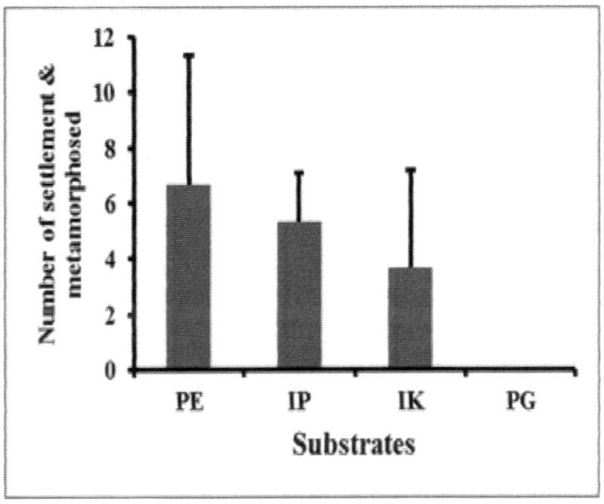

Figure 2. Mean number of settled and metamorphosed pediveligers of *P. maxima* attached on different substrates: PE = polypropylene rope, IP = sugar palm rope, IK = coconut rope, and PG = Pyrex glass (Bars = 95 % Confidence intervals).

3.2. Growth

A pattern of consistently slowly increase in shell length from day 1 until day 4 of pediveligers of *P. maxima* attaching on each type of substrate was observed (Fig. 3).

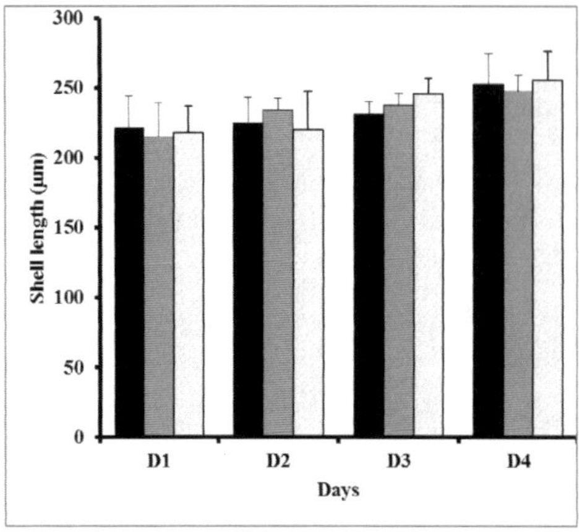

Figure 3. Mean shell length of pediveligers of *P. maxima* settled and metamorphosed on different substrata from day 1 until day 4. D1 = first day, D2 = second day, D3 = third day, and D4 = fourth day. Black bars = coconut fibers, grey bars = sugar palm fibers, and white bars = polypropylene ropes (Bars=are 95 % confidence interval).

The increase in shell length of settled and metamorphosed pediveliger larvae as affected by substratum and time was analyzed using Two-Way ANOVA (Table 2). The results showed that the shell length of settling and metamorphosis of pediveliger larvae were affected by time (P<0.05), but not by substratum (P>0.05).

Table 2. Two-way ANOVA, the shell length of settled and metamorphosed pediveligers of *P. maxima*, with time and substratum as the main effects (**:P<0.01; n.s.: not significant).

Sources	DF	SS	MS	F	P
Time	2	4.570.972	1.523.627	17.353	0.000
Substrates	2	508.667	254.333	2.897	0.075 n.s
Time X Substrates	6	939.778	156.630	1.784	0.145 n.s
Residual	24	2.107.333	87.806		

Furthermore, since shell length was only affected by time, all of the data were combined and analyzed using One-Way ANOVA with time as the main effect. The result showed a significant effect of time on shell length increment (P<0.05) (Tabel 3). Shell length of settled and metamorphosed pediveliger larvae on day 4 was significantly larger than larvae on days 1, 2, and 3 (Fig. 4). However, no difference in shell length among settled and metamorphosed larvae on days 1, 2, and day 3 was determined (P>0.05).

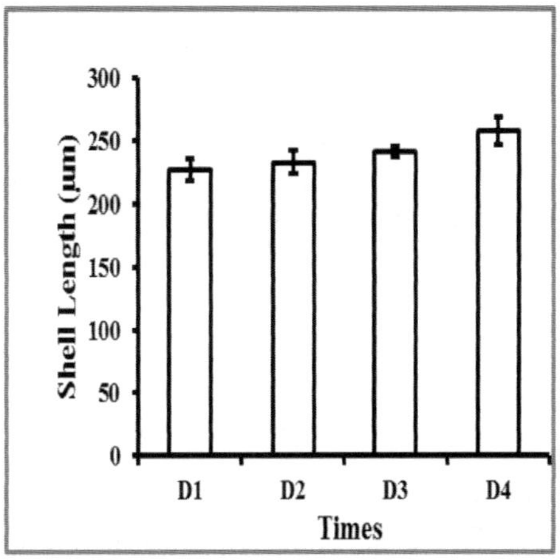

Figure 4. Mean shell length of settled and metamorphosed pediveligers of *P. maxima*. D1 = first day, D2 = second day, D3 = third day, and D4 = fourth day.

Table 3. One-Way ANOVA, shell length of settled and metamorphosed larval *P. maxima* with time as the main effect (**: P<0.01)

Sources	DF	SS	MS	F	P
Time	3	4.570.972	1.523.657	13.712	0.000
Error	32	3.555.778	111.118		

4. Discussion

Settlement for many of marine benthic larvae may be defined as the process of descending from the water column, searching for suitable substrates, and attaching on it, while metamorphosis is defined as a sequence of morphological and organ transformations preparing the larvae for benthic life (Svane & Young 1989; Young *et al* 2002; Carl *et al* 2012).

In the present study, *Pinctada maxima* settled and metamorphosed clearly on polypropylene rope, natural ropes of sugar palm and coconut tree. This might imply that polypropylene and natural ropes of sugar palm, *Arenga pinnata* and coconut three, *Cocos nucifera* were suitable substrates for pediveligers to settle and metamorphose. However, Pyrex glass bottom is apparently not a suitable substrate, since pediveligers might search and test a suitable substrate when crawling continuously at the glass bottom, but eventually swam back into the water column. Behavior of attaching, and crawling, and subsequent swimming back into water column has also been reported for other species such as the coral associated, *Lithopaga simplex* and *L. lessepsiana* (Mokadi et al 1993), *P. margaritifera* (Doroudi et al 1999), *M. edulis* (Bayne 1965), *M. galloprovincialis* (Carl et al 2012; Ompi 2010).

Physical structure of ropes used for substratum could have been a major source of stimuli for pediveligers to settle and metamorphose in this study. Surface pattern of each of the three substrates might increase the surface area and thus substrate complexity as suggested by Carl et al (2012) and

Cristensen et al (2015). Substrate complexity might stimulate metamorphosed pediveligers to penetrate further into the substrate, and at the same time, metamorphosed pediveligers might produce more byssus to attach on the substrate as indication of a strong attachment. A strong attachment on substrata may indicate a new life as a young oyster spat has begun (Kalianasundaram & Ramamoorthi 1987).

Aside from this, all rope surfaces contained very small hairs, which may also stimulate larvae of *P. maxima* to settle and metamorphose in this study. Fine hairs on filamentous materials was reported to be an attractive and suitable substrate for stimulating larvae to settle and metamorphose for other marine bivalves such as larvae of *M. edulis* (Bayne 1965). Furthermore, the organic materials containing palm sugar and coconut trees might also provide additional attraction to pediveligers to be stimulated to settle and metamorphose. Substrate colour may also be a factor to stimulate larvae to settle and metamorphose. For example, in the Pacific Islands, settling larvae of *P. margaritifera* preferred a dark to a light surface of collectors (Braley and Munro 1997). Other examples showed that dark colour of plastic sheets attracted more larvae of *P. martensii* to settle than light colour (Su et al 2007). However, in this study, colour did not seem to have any effect as settling and metamorphosis of pediveligers of *P. maxima* on different substrata with various brown, gold, and dark colours were similar in number.

Growth might not occur in early stages of settling and metamorphosis. In this study, no shell length increment occurred during the first 24 until 72 hours. This could be caused by morphological changes such as loss of velum, reorientation of mouth and foot, including developing of gill filaments such as it was shown for the pediveligers of *Ostrea chilensis* (Videla et al 1998), *Brachidontes rostratus*, *B. erosus*, *Trichomya hirsutus* (Ompi 2010), and *M. galloprovincialis* (Carl et al (2012). During early settlement and metamorphosis, pediveligers might not feed and rely upon stored nutrients and energy while velum and cilia disappear and, at the same time, a new feeding mechanisms develops. Feeding mechanisms might improve at the end and after metamorphosis and, as a result, metamorphosed pediveligers begin to feed, which increases growth activities. In this study, the shell length increment of *P. maxima* pediveligers appeared clearly 72 to 96 hours after introducing substratum.

Overall, this study confirms that polypropylene rope is a good substrate to attract larvae to settle and metamorphos and this substrate has been used as spat collector as reported by Taylor et al (1998). Other artificial materials, such as plastic sheets and ropes, have also been reported as spat collectors (Friedmen et al 1998; Su et al 2007; Libini et al 2013). In this study, natural fiber rope from sugar palm, *Arenga pinnata* and coconut fiber rope from *Cocos nucifera* were also attractive and suitable substrata for pediveligers of *P. maxima* to settle and metamorphose. We recommend that these substrates can be used as spat collectors like other natural spat collectors from bamboo material (Arini & Jaya 2012) and coconut hard shell (Libini et al 2013). A better growth performance after metamorphosis might occur when organ and changing feeding mechanisms have been completed.

5. Conclusion

Pediveligers of *Pinctada maxima* settled and metamorphosed on polypropylene rope, and both natural fiber rope of *Arenga pinnata* and *Cocos nucifera*, but less frequently settle and metamorphose on Pyrex glass. There was no significant preference of artificial or natural fiber ropes for settling and metamorphosis. Increased growth in shell length of pediveligers appeared after settling and metamorphosis.

Acknowledments

We thank PT Arta Samudra Bitung for providing the pearl oyster larvae and for their kind support. We are grateful to the head of the marine biological laboratory, Faculty of Fisheries and Marine Sciences,

IOP Conf. Series: Earth and Environmental Science **162** (2018) 012031 doi:10.1088/1755-1315/162/1/012031

University of Sam Ratulangi, Dr Ir Andry Roeroe, M.Sc., for permitting our use of the microscope. This research was supported by Sam Ratulangi University, BPNP budget, 15 April 2015. Therefore, we are grateful to the Rector of Sam Ratulangi University and the head of the Research and Community Services Institute for supporting us with the allocation of budget.

References

[1] Chae Y and Joo A Y 2017 Effects of micro and nanoplastics on aquatic ecosystems: Current research trends and perspectives *Mar. Poll. Bull.* **115** 20 – 8

[2] Ferita I, Tawarti and Syarif Z 2015 Identifikasi dan karakteristik tanaman enau (*Arenga pinnata*) di Kabupaten Gayo Lues. *Prosiding Seminar Nasional Masyarakat Indonesia* **1 (1)** 31 - 7

[3] Mogea J, Seibert B and Smits W 1991 Multipurpose palms: the sugar palm (*Arenga pinnata* (Wurmb) Merr.) *Agroforestry Systems* **13 (2)** 111- 29.

[4] Friedman F and Bell J D 1996 Effects of different substrata and protective mesh bags on collection of spat of the pearl oyster, *Pinctada margaritifera* (Linnaeus, 1758) and *Pinctada maculate* (Gould, 1850) *Journal of shell fish research* **15 (3)** 535 – 541.

[5] Taylor J J, Southgate P C and Rose R A 1998 Assessment of artificial substrates for collection of hatchery-reared silver-lip pearl oyster (*Pinctada maxima*, Jameson) spat *Aquaculture* **162** 219 – 30.

[6] Libini C L, Manjumol C C, Idu K A A, Kripa V and Mohamed K S 2013 Effect of pediveliger densities and cultch materials on spat settlement of black lip pearl oyster Pinctada margaritifera (Linnaeus, 1758) in hatchery, *Marine Biological Association of India* **55 (1)** 30 - 4.

[7] Karliansyah M R, Waluyo H, Zulhasni, Dwiputra Dj, Ardi T, Rusly A, Bayuadji S and Retnoningtyas H 2010 *Green fins guidelines paduan selam dan snorkeling ramah lingkungan* Kementerian Lingkungan Hidup Jakarta p 71.

[8] Fowler J, Cohen L and Jarvis P 1998 *Practical statistics for field biology* (England John Wiley & Son) p 259.

[9] Sokal R R and Rohlf F J. 1981 *Biometry* (New York Free-man and Company) p 859

[10] Svane I and Young C G 1989 The ecology and behavior of Ascidian larvae *Oceanography Marine Biology Annual Review* **27** 45 – 90.

[11] Young C M, Sewell M A and Rice M E 2002 *Atlas of marine invertebrate larvae* New York.

[12] Carl C, Poole A J, Williams M R, and De Nys R 2012 Where to Settle—Settlement Preferences of *Mytilus galloprovincialis* and Choice of Habitat at a Micro Spatial *Scale PLoS ONE* **7(12)** e52358

[13] Ompi M 2010 Settlement behavior and size of mussel larvae from the family Mytilidae (*Brachidontes erosus* (Lamarck, 1819), *Brachidontes rostratus* (Dunker, 1857), *Trichomya hirsutus* (Lamarck, 1819), and *Mytilus galloprovincialis* Lamarck, 1819 *Journal of Coastal Development* **13 (3)** 215-277.

[14] Mokady O, Bonar D B, Arazi G and Loya Y 1993 Spawning and development of three coral-associated Lithophaga species in the Red Sea, Marine Biology **115** 245 - 52.

[15] Doroudi M, Southgate P C and Mayer R 1999 Growth and survival of blacklip pearl oyster larvae fed different densities of microalgae *Aquaculture International* **7** 179-87.

[16] Bayne B L 1965 Growth and delay of metamorphosis of the larvae of *Mytilus edulis* Ophelia **2** 1 – 47.

[17] Christensen H T, Dolmer P, Hansen B W, Holmer M, Kristensen L D ,Poulsen D LK, Stenberg C, Albertsen C M and Støttrup J G 2015 Aggregation and attachment responses of

blue mussels *Mytilus edulis* impact of substrate Composition, time scale and source of mussel seed *Aquaculture* **435** 245- 51.

[18] Kalyanasundaram M and Ramamoorthi K 1987 Larval development of the oyster *Saccostrea cucullata* (Born) Mahasagar Bulletin of the national institute of oceanography **20 (1)** 53 - 8.

[19] Braley R D and Munro D 1997 Preference for spat collector materials in tanks by larvae of *Pinctada margaritifera* at Penrhyn Atoll Cook Islands SPC Pearl Oyster Bulletin **10** 8 - 11.

[20] Su Z H, Liangmin Y Y and Li H 2007 The effect of different substrates on pearl oyster *Pinctada martensii* (Dunker) larvae settlement *Aquaculture* **271** 377- 383.

[21] Videla J A, Chaparro O R and Thompson R J 1998 Role of biochemical energy serve in the metamorphosis and early juvenile development of the oyster *Ostrea chilensis Mar. Biol.* **132** 635 – 40

[22] Friedman K J, Bell J D and Tiroba G 1998 Availability of wild spat of the blacklip pearl oyster, *Pinctada margaritifera* from open reef systems in Solomon Islands *Aquaculture* **167** 283 - 99.

[23] Arini E and Jaya T S P 2011 The effect of various spat collector materials for spat attachment of pearl oyster (*Pinctada maxima*) *Journal of Coastal Development* **15 (1)** 33 - 44

Spatial assessment of relative resilience potential to support management of coral reef ecosystem in Doreri Bay area, Manokwari Regency, Indonesia

T F Pattiasina[1,2], **A Sartimbul**[2], **B Semedi**[2], **E Y Herawati**[2], **D D Pelasula**[3], **M Krey**[1], **Mulyadi**[4]

[1] Faculty of Fisheries and Marine Science, University of Papua, Manokwari 98314, Indonesia
[2] Postgraduate Program, Faculty of Fisheries and Marine Science, University of Brawijaya, Malang 65145, Indonesia
[3] Research Center for Deep Sea, Indonesian Institute of Science, Ambon 97233, Indonesia
[4] Teluk Cenderawasih National Park, Manokwari 98312, Indonesia

E-mail: th.pattiasina@unipa.ac.id

Abstract. The assessment of ecological resilience is a new approach to support coral reef management, especially in addressing anthropogenic stresses and anticipating the impacts of climate change. This study applies advanced scientific approaches to spatially assess the relative resilience potential and to determine management actions on coral reef ecosystems in Doreri Bay, Manokwari Regency, Indonesia. The combination of underwater photo transect (UPT) method, belt transect, Aqua MODIS satellite data processing, interview, observation and laboratory analysis were applied to collect 11 resilience indicators data, consist of 7 process indicators and 4 stress/pressure indicators. The relative resilience potential and stress were analyzed through several stages including data compilation, normalization, scale setting, and calculation of resilience and stress value, site ranking and site categorization. Furthermore, the relative resilience potential, relative stress values, and the value of individual indicators were queried using criteria to determine target sites and appropriate management actions. The results indicate a spatial variation of the relative resilience potential of coral reefs, where sites with high resilience potential values are located near local community settlements, particularly around Lemon Island and Mansinam Island. Efforts to reduce pollutant sources from land are necessary on sites located near river estuaries and densely populated settlements. Most of the sites meet the criteria for fisheries management/enforcement. Efforts to manage fisheries and law enforcement need to be done on sites at the reefs of Sawaibu and around Raimuti Island.

1. Introduction

Coral reefs are among the richest ecosystems with very high productivity compared to other ecosystems on earth. These ecosystems have important functions both ecologically and economically. In some places, coral reefs even have important cultural values for local communities (Burke et al., 2011; Cinner et al., 2012; Cinner et al., 2013; Weijerman et al., 2015). Nevertheless, in line with the increase in the world population and the increase in development activities that are generally concentrated in coastal areas, the threat to coral reef resources is an important issue today. The combination and sometimes synergies between local pressures and climate change have led to a sharp decline in the percentage of coral cover from about 60% in the past fifty years, to about 20% in current conditions (Hughes et al. 2010; Ateweberhan et al. 2013).In order to prevent further degradation and ensure the sustainability of the world's coral reef systems, serious management efforts need to be undertaken.

The survival of coral reefs is highly dependent on resilience, which is the capacity of coral reefs to survive and recover from degradation and ensure the availability of ecosystem goods and services (Marshall and Schuttenberg, 2006; Mumby et al., 2007; Maynard et al., 2012). If disturbances occur on coral reefs and resilience is at a low level, vulnerability will increase, then there will be a shift from coral-dominated status to a status dominated by algae or other opportunistic organisms (Folke et al., 2004; Norström et al. 2009; Obura and Grimsditch, 2009; Knudby et al., 2014).

Resilience status of coral reef ecosystems is usually measured and monitored using an indicator of abundance or coral cover from important taxonomic groups, whereas those indicators are not sufficient enough to describe resilience. The spectacular decline of coral reefs in many places in the Caribbean is an example of ecological shock, because of the mistake in using only coral abundance as an indicator of resilience. Therefore, experts attempt to assess the development of measures of other indicators to monitor important processes related to coral reef resilience (Hughes et al., 2010; Graham et al., 2013). The development of the concept of assessing the potential for coral reef ecosystem resilience was initiated by Salm et al. (2001). This concept was built in response to a serious coral bleaching event in 1998. This is based on the idea that coral reefs have physical and ecological characteristics that enable some coral reefs to survive or recover from disturbance.

Referring to the above concept Obura and Grimsditch (2009) introduced a guide to the assessment of coral reef ecosystem resilience covering 61 indicators of resilience. However, these guidelines are still constrained in their application, especially in developing countries as they require adequate resources. In order to overcome these constraints, McClanahan et al. (2012) conducted studies to determine important resilience indicators based on three approaches: 1) expert judgment or response to the importance of an indicator compared to other indicators (perceived importance), 2) scientific evidence, and 3) feasibility of each indicator. The results are 11 key indicators affecting coral reef resistance and recovery from local impacts and climate change. Furthermore, their test results support the concept that while high ecological complexity, few but relatively strong indicator variables can affect the dynamics of the ecosystem. Using the framework recommended by McClanahan et al. (2012), coral reef resilience assessments has been implemented by Maynard et al. (2015) in the Commonwealth of the Northern Mariana Islands (CNMI) region.

Although it has been successfully implemented in the CNMI region, this advanced method has never been implemented in Indonesia, especially in new development areas such as Papua region, where the pressure from development activities is quite intense. The rapidly increasing development in Papuaduring the special autonomy era has created chronic pressures on coral reef ecosystems, whereas local communities are still dependent on the services of the ecosystems. In order to support the management of coral reef ecosystems in the region, the assessment of resilience is an urgent need to be implemented.This study aims to spatially assess the relative resilience potential of coral reefs and to identify target areas and management actions in Doreri Bay area, Manokwari Regency, West Papua Province, Indonesia.

2. Study area

Geographically, Doreri Bay is located at the position of 0° 52'43"S - 1° 01' 29" S and 134° 08'06"E - 134° 04'03" E, and is administratively a part of Manokwari District, West Papua Province. Doreri Bay is also part of the Bird's Head Seascape (BHS) region, which has been recognized as one of the largest contributors to the diversity of shallow marine biology in the tropics (Allen and Erdman, 2009). The bay is also thought to be a unique "ecotone" that resembles the transitional boundary between the Gulf of Cenderawasih in the south and the Pacific Ocean. Based on preliminary observations of Allen and Eidman (2008), the butterflyfishes (Chaetodontidae) in Doreri Bay have different characteristics with those in Cenderawasih Bay in the south and the Pacific Ocean in the north. In addition to its ecological uniqueness, coral reef fishery in Doreri Bay serves as a source of food and also a source of income for local communities. Nevertheless, the urgent need for the expansion of Manokwari city as a consequence of the development of the region has caused coral reef ecosystems in Doreri Bay to experience pressure that threatens the sustainability of the ecosystems.

3. Materials and methods

3.1. Resilience and stress indicators

The indicators used in the assessment of coral reef ecosystem resilience are indicators recommended by McClanahan et al. (2012), and belong to two groups according to Maynard et al. (2015), namely: 1) indicators of resilience processes, including variability of sea water temperature, coral diversity, resistant coral species, coral recruitment, macroalgae cover, biomass of herbivorous fish and functional diversity of herbivorous fish; 2) anthropogenic stress/pressureindicators, which includes nutrient level, sedimentation, physical impact and fishing pressure.

3.2. Data collection

Data of resilienceand stress indicators was collected directly in the field and through computer analysis (desktop analysis). Sampling stations were determined on the basis of representation in order to illustrate the overall condition of coral reefs in Doreri Bay. A total of 30 sampling sites for the entire study area were determined according to the distribution patterns of coral reefs based on an initial analysis of Landsat 8 OLI recorded in 2016 and preliminary survey results. The sites distributed at the coral reefs located in the western part of Doreri Bay, in front of Sawaibu Bay, around Lemon Island, around Mansinam Island, and in the eastern part of Doreri Bay. The map of study area showing the location of sampling sites is shown in Figure 1.

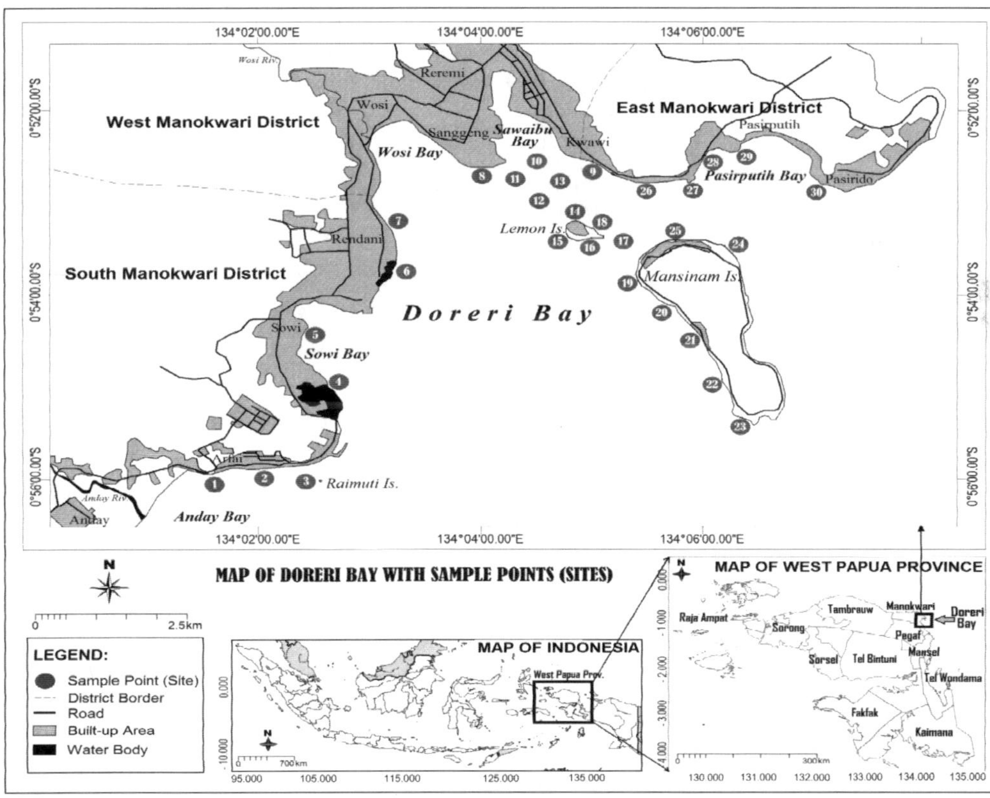

Figure 1. Map of the study area showing the location of Doreri Bay and the sampling sites within the bay.A total of 30 sampling sites were determined according to the distribution patterns of coral reefsin the entire study area.

The underwater photo transect method (Alquezar and Boyd, 2007; Kohler and Gill, 2006) was applied to collect data on coral diversity indicators, resistant coral species, coral recruitment, macroalgae cover and physical impact. Meanwhile, the belt transect method (Jupiter and Egli, 2011) was applied to collect biomass data and functional diversity of herbivorous fish, while interviews and observations to collect data on fishing pressure. Data on temperature variability, nutrient level, and sedimentation/turbidity were collected through satellite image analysis and geographic information system analysis. Methods applied for measuring indicators of ecological resilience is presented in Table 1.

Table 1. Methods applied for measuring indicators of ecological resilience. A total of 11 indicators including 7 indicators of resilience processes (denoted by*) and 4 stress indicators (denoted by**) were measured through field survey, satellite image and GIS analysis.

Indicator	Methods
Sea surface temperature variability *	*Monthly sea surface temperature data from the AquaMODIS satellite period 2002-2017 was obtained from the NEO NASA website. Maximum monthly mean (MMM) was calculated for each pixel. MMM is the month in years with the highest average temperature during the period. The moon with the MMM, together with the month before and the following month is defined as a 12-week warm period (summer). Variability was calculated as a standard deviation from that warm temperature period (Maynard et al., 2015; Sartimbul et al., 2010).*
Coral diversity *	*A list of genera/species was made, then coral diversity was calculated based on the Simpson diversity index for each site, with a range of values from 0 to 1. The higher the value (close to 1), the higher the coral diversity.*
Resistant coral species *	*Each genera/species was assigned a susceptibility value of 1-5 (Marshall and Schuttenberg, 2006; McClanahan et al., 2007, Maynard et al., 2015). Genera/species with a susceptibility value between 1-2 were classified as resistant to bleaching. The proportion (%) of resistant coral species was calculated for each site.*
Coral recruitment *	*Juvenile corals (recruits) are coral with geometric mean size <4 cm. The juvenile coral density (recruit/m^2) was calculated for each site, ie the total number of juveniles divided by the observed area.*
Macroalgal cover *	*Macroalgae cover was calculated as the percentage (%) of the points categorized as macroalgae of the entire points of the benthic cover on each site.*
Funct. groups of herbivorous fish *	*Herbivorous fish were divided into four functional groups: large excavators, browsers, grazer/detritivore and small excavators (Green and Bellwood, 2009). The functional group riches of herbivorous fishes was calculated as the number of functional groups present.*
Biomass of herbivorous fish *	*The average length of the fish size class was applied in the long-weight relationship formula (L-W), ie $W = a \times L^b$, with a and b obtained from FishBase (Froese & Pauly, 2009). The L-W conversion requires a fork length parameter. The conversion factor of total length to fork length (TL-FL) was also taken from FishBase. For each site, "herbivorous biomass" was calculated as the total weight estimate of all herbivorous fish, divided by the area (kg/100 m^2).*
Physical impact **	*The physical effects of human activities and natural phenomena were identified from the coral conditions seen in the photo frames. The proportion (%) of damaged/dead coral was calculated for each site.*
Fishing pressure **	*Based on observations and interviews with local fishermen, it is known that access is the driver of fishing pressure. The multiplication of the standardized values of coastal distance and distance from the fisherman's settlement is the proxy of fishing pressure.*
Pollution/nutrient **	*The multiplication of the standardized value of the distance and the area of the nearest/adjacent watershed to the site is a proxy for pollution/nutrient at each site.*
Sedimentation **	*Composite water sampling was done from bottom to the surface as many as three replicates for each site. Turbidity was measured using a turbidity meter, and the turbidity value (NTU) for each site is the average of all three replications.*

3.3. Data analysis

3.3.1. Assessment of relative resilience potential and relative stress.

Assessment of resilience potential includes several stages: data compilation, normalization, uni-directional scale setting, scale setting in accordance with perceived importance, calculation of average resilience value of each site, calculation of relative resilience value, site ranking and site classification based on value Relative resilience (Maynard et al., 2015). Once compiled, all resilient indicator data is normalized by dividing the value of each indicator from each site to the maximum value of the indicator among all sites. The next step is to set the uni-directional scale for macroalgae cover indicators, as high macroalgae cover scores indicate low resilience conditions. Normalized result values of the macroalgae cover variable are subtracted from the value of "1" (1 - n). Normalized values are then scaled according to the perceived importance of each indicator (in McClanahan et al., 2012). The perceived value of each indicator is divided by the lowest perceived importance (the functional group ofherbivorous fish = 11.00) to obtain scaling multipliers. Furthermore, the normalized value of each indicator is multiplied by the scaling multiplier value and averaged to obtain the crude resilience value. The raw resilience values for each site are then normalized, which is divided by the highest crude resilience value to obtain a standard range from 0 - 1. These values are the final score of resilience for each site.

After calculating the final resilience value of each site, the next process is to rank sites based on the final resilience value of each site. Site ranking is done by sorting the data in the table, where the data is compiled from the highest to the lowest. The ranking of the previous process has not been able to give a clear picture of which sites are high, medium or low resilient. It is, therefore, necessary to group the site so that it is easy to understand how many sites have low, medium or high resilience. Grouping sites are done by calculating the average value (avg) and standard deviation (sd) from the relative resilience value of the entire site. Furthermore, the grouping is done based on the following criteria: High Resilience (value>avg + 1 sd); Medium-high Resilience (value>avg and <avg + 1 sd); Medium-low Resilience (value <avg and>avg - 1 sd); Low Resilience (value <avg - 1 sd).The stresses/pressures of human activitieswere assessed using the same method as the assessment of relative resilience potential. Stress on a given site was expressed relative to the site with the most severe stress due to human activity. High scores indicate high-stress levels and vice versa.

*3.3.2. Identifying target locations and management actions.*The resilience potential values, values of each resilient indicator and stress indicator valueswere tested by criteria for identifying targeted locations for management actions (Maynard et al., 2015). The identification criteria are related to the following management objectives: 1) Conservation; 2) Reduction of pollution sources from the land; 3)Fisheries management and law enforcement; 4) Monitoring bleaching and recovery support; 5) Coral reef restoration/translocation of corals; 6) tourism development. The query name and criteria used to suggest targets for different types of management actions presented in Table 2.

Table 2. Criteria for queriesused to suggesttargets for differenttypes of managementactions on each site (Maynard et al., 2015).

Query Name	Criteria
Conservation (C)	*Sites have high or low resilience potential; currently outside established no-take MPAs*
Land-based Source Pollution Reduction (L)	*Sites have above-average resilience potential & land-based sources of pollution*
Fishery Management and Enforcement (F)	*Sites have above-average resilience potential &fishing access orbelow-average herbivore AFG biomass & above-average fishing access orboth*
Bleaching Monitoring and Supporting Recovery (B)	*Sites have low bleaching resistance & low herbivore AFG biomass*
Reef Restoration/Coral Translocation (R)	*Sites have above-average resilience potential & low coral diversity or coral cover.*
Tourism Outreach and Stewardship (T)	*Sites have above-average coral diversity, fish species richness, and fish biomass*

IOP Conf. Series: Earth and Environmental Science **162** (2018) 012032 doi:10.1088/1755-1315/162/1/012032

4. Results and discussion

4.1. Relative Resilience Potential

The assessment results based on the seven indicators of the resilience process indicate that the relative resilience potential of the reefs varies spatially. The value of the resilience potential between two sites may differ significantly even though they are adjacent to one another. The average value of resilience potential of 30 sites is 0.70 (± 0.12 sd). Resilience potential values ≥ 0.83 were classified as High Resilience Potential, values ≥ 7.0 and <0.83 were classified as Medium-High, values <0.70 and> 0.57 were classified as Medium-Low, and values ≤ 0.57 were classified as Low ResiliencePotential. Among the 30 sites surveyed, 6 sites (20.0%) were classified as relatively high resilience potential, 8 sites (26.7%) were classified as Medium-high, 11 sites (36.7%) were classified as Medium-low, and 5 sites (16.7%) were classified as Low. The highest relative resilience potential value was found at site S19 (Mansinam Island Cemetery), whereas the lowest value was found at site S07 (Rendani Settlement).

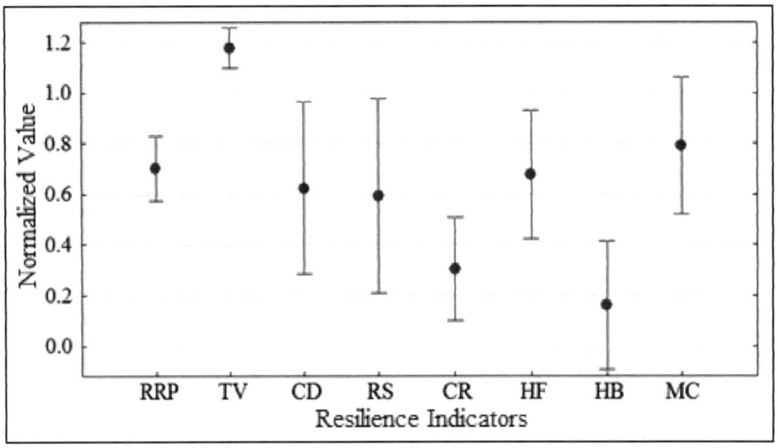

Figure 2. Distribution of normalized values of relative resilience potential (RRP) and resilience indicator values: temperature variability (TV), coral diversity (CD), resistant species (RS), coral recruitment (CR), functionalgroup of herbivorous fish (HF), herbivorous fish biomass (HB) and macroalgae cover (MC). Black circle shape represent average value, while tail equal to "1 sd".

Generally, sites with relatively high resilience potential values are located near local people's settlements, particularly around Lemon Island and the northern part of Mansinam Island, includingsites S19 (Mansinam Island Cemetery), S25 (North Mansinam Island), S15 (Southwest Lemon Island), S12 (South Reef Flat), S29 (Pasirputih Cape), and S16 (South Lemon). These sites generally have high values on indicators of functional groups of herbivorous fish and herbivorous fish biomass. In addition, on these sites, the value of the macroalgae cover indicator and the coral recruitment indicator are constantly low compared to the other six indicators of resilience processes. Even the values of the coral recruitmentindicator at the S15 site (Southwest Lemon Island) and site S16 (South Lemon Island) are low, although both sites belong to a high resilience potential class.

Sites that have the relatively low value of resilience potential are located atthe south of Mansinam Island and western side of Doreri Bay. In addition to site S07 which has the lowest relative resilience potential value, there are four other sites that have relatively low resilience potential value ie, site S21 (MansinamMariculture), S22 (Southwest Mansinam Island), S02 (Arfai-Raimuti), and S23 (Mangewa Cape). Of the five sites, three sites are located in the southwest and south of Mansinam Island, while the other two sites are located on the western side of Doreri Bay. Unlike sites with high resilience potential, the five sites have generally low values in all resilience indicators, particularly functional groups of herbivorous fish, coral diversity, and coral recruitment.Map showing the relative resilience potential of 30 sites presented in Figure 3.

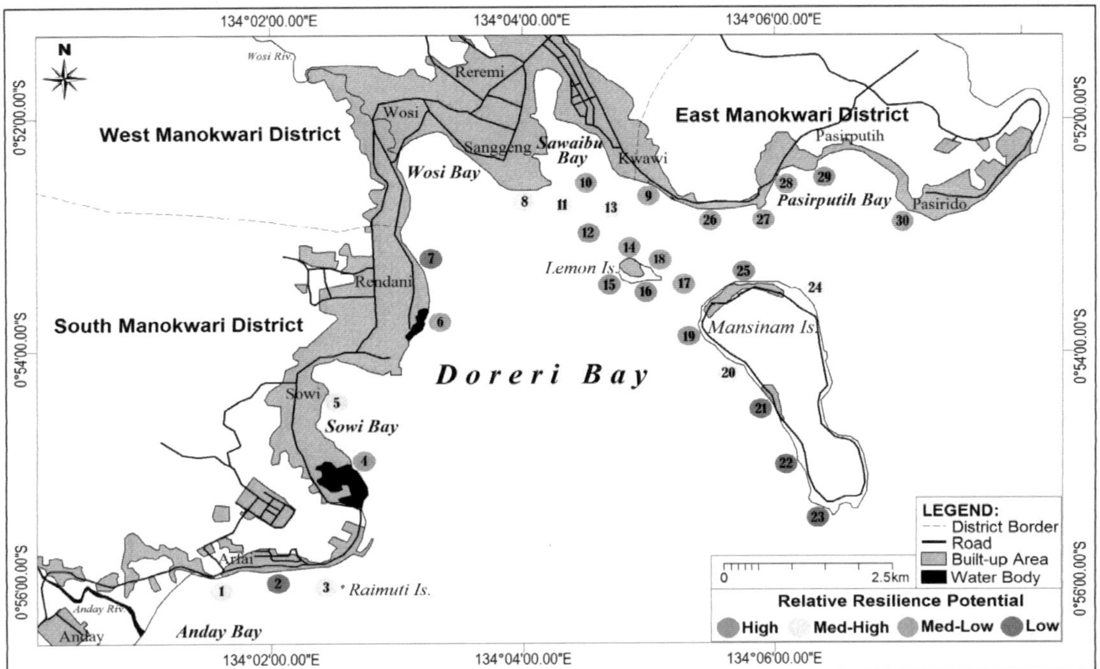

Label	Code	Location	Rank	Label	Code	Location	Rank	Label	Code	Location	Rank
1	S01	Arfai	12	11	S11	West Reef Flat	7	21	S21	Mansinam Mariculture	29
2	S02	Arfai-Raimuti	27	12	S12	South Reef Flat	4	22	S22	South West Mansinam	28
3	S03	Raimuti	8	13	S13	East Reef Flat	14	23	S23	Mangewa Cape	26
4	S04	Telaga Wasti	22	14	S14	North Lemon	24	24	S24	North East Mansinam	11
5	S05	Marampa Port	10	15	S15	South West Lemon	3	25	S25	North Mansinam	2
6	S06	Telaga Rendani	15	16	S16	South Lemon	6	26	S26	Inggandi Beach	18
7	S07	Rendani Settlement	30	17	S17	East Lemon	25	27	S27	Inggandi Cape	23
8	S08	Sanggeng BLK	13	18	S18	North East Lemon	21	28	S28	Pasirputih Cemetery	16
9	S09	Kwawi	20	19	S19	Mansinam Cemetery	1	29	S29	Pasirputih Cape	5
10	S10	North Reef Flat	19	20	S20	Mansinam Bunker	9	30	S30	Pasirido	17

Figure 3. Map showing the relative resilience potential of 30 sites, where 6 sites (20.0%) classified as relatively high, 8 sites (26.7%) Medium-high, 11 sites (36.7%) Medium-low, and 5 sites (16.7%) Low. The highest value found at site S19 (Mansinam Island Cemetery), while the lowest one found at site S07 (Rendani Settlement).

The number of sites classified in the Medium-Lower class is more than the number of sites in other classes. Sites in the Medium-Lower class generally spread from the Pasirputih Bay to the Sawaibu Bay as well as north of Lemon Island. The other two sites are located on the western side of Doreri Bay, in Telaga Wasti and Telaga Rendani. The value of resilience indicators from these sites is generally low, especially the value of functional herbivorous groups and variability of sea surface temperatures, while the macroalgae cover value is quite high. Meanwhile, the sites classified into the Medium-High class are located on the western side of Doreri Bay, particularly at the Arfai-Raimuti Island locations, reef flats in Sawaibu Bay, and Mansinam Island. The values of the resilience indicators of the sites in this class are generally varied, but the values of recruitment indicators of corals are constantly high, and the values of macroalgae cover indicators are generally low.

The results described above indicate that there is a linkage between herbivorous fish indicators (funcional group and biomass), macroalgae cover and resilience potential of coral reefs.The traditional understanding of coral reef ecosystems has focused on three groups of organisms namely coral, algae and fish. Coral and algae are the main benthic producers, but both compete with each other in exploiting space (Pawlik et al., 2016).Herbivores are key processes that support coral reef survival by inhibiting the development of macroalgae which can impact on the living, growth and survival of corals (Heenan and Williams, 2013). Furthermore, increased herbivorous fish have the potential to reduce the competition of algae to corals (Bawole et al., 2014; Bonaldo and Hay, 2014, Stender et al., 2014).Thus, an important thing to support the sustainable management of coral reef ecosystems is a clear understanding of the role by certain types of herbivores in limiting the development of algae as well as supporting coral under certain environmental conditions (Adam et al., 2015).

4.2. Relative Anthropogenic Stress

Based on the calculation of relative anthropogenic stress value from 30 sitesbased on four stress indicators, the relative stress average value is 0.56 (± 0.18 sd). The stress values ≥ 0.74 were classified as High, the values ≥ 0.56 and <0.74 were classified as Medium-high, the values <0.56 and> 0.38 were classified as Medium-Low, and the values ≤ 0.38 were classified as Low. A total of 6 sites (20.0%) were classified as relatively highstress, 7 sites (23.3%) were classified as Medium-high stress, 14 sites (46.7%) were classified as Medium-low stress, and 3 sites (10.0%)were classified as LowStress. The highest relative stress was found on site S01 (Arfai), whereas the lowest relative stress was found on site S11 (West Reef Flat).Map showing the relative anthropogenic stress of 30 sites presented in Figure 4.

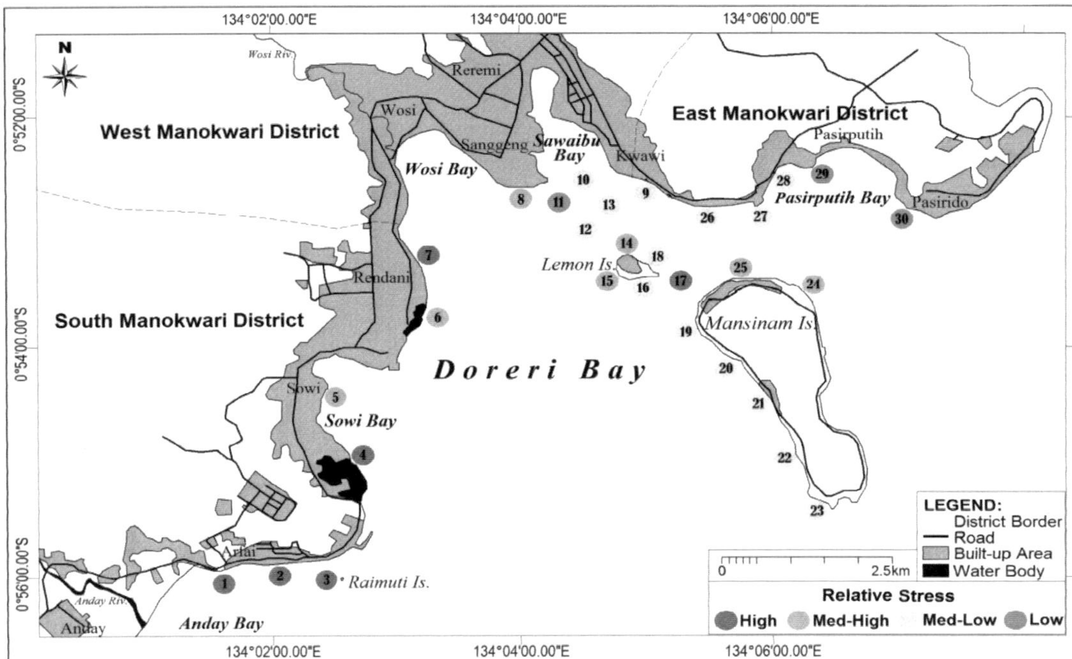

Label	Code	Location	Rank	Label	Code	Location	Rank	Label	Code	Location	Rank
1	S01	Arfai	1	11	S11	West Reef Flat	30	21	S21	Mansinam Mariculture	22
2	S02	Arfai-Raimuti	2	12	S12	South Reef Flat	21	22	S22	South West Mansinam	19
3	S03	Raimuti	4	13	S13	East Reef Flat	17	23	S23	Mangewa Cape	27
4	S04	Telaga Wasti	6	14	S14	North Lemon	12	24	S24	North East Mansinam	7
5	S05	Marampa Port	10	15	S15	South West Lemon	8	25	S25	North Mansinam	11
6	S06	Telaga Rendani	9	16	S16	South Lemon	15	26	S26	Inggandi Beach	26
7	S07	Rendani Settlement	3	17	S17	East Lemon	5	27	S27	Inggandi Cape	20
8	S08	Sanggeng BLK	13	18	S18	North East Lemon	16	28	S28	Pasirputih Cemetery	24
9	S09	Kwawi	18	19	S19	Mansinam Cemetery	25	29	S29	Pasirputih Cape	29
10	S10	North Reef Flat	23	20	S20	Mansinam Bunker	14	30	S30	Pasirido	28

Figure 4. Map showing the relative anthropogenic stress of 30 sites, where of 6 sites (20.0%) were classified as relatively highstress, 7 sites (23.3%) Medium-high, 14 sites (46.7%) Medium-low, and 3 sites (10.0%)Lowstress. The highest relative stress value found at site S01 (Arfai), while the lowest onefoundat site S11 (West Reef Flat).

In general, the relative stress of coral reefs varies spatially among all sites in the study area. However, sites that are classified into relatively high-stress classes are generally located on the western side of Doreri Bay. Sites that have relatively high-stress values are Arfai and Raimuti Island (S01, S02, and S03), TelagaRendani (S04) and settlements in Rendani (S07). Two other sites located between these sites, includingMarampaPort (S05) and TelagaRendani (S06) also have a highstress and classified in the class of Medium-high stress. On the other hand, sites belonging to other classes are located in the middle to the east of Doreri Bay, starting from Sawaibu Bay, Mansinam Island, Lemon Island to Pasirputih Bay. The relative stress value of sites on the western side of Doreri Bay is associated with the generally high value of sedimentation indicator, in addition to the high nutrient/ pollution value, especially the sites in Arfai and Raimuti Island.This relates to the location of these sites, which are relatively close to the river estuaries, where in the upper land there has been conversion from forest to settlement.

Maina et al. (2013) suggests that increased sediment and poor water quality reduce the ability of corals to withstand thermal stress, as well as reduce the ability to recover when experiencing bleaching events. Coastal reef systems generally experience an increase in sediment supply as a result of forest conversion in the upper land. Goatley et al. (2016) also point out that corals near the coast will experience the impact of sediment transport from the mainland, whereas corals far from mainland sedimentary sources will experience in-situ sedimentation disturbances due to physical, biological and chemical processes. In addition to sedimentation problems, nutrification due to human activity often leads not only to the increase of inorganic nutrients such as ammonia, nitrate and phosphate but also to the ratio of the concentrations of these nutrients in the waters. In certain types of phytoplankton the growth becomes chemically unbalanced because the availability of certain nutrient types decreases from the cell's need. This condition is known as nutrient deficiency and results in disturbing effects, such as reduced efficiency of photosynthesis of the symbional algae in coral (Wiedenmann et al., 2012).

4.3. TargetsLocations for Management Actions

The final value of resilience, resilience indicator values and stress indicators have been queried with six criteria for determining appropriate management actions for each site. The number of sites that meet the fisheries management criteria is 19 sites, and the number is the highest compared to the other five criteria. These sites are scattered almost in all parts of Doreri Bay, from the west side to the east side of the bay. The number of sites that meet the criteria for bleaching monitoring and recovery support is 12 sites that are also scattered in almost all parts of Doreri Bay. Meanwhile, a number of 7 sites meet the criteria for the reduction of sources of pollution from the upper land, 5 sites meet the criteria for conservation, 4 sites meet the criteria for tourism development, and there was only 1 site that meets the criteria for coral restoration or coral translocation. The site meets the criteria for coral reef restoration is site S16, located in the southern part of Lemon Island.Map showing the result of queries to identify targets for different management actions presented in Figure 5.

The identification of target locations and management actions indicates that fisheries management needs to be implemented on most sites. It aims to reduce mortality of herbivorous fish as a result of fishing activities. Based on the results of their research on the Hawai'i islands, Weijerman et al. (2013) found a strong association between fish biomass and capture mortality. The high mortality of capture results in changes in fish communities, where there is a decrease in the number of large fish and biomass of piscivores. Further changes in reef fish communities have an effect on the relationship between corals and macroalgae.The results of the study by Stender et al. (2014) in Pelekane Bay shows that coastal access restrictions, efforts to tighten fishing rules, increased oversight of both legal and illegal fishing activities have had an impact on decreasing fishing pressure, and ultimately recovery of coral and reef fish conditions.

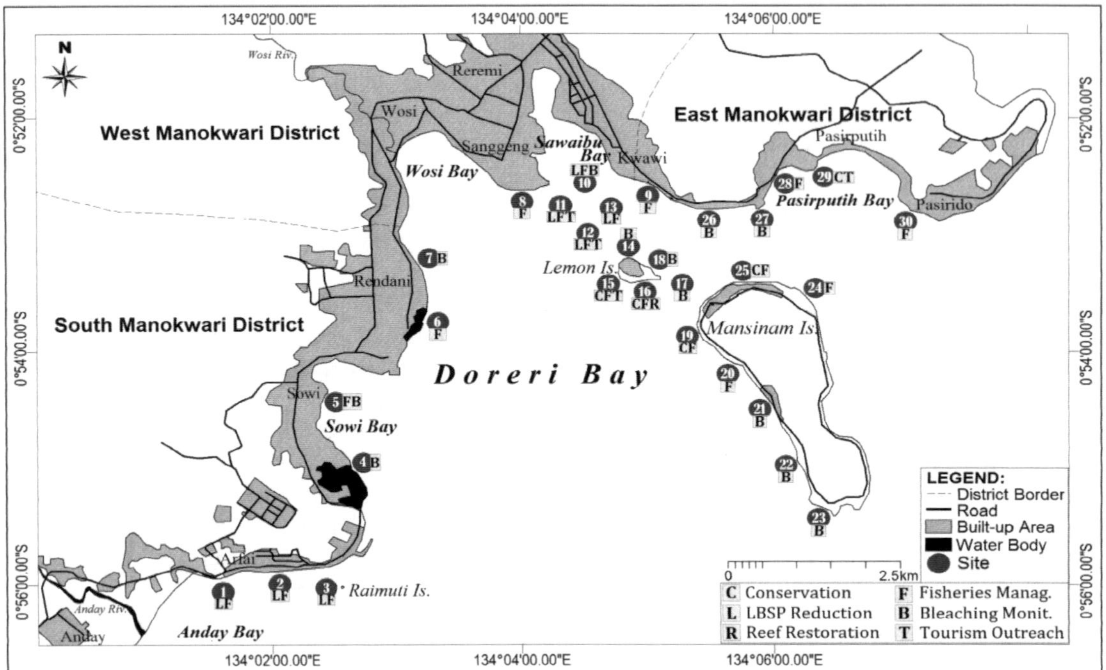

Label	Code	Location	Label	Code	Location	Label	Code	Location
1	S01	Arfai	11	S11	West Reef Flat	21	S21	Mansinam Mariculture
2	S02	Arfai-Raimuti	12	S12	South Reef Flat	22	S22	South West Mansinam
3	S03	Raimuti	13	S13	East Reef Flat	23	S23	Mangewa Cape
4	S04	Telaga Wasti	14	S14	North Lemon	24	S24	North East Mansinam
5	S05	Marampa Port	15	S15	South West Lemon	25	S25	North Mansinam
6	S06	Telaga Rendani	16	S16	South Lemon	26	S26	Inggandi Beach
7	S07	Rendani Settlement	17	S17	East Lemon	27	S27	Inggandi Cape
8	S08	Sanggeng BLK	18	S18	North East Lemon	28	S28	Pasirputih Cemetery
9	S09	Kwawi	19	S19	Mansinam Cemetery	29	S29	Pasirputih Cape
10	S10	North Reef Flat	20	S20	Mansinam Bunker	30	S30	Pasirido

Figure 5. Map showing the result of queries to identify targets for different management actions. A number of 19 sites meet the criteria for fisheries management, 12 sites for bleaching monitoring/recovery, 7 sites for reduction ofpollution from the land, 5 sites for conservation, 4 sites for tourism development, and only 1 site for coral restoration/translocation.

5. Conclusion

The resilience potential and pressure on coral reefs varies spatially. Sites with high resilience potential values are mainly located near local community settlements, particularly around Lemon Island and Mansinam Island. Sites with high-stress values are mainly located on the west side of the bay, particularly in Arfai and around Raimuti Island.In order to reduce stress, local scale management of sedimentation from the upper land is a relevant effort.Most of the sites meet the criteria for fisheries management/enforcement and bleaching monitoring which will contribute to the resilience of coral reef ecoystem in the bay.

Acknowledgement

This work is a part of disertation research of the coresponding author. We would like to express our special gratitude and thanks to the Ministry of Research, Technology and Higher Education of the Republic of Indonesia (Kemenristekdikti), Education Offices of West Papua Province (DinasPendidikanProvinsi Papua Barat),and the Government of Raja Ampat Regency (PemerintahKabupaten Raja Ampat) for the financial support. We are indebted to the Bentara Papua Foundation,Fisheries Science Laboratory and Marine Sciences Laboratory of Papua University for the diving equipment, water sampling equipment and water sample analysis. Finally, We are grateful to

IOP Conf. Series: Earth and Environmental Science **162** (2018) 012032 doi:10.1088/1755-1315/162/1/012032

SCEEPAAR 7000 Diving Team (Rimer Biloro, Syafril, Rizky Karamoy, Ruland Tanati and Freddy Daan) for their hardworking during data collection in the field.

References

[1] Adam T C, Burkepile D E, Ruttenberg B I and Paddack M J 2015 Herbivory and the resilience of Caribbean coral reefs: knowledge gaps and implications for management *Marine Ecology Progress Series* **520** 1-20.

[2] Allen G R and Eidman M 2008 Natural resources in Cenderawasih Bay, Papua. A statement in stadium generale for Fisheries and Marine Science, The State University of Papua Manokwari, Indonesia.

[3] Allen G R and Erdmann M V 2009 Reef fishes of the bird's head peninsula, West Papua, Indonesia *Check List* **5**(3) 587-628.

[4] Alquezar R and Boyd W 2007 Development of rapid, cost effective coral survey techniques: tools for management and conservation planning *Journal of Coastal Conservation* **11**(2) 105-119.

[5] Ateweberhan M, Feary D A, Keshavmurthy S, Chen A, Schleyer M H and Sheppard C R 2013 Climate change impacts on coral reefs: synergies with local effects, possibilities for acclimation, and management implications *Marine pollution bulletin* **74**(2)526-539.

[6] Bawole R, Pattiasina T F and Kawulur E I J 2014 Coral-fish association and its spatial distribution in Cenderawasih Bay National Park Papua, Indonesia. *Aquaculture, Aquarium, Conservation & Legislation-International Journal of the Bioflux Society (AACL Bioflux)*, 7(4) 248-254.

[7] Bonaldo R MandHay M E 2014 Seaweed-coral interactions: variance in seaweed allelopathy, coral susceptibility, and potential effects on coral resilience *PLoS One* **9**(1)e 85786.

[8] Burke L, Reytar K, Spalding M and Perry A 2011 *Reefs at risk revisited* World Resources Institute Washington DC.

[9] Cinner J E, Huchery C, Darling E S, Humphries A T, Graham N A, Hicks C C, Marshall N and McClanahan T R 2013 Evaluating social and ecological vulnerability of coral reef fisheries to climate change *PLoS One* **8**(9)e 74321.

[10] Cinner J E, McClanahan T R, Graham N A J, Daw T M, Maina J, Stead S M, Wamukota A, Brown K and Bodin Ö 2012 Vulnerability of coastal communities to key impacts of climate change on coral reef fisheries *Global Environmental Change* **22**(1) 12-20.

[11] Folke C, Carpenter S, Walker B, Scheffer M, Elmqvist T, Gunderson L and Holling C S 2004 Regime shifts, resilience, and biodiversity in ecosystem management *Annual review of ecology, evolution, and systematics* **35** 557-581.

[12] Froese R and Pauly D 2009 Fishbase Version World Wide Web Electronic Publication Retrieved from http://www.fishbase.org (February 2016).

[13] Goatley C H R, Bonaldo R M, Fox R J and Bellwood D R 2016 Sediments and herbivory as sensitive indicators of coral reef degradation *Ecology and Society* **21**(1) 29.

[14] Graham N A, Bellwood D R, Cinner J E, Hughes T P, Norström A V and Nyström M 2013 Managing resilience to reverse phase shifts in coral reefs *Frontiers in Ecology and the Environment* **11**(10)541-548.

[15] Green A L and Bellwood D R 2009 Monitoring functional groups of herbivorous reef fishes as indicators of coral reef resilience: a practical guide for coral reef managers in the Asia Pacific region (No. 7) IUCN.

[16] Heenan A and Williams I D 2013 Monitoring herbivorous fishes as indicators of coral reef resilience in American Samoa *PloS one* **8**(11)e79604.

[17] Hughes T P, Graham N A, Jackson J B, Mumby P J and Steneck R S 2010 Rising to the challenge of sustaining coral reef resilience *Trends in ecology & evolution* **25**(11) 633-642.

[18] Jupiter S and Egli D P 2011 Ecosystem-based management in Fiji: successes and challenges after five years of implementation *Journal of Marine Biology* **2011**1-14.

[19] Knudby A, Pittman S J, Maina J, and Rowlands G 2014 Remote sensing and modeling In Finkl, C W and Makowski C eds. 2014 *Remote sensing and modeling: Advances in coastal and*

marine resources **9** Springer International PublishingSwitzerland.

[20] Kohler K E and Gill S M 2006 Coral point count with excel extensions (CPCe): A Visual Basic program for the determination of coral and substrate coverage using random point count methodology *Computers & Geosciences* **32**(9) 1259-1269.

[21] Maina J, De Moel H, Zinke J, Madin J, McClanahan T and Vermaat J E 2013 Human deforestation outweighs future climate change impacts of sedimentation on coral reefs *Nature communications* **4** 1-4.

[22] Marshall P A and Schuttenberg H Z 2006 A reef manager's guide to coral bleaching. *Great Barrier Reef Marine Park Authority Australia.*

[23] Maynard J, Mckagan S, Johnson S and Houk P 2012 Coral Reef Resilience to Climate Change in Saipan, CNMI, Field-based Assessments, and Implications for Vulnerability and Future Management. Commonwealth of the Northern Mariana Islands, Division of Environmental Quality.

[24] Maynard J A, Mckagan S, Raymundo L, Johnson S, Ahmadia G N, Johnston L, Houk P, Williams G J, Kendall M, Heron S F and Van Hooidonk R 2015Assessing relative resilience potential of coral reefs to inform management *Biological Conservation* **192** 109-119.

[25] McClanahan T R, Ateweberhan M, Graham N A J, Wilson S K, Sebastián C R, Guillaume M M M and Bruggemann J H 2007 Western Indian Ocean coral communities: bleaching responses and susceptibility to extinction*Marine Ecology Progress Series* **337**1-13.

[26] McClanahan T R, Donner S D, Maynard J A, MacNeil M A, Graham N A, Maina J, Baker A C, Beger M, Campbell S J, Darling E S and Eakin C M 2012 Prioritizing key resilience indicators to support coral reef management in a changing climate *PloS One* **7**(8) e42884.

[27] Mumby P J, Hastings A and Edwards H J 2007 Thresholds and the resilience of Caribbean coral reefs *Nature* **450**(7166) 98-101.

[28] Norström AV, Nyström M, Lokrantz J and Folke C 2009 Alternative states on coral reefs: beyond coral - macroalgal phase shifts *Marine ecology progress series* **376**295-306.

[29] Obura D and Grimsditch G 2009 Resilience assessment of coral reefs: assessment protocol for coral reefs, focusing on coral bleaching and thermal stress Gland: IUCN.

[30] Pawlik J R, Burkepile D E and Thurber R V 2016 A Vicious circle ? Altered carbon and nutrient cycling may xplain the low resilience of Caribbean coral reefs *BioScience* biw 047.

[31] Salm R V, Smith S E and Llewellyn G 2001 Mitigating the impact of coral bleaching through marine protected area design *Coral bleaching: Causes, consequences and response University of Rhode Island:* USA 81-88.

[32] Sartimbul A, Nakata H, Rohadi E, Yusuf B and Kadarisman H P 2010 Variations in chlorophyll-a concentration and the impact on *Sardinella lemuru* catches in Bali Strait, Indonesia. *Progress in Oceanography* **87**(1) 168-174.

[33] Stender Y, Jokiel P L and Rodgers K S 2014 Thirty years of coral reef change in relation to coastal construction and increased sedimentation at Pelekane Bay, Hawai'i *PeerJ* **2**e300.

[34] Weijerman M, Fulton E Aand Parrish F A 2013 Comparison of coral reef ecosystems along a fishing pressure gradient. *PloS One* **8**(5)e 63797.

[35] Weijerman M, Fulton E A, Janssen A B, Kuiper J J, Leemans R, Robson B J, van de Leemput I A and Mooij W M 2015 How models can support ecosystem-based management of coral reefs *Progress In Oceanography* **138** 559-570.

IOP Conf. Series: Earth and Environmental Science **162** (2018) 012033 doi:10.1088/1755-1315/162/1/012033

Depositional environmental evolution of Kalibiuk formation based on paleontological molluscan Study, Cisaat River section, Bumiayu, Central Java, Indonesia

Aswan[1], Elina Sufiati[2], Alfend Rudyawan[1], Desty Kistiani[2], and Thaw Zin Oo[3]

[1] Geology Department, Faculty of earth Sciences and Technology, Institute of Technology Bandung, Jalan Ganesa 10 Bandung, Indonesia
[2] Geological Museum Bandung, Jalan Diponegoro No.57, Bandung, Indonesia
[3] Geology Department, East Yangon University, Thanlyin Township, Yangon Division, Myanmar

E-mail: aswan_gl@gl.itb.ac.id

Abstract. This paper presents new finding from the sedimentary sequences, overlooked by previous researchers, which provide insights to the sea level changes during the deposition of the Middle to Late Pliocene Kalibiuk Formation. Data were acquired from fieldwork via detailed measured section of each sedimentary sequence from the bottom to the upper part of the formation followed by molluscan fossil association analyses. Lithologically, the Cisaat River stratigraphic section is dominated by greenish grey claystone with some light grey fine-grained sandstone intercalation. Based on about 50 m stratigraphic section, 10 molluscan fossils associations were successfully determined. The sequence from bottom to top are: *Corbula–Natica* association, *Nassa* sp. association, *Olividae – Turricula* association, *Pecten – Telina* association, *Telina – Placuna* association, *Corbula – Marginella* association, *Placuna – Natica* association, *Turritella* association, *Turritella – Marginella* association and *Dentalina – Architectonica* association. Changes in vertical succession of the molluscan association indicates a development from shallow marine to terrestrial depositional environment before gradually overlain by terrestrial sediments of the Kaliglagah Formation. At least there are 2 deepening - shallowing sea level cycles during the deposition of Kalibiuk Formation observed from this section.

1. Introduction

The Cisaat River traverse is located to the southwest of Bumiayu, Central Java, Indonesia (Figure 1). Standard geological map of the area referred to the Majenang sheet [1], bounded by the following coordinates: 7° 09' 20.55" S; 108° 57' 13.41" E - 7° 09' 34.53" S; 108° 57' 21.76" E. The traverse, administratively, belong to the Bantarkawung district, Brebes, Central Java Province. This study focuses on the upper part of the Pliocene Kalibiuk Formation [1] and the lower part of the Pleistocene Kaliglagah Formation (Figure 2) with general bedding attitude N060°E/30°SE. This interval is included into the Cheribonian stage in the Neogene stages of Jawa [2] based on the mollusc fossils occurrence or the Zanclian stage in the international geological time scale. This research was carried out due to the abundance and continuous occurrence of marine mollusc fossils across the vertical succession of the formation along the traverse and rare occurrence of benthic forams that are usually used to determine the bathymetric depth of deposition within this sedimentary unit. Therefore a complete account of mollusc fossils analyses will be able to represent the change in depositional bathymetry ranging from tidal to shallow marine environment between the upper Kalibiuk Formation and lower Kaliglagah Formation. Sea level changes study based on molluscs fossils in Indonesia have been carried out by several researchers e.g. [3,4,5,6]. [4] have successfully analyzed the sea level changes using mollusc fossil association in the Nyalindung Formation that shows 9 cycles of shallowing and deepening sequences. This study presents a new observation and analyses of sea level changes on the overlooked sedimentary section of the Kalibiuk Formation in Cisaat River following methods presented in [4].

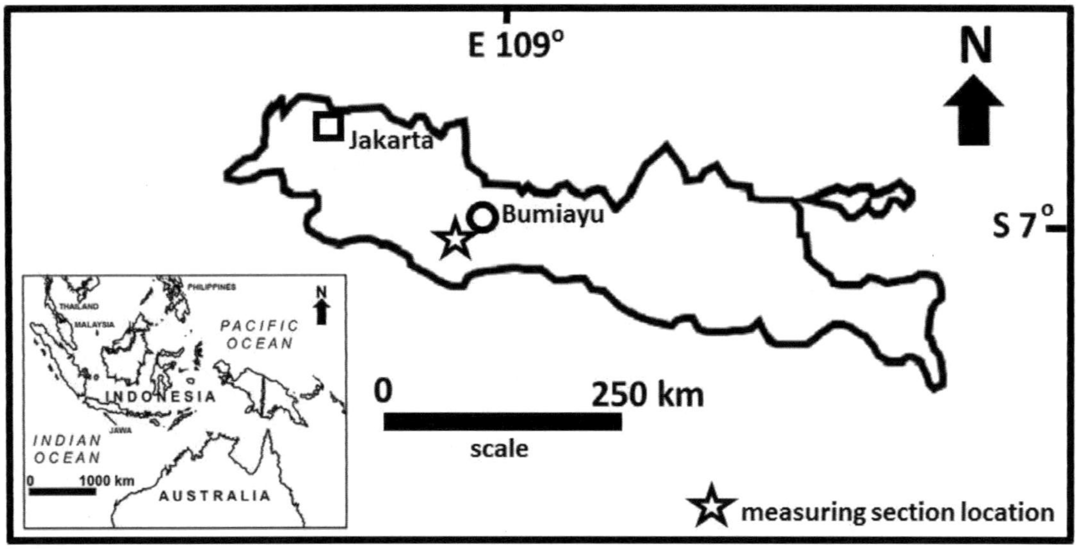

Figure 1. Map showing the location for measuring section (star sign) in Cisaat River to the southwest of Bumiayu city.

2. Geological Setting

The study area is located within the western side of the North Serayu Mountain Range [7]. Stratigraphically, the Kalibiuk Formation was deposited in the middle part of the shallowing upward sedimentary sequence during Pliocene-Pleistocene periods, above the greenish grey, fine-grained, muddy marine sediments of the Tapak Formation and below the terrestrial black shale to conglomerate of Kaliglagah Formation. Some authors suggest that the upper Tapak Formation is in interfinger relationship with the lower Kalibiuk Formation [8,9].

From old to young the Pliocene – Pleistocene stratigraphy [1] are as follows:

- Tapak Formation is characterized by coarse-grained greenish sandstone at the base and gradually fining upward towards the top with occasional yellowish-grey sandy marl intercalation. Brackish-Marine mollusc fossils taken from the upper part of the section suggest Early-Middle Pliocene age deposited in the transition to tidal environment. Total thickness may reach 500 m to the west of study area; the Kalibiuk Formation includes thick conglomerate layers containing mammal fossils (*Merycopotamus nannus* Lydekkery) and fossiliferous calcareous limestone indicating Middle Pliocene age. Lignite layers have been reported in occasional places at the top of the section.
- Kalibiuk Formation was deposited on top of the Tapak Formation. It consists of claystone and fossiliferous blue marl at the base. The middle part contains green sandstone lenses with abundance mollusc fossils while the upper part consists of thin sandstone lenses. It is interpreted to be deposited in the tidal environment. The lower part is interpreted to be interfinger with the upper Tapa Formation. Estimated total thickness of the formation may reach 500 m. The Kalibiuk Formation gradually becomes

the terrestrial Kaliglagah Formation. Balanus limestone often observed at the boundary between the Tapak and Kalibiuk Formation.

- Kaliglagah Formation consists of black claystone, green marl, volcaniclastic sandstone and conglomerate units at the base, the upper part is dominated by coarse sandstone and conglomerate, claystone and marl are rare and completely absent at the top of the section. Lignite intercalation at the base of the unit may reach 1 m thick. Total thickness of this formation is 350 meters. Figure 2 shows the geological map of the area.

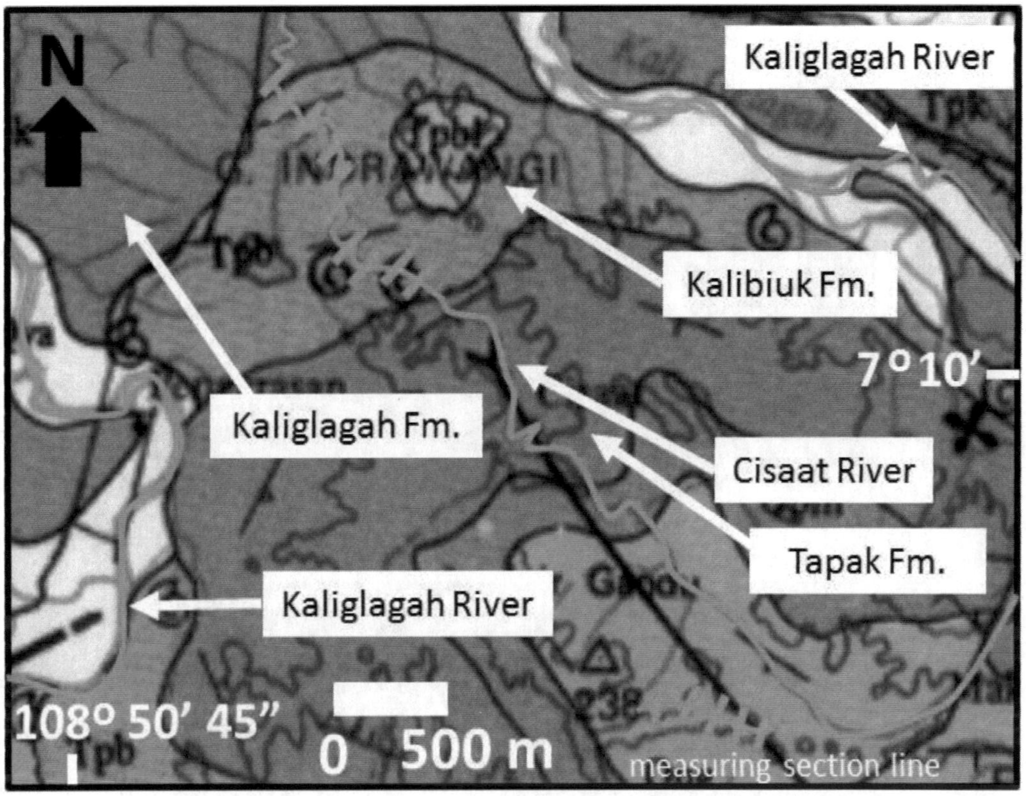

Figure 2. Local geological map of the study area [1]. Blue lines represent river.

3. Methods

The data used in this study is a measured stratigraphic section along the Cisaat River (Figure 2). Some key sections are presented in Figure 3. Lithological boundaries are determined by lithological differences or mollusc fossils content. Useful fossils in this study are those which in their taphonomy are in-situ fossils. Genus and species of each mollusc will be determined following reference from [10,11] and further grouped into mollusc fossil association based on dominance or fossil index at each depositional environment referring to [4,12].The depositional environment interpretation of these fossil associations referring to [13]. Vertical succession of the unit from the bottom to top were then used to construct the local sea level curves in the study area in Pliocene – Pleistocene age.

4. Results

Twelve sedimentary packages of the Kalibiuk Formation yielded 70 mollusc fossils with 24 species each of Pelecypods and Gastropods. Some examples of these fossils are presented in Figure 4. Total measured thickness of the Kalibiuk Formation in Cisaat River is 5.5 meters. It is dominated by claystone with occasional sandstone intercalation. Figure 3 shows the thick claystone outcrop of the Kalibiuk Formation on the riverbank. Examples of mollusc fossils that were used as a basis of each unit

IOP Conf. Series: Earth and Environmental Science **162** (2018) 012033 doi:10.1088/1755-1315/162/1/012033

association acquired from the formations are presented in Figure 4. The stratigraphic records arranged from old (bottom) to young (top) showing ten distinguished fossil associations can be determined from the analyses summarized in Figure 5 and Table 1.

There are five depositional environments determined from the mollusc fossils in this study, they are: upper most intertidal, subtidal, lower intertidal and shallow open marine. These depositional environments are determined from the insitu mollusc fossils. Reworked fossils will not yield an accurate depositional environment interpretation. An accurate environment determination will help to develop better detail of the local sea level curve from the bottom to top part of the stratigraphy.

Figure 3. Kalibiuk Formation outcrop in Cisaat River, Bumiayu, Central Java

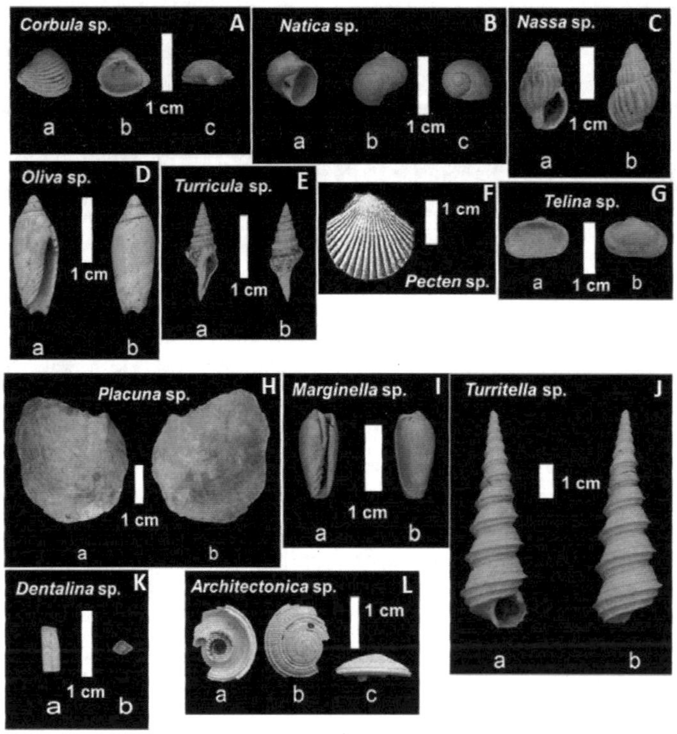

Figure 4. Examples of molluscan fossils found in Kalibiuk Formation, Cisaat River, Bumiayu, Central Java.

IOP Conf. Series: Earth and Environmental Science **162** (2018) 012033 doi:10.1088/1755-1315/162/1/012033

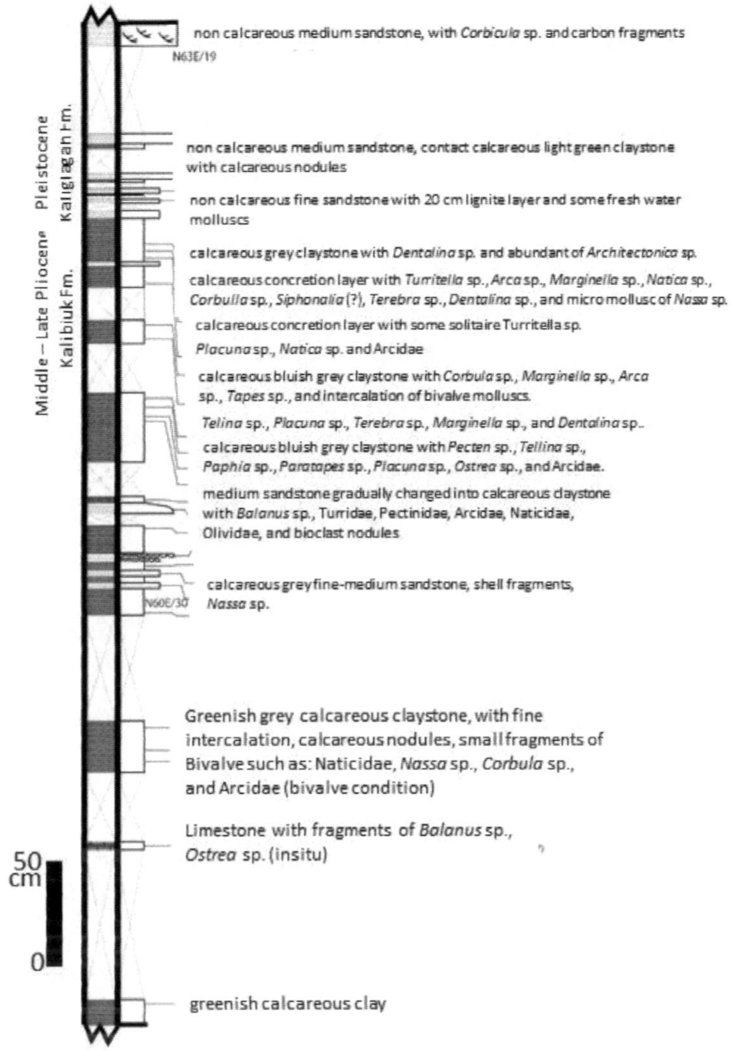

Figure 5. Kalibiuk Formation stratigraphic column, Cisaat River section, Bumiayu, Central Jawa, Indonesia. Note: The coarse clastics dominance towards the top of the section.

Table 1. Vertical succession from top (10) to bottom (1) from the Kalibiuk Formation.

Layers	Fossil Association	Paleodepositional Environment
10	*Dentalina* – *Architectonica* association, *Dentalina* sp. and *Architectonica* sp. found within the grey calcareous claystone and associated with insitu and complete pelecypods fossils.	Quiet uppermost intertidal
9	*Turritella* – *Marginella* association found in the calcareous claystone and associated with other less dominant fossils such as *Arca* sp., *Corbulla* sp., *Siphonalia* sp., *Terebra* sp., *Dentalina* sp., and micro mollusc *Nassa* sp.	Transitional, close to the open marine; lower intertidal

IOP Conf. Series: Earth and Environmental Science **162** (2018) 012033 doi:10.1088/1755-1315/162/1/012033

8	*Turritella* association mostly found in carbonate concretion in the upper part of the bluish claystone. *Turritella* sp. found as solitary fossil species with variations from juvenile to adult species, also found coral and echinoid fragments in the carbonate concretion.	Shallow open marine
7	*Placuna – Natica* association found in claystone and associated with less dominant Arcidae fossils.	Lower intertidal
6	*Corbula – Marginella* association found in calcareous claystone and associated with other less dominant fossils such as *Arca* sp. and *Tapes* sp.	Uppermost intertidal
5	*Telina – Placuna* association found in bluish mudstone and associated with other less molluscs such as *Cytherea* sp., *Arca* sp., *Paphia* sp., *Placuna* sp., *Terebra* sp., *Marginella* sp., and *Dentalina* sp.	Lower intertidal
4	*Pecten – Telina* association found in bluish calcareous mudstone and associated with other less dominant molluscs such as *Placuna* sp., *Ostrea* sp., *Paphia* sp., *Paratapes* sp., broken Arcidae fragments	Quiet open marine
3	*Oliva - Turricula* association found in medium-grained sandstone. Other fossils that are less dominant are Pectinidae, Arcidae and *Balanus* sp..	Shallow open marine
2	*Nassa* association within calcareous sandstone intercalation. *Nassa* sp. is both in complete and broken form while other fossils are mostly broken.	Subtidal
1	*Corbula – Natica* association in complete form within mudstone Generally fossils are small and broken. Samples also contain *Nassa* sp. and Arcidae	Shallow tidal environment; upper most intertidal

5. Discussion

Local sea level curve from the study has been successfully developed from the fossil association (Figure 6). Previous studies [8,9] suggested that towards the top of the Kalibiuk Formation the depositional environment became shallower closer to the terrestrial condition. There were at least 2 deepening – shallowing cycles during the depositional period of the Kalibiuk Formation found in Cisaat River. The base of the Kalibiuk Formation (Unit 1; Table 1) in Cisaat River was deposited in the uppermost intertidal environment. The sediments were gradually deposited at deeper depth towards shallow open marine shown by the fossil association in Unit 2, 3, 4. Shallowing episode started during the deposition of Unit 5 sediments in lower intertidal environment. Maximum drop sea level in this first cycle attained during the deposition of Unit 6 depositional environment returned to the uppermost intertidal. Unit 7 and 8 represent the second deepening episodes and unit 9 and 10 indicate the final shallowing episodes before the Kalibiuk Formation was gradually overlain by Kaliglagah Formation that was deposited in the terrestrial setting.

Turritella – Marginella association (Unit 6) represents the most optimum condition for molluscs to live. This is due to the observation that shows there are more variations in fossil genera within the beds carrying this association; the fossils are also commonly found in complete forms.

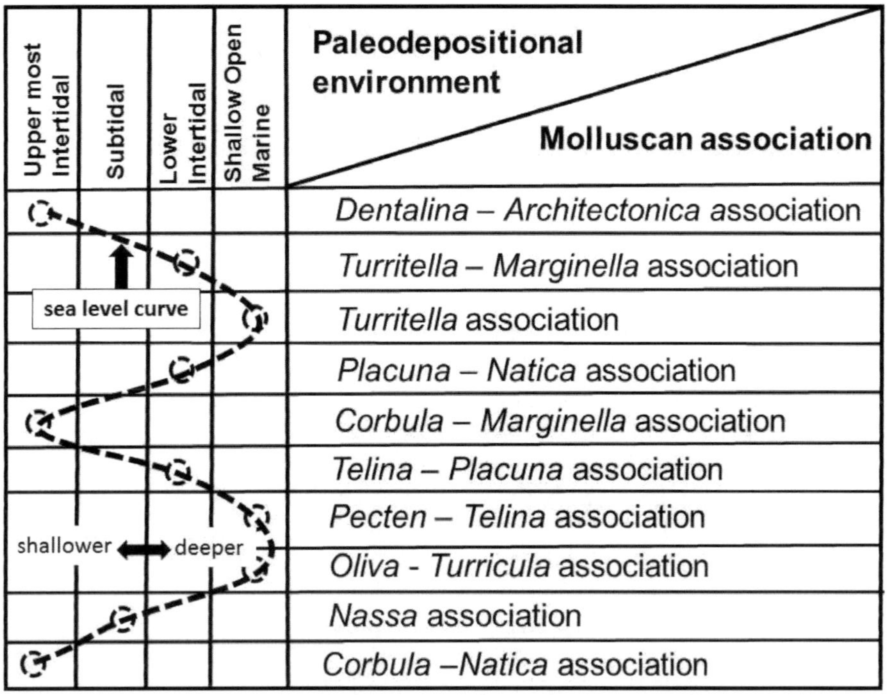

Figure 6. Local sea level change during the deposition of the Kalibiuk Formation in the Middle-Late Pliocene.

6. Conclusions

The study presents local sea level curve during the deposition of Middle-Late Pliocene Kalibiuk Formation generated from 2 deepening and shallowing cycles demonstrated by 10 mollusc fossils association.

From bottom to top the depositional environments are as follow: (1) upper most intertidal (*Corbula – Natica* association), (2) subtidal (based on *Nassa* association), (3) shallow open marine (marked by *Oliva - Turricula* association, (4) quiet open marine (*Pecten – Telina* association), (5) lower intertidal (*Telina – Placuna* association), (6) upper most intertidal (*Corbula – Marginella* association), (7) lower intertidal (*Placuna – Natica* association), (8) shallow open marine (*Turritella* association), (9) lower intertidal (*Turritella – Marginella* association), and (10) upper most intertidal (*Architectonica - Dentalina* association). Optimum condition for molluscs development in the Kalibiuk Formation is represented by the beds carrying *Turritella-Marginella* association.

Acknowledgements

We would like to express our gratitude to the Geology Study Program at ITB for the support and opportunity to use the facility during the work of this study. We would also like to thanks to the Dean and all staff of Faculty of Earth Sciences and Technology (FITB) – ITB and Head of Geology Expertise Group and staff, who helped the author by giving the permission to do this collaboration research. Our thanks also to Irman Yudi Abdurrahman from Bandung Geological Museum for provided geological map of research area. The field research was funded by Bandung Geological Museum.

References

[1] Kastowo 1975 *Geological Map of Majenang Quadrangles, Jawa.* Bandung: Department of Mines and Energy – Directorate General of Geology and Mineral Resources – Geological Research and Development Centre.

IOP Conf. Series: Earth and Environmental Science **162** (2018) 012033 doi:10.1088/1755-1315/162/1/012033

[2] Oostingh C H 1938 *Mollusken Als Gidsfossielen Voor Het Neogeen in Nederlandsh Indie* (Surabaya: Handelingen Van Het Achtste Nederlandsch – Indisch Natuur Wetenschappelijk Congres).

[3] Kitamura A, Ubukata T 2003 The sequence of local recolonization of warm-water marine molluscan species during a deglacial warming climate phase: a case study from the early Pleistocene of the Sea of Japan. *Palaeogeography, Palaeoclimatology, Palaeoecology*, vol. 199, pp. 83 – 94.

[4] Aswan, Ozawa T 2006 Milankovitch 41000-year Cycles in Lithofacies and Molluscan Content in The Tropical Middle Miocene Nyalindung Formation, Jawa, Indonesia. *Palaeogeography, Palaeoclimatology, Palaeoecology* vol. 235, pp. 382 – 405.

[5] Lopes R P, Simone LR L, Dillenburg S R, Schultz C L, Pereira J C 2013 A middle pleistocene marine molluscan assemblage from the southern coastal plain of Rio Grande do Sul State, Brazil. *Revista Brasileira de Paleontologia*, vol. 16, no. 3, pp. 343-360.

[6] White T S, Preece R C, Whittaker J E 2013 Molluscan and ostracod successions from Dierden's Pit, Swanscombe: Insights into the fluvial history, sea-level record and human occupation of the Hoxnian Thames. *Quaternary Science Reviews*, vol. 70, pp. 73-90.

[7] Van Bemmelen R W 1949 *The Geology of Indonesia*. Vol IA, 1st Edition. Netherland: Government Printing Office. The Hague.

[8] Alfian A, Aswan, Gentur W 2013 Studi Formasi Tapak dan Formasi Kalibiuk Daerah Sumampir dan sekitarnya, Kecamatan Rembang, Kabupaten Purbalingga, Jawa Tengah. (Tapak and Kalibiuk Formation Study in Sumampir area and surroundings, Rembang District, Purbalingga Region, Central Jawa). *JTM ITB*, vol. 20, no. 2, pp. 55-66.

[9] Arif H H, Aswan, Gentur W 2014 Lingkungan Pengendapan Formasi Tapak dan Formasi Kalibiuk, Daerah Karanganyar dan sekitarnya, Kabupaten Purbalingga, Jawa Tengah (Paleodepositional Study of Tapak and Kalibiuk Formations in Karanganyar area and surroundings, Purbalingga Region, Central Jawa). *Buletin Geologi ITB*, vol. 41, no. 2, pp. 160-174.

[10] Martin K 1879–1880 *Die tertiärschicten auf Java. Nach den Entdeckungen von F. Junghuhn. Paleontologischer Theil.* Geolo- gische-Reichsmuseum, Leiden. IX +164+51+6 pp, 28 pls.

[11] Sufiati E 2013 *Koleksi Museum Geologi: Fosil Moluska Holotype dari Bumiayu dan Cirebon, Jawa* (*Geological Museum Collection: Holotype Molluscs Fossils from Bumiayu and Cirebon, Jawa*). Bandung Geological Museum, Geological Agency, Ministry of Energy and Mineral Resources, Bandung.

[12] Okutani T 2000 *Marine Mollusks in Japan*. Tokai University Press, Tokyo.

[13] Fairbridge RW, Bourgeois J 1978 *Encyclopedia of Sedimentology* (Encyclopedia of Earth Sciences Vol. VI). Hutchinson and Ross Inc., Stroudsburg.

IOP Conf. Series: Earth and Environmental Science **162** (2018) 012034 doi:10.1088/1755-1315/162/1/012034

Ichnofossils study of paleocene sediment source rock cores from Bintuni basin, West Papua, Eastern Indonesia

Aswan[1], Alfend Rudyawan[1], and Thaw Zin Oo[3]

[1] Geology Department, Faculty of Earth Sciences and Technology, Institut Teknologi Bandung, Jalan Ganesa 10 Bandung, Indonesia
[3] Geology Department, East Yangon University, Thanlyin Township, Yangon Division, Myanmar

E-mail: aswan_gl@gc.itb.ac.id

Abstract. Ichnofossils content of Paleocene source rock sediment were examined based on two wells cores from Bintuni Basin, West Papua, Indonesia. The cores are from Paleocene interval of Waripi - Daram Formations and known as the potential source rock sediment interval. This study aims to determine the sea level changes during the depositional of the Waripi - Daram Formations. Representative of 18 ichnogeneras including: *Arenicolites*, *Asterosoma*, *Chondrites*, cryptic bioturbation, *Glossifungites*, *Helminthoida*, *Nereites*, *Ophiomorpha rudis*, *Palaeophycus*, *Phycosiphon*, *Planolites*, *Rosselia*, *Scolicia*, *Skolithos*, *Spirophyton*, *Teichicnus*, *Thalassinoides*, and *Zoophycos*. The sea level curve generated from both wells based on ichnofossils association, suggest that there were at least three deepening – shallowing cycles. The cycles generally begins by inner to middle neritic interval that sign by appearance of *Ophiomorpha* and *Chondrites* with few *Phycosiphon*, while the deeper interval of outer neritic, upper bathyal into abyssal usually indicated by *Zoophycos*, *Helminthoida*, *Nereites*, *Spirophyton* and abundant of *Phycosiphon*. Sea level curve reconstructions based on the depositional environment suggest that during Paleocene Well-2 was located closer to the land compared to Well-1. Rapid shift in the depositional environment could indicate an initiation of regional uplift in the area caused by a new subduction system to the north of the Basin.

1. Introduction

This study aims to determine the sea level changes during the depositional of the Waripi and Daram Formations in the Paleocene age utilizing the trace fossils from core data from two wells. Ichnofossils studies in Indonesia have been carried out by some researcher i.e. [1,2,3,4,5]. This study presents the first ever result of the ichnofossils studies focused on the sea level changes by using core data from two wells in the hydrocarbon prolific Bintuni Basin, West Papua, Indonesia (Figure 1). The sedimentary units in this study consist of slightly calcareous grained muddy sandstone, claystone, siltstone and shale part of the Waripi – Daram Formation of Paleocene age which are considered to be the source rock for the petroleum system in the basin. Foraminifera studies have been carried out but proved to be inconclusive. Observation on the ichnofossils at a particular section from these two wells are utilised to interpret the depositional environment. Vertical succession of the depositional environment is then compiled to generate the local sea level curve during the Waripi – Daram Formation deposition period. The sea level curve can be used to support the Cenozoic tectonic evolution studies of the Bird's Head area in Papua.

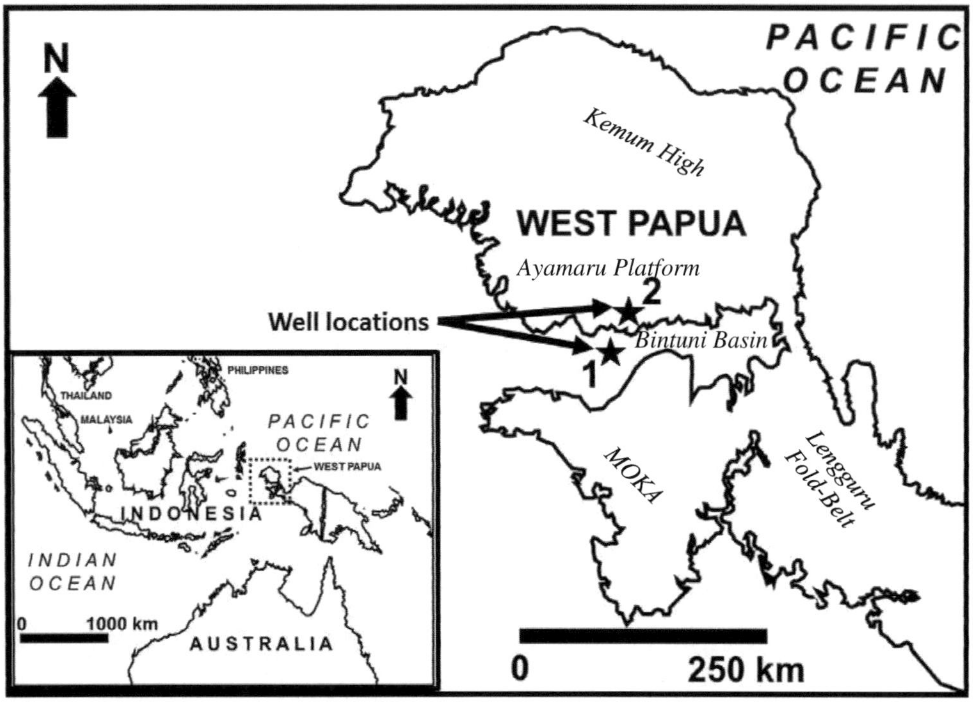

Figure 1. Star symbols mark the location of Well-1 (1) and Well-2 (1), Bintuni Basin, West Papua.

2. Geological Setting

The Bintuni Basin is located in the Bird's Head region of Papua that was formed by compressional tectonic between the Pacific and Australian Plate since the Paleogene [6]. The well is located in the northwest shoulder of the Basin. The Paleocene interval, as a focus of this study, is part of the Waripi – Daram Formation that was deposited conformably above the Upper Kembelangan Formation. [7] interpreted this rock unit age as Paleocene due to contains no age diagnostics fossils but lies with gradational contacts between the Eocene Yawee Formation and Late Cretaceous Ekmai sandstone, but may partly be a lateral equivalent of the Ekmai sandstone. This formation consists of sandstone, claystone and shale which collectively indicate deep marine sedimentary deposit. This rock unit crops out in the western mountains of the Central Range from where it extends westwards into the southern edge of the Birds Neck [7].

3. Methods

Ichnofossils types and sedimentary structures are examined and documented during core observation. Representative core leg is shown in Figure 2. Core section from Well-1 is acquired from 8.877 ft. - 8.602 ft. deep, while from Well-2 the similar interval was acquired between 8.811 ft. - 7.590 ft. deep. However, there are some intervals missing from the core. Changes in depositional environment in the study will be based on "changes in the characteristics of ichnofossils content" or "changes of ichnofossils association". Ichnofossils determination and depositional paleoenvironmental interpretation in general are based on [8,9], while depositional environmental zonations were based on [10]. Ichnofossils other than those contained in the literatures as mentioned above such as *Phycosiphon* based on [11] and *Nereites* were determined based on [12] while *Spirophyton* were based on [13]. The bathymetry of depositional environment is determined based on Bathymetric Zone classification from [14].

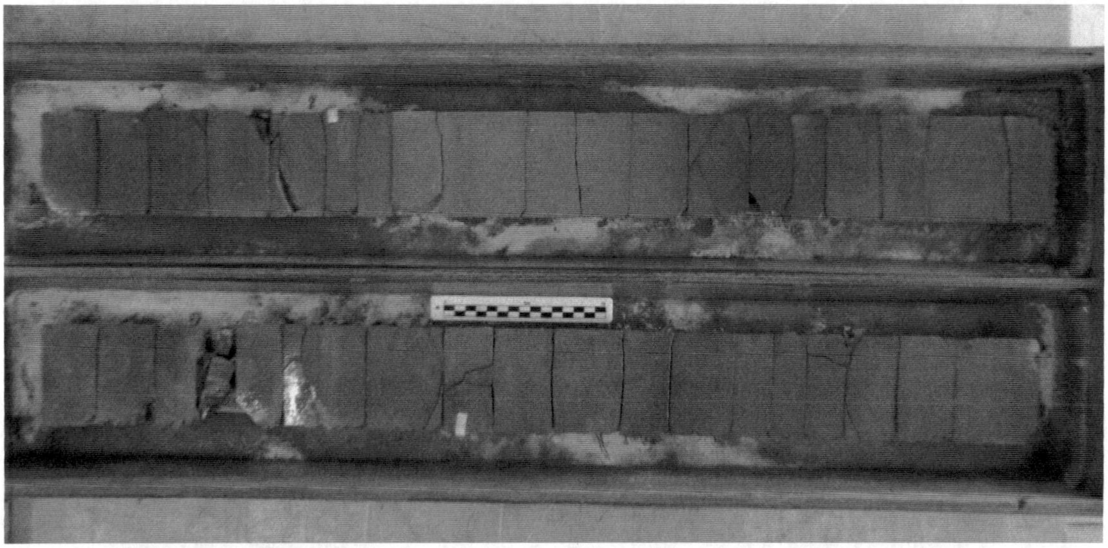

Figure 2. Representative core leg from well number 1 (Well-1).

Determination of ichnofossils association in each interval was carried out based on the predominant ichno taxa, the presence of ichnofossils index, or the presence of ichnofossils that appears only in specific intervals. Interpretation of depositional environments for each interval is based on the change in ichnofossils association, or their characteristics. In case of discrepancy in the depositional environment bathymetry of the studied wells, the "characteristic ichnofossils" or "characteristic ichnofossils association" from each well should be compared.

Seven depositional bathymetric zones were recognized from the ichnofossils in the section, they are: tidal, shallow neritic (inner shelf), middle neritic, outer neritic, upper bathyal, lower bathyal and abyssal. Using the available data and interpretation, vertical succession of the sea level curve during the deposition period of the Waripi – Daram Formation can be determined from the bottom to the core of each core section.

4. Results

The lithology association from the section is dominated by slightly calcareous fine-grained muddy sandstone, claystone, siltstone and brown-grey shale. Ichnofossils determination from these two wells suggest 19 ichnogenera and floral trace including: *Arenicolites*, *Asterosoma*, *Chondrites*, cryptic bioturbation, *Glossifungites*, *Helminthoida*, *Nereites*, *Ophiomorpha rudis*, *Palaeophycus*, *Phycosiphon*, *Planolites*, *Rosselia*, root structure, *Scolicia*, *Skolithos*, *Spirophyton*, *Teichicnus*, *Thalassinoides*, and *Zoophycos*. Eight ichnofossils association can be determined from each well (Table 1 and 2).

Each ichnogenera that are used as a basis for ichnofossils association are shown in Figure 3 (Well-1) and Figure 4 (Well-2). From these ichnofossils association, depositional environment for the particular section can be determined.

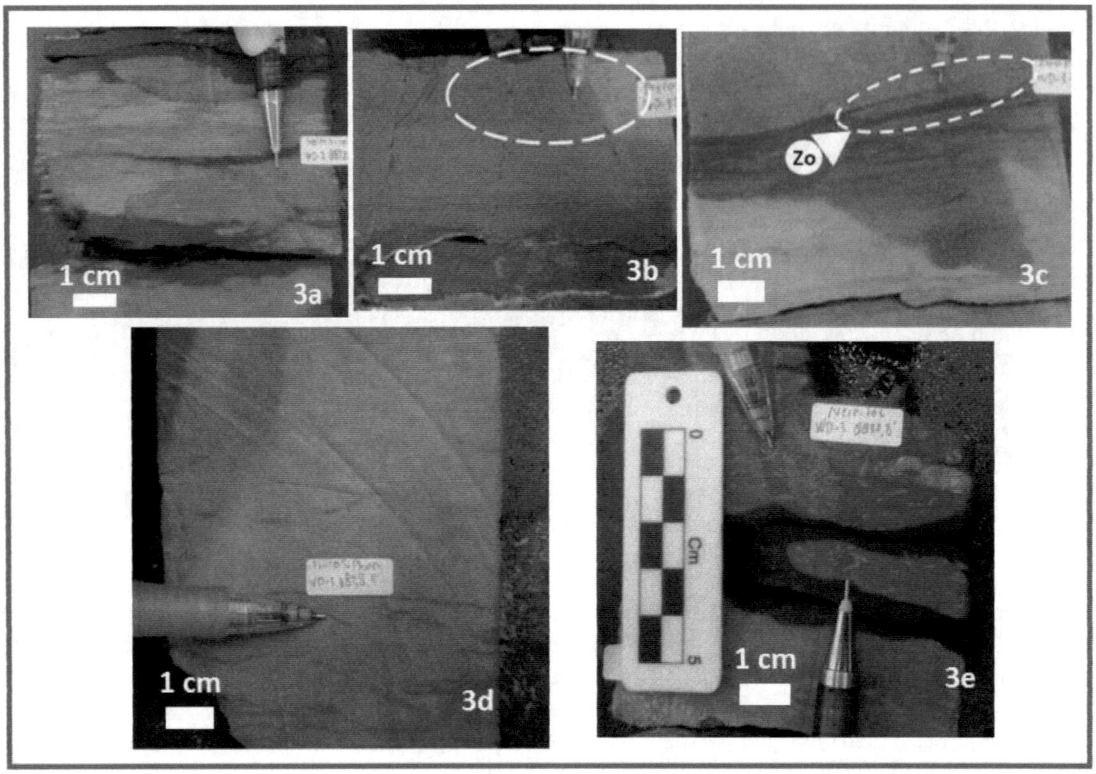

Figure 3. Ichnogenera found in Well-1; *Helminthoida* (3a; depth: 8.872.1 ft), *Phycosiphon* (3b; depth: 8.843 ft), *Zoophycos* (3c; Zo; depth: 8.853 ft), large *Phycosiphon* (3d; depth: 8.838 ft), *Nereites* (3e; depth: 8.837.8 ft).

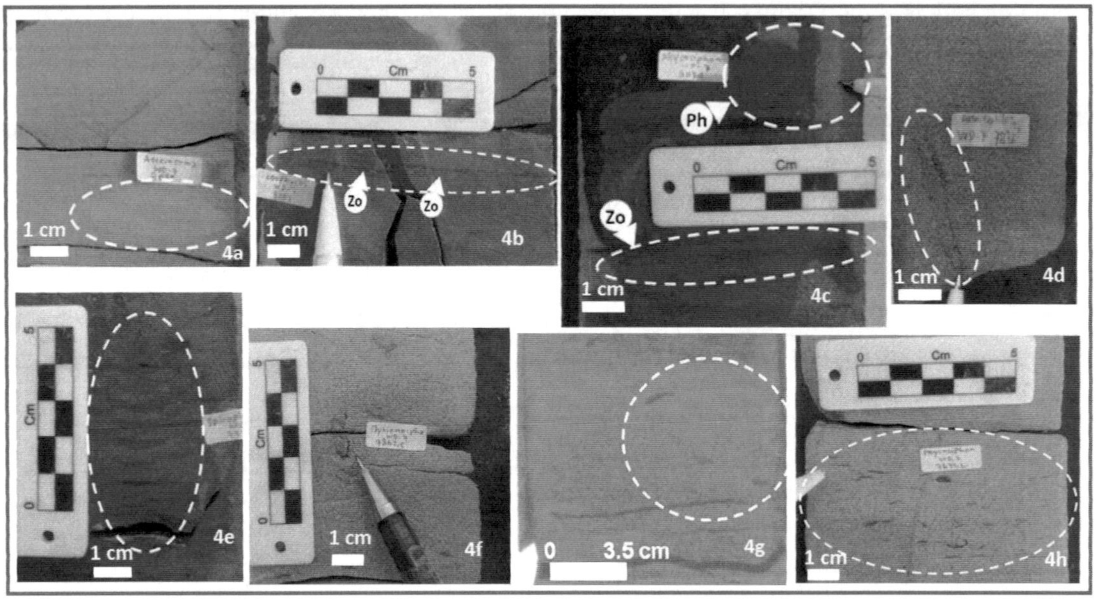

Figure 4. Ichnogenera found in Well-2; *Asterosoma* (4a; depth: 8.085 ft), *Zoophycos* (Zo) (4b; depth: 8.082 ft), *Phycosiphon* (Ph) and *Zoophycos* (Zo) (4c; depth: 8.034 ft), *Arenicolites* (4d; depth: 7.872 ft), *Spirophyton* (4e; depth: 7.777.4 ft), *Ophiomorpha* (4f; depth: 7.862.5 ft), *Chondrites* (4g; depth: 7.602 ft), *Phycosiphon* (4h; depth: 7.633 ft).

Table 1. Ichnofossils association acquired from particular depth within Well-1, Bintuni Basin.

Ichnofossils association	Acquired depth	Depositional bathymetry
Ophiomorpha - *Chondrites* with few-rare of *Phycosiphon*	8.628 – 8.602 ft.	transition zone between middle neritic and shallow neritic (inner shelf)
Phycosiphon – rare of *Ophiomorpha* and *Chondrites* disappearance	8.657.17 – 8.628 ft.	middle neritic bathymetric zone
Ophiomorpha - *Chondrites*	8.702.3 – 8.657.17 ft.	transition zone between middle neritic and shallow neritic (inner shelf)
decreasing of *Phycosiphon*	8.731.4 – 8.702.3 ft.	outer neritic bathymetric zone
Abundant of *Phycosiphon* and appearance of *Zoophicos*	8.744.8 – 8.731.4 ft.	transition zone between upper bathyal and outer neritic.
Ophiomorpha with highly decreasing of *Phycosiphon*	8.770.8 – 8.744.8 ft.	transition zone between middle neritic and shallow neritic (inner shelf)
Helminthoida – abundant of *Phycosiphon* without *Zoophicos*	8.842.3 – 8.770.8 ft.	bathyal bathymetric zone
Helminthoida – *Phycosiphon* with *Zoophicos*	8.877 – 8.842.3 ft.	transition zone between upper bathyal and outer neritic

Table 2. Ichnofossils association acquired from particular depth within Well-2, Bintuni Basin.

Ichnofossils association	Acquired depth	Depositional bathymetry
Few *Chondrites* – *Thalassinoides* associated with mud drapes	7.601.3 – 7.590 ft.	tidal depositional area
Ophiomorpha - *Chondrites*	7.689 – 7.601 ft.	transition zone between middle neritic and shallow neritic (inner shelf)
abundant of *Phycosiphon* - *Zoophycos* – with some *Chondrites* and *Ophiomorpha*	7.869.71 – 7.689 ft.	transition zone between outer neritic and middle neritic
Arenicolites - *Chondrites*	7.893 – 7.869.71 ft.	shallow neritic bathymetric zone
Phycosiphon - *Zoophycos*	7.975.5 – 7.893 ft.	transition zone between upper bathyal and outer neritic
Rosselia - *Skolithos*	8.002.4 – 7.975.5 ft.	transition zone between middle neritic and shallow neritic (inner shelf)
Zoophycos – abundant of *Phycosiphon*	8.065.5 – 8.002.4 ft.	transition zone between upper bathyal and outer neritic
Scolicia – *Zoophycos* with few *Asterosoma*	*8.811 – 8.065.5 ft.*	transition zone between outer neritic and middle neritic

5. Discussion

Regionally, Paleocene deposits represent passive continental margin sequences dominated by deep marine limestone and mudstone deposits. At the present day, Well-1 and Well-2 is located in the Bintuni Bay and at the edge of the Ayamaru Platform. However, it is evident from the ichnofossils association that contain outer neritic, upper bathyal and abyssal, sedimentary sequences in Well-1 are interpreted to

IOP Conf. Series: Earth and Environmental Science **162** (2018) 012034 doi:10.1088/1755-1315/162/1/012034

be deposited in deeper environment compared to Well-2 which contain tidal to outer neritic sequence. This suggests that land is probably emerged near the Well-2 and Well-1 is more of basinal area during the deposition of the Waripi – Daram Formation. The sea level curve generated from both wells (Figure 5 and 6), suggest that there was at least three deepening – shallowing cycles in the Bintuni Basin in the Paleocene age.

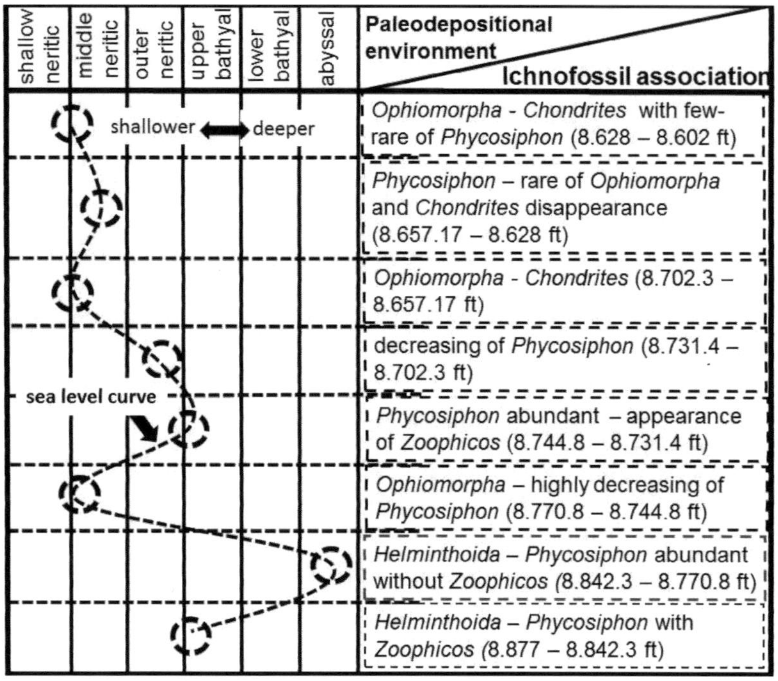

Figure 5. Paleocene sea level curve in the Bintuni Basin determined from the Waripi – Daram Formation in Well-1.

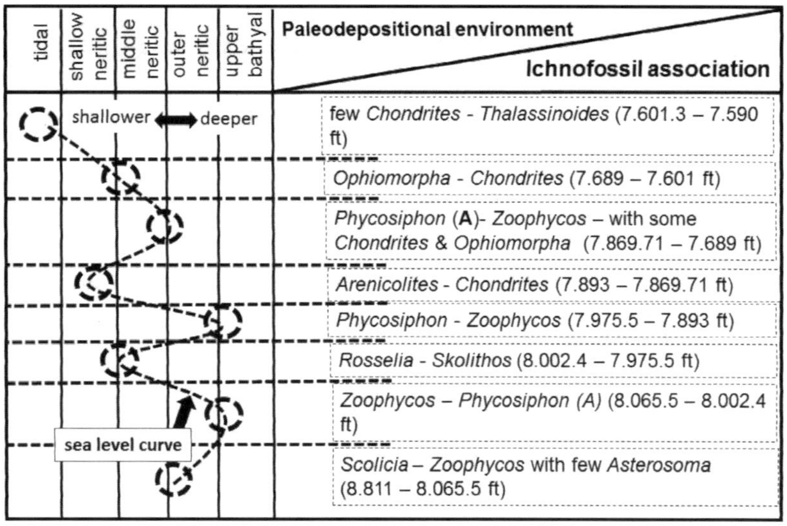

Figure 6. Paleocene sea level curve in the Bintuni Basin determined from the Waripi – Daram Formation in Well-2.

There is a notable jump in bathymetry from abyssal to middle neritic in the lower sequence of Waripi – Daram Formation in Well-1 but not so clear in Well-2. This could suggest that the start of the jump in bathymetry marks the initial regional uplift in the Bintuni Basin. Towards the top of the sequence, sedimentary units in both wells indicating shallower depositional environment. This is interpreted to be caused by a change in the geodynamic system to the north of the basin. For the last part of the Mesozoic period the Bintuni Basin was a part of Australian passive margin system. In the early Cenozoic there was a south-directed subduction system from the Pacific Plate under the Australian Plate [15]. This caused regional uplift in the overriding plate thus the rapid shallowing episodes of the basin.

The lack of limestone and the occurrence of clastics sequences within the sequence suggest that the Bintuni Basin was in a relatively close proximity to the exposed terrain as the source for its sediments in the Paleocene; they are probably sourced from the Kemum High. To the south of Well-1 is the well-known MOKA (Misool-Onin-Kumawa Anticlinorium), deeper environment of sedimentary units in Well-1 confirms that the anticlinorium was caused by a tectonic event post Paleocene.

6. Conclusion

Paleocene sea level curve has successfully been determined using ichnofossils and ichnofossils association from core data of Waripi – Daram Formation acquired from two wells in the Bintuni Basin. This research also suggests three significant sea-level fluctuations between shallow neritic, middle neritic, outer neritic, upper bathyal, lower bathyal and abyssal marine conditions in the Well-1 and between tidal, shallow neritic, middle neritic, outer neritic and upper part of upper bathyal in the Well-2. Three episodes of sea level fluctuations were recorded from the sediments; the most notable jump can be seen at the lower section of the formation. Sedimentary units in Well-1 were deposited in the deeper environment compared to those in Well-2. This suggests that land was probably emerged in the north sending sediments towards the basin in the south.

Acknowledgements

We would like to express our gratitude to the Dean of Faculty of Earth Sciences and Technology (FITB), the Head of Geology Expertise Group, the Geology Study Program – ITB and all their staff for the permission and support to do this study. We are extremely grateful to Astuti Purwati, Ruly Mardani and Erick Victor Yudhanto from BP Indonesia for giving us the opportunity to study the core samples. Our thank also to all SIGMA Serpong staff who helped us to prepare core samples for ichnofossil observation. We also thank Dr. Indra Gunawan from Geology Department – ITB, Professor George Pemberton and Professor John-Paul Zonneveld from the Alberta University, Canada, for the discussion and their help for the improvement of this paper.

References

[1] Savrda C E 1995 Ichnologic Applications in Paleoceanographic, Paleoclimatic, and Sea-Level Studies. *PALAIOS* **10**, p 565–577

[2] Wilson M A, Curran H A and White B 1998 Paleontological Evidence of a Brief Global Sea-Level Event during the Last Interglacial. *LETHAIA* **31**, p 241-250

[3] Savrda C E, Krawinkel H, McCarthy F M G, McHugh C M G, Olson H Cand Mountain G 2001 Ichnofabrics of a Pleistocene slope succession, New Jersey margin: relations to climate and sea-level dynamics. *Palaeogeography, Palaeoclimatology, Palaeoecology*, **171**, p 41-61

[4] Aswan 2017 Jurassic Deposit Ichnofossils Study of Cores from Bintuni basin, Eastern Indonesia. Proceeding World Academy of Science, Engineering and Technology Conference, Phuket, Thailand.

[5] Aswan, Nurcahyo I B and Thaw Z O 2017 Jurassic and Paleocene Deposit Ichnofossil Study of Cores from Bintuni Basin- Eastern Indonesia Comparison between Shallow and Deep Marine Ichnofossil Associations. Proceeding Joint Convention Malang.

[6] Sapiie B 1998 Strike-slip faulting, breccia formation and porphyry Cu-Au mineralization in the Gunung Bijih (Ertsberg) mining district, Irian Jaya, Indonesia (PhD dissertation, University of Texas at Austin, Austin, Texas, USA).
[7] Pieters P E, Ratman N, Sukamto R, Pigram C J, Trail D S and Dow D B 1983 The Stratigraphy of Western Irian Jaya. *Bulletin Geological Research and Development Centre* **8**
[8] Pemberton S G, MacEachern J A and Frey, R W 1992 Trace Fossil Facies Models: Environmental and Allostratigraphic Significance in: Facies Models Response to Sea Level Change 1st, *Geological Association of Canada*, p 47-72
[9] Ichnology Research Group and Pemberton S G 1999 *ICHNOLOGY ATLAS* (Alberta, Canada Ichnology Research Group)
[10] Krause F F, Deutsch K B, Joiner S D, Barclay J E, Hall R Land Hills, L V 2012 *Cretaceous Cardium Formation of the Western Canada Sedimentary Basin in: Geological Atlas of the Western Canada Sedimentary Basin* (Alberta Geological Survey, Alberta)
[11] Wetzel A and Bromley R G 1994 Phycosiphon incertum revisited: *Anconichnus horizontalis* is Its Junior Subjective Synonym. *Journal Paleontology* **68**, p 1396-1402
[12] Bendella M and Mehadji A O 2014 Depositional Environment and Ichnology (*Nereites* ichnofacies) of the Late Devonian Sahara Region (SW Algeria) (Arab Journal Geoscience in press)
[13] Benton M J and Harper D A T 1997 Basic Paleontology (Essex, Addison Wesley Longman) p 342
[14] Tipsword H L, Setzer F M, and Smith L F 1966 Interpretation of Depositional Environment in Gulf Coast Petroleum Exploration from Paleoecology and Related Stratigraphy (Houston: *Gulf Coast Association of Geological Societies* **XVI**)
[15] Hall R 2011 Australia-SE Asia collision: plate tectonics and crustal flow. In: The SE Asian gateway: history and tectonics of Australia-Asia collision. (London: Geological Society of London) 355, pp 75-109

IOP Conf. Series: Earth and Environmental Science **162** (2018) 012035 doi:10.1088/1755-1315/162/1/012035

Chlorophyll-A variability in the southern coast of Java Island, Indian Ocean: corresponding to the tropical cyclone of Ernie

U Efendi[1], A Fadlan[2], A M Hidayat[1]

[1] Under Graduate Program of Meteorology, State College of Meteorology Climatology and Geophysics, Indonesia

[2] Lecturer, Department of Meteorology, State College of Meteorology Climatology and Geophysics, Indonesia

E-mail: usman.efendi@bmkg.go.id

Abstract. Tropical cyclone is an atmospheric phenomenon formed above warm ocean in a tropical region. Generally, tropical cyclone gives negative impact, such as extreme rainfall, strong winds, and extreme wave height. On the other hand, however, tropical cyclone gives positive impact on triggering the ocean fertility by induced chlorophyll-a concentration caused by vertical mixing and upwelling in the ocean. Related to this, a study has been conducted to determine the concentration of chlorophyll-a after the occurrence of Ernie cyclone in Indian ocean near western part of Australia and along the southern coast of Java island. The data that used in this study is daily wind speed data from 5 - 11 April 2017, daily average data of SST from 5 - 11 April 2017, and 8-day composite data of chlorophyll-a data from Aqua-MODIS satellite. The analysis of chlorophyll-a concentration showed a slow response of ocean fertility increment in the southern coast of Java island. Chlorophyll-a concentration along the cylone trajectory prior to the cyclone was in the range of 0.08 mg/m^3 – 0.15 mg/m^3 and significantly increased up to 88%–100% in a week after the occurrence of cyclone. On the next third and fourth weeks after cyclone, the concentration of chlorophyll-a along the cyclone trajectory decreases gradually. Meanwhile its sea surface temperature decreased significantly, especially in areas where cyclone had passed.

1. Introduction

Tropical cyclone is one of the most destructive meteorological activities on earth which commonly associated with strong wind, heavy rain, and induce large wave in the ocean. Besides numerous calamities caused by the activity of cyclone, studies have shown that cyclone events lead to the increase of nutrient and phytoplankton increment in the water column, which depend on its translation speed and intensity [1]. The energy of cyclone penetrates into the ocean cause strong entrainment and mixing [2][3][4]. This energy play a substantial role on the changes of upper ocean environment and its physical characteristic [5][6][7][8][9][10][11].

The maximum cooling observed on the right of the cyclone track in northern hemisphere and to the left of the track in southern hemisphere when the cyclone moves rapidly, while the maximum cooling occurs well near or on the track for slowly moving cyclone [12]. Suetsugu et al. [13] revealed that atmosphere-ocean parameter dependence, such as storm strength level, storm translation speed, and upper ocean thermal structure, influence the cooling process of sea surface temperature. As conclusion, they described that the strong and slow moving cyclone causes maximum cooling of 5 °C in the cyclone wake.

According to Lin and Wu [14], the emergence of cyclones, which decrease sea surface temperature (SST), lead to subsequent changes in nutrient concentrations and the uptake of nutrients (upwelling) by phytoplankton, which significantly affects chlorophyll-a (Chl-a) and depth-integrated primary productivity (IPP). Upwelling brings rich nutrient water from sub surface to the euphotic zone where there is abundant light for photosynthesis but often lack of nutrients [15]. The chlorophyll-a increment may occur along the cyclone track or in nearby coastal region. Hung et al. [16] observed that the increase of chlorophyll-a concentration not only at the surface, but also above thermocline. After the occurence of typhoon Fengwong, they found that the in situ measured on the subsuface chlorophyll-a concentration was 50% higher than the imagery analysis using moderate-resolution imaging spectroradiometer (MODIS) data, therefore the depth of euphotic zone also increased.

Hong and Sohn [17] and Kim et al. [18] declares that upwelling, caused by Ekman pumping, was active when cyclone passes over a specific sea surface and the cold-water injection into the surface layer from a deeper layer continues to spread horizontally outwards. This mechanism considered as one of the main factors supporting vertical mixing, and, as a result changes on various factors which may last for quiet long period [19]. On April 2017, a tropical cyclone, named Ernie, was form at latitude 13.30°S and longitude 110.8°E in the Indian ocean. This cyclone grew intensively until it was declared as tropical cyclone category 4 based on Saffir-Simpson hurricane scale with a maximum wind speed of 130 knots. Although Ernie cyclone did not pass Indonesia but the distance between this cyclone and southern coast of Java was quite close.

During cyclone period, the measurement of biological and physical parameter onboard ship is very arduous due to severe weather conditions as one of the cyclones impact. Krishna and Song [20] stated that the development of multi satellite remote sensing, especially microwave remote sensing, in combination with numerical simulation, provide powerful tools for further understanding the upper ocean changes induced by cyclone. Hence, the development of multi-satellite remote sensing provides worthwhile equipment to analyze the upper ocean dynamic changes. Most studies of cyclone effects on chlorophyll-a have focused on the concentration of uptake nutrient around the cyclone treajectory. However, the impacts of cyclone on the coastal region remain unclear. Therefore, this study examines the effect of Ernie cyclone on the concentration of chlorophyll-a on the southern coast of Java island and along the cyclone track using very high-resolution satellite imagery.

2. Data and Methodology

2.1 Research Site

The study was conducted along the track of Ernie cyclone and along the southern coast of Java island (Fig. 1). Analysis of the impact of tropical cyclones on aquatic area was carry out in two main areas: along the emergence of Ernie cyclone (cyclone track, A1) and along the coastal region in the southern coast of Java (A2), which precisely located at the coordinates of 7° 58’ 45’’ S 109° 43’ 15’’ E. Coastal region (A2) put into important consideration because it is assumed as the most potential region of fishing.

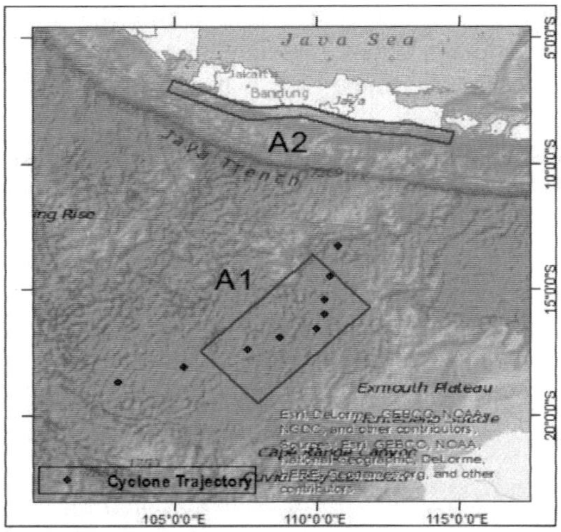

Figure 1. The research site focused on 2 regions depicted as A1 and A2. A1 showed an area of cyclone trajectory, while A2 was a coastal region on the southern part of Java.

2.2 Data Sources

2.2.1 Cyclone information for Southern Indian Ocean. The tropical cyclone data that used in this study is provided by the Joint Typhoon Warning Center (JTWC), from 6 - 10 April 2017, in the geographical range of 100°E-115°E and 5°-20°S (Fig. 1). Ernie cyclone was active in the southern coast of Java island, Indian ocean for about 5 days (Table 1). The table contain several information related to the cyclone, such as the name of tropical cyclone, the coordinates location of its center, observed time (about 6-h interval), and maximum sustained wind speed. Tropical cyclone Ernie is the third tropical cyclone occurs in the southern Indian ocean in 2017. The maximum wind velocity of tropical cyclone Ernie is considered to be the highest amongst other tropical cyclones, reached up to 130 knots. Ernie cyclone was declared as a tropical cyclone category 4 on the Saffir-Simpson hurricane scale. Based on tropical cyclone data obtained from www.weather.unisys.com, the coordinates and wind velocity of tropical cyclone Ernie can be seen in the following table 1.

Tabel 1. The detailed data about the condition at pre-cyclone periods, at the cyclone event, and post-cyclone. The data contains information about times of occurrence, coordinate location of the cyclone, maximum speed, and also its category based on Saffir-Simpson hurricane scale.

Dates	Hour (UTC)	Latitude	Longitude	Maximum wind speed (knots)	Category
6/April/2017	06	-13.3	110.8	35	Tropical storm
6/April/2017	18	-14.5	110.5	40	Tropical storm
7/April/2017	06	-15.4	110.3	95	Tropical cyclone (2)
7/April/2017	18	-16	110.3	130	Tropical cyclone (4)
8/April/2017	06	-16.6	110	95	Tropical cyclone (2)
8/April/2017	18	-16.9	108.7	90	Tropical cyclone (2)
9/April/2017	06	-17.4	107.6	85	Tropical cyclone (2)
9/April/2017	18	-18.1	105.3	55	Tropical storm
10/April/2017	06	-18.7	103	40	Tropical storm

Source: http://weather.unisys.com/hurricane/

2.2.2. Sea surface temperature (SST). The value of daily average SST is derived from NASA Jet Propulsion Laboratory (JPL) Multi-scale Ultra-High-Resolution Sea Surface Temperature (MUR SST) with specific resolution of 1 km. SST data can be obtained online through https://mur.jpl.nasa.gov/DownloadDataText.php. The daily average of sea surface temperature was displayed in the form of spatial distribution map from 5 - 11 April 2017.

2.2.3. Sea surface wind. The blended sea winds contain globally gridded, high resolution ocean surface vector winds and wind stresses on a global 0.25° grid. The spatial variation of wind force produces a consequent variation in the corresponding with Ekman transport in the surface layer, causing convergence and divergence in different areas. This leads to "Ekman pumping" [21]. Wind data is obtained from the Advanced Scatterometer (ASCAT) sensor on the METOP satellite. The wind data used in this research is daily average data of surface wind. The data measured at a height of 10 meters above sea level and has a grid resolution of 0.25° x 0.25°. The data can obtain online through ERDDAP server website (http://coastwatch.pfeg.noaa.gov/erddap/index.html). Furthermore, the wind speed and direction were calculated according to the following formula:

$$Wind\ speed = \sqrt{x\ wind^2 + y\ wind^2} \quad (1)$$

$$wind\ direction = 90 - \left(\left(\frac{180}{\pi} * ATAN2\left(\frac{x_{wind}}{y_{wind}}\right)\right)\right) \quad (2)$$

Where x wind is zonal wind (knots) and y wind is meridional wind (knots) [22]. The calculated data were displayed in the form of spatial distribution of daily average wind speed and direction.

2.2.4. Chlorophyll-a concentration. There are several ways to analyzed the variability of chlorophyll-a concentration in the ocean. Susilo and Hardianti [22] used sensor of the Visible Infrared Imaging Radiometer Suite (VIIRS) on the Suomi National Polar-orbiting Partnership or Suomi NPP satellite. Nuris et al. [23] declared that the results of VIIRS - NPP sensor and Moderate Resolution Imaging Spectroradiometer (MODIS) had the same pattern, however the estimation of chlorophyll-a concentration from the MODIS sensor was higher than VIIRS -NPP sensor. Therefore, this study used chlorophyll-a concentration data from MODIS sensor on the Terra satellite /Aqua-MODIS. Spatial distribution and temporal variation of chlorophyll-a concentration is based on the 8-day composite data. The used of this method is due to unavailability data of chlorophyll-a concentrations since the cyclone trajectory area is mostly covered by cloud. Chlorophyll data can be obtained online through ERDDAP server website (http://coastwatch.pfeg.noaa.gov/erddap/griddap/erdMH1chla8day.html). The chlorophyll data were presented in the form of spatial distribution and hovmoller diagram. The spatial map of chlorophyll-a was made within 3 times period, 30 March - 6 April 2017, 7 – 14 April 2017, and 15 – 22 April 2017. The hovmoller diagram was created to determine the spatio-temporal variation of chlorophyll-a along the southern coast of Java Island beginning at early February 2017 to early June 2017.

2.3 Data Analysis

Distribution of sea surface temperature, daily average wind, and 8-day composite data of chlorohyll-a was analyzed using GrADS 2.02.oga.2 software. The analysis was conducted to determine the pattern of sea surface temperature, wind speed and direction, and also the variability of chlorohyll-a during pre-cyclone period, at cyclone event, and post cyclone period. The pattern of each data then analyzed in two research areas, A1 and A2, in order to comprehend the effects of ernie cyclone toward the variability of chlorophyll-a concentration along the southern coast of Java island. The variability of chlorophyll-a on the southern coast of Java island (A2) before and after the occurrence of Ernie cyclone then analyzed using Ocean Data View (ODV) in order to create a hovmoller diagram.

3. Result and Discussion

3.1. Computation of daily-average surface wind from METOP-ASCAT

Surface wind, which blow over the ocean, is parameter that has considerable impact due to the activity of tropical cyclone Ernie. Significant pressure difference between the cyclone center and its surrounding areas made the wind speed in that area was quite high. When cyclone happens, wind velocity generally increases along the cyclone trajectory, but tends to be weak at the center of the cyclone [22]. Tropical cyclone also leads to the increase of wind speed and retard the current movement [1]. Surface wind conditions around the area of Ernie cyclone and southern coast of Java island were represented in Fig. 2. Wind speed shown in the daily-average value from ASCAT sensor so that the value was lower when compared to the maximum value of wind velocity stated in table 1. This different occurred because the wind speed value shown by ASCAT sensor is the average value not the maximum value as what stated on table 1.

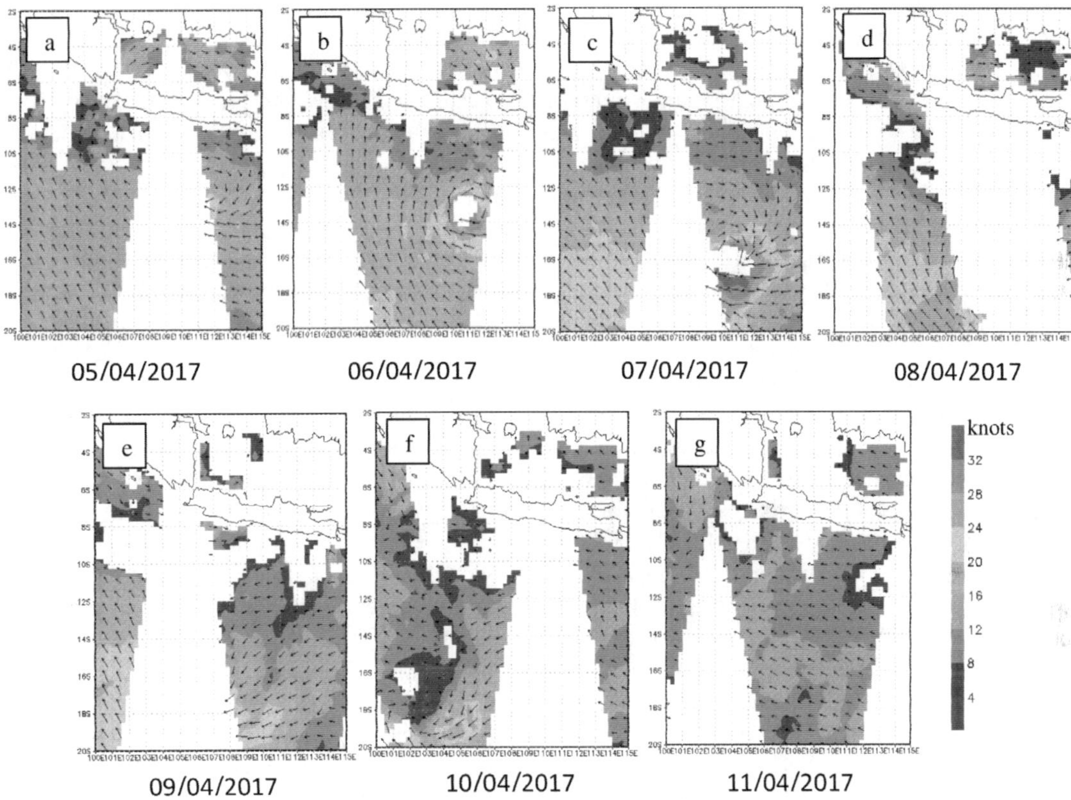

Figure 2. The daily-average data of surface wind from pre-cyclone period, a) 5 April 2017, to the post-cyclone period of tropical storm, g) 11April 2017, in the southern coast of Java island. Wind speed is expressed in knots.

Generally, the daily-average of surface wind direction in April is dominated by easterly wind pattern [24]. However, surface wind direction at the initial phase location of cyclone-formation until the occurrence of Ernie cyclone was dominated by westerly wind patterns. This indicated that the tropical cyclone phenomenon changed the daily-average wind direction in April. As consequences, the changes in the surface wind direction will also impact on the changes of the sea surface current direction [25][26].

The daily-average wind speed on 6 April 2017 in the southern coast of Java (A2 region) was quite high, within the range of 8 - 20 knots. At that time, the location of Ernie storm was quite close to the A2 region, therefore it induced large pressure gradient between A1 and A2. The bottom line is, the higher value of air pressure gradient, the greater the wind speed occured as the impact of cyclone. Tropical cyclone activity continued to develop and grow, reached its peak, and classified into 4th scale tropical cyclone in the night of 7 April 2017. In this condition, the surface wind speed in A2 region decreased slightly, however the value of wind speed in A1 region increased up to more than 32 knots. Analysis on the streamline showed there was a divergence in the southern region of Banten province caused a decrease in wind speed in A2 region.

A day after Ernie cyclone reached its peak intensity, the condition of the surface wind could not be observed, regarding the inability of ASCAT sensors to record the conditions in A2 region due to the dense cloud coverage. But in general, the condition of surface wind decreased in the range of 6-16 knots along with weakening activity of Ernie cyclone until it was started to decay on 10 April 2017. The weakening of surface wind speed continued because the distance of cyclone track to the A2 region was getting further. As time goes by, the activity of Ernie cyclone became weaker and therefore triggered the weakening of air pressure gradients between the two areas.

Meanwhile, the wind direction after the occurence of Ernie cyclone experienced a significant change in the A2 region. On 8 - 9 April 2017, the southern coast of Java island was still dominated by westerly wind pattern as the impact of Ernie cyclone, but on 10 - 11 April 2017 when the cyclone started to decay, the daily-average wind direction changed again into easterly wind pattern. As the wind blows from east to west, the direction of the current will flow to the south by the Ekman transport, so that the sea-surface current would flow away from A2 region and there would be a vacuum of sea water along that region. This condition triggered entrainment of colder water from below which is accompanied by an enhancement of primary production moved to the upper oceanic layer, commonly known as upwelling mechanism. Generally, sea water temperatures formed below tend to be cooler than that on the surface layer, therefore the changes in surface wind patterns will also affect sea surface temperature patterns.

3.2. Response of sea surface temperature

Temporal variation and spatial distribution of sea surface temperature represent the water column conditions before the decayed of tropical cyclone (Fig. 3). One of the factors that triggered the occurrence of tropical cyclone in a certain region is the warmer of Sea Surface Temperature (SST) on that region compare to its surrounding. The formation of tropical cyclones around the equator generally occurs at temperatures above 26.5°C [27][28][29] and in the region with latitude of 35°, cyclone also occurred at the same temperature [30]. Sea surface temperatures in A1 and its surrounding areas prior to the occurrence of tropical cyclones (April, 5 – 6 2017) showed a relatively high value of SST between the range of 30 - 32°C. The high sea surface temperatures led to the formation of a low-pressure area which triggered the occurrence of tropical cyclone Ernie [22].

A1 region showed a significant decrease in sea surface temperature during the occurrence of tropical cyclone events which was 7°C lower than the previous day. The sharp drop in the sea surface temperature on the tropical cyclone trajectory caused by Ekman pumping mechanism induced upwelling and the vertical mixing process. This mechanism caused the cold water injection into the surface layer from a deeper layer continues to spread horizontally outwards. This horizontally spreading is considered as one of the main factors supporting vertical mixing, and, as a result, continues changes in various factors may last for some time [17].

Tropical cyclone creates a whirlpool that rotates clockwise in the southern hemisphere and anti-clockwise in the northern hemisphere. This vortex caused a change in ocean surface currents. In southern hemisphere, the direction of the ocean currents rotated to the left (stayed away from the center of the cyclone) due to the Ekman transport mechanism, resulting an emptiness of sea water at the center of the cyclone. This emptiness replaced by the rising of cold water mass from the deeper layer of the sea. Therefore, sea surface temperature showed a relatively low value between the range of 24 - 27°C during the cyclone period and maximum cooling happened when Ernie cyclone was on its peak intensity. This cold pool slowly decreased after the passage of Ernie cyclone which accompanied by the weakening of the cyclone activity. This clearly indicated that the maximum cooling and hence the cold pool, at the

IOP Conf. Series: Earth and Environmental Science **162** (2018) 012035 doi:10.1088/1755-1315/162/1/012035

left side of cyclone track, was generated by the cyclonic wind-stress curl and the associated upward Ekman pumping.

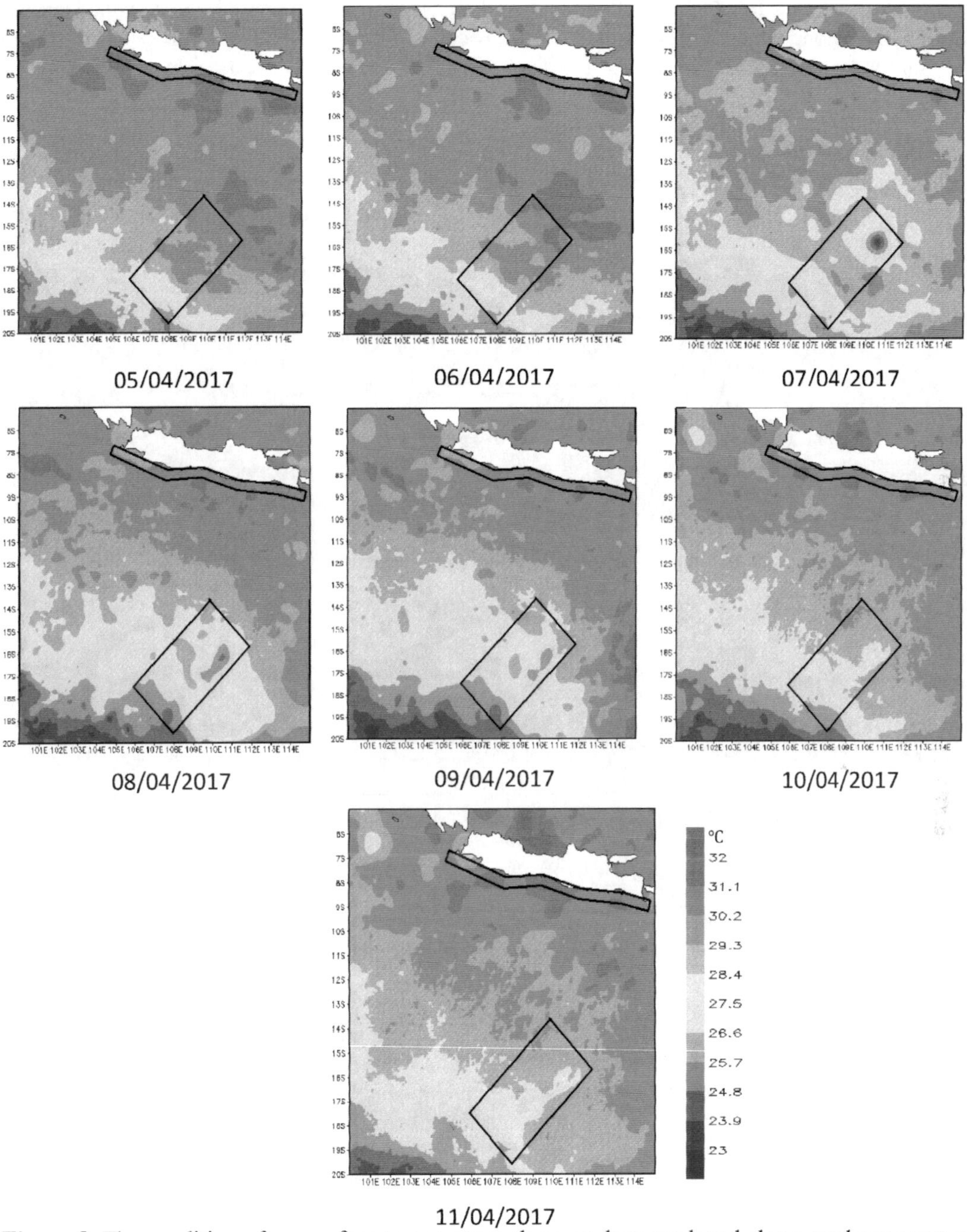

Figure 3. The condition of sea surface temperature along cyclone track and along southern coast area of Java island before the occurrence of tropical cyclone Ernie (5 April 2017) to the decayed phase of Ernie cyclone (11 April 2017). The condition of sea surface temperature expressed in Celsius (°C).

The condition of sea surface temperature in A2 region gave different response compared to A1 region. In general, the spatial distribution of sea surface temperature in A2 region tended to warm within the range of 30°C to 32°C and lasted for quiet long period since the occurrence of cyclone event to the end of the Ernie cyclone formation. The high value of sea surface temperatures in this region indicated a downwelling process. The wind that blows eastward along the southern coast of Java evoked surface currents leading to land which caused the mass of sea water along the coast pushed into its below deeper layers. This condition causes the mass of warm sea water to fill up the empty space in coastal areas [26].

3.3. The variability of chlorophyll-a concentration

Tropical cyclone activity contributes to the enhancement of vertical mixing and upwelling processes along the cyclone track. One of the parameters which closely related to this process is the variability of chlorophyll-a. Tropical cyclones could trigger the formation of eddy and upwelling processes that bring nutrient elements from the deeper layer and inject it into the surface layer [31]. The response of chlorophyll-a concentration on the A1 and A2 region due to the activity of Ernie cyclone can be seen in Fig. 4.

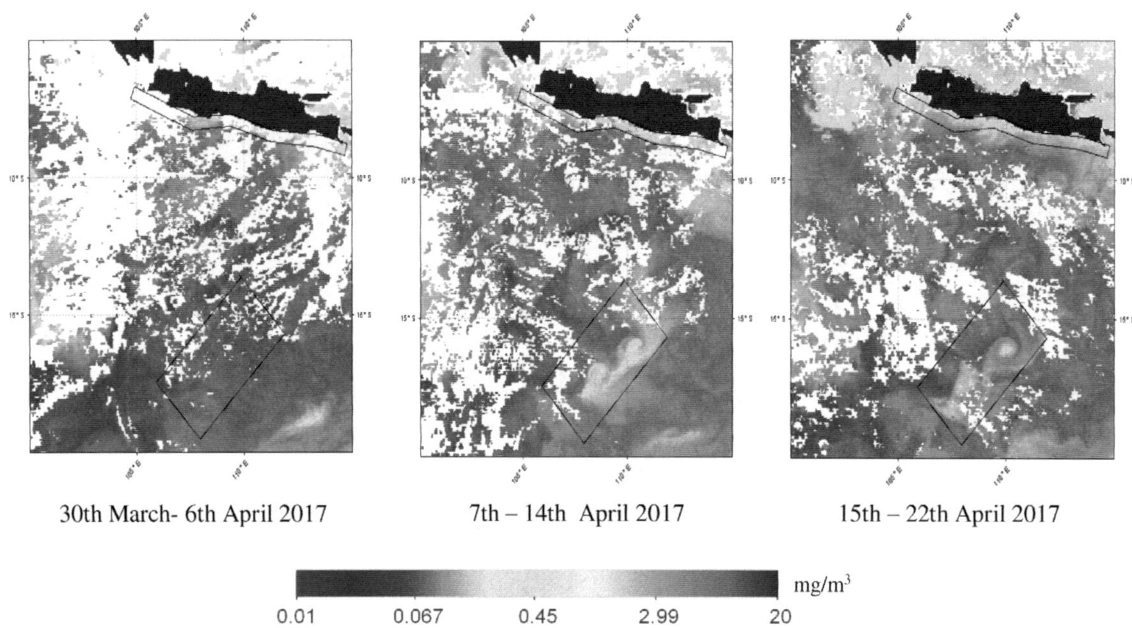

Figure 4. The 8-day composite data of chlorophyll-a variability in A1 and A2 region from 30 March – 22 April 2017.

This study used 8-day composite data of chlorophyll-a concentration. This is due to the incomplete data of daily chlorophyll-a concentration since there was a dense cloud cover in the tropical cyclone region. At the beginning period of the storm formation (30 March – 6 April 2017) based on Unisys Weather (table 1), it was clearly seen that the concentration of chlorophyll-a in A1 region was relatively low within the range of 0.08 mg/m^3 - 0.15 mg/m^3. The remote location of the tropical cyclone area from the coastal region and river estuaries made the contribution of soluble nutrient, usually produced by the river streams, was very low so that the chlorophyll-a concentration in this region was relatively low throughout the years.

During the cyclone period (7 - 14 April 2017), there was a significant increase in chlorophyll-a concentration in A1 region ranged between 0.15 mg/m^3 - 0.30 mg/m^3 and persisted for around 2 weeks throughout its post-cyclone period. The Ekman pumping mechanism allowed the water mass of rich-nutrient from deeper ocean layer to rise into the surface of tropical cyclones. Therefore, there was an increase of chlorophyll-a concentration in A1 region after the occurrence of Ernie cyclone.

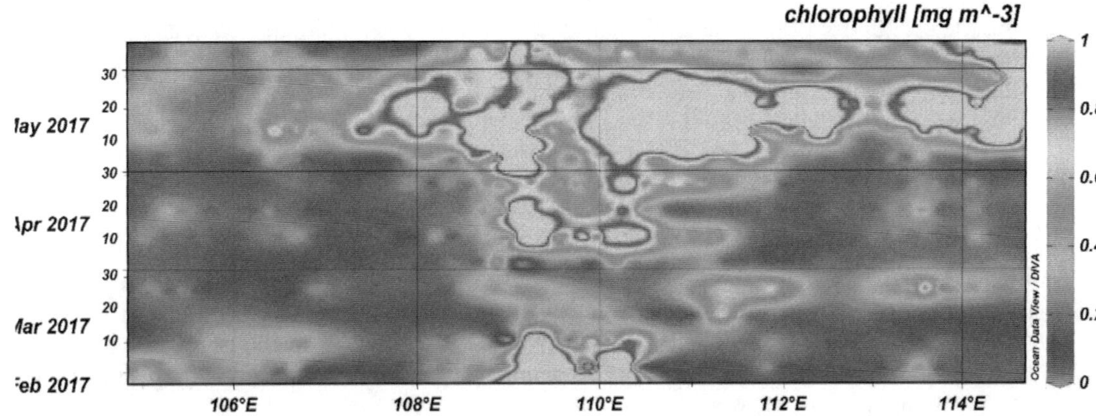

Figure 5. Variations of chlorophyll-a distribution on the southern coast of Java island (A2) before and after the occurrence of Ernie cyclone showed on the hovmoller diagram. The chlorophyll-a concentration expressed in mg/m^3.

Several related studies also showed that there is an increase of chlorophyll-a concentration along the cyclone trajectory. For instance, the concentration of chlorophyll-a was rose fivefold compared to the condition before the tropical cyclone LAM happened [22]. Meanwhile, the results of Ye et al., [32] showed that Nuri tropical cyclone was able to increase chlorophyll-a concentration five times higher than normal levels on the second day and continued nearly a week later.

However, the activity of Ernie cyclone gave different response to the variability of chlorophyll-a concentration in A2 region. The spatial and temporal distribution of chlorophyll-a in A2 region could be seen in the hovmoller diagram (Fig. 5). Before the occurrence of tropical cyclone, the overall chlorophyll-a concentration was quite low within the range of 0.1 mg/m^3 to 0.4 mg/m^3. The low concentration of chlorophyll-a was the impact of the changes in wind direction and ocean current patterns in A2 region which induced downwelling mechanism.

The concentration of chlorophyll-a in A2 region experienced a significant increase after the occurrence of tropical cyclone Ernie. Significant increase on the chlorophyll-a concentration after the occurence of Ernie cyclone due to the direction of sea surface wind which turned back into its normal condition (surface wind blow westwards in April), generated an upwelling process and triggered injection of colder sea water from deeper layer to the surface layer of the ocean [26]. Fig. 5 showed that the response of chlorophyll-a concentration in A2 region was quite slowly than that on the A1 region. It required considerable time (approximately one month after the Ernie cyclone happened) to induce significant increase in chlorophyll-a concentration in A2 region. This is because the wind speed in A2 region was much slower compared to the wind speed around the cyclone area (A1 region), therefore water mass transport mechanism in A2 region took longer times than that in A1 region.

4. Conclusions

The Ernie cyclone gave different effects on the area along the southern coast of Java island (A1) and along the cyclone trajectory (A2) area. The surface wind speed gradually increase along the cyclone track and decrease after the occurrence of cyclone. The increase in wind speed induced an Ekman pumping mechanism that caused the occurence of upwelling in A1 region. This upwelling process was characterized by a decrease in sea surface temperature, accompanied by a significant increase in chlorophyll-a concentration in A1 region. The activity of tropical cyclone Ernie caused westerly wind pattern along the southern coast of Java. The eastwards wind pattern (wind flows from west to east) caused downwelling mechanism in the A2 region. This downwelling process was clearly showed by the high value of sea surface temperature and low chlorophyll-a concentration during the occurrence of tropical cyclone. Meanwhile, the wind direction pattern turned back into its normal, westwards wind pattern, after the occurrence of tropical cyclone Ernie. The westwards wind pattern (wind flows from east to west) led to the occurence of upwelling mechanism which can be identified by the increase of chlorophyll-a concentration in the A2 region after the occurrence of Ernie cyclone. This increase in chlorophyll-a concentration required longer time because the wind speed in A2 region was much more slower than that in A1 region, so that water mass transport mechanism in A2 region took longer time compared to that in A1 region.

Acknowledgments
We acknowledge Joint Typhoon Warning Center (JTWC) for providing basic data on the tropical cyclone events, NASA Jet Propulsion Laboratory (JPL) for providing Multi-scale Ultra-high Resolution Sea Surface Temperature (MUR SST) data, and NOAA for providing daily-average data of sea surface wind and 8-day composite data of chlorophyll-a concentration. Lastly, we thank Dr. *Endarwin*, S.Si, M.Si, Reynhard Syatauw, and Erwin Hayanata S.Sos for technical support and advice.

References
[1] Zhao, H., Tang, D., Wang, Y., 2008. *Comparison of phytoplankton blooms triggered by two typhoons with different intensities and translation speeds in the South China Sea*. Mar. Ecol. Prog. Ser. 365, 57e65.
[2] Shang, S., Li, L., Sun, F., Wu, J., Hu, C., Chen, D., Shang, S., 2008. Changes of temperature and bio-optical properties in the south China sea in response to typhoon lingling, 2001. *Geophys. Res. Lett.* 35 (**10**)
[3] Pan, J. and Sun, Y., 2013. Estimate of ocean mixed layer deepening after a typhoon passage over the south China sea by using satellite data. *J. Phys. Oceanogr.* 43 (**3**), 498-506.
[4] Zhao, H., Shao, J., Han, G., Yang, D., Lv, J., 2015. *Influence of typhoon matsa on phytoplankton chlorophyll-a off East China.* PLoS One 10 (9), e0137863
[5] Black, P.G., 1983. Ocean Temperature Changes Induced by Tropical Cyclones. Ph.D. *dissertation*. Pennsylvania State University, 278.
[6] Gopala Krishna, V.V., Murty, V.S.N., Sarma, M.S.S., Sastry, J.S., 1993. Thermal response of upper layers of Bay of Bengal to forcing of a severe cyclonic storm: a case study. *Indian J. Mar. Sci.* **22**, 8–11.
[7] Chinthalu, G.R., Seetaramayya, P., Ravichandran, M., Mahajan, P.N., 2001. Response of the Bay of Bengal to Gopalpur and Paradip super cyclone during 15–31 October 1999. *Curr. Sci.* 81 (**3**), 283–291.
[8] Babin, S.M., Carton, J.A., Dickey, T.D., Wiggert, J.D., 2004. Satellite evidence of hurricane-induced phytoplankton blooms in an oceanic desert. *J. Geophys. Res.* **109**, C03043.
[9] Chang, Y., Liao, H.-T., Lee, M.-A., Chan, J.-W., Shieh, W.-J., Lee, K.-T., Wang, G.-H., Lan, Y.-C., 2008. Multi-satellite observation on upwelling after the passage of Typhoon Hai-Tang in the southern East China Sea. *Geophys. Res. Lett.* **35**, L03612. http://dx.doi.org/10.1029/2007GL032858.
[10] Gierach, M.M., Subrahmanyam, Bulusu, Prasad, T.G., 2009. Physical and biological responses to Hurricane Katrina (2005) in a 1/25_ nested Gulf of Mexico HYCOM. *J. Mar. Syst.* **78**, 168–179.
[11] Lee, Chung I, Park, Mi-Ok, 2010. Time series changes in sea-surface temperature, chlorophyll a, nutrients, and sea-wind in the East/Japan Sea on the left- and right-hand sides of Typhoon Shanshan's track. *Ocean Sci. J.* 45 (**4**), 253–265.
[12] Stramma, L., Cornillon, P., 1986. Satellite observation of sea surface cooling by hurricanes. *J. Geophys. Res.* ***91***, 5031–5035.

[13] Suetsugu, M., Kawamura, H., Nishihama, S., 2000. *Sea Surface Cooling Caused by Typhoons in the Western North Pacific Ocean*. In: PORSEC proceedings, vol. 1. Goa, India, pp. 258–262.
[14] Lin, I.I., Wu, C.C., 2007. Typhoon-ocean interactions inferred by multisensor observations. *Recent Prog. Atmos. Sci.* 358.
[15] Babin, S. M., J. A. Carton, T. D. Dickey, and J. D. Wiggert. 2004. Satellite evidence of hurricane-induced phytoplankton blooms in an oceanic desert, *J. Geophys. Res.*, **109**, C03043, doi:10.1029/2003JC001938.
[16] Oceans 109 (C3). Hung, C.C., Gong, G.C., Chou, W.C., Chuang, C.C., Lee, M.A., Chang, Y., Chuang, W.C., 2010. The effect of typhoon on particulate organic carbon flux in the southern East China Sea. *Biogeosciences J.* 7 (**10**), 3007e3018.
[17] Hong, C.H., Sohn, I.S., 2004. Sea surface in the East Sea with the passage of typhoons. *J. Korean Fish Soc*. 37 (**2**), 137–147.
[18] Kim, S.W., Yamad, K., Jang, L.H., Hong, C.H., Go, W.J., Suh, Y.S., Lee, C., Lee, G.H., 2007. Short-term variation of sea surface temperature caused by Typhoon Nabi in the Eastern Sea of Korean peninsula derived from satellite data. *J. Korean Fish Soc*. **40** (2), 102–107.
[19] D. Chen, L. He, F. Liu, and K. Yin, Effects of typhoon events on chlorophyll and carbon fixation in different regions of the East China Sea. Estuar. *Coast. Shelf Sci.*, vol. **194**, pp. 229–239, 2017.
[20] K. Muni Krishna and G. Song, Physical and biological changes in the south Bay of Bengal due to the Baaz cyclone. *Adv. Sp. Res.,* vol. **56**, no. 8, pp. 1658–1666, 2015.
[21] Enriquez, A.G., Friehe, C.A., 1995. Effects of wind stress and wind stress curl variability on coastal upwelling. *J. Phys. Oceanogr.* 25 (**7**), 1651–1671.
[22] Susilo, Eko., Hadianti, Sri., 2016. *Dapatkah Siklon Tropis Picu Peningkatan Konsentrasi Klorofil-A? (Studi Kasus : Siklon Tropis LAM).* Seminar Nasional Penginderaan Jauh, 972 – 980.
[23] Rayhan Nuris, J. L. 2015. Chlorophyll-a concentrations estimation from AQUAMODIS and VIIRS-NPP satellite sensors in south Java sea waters. *Int. J. of Remote Sensing and Earth Sciences Vol. **12** No. 1*, 63 - 70.
[24] Fadlan, Ahmad. Pengaruh Fenomena Monsun, El Nino Southern Oscillation (ENSO) dan Indian Ocean Dipole (IOD) terhadap Variabilitas Tinggi Muka Laut di Perairan Utara dan Selatan Pulau Jawa. M.S. *thesis*, Marine Science Dept., Diponegoro Univ., Semarang, Indonesia, 2017.
[25] Dietze, H., Löptien, U. 2016. Effect of surface current-wind interaction in eddy-rich general ocean circulation simulation of the Baltic sea. *Ocean Sci. J.* **12**: 977-986.
[26] Yoga, Bima, Raden. 2014. Dinamika *upwelling* dan *downwelling* berdasarkan variabilitas suhu permukaan laut dan klorofil-a di perairan selatan Jawa. *J. Oseanografi* (**3**):57-66
[27] Laing, A., and J. L. Evans, 2011: *Introduction to Tropical Meteorology. 2nd ed. University Corporation for Atmospheric Research.* [Available online at www.goes-r.gov/users/comet/tropical/textbook_ 2nd_edition/index.htm.]
[28] Ackerman, S. A., and J. A. Knox, 2015: *Meteorology: Understanding the Atmosphere. 4th ed. Jones and Bartlett Learning*, 575 pp.
[29] Cowan, R. M., Davies, E. L., Fairman Jr, J. G., Galarneau Jr, T. J., Schultz, D. M. 2015. *Revisting the 26.5°C Sea Surface Temperature Threshold for Tropical Cyclone Development*. American Meteorolgy Society. 1929-1943.
[30] Dare, R., dan Mcbride, J.L., (2011). The threshold sea surface temperature condition for tropical cyclogenesis. *Journal of Climate,* 24 (**17**):4570-4576.
[31] Wang, G., Ling, Z., dan Wang, C. 2009. Influence of tropical cyclones on seasonal ocean circulation in the South China Sea. *Journal of Geophysical Research*, **114**:1-9
[32] Ye, H.J., Sui, Y., Tang, D.L., dan Afanasyev, Y.D. 2013. A subsurface chlorophyll a bloom induced by typhoon in the South China Sea. *Journal of Marine Systems*, **128**:138145.

IOP Conf. Series: Earth and Environmental Science **162** (2018) 012036 doi:10.1088/1755-1315/162/1/012036

The readiness of small island community to the development of potential renewable energy: case study of Bulang Kebam Island, Riau Island province, Indonesia

P A Soejarwo[1], W P Fitriyanny[2], A R Hakim[3] and A Heriati[4]

[1] Research Center for Social Economic Marine and Fisheries, Gedung Balitbang KP I Lt. 4 Jl Pasir Putih No 1 Ancol Timur, Jakarta Utara, Indonesia
[2] Center for Integrated Research in Environmental Geoscience, Bandung Ocean Technology Research and Management, Jl. Bukit Dago Utara II No 11, Bandung, 40135, Indonesia
[3] Center for Ocean Modelling Research and Operational, Bandung Ocean Technology Research and Management, Jl. Bukit Dago Utara II No 11, Bandung, 40135, Indonesia
[4] Research Center of Marine, Gedung Balitbang KP II Lt. 4 ,Jl Pasir Putih No 1 Ancol Timur, Jakarta Utara, Indonesia

E-mail: ari_permana008@yahoo.com

Abstract. Bulang Kebam Island is one of the small islands located in the west part of Batam City, Riau Islands Province. The region has a deficiency of energy because both central and local governments have not been able to build electrical grid connections for the region, hence the electricity demand can only be met for six hours per day using a diesel generator. In order to provide a more continuous supply, an alternative source of energy by utilizing the potential of tidal energy resources around Bulang Kebam Island is needed. In support of an effective and efficient renewable energy project development, a study of the community readiness level in Bulang Kebam Island is conducted. *Rapid Appraisal for Community Readiness* (RAPCR) method is used for this purpose, which consists of social, economic, environment, technology, and institutional factors. The results showed that the community readiness index were categorized as sufficiently ready with values of 61.84, 64.15 and 72.55 for the environment, technology, and institutional factors respectively. While from the economic and social factors, community readiness index were categorized as ready with values of 82.94 and 84.93. Furthermore, the dominant attributes from each factor were analyzed and ranked by using Leverage analysis. The most dominant attribute from economic, environment, technology, social, and institutional factors respectively are potential for renewable energy funding from the community, level of environmental difficulty on the implementation of renewable energy, area availability for renewable energy technology, a clear roles in the implementation of renewable energy, and commitment of local government to renewable energy. The results of this study is expected to be a reference and recommendations for the development of appropriate target and advantage of renewable energy project development in Bulang Kebam Island.

IOP Conf. Series: Earth and Environmental Science **162** (2018) 012036 doi:10.1088/1755-1315/162/1/012036

1. Introduction

Indonesia consists of many small islands and most of the islands had been recorded in the United Nation and about 2.509 new verified islands had been reported to the United Nation in the 11th United Nations Conference on the Standardization of Geographical Names (UNSCSGN) on August 2017. Remote location, distance form main islands, and accessibility are main obstacles for the development of small islands. Therefore the small islands are left behind compare with other.

Figure 1. Research Locations is located at Bulang Kebam Island, Riau Island Province Indonesia.

On the other hands, the potential of marine resources in small islands is quite large, such as the potential of fisheries, minerals, sea currents and tides. Abundant marine resources in Indonesia are still not utilized optimally, especially in the potential of currents and tides that can be used as a source of renewable energy. The utilization of the current and tidal currents energy can be converted as the fulfillment of electricity needs in the small island in Indonesia. However, at this time the fulfillment of electric energy sources contained in small islands is still relying from fossil energy sources by using oil fuel. Fossil energy sources certainly have limited amounts. In term of that someday in the future scarcity of the fossil energy will occur and even running out. In Bulang Kebam Island the fossil energy only support six hours per day and that was not enough to fullfil the need of electrical energy in this island. Therefore, it needs a supply of renewable and environmentally friendly power source by considering that the source of the fossil energy is running low and also challenge to distribute fossil energy in small island [1].

Implementation of renewable energy in Indonesia has its own challenge where every area has different characteristic of resource potential, weather and climate variability as well as community characteristic. These complexities well recognized so that the complete image of area readiness in the implementation of renewable energy can be achieved.

The successful of renewable energy development in Bulang Kebam Island is influenced by environment, technology, economic, social and institutional factors. Community norms and values must be considered as important factors affecting the support of community-based development [2]. This research aim to assess the community readiness based on the factors above in the development of renewable energy in Bulang Kebam Island

2. Methods

2.1. Location and time research

This research was conducted on February until March 2017 in Bulang Kebam island which is administratively located in Bulang District, Riau Island Province (Figure 1). The geographical location is at 00° 51' - 1° 06' North Latitude and 103°48' - 104°06'East Longitude. Primary data were attained from questionnaire of selected respondents in Bulang Kebam Island. Secondary data were obtained from the Central Bureau of Statistics (BPS), Agency of Marine and Fisheries or Ministry of Maritime Affairs and Fisheries (KKP), Bulang District, Riau Island Province and other publications.

2.2 RAPCR method

RAPCR (Rapid Appraisal for Community Readiness) is a rapid appraisal technique that conducted to evaluate the community readiness in order to prepare communites to be more successful in their effort to address a variety of issues. It was adopted from Pitcher and Preikshot, 2001. This allows us to describe the developmental level of a community to spesific issue [3]. RAPCR steps can be outlined as follows:

(i) **Identification and scoring of each attribute**. Identification conducted with review and determine the attribute in 5 factors: environment, technology, economic, social and institutional. Each factors has 5 attributes so that the total of attribute is 25. Determination of attribute for each factor is acquired through the study literature that is relevant with this research. The field survey and interview with the communities, expert and related institution were conducted to receive more accurate information. At the final step is scoring of each attribute which reffers to technique from Pittcher T.J and Preikshot, 2001 and Susilo, 2003 ranging from the lowest score 0 for the most unprofitable condition and the highest score 2 for the most profitable condition. Scoring is also determined by primary data that resulted from questionnaire and supported with literature and competent experts. Attribute and scoring of 5 factors can be seen in the Table 1.

(ii) **Ordination process**. Process to visualize the position of seaweed farming sustainability point in two dimentional based on the scoring of each attribute which is conducted by using Multidimensional Scaling method [4] and [5]. The distance calculation between attribute points, horizontal and vertical reference points, and anchor reference points are conducted by using euclidian distance formula below :

$$d = \sqrt{|\, x1 - x2\ ^2| + \ |y1 - y2^2| + \cdots} \tag{1}$$

Where d is distance between attribute, x is attribute of factor 1, y is attribute of factor 2,...

(iii) **Rotation process and arrangement of readiness scale in RAPCR**. This process is to align the quality (good or bad) position of horizontal reference points with the horizontal axis x. Reference point up is located above the horizontal axis x, whereas reference point down is located below the horizontal axis x. Position of environment, technology, economic, social and institutional points which is located in the scale 0 – 100 describe the community readiness status of potential renewable energy development in Bulang Kebam Island. Category of community readiness status adopted from [5] and [6] shown in Table 2. Meanwhile to know the correlation of continuity index within each factors, the kite diagram was built up, therefore the integrated review from all related factors can be conducted [4] and [7].

(iv) **Leverage analysis**. This process is conducted to find the most dominant to the community readiness status of potential renewable energy development [4] and [7]. The leverage analysis is the calculation of Root Means Square (RMS) from each attributes in five factors by using formula below

$$Total\ RMS = \sqrt{\frac{\sum_{j=1}^{n}((Vf(i,1) - Vf(cent,1))^2}{n}} \tag{2}$$

Where Vf (i,1) is position of attribute point matrix V in line i column 1 (x axis), Vf (cent, 1) is center value of position of attribute point matrix V in line i column 1 (x axis) and n is number of attributes

IOP Conf. Series: Earth and Environmental Science **162** (2018) 012036 doi:10.1088/1755-1315/162/1/012036

Table 1. Factor and Attribute of Renewable Energy Development

Factor	Attributes	Score (min 0 – max 2)
Environment	1. Potential resource of renewable energy	2 (available)
	2. Impact of renewable energy utilization to environment condition	1 (netral)
	3. Level of environment difficulties to the renewable energy implementation	2 (difficult)
	4. Sea water suitability to the renewable energy	1 (sufficient)
	5. Affordability of renewable energy location	1 (sufficient)
Technology	1. Availability of power generator technology	2 (available)
	2. Availability of electrical distribution technology	2 (available)
	3. Technology suitability for electrification supply and demand	1 (sufficient)
	4. Area availability for technology of renewable energy	1 (sufficient)
	5. Infrastructuire availability to support technology of renewable energy	0 (not available)
Economic	1. Potential funding of renewable energy from government	1 (few)
	2. Potential funding of renewable energy from private sectors	2 (available)
	3. Potential funding of renewable energy from community	2 (available)
	4. Absorption of employee	2 (many)
	5. Existence of formal finance institution	0(not available)
Social	1. Influence the exsitence of electricity resource from renewable energy to the socio cultural values	2 (high)
	2. Community awareness to the resources and benefits of renewable energy	2 (high)
	3. Consumer users potential of renewable energy	2 (high)
	4. Clear roles distribution in the renewable energy implementation	1 (sufficient)
	5. Impact of renewable energy to the community welfare	2 (high)
Institutional	1. Maintenance institution existence of renewable energy technology	2 (available)
	2. Local institution existence for electricity management from renewable energy	2 (available)
	3. Commitment of local government to renewable energy	1 (sufficient)
	4. Community group involvement to manage renewable energy technology	2 (available)
	5. Monitoring of related institution to the power plant company	1 (sufficient)

Table 2. Status Category of Community Readiness

No	Index	Category
1	0 - 25	Not Ready
2	26 - 50	Less Ready
3	51 - 75	Sufficiently Ready
4	75 - 100	Ready

3. Result and discussion

3.1. Community Readiness Index for Renewable Energy Development

Community readiness index from environment, technology, economic, social and institutional factor can be seen from Figure 2 below.

3.1.1. Environment

Index of community readiness based on RAPCR analysis in environment factor is 68.14 that categorized in sufficiently ready category. This show that community readiness level in environment factor is sufficiently ready to support renewable energy development. This condition should have more attention since the index is closer to 50 which can be easily shift into less ready category. Environment factor is the important factor for renewable energy development because the natural environment especially from the ocean can provide potential natural resources that can be conversed as new source of renewable energy. For instance, the location of Bulang Kebam island is between small islands in Batam area, this condition leads to the tidal current energy potency from its surrounding straits.

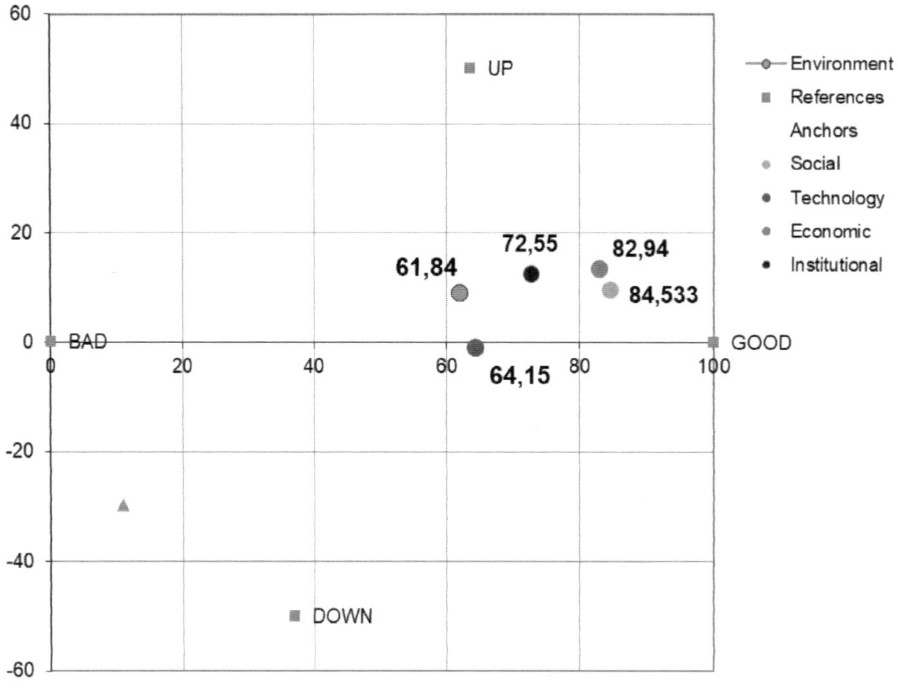

Figure 2. Community Readiness Index for Renewable Energy Development

3.1.2. Technology

Index of community readiness based on RAPCR analysis in technology factor is 64.15 that categorized in sufficiently ready category. Proper technology can support sustainability of renewable energy development but on the other hand, some of inproper technology might be the hardest obstacles to find, since the price of renewable energy cannot compete with the price of subsidized fossil energy. The circumstance conducts the renewable energy development running slow. In case of that, the selection of renewable energy technology must be done properly.

3.1.3. Institutional

Index of community readiness based on RAPCR analysis in institutional factor is 72.55 that categorized in sufficiently ready category. Readiness criteria in this factors focus on the integration of

IOP Conf. Series: Earth and Environmental Science **162** (2018) 012036 doi:10.1088/1755-1315/162/1/012036

involvement, monitoring and existence of institutional from community, government and also related stakeholders in renewable energy development.

3.1.4. Economic
Index of community readiness based on RAPCR analysis in economic factor is 82.94 that categorized in ready category. Funding for the implementation, operational and maintenance of renewable energy should be integrated properly in order to achieve the sustainability of renewable energy development. The contribution should come from government, private and community institutional.

3.1.5. Social
Index of community readiness based on RAPCR analysis in social factor is 84.53 that categorized in ready category. In social factor, community in Bulang Kebam island had already expected another source of energy form several years ago. Lack of electricity supply became a main problem in this small island and the community really want the solution as soon as possible. Social aspect should integrate the impact of renewable energy, community assessment, clear distribution from roles of renewable energy development.

3.2. Leverage Analysis
Leverage analysis was conducted by calculating root mean square of each attribute from environment, technology, economic, social and institutional factors by using equation (2). The most dominant attribute for each factors can be seen from Figure 3 to Figure 7 below.

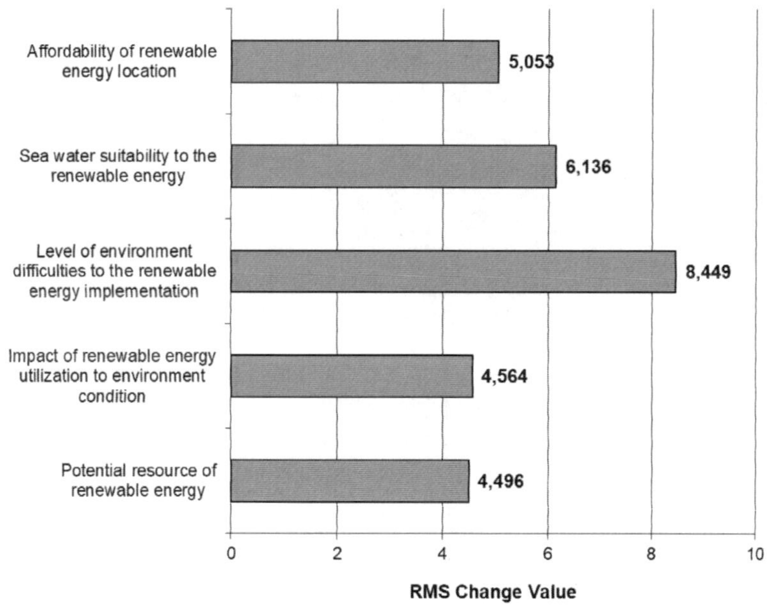

Figure 3. Leverage analysis of environment factor

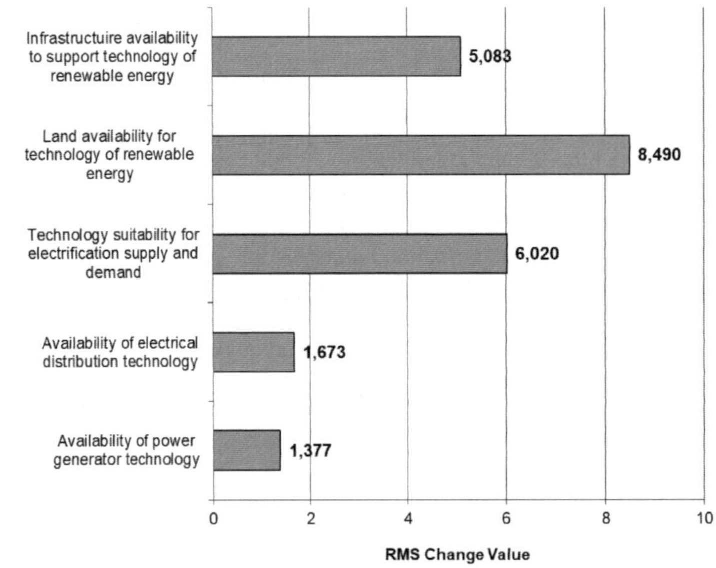

Figure 4. Leverage analysis of technology factor

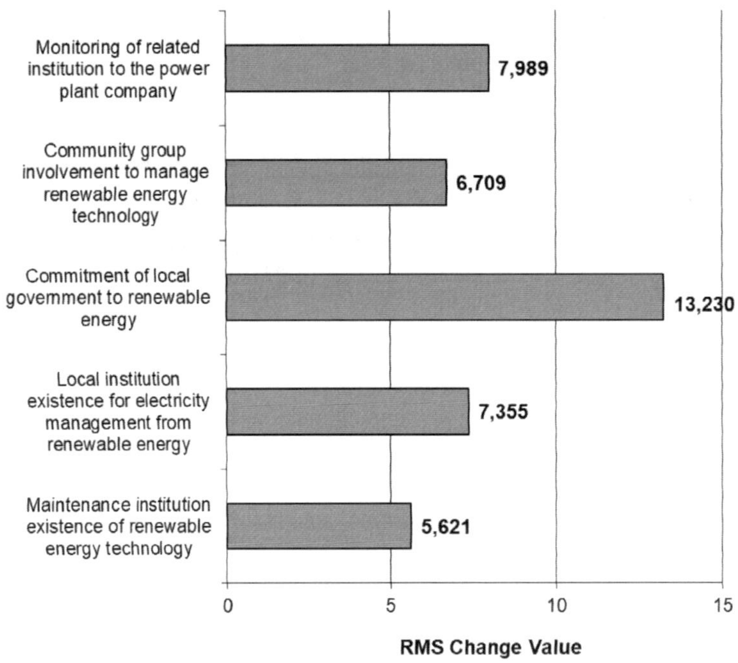

Figure 5. Leverage analysis of institutional factor

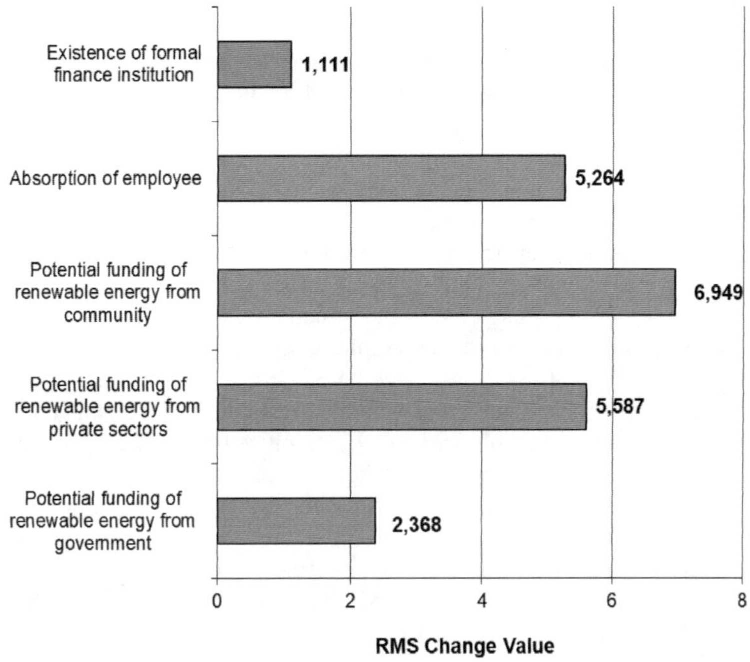

Figure 6. Leverage analysis of economic factor

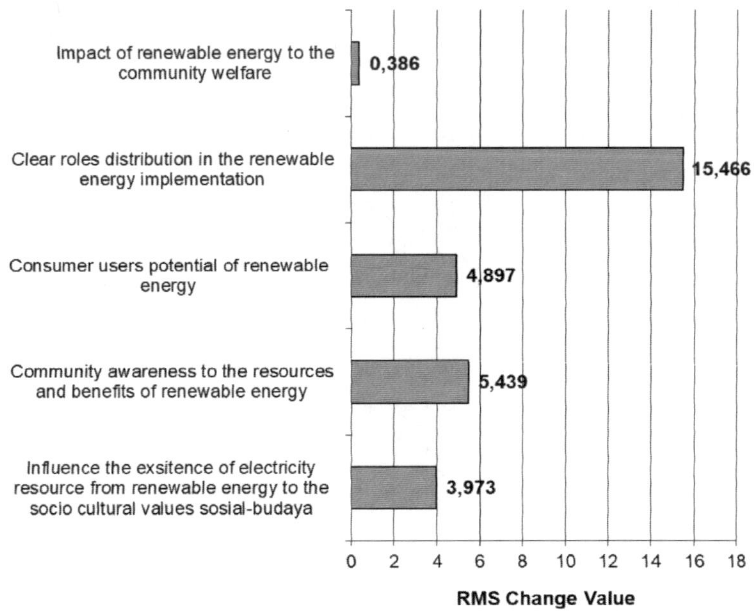

Figure 7. Leverage analysis of social factor

IOP Conf. Series: Earth and Environmental Science **162** (2018) 012036 doi:10.1088/1755-1315/162/1/012036

The most dominant atrribute in environment factor analysis based on RMS changed value is the level of environment difficulties to the renewable energy implementation within value of 8.44 as the result of total RMS calculation by using equation (2). Difficulties level from the location of renewable energy implemetation must be noticed properly since it can affect affect the succesful of renewable energy development in Bulang Kebam island. Including the distance, characteristic of location, geomorphology, oceanographic and weather condition, and road condition.

From technology factor analysis, the most dominant attribute as the result from calculation of equation (2) is area availability for technology of renewable energy within value of 8.49. In some cases, the equipment of renewable energy technology need some spaces to be installed. This area of installation should be well prepared to support the equipment of renewable energy technology.

Furthermore, the most dominant attribute in institutional factor analysis is commitment of local government to renewable energy within value of 13.23 as the result of calculation by using equation (2). Local government plays an important role to the sustainability of renewable energy development in Pulau Panjang. Local government should taken into account the annual budget in its local budget planning to support the development.

Based on economic factor analysis, the most dominant attribute is potential funding of renewable energy from community within value of 6.94 as the result of calculation by using equation (2). Community participation in renewable energy development in Bulang Kebam island should also be supported by the funding from community itself. Nevertheless the funding implementation must be integrated properly with funding from government and also private sectors. So that the roles from each stakeholders can be clearly implemented. In this case, community in Pulau Bulang Kebang are willing to pay cost for energy availability. This case was similar with Madura Island where community in Madura had a high williness to pay for electricity that reach 486.38 IDR per kwH. This rate was higher than State Electricity Company (PLN) rate [8].

From social factor analysis, the most dominant attribute is clear roles distribution in the renewable energy implementation within value of 15.46. Each stakeholders need to now and understand their roles in renewable energy development, so that the development can run smoothly and integrated each other.

3.3. Kite Diagram

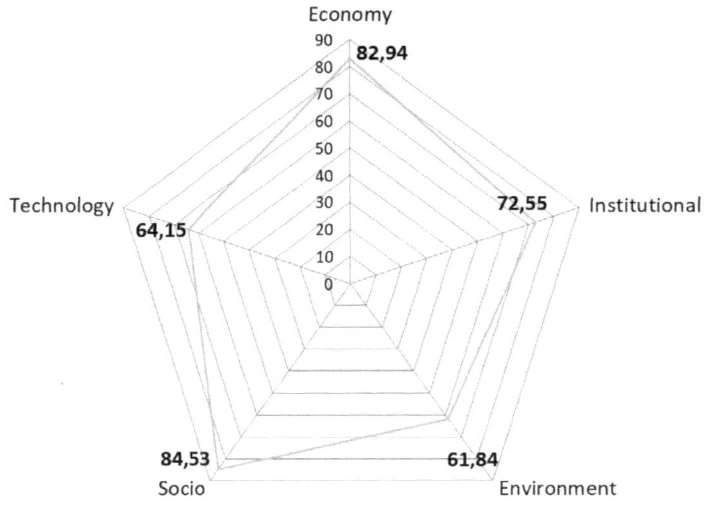

Figure 8. Kite Diagram of Community Readiness in Bulang Kebam Island

From the kite diagram above, it can be seen that community readiness from social factor has the highest value that describe Bulang Kebam community is very ready to support renewable energy development in view of the fact that electricity is the most important things that must be fulfilled immediately and it will bring impact to the social activity sectors in Bulang Kebam island. Economic factor as the second high value (82.94) from kite diagram emphasizes that most of community in Bulang Kebam island is willing to pay more for the electricity availability. They really need the availability of continuous electrical energy expected from renewable energy.

4. Conclusion

Result of community readiness in Bulang Kebam Island to the development of renewable energy by using RAPCR method shows that factor of environment, technology and institutional are categorized in sufficiently ready category within value RAPCR of 68.14, 64.15 and 72.55. Other factors such as economic and social are categorized in ready category with value RAPCR of 82.94 and 84.53.

Findings on leverage analysis show that the most dominant attribute in environment factor is level of environment difficulties to the renewable energy implementation within value of 8.44. From technology factor is area availability for technology of renewable energy within value of 8.49. From institutional factor is commitment of local government to renewable energy within value of 13.23. While from economic factor is potential funding of renewable energy from community within value of 6.94. And social factor is clear roles distribution in the renewable energy implementation with value of 15.46.

Kite diagram shows social factor has the highest value which emphasizes that community in Bulang Kebam Island is very ready to support renewable energy development. Development since electricity is the most important things that must be fulfilled immediately and it will bring impact to the social activity sectors in Bulang Kebam Island. From economic factor, most of community in Bulang Kebam Island is willing to pay cost of electricity availability. They are really looking forward to the availability of continuous electrical energy that is expected from renewable energy.

Acknowledgement

This research is funded by Botram Science Foundation, award number 2017001-A5.

References

[1] Veldhuis A J, Reinders A H M E 2013 *Reviewing the potential and cost-effectivess of grid-connected solar PV in Indonesia on a provincial level*. J. *Renewable and Sustainable Energy Reviews. Appl.* **27**, 315–324

[2] Donnermeyer J F, Oetting E R, Plested B A, Edwards R W, Jumper-Thurman P & Littlehunder L. 1997. *Community readiness and prevention programs. J. Community Development. Appl.* **28(1)**, 65-83

[3] Thurman P J, Plested B A, Edwards R W, Foley R, Burnside M. 2003. *Community readiness: the journey to community healing. J. Psychoactive Drugs. Appl.* **35 (1)**, 27-31

[4] Pitcher T J, Preikshot D. 2001. *RAPFISH : a rapid appraisal technique to evaluate the sustainability status of fisheries. J. Fisheries Research. Appl.* **49**, 255-270

[5] Susilo S B. 2003. *Keberlanjutan pembangunan pulau-pulau kecil: Studi kasus Kelurahan Pulau Panggang dan Pulau Pari, Kepulauan Seribu, DKI Jakarta*. Disertasi. Program Pascasarjana IPB, Bogor.

[6] Marzuki M. 2014. *Desain pengelolaan budidaya laut berkelanjutan di teluk saleh kabupaten Sumbawa*. Disertasi, Program Pascasarjana IPB.Bogor

[7] Kavanagh P, Pitcher T J. 2004. *Implementing microsoft excel software for RAPFISH: a technique for the rapid appraisal of fisheries status. Fisheries Center Research Reports.* **12(2)**

[8] Firdaus A M, Kusumastanto T, Nurjaya I W. 2015. *Analisis ekonomi dan kebijakan pengembangan energi arus laut di Selat Madura, Provinsi Jawa Timur*. Warta KIML Pusat Penelitian Perkembangan Iptek LIPI. **13(1)**

Study of ocean thermal energy resources in Para'baya, West Sulawesi

Y B Wicaksono[1,2] and M R Putri[1]

[1] Department of Oceanography, Faculty of Earth Science and Technology, Bandung Institute of Technology, Jl. Ganesha 10 Bandung, 40132, Indonesia.

[2] Energy Engineering, School of Engineering, University College Cork, College Road, Cork, T12 K8AF, Ireland

Email: y.wicaksono@umail.ucc.ie

Abstract: Ocean thermal is one of the renewable energy resources. In 1981, a design of Ocean Thermal Energy Conversion (OTEC) pilot power plant was proven could produce electricity. The performance of OTEC depends on the possible temperature differences of the warm and cold seawater, at least 20°C. In the coastal of Para'baya, this requirement can be satisfied since the distance to reach the 1000 m depth is less than 10,000 m from the shore. The result of ocean thermal power calculation in Makassar Strait shows that Para'baya could produce higher power than any other place in west coast of Sulawesi, with an average power output of 120.35 kW. This ocean thermal study, in the coast of Para'baya, used sea surface temperatures (SST) data from the result of Long-term Indonesian Throughflow Model Simulation (LITHMOS) over 24 years (1982 – 2006), and sea temperature data at 1000 m depth from the World Ocean Atlas (WOA) 2009. The result shows that ocean thermal energy distributions in the Makassar Strait were affected by the combination of gust, wind direction, and sun position which varies in each season. Maximum ocean thermal power is reached during the first transitional season (March, April, May) with output power of 128 kW, and the minimum power is achieved during the dry season (June, July, August) with an output power of 114 kW.

1. Introduction

The ocean is a vast resource of energy. Ocean thermal is one of energy source that can be harvested from the ocean. This energy comes from the absorbed solar energy by the ocean. Nowadays, a method to utilize thermal energy from the ocean to produce electricity has been invented, namely Ocean Thermal Energy Conversion (OTEC). This system uses warm sea surface water and cold deep sea water with minimum 20°C of temperature different [1–3]. In the tropical area, the 20°C temperature differences mostly can be acquired from sea surface water and 1000 m depth deep water [1].

Para'baya is a coastal region in West Sulawesi Province, Indonesia, located in coordinate 3.16°S; 118.82°E (Figure 1.a). It is one of the potential areas for ocean thermal energy exploration. It is located on

IOP Conf. Series: Earth and Environmental Science **162** (2018) 012037 doi:10.1088/1755-1315/162/1/012037

the equator where the sea surface temperature (SST) is always higher than 27.5°C. Another reason is the steepness of the bathymetry. A thousand meters depth can be reached within 9,300 m from the shoreline (Figure 1.b).

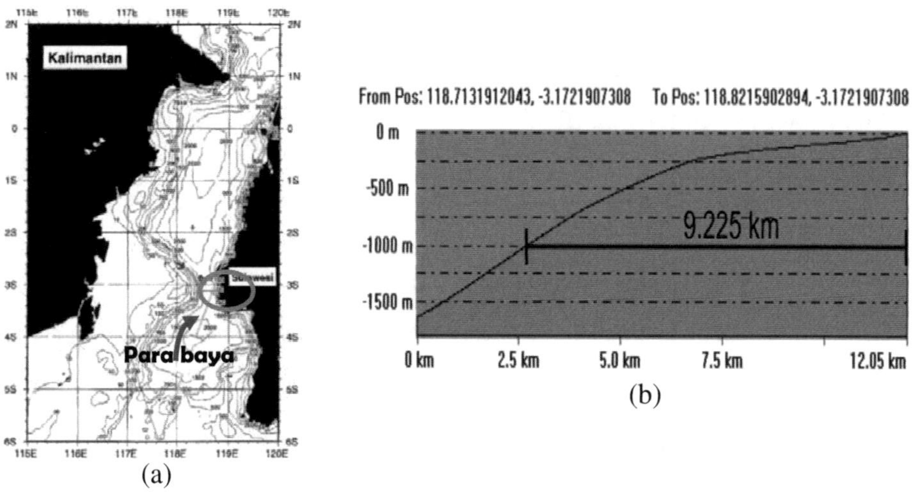

Figure 1. (a) Bathymetry Map of Para'baya [4], **(b)** Section View of Para'baya Coastal Area

This study calculates the ocean thermal energy potential of Para'baya, by simulating an OTEC design and calculate the heat exchange within the system. The simulation of the OTEC design use the SST output of Long-term Indonesia Throughflow Model Simulation (LITHMOS) from 1982 to 2006 [5], and World Ocean Atlas 2009 [6] water temperature of 1000 m depth as the input variable. The purpose of this study is to analyze the monthly variation of ocean thermal energy in Para'baya coastal area.

2. Literature Review

Since the early development of OTEC system, several variations of method had been proposed. The most common varieties are classified based on its thermodynamic cycle; closed cycle, open cycle, and hybrid cycle [1].

In 1981, an OTEC pilot power plant was built in the Republic of Nauru. This power plant applied the OTEC closed cycle system [7]. This OTEC system works by evaporating the working fluid into the steam and let it flow through the turbine to move the turbine. As the turbine running, it generates electricity. After the working fluid pass through the turbine, it is condensed back to the liquid state in the condenser and pumped back to the evaporator for the next cycle. The heat source to evaporate working fluid is extracted from the warm sea surface, and the heat is dumped into the cold deep water to condense the working fluid [1]. Table.1 shows the specification of the Nauru's OTEC pilot power plant. The average sea surface temperature in Nauru is around 29.8°C, while the temperature in 600-meter depth is 8.1°C. The pilot power plant project in Nauru, on average could produce 100.5 kW electricity with a maximum power output of 120 kW electricity [7].

Table 1. Specification of OTEC Pilot Power Plant, Republic of Nauru [1]

	Evaporator	Kondensor
Shell ID (Inner Diameter) (m)	1.9	1.5
Length of Shell (m)	8	9.1
Number of Tube	870 (*enhanced*)	914 (*grooved*)
Tube OD (Outer Diameter) (mm)	25.4	25.4
Thickness of Tube (mm)	0.7	0.6
Length of Tube (m)	5.35	6
Seawater Flow Rate (kg/s)	402.8	391.7
Seawater Flow Velocity (m/s)	2	1.97
Pressure drop (m)	2.9	3.71
Freon Flow Rate (kg/s)	20.6	20.6
Effective Area Heat exchanger (m^2)	371.4	437.6
Overall U (Watt/m^2K)	3000	2560

Note: Rankine Closed Cycle; Working Fluid: Freon R-22;
Heat Exchanger Type: Shell & Tube HXs; Tube Material: Titanium;
Horizontal Evaporator; Vertical Condenser

Figure 1 shows the location of Parabaya, and the section view of the steep bathymetry of Para'baya's coast. It makes this area suitable for OTEC system application, where it needs the least energy to pump up the cold deep sea water compared to any other location on the west coast of Sulawesi. The proposed coordinate is also close to residential area. Therefore, the generated electricity can be used to supply the electricity demand load of this region.

3. Power Calculation of OTEC Closed Cycle

OTEC works by utilizing warm sea surface water and cold deep layer water. Therefore, Sea surface temperature and deep sea water temperature data are needed to calculate the produced energy. The type of OTEC design also determines the process within the system, hence influences the system power generation. LITHMOS temperature output at the surface is used as the sea surface temperature in the data, and WOA 2009 temperature data of Makassar Strait at 1000 m depth is used as the deep-water temperature data in the calculation. The power generation system is OTEC closed cycle as specified in Table 1. The schematic of the OTEC closed cycle is illustrated in Figure 2.

The power calculated in this study is only gross power, without taking account of power required for the pump and any other parasitic power demand. The power produced by the turbine is assumed to be the heat rate difference between evaporator and condenser. The power generation by the OTEC system is formulated in equation (1) [1,8]:

$$\dot{W}_t = \dot{Q}_e - \dot{Q}_k \qquad (1)$$

Where $\dot{W}_t$ is the power produced by the Turbine, $\dot{Q}_e$ is the heat extracted from the sea surface water in evaporator, and $\dot{Q}_k$ is the heat absorbed by the cold deep water in the condenser.

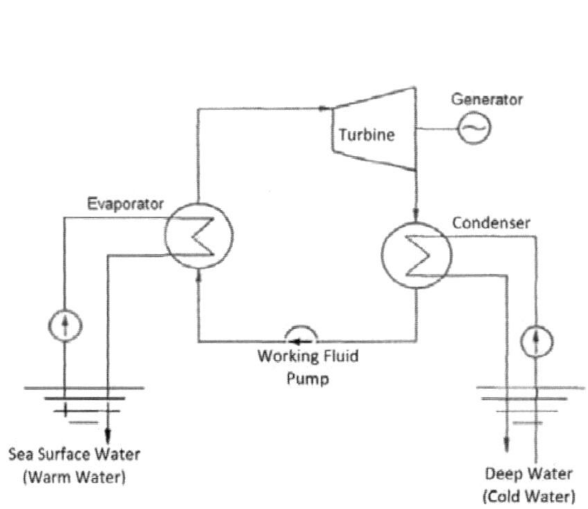

Figure 2. OTEC Closed Cycle schematic [7]

The heat rate, $\dot{Q}_e$ and $\dot{Q}_k$ are calculated with equation (2):

$$\dot{Q} = \Delta T_m UA = \frac{UA(\Delta T_i - \Delta T_o)}{\ln(\Delta T_i/\Delta T_o)} \quad (2)$$

Where $\dot{Q}$ is the heat rate between sea water and working fluid in each heat exchanger, U is the thermal conductance of the heat exchanger, A is the effective area of the heat exchanger, ΔT_i is the temperature differences between incoming seawater and incoming working fluid in each heat exchanger, ΔT_o is the temperature differences between outgoing seawater and outgoing working fluid in each Heat Exchanger.

The outlet temperature of sea water and working fluid temperature influenced by the heat flow rate in each heat exchanger. Hence, the outlet temperature should be calculated using the derivation of heat transfer formula (3):

$$T_{out} = \frac{\dot{m}_1 C_H T_{in1} + \dot{m}_2 C_2 T_{in2}}{\dot{m}_1 C_H + \dot{m}_2 C_2} \quad (3)$$

For the evaporator part, the index 1 indicates the reference fluid, while index 2 indicates the target fluid. For the condenser part, the index 1 indicates the target fluid, while index 2 indicates the reference fluid. And index H show the heat exchanger material. Where $\dot{m}$ is the flow rate of the fluid, T_{in} is the incoming fluid temperature, C_2 is the fluid specific heat, either sea water or working fluid, and C_H is the heat exchanger specific heat.

The working fluid of this study is Freon R-22. The specific heat of Freon is 1.22 kJ/kgK, and the specific heat of seawater is 3.93 kJ/kgK. The heat exchanger material is titanium, and its specific heat is 0.47 kJ/kgK.

To solve $\dot{Q}_e$ and $\dot{Q}_k$ using equation (2), ΔT_i and ΔT_o need to be solved first using equation (3). However, the incoming working fluid from evaporator is the outward working fluid from condenser, and the incoming working fluid of condenser is the outward working fluid from evaporator. Hence, for this system, equation (3) is an implicit equation and need to be solved using iterative method.

4. OTEC Power Output

The average power output of the OTEC simulation at Makassar Strait is shown in Figure 3. The simulation shows different power output between southern and northern part of Makassar Strait. The southern part has higher potential compared to the north of Makassar Strait (Figure.3a). However, the region with the highest ocean thermal energy potential is quite far from the shore. The Capital Expenditure (CapEx) to build the floating platform for the OTEC system in that area is expected to be more expensive compare to the onshore OTEC platform, due to the technological complexity.

Para'baya is the closest shore with the most exceptional ocean thermal energy potential along the West Sulawesi Coast (Figure 3.a). The average power output is 120.39 kW. However, the power output is seasonally varied in a year basis (Figure 3.b). During the wet season (December, January, February) and dry season (June, July, August) the average power output decreased. While during first transitional season (March, April, May) and second transitional season (September, October, November), the average power output increased.

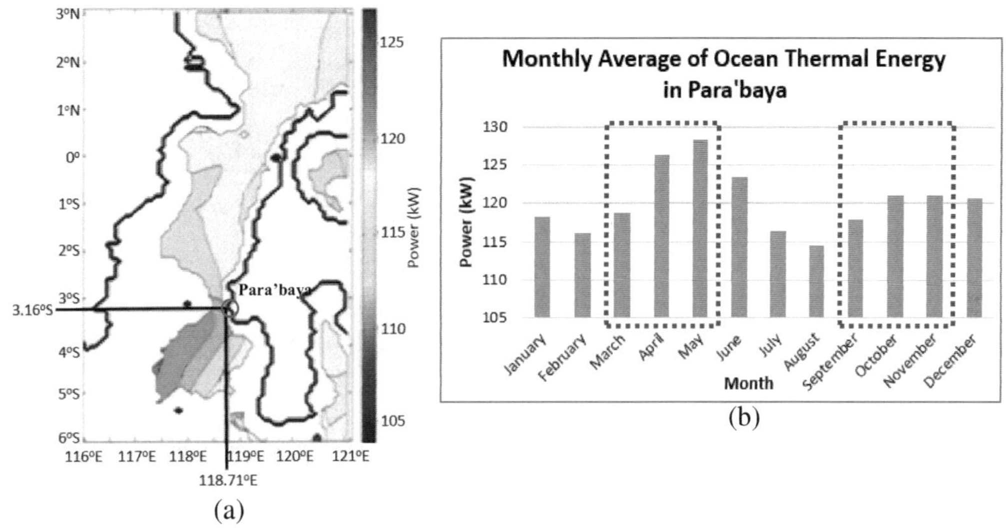

Figure 3. (a) Average Ocean Thermal Energy Distribution Map in Makassar Strait, **(b)** Monthly Average of Ocean Thermal Energy in Para'baya

In the wet season, the maximum ocean thermal power is 120.91 kW, which is acquired in December, while the lowest output is in February with a power output of 116.33 kW. The maximum power output during the first transitional season is in May with a power output of 128.36 kW, and the minimum output is in March with 118.83 kW. During the dry season, the highest ocean thermal output is around 123.29 kW which is acquired in June, and the lowest ocean thermal output is in August with 114.7 kW power output. In the second transitional season, the power output does not have high variation. The highest ocean thermal is in November with power output of 120.16 kW, and the smallest is in September with 118.27 kW output.

5. Sea Surface Temperature (SST) Variability

The SST variability is the primary cause of the ocean thermal energy variation. The correlation between sea surface temperature and the ocean thermal energy in Para'baya is very high, 0.86. The ocean thermal power increases when the SST jumped and it decreases when the SST droped. SST varies, as the weather changing. Meanwhile, the deep layer water below 1,000 m depth temperature is hardly had direct influenced from the weather. Hence, the temperature variation is low and doesn't give significant impact to the ocean thermal power. Figure 4 shows the vertical variation of temperature over the time in Para'baya based on the output of LITHMOS for 1st of January 1982 to 31st of December 1983. The temperature in the surface is fluctuative, while the temperature bellow 900 m is relatively uniform.

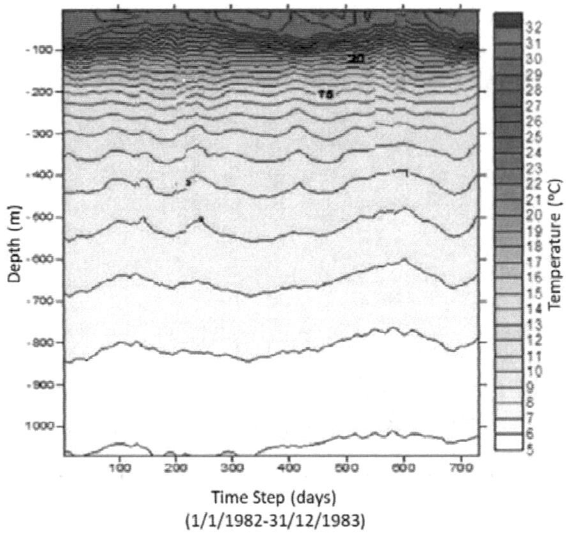

Figure 4. Para'baya Vertical Variation of Temperature from 1st of January 1982 to 31st of December 1983 of LITHMOS output.

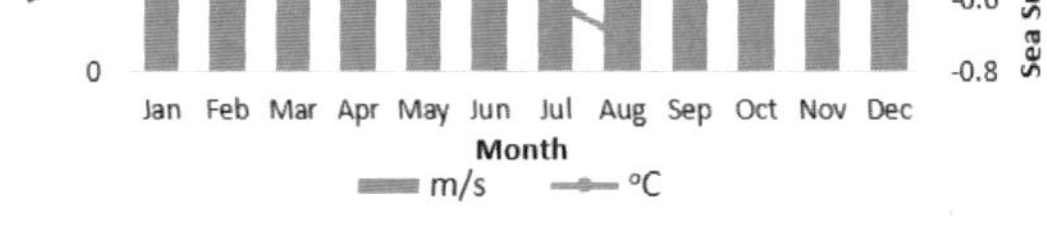

Figure 5. Monthly Average of Wind Speed at Coordinate 2.5°S;117.5°E [9] and Sea Surface Temperature Anomaly at Para'baya.

The intensity of solar radiation and the wind speed influence the SST. The higher solar irradiation received by the ocean then, the higher sea surface temperature would be. The water particle absorbs more heat if the solar irradiation is greater. However, the SST decreases when the gust of wind is greater. Figure 5 shows the monthly average wind gust in the southern part of Makassar Strait and the monthly average of SST variation. When the wind blows hard, it causes turbulent in the sea water. The turbulence mixes the water and distributes the heat evenly. However, the turbulence also blocks the solar ray; hence, the surface temperature is low when the wind is strong. And vice versa, the sea surface temperature is high when the wind is weak.

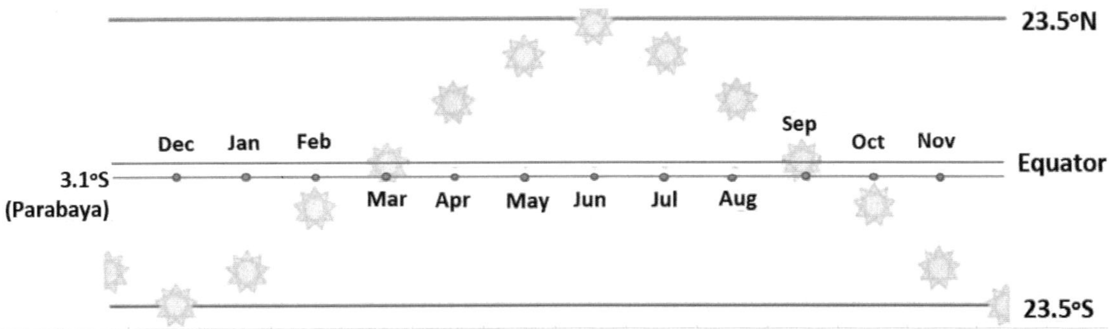

Figure 6. Relative sun position over Para'baya's latitude (3.1°S)

Figure 6 shows the relative position of the sun to the latitude of Para'baya. The sun is above Para'baya around March; thus, the solar irradiation reached the maximum point. However, the lowest wind speed in Makassar Strait is around April – May; hence the sea surface temperature in May is the highest. On average, the temperature anomaly in May is 0.6°C above average. In July, the sun position is in the farthest north from Para'baya. The sea surface temperature is low, and it gets even lower in August due to the strengthening of the wind. The strong wind to the north during August is causing an upwelling effect on the west coast of Sulawesi, and it also contributes to the sea surface temperature drop. On average, the sea surface temperature in August drops by -0.74°C compare to the average temperature.

In December, the sun reaches the farthest south from Para'baya, the wind speed also relatively high, but the wind is blowing to the south and doesn't cause an upwelling effect on the west coast of Sulawesi. Therefore, the temperature drop is not as much as in August. The average sea surface temperature in December is still 0.17°C higher than overall average sea surface temperature.

Meanwhile, in January – February, the sun is getting closer to Para'baya. But the wind is stronger than the wind magnitude in December. Therefore, the sea surface temperature keeps getting lower until February. The average sea surface temperature of Para'baya water is -0.17°C lower than the overall average temperature.

6. Conclusion

Para'baya is one of the potential locations for the ocean thermal energy utilization. The ocean thermal power simulation shows Para'baya coast could produce quite significant power compared to others region in the west coast of Sulawesi. Another reason is the bathymetry of Para'baya is steep enough to reach the 1000 m depth. The water intake from the deep layer would not be very far from the shoreline. Therefore, it would minimize the power requirement of the pump; hence decrease parasitic power demand and increase the net power output.

The ocean thermal power in Para'baya varies over the time as the SST changes. The SST has strong influence on the OTEC closed cycle power generation. Meanwhile, the SST is varied in season. The combination of seasonal solar radiation intensity and wind magnitude determine the SST variation. The highest ocean thermal power in Para'baya is in May when the wind magnitude is at the lowest, and the sun is not in its farthest position to Para'baya. The average gross ocean thermal power in May is 128.36 kW. The smallest ocean thermal power is in August when the wind magnitude is high, and there is an indication of upwelling effect in the west coast of Sulawesi. The average gross ocean thermal power in August is 114.7 kW.

IOP Conf. Series: Earth and Environmental Science **162** (2018) 012037 doi:10.1088/1755-1315/162/1/012037

Acknowledgement
Biggest regards to Ms. Ivonne Radjawanne and Ms. Susanna Nurdjaman for all of the feedback on this ocean thermal energy resources study. Gratitude to Mr. Benhard Mayer and Mr. Thomas Pohlmann for the LITHMOS temperature output data. And also to Mr. Delyuzar Ilahude and all of the Geomarine 3 Tarakan Research Cruise for the chances and experiences to directly measure the field temperature data in Makassar Strait. This research paper is funded by Indonesian Endowment Fund for Education (LPDP).

References
[1] Avery W H and Wu C 1994 *Renewable Energy from the Ocean: A Guide to OTEC* (New York: Oxford University Press)
[2] Yeh R-H, Su T-Z and Yang M-S 2005 Maximum output of an OTEC power plant *Ocean Eng* **32** 685–700
[3] Vega L A 2010 Economics of ocean thermal energy conversion (OTEC): an update Offshore Technology Conference (Offshore Technology Conference)
[4] Waworuntu M, Garzoli S L and Olson D B 2001 Dynamics of the Makassar Strait *J. Mar. Res.* **59** 313–25
[5] Mayer B, Damm P E, Pohlmann T and Rizal S 2010 What is driving the ITF? An illumination of the Indonesian throughflow with a numerical nested model system *Model. Obs. Indonesia. Throughflow* **50** 301–12
[6] NOAA World Ocean Atlas 2009 http://www.nodc.noaa.gov/OC5/WOA09/pr_woa09.html access 9th of June 2017 14.02 GMT
[7] Mitsui T, Ito F, Seya Y and Nakamoto Y 1983 Outline of the 100 kW OTEC Pilot Plant in the Republic of Nauru *IEEE Power Eng. Rev.* **PER-3** 47–8
[8] Incropera F P 2007 *Introduction to heat transfer* (Hobokenm NJ: Wiley)
[9] NOAA ESRL : PSD : Monthly Mean Timeseries http://www.esrl.noaa.gov/psd/cgi-bin/data/timeseries/timeseries1.pl access 11th of June 2017 18.25 GMT

IOP Conf. Series: Earth and Environmental Science **162** (2018) 012038 doi:10.1088/1755-1315/162/1/012038

Estimation of marine gravity anomaly model from satellite altimetry data (Case Study : Kalimantan and Sulawesi Waters-Indonesia)

P Hartanto[1], S Huda[1], W Putra[1], E D Variandy[1], B Triarahmadhana[1], D Pangastuti[1], A M Pahlevi[1] and C Hwang[2]

[1] Geospatial Information Agency of Indonesia, Jl. Raya Jakarta-Bogor Km. 46, Cibinong 16911, West Java, Indonesia
[2] National Chiao Tung University, 1001 University Road, Hsinchu, Taiwan 300, ROC

E-mail: hartantoprayudha@gmail.com

Abstract. Nowadays satellite altimetry has become an advanced instrument to observe many natural physical phenomena, such as sea-level rise, ocean circulation, water mass changes, and marine gravity anomaly. The use of satellite altimetry data to compute marine gravity anomaly provides good results and costs relatively low. Those advantages make geodesists utilize this method as an alternative in geoid determination, especially over the seas. Several sets of satellite altimetry data from Cryosat 2, Jason 1 phase C, Geosat and ERS1 were used to compute gravity anomaly over the surrounding waters of Kalimantan and Sulawesi Island in Indonesia. The study area spans between -7^0-7^0 N and 108^0-127^0 E with a spatial resolution of 1'x1'. In the pre-processing step, the altimetry data especially Geosat and ERS1, were retracked to reduce errors due to the land influence. The main computation step was done by using two different methods, least square collocation (LSC) and Inverse Vening-Meinesz (IVM). The computed gravity anomaly models then assessed with the in-situ marine gravity data from the National Geophysical Data Center (NGDC). The best model in term of RMS error is the 10 km Gaussian filtered LSC with an RMS error of 15.042 mgal. The least accurate model is the non-filtered IVM with an RMS of 16.704 mgal.

1. Introduction

The Kalimantan and Sulawesi are two main islands of Indonesian Archipelago. In order to determine the regional geoid, the government of Indonesia in collaboration with Technical University of Denmark (DTU) ran airborne gravity surveys over those two Islands in 2008-2009 [1]. However, the surveys did not cover the waters surrounding those two islands [2]. In order to fill those gaps, some geodetic mission altimetry datasets were used to determine the gravity anomalies over an area between 7°S - 7° N and 108°E - 127°E. The datasets consist of Sea Surface Height (SSH) measured by Cryosat-2, Jason-1 phase C, Geosat and ERS-1 satellite altimetry missions. By definition, SSH is the difference of altimeter range from the satellite altitude above the reference ellipsoid [3], or in another word SSH is the height of instantaneous sea surface above the reference ellipsoid.

To make the SSH data usable in gravity anomaly determination, some accuracy defects related to land occurrence should be reduced by performing waveform retracking. Waveform retracking was performed to fit a model or functional form to the measured waveforms, and retrieve geophysical parameters such as range, echo power, etc [4]. Four retracking methods are available to use; threshold [5], subwaveform threshold [6], improved threshold [7], and beta-5 parameter [8]. The past research by Hsiao et al., shows that the subwaveform retracker gave the best result for SSH and altimeter-derived gravity values [9].The corrected altimetry range then used in SSH determination. Further explanation of the altimetry data handling will be discussed in Section 2 and 3.

Next, the obtained SSH of those altimetry missions are then combined together to be used as the main input in gravity anomaly determination. Moreover, supplementary data are required, those are tide, reference geoid and sea surface topography (SST) models. This research objective is to make the best altimetry-only gravity model with spatial resolution of 1' x 1'.

2. Data

2.1. Satellite Altimetry Data

There were four satellite altimetry missions used as the main input for this research, Geosat/GM, ERS-1/GM, Jason-1/GM and Cryosat-2. The "GM" phrase defines that an altimetry mission is a geodetic mission or in other word a non-repeat mission. The Geosat/GM is a sun-synchronous mission which orbited earth in the period of March 31st 1985 to September 30th 1986, with the sampling frequency of 10 Hz. The ERS-1/GM was gathered from the European Space Agency (ESA), which consisted of two 168 days data cycles with a sampling frequency of 20 Hz. The Jason-1/GM data was downloaded from AVISO data center or Jet Propulsion Laboratory (JPL) alternatively. This mission is a drifted orbit of Jason-1 repeat mission, with a repeat period of 406 days and sub cycles of 3.9, 10.9, 47.5, and 179.5 days [10]. The Cryosat-2 data used are the level 1B data with a sampling frequency of 20Hz [11]. Because of the different sampling rates of each satellite mission, the SSH datasets should be resampled into 2Hz. Besides, the datasets also needed to be freed from outlier by applying a Gaussian filter with a 10 km window size.

2.2. NAO99 Ocean Tide Model

Before taken into further process, the altimetry SSH data had to be corrected from the tidal effect. This effect was removed by using a global ocean tide model. In this research we used the NAO99b ocean tide model. This model has a spatial resolution of 0.5^o (global version) and 5' (Japan regional version) and predicts the 16 major tide constituents [12].

2.3. RIO05 Mean Dynamic Topography Model

The RIO05 MDT model was generated from the combination of altimetry data (T/P and ERS1), in-situ measurements (buoys velocities, XBT, CTD) from 1993-2002, and refers to EIGEN-GRACE03S geoid model on a 30' x 30' regular grid [13]. The standard deviation of this model is 0.713 m according to Marchenko et. al [14].

3. Method

3.1. Waveform Retracking

After preparing all the data needed (see sub-section 2.1), some corrections would be performed. The altimetry range measurement had to be corrected from the coastal and shallow water effects or well known as the waveform retracking.

Waveform retracking is an improvement method of altimetry range measurement by determining the location of epoch on the leading edge. This epoch, commonly called as tracking gate, is a pre-defined value which computed by the satellite data provider. However, in some cases, especially in near coastal and shallow waters, the given value is not representing the actual state of water surface. Several retracking methods, either statistic or deterministic have been developed. In this research, the applied retracker is subwaveform threshold. This retracker first identifies the leading edge based on subwaveform correlation analysis, and then computes the retracking gate by using a threshold method [6]. All altimetry data but the Cryosat-2 had been retracked by those aforementioned retrackers. The

Cryosat-2 altimeter uses the latest technique of delay/Doppler [15] which improves the determination of altimeter range near coastal area [9].

3.2. Gravity Anomaly Determination

The remove-compute-restore (RCR) technique was applied to determine the gravity anomalies. This technique is based on the separation of gravity anomaly signal into three different spectral components, the long wavelength, the short wavelength and the residual part (see figure 1 and eq. 1). Figure 1 shows that the addition of three different undulation (N) wavelengths gives the most realistic geoid model. Each component of N is computed from corresponding gravity anomaly (Δg) by the Stokes integration formula [16]. The geoid determination by using gravity data is called the gravimetric geoid modelling.

The first step in RCR technique is computing the along-track gradient. The second is removing the reference gradient from global gravity model and followed by removing outliers. The outlier removal was done by applying τ test as mentioned in the theory of Pope [17]. The next step is computing the residual gravity anomalies. There were two different methods in this computation step, Least-Square Collocation (LSC) and Inverse Vening-Meinesz (IVM). The last step is restoring gravity anomalies from the global gravity model. Eight different models were made and would be assessed with the in-situ data. Those models were 7 LSCs and an IVM. The LSC models were differed from each other by the type and size of the filter applied. Filters were applied to overcome the remains of high frequency noise. In addition, we also assessed Sandwell v23.1 altimeter-only global gravity model. This model has RMS error of 2.6 mgal when compared to National Geospatial Intelligence Agency (NGA) shipborne data [18]. The list of models are shown in table 1. Furthermore, the defined grid-size of the models is 1' x 1' due to the cross-track spacing of the altimetry missions are about 1-2 km.

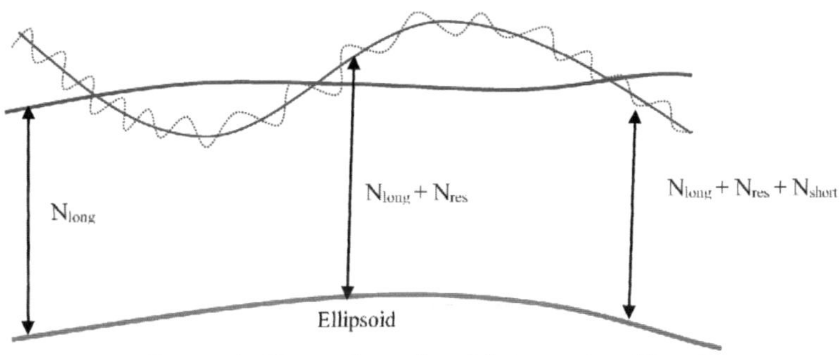

Figure 1. Illustration of geoid above the ellipsoid

$$\begin{cases} N = N_{long} + N_{res} + N_{short} \\ \Delta g = \Delta g_{long} + \Delta g_{res} + \Delta g_{short} \end{cases} \quad (1)$$

with :

N	= geoid undulation	Δg	= gravity anomaly
N_{long}	= long wavelength component of N	Δg_{long}	= long wavelength component of Δg
N_{res}	= short component of N	Δg_{res}	= residual component of Δg
N_{short}	= short wavelength component of N	Δg_{short}	= short wavelength component of Δg

Table 1. List of gravity anomaly models

Model	Method	Filter
Kalsul 1	LSC	Non filtered
Kalsul 2	LSC	Gaussian 5km
Kalsul 3	LSC	Gaussian 10km
Kalsul 4	LSC	Gaussian 15km
Kalsul 5	LSC	Median 5km
Kalsul 6	LSC	Median 10km
Kalsul 7	LSC	Median 15km
Kalsul 8	IVM	Non filtered
Sandwell v23.1	-	-

3.3. Least-square collocation (LSC)

The LSC method computes final gravity anomaly by summing the residual gravity anomalies (Δg_{res}) and those computed from a reference gravity field (Δg_{ref}) [9]. Beside SSHs from altimetry, a global geopotential model was needed as the reference gravity field. The selected reference was the EGM08 to degree and order 2190. Along with the SSHs, EGM08 gave the value of residual geoid gradient (ε_{res}). The computed residual geoid gradients were used to calculate the residual gravity anomalies by equation 2 [19]:

$$\Delta g_{res} = C^S_{\Delta g\varepsilon} C^S_{\Delta g\Delta g} \begin{pmatrix} C^o_{\varepsilon\varepsilon} + D_\varepsilon & C^o_{\varepsilon\Delta g} \\ C^o_{\Delta g\varepsilon} & C^o_{\Delta g\Delta g} + D_{\Delta g} \end{pmatrix}^{-1} \begin{pmatrix} \varepsilon \\ \Delta g \end{pmatrix} + \Delta g_{ref} \qquad (2)$$

with

Δg_{res} = vectors of residual gravity anomalies
ε_{res} = residual geoid gradients
$C_{\Delta g\varepsilon}$ = gravity anomaly-residual geoid gradients covariance matrices
$C_{\varepsilon\varepsilon}$ = residual geoid gradient-residual geoid gradient covariance matrices
$C_{\Delta g\Delta g}$ = gravity anomaly- gravity anomaly covariance matrices for
D_ε = noise of residual geoid gradient
$D_{\Delta g}$ = noise of residual gravity anomaly
Δg_{ref} = reference gravity anomaly

The final gravity anomaly was obtained by adding the residual gravity anomalies into those computed from EGM08 [9]. A brief explanation of this LSC method can be found in [9] and [20].

3.4. Inverse Vening-Meinesz (IVM)

This method computes gravity anomaly from deflections of the vertical [21]. The deflections of the vertical (DOV) is defined as the spatial angle between the normal gravity vector on the reference ellipsoid and the actual gravity vector on the geoid [22]. The first three steps in this technique are similar as in LSC; get along-track gradient, remove reference gradient of global gravity model, and remove outliers. However, an intermediate step is needed in this technique, which is the gridding of north and east gradient components as written in equation 3.

$$\begin{pmatrix} \xi \\ \eta \end{pmatrix} = \hat{S} = C_{sl}(C_{ll} + C_n)^{-1} l \qquad (3)$$

with:

$\hat{S}$ = prediction vector (with ξ, η)
ξ = north component of DOV
η = east component of DOV
l = observation vector, the geoid gradient from altimetry
C_{ll} = covariance matrices for l and l

IOP Conf. Series: Earth and Environmental Science **162** (2018) 012038 doi:10.1088/1755-1315/162/1/012038

C_n = covariance matrices for the noise of *l*
C_{sl} = covariance matrices for *s* and *l*

Vector l is an observation vector contains the geoid gradient derived from altimetry SSHs and EGM08 as the reference gravity field (ε_{res}, see sub-section 3.3).
The next step is converting DOV to gravity anomaly by using the IVM formula, as written in equation 4.

$$\begin{Bmatrix} N_p \\ \Delta g_p \end{Bmatrix} = \frac{1}{4\pi} \begin{Bmatrix} R \\ \gamma \end{Bmatrix} \iint_\sigma (\xi_q \cos\alpha_{qp} + \eta_q \sin\alpha_{qp}) \begin{Bmatrix} C' \\ H' \end{Bmatrix} d\sigma_q \quad (4)$$

where

N_p	= geoidal height at p	$\xi_q \eta_q$	= north and east component of DOV at q
Δg_p	= gravity anomaly at p	α_{qp}	= azimuth from q to p
R	= mean Earth radius	$d\sigma_q$	= surface element = $cos\phi_q d\phi_q d\lambda$
γ	= normal gravity	$\phi_q \lambda_q$	= latitude and longitude of q
C'H'	= Kernel function		

Because of the singularity of the Kernel function C' and H' at zero spherical distance, the innermost zone effects on geoid and gravity anomaly must be taken into account and were computed by equation 5:

$$\begin{Bmatrix} N_p \\ \Delta g_p \end{Bmatrix} = \frac{1}{4} (\xi_y + \eta_x) \begin{Bmatrix} s_0^2 \\ 2s_0\gamma \end{Bmatrix} \quad (5)$$

where

$$\xi_y = {}^{\delta\xi}/_{\delta y} \qquad \eta_x = {}^{\delta\eta}/_{\delta x} \qquad s_0 = \sqrt{\frac{\Delta x \Delta y}{\pi}} = the\ size\ of\ the\ innermost\ zone$$

When using the RCR procedure, the error in using spherical approximations should be very small compared to data noise. Consider the formula of error-free LSC in the case of using ellipsoidal correction [23] as written in equation 6:

$$S = C_{sl} C_{ll}^{-1} (l - e^2 l^l) + e^2 s^l \quad (6)$$

e^2 is the squared eccentricity of a reference ellipsoid, which is about 0.006694 for the GRS80 ellipsoid. If the largest element of DOV in l^l was assumed as 100 *µrad*, then the largest element in $e^2 l^l$ would be 0.66 *µrad*. This value was far smaller than the noise of DOV from the multi-mission satellite altimetry. In conclusion, the value of ellipsoidal correction could be neglected.

3.5. NGDC Marine Gravity Data
The NGDC datasets were used as the assessment values for the modelled gravity anomalies. The datasets consist of several data types such as gravity, magnetics and bathymetry, but in this research we only utilized the gravity data. Over the research area, the gravity data was observed by many shipborne survey missions dated back from 1963 to 1991. The data were available in MGD77T format [24].

As the models were in the grid format, an interpolation should be made to the NGDC data. We used a two dimensional polynomial interpolation to define an NGDC gravity value on the desired cell grid. This value then subtracted from the models corresponding value to calculate the difference. Figures 5, 6 and 7 below show the map of the differences while table 2 presents the statistics.

IOP Conf. Series: Earth and Environmental Science **162** (2018) 012038 doi:10.1088/1755-1315/162/1/012038

4. Results and Analysis

In figure 2, Kalsul 1 model, it can be seen that the spatial distribution of gravity anomalies over the area is similar as in figure 3 (Kalsul 8) and figure 4 (Sandwell). All figures show gravity anomalies ranged from about -300 to 300 mgal with a pattern of negative anomalies on the waters east-side of Sulawesi Island (-2° - 0° N and 122° – 127° E). This pattern represents an area of deep waters.

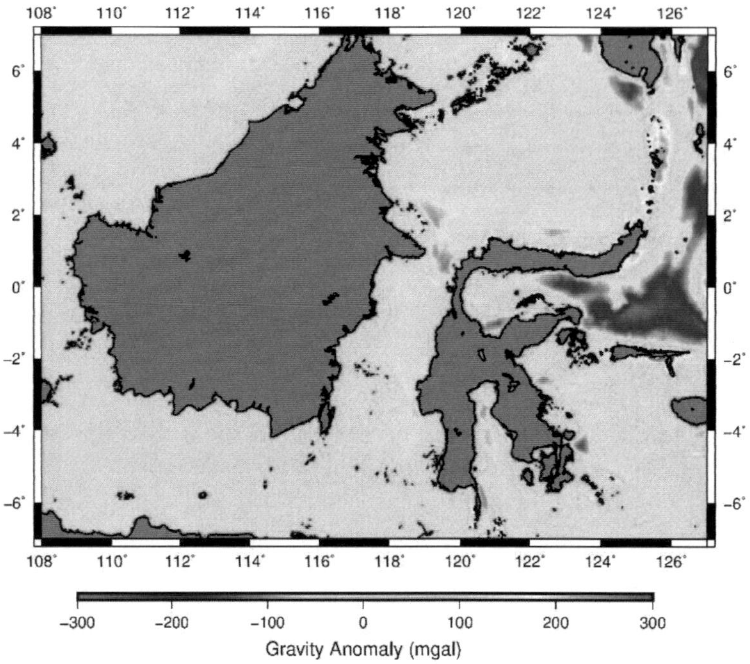

Figure 2. Gravity anomaly as computed by LSC technique (Kalsul 1)

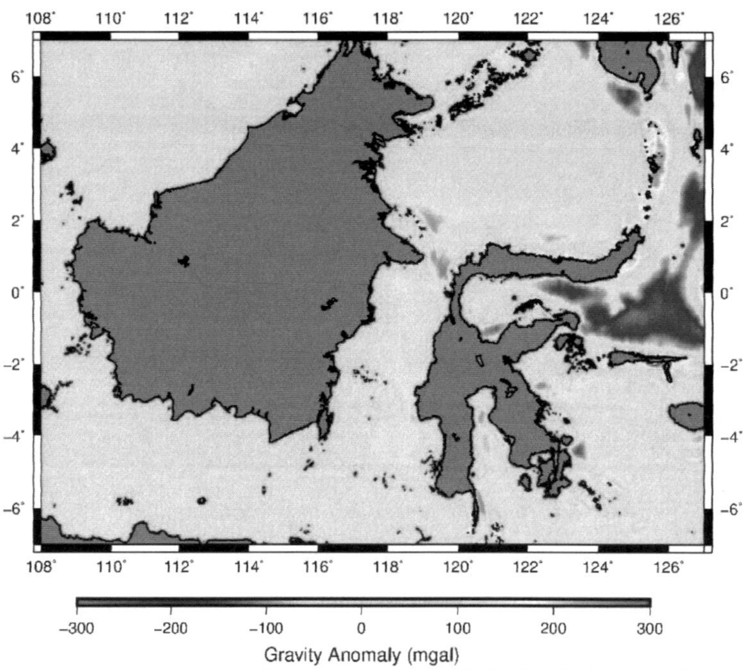

Figure 3. Gravity anomaly as computed by IVM Technique (Kalsul 8)

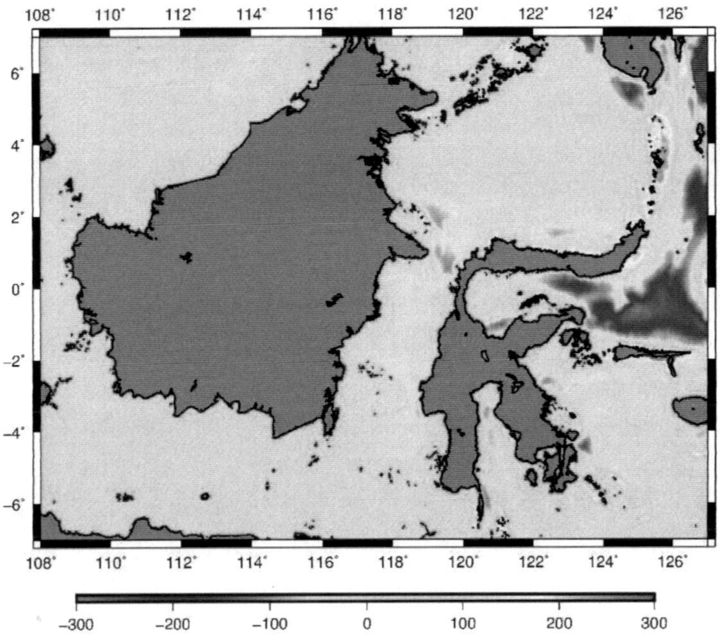

Figure 4. Gravity Anomaly from Sandwell v 23.1

To get a better understanding of the models quality, the gravity anomaly differences were plotted on the NGDC shipborne tracks (figure 5, 6 and 7). Figure 5 shows the differences of the non-filtered LSC model (Kalsul 1), while figure 6 and 7 display the differences of IVM (Kalsul 8) and Sandwell models respectively. Kalsul 1 and Kalsul 8 show a similar pattern with the biggest difference occurs in around latitude 6^0 N. Furthermore, the Kalsul 1 model give a better accuracy than Kalsul 8 by 1.62 mgal. The application of Gaussian filter with a 10 km window size (Kalsul 3) had improved the accuracy of LSC model by 0.2% (Table 2). On the other side, the application of median filter with the same window size only gave 0.17% improvement. Overall, the LSC models outperformed the IVM one.

Compared to Sandwell, our models are better in term of RMS error by almost 6 times. This result is not only consistent with the previous research by Hsiao et al [9], but also indicate that the Sandwell model is not accurate enough in the research area. However, the Sandwell model has a better mean deviation (0.411 mgal) than the Kalsul 8 (0.581 mgal). In term of mean deviation, the best computed model is Kalsul 7, with mean deviation of -0.019 mgal, while Kalsul 3 is the best model in term of RMS error (15.042 mgal) and STD of error (15.041 mgal). The complete results shown in table 2 below.

Table 2. Statistics of the gravity anomaly models

Model	Max (mgal)	Min (mgal)	Mean (mgal)	RMS (mgal)	STD (mgal)
Kalsul 1	277.636	-227.900	0.084	15.077	15.077
Kalsul 2	277.712	-227.072	0.085	15.064	15.064
Kalsul 3	277.889	-223.814	0.094	15.042	15.041
Kalsul 4	277.866	-221.080	0.115	15.072	15.071
Kalsul 5	277.534	-227.403	0.076	15.064	15.064
Kalsul 6	277.707	-224.629	0.031	15.051	15.051
Kalsul 7	278.113	-221.516	-0.019	15.140	15.140
Kalsul 8	284.513	-230.813	0.581	16.704	16.694
Sandwell	588.134	-451.830	0.411	93.923	93.923

a) b)

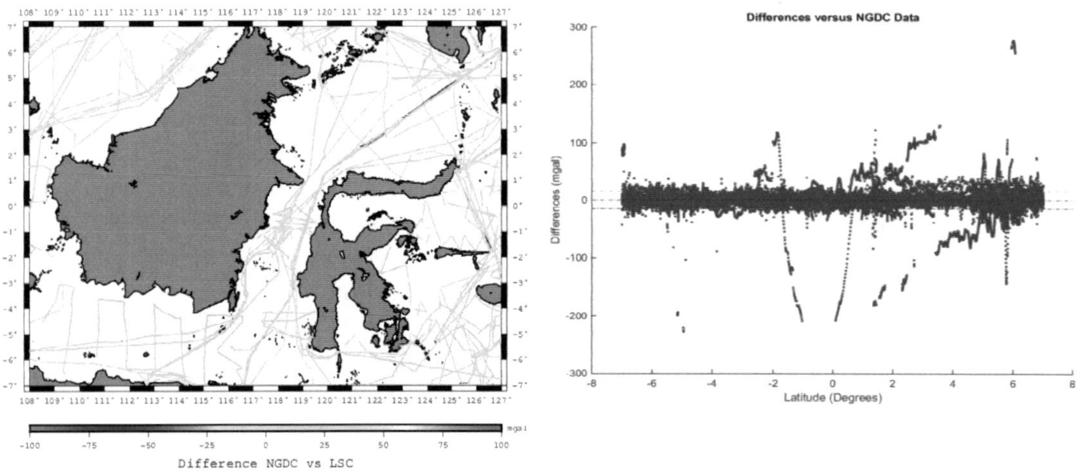

Figure 5. a) Differences between LSC (Kalsul 1) and NGDC and b) its latitudal profile

Figure 5a and 6a show green lines all over the research area. The green lines indicate near zero difference to the NGDC marine gravity data. However, in the northern part of Sulawesi waters, a pattern with blue color appeared, this is an indication of negative differences with magnitude -50 to -75 mgal. The latitudal profiles of the anomaly differences show the existence of major errors (figure 5b and 6b), especially in latitude 2^0 - 1^0 S, 0^0 - 1^0 N, and 1.5^0 - 6^0 N. Those phenomena appear because of some gross errors in at least three NGDC datasets.

On the other hand, figure 7 does not show a similar pattern. Figure 7a shows that the difference between Sandwell 23.1 and in-situ data vary all over the area, indicated by multi-colored lines. However, there is a small region in the middle of research area which gives errors ±250 mgal, shown in blue/purple lines. Those errors possibly caused by high frequency noises affected by the land influence on that narrow waters region. Figure 7b shows major errors in almost all latitude. This is an indication that Sandwell v 23.1 model still has some systematical errors in the research area. The most probable systematical errors source is aliasing which occurred when cutting the global model into smaller regional parts.

a) b)

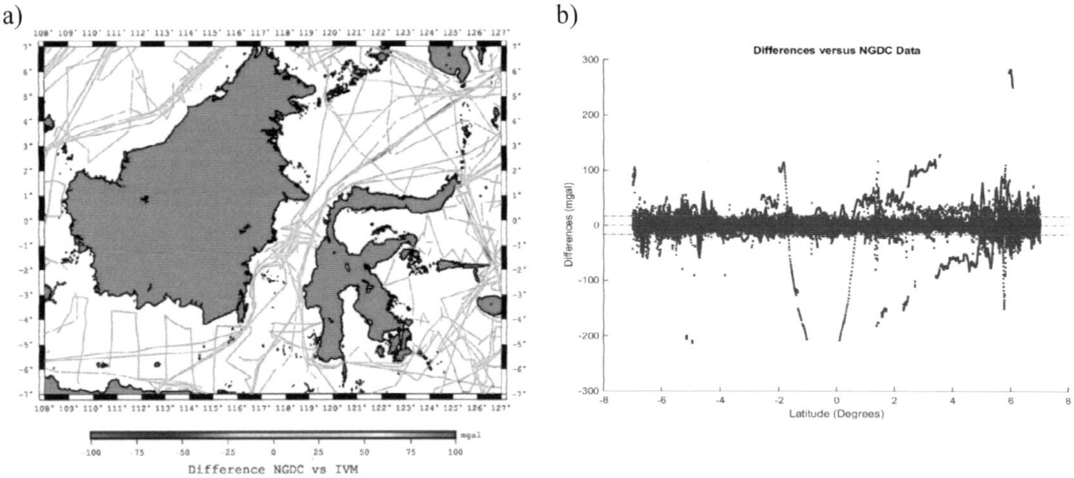

Figure 6. a) Differences between IVM (Kalsul 8) and NGDC and b) its latitudal profile

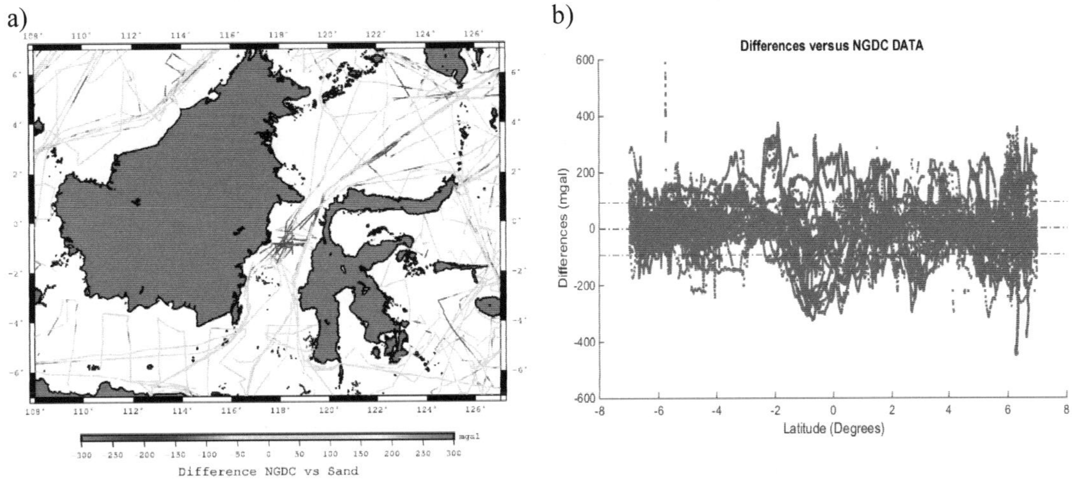

Figure 7. a) Differences between Sandwell 23.1 and NGDC and b) its latitudal profile

5. Conclusion

This research has produced eight gravity anomaly models with mean deviation ranged from -0.019 to 0.581 mgal, and RMS error ranged from 15.042 to 16.704 mgal. Those values were obtained by using NGDC marine gravity data as the assessment values. The best model in term of RMS error is Kalsul 3, an LSC gravity anomaly model with 10 km Gaussian filter applied, while the least accurate one is Kalsul 8, an IVM gravity anomaly model with no filter applied. Kalsul 3 might not be the most optimum model, yet it is suitable enough to determine the regional geoid model of Indonesia, especially on waters area. To improve the models quality, a list of supplementary data could substitute the data used, such as Indonesian tide gauges assimilated tidal model as an alternative to NAO99, CNES-CLS09 as an alternative to RIO05 MDT model, and the incoming EGM as the substitution to EGM08 geoid reference. The use of most recent altimetry missions, such as ICESat and Saral-Altika, could also enhance the quality of the model, as it densifies the altimetry footprints over the research area.

Acknowledgements

This paper is a product of a collaborative short course held by Geospatial Information Agency of Indonesia (BIG) and National Chiao Tung University (NCTU) of Taiwan. The course was funded by the non-degree RISET-PRO scholarship managed by the Indonesian Ministry of Research, Technology and Higher Education. The softwares used were developed by the Civil Engineering Department of NCTU.

References

[1] Kasenda A, Pangastuti D and Prijatna K 2015 Determination of gravimetric geoid model in kalimantan and sulawesi indonesia *FIG Working Week 2015 Proc. (Sofia)* 7706

[2] Pahlevi A, Pangastuti D, Sofia N, Kasenda A and Prijatna K 2015 Determination of gravimetric geoid model in sulawesi indonesia *FIG Working Week Proc. (Sofia)* 7706

[3] Dumont J, Rosmorduc V, Lillibridge J, Scharoo R, Desai S, Bonekamp H, Figa J, Bronner E and Picot N 2017 OSTM/Jason-2 Products Handbook (NOAA/NESDIS: Polar Series/OSTM) J400

[4] Gommenginger C, Thibaut P, Fenoglio-Marc L, Quartly G, Deng X, Gomez-Enri J, Challenor P and Gao Y 2011 Retracking altimeter waveforms near the coasts *Coastal Altimetry* ed S. Vignudelli *et al* (Berlin: Springer-Verlag) pp. 61-101

IOP Conf. Series: Earth and Environmental Science **162** (2018) 012038 doi:10.1088/1755-1315/162/1/012038

[5] Davis C 1997 A robust threshold retracking algorithm for measuring ice-sheet surface elevation change from satellite radar altimetry *IEEE Transactions on Geoscience* **35** 974-79

[6] Yang Y, Hwang C, Hsu H, Er D and Wang H 2012 A subwaveform threshold retracker for ers-1 altimetry: a case study in the antartic ocean *Computer and Geosciences* **41** 88-98

[7] Hwang C, Guo J, Deng X, Hsu H and Liu Y 2006 Coastal gravity anomaly from retracked geosat/gm altimetry: improvement, limitation and the role of airborne gravity data *J. of Geodesy* **80** 204-16

[8] Martin T, Zwally H, Brenner A and Bindschadler R 1983 Analysis and retracking of continental ice sheet radar altimeter waveforms *J. of Geophysical Res.* **88** 1608-16

[9] Hsiao Y, Hwang C, Cheng Y, Chen L, Hsu H, Tsai J, Liu C, Wang C, Liu Y and Kao Y 2016 High-resolution depth and coastline over major atolls of south china sea from satellite altimetry and imagery *Remote Sensing of Environment* **176** 69-83

[10] Picot N *et al* 2016 Jason 1 Product Handbook (CNES/AVISO+ : SALP-MU-M5-OP-13184-CN)

[11] Bouzinac C 2012 *CryoSat Product Handbook* (London : ESRIN-ESA and Mullard Space Science Laboratory-University College London) p 55

[12] Matsumoto K, Takenazawa T and Ooe M 2000 Ocean tide models developed by assimilating topex/poseidon altimeter data into hydrodynamical model: a global model and a regional model around japan *J. of Oceanography* **56** 567-81

[13] Rio M, Schaeffer P, Hernandez F and Lemoine J 2005 The estimation of the ocean mean dynamic topography through the combination of altimetric data, in-situ measurements and grace geoid: from global to regional studies *GOCINA International Workshop (Luxembourg)*

[14] Marchenko A, Tretyak K, Lopyshansky A and Pavliv T 2010 Recent dynamic ocean topography models and their comparison *Infrastructure and ecology of Rural Areas* **11** 151-158

[15] Raney R 1998 The delay/doppler radar altimeter *IEEE Transactions on Geoscience and Remote Sensing* **36** 1578-88

[16] Heiskanen W and Moritz H 1967 *Physical Geodesy* (San Fransisco: W.H. Freeman and Co) p 92

[17] Pope A 1976 The statistics of residuals and the detection of outliers *NOAA Technical Report U.S. Dept. of Commerce, National Oceanic and Atmospheric Administration, National Ocean Survey* (Rockville, MD : Geodetic Research and Development Laboratory)

[18] Sandwell D, Muller R, Smith W, Garcia E and Francis R 2014 New global marine gravity model from cryosat-2 and jason-1 reveals buried tectonic structure *Science* **346** 65-67

[19] Hwang C and Parsons B 1995 Gravity anomalies derived from Seasat, Geosat, ERS-1 and Topex/Poseidon altimeter and ship gravity: A case study over the Reykjanes Ridge *Geophysical J. Int.* **122** 551-68

[20] Rummel R, Sjoberg L and Rapp R 1978 The determination of gravity anomalies from geoid heights using the inverse Stokes' formula, Fourier transforms, and least squares collocation (NASA: Columbus, OH)

[21] Hwang C 1998 Inverse vening meinesz formula and deflection-geoid formula: application to prediction of gravity and geoid over the south china sea *J. of Geodesy* **72** 304-12

[22] Ali A, Nan L and Peng J 2007 Determination of Gravity Anomalies over the Arabian Sea *J. of Applied Sciences* **7** 877-82

IOP Conf. Series: Earth and Environmental Science **162** (2018) 012038 doi:10.1088/1755-1315/162/1/012038

[23] Moritz H 1980 *Advanced Physical Geodesy* (Michigan: Wichman) p. 328

[24] Hittelman A, Groman R, Haworth R, Holcombe T, McHendrie G and Smith S 2010 *The Marine Geophysical Data Exchange Format - MGD77T/Legacy MGD 77* (Boulder, Colorado : National Geophysical Data Center)

IOP Conf. Series: Earth and Environmental Science **162** (2018) 012039 doi:10.1088/1755-1315/162/1/012039

GEOBIA an (Geographic) Object-Based Image Analysis for coastal mapping in Indonesia: *A Review*

F Hidayat, A W Rudiastuti, and N Purwono

Geospatial Information Agency, Jalan Raya Jakarta – Bogor Km 46, Cibinong, Bogor, Indonesia

E-mail: pahrulhdyt@gmail.com

Abstract. The traditionally human-visual or manual interpretation is generally chosen method for spatial feature extraction on map producing process in Indonesia, although the pixel-based feature extraction is used in particular step. The main problem of human-visual method is the dependency to operator capacity. Subjectivity may occur because each operator has different ability. In early 20th century, the image segmentation method was still developing and now Geographic Object-based Image Analysis (GEOBIA) method is introduced to mapping process. Researchers are still developing and applying GEOBIA for many mapping purposes. The advantage of GEOBIA is object-based segmentation which contextually influenced by spatial feature characteristics. GEOBIA provides more meaningful information than traditional pixel-based image analysis by allowing less well-defined edges or borders between different classes. On the resulting maps, there are divisions between different types of object classification, for example where some species of mangroves meet are not generally represented by a single object. The developing of remotely sensed imagery data quality requires the developing of extraction method as well. This article discusses the implementation of GEOBIA for coastal mapping in Indonesia. Some research combined pixel-based and GEOBIA method, and some get the overall accuracy > 70% of GEOBIA method. By developing a new segmentation accuracy measure, we evaluated that segmentation accuracies decrease with increasing segmentation scales and the negative impact of under-segmentation errors become significantly rise at large scale. The use of GEOBIA for coastal mapping in Indonesia are still lacking, locally applicable and still developing the transferability rule sets. The conclusion is that the innovation is needed to make GEOBIA method applicable for mass production of Indonesian coastal map.

1. Introduction

Visual interpretation and digital image processing is a kind of method to extract image information in remote sensing technique. The extraction of image information is a process to recognize objects of an aerial or satellite imagery, while the objects involve some man-made or natural objects such as land cover, buildings, roads, and others [1]. There are many advantages of image extraction such as object identification or object changes detection. The object changes detection is useful for monitoring as land cover, deforestation estimation, and others [2].

Visual or manual image interpretation still a common way that used in the mapping as well as in Indonesia, for example is topographic mapping. One of the advantages of visual interpretation is efficient at cost because that process is simpler rather than digital image processing, even though visual interpretation is limited to small mapping coverage relatively [3] [4]. Both visual interpretation and digital image processing have their advantages and disadvantages. The subjective process (matter) is one of disadvantages of visual interpretation, because it has multivarious results depending on the image interpreter [5].

In line with technological development, digital image processing also has been developed, for example in terms of image information extraction such as pixel-based and object-based (still going to be developing). In the 1970's, the image processing concept that mostly developed is pixel-based. While in the earth modeling by computer-based usually has the problem because the representation of objects is very complex. Whereas the purpose of remote sensing technique is to obtain information as accurately as possible without override the quantitative approach. The concept of object-based information extraction is accommodating the concept of image segmentation. It using the homogeneity principle of the pixel group, and it being one of the methods of remote sensing that offers more information extraction accurately [6].

Object-Based Image Analysis (OBIA) is specifically intended to extract the imagery into image-objects and assess by its spatial characteristics that taken from spectral, temporal and other values [7] [8]. One of the main processes in OBIA is image segmentation into specific image [7] [1] [9]. As no longer new, the topic about image segmentation was developing in 1950's (according to Scopus Database, the oldest research with keyword image segmentation identified in 1954). One of image-segmentation study is use a stepwise optimization approach that published in 1989 [10]. A review of image-segmentation applications using different algorithms has also been done in 1993 [11].

Strengths, Weaknesses, Opportunities, and Threats (SWOT) analysis of OBIA was formally presented at the first OBIA International Conference in 2006 [7]. In its development, the term of OBIA is devoted and determined into Geographic Object-Based Image Analysis (GEOBIA), and proposed to be a part of remote sensing discipline [12]. Several studies that apply GEOBIA use various types of the data, ranging from very low to very high resolution such as ASTER, JERS, Landsat ETM, RADARSAT, SPOT, Quickbird, IKONOS, Envisat, etc [8]. The purpose of using the GEOBIA are multivarious, e.g. to extract the road objects, buildings, land cover, mangrove distribution and etc. [13] [14] [15] [16] [17]. One of the application of GEOBIA is for coastal studies.

Based on the principles stated in the United Nations Convention on The Law of The Sea (UNCLOS) Indonesia is referred to as the archipelago country (Indonesia has ratified UNCLOS through The Indonesian Law No. 17 of 1985) [18] [19]. As the archipelagic country, Indonesia has coastline as along 99.093 KM, and it is the second longest of world [20] [21], the GEOBIA method should be used as an alternative to coastal mapping. The application of GEOBIA for coastal mapping is to be known whether the GEOBIA method can be used to mass production particularly for coastal mapping in Indonesia. This paper aims to review related on GEOBIA and its application for coastal mapping particularly in Indonesia. The result of this review is expected to know the role and position of Indonesia in the application and development of OBIA method.

2. GEOBIA today

GEOBIA is always associated to the several processes related with image processing. Some previous studies have developed digital image processing for various purposes such as health study, mapping and more. The paper summarizes some of the processes related to GEOBIA method, which sourced by some of previous studies (see Table 1.)

Table 1. Process related to GEOBIA.

	References
Image segmentation	[10] [11] [22] [1] [23] [9]
Feature extraction	[24] [25] [26] [13] [14] [27] [28]
Object classification	[29] [30] [15]
Developing automated method	[8] [12] [29] [16] [31]

IOP Conf. Series: Earth and Environmental Science **162** (2018) 012039 doi:10.1088/1755-1315/162/1/012039

This paper tries to prove that the image segmentation has been existed before the emergence of OBIA term in 2006 and GEOBIA in 2008 by searching from references database i.e. Scopus database [32]. The first search, by using keywords "image segmentation" with categories of "document article" and set the year of search for "all year", results 53901 articles distributes from 1954 - 2017 (Figure 1).

The second search, by using keywords "object-based image analysis" or "object-based image analysis" or "object-based classification" or "OBIA" or "GEOBIA" and using categories of "documents article" and set the years for "all year", results 1176 articles distributed from 1954 to 2017 (Figure 2). Based on the results of the search, an article was identified with the title" Object-oriented image analysis "published in 1987. In it article is written that “An image analysis workstation is described which supports the object-oriented representation and manipulation of grayscale sensed image objects” [33]. There are top of 20 journals related to Object-based Supervised Classification as shown as Figure 3 [34]. Further it proves that OBIA and GEOBIA is a basic to take into consideration in the development and application of image processing.

Formally, OBIA has a conference held every two years or biennial. The first OBIA international conference was held in 2006 at Salzburg University, Austria with the topic "Bridging Remote Sensing and GIS". Further, in its conference begin with appoint the sub-topic "automated", which is intended to develop automation trends in mapping [35]. The second conference was held in 2008 in Calgary, Alberta, Canada with the topic "GEOBIA - Pixels, Objects, Intelligence: GEOgraphic Object Based Image Analysis for the 21st Century". The sub-topic is consisted of such as "Comparative studies: object-based and pixel-based methods", "Evaluation and comparison of segmentation methods", "New classification methods", "Requirements for object-based accuracy assessment", etc. In 2008, it began to specified the scope of OBIA only into Geographical aspect, and it transform to term of GEOBIA [36]. The third conference was held in 2010 in Ghent, Belgium. Several articles discussed about the use of GEOBIA methods for the development of "Free and Open Source Geomatics solution for residential waste", "Classification of High Resolution Optical and SAR Fusion Image Using Fuzzy Knowledge and Object-Oriented Paradigm", and "OBIA for change detection of multi-temporal image"[37]. The fourth conference was held in 2012 in Rio de Janeiro, Brazil. The sub-topic is consisted of such as "Accuracy assessment, aquatic and coastal ecosystems, change detection, classification, feature extraction, LiDAR and SAR application", etc. [38]. The fifth conference was held in 2014 in Thessaloniki, Greece with the topic "Advancements, trends and challenges". This indicates that GEOBIA's development is being prepared to more advanced level with possible emerging issue is the Modifiable Area Unit Problem (MAUP) [39]. Sixth Conference was held in 2016 in Enschede, The Netherlands with the topic “Solutions and Synergies”. The sixth conference discuss about how to develop optimized, generic and transferable methods for solving standard problems in GEOBIA applications. Transferable is one of the keys of GEOBIA's success [40] The number of conference papers in each OBIA / GEOBIA conference presented in Figure 4.

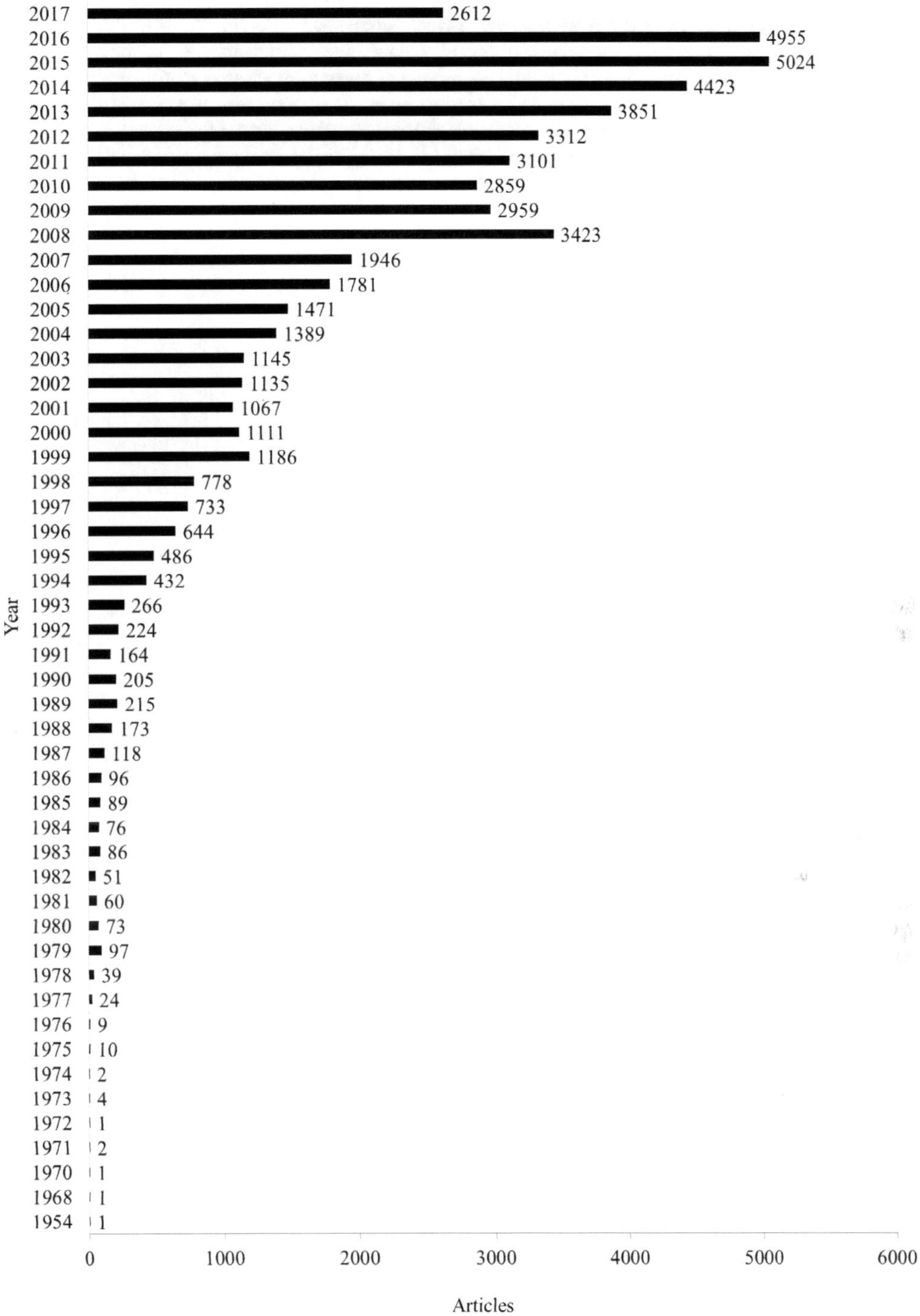

Figure 1. Number of Articles related to "image segmentation" keywords in Scopus Database (1954 – 2017)

IOP Conf. Series: Earth and Environmental Science **162** (2018) 012039 doi:10.1088/1755-1315/162/1/012039

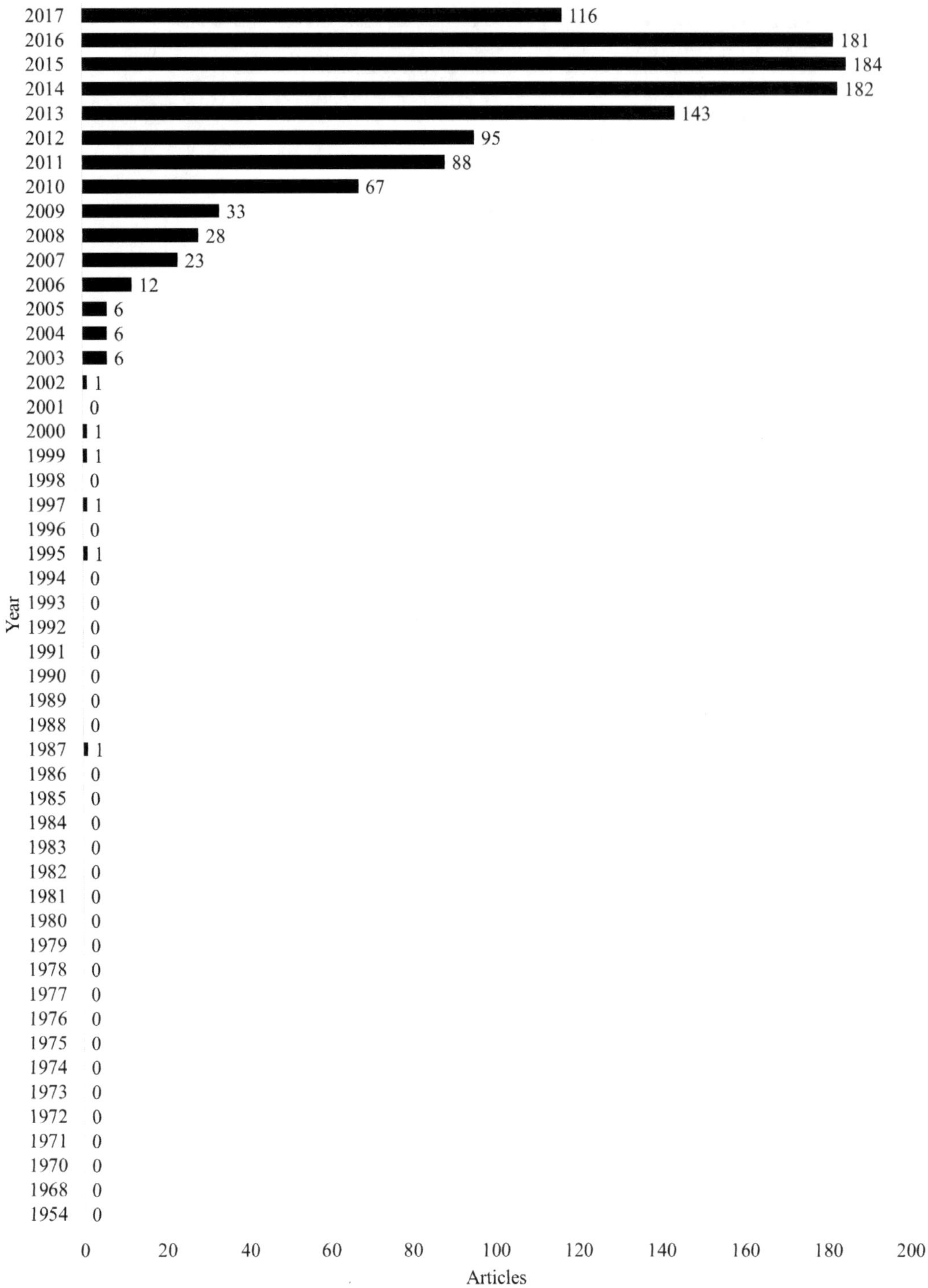

Figure 2. Number of Articles related to "object-based image analysis" OR "object-oriented image analysis" OR "geographic object-based image analysis" OR "object-based classification" OR "OBIA" OR "GEOBIA" keywords in Scopus Database (1954 – 2017)

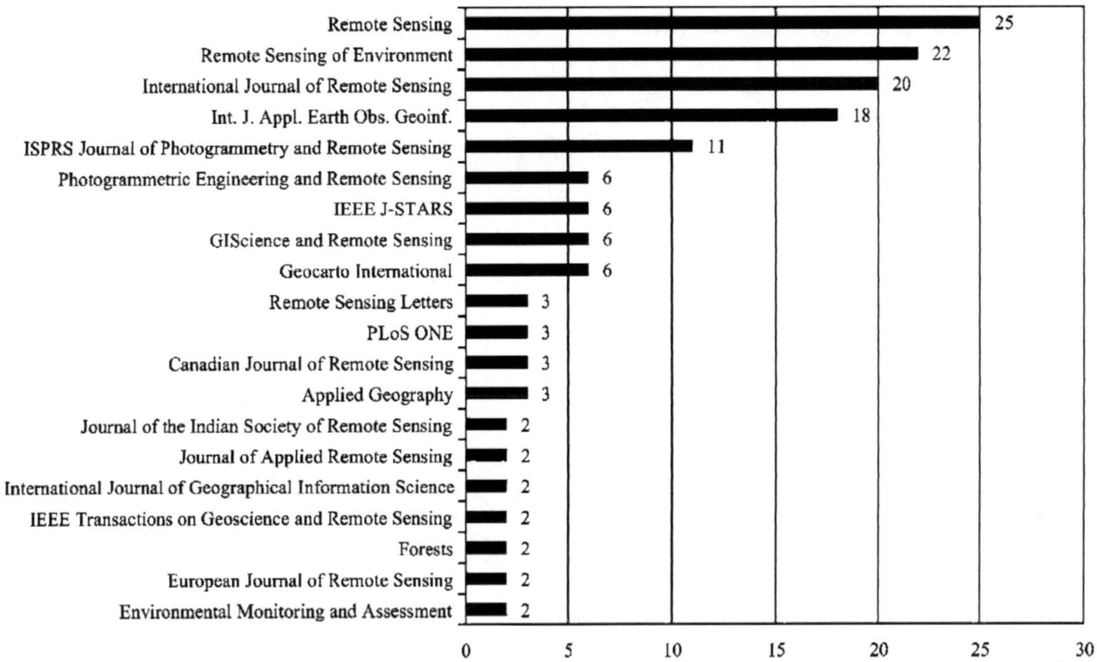

Figure 3. Top 20 journals for articles related to object-based supervised classification [34]

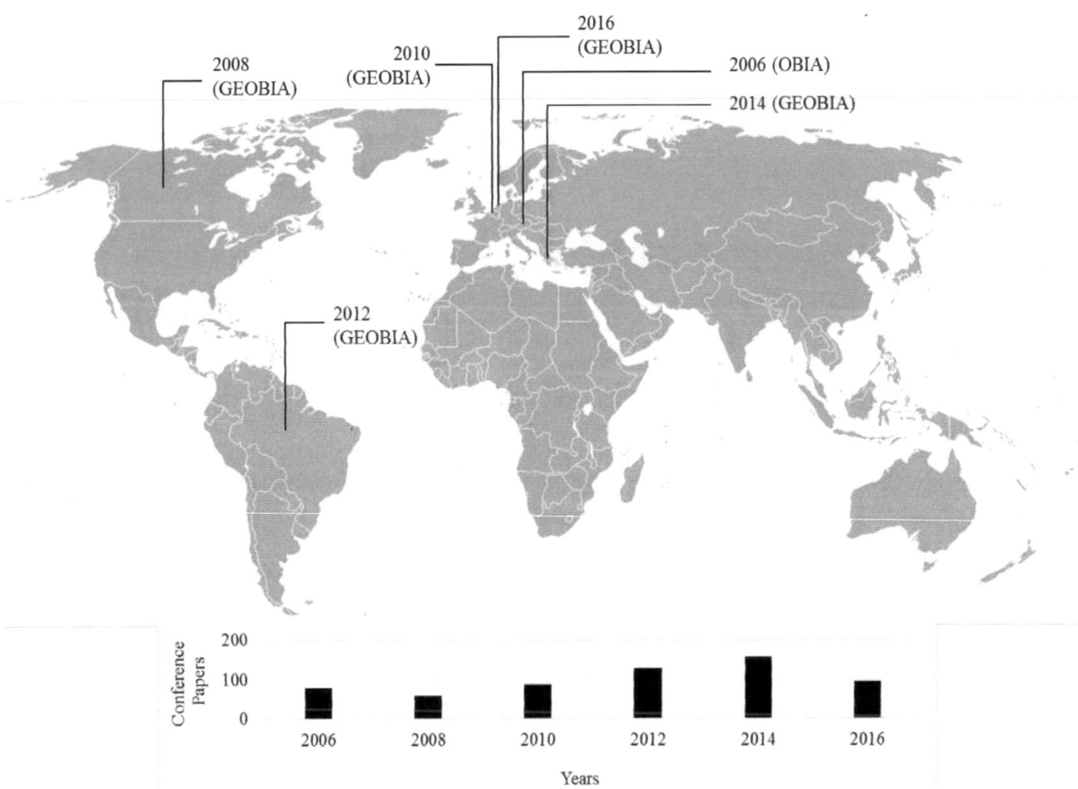

Figure 4. OBIA to GEOBIA conference trends [35] [36] [37] [38] [39] [40]

3. GEOBIA for coastal mapping research

Coastal refers to regional mosaics of habitats including intertidal habitats (mangrove, marshes, rocky, etc.), semi-enclosed bodies of water (bays, estuaries, gulf, etc.), benthic habitats (coral reefs, sea-grass bed, etc.) [41]. Researches that utilize OBIA for coral reef ecosystem has been done in decades [42] [43] [44] [45]. The fundamental difference of OBIA is in analysis unit which based on image object not a single pixel. OBIA was used for geomorphology and coral reef ecosystem mapping in three different waters and get higher accuracy [42]. Based on various sources, authors have summarizes the example of GEOBIA application for coastal mapping researches (Table 2).

Table 2. Researches about the use of GEOBIA for coastal mapping

Classification	Title	Result	Reference
Benthic habitats	Mapping coral reefs at reef to reef system scales, 10s–1000s km^2, using object-based image analysis	- Hierarchical levels: reef, reef type, geomorphic zone, and benthic community - The rule sets increased in complexity following the hierarchical level increasing - Overall accuracy ranging from 76% to 82% of 'geomorphic zone' hierarchical maps - Overall accuracy ranging from 52% to 75% of 'benthic community' hierarchical maps - These sets of hierarchical maps are an important outcome considering the differences that arise due to the extent and composition of a reef, satellite image quality, and environmental conditions.	[43]
	The Application Of Object Based Analysis Of High Spatial Resolution Imagery For Mapping Large Coral Reef Systems In The West Pacific At Geomorphic And Benthic Community Spatial Scales	- Overall accuracy of 'geomorphic zone' spatial scales map 82.1% - Overall accuracy of 'benthic community' spatial scales map 66.6% - The rule sets developed for one site could be implemented on other reefs that had similar characteristics after adjusting the thresholds for some of the mapping categories - The study demonstrates that object based image analysis is robust approach that can be applied on a variety of reef systems to create habitat maps at different spatial scales:1) reef, 2) reef type; 3), geomorphic zone, and 4) benthic community.	[46]
	Semi-Automated Object-Based Classification of Coral Reef Habitat using Discrete Choice Models	- Using a multinomial logistic model to perform an initial automated pass at classifying coral reef habitat from satellite imagery was able to attain assignment accuracies exceeding 85% - Quantifying uncertainty is important to the end-user when developing marine spatial planning scenarios and populating spatial models from reef habitat maps	[47]
	Multi-scale, object-based image analysis for mapping geomorphic and ecological zones on coral reefs	- Overall accuracies for 'geomorphic zones' of Heron (Great Barrier Reef), Ngderack (Palau), and Navakavu (Fiji) about ±80%, ±80% and ±80% respectively using GEOBIA - Overall accuracies for 'benthic community' of Heron (Great Barrier Reef), Ngderack (Palau),	[42]

Classification	Title	Result	Reference
		and Navakavu (Fiji) consecutively 77.9%, 51.6%, and 65.4% using GEOBIA	
Intertidal habitats	The Influence of Polarimetric Parameters and an Object-Based Approach on Land Cover Classification in Coastal Wetlands	- Overall accuracy of Proposed method is 87.3% - Overall accuracy of Wishart supervised classification is 66.6% - Overall accuracy of Proposed method without polarimetric parameters is 74.0% - Overall accuracy of Proposed method without an object-based analysis is 71.1% - Overall accuracy of Proposed method without textural and geometric information is 73.2 - Overall accuracy of Proposed method using the nearest-neighbor classifier is 80.5%	[48]
	Mangrove Species Identification: Comparing WorldView-2 with Aerial Photographs	- PS-WV2-VIS (pan-sharpening, visible spectral i.e., blue, green, yellow red and red-edge bands) has 89% of overall accuracy - PS-WV2-R/NIR1 (pan-sharpening, red, red-edge and NIR1 bands) has 87% of overall accuracy - AP0.14M (Aerial Photograph) has 68% of overall accuracy - AP0.5M (Aerial Photograph) has 68% of overall accuracy - WV2-VIS (without pan-sharpening, visible spectral i.e., blue, green, yellow red and red-edge bands) has 58% of overall accuracy - WV2-R/NIR1 (red, red-edge and NIR1 bands) has 42% of overall accuracy - The combination of high spatial resolution remote sensing data using a relatively large number of spectral bands within the visible, NIR region and the SVM non-linear machine learning classification technique can be a powerful way for mixed environments such as mangroves.	[17]
	An Object-Based Classification of Mangroves Using a Hybrid Decision Tree—Support Vector Machine Approach	- Using a hybrid decision-tree and SVM approach, true mangrove species and associate mangrove species were classified with an accuracy of 94% at the object level. - There is a necessity for greater spectral resolution to distinguish between the subtle differences between individual species while combine non-linear machine-learning classification techniques such as support vector machine with OBIA using very-high spatial resolution data in order to provide marked improvement in the classification of vegetation types	[49]

Classification	Title	Result	Reference
	Characterizing the Spatial Structure of Mangrove Features for Optimizing Image-Based Mangrove Mapping	- While using original and resampled WorldView-2 images, the author found that a pixel size ≤2 m was suitable for mapping canopy and inter-canopy related features within mangrove objects; - Pixel size of ≥4 m was more appropriate for mangrove vegetation formation, communities and larger mangrove features mapping; - The green and red-edge bands were optimum for discriminating smaller sized mangrove features (<8 m); - The near infrared1 band was more suitable for identifying feature ≥8 m; and - The NDVI image was suitable for mapping all targeted features. - The study site limited in Moreton Bay, Australia, with sub-tropical mangroves.	[50]
	Transferability of object-based rule sets for mapping coastal high marsh habitat among different regions in Georgian Bay, Canada	- To examine the trend (the accuracy associated with each scene was generally dependent on the origin of the training set): how accurate each rule set is with regard to one scene/region and how accurate one rule set is with regard to all scenes/regions. - Significant differences among rule sets existed at the scene level, but no single rule set emerged as being superior while applied to the other two scenes.	[51]
Coastline	Extraction of Coastline in Aquaculture Coast from Multispectral Remote Sensing Images: Object-Based Region Growing Integrating Edge Detection	- Proposed automated method (object-based region growing integrating edge detection, OBRGIE) for aquaculture coastline mapping using multispectral imagery data; - The method is robust to segmentation scale parameter; and - The method could derive coastlines with pixel level accuracy in an automated way from relatively fine-to-moderate spatial resolution imagery	[52]
	Coastal Changes and Movements in the Wider Vlora (Albania) Area	- Change Vector Analysis (CVA) is more exact result than GEOBIA approach while using images with 30m of spatial resolution; - The situation will be different when dealing with high-resolution data (10 m or higher spatial resolution).	[53]

In remote sensing aspect, coastal mapping holds onto two basic structures, geomorphologic zone, and benthic habitat which both determine classification. Besides, the moderate resolution imagery is more appropriate for studying reef geomorphology than reef biology [54]. Landscape, form, spread, compactness, and depth are features that compromise the geomorphologic zones. As well the benthic habitat assigned by the percent coverage of each benthic habitat class [55]. Research combining much information comprehends geomorphologic and biotic profusion has been done by Green *et al* [56].

Also in another benthic habitat, OBIA also implemented on sea-grass mapping projects. Baumstark *et al* [45] processed IKONOS imagery using a combination of water column correction, pixel classification, and image segmentation techniques that yield a sea-grass density index map with high spatial detail and gain overall accuracy around 77% [57]. Thus, water column correction play an important role to gain more accuracy in mapping. However, for sea-grass mapping, depth and distance from shore became significant parameters for discriminating spectrally between sea-grass and colonized hard bottom. In this case, Baumstark *et al* [45] achieved 78% overall accuracy using OBIA for Sea-grass, colonized hard bottom and sediment compared to 71% overall accuracy using the photo-interpretation method. This study revealed that OBIA as an alternative technique for offshore mapping with capability of producing higher thematic and spatial resolution maps.

In OBIA, appropriate threshold values are a major step to provide better classification results. The feature attribute used in this research consists of two; features related to objects and features related to the class [58]. Object-related features are obtained by evaluating the segmentation image objects. The features used are layer/ spectral value (average and ratio), hierarchy, thematic attributes, NDVI (normalize different vegetation index). Functions that related to the object class generated and consider the proximity of inter-class existence. The feature is a relation with neighbour objects that refer to objects that already have classes at the same level into the image object hierarchy class. Another feature used is the distance (unit pixels) to indicate centre point distance of an object reviewed with another central point that is nearby. All features employed in OBIA greatly simplified classification process of each class based on its characteristics and yielded better accuracy. By developing a new segmentation accuracy measure, we evaluated that segmentation accuracies decrease with increasing segmentation scales and the negative impact of under-segmentation errors become significantly rise at large scale.

4. GEOBIA application in Indonesia

GEOBIA application in Indonesia is still limited on the scale of research, and this is possible because GEOBIA is a new paradigm in image classification [59]. The implementations of GEOBIA method in Indonesia are not only for the coastal area but also covers of land and another theme. The classifying process of this method uses a hierarchical process of segmentation at the pre-processing stage so that the characteristics of the object can be added with some additional information such as text, texture, context, and other information related to the object [9].

The reef level and geomorphological zone level mapping using WorldView-2 with object-based classification method (OBIA) located in Pari Island able to gain accuracy respectively 97% and 87%. [44]. For benthic habitat mapping study, GEOBIA usage is produced higher accuracy (73.3% while using sun glint corrected image) compared to the per-pixel classification (64.3%) and is comparable to the result of the original image [60]. Wahidin *et al* [61] had tried classified Landsat 8 OLI image coral for reef benthic habitat mapping employing OBIA within seven classes (sand-rubble, rubble, coral, sand, sand-algae, sea-grass, and rubble-sand). The object-based classification was executed using several hybrid classification algorithms of the learning machine. The highest overall accuracy and Kappa on coral reef habitat mapping were produced by Support Vector Machine (SVM) algorithm (73% and 0.64). Then it is followed by Random Tree (RT), Decision Tree (DT), Bayesian, and k-Nearest Neighbour (KNN) algorithms of 68% & 0.59, 67% & 0.56, 66% & 0.53, and 56% & 0.46, respectively. It claimed the classification methods based OBIA approach within > 6 classes produced better accuracy than the pixel based classification technique. The other research compares the suitability of high (QuickBird 2, IKONOS, GeoEye-1, WorldView-1) and moderate (Landsat 7 ETM+, Landsat TM, ALOS) spatial resolution imageries for coral reef mapping in Indonesia. It concludes that both moderate and high spatial resolution imageries are suitable for benthic community and geomorphic zone mapping with an area more than 6.39 km^2 [62]. Kamal *et al* [63] focused on mangrove mapping research using GEOBIA and compare various imageries such as Landsat TM, ALOS AVNIR-2, WorldView-2, WorldView-2+LiDAR to get multi-scale mangrove composition. The locations are different while develop the classification rule sets and test the applicability of the rule sets. Moreton Bay in Australia as the primary site to develop the classification rule sets and Karimunjawa Island in Indonesia to test the developed classification rule sets. The use of Landsat TM, AVNIR-2, WorldView-2, WorldView-2+LiDAR data provided overall accuracies of 82%, 82%, 85%, 94% respectively for mangrove and non-mangrove mapping.

Out of coastal topics, several studies are shown OBIA for inland mapping. Thouret *et al* [64] applied OBIA for disaster, risk, and hazard. Object-oriented classification (OOC) and normalized difference spectral indices of vegetation, moisture and soil redness (NDVI, NDWI, and NDRSI) to identify and map pyroclastic and lahar deposits from the 2010 eruption of Merapi Volcano, Indonesia. This study shows that OBIA is ready to be embedded in GIS platform which can contribute to hazard and risk assessment in the area after a disaster. As well Zylshal *et al* [65] use a stereo using an object-based classification method. ESP tool was then used to determine the optimal segmentation parameter values. For further classification, decision tree algorithms were then performed to derive 11 land use/land cover classes. This gives GEOBIA an opportunity and challenges in the mapping and monitoring of land use. In the other location, Darmawan *et al* [66] use object-based approaches for land use/land cover changes identification. The process is Landsat imagery pre-processing, field survey, classification. To get an optimal segmentation, Darmawan *et al* [66] use trial and error process. That study indicates that cover type of artificial surface such as urban area, industrial and bare areas are increasing. Alrassi *et al* [67] tried to assess the GEOBIA classification method for generating the classification of land use in coastal areas of Banyuwangi, and also validated by field data. It yielded that Banyuwangi Regency is dominated by mangrove, agricultural fields, mixed farms, settlements, and ponds. Nonetheless Puturuhu *et al* [68] compared the accuracy of interpretation methods and determine the location of landslide events using Landsat 8, QuickBird, and SRTM images. Landslide candidates determined by several methods, i.e., layered visual interpretations, digital transformation with NDVI, OBIA, Toposhape, and NDVI-OBIA combinations, and NDVI-OBIA-Toposhape. The various accuracy achieved such as 47% using digital transformation with NDVI, 45% using OBIA, 47% using Toposhape, and 53% for the combination of NDVI, and the combination of NDVI-OBIA-Toposhape 53%. Landslide type found from the layered visual interpretation combined with field observations, i.e., soil rotational slide types found in 7 points (38.9%), soil creep, debris flow, earth slide, and multiple rotational slides. In this case, the best interpretation method of landslide occurrence is layered visual interpretation with 90% of accuracy.

5. GEOBIA for national map production

National coastal map producer is refers to the custodianship regulation. Based on the existing regulation [69], mangrove ecosystem, sea-grass, and coral reef map custodian in Indonesia respectively are Ministry of Environment and Forestry, Indonesian Institute of Sciences and Indonesian Institute of Sciences. There are several regulations for mapping procedure of national coastal mapping. First regulation is Head of Geospatial Information Agency Regulation No 3/2014, about the technical guidance of mangrove geospatial data collecting and processing [70]. Then Head of Geospatial Information Agency Regulation No 8/2014, about the technical guidance of collecting and processing of shallow waters habitat geospatial data [71]. On 2014, Indonesian Institute of Sciences (LIPI) has published the manual book "Technical Guidance for Sallow Waters Habitat Mapping" of which references are National Standard of Indonesia documents (SNI) and Head of Geospatial Information Agency Regulation [72]. Both of Head of Geospatial Information Agency Regulation and manual book allowed the use of OBIA/GEOBIA method for mangrove and shallow water habitat mapping.

Even OBIA/GEOBIA usage has been allowed, Indonesian Institute of Sciences (LIPI) did not use it for massive or national coastal map production. Authors argue that one of the reasons is rule sets transferability problem. For national map of coral reef and sea-grass, LIPI use the unsupervised classification on Landsat ETM+, Landsat 8 OLI, SPOT-5 and through ground truth for validation [73] [74]. The result of coral reef and sea-grass mapping using that method are displayed in figure 5 to figure 8.

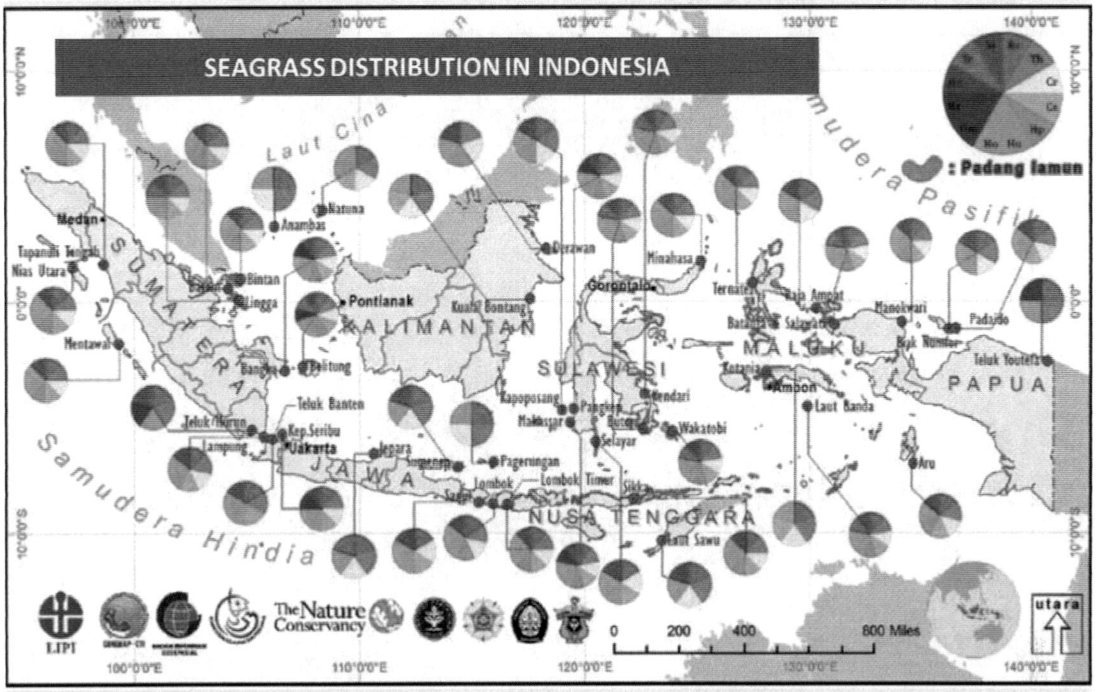

Figure 5. Seagrass distribution in Indonesia [73]

Figure 6. Seagrass status in Indonesia, 2017 [73]

Figure 7. Coral reefs distribution in Indonesia [74]

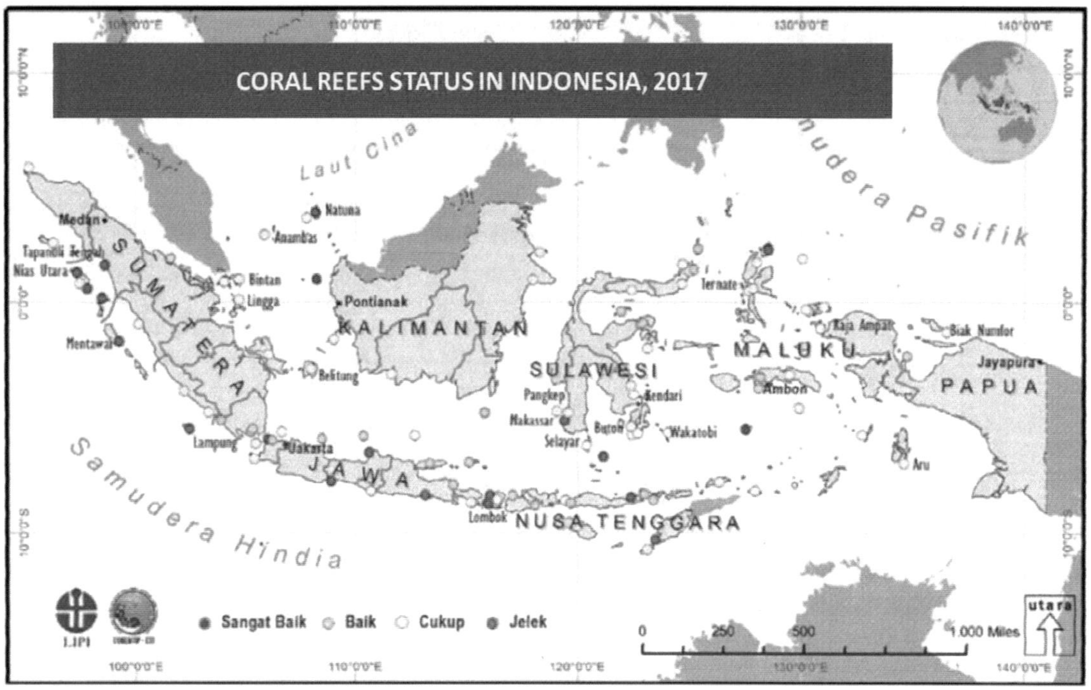

Figure 8. Coral reefs status in Indonesia, 2017 [74]

Many aspects are needed to make GEOBIA could be used operationally for national map production. Authors argue that even GEOBIA have many advantages e.g., GEOBIA could significantly contribute through raster images to vector products conversion [75], GEOBIA is still 'new paradigm' which is needs more innovative ideas and efforts to make it applicable for general conditions [59]. In addition, segmentation, sampling, feature selection, classification and accuracy assessment are kind of uncertainty factors [34]. Moreover, limited access of suitable software is become the other problems. Then the fully transferability become the main question while applying GEOBIA for massive production.

6. Conclusion

In remote sensing especially for feature extraction, manual interpretation has some weakness *e.g.* subjectivity. OBIA and GEOBIA is a basic to take into consideration in the development and application of image processing. GEOBIA has been developed for automation process and increase geometry detail. One of GEOBIA development target is mapping problem which is related to inefficiencies could be solved. Generally, GEOBIA applications in the whole of the world are still limited on the scale of research. The number use of GEOBIA for coastal mapping studies in Indonesia are still lacking, locally applicable and still developing the transferability rule sets.

National coastal map producer in Indonesia is refers to the custodianship regulation. The producer means legal institutions did not use GEOBIA for massive or national coastal map production. Authors argue that one of the reasons is rule sets transferability problem. Many aspects are needed to make GEOBIA could be used operationally for national map production. Segmentation, sampling, feature selection, classification and accuracy assessment are kind of uncertainty factors. Moreover, limited access of suitable software is become the other problems. Then the fully transferability become the main question. The innovation and effort are needed to make GEOBIA method applicable for massive production of Indonesian coastal map. The richness of Indonesian coastal resources should be the main factors why Indonesia must be involved in developing GEOBIA method.

Acknowledgments

The authors very pleased and gratefully acknowledge Geospatial Information Agency especially Centre of Research, Promotion and Cooperation for the financial support in the process of research.

References

[1] Cheng G and Han J A 2016 Survey on object detection in optical remote sensing images *ISPRS J. of Photogrammetry and Remote Sensing*. **117** pp 11-28

[2] Singh A 1989 Digital change detection techniques using remotely-sensed data *Int. J. Remote Sensing*. **10(6)** pp 989-1003

[3] Snook P W, Winterbenrger K C, Merrit N E, and Czaplewski R L 1987 Comparison between Digital and Manual Interpretation of High Altitude Aerial Photographs *Photogrammetric Engineering and Remote Sensing*. **53(5)** pp 531-534

[4] Husson E, Ecke F, and Reese H 2016 Comparison of Manual Mapping and Automated Object-Based Image Analysis of Non-Submerged Aquatic Vegetation from Very-High-Resolution UAS Images *Remote Sensing*. **8(9)** pp 724

[5] Natural Resource Canada 2015 [Online] cited 2017. Available from: http://www.nrcan.gc.ca/node/9303

[6] Blaschke T and Strobl J 2001 What's wrong with pixels? Some recent developments interfacing remote sensing. *Zeitschrift fur Geoinformationssysteme*. 6 pp 12-17

[7] Hay G J and Castilla G 2006 Object-Based Image Analysis: Strengths, Weaknesses, Opportunities and Threats (SWOT). In *Bridging Remote Sensing and GIS: 1st Int. Conf. on Object-based Image Analysis (OBIA 2006)*. ed Lang S, Blaschke T, and Schöpfer E (Salzburg: ISPRS)

[8] Blaschke T and Lang S 2006 Object Based Image Analysis for Automated Information Extraction - A Synthesis. In *Measuring The Earth II: Latest Developments with Digital Surface Modeling and Automated Feature Extraction*. (San Antonio: ASPRS)

[9] Blaschke T 2010 Object based image analysis for remote sensing. ISPRS *J. of Photogrammetry and Remote Sensing*. **65(1)** pp 2-16
[10] Beaulieu J and Goldberg M 1989 Hierarchy in Picture Segmentation: A Stepwise Optimization Approach *IEEE Transactions on Pattern Analysis and Machine Intelligence.* **11(2)** pp 150-163
[11] Pal N R and Pal S K 1993 A Review on Image Segmentation Techniques *Pattern Recognition.* **26(9)** pp 1227-1294
[12] Hay G J and Castilla G 2008 Geographic Object-Based Image Analysis (GEOBIA): A new name for a new discipline. *Object-Based Image Analysis. Lecture Notes in Geoinformation and Cartography*. ed Blaschke T, Lang S, Hay G J (Berlin, Heidelberg: Springer) pp 75-89
[13] Tomljenovic I, Tiede D and Blaschke T 2016 A building extraction approach for Airborne Laser Scanner data utilizing the Object Based Image Analysis paradigm *Int. J. of Applied Earth Observation and Geoinformation.* **52** pp 137-148
[14] Wang Y, Wang J and Wang Q 2009 An Expert System for Road Extraction from Remote Sensing Image. *In 3rd Int. Symp. on Intelligent Information Technology Application Workshops.* (NanChang: IEEE Computer Society) pp 125-128.
[15] Arvor D, Saint-Geours N, Dupuy S, Andrés S and Durieux L 2013 Identifying optimal classification rules for geographic object-based image analysis. In *XVI Simpósio Brasileiro de Sensoriamento Remoto*. (Iguaçu) pp 2290-2297
[16] Hay G J, Castilla G, Wulder M A and Ruiz J R 2005 An automated object-based approach for the multiscale image segmentation of forest scenes. *Int. J. of Applied Earth Observation and Geoinformation.* **7(4)** pp 339-359
[17] Heenkenda M K, Joyce K E, Maier S W and Bartolo R 2014 Mangrove Species Identification: Comparing WorldView-2 with Aerial Photographs *Remote Sensing* **6(7)** pp 6064-6088
[18] United Nations 1982 *United Nations Convention on the Law of the Sea* (Montego Bay). pp 36
[19] Republik Indonesia 1985 *Undang Undang No. 17 Tahun 1985 Tentang : Pengesahan United Nations Convention On The Law Of The Sea (Konvensi Perserikatan Bangsa Bangsa Tentang Hukum Laut).* (Jakarta)
[20] Badan Informasi Geospasial 2015 [Online] cited 2017 July 05. Available from: http://www.bakosurtanal.go.id/berita-surta/show/pentingnya-informasi-geospasial-untuk-menata-laut-indonesia
[21] Nag O S 2017 *World Atlas.* [Online] cited 2017 July 05. Available from: http://www.worldatlas.com/articles/countries-with-the-most-coastline.html
[22] Drăguţ L, Csillik O, Eisank C and Tiede D 2014 Automated parameterisation for multi-scale image segmentation on multiple layers *ISPRS J. of Photogrammetry and Remote Sensing.* **88** pp 119-127
[23] Drăguţ L, Tiede D, and Levick S R 2010 ESP: a tool to estimate scale parameter for multiresolution image segmentation of remotely sensed data *Int. J. of Geographical Information Science.* **24(6)** pp 859-871
[24] Ruiz L A, Reico J A, Sarría F, and Hermosilla T 2011 A feature extraction software tool for agricultural object-based image analysis *Computers and Electronics in Agriculture*. **76** pp 284-296.
[25] Wang J, Treitz P M, and Howarth P J 2007 Road network detection from SPOT imagery forupdating geographical information system in the rural-urban fringe. *Int. J. of Geographical Information System.* 2007 **6(2)** pp 141-157
[26] Nussbaum S, Niemeyer I, and Canty M J 2006 SEATH - A New Tool for Automated Feature Extraction in the Context of Object-Based Image Analysis. In *Bridging Remote Sensing and GIS: 1st Int. Conf. on Object-based Image Analysis (OBIA 2006).* (Salzburg)
[27] Kurtz C, Passat N, Gançarski P, and Puissant A 2012 Extraction of complex patterns from multiresolution remote sensing images: A hierarchical top-down methodology *Pattern Recognition*. **45(2)** pp 685-706

[28] Ghassemian H and Landgrebe D A 1988 Object-Oriented Feature Extraction Method for Image Data Compaction *IEEE Control System Magazine*. pp 42-48
[29] Castilla G and Hay G J 2007 An Automated Delineation Tool for Assisted Interpretation of Digital Imagery. In *ASPRS 2007 Annual Conf. : Identifying Geospatial Solutions.* (Tampa: ASPRS). pp 137-146
[30] Walter V 2004 Object-based classification of remote sensing data for change detection *ISPRS J. of Photogrammetry & Remote Sensing.* **58(3-4)** pp 225-238
[31] Sebari I and He D 2013Automatic fuzzy object-based analysis of VHSR images for urban objects extraction *ISPRS J. of Photogrammetry and Remote Sensing*. 79 pp 171-184
[32] Elsevier B V 2017 *Scopus.* [Online] cited 2017 July 04. Available from: https://www.scopus.com/home.uri
[33] Rearick T C 1987 Object-Oriented Image Analysis. In *Image Understanding and the Man-Machine Interface*. (Los Angeles: SPIE) pp 177-180
[34] Ma L, Li M, Ma X, Cheng L, Du P, and Liu Y 2017 A review of supervised object-based land-cover image classification *ISPRS J. of Photogrammetry and Remote Sensing*. **130** pp 277-293
[35] Lang S, Blaschke T, and Schöpfer E ed 2006 Bridging Remote Sensing and GIS. *1st Int. Conf. on Object-based Image Analysis (OBIA 2006).* (Salzburg, Austria: ISPRS).
[36] Hay G J, Blaschke T, and Marceau D ed 2008 *GEOBIA 2008 - Pixels, Objects, Intelligence: GEOgraphic Object Based Image Analysis for the 21st Century.* (Calgary, Canada: ISPRS)
[37] Addink E A and Van Coillie F M B ed 2010 GEOBIA 2010: Geographic Object-Based Image Analysis. (Ghent, Belgium: ISPRS)
[38] Feitosa R Q, Pedro da Costa G A O, Maria de Almeida C, Fonseca L M G, and Kux H J H ed 2012 *Int. Conf. on Geographic Object-Based Image Analysis, 4. (GEOBIA).* itors. (Rio de Janeiro, Brazil: The Brazilian National Institute for Space Research (INPE))
[39] Gitas I, Mallinis G, Patias P, Stathakis D, and Zalidis G ed 2014 GEOBIA 2014 Advancements, trends and challenges, 5th Geographic Object-Based Image Analysis Conf. *South-Eastern European J. of Earth Observation and Geomatics*. **3(2S)**. (Thessaloniki, Greece)
[40] Kerle N, Gerke M, and Lefevre S ed 2016 *GEOBIA 2016 : Solutions and Synergies.* (Enschede: University of Twente Faculty of Geo-Information and Earth Observation (ITC)).
[41] UNESCO, IOC, WMO, UNEP, ICSU, FAO, and IGBP 2003. *The integrated Strategic Design Plan for the Coastal Ocean Observations Module of the Global Ocean Observing System.* (Paris:UNESCO). pp 35
[42] Phinn S R, Roelfsema C M, and Mumby P J 2012 Multi-scale, object-based image analysis for mapping geomorphic and ecological zones on coral reefs *Int. J. of Remote Sensing.* **33(12)** pp 3768-3797
[43] Roelfsema C, Phinn S, Jupiter S, Comley J, and Albert S 2013 Mapping coral reefs at reef to reef-system scales, 10s–1000s km2, using object-based *Int. J. of Remote Sensing.* **34(18)** pp 6367-6388
[44] Anggoro A, Siregar V P, and Agus S B 2015 Pemetaan Zona Geomorfologi Ekosistem Terumbu Karang Menggunakan Metode OBIA, Studi Kasus di Pulau Pari *J. Penginderaan Jauh.* **12(1)** pp 1-12
[45] Baumstark R, Duffey R, and Pu R 2016 Mapping seagrass and colonized hard bottom in Springs Coast, Florida using WorldView-2 satellite imagery *Int. J. of Remote Sensing.* **181** pp 83-92.
[46] Roelfsema C, Phinn S, Jupiter S, Comley J, Beger M, and Paterson E 2010 The Application of Object Based Analysis of High Spatial Resolution Imagery for Mapping Large Coral Reef System in The West Pacific at Geomorphic and Benthic Community Spatial Scales. In *IGARSS 2010 - 2010 IEEE Int. Geoscience and Remote Sensing Symp*. (Honolulu) pp 4346-4349.
[47] Saul S and Purkis S 2015 Semi-Automated Object-Based Classification of Coral Reef Habitat using Discrete Choice Models *Remote Sensing*. **7(12)** pp 15894-15916

[48] Chen Y, He X, Jing W, and Xiao R 2014 The Influence of Polarimetric Parameters and an Object-Based Approach on Land Cover Classification in Coastal Wetlands *Remote Sensing*. **6(12)** pp 12575-12592

[49] Heumann B W 2011 An Object-Based Classification of Mangroves Using a Hybrid Decision Tree—Support Vector Machine Approach *Remote Sensing*. **3(11)** pp 2440-2460

[50] Kamal M, Phinn S, and Jogansen K 2014 Characterizing the Spatial Structure of Mangrove Features for Optimizing Image-Based Mangrove Mapping *Remote Sensing*. **6(2)** pp 984-1006

[51] Rokitnicki-Wojcik D and Wei A 2011Transferability of object-based rule sets for mapping coastal high marsh habitat among different regions in Georgian Bay, Canada *Wetlands Ecology and Management*. **19** pp 223-236

[52] Zhang T, Yang X, Hu S, and Su F 2013 Extraction of Coastline in Aquaculture Coast from Multispectral Remote Sensing Images: Object-Based Region Growing Integrating Edge Detection *Remote Sensing*. **5(9)** pp 4470-4487

[53] Kanjir U and Gregorič Bon N 2016 Coastal Changes and Movements in the Wider Vlora (Albania) Area. *GEOBIA 2016 : Solutions and Synergies*. Kerle N, Gerke M, and Levefre S, ed. (Enschede)

[54] Bainbridge S J and Reuchelt R E 1988 An assessment of ground truth methods for coral reef remote sensing. *Proceedings of the 6th International Coral Reef Symposium*. Choat J H et al ed. (Townsville) pp 439-444

[55] Andrefouet S *et al* 2003 Multi-site evaluation of IKONOS data for classification of tropical coral reef environments *Remote Sensing of Environment*. **88** pp 128-143

[56] Green E P, Mumby P J, Edwards A J, and Clark C D 2000 *Remote Sensing Handbook for Tropical Coastal Management: Coastal Management Sourcebooks 3rd* edition. Edwards A J ed (Paris: UNESCO)

[57] Baumstark R, Dixon B, Carlson P, Palandro D, and Kolasa K 2013 Alternative spatially enhanced integrative techniques for mapping seagrass in Florida's marine ecosystem *Int. J. of Remote Sensing*. **34(4)** pp 1248-1264

[58] Trimble 2014. *Ecognition developer: User guide*

[59] Blaschke T *et al* 2014. Geographic Object-Based Image Analysis – Towards a new paradigm *ISPRS J. of Photogrammetry and Remote Sensing*. 87 pp 180-191

[60] Wicaksono P and Adhimah F 2017 Accuracy Assessments of Pan-Sharpened Image for Benthic Habitats Mapping *J. of Geomatics and Planning*. **4(1)** pp 27-40

[61] Wahidin N, Siregar V P, Nababan B, Jaya I, and Wouthuyzen S 2015 Object-based image analysis for coral reef benthic habitat mapping with several classification algorithms. In *The 1st Int. Symp. on LAPAN-IPB Satellite for Food Security and Environmental*. (Elsevier B.V) pp 222-227

[62] Adji A S 2014 Suitability Analysis of Multispectral Satellite Sensors for Mapping Coral Reefs in Indonesia, Case Study: Wakatobi Marine National Park *Mar. Res. in Indonesia*. **39(2)** pp 73-78

[63] Kamal M, Phinn S, and Johansen K 2015 Object-Based Approach for Multi-Scale Mangrove Composition Mapping Using Multi-Resolution Image Datasets *Remote Sensing*. **7(4)** pp 4753-4783

[64] Thouret J C, Kassouk Z, Gupta A, Liew S C, and Solikhin A 2015 Tracing the evolution of 2010 Merapi volcanic deposits (Indonesia) based on object-oriented classification and analysis of multi-temporal, very high resolution images *Remote Sensing of Environment*. **170** pp 350-371

[65] Zylshal, Yulianto F, Pasaribu J M, and Prasasti I 2015 Landuse/Landcover Extraction From SPOT 6 Imagery Using Object Based Image Analysis Approach: A Case Study of Jakarta. In *The 36th Asian Conference of Remote Sensing*. (Quezon City, Metro Manila, Philippines)

[66] Darmawan S, Sari D K, and Nugroho H 2013 Object Based Approaches for Identification of Land Use/Land Cover Changes Using Multitemporal Landsat Imagery in Bandung, West Java,

IOP Conf. Series: Earth and Environmental Science **162** (2018) 012039 doi:10.1088/1755-1315/162/1/012039

Indonesia. *The 34th Asian Conf. on Remote Sensing.* Pramono G H, Ramdani D, Barus B, Ariansyah R M ed. (Bali: Asian Association on Remote Sensing). pp 702-709

[67] Alrassi F, Salim E, Nina A, Alwi L, Danoedoro P, and Kamal M 2016 GEOBIA for Land Use Mapping Using WorldView-2 Image in Bengkak Village Coastal, Banyuwangi Regency, East Java. In *2nd Int. Conf. of Indonesian Society of Remote Sensing (ICOIRS).* (IOP Publishing)

[68] Puturuhu F, Danoedoro P, Sartohadi J, and Srihadmoko D 2017 The Development of Interpretation Method for Remote Sensing Imagery in Determining the Candidate of Landslide in Leitimur Paninsula, Ambon Island *Jurnal Ilmu Lingkungan.* **15(1)** pp 20-34

[69] Badan Informasi Geospasial 2015 *Keputusan Kepala Badan Informasi Geospasial No 54 Tahun 2015 tentang Wali Data Informasi Geospasial Tematik.* (Bogor)

[70] Badan Informasi Geospasial 2014 *Peraturan Kepala Badan Informasi Geospasial No 3 Tahun 2014 tentang Pedoman Teknis Pengumpulan dan Pengolahan Data Geospasial Mangrove.* (Bogor)

[71] Badan Informasi Geospasial 2014 *Peraturan Kepala Badan Informasi Geospasial No 8 Tahun 2014 tentang Pedoman Teknis Pengumpulan dan Pengolahan Data Geospasial Habitat Dasar Perairan Laut Dangkal.* (Bogor)

[72] Prayuda B 2014 *Panduan Teknis Pemetaan Habitat Dasar Perairan Laut Dangkal.* Suyarso ed. (Jakarta: CRITC COREMAP II LIPI)

[73] Hermawan U E, Sjafrie N D M, Supriyadi I H, Suyarso, Iswari M Y, Anggraini K, and Rahmat 2017 *Status Padang Lamun Indonesia 2017.* (Jakarta: Puslit Oseanografi-LIPI)

[74] Giyanto, Abrar M, Hadi T A, Budiyanto A, Hafizt M, Salatalohy A, and Iswari M Y 2017 *Status Terumbu Karang Indonesia 2017.* Suharsono ed. (Jakarta: Puslit Oseanografi – LIPI)

[75] Johansen K, Bartolo R, and Phinn S 2010 Special Feature - Geographic Object-Based Image Analysis *J. of Spatial Science.* **55(1)** pp 3-7

IOP Conf. Series: Earth and Environmental Science **162** (2018) 012040 doi:10.1088/1755-1315/162/1/012040

Utilization of AIS (Automatic Identification System) sensor for mobility monitoring of fishing vessel based on PFZ (Potential Fishing Zones) distribution

S M Purnama, D Y Hutapea and R Permala

Remote Sensing Technology and Data Center, LAPAN, Indonesia

E-mail: sylmuflichin@gmail.com

Abstract. Indonesia is known worldwide as an archipelagic country with a vast territory and a country consisting of many islands. The length of the coastline reaches about 81,000 km2, so Indonesia is the largest archipelagic country in the world with two thirds of its territory consisting of oceans. One of the efforts that need to be conducted is to improve the economy and accelerate the development of the marine sector. With this policy, the security of marine and marine resource management is very important. Exclusive Economic Zone (EEZ) and continental foundation are vital areas that need to be guarded and guaranteed for the benefit of maritime sovereignty of the Republic of Indonesia.The development of science and technology of remote sensing, LAPAN in cooperation with IPB to make satellite LAPAN / IPB-A3 or A3LISat. These satellites serve as monitoring of paddy field resources. These satellites are orbiting 14 times a day in the territory of Indonesia is equipped with AIS (Automatic Identification System) sensor where the sensor is to monitor the mobility of the ship. The AIS sensor can provide location coordinate information and temporarily obtain ships vessel information in Indonesian seas. This research tried to utilize the AIS sensor and combined with PFZ (Potential Fishing Zone) information result of remote sensing data and analyzes the mobility of fishing boats recorded by A3 satellite whether it is in accordance with the PFZ information extracted from remote sensing satellite LISat data.

1. Introduction

[1] National Institute of Aeronautics and Space (LAPAN) is always working to improve product quality and information technology in the field of aviation and space to create an advanced and independent Indonesia. LAPAN effort is to bring technology is satellite LAPAN A2 and A3 (LISAT) LAPAN IPB Satellite. The satellite is fully the result of work of the nation and made in Indonesia. The objective of LAPAN-A2 satellite project is to achieve the design, integration and operation of micro satellite in Indonesia. LAPAN-A2 main mission is Earth Observation (video/matrix RGB camera surveillance). In addition to that the satellite carry AIS (Automatic Identification System) receiver to monitor maritime traffic, a reaction wheel made by LAPAN (LPN-001) for space proofing, and an amateur radio Automatic Packet Reporting System (APRS), as well as amateur radio voice repeater, for Indonesian Amateur Radio Organization (ORARI).

[2] The LAPAN-A3/IPB, also referred to as LISat, is a cooperative remote sensing microsatellite project between LAPAN (National Institute of Aeronautics and Space of Indonesia) Jakarta and IPB (Bogor Agricultural University or Institute Pertanian Bogor) located in Bogor, Indonesia. LAPAN-

IOP Conf. Series: Earth and Environmental Science **162** (2018) 012040 doi:10.1088/1755-1315/162/1/012040

A3/IPB was launched on June 22, 2016 (03:56 UTC) as a secondary payload (piggy back) to ISRO's CartoSat-2C spacecraft. The launch site was SDSC (Satish Dhawan Space Center) in India and the launch vehicle is PSLV-C34. Orbit of LAPAN A3 satellite are Sun-synchronous circular orbit, altitude = 515 km, inclination = 97.5°, local time on descending node (LTDN) = 9:30 hours. The LAPAN-A3/IPB mission, this polar sun-synchronous orbit will make about two times contact per day (day and night) of about 11 minutes average. In this limited contact scenario, much of data will be downlinked in near real time to the ground station through - band communication link that contain the information of data imagery and shipping monitoring data as well.

[3] LAPAN-A3 satellite brings a multispectral imager as main payload which has four color channel, i.e.: red, green, blue and near-infrared. The imager optics system uses beam-splitter mechanism, which divides the incoming light passing lens system into each color channel detector.

This research is to utilize the information of fishing vessel data obtained from AIS sensor and combined with PFZ information resulted from remote sensing data utilization. This study analyzes the mobility of fishing vessels recorded by LAPAN A2 and A3 satellite. The mobility of the fishing vessel is appropriate, knowing and utilizing the PFZ (Potential Fishing Zone) information.

2. Research Background

[4] The importance of the maritime sector in Indonesia has been known for a long time, even long before the Republic of Indonesia was formed, when it was still in the form of small kingdoms with their respective interests in each region. In development in the marine sector in various aspects, one of which is increasing the catch of fish with fishing vessel . [5] The main problem of optimizing fish catch especially pelagic fish is the very limited data and information about oceanographic condition which is closely related to fishing potential area. Identification of potential fishing areas using remote sensing technology is an indirect means of identification. From remote sensing data, identification of oceanographic parameters that are closely related to fish habitat or areas that are suspected of potential place for fish, such as the occurrence of thermal front or upwelling. This technology is known as PFZ (Potential Fishing Zone).

[6] The Automatic Identification System (AIS) was developed to provide identification and location information to vessels and shore stations with the aim of exchanging different types of data including position, identification, course, speed and others. On one hand this allows vessels to anticipate and thus avoid collisions at sea by means of continuous traffic monitoring with several navigation aids, while on the other hand, AIS also offers important ship monitoring services to coastal guards as well as search and rescue organizations.

One of the LAPAN-A2 and LAPAN-A3 missions is maritime traffic monitoring using AIS. AIS information can be used to analyze the distribution of fishing vessel and potential fishing zone. [8]The benefits of AIS (Automatic Identification System) can be used for An Oil Spill Tracking Application.

3. Research Method

To utilization of AIS sensor for mobility monitoring of fishing vessel based on PFZ distribution, there are two kinds of material. Primary data for LAPAN A2 and A3 satellite data with our AIS and secondary data information using data vector for administrative Boundary of Indonesia.

Table 1. Material Research

	Data	Period	Location
Primary Data	LAPAN A2 and A3	Jan-Des 2016 and Jan-May 2017	Indonesia
Secondary Data	Vector Data: Administrative Boundary	2015	Indonesia

The selection of data period for LAPANA2 and A3 satellites is the last 1.5 years from this research period, expected to know the condition of Indonesian waters at this time.

[6]For LAPAN-A2 Satellite, AIS data is downloaded using Biak Ground Station, which is equipped with 11 meter S-Band (2220MHz) antenna. The downloaded file is stored in a binary format and uploaded to a cloud server. The Bogor Ground Station downloaded the file for extraction and processing. The processed data produce ship information such as geolocation, time, and other properties, presented in a text file. Workflow of the system LAPAN A2 can be described as shown in the picture below in figure 1.

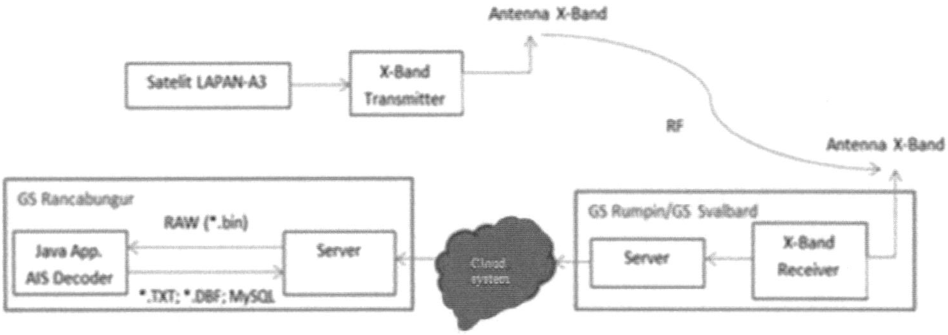

Figure 1. Workflow Process Downloaded LAPAN-A2 Information

The operations are carried out in the development includes:

a. Modules to read the AIS raw file. It is tasked to locate the frame starting point, frame ending point, and calculate the size of the frame. Each frame contains dozens of AIS data.
b. Modules to locate the code word that represents single data lines, which identify the AIS data.
c. Modules to perform CRC16-CCITT calculation and distinguish between bad CRC16-CCITT (represented by 1) and good CRC16-CCITT (represented by 0).
d. Modules to perform decoding of all the data according to its respective code word.

[7] For LAPAN-A3 Satellite, AIS data is downloaded using Svalbard Ground Station in Norway. The station is also equipped with 11 meter X-Band (8200 MHz) antenna. The file is stored in binary format and uploaded to a cloud server. The Bogor Ground Station downloaded the file for extraction and processing. The processed data produce ship information such as geolocation, time, and other properties, presented in text file. Workflow of the system LAPAN A3 can be described as shown in the picture below in figure 2.

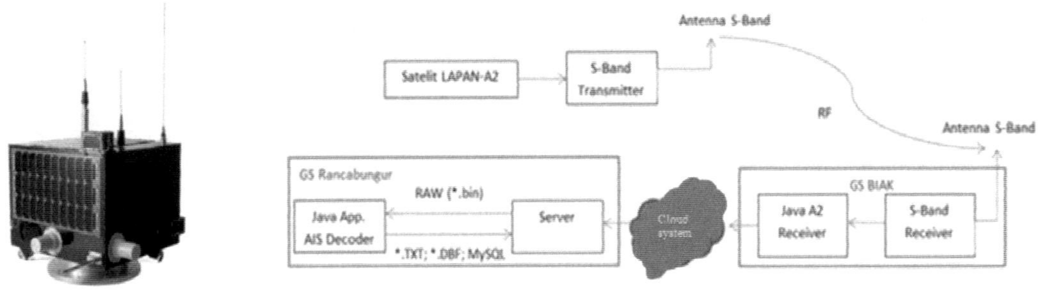

Figure 2. Workflow Process Downloaded LAPAN-A3 Information

The operations are carried out in the development includes:

a. Modules to read the AIS raw file and remove the CCSDS format (header and parity). It is roled to locate the frame starting point, frame ending point, and calculate the size of the frame. Each frame contains dozens of AIS data.
b. Modules to locate the code word that represents single data lines, which identify the AIS data.
c. Modules to perform CRC16-CCITT calculation and distinguish between bad CRC16-CCITT (represented by 1) and good CRC16-CCITT (represented by 0).
d. Modules to perform decoding of all the data according to its respective code word.

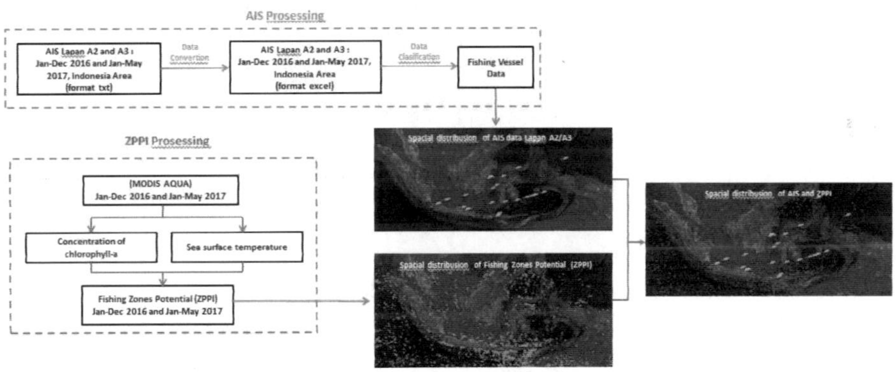

Figure 3. Mapping Flow AIS and PFZ

In Figure 3 mapping flow AIS with PFZ. The AIS data on the classification of the fishing vessel is subsequently displayed spatially so that the distribution is visible. The results are then combined with potential fishing zone (PFZ) data to determine the distribution of fishing vessels are appropriate or not appropriate to PFZ. Data PFZ is extraction of chlorophyll-a concentration and sea surface temperature. PFZ data obtained from weather satellites MODIS AQUA from January to December 2016 and January to May 2017.

4. Result

AIS LAPAN A2 / A3 data provides information on recording time, MMSI (Maritime Mobile Service Identity), navigation, and location. Information MMSI is a series of 6-9 digit numbers sent in digital data through a radio frequency channel with the purpose of being the unique (unique) identity of a ship to the ship station. MMSI information describes ship name information and ship size according to MSSI database. Navigation information describes ship activity and ship type. Location information is the coordinate location information when the data is recorded by satellite. This data is very important because as a reference in knowing the distribution of ships in Indian Ocean.

Based on AIS LAPAN A2/A3 data of 2016 and 2017, the distribution of fishing vessels tends to focus on Indonesian eastern waters, around Banda Sea and Timor Sea. Fishing vessels spread across the Java Sea, Makassar Strait, Celebes Sea, Molucca Sea, Banda Sea, Flores Sea and Halmahera Sea.

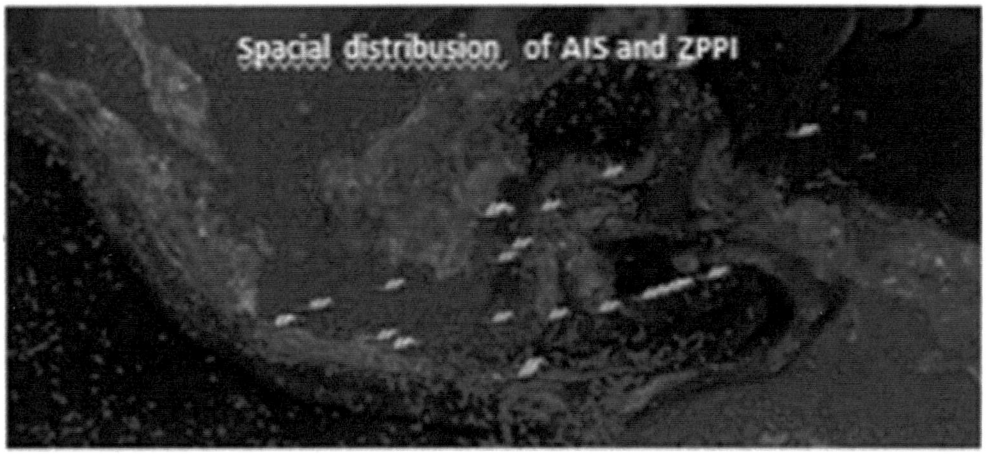

Figure 4. Result Mapping AIS and PFZ

5. Conclusion

After plotting with PFZ data, the analysis showed that not all fishing vessels are in fish potential location. There are several ships on the high seas and there are also ships that are leaning on the harbour. AIS LAPAN A2/A3 data is very helpful in identifying and monitoring the fishing vessel. Fishing vessel information can be determined using this AIS data, but only on the level of ship type identification. AIS data has not been able to provide information up to the level of ship activity when at a recording location, such as loading and unloading of goods, illegal fishing and illegal oil transactions. To be more accurate the activity information can be supported with ground check data.

Acknowledgments

We thank Mr. Dedi Irawadi, Chief of Remote Sensing Technology and Data Center LAPAN, for supporting us. We would also like to show our gratitude to Mr. Rubini Jusuf for the permission to attend to this Seminar.

References

[1] Wuryandani, et all. 2015 *Mewujudkan Agenda Prioritas NAWACITA*. (Jakarta: P3DI Setjen DPR RI dan AZZA Grafika)
[2] LAPAN. 2015. *Roadmap Reformasi Birokrasi LAPAN 2015-2019*
[3] https://directory.eoportal.org/web/eoportal/satellite-missions/l/LAPAN-a2
[4] Tahir AM, Hakim PR, Syafruddin AH 2016 Peningkatan kualitas focus citra imager multispectral satelit LAPAN-A3 *Jurnal Teknologi Dirgantara* **14** 37-50

IOP Conf. Series: Earth and Environmental Science **162** (2018) 012040 doi:10.1088/1755-1315/162/1/012040

[5] Nainggolan PP 2015 Kebijakan poros maritime dunia Joko Widodo dan implikasi internasionalnya *Politica* **Vol.6** 167-190
[6] Hasyim B 2004 Penerapan informasi zona potensi penangkapan ikan (PFZ) untuk mendukung usaha peningkatan produksi dan efisiensi operasi penangkapan ikan *Pengantar ke falsafah sains* 2-14
[7] Cervera M, Ginesi A and Eckstein K 2009 Satellite-based vessel automatic identification system: a feasibility and performance analysis *Int. J. Satell. Commun. Network* **29** 117-142
[8] Schwehr K, McGilivary PA 2007 Marine ship automatic identification system(AIS) for e nhanced coastal security capabilities: an oil spill tracking application *Proceedings of the 2007 OCEANS conference* 1131-39
[9] Satellite Technology Center LAPAN 2012 Assembly, Integration and Test Documentation.
[10] Satellite Technology Center LAPAN 2016 Assembly, Integration and Test Documentation.

Salt pond analysis using ALOS PALSAR case study Sampang, Madura-Indonesia

S M Purnama, A Cahyawati and D Y Hutapea

Remote Sensing Technology and Data Center, LAPAN, Indonesia

E-mail: sylmuflichin@gmail.com

Abstract. Indonesia is a country that has a large sea area of about 96,079.15 km2. With this condition we can utilize the natural resources that are infested in the sea, such as fish, and salt. Indonesia has exceptional salt pond commodities from various areas such as Cirebon, Sampang, Indramayu Madura, and Rembang. These resources need to be monitor and inventory in proper way. The development of science and technology, there is an effective and efficient way to identify, monitoring and inventory of salt ponds using remote sensing technology. In this last period the saltwater pond infiltration is still inclined to utilize optical sensor system to monitor and infuse. The method used is visual interpretation through multipolarization composite radar image by using interpretation key. The weakness of passive sensors is affected by the presence of cloud cover. The location of salt ponds in Indonesia is mostly located on the coastal regions that is visually affected cloud cover. In addition the use of active sensor is become a challenging for this aspect. This study was conducted to analysis of hues, patterns, and textures in order to recognize the salt pond object using Alos PALSAR Image (Active Sensor). The salt pond object in multi-polarized composite HH-HH / HV composite radar images is recognizable in dark blue with regular pattern plots with characterized by brighter / clear / bright pink ridges.

1. Introduction

Indonesia has a long coastline, potentially for salt pond. Salt pond is one of superior products in marine field. In recent years salt has become one of the national strategic commodities. Salt Pond is an artificial shallow pond designed to produce salt from sea water or salt water. People's salt production is almost spread throughout Indonesia. Based on the distribution map of people's salt production site In 2015 the land area reaches 25,830.34 Ha [1]. With the area of salt ponds that are so widespread and scattered throughout Indonesia, so monitoring activities is necessary. The most effective and efficient for monitoring is by using remote sensing technology. Remote sensing in the broader sense, measurement or information acquisition of some nature of the object or phenomenon, using a recorder that is physically not in direct contact or intersection with the object or phenomenon studied [2]. Remote sensing system consists of several components [3]. The series of components includes: 1) power source, 2) atmosphere, 3) objects, 4) sensors, and 5) data acquisition and data usage.

Remote sensing technology is divided into two sensor systems, passive sensors and active sensors. Passive sensor is an optical sensor that utilizes sunlight as energy to record an object on the surface of the earth. Active sensor is a radar sensor with its own lighting system to be able to record objects on the surface of the earth. In the case of monitoring and calculation of salt ponds in coastal areas, optical sensors have a weakness that is cloud cover and foggy compared to active / radar sensors. Remote sensing images with active / radar sensors are helpful in monitoring activities of salt ponds because they have the ability to penetrate clouds and fog.

This research uses ALOS PALSAR radar image to perform the identification of salt pond object. The Phased Array L-Band type Synthetic Aperture Radar (PALSAR) is an active microwave sensor that uses L-Band frequencies to achieve groundless observations on a day and night basis. Frequency L-Band Alos PALSAR 1.3 GHz with 10m spatial resolution.

Sampang Madura is selected as a study area because the area has 3,064,55 Ha with production of 398,983.61 ton. Based on data of salt people productivity Year 2015, Sampang is the second highest production in Indonesia after Cirebon. The purpose of this research is to identify and analyze the object of salt pond based on color, hue, and texture using ALOS PALSAR image for Sampang region, Madura, East Java Province.

2. Material and Method

2.1 Time and Research Location

This research was conducted in February 2017 located in Sampang, Madura, East Java, map of research location:

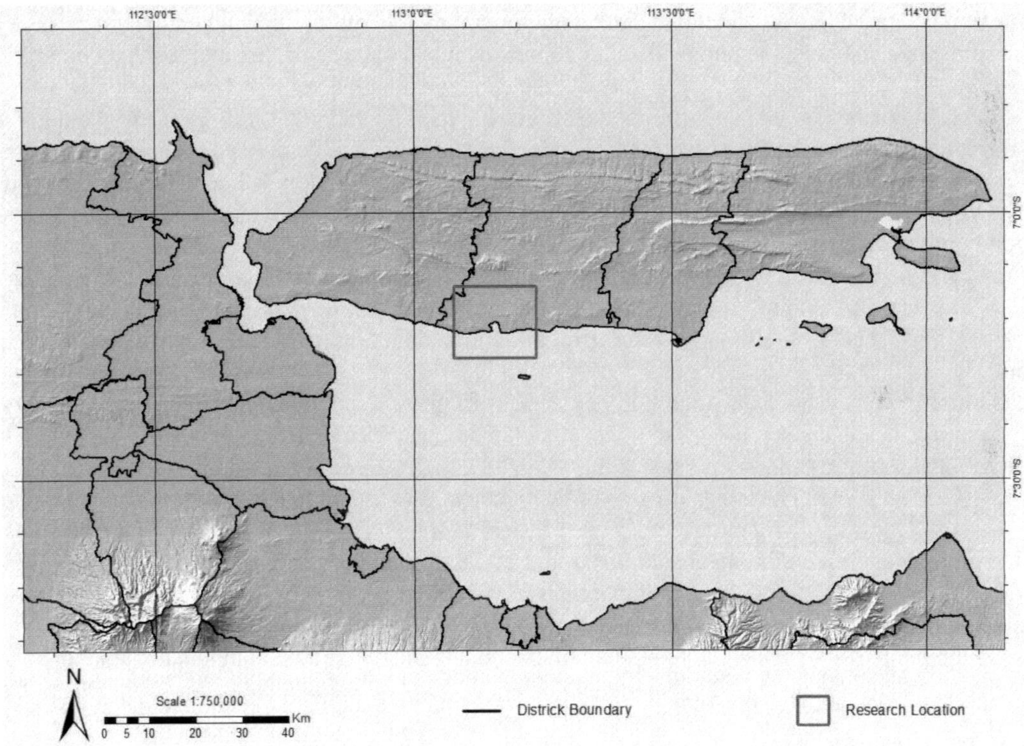

Figure 1. Map Location

2.2 Data Collection

There are of material in this research, namely:

a) Primary data

Raster data : ALOS PALSAR FBD level RT1, Frame 7040, Path 426

Resolution Data : 12,5 meter

Time data : June 20, 2009

Data Location : Sampang, Madura, East Java

2.3 Software and Hardware

There are two software in this research, namely :

a) ENVI 4.5
b) ArcGIS 10.3

There are hardware in this research :

a) Laptop Dell Latitude E7440
   - RAM 8 GB
   - Storage SSD 256 GB

2.4 Analisis Data

ALOS PALSAR RT1 level is remote sensing data that has been corrected terrain level that has been calibrated gamma naught so that satellite image of study area is not affected by topography effect such as layouver (inward fall effect), foreshortening (shortening of the front slope), and shadow (shadow effect).

Radar image processing takes into account the conversion of the slant range to ground range. Varying qualities will be obtained from an independent looks great with a different azimuth. Image taken have variations in pixel size and data processing can be done in a Geographic Information System (GIS). Digital radar processing includes conversion and compression of radar signals in both directions. Azimuth and distance into a format that can be interpreted by the interpreter. In the processing there are steps that include compression range, azimuth compression, signal detection, multilook processing, and data storage [4].

Geometric correction can be grouped into two types namely orbital geometry model and transformation based on field control points [5]. Geometry models can be divided menajdi orbital aspect ratio correction, correction skewness, and correction of the Earth's rotation. While the transformation by GCP (Ground Control Point) consists of geometric correction by rectifying imagery to map and correction of geometric rectification of the image to the image [5]. In general, geometry correction aims to correct the image that has coordinates that correspond to geographical coordinates, matching the position of the image with the image or map that have been corrected or kooordinat transforming the image, and then register the image to the map so as to produce images with a particular projection system.

Radiometric correction radar image is done using filtering techniques. This filtering is a way to mitigate or reduce the appearance of speckle on a radar image. Speckle trlihat as black spots and white or commonly known as 'salt and pepper'. Filters can be done in a variety of window sizes such as 3x3, 5x5, 7x7, 9x9, or 11x11. The larger the size of the window, then the image looks more refined or 'blur'. Here are some techniques filters are often used in research are: (1) Filter Lee used to refine spots or speckle on image to maintain image sharpness and of detail while reducing noise in the image. Pixels that do filter is replaced by the pixel values of the surrounding nearby; (2) Filter Frost serves to reduce noise while maintaining edge on radar imagery [6].

2.5 Metode and Flow Chart

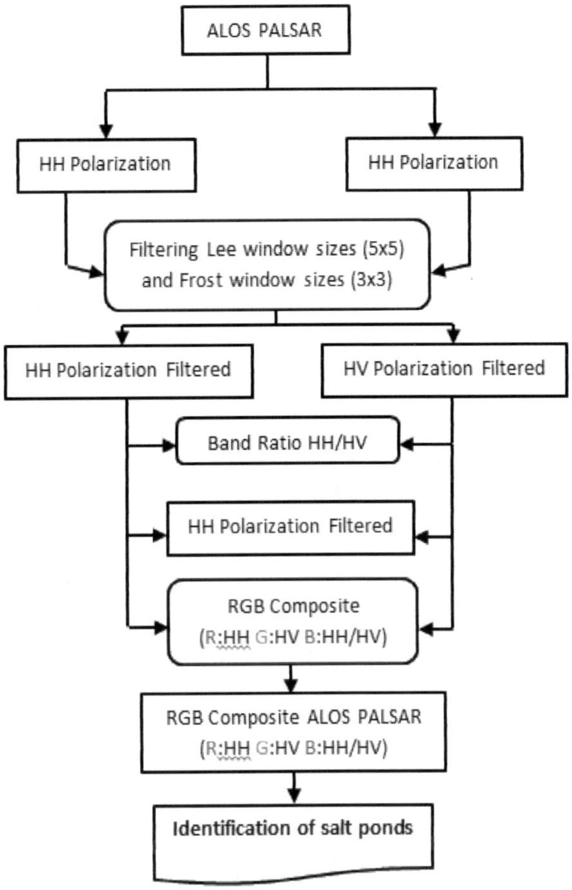

Figure 2. Research flow chart.

3. Processing Data

SAR images are calibrated using a logarithmic scale decibel (dB) and should pay attention to the logarithmic nature of these values when doing statistical calculations. To determine the average value of each part of the image, the calculation should be based on the values of the power scale. The scale power value can then be converted back into a logarithmic scale that represents the values of each pixel dB SAR data [7]. There are three kinds kalibrasipada SAR data and is described in the figure below.

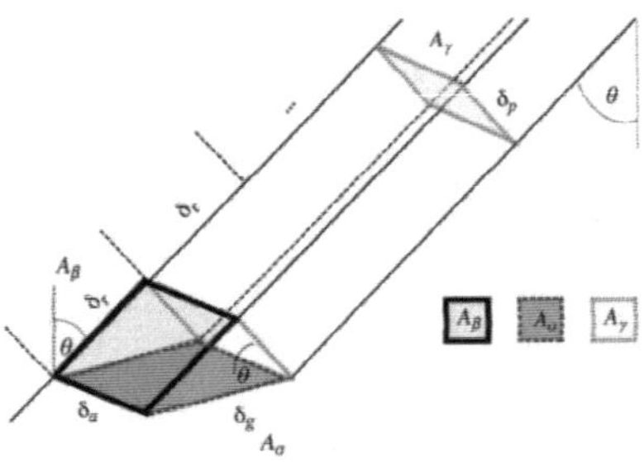

Figure 3. Illustrated of the reference areas for the three radar backscatter beta, sigma, and gamma naught by [8].

To explain the actual equations in each calibration, it must use logarithmic scale with decibels (dB) is 10log10. Thus the equations of the three calibration types in dB units become by [9]:

$$\text{Sigma naught } (\sigma o) = 10 * 10\log 10\ (DN2) + CF$$

With a DN is a Digital Number and CF is a predetermined calibration factor of -83 for HH and HV polarization, then the result of sigma naught on the ellipsoid earth model or the Geocoded Terrain Corrected product (GTC), as well as be developed using gamma naught into Radiometric Terrain Corrected (RTC).

$$Gamma\ naught\ (\gamma^o) = \sigma^o \frac{A_{flat}}{A_{slope}} \left(\frac{Cos\ \alpha_{ref}}{Cos\ \alpha_{loc}}\right)^n$$

Aflat is the value of the image pixel size under flat earth conditions, Aslope is the image pixel size based on actual field conditions (topographic variations). Aref is the incident angle calculated at the midpoint of the image scene for ALOS PALSAR image with a large nadir-off angle of 34.3 obtained values for αref = 38.7° whereas for αloc is the local incidence angle of each pixel based on topographic variation. The value of n is a constant adjusted to the depth or thickness of an object ranging from 0 to 1, for the study of vegetation is usually the value of n = 1 because the thickness of the canopy is difficult to determine the depth or thickness of an object ranging from 0 to 1, for the study of vegetation is usually Value of n = 1 because the thickness of the canopy is difficult to determine.

However, ALOS PALSAR RT1 level still influenced by noise (speckle), to overcome this, then do the filtering on each polarization, lee filter with window size 5x5 and frost 3x3. Lee filters are used to smooth the spots or speckle on the image by maintaining the sharpness of the image and detail while reducing the noise in the image. The pixel that the filter does is replaced by nearby pixel values; While Frost Filter serves to reduce noise while maintaining the edge of the radar image.

4. Result

The identification of salt ponds on ALOS PALSAR imagery is more focused on visual interpretation through the ecological spatial approach associated with the tidal coast. The appearance of the salt pond area is more easily identified by looking at the embankment pattern of the pond, in indicating the firm appearance of embankment ponds, the polarization is better than the HH polarization. This is because HH polarization is more sensitive to angular reflections.

Comparison of backscatter scattering values on HH (Top) and HV (Middle) polarization in pond embankment objects is significant, -1db to 2db for HH and -4db polarization to -5db for HV polarization. The image used has been corrected to the terrain level (RTC / radiometric terrain corrected) and calibrated gammanaught. HH polarization is also more sensitive in distinguishing areas of salt ponds that are still in the form of water and ponds that have been dried (no salt).

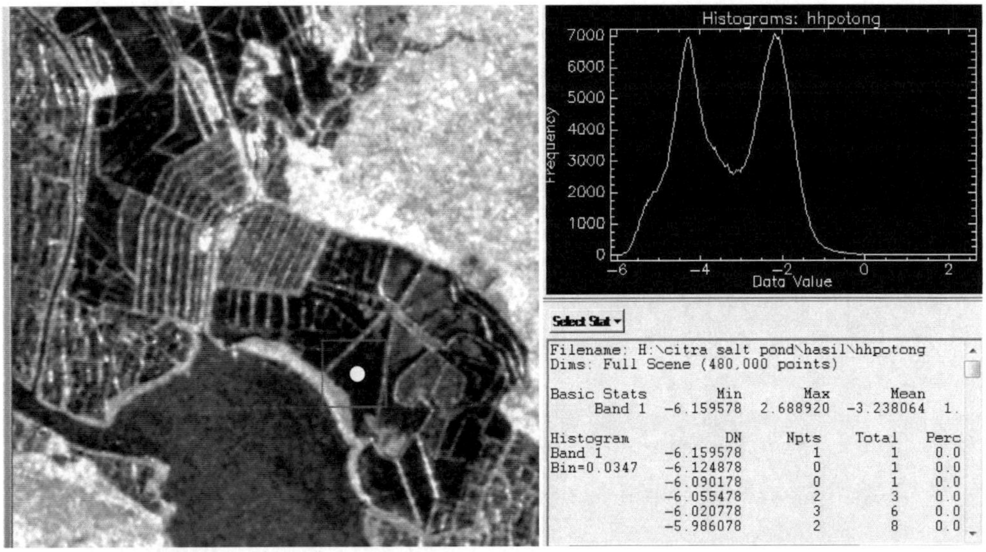

Figure 4. Illustration of HH polarization and histogram digital number

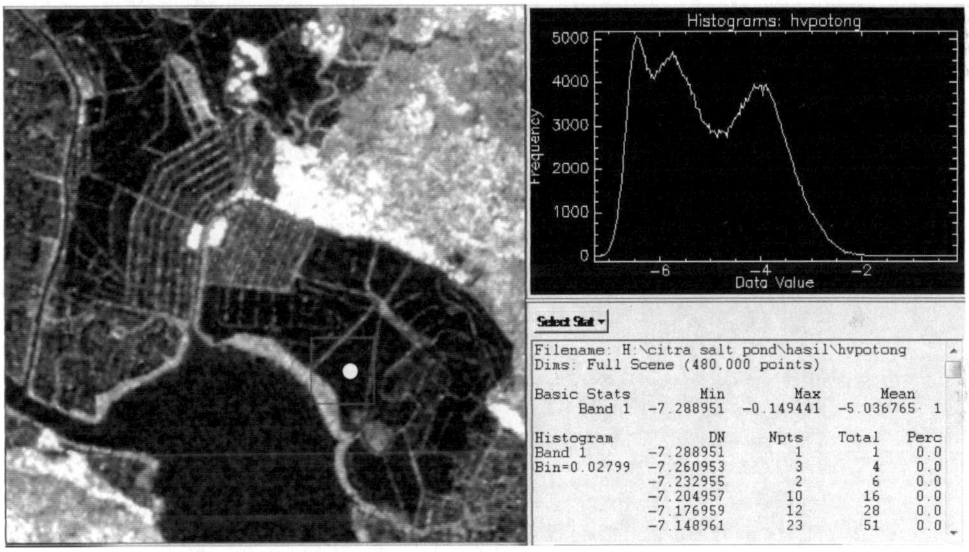

Figure 5. Illustration of HV polarization and histogram digital number

IOP Conf. Series: Earth and Environmental Science **162** (2018) 012041 doi:10.1088/1755-1315/162/1/012041

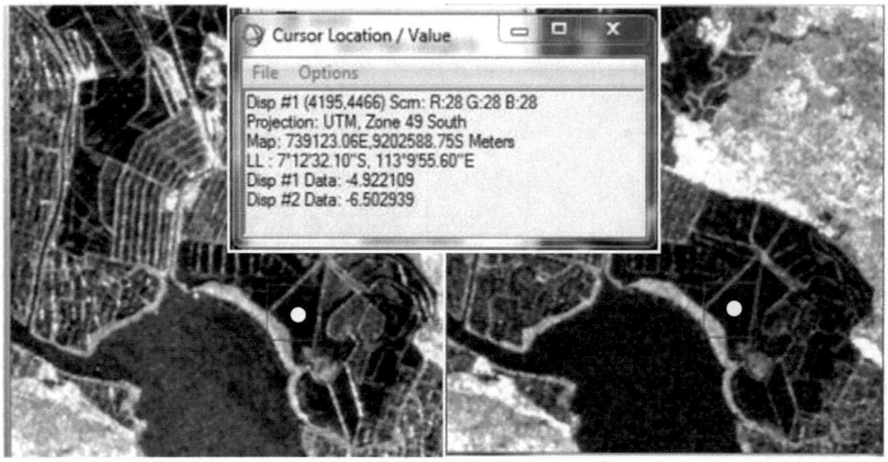

Figure 6. Comparison of digital values (object at the yellow point).

The results of the research are 3 polarization bands R:HH, G:HV and B:HH/HV. ALOS PALSAR with 3 bands then performed band ratio with HH polarization and HV polarization to get the new band. The new band is a ratio of HH/HV, the band was used in this study for the identification of salt ponds.

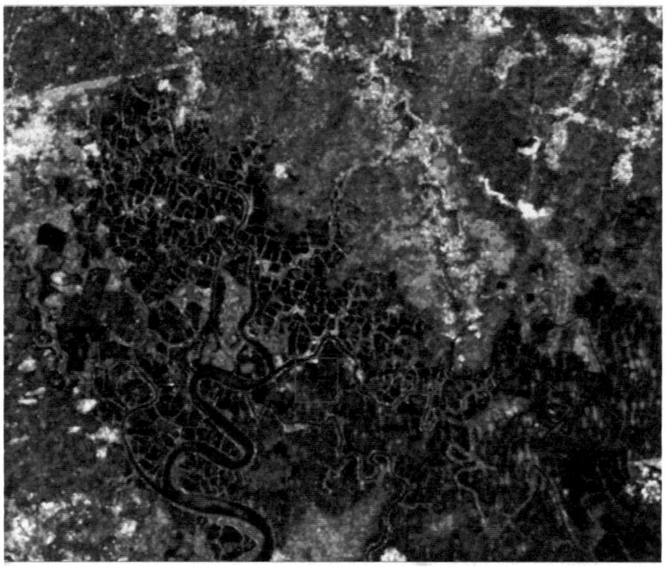

Figure 7. RGB Composite ALOS PALSAR

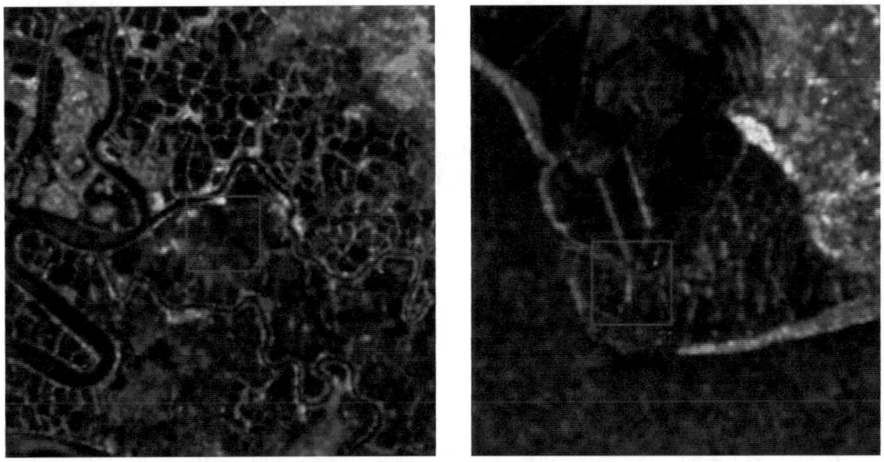

Figure 8. Paddy Field (left) and Salt Pond (right)

HH polarization is colored red, HV is colored green, and HH/HV is colored blue. The result of mixing of 3 colors / bands is obtained, vegetation objects especially forest (dominated scattering volume) will be green. The green color is because HV polarization is sensitive to volume scattering.
The object of the path, or the settlement (dominated by angular reflections) on the blue composite image, the greater the angular reflection that occurs (eg on objects with sharp corners such as bridges, or building corners) the object will be purple. For a water object will display a blue color (reflection mirror), HH polarization is also sensitive to mirror reflections[6].

Salt ponds on composite images have patterns and shapes similar to paddy fields, with plot patterns. What distinguishes is a salt pond in it containing water and open land (when drying) so that the radar image looks blue (mirror reflection). While bunds (dividers between ponds) appear on a purple image, different for paddy fields (farmland) where paddy fields are shrubs or dirt roads interspersed with shrubs and trees (radar images appear green).

5. Conclusion

The conclusion is a visual interpretation of spatial ecological approach ALOS PALSAR can identify salt ponds. Own ecological approach focuses on the shape, pattern and spatial association to the coastal ecology. ALOS PALSAR has featured can penetrate clouds so it is appropriate for the analysis of coastal areas. So ALOS PALSAR can be used for identification, monitoring and calculation of the salt pond area.

Acknowledgements
We thank Mr. Dedi Irawadi, Chief of Remote Sensing Technology and Data Center LAPAN, for supporting us. We would also like to show our gratitude to Mr. Rubini Jusuf for the permission to attend to this Seminar.

References

[1] Direktorat Jenderal Pengelolaan Ruang Laut 2015
[2] Howard, John A 1996 *Remote Sensing of Forest Resources: Theory and Application* (Diterjemahkan oleh Hartono, Dulbahri, Suharyadi, Danoedoro, dan Jatmiko) London: Chapman & Hall
[3] Sutanto 1987 *Penginderaan Jauh Jilid* . Yogyakarta Gadjah Mada University Press
[4] Bakosurtnal 1998 *Penginderaan Jauh Radar* Proyek Evaluasi dan Perencanaan Sumberdaya Laut
[5] Danoedoro P 2012 *Pengantar Penginderaan Jauh Digital,* Yogyakarta: Andi Offset.
[6] Leeuw Carvalho 2009 *Performance Evaluation of Several Adaptive Speckle Filters for SAR*

IOP Conf. Series: Earth and Environmental Science **162** (2018) 012041 doi:10.1088/1755-1315/162/1/012041

Imaging. Anais XIV SImposio Brasileiro de Sensoriamento Remoto, Natal, Brasil, 25 – 30 April 2009. INPE, p. 7299 – 7305

[7] Lavalle, Marco., Wright, T. 2009. *Absolute Radiometric and Polarimetric Calibration of ALOS PALSAR Products*. Calibration_palsar_products_**v13**.doc

[8] Atwood et al 2012 *Improving PolSAR Land Cover Classification With Radiometric Correction of the Coherency Matrix* IEEE Journal of Selected Topics in Applied Earth Observation and Remote ASensing p 1939 – 1404

[9] Imhoff M L 2002 *Radar backscatter and biomass saturation: ramifications for global biomass inventory* Geoscience and Remote Sensing: **Vol 33** p 511-518

[10] Cahyawati, Atika 2017 *Analisis citra alos palsar untuk estimasi stok karbon atas permukaan pada tegakan di tiap ekosistem hutan di SPTN I Sukadana, Taman Nasional Gunung Palung* Undergraduate thesis

IOP Conf. Series: Earth and Environmental Science **162** (2018) 012042 doi:10.1088/1755-1315/162/1/012042

Identification of flood area in the coastal region using remote sensing in Karawang Regency, West Java

Bambang Riadi[1], Baba Barus[2], Widiatmaka[2], Moh Yanuar[3], Bambang Pramudya[4]

[1]Study Natural Resources and Environmental Management
[2]Department of Soil Science and Land Resources
[3]Department of Civil and Environmental Engineering
[4]Department of Agroindustrial Technology
Bogor Agricultural University, Jalan Raya Dramaga, Babakan Dramaga, Bogor - Indonesia

E-mail: bambang.riadi@big.go.id, briadi_jasinfo@yahoo.com

Abstract. Flooding often occurs in the lowlands, while the area is usually a rice production centres which affects the rice crop. Rice fields is generally located on fluvial plains and usually located in coastal areas, with relatively lower altitudes than roads or settlements. Physically wetland is an area flooding hazard and wetlands tend to longer floods than surrounding areas.. Flood-hazard areas are formed due to sedimentation and usually at the confluence of river. This area is a fertile area for agriculture, has shallow ground water which is well suited for residential and urban areas. Karawang regency is one of the largest producers of rice commodities in West Java, therefore the Karawang regency government faces a challenge in maintaining the level of rice production due to global climate change that worsen the threat of flood hazard, in addition to land use change and increasing population. Based on the results of the spatial analysis from period of 2000 to 2006 the wetland area have been converted up to 7,493 ha. For the period of 2006 to 2015 wetland area have been converted up to 14,416 ha. Utilization of Geographic Information System with a model of Cellular Automata - Marcov Chain is an appropriate method for studying the phenomenon of expansion of build ups land in coastal region. The primary data used in the research is a land cover map of 2000, 2006 and 2015, which were the results of the multispectral classification of Landsat. The prediction of land cover change for the next fifteen years shows a continuous increase in built up area. The expansion of built up area and the possibility of widespread flood hazard areas, have inverse relationship with the decreasing of agricultural land area.

1. Introduction

Flood disasters often occur in the rainy season, especially in coastal areas, this incident is a major issue in Indonesia. Flooding is an event where normally dry land becomes flooded by water, this is caused by high rainfall and lowland upland conditions and concave areas. Geomorphological phenomenon provides information that any landform of flood-formed can inform about flood vulnerability and its characteristic. Determining the zone of flood hazard areas using remote sensing satellite imagery can be done by combining the flood phenomenon and the ability of satellite data.

IOP Conf. Series: Earth and Environmental Science **162** (2018) 012042 doi:10.1088/1755-1315/162/1/012042

Delineation of flooded areas needs to be done quickly, as it is an important spatial information to support emergency response. The level of vulnerability of flood affected areas is identified from the character of its territory such as landform, river-side slopes, meandering, natural containment, and the presence of flood control building. Landforms such as alluvial plains, alluvial valleys, river bends, and swamps are areas susceptible to flooding because it is a low or hollow region with a slope of <2% [8]. The development of information technology presents a challenge in the distribution of spatial information of flood areas that can be determined more quickly, precisely and accurately. The parameters used are land form, land cover relief, slope, soil type and rock type which is analyzed by Geographic Information System and validated by using secondary data such as data of inundation area and field survey result data. The indicator variable that has a large influence on the occurrence of floods is land cover, which is an open field land conversion from the land used for recharge becomes built up area. The purpose of this study is to describe and map the flood-hazard areas, as well as the assessment of land cover changes using Landsat-8 satellite images of 2015 in Karawang regency.

Methods for identifying flood hazard areas using satellite imagery data can be done by identifying areas that have response to flooding. The Landsat-8 image is a multispectral image that can present the earth's surface physical information from a region, the information identified from the image using the regional wetness level parameter. Band 5 (NIR) is a useful channel in the manufacture of watersheds with land that produce very low reflections. Band 5 (NIR) and Band 7 (SWIR 2) are efficient in distinguishing wet and dry soil so that wetland boundaries can be delineated. Tasseled Cap Transformation (TCT) is a mathematical formula for calculating brightness, greenness, and wetness of digital numbers in each band (band 1 to band 5 and band 7) on Landsat images [6]. Calculation of greenness level by TCT method can only be applied using Landsat satellite image, TCT can also be used to analyze dry land, wetlands. The use of spectral value transformation is an attempt to refine objects in digital data satellite imagery that will generate new information. Transformation of spectral values can be divided into two, namely (1) sharpen information, (2) summarize information from the number of existing channels [7].

Utilization of satellite imagery for flood identification has advantages compared with conventional methods, because by using the image of early delineation activity of flood and puddle areas can be done before it is validated. The wet / flooded region has a spectral response that looks darker than the unflooded area that looks bright / red. This flooded and puddle-hazard map when overlaid with river hydrological network maps will be more attractive because it can therefore be monitored for areas with the potential to experience subsequent flooding if river flow or rainfall continues to increase. The accuracy of land cover information will facilitate the analysis of the planning and development of a region. The Landsat Image 8 of 2015 in this study is used to identify current land cover conditions and wet areas indicated as flooded areas, has a resolution of 30 m and 12 bit pixels, this image data can be accessed for free, and is an opportunity for maximum utilization.

1.1. Landsat-8 Imagery

The Landsat-8 satellite (Table.1.) used on this project, has an On-board Operational Land Imager (OLI) and Thermal Infrared Sensor (TIRS) sensor with 11 bands, among these bands, 9 bands (bands 1-9) are in OLI and 2 others (Bands 10 and 11) on TIRS [1].

Table 1. Specifications spectral sensor canals imagers LDCM (Landsat-8)

Sensor	Band	Wavelength (μm)	Resolution (m)	Sensor
1. Operational Land Imager (OLI)	*Band 1*	*0.43 - 0.45*	*30*	*Visible (Coastal aerosol)*
	Band 2	*0.45 - 0.51*	*30*	*Visible (Blue)*
	Band 3	*0.53 - 0.59*	*30*	*Visible (Green)*
	Band 4	*0.64 - 0.67*	*30*	*Visible Red (Red)*
	Band 5	*0.85 - 0.88*	*30*	*Near Infrared*
	Band 6	*1.57 - 1.65*	*30*	*SWIR 1*

	Band 7	*2.11 - 2.29*	*30*	*SWIR 2*
	Band 8	*0.50 - 0.68*	*15*	*Panchromatic*
2. Thermal	*Band 9*	*1.36 - 1.38*	*30*	*Cirrus*
Infra-red	*Band 10*	*10.60 - 11.19*	*100*	*TIRS 1 (Thermal Infrared)*
Sensor (TIRS)	*Band 11*	*11.50 - 12.51*	*100*	*TIRS 2 (Thermal Infrared)*

Source : Butler,2013 (*Obtained from USGS website)*

2. Method

This research applies descriptive qualitative method based on the evaluation of the existing condition, research variables include the spatial extent obtained from the results of spatial analysis with multi time. Classification of spatial analysis of land use change using topography map of year 2000, satellite image of Landsat year 2006 and year 2015. Result of extraction of topographic map of 2000 scale 1: 25.000 as reference in analysis of land use change. The steps to be implemented in this study are presented in the flow chart of the research method attached to (Figure.1).

For the purposes of identifying flood-hazard areas the Tasseled Cap Transformation (TCT) method is used, the transformation of spectral values is an attempt to sharpen objects in digital data satellite imagery that will generate new information. The new coefficients are analyzed in detail to confirm Landsat-based imagery 8, TCT indicating that the new set comes from the Landsat 8 OLI coefficient. This process as a continuation of data continuity through TCT since 1972 re-formulated TCT for OLI on Landsat 8, this technique will be very useful to study the estimated land cover for the future [2]. Land surface objects have varied spectral responses in wet / waterlogged and dry / unfertilized areas. In the vegetated areas, during floods there is an increase in reflectance values in visible channels (band 1 - 4 OLI). In contrast to Infrared channels (NIR, SWIR, MIR) decreased. This condition is different for soil objects (open land), where both the visible and infrared spectrum decreases.

The Tasseled Cap Transformation counting procedure based on [2] is
i1*0.1511+i2*0.1973+i3*0.3283 +i4*0.3407+i5*-0.7117+i6*-0.4559 (wetness)
i1*-0.2941+i2*-0.2430+i3*-0.5424+i4*0.7276+i5*0.0713+i6*-0.1608 (greenness)
i1*0.3029+i2*0.2786+i3*0.4733+i4*0.5599+i5*0.5080+i6*0.1872 (brightness)

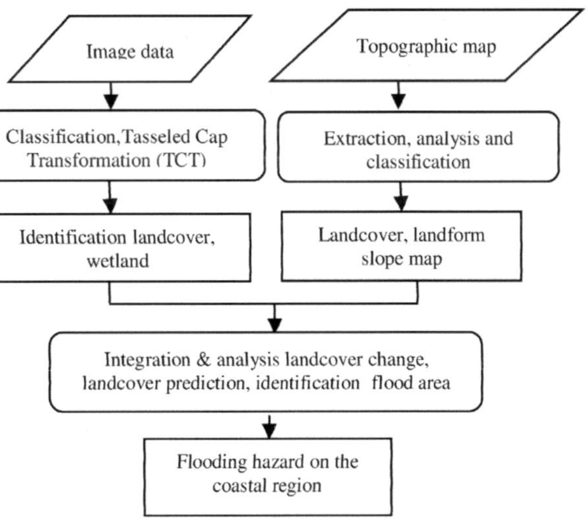

Figure 1. Flow chart of research methods

The spectral that can be used to assess the temperature conditions in an object on the Earth's surface is the thermal spectral. This thermal spectral usage can be done with brightness temperature analysis. Brightness and Greenness are considered to be vegetation visions, whereas between Brightness and Wetness is considered a plot of land for Landsat TM [4]. The vegetation and water indexes derived from satellite reflection data are the two main sources of information for monitoring land cover. Land surface objects have varied spectral responses in wet / waterlogged and dry / unfertilized areas [4]. In the vegetated areas, during floods there is an increase in reflectance values in visible channels (band 1 - 4 OLI). In contrast to Infrared channels (NIR, SWIR, MIR) decreased. This condition is different for the object of land (open land), where the visible spectrum and infrared decrease [2]. From field test to NDVI spectral transformation process result gives 68,17%; The process of transformation of Principal Component Analysis (PCA) spectral value of 96.13%; And Tasseled Cap Transformation (TCT) provide an accuracy of 92.44% [7]. The result of accuracy of the process of merging the transformation of spectral values (Principal Component Analysis + NDVI + Tasseled Cap Transformation) then done the process of supervised classification gives accuracy value 92,61%, not proven to give addition of better accuracy [2].

Multicriteria analysis for flood hazard mapping, in determining flood hazard area using rainfall parameter, altitude data, slope map, soil type, geological data, land cover data, and landform. Rainfall is the main source causing flooding, areas with high rainfall have the possibility of greater flood events compared to areas with lower rainfall. In this study, rainfall data used obtained from the Meteorology Climatology Agency and Geophysics Agency. Flood events are more common in lowland areas and have flat slopes because the water tends to be slow to flow, compared to areas that are at high altitudes. Slope and elevation parameters are made using contour data from topographic maps. Different types of soils have different levels of infiltration and water holding capacity, soils with coarse textures have higher infiltration capacity than fine textured soils. But in this study, the classification of land used in the Order category [10]. Geological conditions directly affect the rate of infiltration of water in the form of deep permeation, geological data for this study obtained from geological map on a scale of 1: 100,000 Karawang Sheet, which will be used as secondary data..

Geomorphology, type of main rock, soil type, slope and rainfall as the main inputs, to recognize flood-hazard areas can be analyzed by overlaying various parameters that result in flooding. Physical characteristics of the land as inputs play a role in determining the level of flood vulnerability, therefore the form of land has an important role in this research. Because the form of land is one place where the process of running water that comes from the input of rain to the sea. A review of landforms can also represent slope conditions, drainage conditions and in general can also concern existing soil conditions. Another factor that also gives a great influence on the occurrence of flooding is the land cover that serves as a parameter against water to flow. In making the map of flood hazard to implement the overlay of the five parameters trigger the occurrence of floods, of the five factors is a big enough role is on the level of rainfall that exists.

3. Result and discussion

The result of transformation process of spectral value with Tasseled Cap Transformation (TCT) gives three bands created are commonly held to represent Brightness, Greenness and Wetness. The result of transformation process of multi-temporal colour composite spectroscopy (RGB) value can separate the appearance of rice field / wetland area with non-rice field can be seen clearly. Blue colour associated with water, red colour associated with open area / settlement / dry land / open land . For green colour associated with vegetation (Figure.2.a).

The analysis of spectral pattern can be explained briefly in (Figure.2.a.), analysis of satellite data of Landsat colour composite with RGB TCB (Tasslecap brightness) -TCW (tasslecap wetness) -TCW (tasslecap greenness) composition, giving red appearance to non- vegetated, green on areas with vegetation, and blue or dark on the appearance of waters or wet objects. The gradations of redness depend on the level of vegetation and water shortages, the less the vegetation gets wetter and the TCW gets wetter the blue or darker the colour.

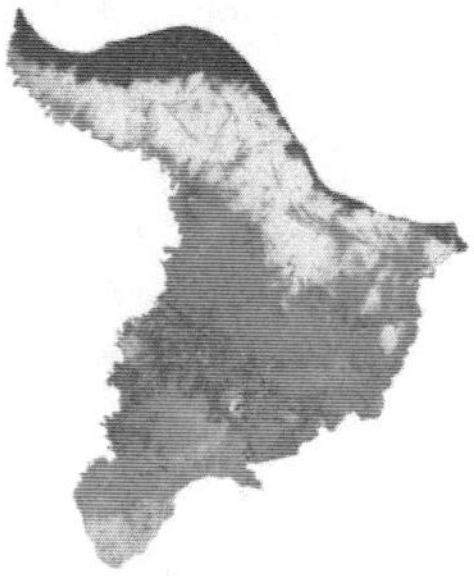

Figure 2. (a) TCT RGB

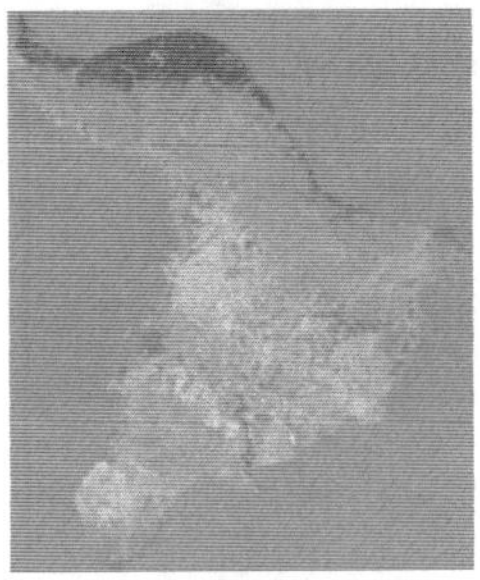

Figure 2. (b) TCT Brightness

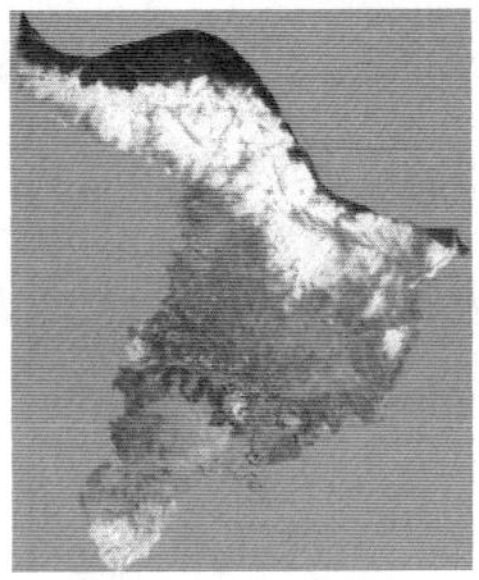

Figure 2. (c) TCT. Greenness

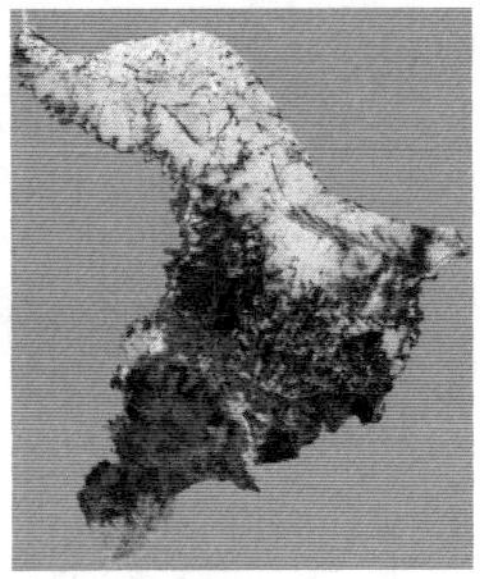

Figure 2. (d) TCT. Wetness

Result of brightness, light / white colour is dry land / open area / roof / road etc. Water gives a dark colour, while vegetation follows the density of plants with grey colour (Figure.2.b). The greenness results show the light / white colour of the vegetation, for dark colour is the wet / water area (Figure.2.c.). The result of wetness of water appearance is white / bright. The darker the grey scale, for vegetation following the density if the vegetation is denser lighter if it is wet. The open area gives a darker spectral display (Figure.2.d).

The results of the analysis of flood hazard areas with multicriteria method referring to [9] (Table.1). The coastal flood hazard area can be calculated using the formula in Table 2 using the formulation of parameter 1 and parameter 4 in Table.1.

Table 2. Calculation of flood hazard class

No.	Parameter	Scoring	Weighting	Value
1.	Calculation of flood hazard class for rainfall (30%)			
	≥ 200 mm	*3*	*30*	*0.90*
	50 – 200 mm	*2*	*30*	*0.60*
	≤ 50 mm	*1*	*30*	*0.30*
2.	Calculation of flood hazard classes for landcover (70%)			
	Settlements	*3*	*50*	*1.05*
	Shrub/Farm	*2*	*50*	*0.70*
	Rice field/ forest	*1*	*50*	*0.35*
3.	Calculation of flood hazard characteristic of slope land (70%)			
	0% - 2%	*3*	*50*	*1.05*
	2% - 4%	*2*	*50*	*0.70*
	>4%	*1*	*50*	*0.35*
4.	Calculation of flood hazard coastal of slope land (70%)			
	0% - 2%	3	70	2.10
	2% - 4%	2	70	1.40
	>4%	1	70	0.70

Source : SNI 8197, 2015 [9]

Slope maps are processed using contour altitude and spot height data on a 1: 25,000 scale topography map. That research area has slope <2% with area 75.61%, slope 16-25% with wide 13,43%, slope 2-8% with wide 5,24% and slope> 40% with wide 5,61%. Landforms in areas susceptible to flooding due to low areas or basins with slopes <2%, or having high flood vulnerability (Table.2.), (Figure.3.) are maps of spatial assessment of flood hazard areas [9]. The average rainfall condition is 1,100 - 3,200 mm / year with the highest rainfall in January to April and December with rainfall of ≥200 mm [3].

IOP Conf. Series: Earth and Environmental Science **162** (2018) 012042 doi:10.1088/1755-1315/162/1/012042

Table 3. Calculation of coastal flood hazard class

Weighting value (1+4)	Class interval	Hazard classes
0.90+2.10= 3.0	*2.10 – 3.00*	*High*
0.60+1.40=2.0	*1.10 – 2.00*	*Medium*
0.30+0.70=1.0	*0.10 – 1.00*	*low*

Source : SNI 8197, 2015 [9]

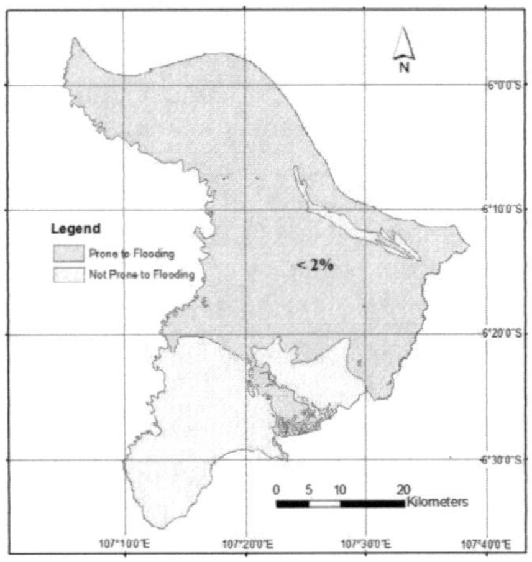

Figure 3. Flood hazard map

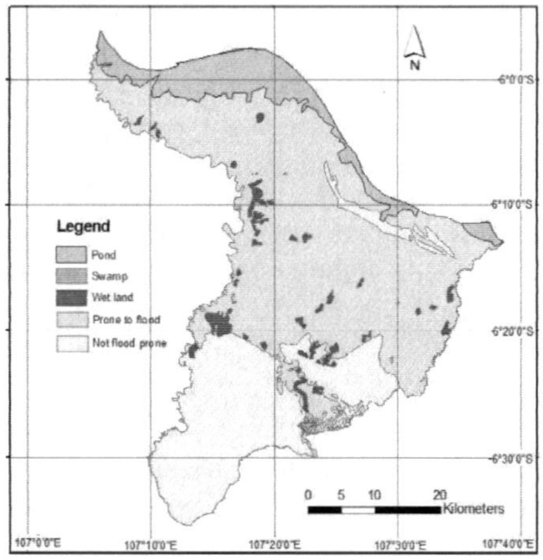

Figure 4. TCT Analysis overlay with Flood hazard map

Landsat 8 Image Data Recordings dated October 18, 2015, transformed spectral Tasseled Cap Transformation (TCT) spectral values that can provide information on wetland conditions as an early indication of low areas and will often be inundated (Figure.4).

Projected future land cover changes will be made with the predicted changes in the present approach through spatial modelling. The prediction model uses CA-Markov, to predict future land cover changes of 16 years or in 2031, this software has the ability to simulate land cover changes with Markov chain procedures [5]. The CA-Markov model is run by using IDRISI software, the 2000 data input, 2006 data is used as the basis for projected 2015. The projected land cover change for 2031 uses 2000 data, 2006 data and 2015 data. Data used in This process is raster format data with pixel size 30x30 meter.

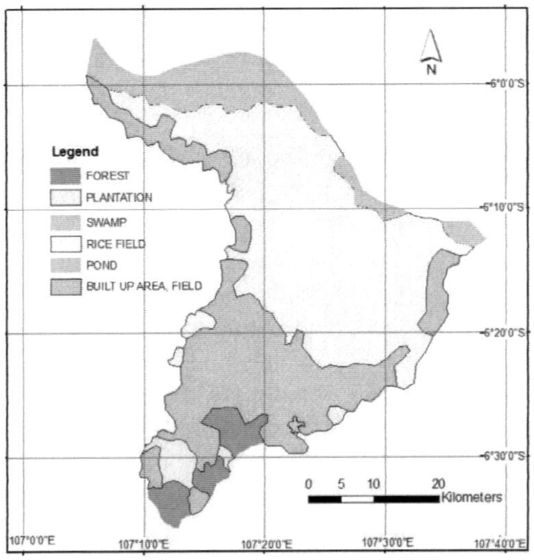

Figure 5. Land cover in 2015

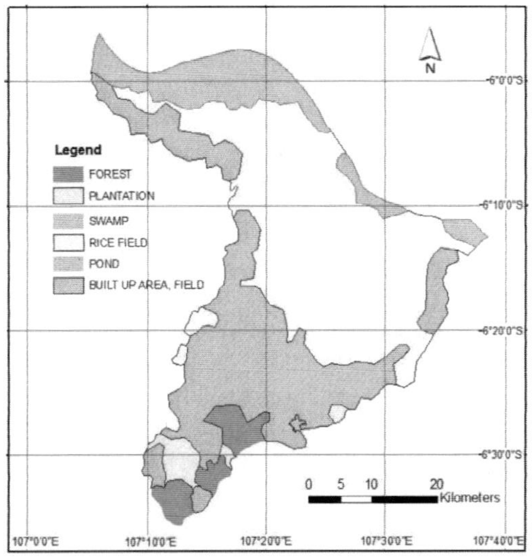

Figure 6. Projected land cover in 2031

Indication of the result of the analysis is the decrease of wetland area and the widening of the developed land, the widened land expansion has reduced the wetland area from 51,41% of area by 2015 (Figure.5) to 45,81% from the area in 2031 (Figure.6.) And the expansion of wake land is in flood-hazard areas.

4. Conclusion

The benefits of identification of flooded and inundated areas can be used to alert communities living in the region about the dangers of flooding and puddles. Transformation of Tasseled Cap Transformation (TCT) spectral values can provide information on wetland conditions as an early indication of low areas and will often flooded. The analysis of land use change using the land transition change matrix in 2000-2015 to predict land use change in 2031 suggests that there is a significant change from converted agricultural land to built up area. Projected land use changes in 2031 indicate an increase in the build up area that converts rice fields, and in general the expansion of build up areas occupies flood-hazard areas.

Acknowledgment

Thanks to the Government of Karawang Regency, Central Bureau of Statistics Karawang Regency, Geospatial Information Agency which has provided easiness in obtaining data to support this research.

References

[1] Butler, K 2013 Band Combination For Landsat 8 http://www.georeferencial.com.br/UNIFESP/DVD_Curso_2015/Band/Combinations/forLandsat/08_melhor/organizado.pdf

[2] Baig M H A, Zhang L, Shuai T, Tong Q 2015 Derivation of A Tasseelled Cap Transformation Based on Landsat 8 at- Satellite Reflectance *Remote Sensing Letters* **5(5)** 423-431

[3] BPS (Badan Pusat Statistik /Central Bureau of Statistics) 2015 National Socio-Economic Survey and Karawang In Figures

[4] Crist E P and Cicone R C 1984b A Physically-Based Transformation of Thematic Mapper Data- the TM Tasseled Cap *IEEE Transactions on Geoscience and Remote Sensing*. GE–**22 (3)** 256–263

[5] Eastman J R 2009 IDRISI Taiga: guide to GIS and image processing Worcester: Clark Labs.Foody, G. M 2002 Status Of Land Cover Classification Accuracy Assessment *Remote Sensing of Environment* **80** 185–201

[6] Kauth R J and Thomas G S 1976 The tasseled Cap – A Graphic Description of the Spectral-Temporal Development of Agricultural Crops as Seen by LANDSAT *Proceedings of the Symposium on Machine Processing of Remotely Sensed Data, Purdue University of West Lafayette, Indiana*. pp. 4B-41 to 4B-51

[7] Murdiyanti S R 2010 Integrasi Transformasi Spektral Citra Landsat ETM Dan SIG Untuk Pemetaan Pola Rotasi Tanam Lahan Sawah Kabupaten dan Kota Semarang Serta Daerah Sekitarnya di Jawa Tengah *Majalah Geografi Indonesia (MGI)* **24(2)** 121-141

[8] Paimin, Sukresno, Pramono I P 2009 Teknik Mitigasi Banjir dan Tanah Longsor *Penerbit Tropenbos International Indonesia Programme, Balikpapan*

[9] SNI 8197 (Indonesia National Standart) 2015 Method of flood mapping on a scale of 1: 50000 and scale 1: 25000

[10] Widiatmaka W, Ambarwulan I, Firmansyah K, Munibah P B K, Santosa, Murdaningsih W Y, Utomo C 2015 Multicriteria Land Evaluation and Remote Sensing for Flood Hazard Delineation in Lower Citarum Sub- Watershed: Implication to Paddy Field Area The 1st International Conference of Indonesian Society for Remote Sensing 2015 " Harnessing Earth Information from Space" At October 27-28th 2015 Geomatic Engineering

[11] Widiatmaka W, Ambarwulan I, Firmansyah K, Munibah P B K, Santoso, Sudarsono 2014 Land Suitability and Dynamic System Modelling to Define Priority Areas of Soybean Plantation in Paddy Fields in Karawang, West Java. *Agrivita* **36(3)**

IOP Conf. Series: Earth and Environmental Science **162** (2018) 012043 doi:10.1088/1755-1315/162/1/012043

Long term variation of sea level anomaly (September 1992-January 2017) in the Indonesian sea from multi-mission satellite altimetry data

Dina A. Sarsito[1], Kosasih Prijatna[1], Dudy D. Wijaya[1], Nur Fajar T[1], Ivonne M. Radjawane[2], Wiwin Windupranata[3] and Brian Bramanto[1]

[1] Geodesy Research Group, Faculty of Earth Sciences and Technology, Bandung Institute of Technology, Jl. Ganesha 10 Bandung, Indonesia
[2] Oceanography Research Group, Faculty of Earth Sciences and Technology, Bandung Institute of Technology, Jl. Ganesha 10 Bandung, Indonesia
[3] SSKWPL, Faculty of Earth Sciences and Technology, Bandung Institute of Technology, Jl. Ganesha 10 Bandung, Indonesia

E-mail: dsarsito@gd.itb.ac.id

Abstract. Sea level anomaly (SLA) is one of the oceanic parameters that play a crucial role in the ocean dynamics. In the Indonesian region, spatio-temporal characteristic of SLA is significantly affected by local topographic features of the region. In this research, we investigate long-term variation of SLA over the region based on Topex/Poseidon, Jason-1 and Jason-2 data, during the period of Sept 1992 to January 2017. After removing some geophysical effects, time series of SLA values at each point of the collinear tracks are analyzed. The result show the long term MSSH in Indonesia area have increase from western part of Indonesia to eastern part with range -40 m to 80 m. The long term SST in Indonesia have almost same trend with the deviation from geoid undulation full degree with short wavelength type, and this phenomena represent the local effect such as shallow bathymetry and close sea. And for long term SLA, Indonesia shows positive value that the anomaly increasing with variable from western part to eastern part of Indonesia from 2.5 mm/year until 7 mm/yr.

1. Introduction

Sea level anomaly (SLA) is deviation between time independent sea surface height (SSH) with respect to the mean sea surface height (MSSH) using ellipsoid as a mathematical reference surface. The variability of SLA is affected by geographical setting each area and play a crucial role in dynamic of ocean and can give important information for geoid computation in geodetic research and ocean circulation-sea level rise study [1] tidal modeling in physical oceanography and environmental research. In this research, we will use altimetry data to study SLA, to get better understanding of the SLA phenomena in Indonesia, an archipelagic zone, that lay on equatorial area that have many type of ocean circulations. Since 1992 until 2017, SLA is derived using various type of altimetry data and gave promising accuracy. Archiving, Validation and Interpretation of Satellites Oceanographic data (AVISO), Physical Oceanography Distributed Active Archive Center (PO.DAAC), European Meteorological Satellite (EUMETSAT), and National Oceanic and Atmospheric Administration (NOAA) are the agencies that distribute altimetry data. The mirror of all satellite altimeter missions data can also derived from Radar Altimeter Database System (RADS) [2] [3] that is developed by Delft Institute for Earth-Oriented Space Research (DEOS) and the NOAA Laboratory for Satellite Altimetry, that consist of validated and cross-calibrated sea level data. To get better understanding about SLA implications to geodetic and oceanographic studies in the Indonesian area, in this research, we study long-term SLA based on 30 years of sea-level data (more than 1 full period of 18.6 year) derived from multi-mission of satellite Altimetry collected in Radar Altimeter in the research coverage area -15S/90E to 15N/150E. The estimation method to determine the SLA in here is purely a geometric-geodetic strategy by estimate the cross over value and interpolate it in radius 5 km. The investigation area for full degree (n = 2190) and short wavelength (with degree n = 180) period of signal is chosen in Indonesia area.

2. Methods

2.1. Sea Level Surface and Satellite Altimetry

In studying sea surface, there are several types of surfaces that are used as reference (Figure 1). The choice of surface type used depends on the usage requirements as well as the necessary analysis. The reference datum used there are 2 types, namely geoid as the real potential field and ellipsoid as the mathematical reference field of the earth's surface. When using geoid as reference surface, it can be obtained Instantaneous Sea Surface Height (ISSH) or often called Sea Surface Height (SSH) and MSSH (Mean Sea surface Height). The difference between ISSH and MSSH is referred to as Sea Level Anomaly (SLA) (Equation 1). Whereas if the ellipsoid surface is used as a reference it will be obtained Dynamic Topography (DT) and Mean Dynamic Topography (MDT). In addition to the above, there is a type of sea surface that is also used as a reference that is Sea Surface Topography (SST) which is the difference between ISSH with geoid undulation (N) (Equation 2).

$$SLA = SSH - MSS \tag{1}$$

$$SST = SSH - N \tag{2}$$

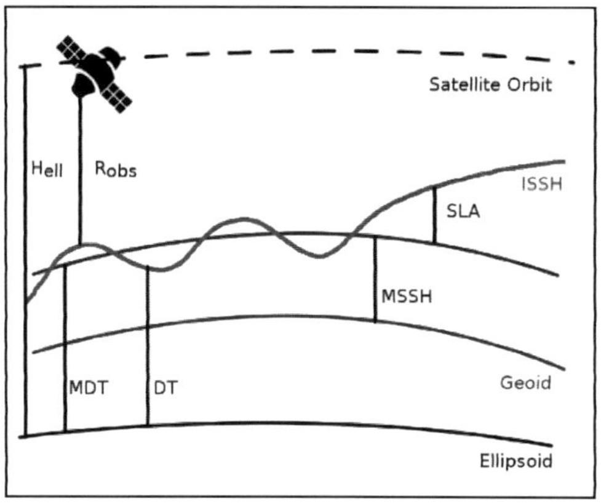

Figure 1. Sea surface types and principle of satellite altimetry (modified from [5])

N values in the ocean can be obtained in various ways, i.e. marine/sub-bottom gravimetry, airborne gravimetry, satellite gravity or Global Geopotential Model (GGM). For the area of research that has not been organized geoid model local/regional, then N obtained from GGM by using Equation 3.

$$N(\lambda,\theta) = \frac{GM}{r\gamma(\lambda,\theta)} \sum_{n=2}^{N_{max}} \sum_{m=0}^{n} \left(\overline{C_{nm}^{*}} \cos m\lambda + \overline{S_{nm}^{*}} \sin m\lambda\right) \overline{P_{nm}^{*}} \sin\theta \tag{3}$$

Where (r, λ, θ) is the radius-longitude in the sphere system, G is the gravitational constant, M is the mass of the globe of the fingers radius R, and $\overline{C^*}$, $\overline{S^*}$ dan $\overline{P^*}$ is a fully normalized coefficient and associated Legendre's polynomials with degree n and orde m.

The combination of altimetry satellite technology (dedicated to determining sea-water dynamics) with positioning satellites (GNSS) allows for the measurement of the height difference (vertical distance) of the satellite up to the surface of seawater (Figure 1). The vertical distance (R_{obs}) is obtained by measuring the travel time back (t) of the electromagnetic waves emitted by the satellite based on Equation 4

$$R_{obs} = c * 0.5t \quad (4)$$

Where c is the speed of light in a vacuum. R_{obs} still contains errors as well as bias that must be corrected first. The correction is adjusted to the source of the orbit (H_{ell}), the satellite instrument ($corr_{ins}$), the propagating media (ionosphere c_{ion} – wet troposphere c_{tw} – dry troposphere c_{td}) and the reflecting medium of sea state bias c_{ssb} , while the geophysical bias derived from ocean tide c_{ot} loading tide c_{lt}, solid earth tide c_{set}, pole tide c_{pt} and atmospheric tidec_{ib} . Furthermore, the SSH value is obtained by reducing the satellite heights to the H_{ell} ellipsoid with the vertical distance R_{obs} along with the error correction and the bias in equation 5

$$SSH = H_{ell} - R_{obs} - c_{ion} - c_{tw} - c_{td} - c_{ins} - c_{ssb} - c_{ot} - c_{lt} - c_{set} - c_{pt} - c_{ib} \quad (5)$$

2.2. Radar Altimeter Database System (RADS) and strategy of altimetry data processing

Radar Altimeter Database System (RADS) where carried out in here for data retrieval and reduction as the archiving and processing system initiative by TU-Delft, NOAA and Altimetrics LLC [6] [8]. This system was re installed in ITB at 2016 in the frame of ITB-UTM-DEOS collaboration as the continuation the 2005 SEAMERGES project (an EU funded project (AUNP) that aimed for knowledge, methods and data exchange related to satellite altimetry, InSAR and GPS (www.deos.tudelft.nl/seamerges)). The altimetry data base are consist of all radar altimetry mission data in the timeline from 1992 until early 2017 that portrayed in Figure 2 and RADS system layout is in Figure 3 below:

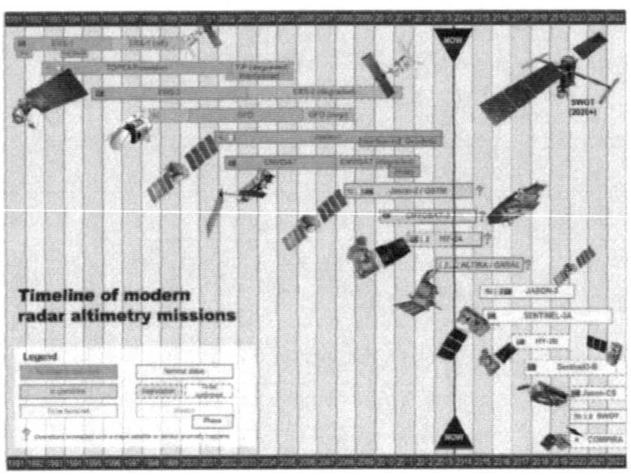

Figure 2. Timeline of modern radar altimetry mission [5] [6]

IOP Conf. Series: Earth and Environmental Science **162** (2018) 012043 doi:10.1088/1755-1315/162/1/012043

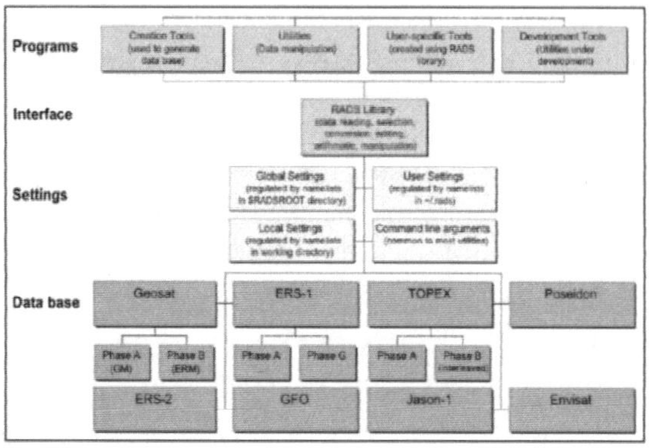

Figure 3. Radar Altimeter Database System (RADS) [2]

Strategy for altimetry data processing in here are by merging all choices satellite altimeter missions using collinear track to normal reference point and every footprint is calculated before interpolated for nearby points in radius 5 km surrounding cross over points (Figure 4) that modified from [7],[8] and [9]. The value are calculated on daily solution and collected per month to get monthly solution.

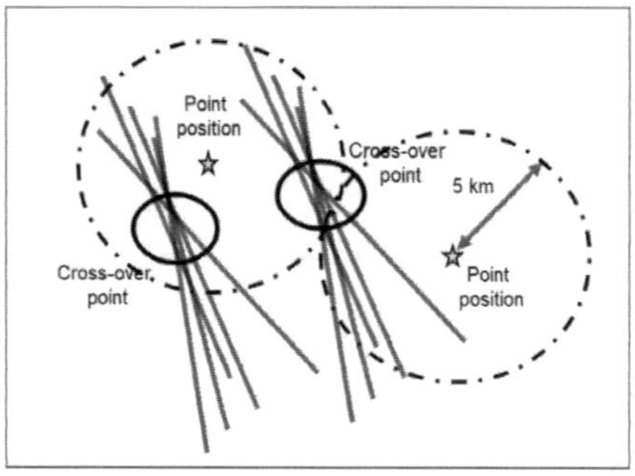

Figure 4. Strategy for altimetry data processing

Table 1 is summary of corrections and models that applied at RADS for this research, in which specific models are used for correcting each satellite altimetry missions in RADS and an updated geophysical and environmental corrections were also applied. The sea level data have been corrected from error and bias as mention before.

Table 1. Corrections and models applied for altimeter processing in RADS

Correction / Model	Description
Orbit/Gravity Field	EIGEN GL04C
Dry Troposhere	Atmospheric pressure grids; ECMWF (Model)
Wet Troposphere	All satellites
Ionosphere	NIC08
Inverse Barometer	IB (Model, Local Pressure)
Dynamic atmosphere	All satellites: MOG2D
Solid Earth Tide	Applied (Elastic response to tidal potential)
Ocean Tide	GOT4.8
Load Tide	GOT4.8

Pole Tide	Polar Wobble
Sea State Bias	CLS non parametric ; ERS: BM3/BM4 parametric
Reference	DTU10
Engineering flag	Applied

3. Results and Discussion

The research coverage area is between -15S/90E to 15N/150E with 110 points of observation are chosen nearby the crossover position between satellite altimetry tracks in Figure 5, with Topex/Poseidon, Jason-1 and Jason-2 data from 1992-2017.

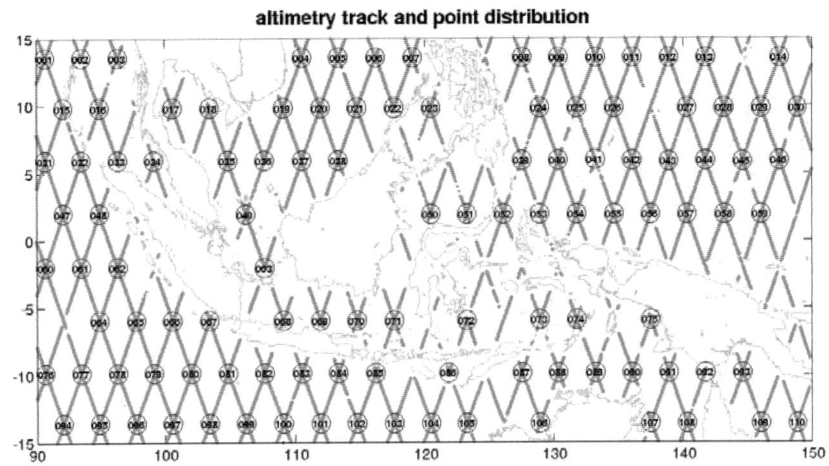

Figure 5. Altimetry tracks and point distribution

The next step is to calculate MSSH at all points of observation as shown in Figure 6.

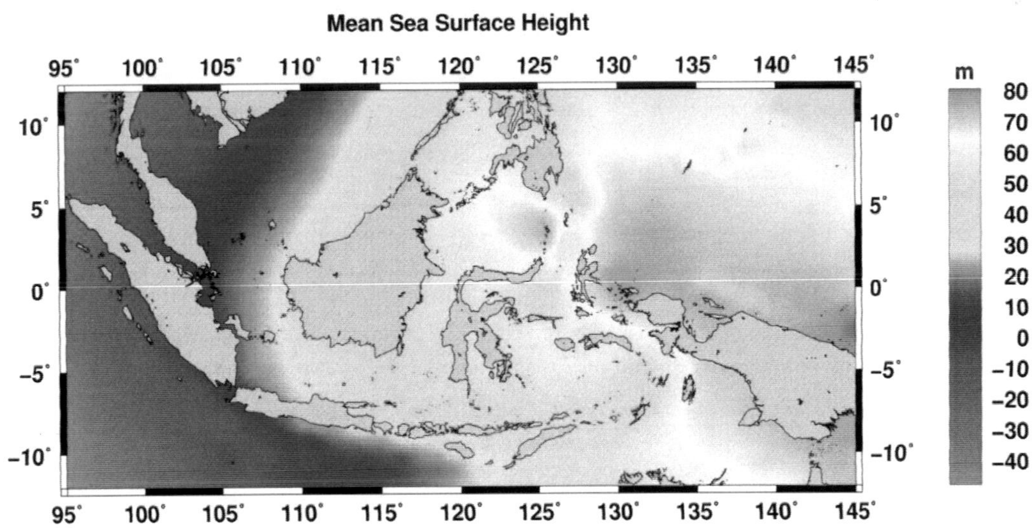

Figure 6. Indonesian MSSH from 1992-2017

The observed pattern is that in the western part of Indonesia it is worth -40 meters which rises gradually eastward to the highest value close to 80 meters for the northern region of Papua. For the central part of Indonesia the value ranges from 30 meters to 60 meters, where there is a significant increase of 70 meters for the northern part of Sulawesi which is interspersed with a lower MSSH gap of 50 meters in Halmahera double subduction area and rises again next to east after the Halmahera archipelago.

To get a clearer picture of the phenomenon of MSSH, we use the geo-undulation value of GGM EGM-2008 using 2 degree types, i.e. full wavelength n = 2190 and long wavelength n = 180 and calculate the SST value for both types of wavelengths used. Figure 7 shows a cross-section at latitude 0^0 that passes through various islands in the study area and at latitude -10^0 over more open oceans. These latitudes are selected to obtain a clear picture of the effect of land crossing the altimetry satellite path. The disconnected line for MSSH and SST indicates that the data passes through the land area, whereas the GGM undulation data does not experience a data gap as it passes through the mainland. For both cross sections, it is apparent that the MSSH value has a value that is not significantly different from the geoid undulation value of both full wavelength and long wavelength types with a maximum value of 5 meters, where this deviation appears in the transition region between the sea-land. For SST both for full wavelength and long wavelength types are seen to have a pattern that is close to the difference in GGM geoid undulation values of both wavelength types. This shows the effect of the dominant local effect (high frequency/short wavelength local phenomenon). Visible deviation of GGM difference model can reach 10 meters, especially for narrow/closed seawater area.

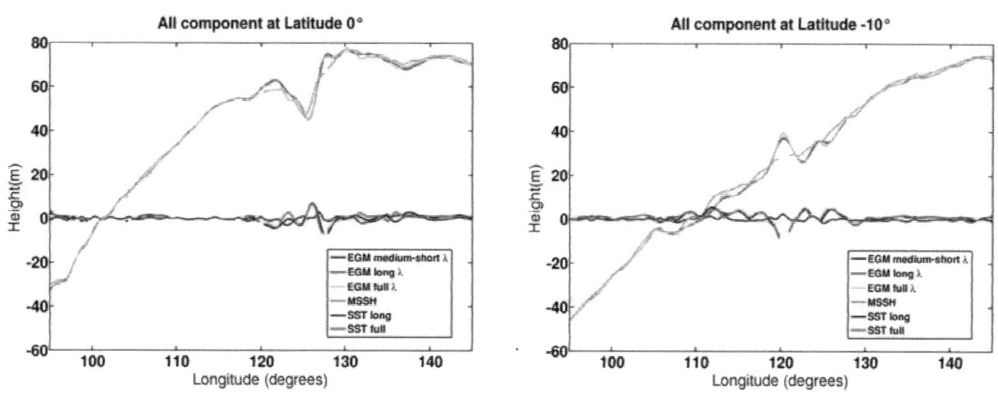

Figure 7. MSSH, SST and GGM EGM-2008 (n = 2190 and n = 180)

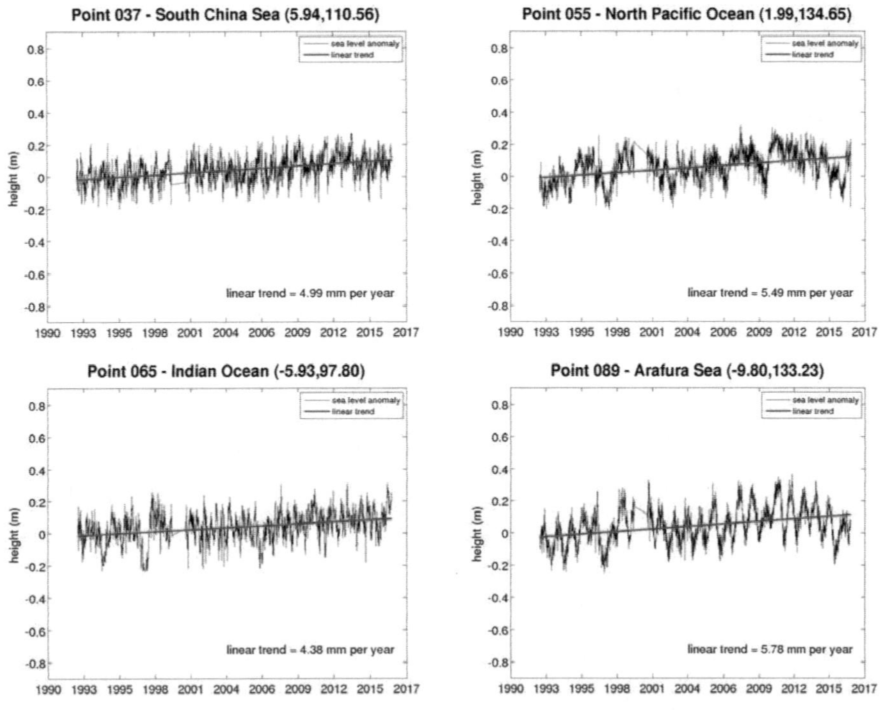

Figure 8. SLA and its trend at observation points

The next step is to know the SLA for all observation points, with sample stations in Figure 8 representing 110 observation stations. The point of 037 South China Sea representing the western part of western Indonesia has a positive linear trend of 4.99 mm/year, while for the southern latitude 065 point (Indian Ocean) also shows a linear trend of 4.38 mm/yr. For the eastern part of Indonesia the northern latitude is represented by the 055 (North Pacific Ocean) point which has a positive linear trend of 5.49 mm/year and the point 089 (Arafura Sea) in southern latitude with a linear trend value of 5.78 mm/year. In general, the Indonesian territory experienced a rising sea level trend represented by a positive SLA with variations in each region.

SLAs for all observation stations can be seen in Figure 9 and the SLA range (difference between maximum and minimum SLA values in the same station) in figure 10.

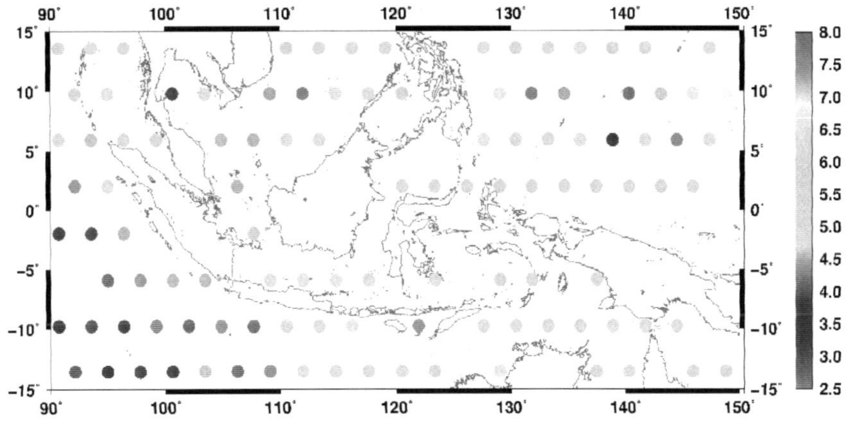

Figure 9. Linear trend of SLA in mm/year

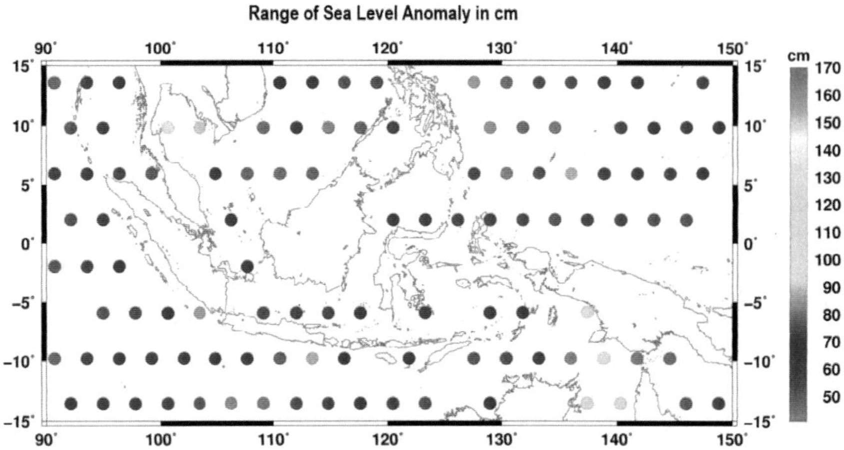

Figure 10. Range of SLA in cm

Indonesia's territory as a whole has a positive trend value of about 2.5 mm /year in the west (station : 076 and 094) and eastward increments with gradually increased values up to 7 mm/year in the northern region of the Pacific Ocean. The highest SLA is located in this section (station: 025, 026, 027 and 028) which is the open sea area. For shallow sea or closed sea areas, the SLA value changes are relatively small, this is probably due to the quality of altimetry data coverage is not as good for open sea areas, where much noise is still present in the data. Phenomenon SLA can also be seen from the range in figure 10, the average for all regions with data hose from 1992-2016 is seen on average around 50-75 cm, except for southern Java (station 084) and western part of Sumatra (station 067) with a value of 90 cm and a higher value reaching above 100 cm between Papua-Australia (stations: 075, 090, 091, 107 and 108). This anomaly happens due to the effect of the land being cut off by the narrow and not too deep so it affects the altimetry data when reflected back to the satellite.

4. Conclusion

In this study, we can recognize that the long term MSSH in Indonesia area have increase from western part of Indonesia to eastern part with range -40 m to 80 m. This MSSH majority pattern have almost same pattern with geoid undulation from GGM with full degree n = 2190 and degree n =180, except for the transition area between land and sea. The long term SST in Indonesia have almost same trend with the deviation from geoid undulation full degree with short wavelength type, and this phenomena represent the local effect such as shallow bathymetry and close sea. And for long term SLA, Indonesia shows positive value that the anomaly increasing with variable from western part to eastern part of Indonesia from 2.5 mm/year until 7 mm/yr. The range between the highest and the lowest SLA in each station shows an 50-75cm in average for Indonesia. Several SLA - anomaly range happened in close sea and shallow sea. The long term MSSH, SST and SLA in Indonesia shows variability that indicate not only the geographical variation but also the other oceanography phenomena, that still need more investigation to get better understanding the causes and impacts of SLA.

Acknowledgements

The authors would like to thank to TU Delft, NOAA and Altimetrics Llc for providing altimetry, and to Marc Naije (TU Delft) and Ami Hassan Md Din (UTM – Skudai Johor Bahru) for altimetry state of the art.

References

[1] Church J A and White N J Sea-level rise from the late 19th to the early 21st Century, Surv. Geophys. **32** 585–602, doi:10.1007/s10712-011-9119-1, 2011

IOP Conf. Series: Earth and Environmental Science **162** (2018) 012043 doi:10.1088/1755-1315/162/1/012043

[2] Naeije M, Scharroo R, Doornbos E and Schrama E 2008. Global Altimetry Sea Level Service: GLASS. NIVR/SRON GO project: GO 52320 DEO

[3] Scharroo R 2016 RADS User Manual - Version 4.2.4 - 6 June 2016

[4] Fu L L and Cazenave A 2001 Altimetry and Earth Science, A Handbook of Techniques and Applications. Vol. 69 of International Geophysics Series. Academic Press, London

[5] P -Y Le Traon, D Antoine, A Bentamy, H Bonekamp, L A Breivik, B Chapron, G Corlett, G Dibarboure, P DiGiacomo, C Donlon, Y Faugère, J Font, F Girard-Ardhuin, F Gohin, J A Johannessen, M Kamachi, G Lagerloef, J Lambin, G Larnicol, P Le Borgne, E Leuliette, E Lindstrom, M J Martin, E Maturi, L Miller, L Mingsen, R Morrow, N Reul, M H Rio, H Roquet, R Santoleri and J Wilkin 2015 Use of satellite observations for operational oceanography: recent achievements and future prospects, *Journal of Operational Oceanography*, **8**:sup1, s12-s27, DOI: 10.1080/1755876X.2015.1022050

[6] Aviso, Website: http://www.aviso.oceanobs.com/en/

[7] Andersen O B and Scharroo R 2011 Range and Geophysical Corrections in Coastal Regions: And Implications for Mean Sea Surface Determination *In Coastal Altimetry*. Springer. doi:10.1007/978-3-642-12796-0.

[8] Bertiger W, Desai S, Dorsey A, Haines B, Harvey N, Kuang D, Lane C, Weiss J and Sibores A 2008. Jason-2. *Precision orbit determination status, OSTST workshop, Nice*. Nov 2008

[9] Trisirisatayawong I, Naeije M, Simons W and Fenoglio-Marc L 2011 Sea level change in the Gulf of Thailand from GPS corrected tide gauge data and multi-satellite altimetry. Global Planet. *Change*. **76**: 137–151

IOP Conf. Series: Earth and Environmental Science **162** (2018) 012044 doi:10.1088/1755-1315/162/1/012044

Oceanographic factors related to Eastern Little Tuna (*Euthynnus affinis*) catches in the west Java Sea

Mega Syamsuddin, Sunarto, and Lintang Yuliadi

Department of Marine Sciences, Faculty of Fishery and Marine Science
Universitas Padjadjaran, Jl. Raya Bandung–Sumedang KM. 21 Jatinangor 45363
Indonesia

E-mail: vegha16@gmail.com

Abstract. The satellite data included sea surface height anomaly (SSHA) from AVISO, sea surface temperature (SST) and chlorophyll-*a* (Chl-*a*) from Aqua MODIS and Eastern Little Tuna (*Euthynnus affinis*) catches were used as a combined dataset to understand the ocean variations and further addresses their relations with the Eastern Little Tuna catches in the Java Sea. The fish catches and remotely sensed data were analyzed for the 5 years datasets from 2010-2014. The relationships of oceanographic factors and catch distribution were explored with a generalized additive model (GAM). Catch rates varied temporally relatively significant over year-round. The Eastern Little Tuna catch rates have the peak season in March and October (900 ton to 1000 ton). The ELT catch peaks during the transition season from southeast to northwest monsoon (September to November) and decreases during southeast monsoon (June to August). The GAM results showed that the 3-oceanographic parameter combination models explained the highest deviance (41.4%) with Chl-*a* explained the highest deviance (23.3%). High probabilities of Eastern Little Tuna catches corresponded to marine productivity of Chl-*a* concentration ranging from 0.3-0.5 mg/m^3, for SSHA ranging from 0-8 cm and SST ranging from 28-29°C. We recommend to have further investigations through the use of long-term historical time series to predict fishing ground locations and an emerging need to improve our climate understanding and forecast skill to conserve small pelagic catches.

1. Introduction

The Eastern Little Tuna (*Euthynnus affinis*) represents a dominant catch in the Java Sea. The Eastern Little Tuna is an epipelagic, neritic species inhabiting water temperatures ranging from 18 °C to 29 °C and tend to form multispecies schools by size. The Eastern Little Tuna is largely confined to continental shelves and islands of the western Pacific and the Indian Ocean [1]. Although also inhabiting ocean waters, this species prefer to stay close to the coast and juvenils are even found in the bays and harbours. It is a highly migratory species and frequently forms large schools which are often mixed with other scombrid species. Although sexually mature fish may be encountered throughout the year, there are seasonal spawning peaks varying according to regions: i.e. March to May in Philippine waters; from the middle of the North West monsoon period to the beginning of the South East monsoon (January to July) off East Africa; and probably from August to October off Indonesia [2].

The behaviour and survival of Eastern Little Tuna have been related to a range of environmental and ecological factors, necessitating the need to combine several oceanographic factors to investigate the effects of oceanographic factors on the Eastern Little Tuna catch in the west Java Sea. As proxies of potential tuna fishing grounds, sea surface height (SSH) measurement which could indicate oceanic features, as well as measurements of sea-surface temperature (SST), have been used to investigate productive frontal zones [3]. Thermal or color gradients in satellite images that arise from the circulation of water masses often indicate areas of high productivity [4]. Chlorophyll-*a* (Chl-*a*) data can also be used as a valuable proxy for water mass boundaries that can influence pelagic fish distribution in a region.

On the other hand, there seems to be only a few studies of the oceanographic variability impacts on Eastern Little Tuna distributions in the Java Sea. The satellite data included SSHA, SST and Chl-*a* and Eastern Little Tuna catches were used as a combined dataset to understand the ocean variations and further addresses their relations with the Eastern Little Tuna catches in the Java Sea. Therefore, understanding the effect oceanographic conditions and Eastern Little Tuna catches is an essential step towards the sustainable management of Eastern Little Tuna resources in the Java Sea.

2. Study area

The study area was in the western of Java Sea in the Indonesian Seas region focusing on geographical coordinates of 3 °S to 7 °S and 108 °E to 110 °E, as shown in Figure 1. The oceanographic condition in the study area is affected by the monsoon wind reversal drives the surface layer of the Java Sea southeastward, bringing relatively cool, fresh South China Sea water into the region [5].

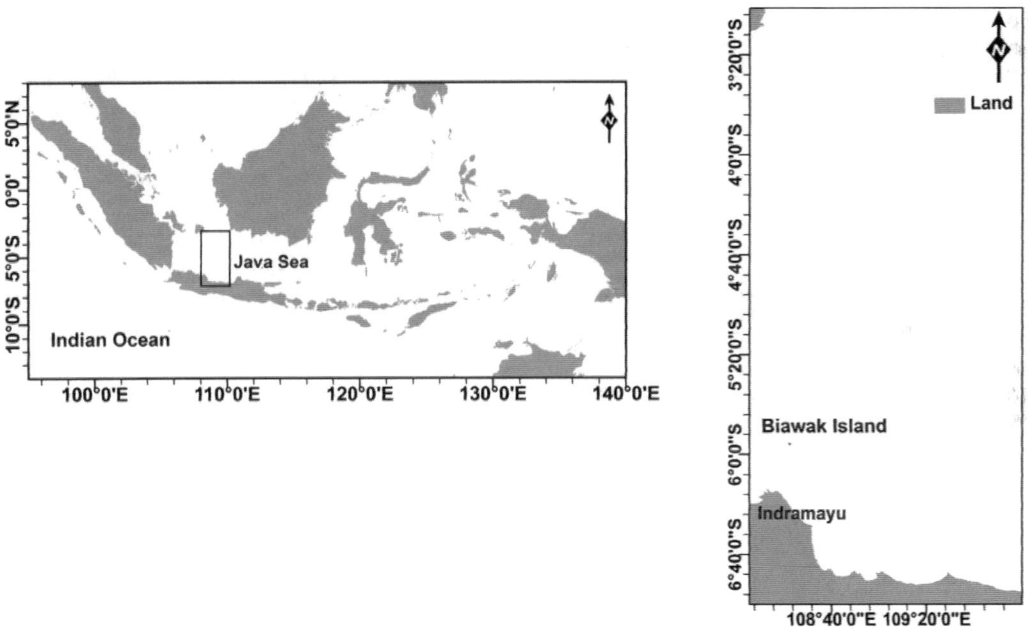

Figure 1. Map of Indonesian Seas with the inset box representing the study area in the west Java Sea

3. Data and methods

3.1 Data

This study utilized a time series of Eastern Little Tuna catch data that were collected from fishing logbooks provided by the Fishing Port of Indramayu, West Java and the Ministry of Marine Affairs and Fisheries Indonesia. The remotely sensed data consist of sea-surface temperature (SST), sea surface height anomaly (SSHA) and Chlorophyll-*a* (Chl-*a*). The SST and Chl-*a* were derived from satellite imagery of Aqua Modis and downloaded from http://oceancolor.gsfc.nasa.gov. The SST and Chl-*a* data had spatial and temporal resolutions of 4 km and monthly, respectively. The SSHA data derived from the TOPEX and Poseidon ERS-1/2 altimeters which are produced and distributed by Archived Validation and Interpretation of Satellite Oceanographic Data (AVISO). For data analysis, all SSHA, SST, and Chl-*a* were composed into monthly data and resampled into 9 km spatial resolution. In this study, the fish catch and satellite remotely sensed data were analyzed for the 5 years datasets from January 2010-December 2014.

3.2 Methods

3.2.1 Wavelet Spectrum Analysis

In this study, the wavelet analysis was applied on the monthly Eastern Little Tuna catch data, SSHA, SST and Chl-*a* data. This study applied the cross wavelet spectrum (XWT) and wavelet coherence (WTC). The XWT finds region in time frequency space where the time series show high common power and the WTC analysis can provide further information about time-scale-dependent phase shifts and correlations between two different time signals. In the other words, wavelet coherence analysis provides a quantitative way to describe the correlations between two time signals as a function of both time scale and phase shift.

The spectra are significant at the 95% confidence level assuming a background red noise. High energy variances are represented in red and low values are in blue as shown in the colorbars that represent different wavelet amplitude. The black line (solid contour) shows the 95% confidence level and a thin black line shows the cone of influence (COI) and are the region of the wavelet spectrum in which edge effects become important [6].

3.2.2 Generalized Additive Model (GAM)

This study examined the relationship between Eastern Little Tuna and oceanographic variables using generalized additive model (GAM). The GAM has been utilized in the previous fishery studies related to the oceanographic factors [7,8]. The advantages in using GAM are its simplicity beside the predictor variables having non-linear effects upon the response variable. The GAM was applied to Eastern Little Tuna catches in order to explore the spatial trends in distribution influenced by the SSHA, SST, and Chl-*a*.

All explanatory model terms were treated as continuous variables with the spline smoothers initially fitted to each term in the model [9]. A step-wise GAM was performed to determine the best fitting model prior to applying the final GAM to the entire data set. Akaike's information criterion (AIC) were used to determine the optimal set of explanatory variables. The model with the smallest AIC can be selected as the optimal model. GAMs were constructed in the R program (version 2.14.0) software using the *gam*.function of the *mgcv* package [10]. The GAMs were fitted in the formula:

$$g(u_i) = \alpha_0 + s_1(x_{1i}) + s2(x_{2i}) + s3(x_{3i}) + \ldots . s_n(x_{ni}), \quad (1)$$

where,

g = the link function,

u_i = the expected value of the dependent variable (HR of bigeye tuna),

α_0 = the model constant,

s_n = a smoothing function for each of the model covariates x_n.

4. Result and Discussion

4.1 Oceanographic factors associated with Eastern Little Tuna cathes

This study has shown the effect of monsoon inducing oceanographic condition in the west Java Sea. Seasonal change features were dominant for all the selected oceanographic parameters of SSHA, SST and Chl-*a*, and also Eastern Little Tuna catches, respectively. The time series plots on Figure 2 showed that the Eastern Little Tuna catch rates have the peak season in March during transition period from northwest monsoon to southeast monsoon and October during transition period from southeast monsoon to west monsoon (900 ton to 1000 ton).

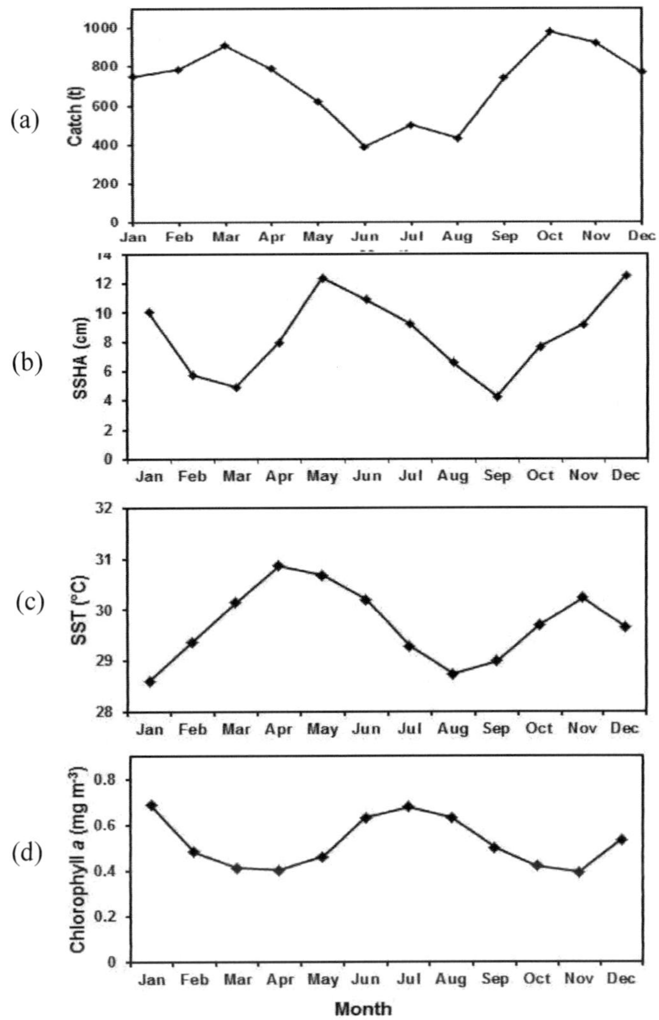

Figure 2. Mean monthly of (a) catch rate of Eastern Little Tuna, (b) SSHA, (c) SST, (d) Chl-*a* concentration during 2010-2014. The x-axis represents the month, and y-axis shows the variable value.

The peak catches corresponded with the value of SSHA ranging from 4-8 cm, SST ranging from 29 °C to 30 °C following the decreasing Chl-*a* concentrations in September to November (0.4 to 0.5) mg · m^{-3}. These results are consistent with previous studies who found that Eastern Little Tuna peaks during transition period from northwest monsoon to southeast monsoon with the preference habitat of warmer SST from 29.03-31.58°C and lower nutrient [11]. Other research revealed that Eastern Little Tuna is a typical warm water species that does not a preference for low temperature, thus ocean temperature plays an important role in the distribution of *Euthynnus affinis* [12].

4.2 Fishing ground modeling of GAM

The results of the GAM are presented as 1-parameter, 2-parameter, and 3-parameter models. The GAM analysis showed all explanatory variables used in the models to be statistically highly significant ($P<0.0001$) for SSHA, SST, Chl-*a* (Table 1). The addition of predictor variables at different levels gave an increased result in the deviance explained. GAM models explained 23.3%, 14.2%, and 0.0024% of the deviance explained in the single parameter models for Chl-*a*, SSHA, and SST, respectively. The three-parameter combination models explained the highest deviance (41.4%).

Table 1. Results from general additive models (GAMs) derived from catch rates of Eastern Little Tuna in 2010–2014 as a function of the oceanographic parameters (N=61 samples). The best model was selected on the basis of the significance of predictor terms, and increase in cumulative deviance explained (CDE).

Model	Parameter	Deviance explained (%)
catch ~ s(chla)	Chl-*a*	23.3***
catch ~ s(ssha)	SSHA	14.2**
catch ~ s(sst)	SST	0.0024
catch ~ s(chla) + s(ssha) + s(sst)	Chl-*a*, SSHA, SST	41.4***

***indicates statistical significance at the 0.0001 level

Figure 3 showed the individual effects of predictor variables associated with each variable of SSHA, SST, and Chl-*a* on Eastern Little Tuna catch. The model results showed the most favorable oceanographic condition for Eastern Little Tuna as observed for SSHA ranging from 0 to 8 cm, for SST between 28ºC and 29.5ºC, for Chl-*a* from 0.3 to 0.5 mg m^{-3}. The declined catches of bigeye tuna were observed for higher SSHA 8 -20 cm; SST values from 29.6 to 31.5ºC; higher Chl-*a* values above 0.5 mg m^{-3}.

Our finding is in agreement with other research indicating that *Euthynnus affinis* prefer to live in warmer surface water with the average value of SST 29.46 ºC and lower Chl-*a* concentration of 0.3 mg m^{-3} [13].

The GAM results confirmed that Chl-*a* seemed to be one of the dominant factors to explain the variability in the study area. Thermal or Chl-*a* fronts often indicate areas of high biological productivity, and hence a high probability finding fish. Primary production provides an attractive habitat for small pelagic fish species [4,14]. The SSHA and SST were the second and third significant predictor of Eastern Little Tuna. Predominantly positive SSHA and warm SST have a greater effect on catches implying preference for habitat optimum of Eastern Little Tuna.

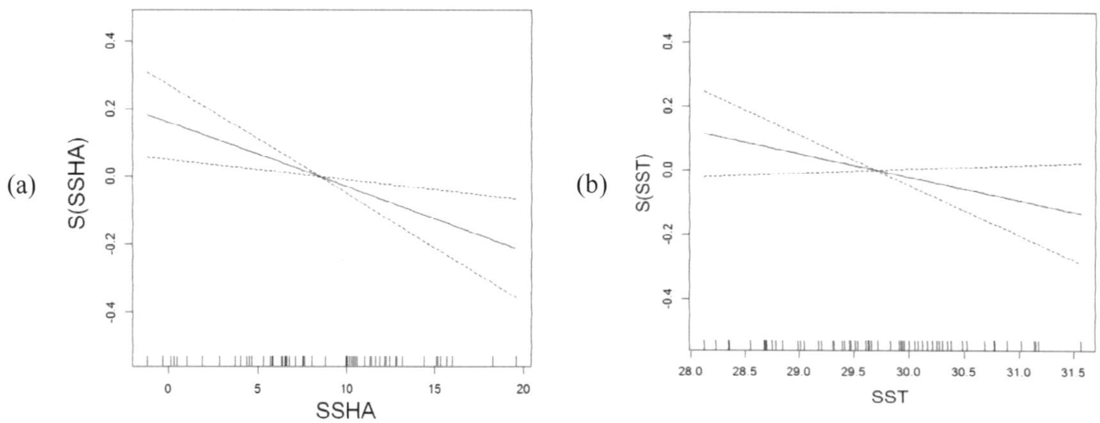

(c)

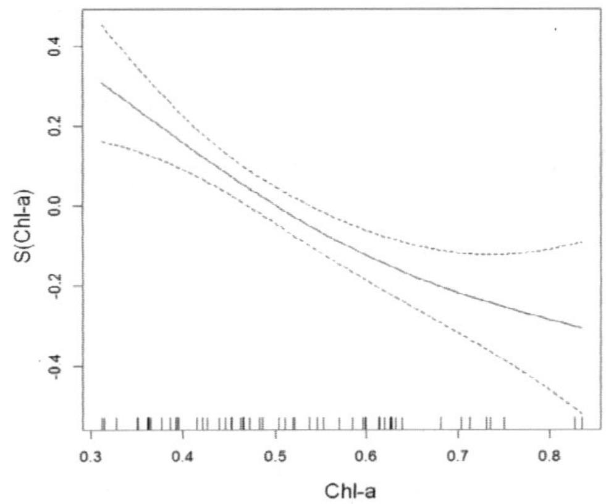

Figure 3. Effect of the 3 oceanographic variables (a) sea-surface-height anomaly (SSHA), (b) sea-surface temperature (SST), and (c) chlorophyll-*a* (Chl-*a*) concentrations—on Eastern Little Tuna catches derived from generalized additive model (GAM) during 2010-2014 (N=61 samples). The x-axis shows the values of the explanatory variables, and the y-axis shows the contribution of the smoother to the fitted values. The tick marks on the horizontal axis represent the values of the observed data points; the thick line indicates the fitted function. Dashed lines represent 95% confidence intervals

4.3 Cross wavelet spectrum analysis (XWT) and Wavelet Coherence (WTC)

In order to clarify the physical characteristics of the seasonal change, we performed the wavelet spectrum analysis. The Cross Wavelet Transform (XWT) finds regions in time frequency space where the time series show high common power. The XWT showed correlations of Eastern Little Tuna catch with SSHA, SST, and Chl-*a* from 2010 until 2014. The XWT signal revealed the interannual signal with the period of 16 months for all parameters (Figure 4). The XWTs between Catch and Chl-*a* indicated correlations within interannual signal of period 16 months during middle of the year of 2011-2014. The XWT between Catch and SST, Catch and SSHA were also showed that there was a strong correlation within seasonal signal of period 4-8 months and interannual signal of 16 months during the year of 2012-2014.

IOP Conf. Series: Earth and Environmental Science **162** (2018) 012044 doi:10.1088/1755-1315/162/1/012044

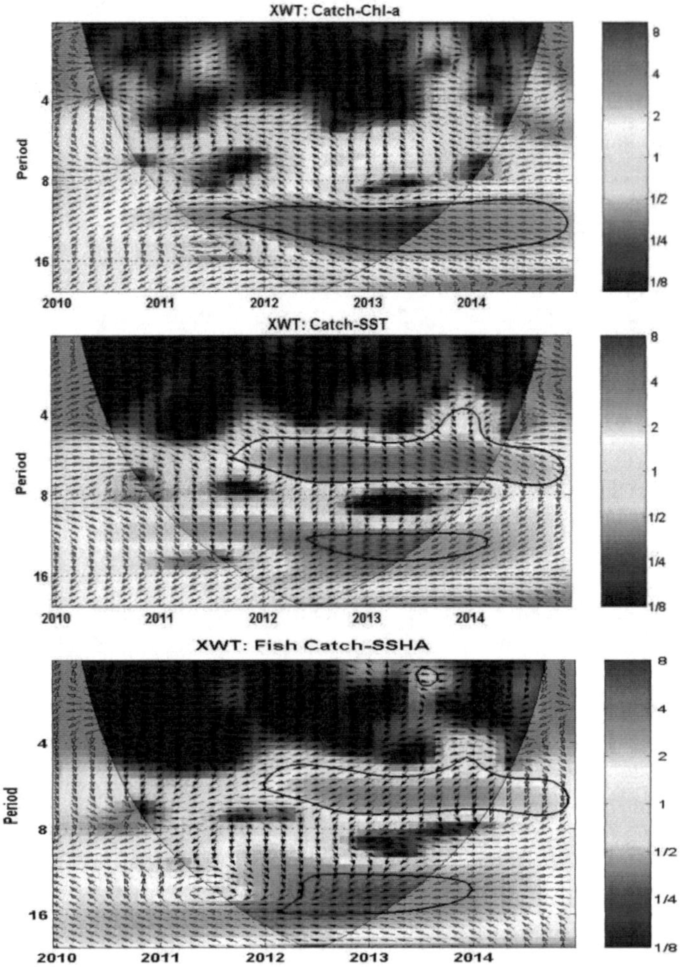

Figure 4. The cross wavelet spectrum (XWT) for Eastern Little Tuna catch in relation to SSHA, SST and Chl-*a* during 2010-2014. The y-axis is represented a period (monthly) and x-axis showed the time (year). Colorbars represented different wavelet amplitude, the black line (solid contour) showed the 95% confidence level and a thin black line showed the cone of influence

The wavelet coherence (WTC) is especially useful in highlighting the time and frequency intervals where two phenomena have a strong interaction. The coherence is defined as the cross-spectrum normalized to an individual power spectrum. It is a number between 0 and 1, and gives a measurement of the cross-correlation between two time-series and a frequency function. If the coherence between two series is high, the arrows in the coherence spectra show the phase between the phenomena: arrows at 0° (horizontal right) indicate that both phenomena are in phase and arrows at 180° (horizontal left) indicating that they are in anti phase. The WTCs between the parameters are shown in Figure 5.

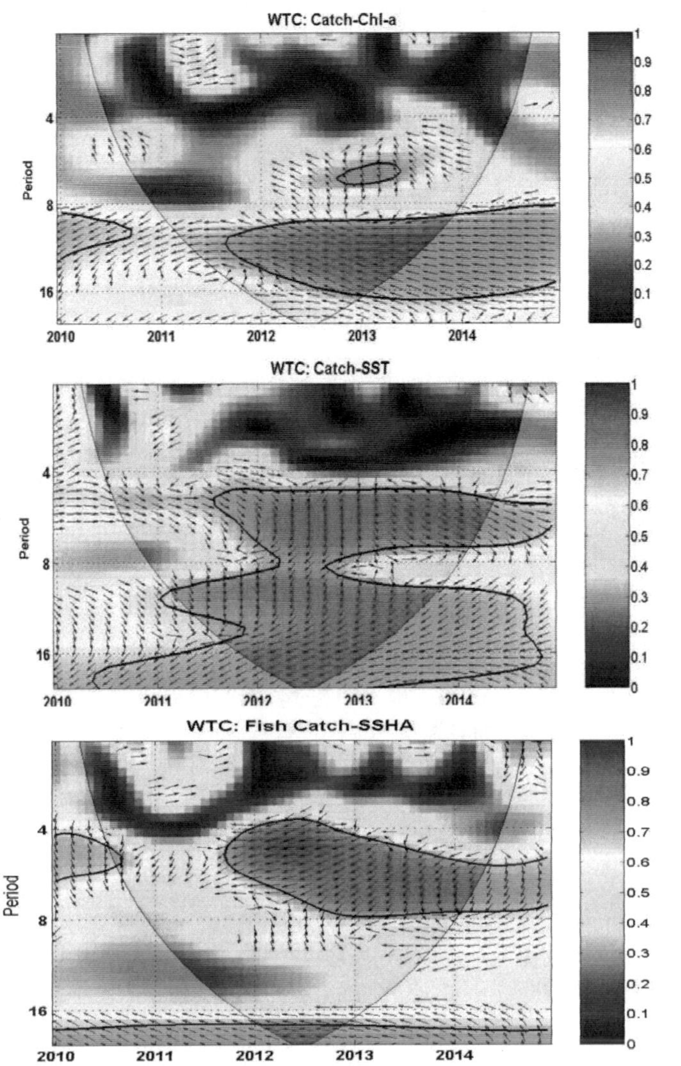

Figure 5. The wavelet coherence (WTC) for Eastern Little Tuna in relation to SSHA, SST and Chl-*a* during 2010-2014. The y-axis is represented a period (monthly) and x-axis showed the time (year). Colorbars represented different wavelet amplitude, the black line (solid contour) showed the 95% confidence level and a thin black line showed the cone of influence

The WTC between Eastern Little Tuna catches and all oceanographic parameters had a coherence of 0.8 (80%) within interannual signal of period 16 months during middle of 2011-2014. The coherence of 0.8 between catches and chl, ssha at the 4 and 16 months frequency in the interval 2012-2014 with tendency, anti phase for catches, SST, Chl-*a*, and SSHA.

The results of XWT indicated that the strong correlation of among catches, Chl-*a*, SST and SSHA within interannual and seasonal pattern were consistent with the strong coherence of 80% from the WTC analysis. This means that all oceanographic parameters have strong correlation with the Eastern Little Tuna catches. Furthermore, the results of GAM and wavelet analysis could be used to identify the correlation among the catches and all oceanographic parameters. The wavelet results emphasize how oceanographic factors influence the Eastern Little Tuna catches and identify the peak catches based on the season changing. This is very important precursor to know when is the best fishing season and the information will be useful for the fisherman and policy maker to reduce risk and to have sustainable fishing management.

5. Conclusion

The present study could identify the favourable oceanographic condition as a habitat preferences for Eastern Little Tuna based on the changing of oceanographic factors. High probabilities of Eastern Little Tuna catch were observed for Chl-*a* concentration ranging from 0.3-0.5 mg.m^{-3}, for SSHA of 0-8 cm and SST values of 28-29°C. Statistical analysis showed very clear relationship between Eastern Little Tuna catches and all oceanographic parameters to seasonal change with most predominant signal in the interannual time scale. The GAM results for the period of 2010-2014 indicated that 3-oceanographic parameter combination models explained the highest deviance (41.4%) in which the Chl-*a* is a more important contributing factor than the other environmental variables (SSHA and SST) in this region. The wavelet coherence confirmed that there were a strong correlation between catches and all parameters (0.8). This information will be useful for the fishermen and policy maker to have sustainable fishing management. We recommend to have further investigations through the use of long-term historical time series to predict fishing ground locations and an emerging need to improve our climate understanding and forecast skill to conserve small pelagic catches.

Acknowledgements

The authors wish to thank the Ministry of Research, Technology and Higher Education of the Republic of Indonesia for the research grant. We thank Padjadjaran University, Marine Affairs and Fisheries Indramayu, the Goddard Space Flight Center/NASA, Physical Oceanography Distributed Active Archive Center, NOAA Climate Prediction Center, MSAT Committee and also CSPACE Project Team.

References

[1] Pepperell, J 2010 *Fishes of the open ocean*. The University of Chicago Press, Chicago. 75-80 pp
[2] FAO 2017 Species Fact Sheet Euthynnus affinis. Food and Agriculture Organization of the United Nations. FAO Fisheries and Aquaculture Department.
[3] Zainuddin M, Saitoh S, and Saitoh K 2004 Detection of potential fishing ground for albacore tuna using synoptic measurements of ocean color and thermal remote sensing in the northwestern North Pacific. *Geophysical Research Letter*.
[4] Saitoh S, Chassot E, Dwivedi R, Fonteneau A, Kiyofuji H, Kumari B, Kuno M, Matsumura S, Platt T, Raman M, Sathyendranath S, Solanki H, and Takahashi F 2009 Remote sensing applications to fish harvesting. *In* Remote sensing in fisheries and aquaculture. *Reports of the International Ocean-Colour Coordinating Group (IOCCG)*, No. 8 (M.-H. Forget, V. Stuart, and T. Platt, eds.), IOCCG, Dartmouth, Canada. 57–76pp.
[5] Qu, T., Y. Du, J. Strachan, G. Meyers, and J. Slingo 2005 Sea Surface Temperature and Its Variability in the Indonesian Region. *Oceanography*. **18**: 50–123.
[6] Torrence, C., and Compo, G.P 1998 A practical guide to wavelet analysis. *Bulletin of the American Meteorological Society*, **79**: 61-78.
[7] Howell and Kobayashi 2006 El Nino effects in the Palmyra Atoll region: oceanographic changes and bigeye tuna (*Thunnus obesus*) catch rate variability. *Fisheries Oceanography*. **15**: 477-489.
[8] Sabates, A., Martin, P., and Raya, V 2012 Changes in life-history traits in relation to climate change:bluefish (*Pomatomus saltatrix*) in the northwestern Mediterranean. *ICES Journal of Marine Science*, **69**(6): 1000-1009.
[9] Zuur A F, Ieno E N, Walker N J, Saveliev A A, and Smith G M 2009 *Mixed effects models and extensions in ecology with R*. Springer, New York. 574 p.

IOP Conf. Series: Earth and Environmental Science **162** (2018) 012044 doi:10.1088/1755-1315/162/1/012044

[10] Wood S N 2006 *Generalized additive models, an introduction with R.* Chapman and Hall, London. 392 p.
[11] Amri K dan Satria F 2013 Impact of climate anomaly on catch composition of neritic tuna in Sunda Strait. Ind. Fish. Res. J, **19**(2):61-72
[12] Chiou W D, Lee, L K 2004 Migration of kawakawa Euthynnus affinis in the waters near Taiwan. Fisheries Science, **76**:746-737
[13] Hidayat T, Febrianti E, Restianingsih, Y H 2016 Pattern and Spawning Season of Kawakawa (*Euthynnus affinis*, Cantor 1850). Bawal, **8**(2):101-108
[14] Syamsuddin M, Saitoh S, Hirawake T, Syamsudin F, and Zainuddin M 2016 Interannual variation of bigeye tuna (Thunnus obesus) hotspots in the eastern Indian Ocan off Java *International Journal Of Remote Sensing*

Variabilities of the sea level anomalies in the upstream areas of the Kuroshio Current

Y J Wijaya[1,2], S Sugitani[3], D N Sugianto[4,5], D P Wijayanti[2] and Y Hisaki[1]

[1] Faculty of Science, University of The Ryukyus, Nishihara, 903-0213 Japan
[2] Faculty of Fisheries and Marine Science, Diponegoro University, Semarang, 50275 Indonesia
[3] National Institute of Information and Communications Technology, 4484 Onna, Kunigami, 904- 0411 Japan
[4] Oceanography Department, Diponegoro University, Semarang, 50275 Indonesia
[5] Centre for Disaster Mitigation and Rehabilitation Studies, Diponegoro University, Semarang, 50275 Indonesia

E-mail: yusufjatiwijaya@gmail.com

Abstract. The relationship between The Kuroshio velocities and Sea Level Anomalies (SLA) in the upstream areas (between Taiwan and Yonaguni Island) is investigated based on satellite altimetry observation. Mean of the SLA data from 2005-2008 show that in the northeast of Taiwan tends to form strong eddy activity. Furthermore, the type of eddy that forms in the upstream areas is difference every season, cold (warm) eddy more exists in summer (winter). The speed of the Kuroshio in the upstream areas is determined by combination of High-Frequency (HF) radar and Copernicus Marine Environment Monitoring Service (CMEMS) data. These two are exhibited that Kuroshio speed become faster (slower) in summer (winter). High positive correlation of the Kuroshio speed and SLA are found in between Yonaguni and Iriomote Island. When the Kuroshio speed in the middle of HF radar areas (123.375° N) became faster, it coincided with transition of warm eddy to cold eddy in the northeast of Taiwan and vice versa.

1. Introduction

Kuroshio also called Japan Current, is the strong surface current of the North Pacific Ocean gyre, flowing past Taiwan and Ryukyu Island in Japan. Lately, satellite altimetry data with high resolution used frequently to improve the understanding of this phenomenon. In the East of Taiwan, Sea Surface Height Anomaly (SSHA) observed by satellite altimetry are found to be well correlated with the Kuroshio transport, both offshore and onshore Kuroshio meandering could be generated by the negative anomaly which are simultaneously with the low Kuroshio transport based on the World Ocean Circulation Experiment (WOCE) PCM-1 [1]. Prior study recognized mesoscale eddy southeast of Taiwan later on would encourage seaward (or coastward) of the Kuroshio east of Taiwan 40 days later [2]. Westward propagating warm eddies from 18°N to 23°N may form a large Kuroshio meander and then, the meanders led to a large 100-days transport fluctuation in the East Taiwan Channel [3]. El Niño/Southern Oscillation (ENSO) and Pacific decadal oscillation (PDO) have an influence on these variabilities of the Kuroshio [4,5].

Furthermore, ocean circulation in the upstream areas of the Kuroshio have been widely recorded in another several observations, such as variability of the SLA around northeast of Taiwan using short and long term datasets [6,7], study of the Kuroshio axis by High-Frequency (HF) radar [8] and the Kuroshio intrusion role in atmospheric variability [9]. Northeast of Taiwan has a unique geography setting that consist of the continental shelf and slope, where eddies are frequently forms. In northeast of Taiwan, cyclonic eddy found more exist in summer than winter [10]. Low sea level anomaly is one indicator that represents cyclonic eddy [11]. This eddy associated with cold dome, and would be disappeared caused by seasonal migration of the Kuroshio axis [12].

Enhancement of quality on satellite altimetry resolution in recent decades, make accuracy of the Kuroshio's observations also increased. For example, SLA reanalysis data from Copernicus Marine Environment Monitoring Service (CMEMS) derived from Mean Sea Surface (MSS) which already filtered from residual noise and small-scale signal. Since May 2017, geostrophic velocities data (MADT-UV) delayed time from AVISO processed and distributed by CMEMS. Geostrophic velocities are one of important product of satellite altimetry which obtained from the first derivative of sea surface height [13]. But it should be noted, in order to improve accuracy much better, it should not only rely on satellite altimetry data but also another instrument with higher spatial resolution. High-Frequency radar is one of reliable instrument which provide ocean currents data by send radio waves and measure scattered signal from the surface ocean [14]. In general, two HF radars needed and would be placed at a certain distance to produce the current velocity map. In Yonaguni and Ishigaki Island, there is a pair of High-Frequency radar that have been observing sea surface current in the upstream areas since 2001. Combination of these two-instrument expected to providing comprehensive explanation of variability of the SLA and Kuroshio.

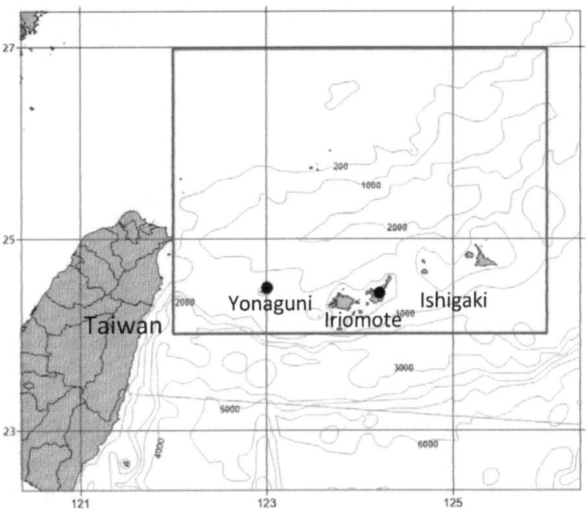

Figure 1. Observation location map with bathymetry contours (200, 1000, 2000, 3000, 5000 and 6000 m), black dots represent the location of HF radar at Yonaguni and Ishigaki Island. Red box is northern and eastern boundary for HF radar observation.

In this paper, we determine speed of the Kuroshio and SLA in summer and winter. Seasonal averaged of the SLA maps present the tendency of eddy that to be formed, particularly in the northeast of Taiwan. The Kuroshio speed data are obtained by High-Frequency radar and satellite altimetry. Meanwhile, seasonal mean of the Kuroshio reveals difference feature that will be occurred during the season. Finally, correlation map of these parameter also shown, both these parameters affected each other and led to the formation of eddy northeast of Taiwan.

IOP Conf. Series: Earth and Environmental Science **162** (2018) 012045 doi:10.1088/1755-1315/162/1/012045

2. Materials and Methods

Table 1 describes the data that used in this paper in brief, one of them is satellite altimetry data (i.e. geostrophic velocity and SLA). Copernicus Marine Environment Monitoring Service (CMEMS) distributes delayed time SLA data which is generated by combination of several satellite altimetry mission (JASON-1 and 2, ENVISAT, ERS-1 and 2). Computation of raw SLA, cross validation, filtering and sub-sampling and generation of by-products process, all

Table 1. HF Radar and Satellite Altimetry Data Specifications.

Data	Spatial Resolution	Temporal resolution
HF Radar	7 km x 7 km	30 minutes
Satellite (Geostrophic Current)	± 27 km x 27 km	Daily
Satellite (SLA)	± 27 km x 27 km	Daily

have been done in order to generating the SLA product [15]. Spatial resolution of SLA data from January 2005 to December 2008 are 0.25° x 0.25° (± 27 km x 27 km). The SLA was derived from the difference between sea surface height (SSH) and mean sea surface height (MSSH).

$$SLA = H - R_{obs} - \Delta RGeo_{corrs} - MSSH \quad (1)$$

Where H is representative of absolute sea surface height, R_{obs} is the distance from spacecraft antenna to sea surface that already corrected, $\Delta RGeo_{corrs}$ is geophysical corrections (e.g. tides, atmospheric delay and sea state bias) and MSSH is the ocean surface height above satellite altimetry reference ellipsoid over a certain time.

Geostrophic velocity was obtained from first derivative of the sea surface height (SSH). The data are used as representation of the Kuroshio speed, the zonal (u) and meridional (v) component of geostrophic velocity can be derived from equation 2. In which, g is gravitational constant, the Coriolis parameter $f = 2\Omega sin\varphi f$, where Ω is the angular speed of the earth rotation and φ is the latitude, h is SSH that obtained from altimeter.

$$u_s = \frac{g}{f}\frac{\partial h}{\partial y}, v_s = \frac{g}{f}\frac{\partial h}{\partial x} \quad (2)$$

A pair of Long-Range High-Frequency Ocean Radar are operated by The National Institute of Information and Communication (NICT) Japan, placed on Yonaguni and Ishigaki Island. The Long-Range High-Frequency Ocean Radar (hereafter HF radar) is a kind of a doppler radar which are aimed at monitoring surface current in the upstream areas. The HF radar produced surface current components with 7 km spatial interval and covered approximately 200 km. Eastward and northward current component every 30-minutes temporal interval from April 2005 to August 2008 use in this paper. Strong typhoon was happened in the late of September 2006 and led to interference in HF radar result, thus we omitted the data in that month.

The investigation of seasonal variability of the Kuroshio current and SLA in the present study are performed by computing the average value from December to February for winter and June to August for summer season. Seasonal variability of the Kuroshio from HF radar and satellite altimetry will be compared and expected to show variabilities more clearly. The SLA were used to detect eddies, in particular northeast of Taiwan.

IOP Conf. Series: Earth and Environmental Science **162** (2018) 012045 doi:10.1088/1755-1315/162/1/012045

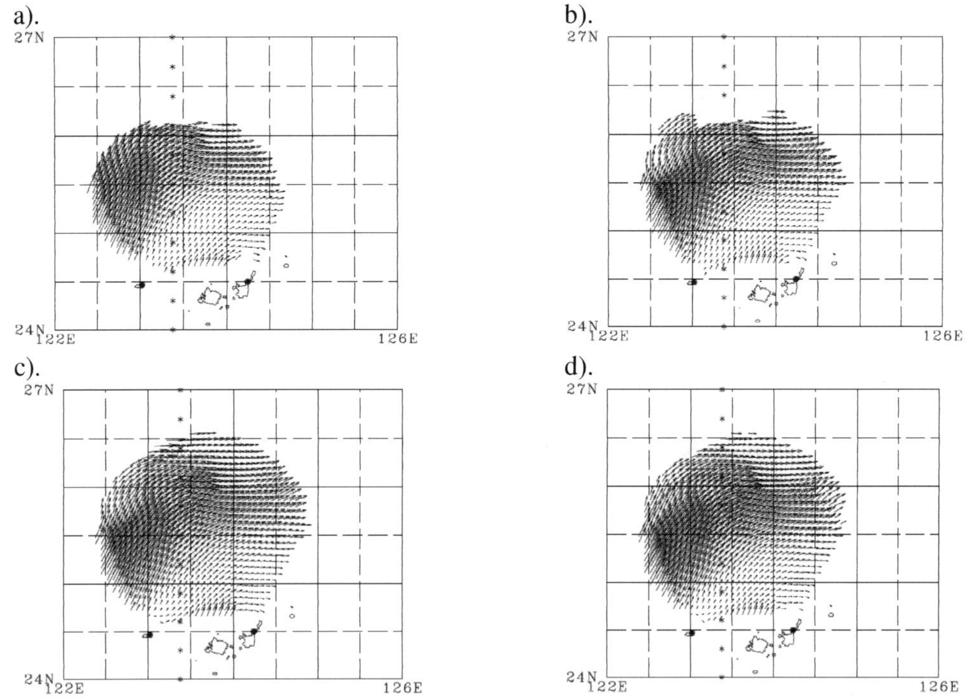

Figure 2. Mean annual of the Kuroshio speed in 2005 (a), 2006 (b), 2007 (c), 2008 (d). The point marks indicate the position of HF radar. The meridional line at 123.375°E as reference for observation of the Kuroshio is also plotted by a star line.

3. Results

3.1. Variabilities of the Kuroshio by HF Radar and Satellite Altimetry

In order to facilitate within monitoring of variabilities of the Kuroshio, we need to determine specific point or line that passed by the Kuroshio consistently. Based on HF radar and altimetry data, meridional line at 123.375°E assumed as specific line for the Kuroshio observation. Figure 2 indicates the Kuroshio continuously crosses at 123.375°E which is represented by the length of vector. The latitude for the Kuroshio observation boundaries are confirmed by Fig. 5, in which these boundaries have a purpose to perceive variabilities of the Kuroshio by satellite altimetry.

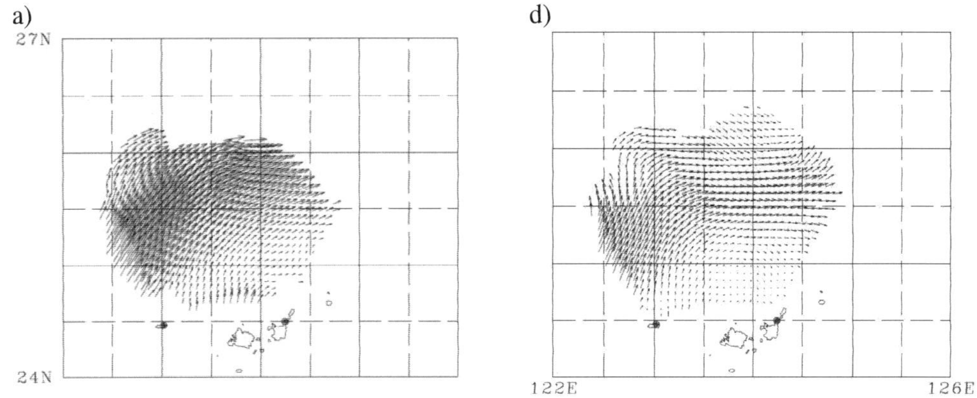

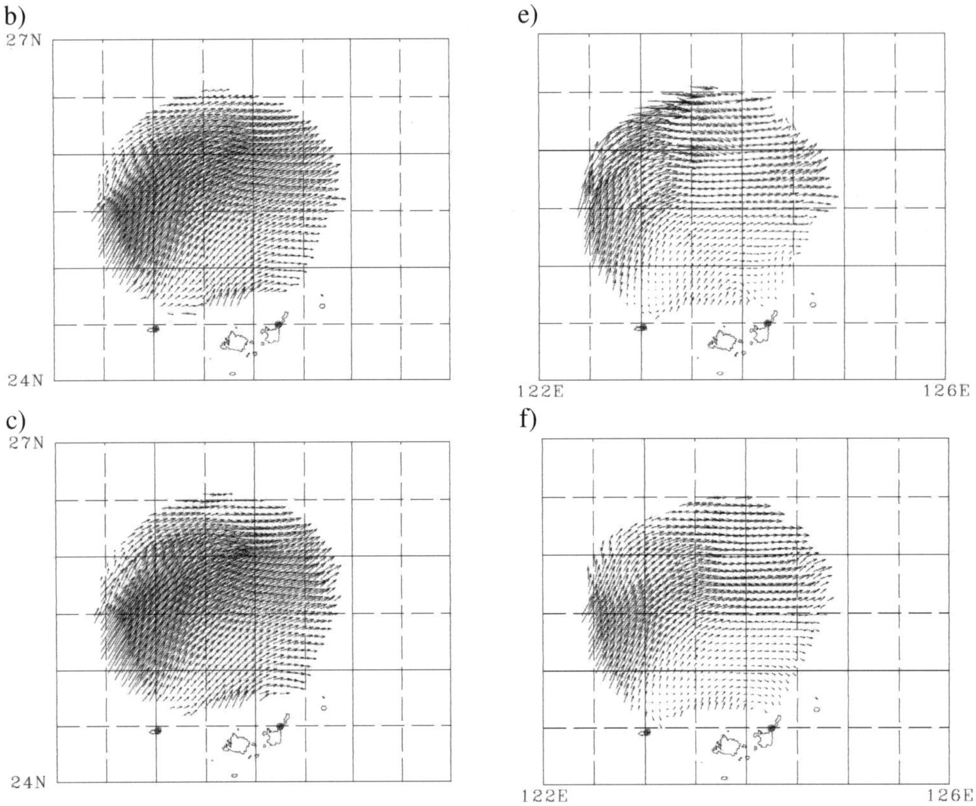

Figure 3. Mean seasonal variabilities of the Kuroshio by HF Radar in summer 2006 (a), 2007 (b), 2008 (c) and in winter 2006 (d), 2007 (e) and 2008 (f). The dotted marks are the same as Fig. 1.

The seasonal variabilities of the Kuroshio by HF Radar are exhibited in Fig. 3. Vector in summer seem more length and crowded than in winter, it means the Kuroshio speed shows a tendency to be faster in summer and slower in winter. In winter, the vectors that represent strong current are seen shift northward, meanwhile in summer move to southward. It indicates when summer (or winter) the Kuroshio shifted southward (or northward), These have been reported in prior studies in which the Kuroshio led northward on to shelf of the East China Sea in winter [16]. Actually, the shifting is clearest in 2007, but unclear in other years. Although shifts of the Kuroshio are unclear in 2006 and 2008, stronger (or weaker) remained clear.

Variabilities of the Kuroshio with magnitude above 0.3 by satellite altimetry data are extracted in Fig. 4. The results look exactly the same as maps that are produced by HF radar. The Kuroshio tends to be faster (slower) in summer (winter), but with the advantages northward and southward shifts of the Kuroshio are more clearly visible. All of figure in winter (bottom panel) indicates that the Kuroshio moves northward, meantime in summer (upper panel) moves southward. The Kuroshio intrusion reportedly increased as the Kuroshio shifted northward in winter [6].

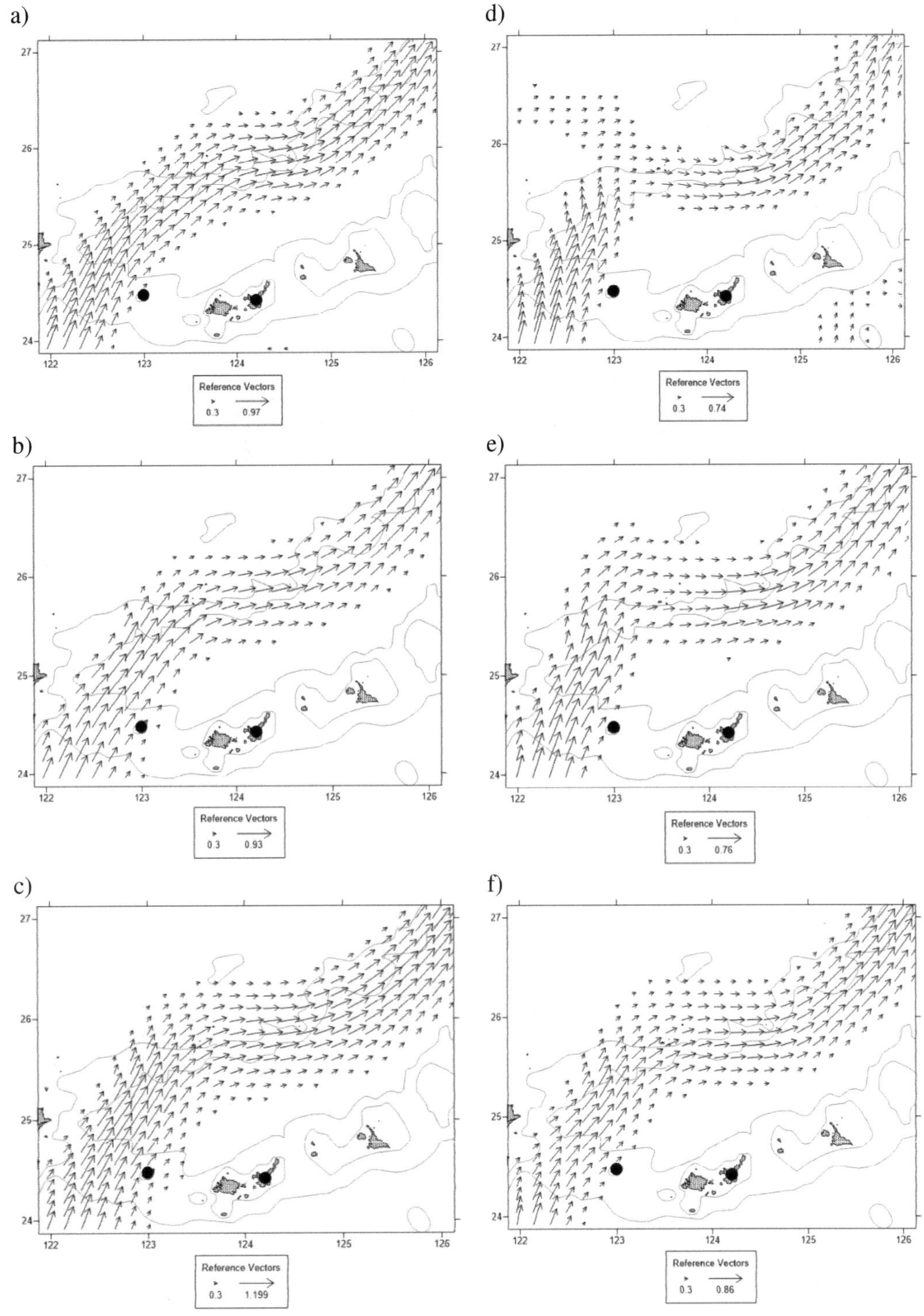

Figure 4. Mean seasonal variabilities of the Kuroshio by satellite altimetry in summer 2006 (a), 2007 (b), 2008 (c) and in winter 2006 (d), 2007 (e) and 2008 (f), superimposed on the bathymetry contours (200 and 1000 m). The magnitudes under 0.3 are removed from data for convenience. The dotted marks are the same as Fig. 1.

Since variability of the Kuroshio was shown clearer by satellite altimetry, therefore in order to better understand about the Kuroshio, we have constructed Hovmöller (latitude-time) diagram at meridional line 123.375°N using time series data of geostrophic velocity from satellite altimetry. Hovmöller diagram is a common ways of plotting oceanography data, in order to display the pattern. Hovmöller diagram in Fig. 5 reveals variabilities of the Kuroshio by satellite altimetry for 4 years. As well as Fig. 3 and Fig. 4, variabilities of the Kuroshio tend likewise to be moved northward (or southward) in winter (or summer). However, the shifts of the Kuroshio in Fig. 5 described more clearly. When the Kuroshio speed is faster (or slower), the Kuroshio shifts southward (or northward). Hovmöller diagram Fig. 5 shows the Kuroshio can reach out speeds over 0.9 m s^{-1} in summer, and weakened to less than 0.3 m s^{-1} in spring or winter. Nevertheless, there are exception which the Kuroshio speed exceeds 0.6 m s^{-1} in January 2008. It will be discussed in the next chapter.

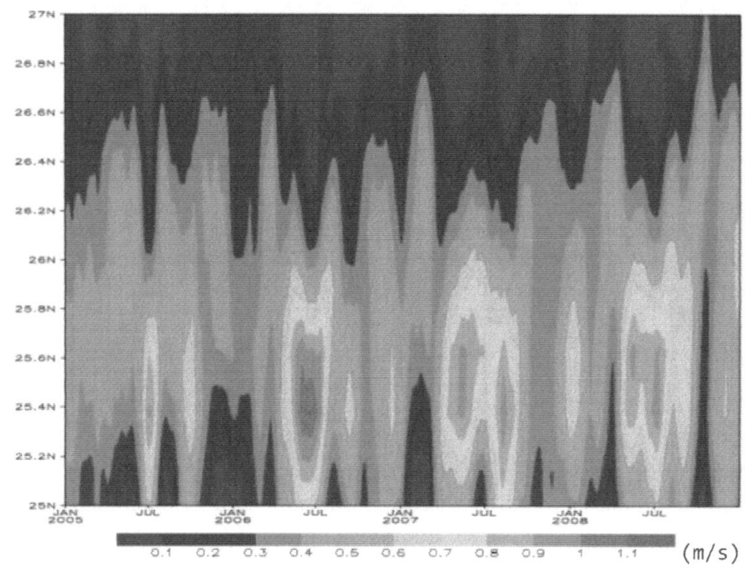

Figure 5. Hovmöller (latitude-time) diagram of variabilities of the Kuroshio speed by satellite altimetry data from 2005-2008 at meridional line 123.375°E with 35-days low pass filtered. Latitude (time) is indicated by X-axis (Y-axis).

3.2. Variabilities of the SLA Northeast of Taiwan

Figure 6 exhibits mean seasonal of the SLA during winter and summer from 2005-2008. The SLA northeast of Taiwan in continental slope indicates lower (or higher) in summer (or winter). This simultaneity also implies that there is a tendency to form cold dome in summer (Fig. 6b), especially at meridional line 122.625°E. The SLA inside the ring is higher (or lower) than surrounding in winter (or summer), denotes the emergence of warm (or cold) eddy. Meantime, in the centre of HF radar observation area (between Yonaguni and Ishigaki Island) is surrounded by the negative SLA in winter (Fig. 6b) and the inverse is the positive SLA in summer.

IOP Conf. Series: Earth and Environmental Science **162** (2018) 012045 doi:10.1088/1755-1315/162/1/012045

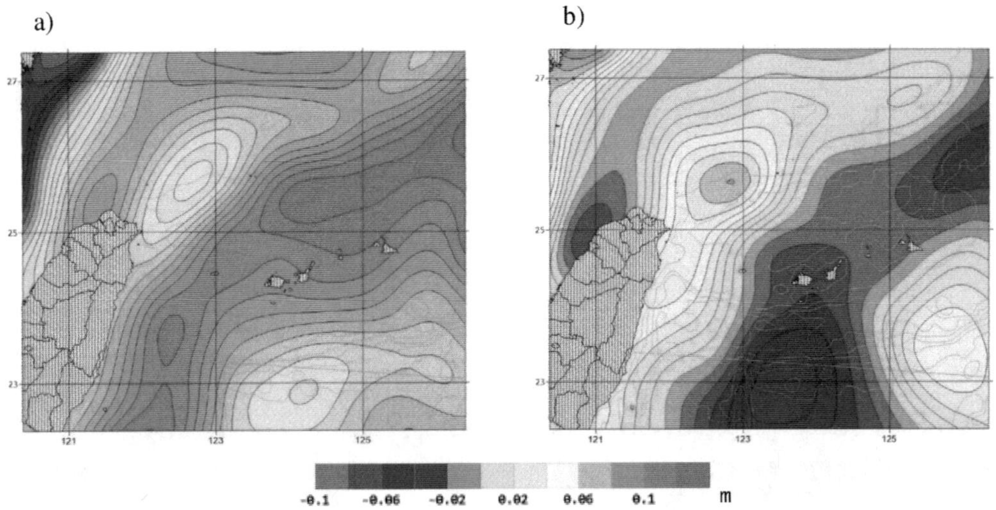

Figure 6. Mean of the SLA in summer (a) and in winter (b) from 2005-2008, contour intervals are 0.01 m.

Mean of the Kuroshio speed for one week is displayed by the upper panels of Fig. 7. Fig. 7a describes the surface current speed when the positive SLA appears in northeast of Taiwan in Fig. 7c and 7d, whereas Fig. 7b is for Fig. 7e and 7f. The Kuroshio moved northward in March 2008, it is quite different with the Kuroshio speed in May 2008 which shifted southward. In May 2008 (or March 2008), the Kuroshio speed became faster (or slower) when shifted southward (or northward) as mentioned earlier. That motion occurred simultaneously with the positive and negative SLA northeast of Taiwan.

To better understanding about SLA and Kuroshio correlation. We choose occurrence on spring 2008 to observe, when transition from weak to strong Kuroshio took place. The transition from warm to cold core in northeast of Taiwan is shown by Fig. 7c to 7f (middle and bottom panel), positive (or negative) SLA are appeared on west of the red boxes that represent Fig. 7a and 7b boundaries. These SLA existed at meridional line 122.625°E in which the core of the SLA located along the continental slope. As that reason, meridional line at 122.625°E determined as the SLA observation line for Fig. 8. Beside the SLA transition northeast of Taiwan, from 18th to 19th of March 2008 seen that the SLA located at 22.375°N joined the SLA east of Taiwan and towards HF radar observation areas. Furthermore, at 19th of March the SLA northeast of Taiwan are associated with these SLA (Fig. 7d). Like the prior statement, previous study mentioned that warm eddies southeast of Taiwan related to other eddies within the HF radar observation areas [2]. Meanwhile, from 8th to 9th of May 2008 shown by Fig. 7e and 7f, the SLA from southeast of Taiwan are moved out and merged with the SLA southwest of the red boxes. When the Kuroshio speed is slower (or faster) in Fig. 7a (or Fig. 7b), the SLA between Iriomote and Ishigaki Island turned out higher (or lower). Figure 7 overall implies that southern and stronger (or northern and weaker) Kuroshio triggered the negative (or positive) SLA northeast of Taiwan.

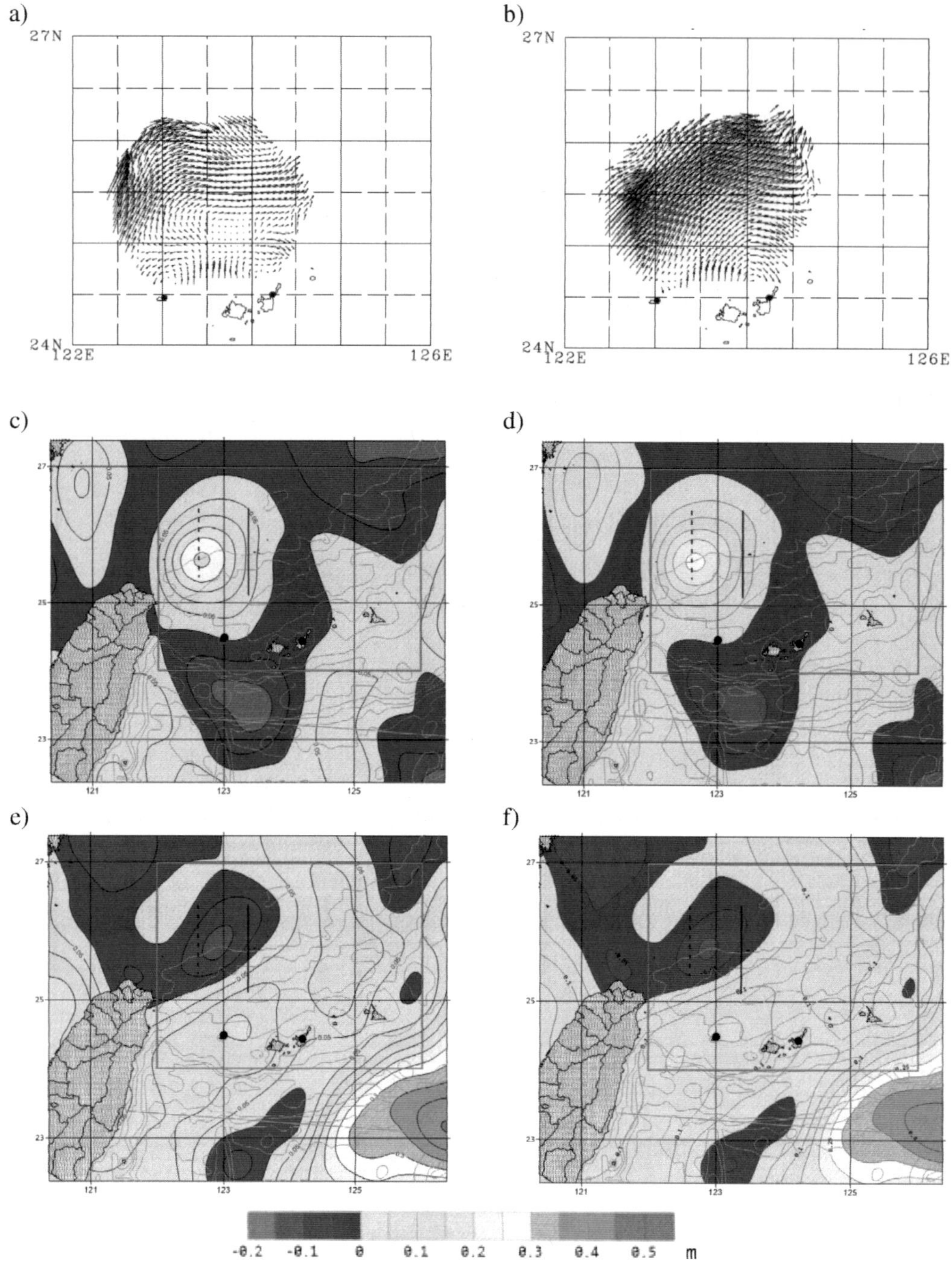

Figure 7. Mean of the surface current speed from 14-21 March 2008 (a), and 7-14 May 2008 (b). The SLA maps on 18 March 2008 (c), 19 March 2008 (d), 8 May 2008 (e) and 9 May 2008 (f). Black dots and red boxes in the lower panels are the same as Fig. 1. The contour intervals are 0.05 m.

In order to support the above paragraph statement, the Kuroshio speed and SLA from 2005-2008 are plotted as Fig. 8. In order to reduce noise from satellite altimetry, all geostrophic components and SLA data were smoothed by 35-days low-passed filtered. These two are found to be well correlated, as the

SLA is higher (or lower) when the Kuroshio speed become slower (or faster). Namely, the SLA northeast of Taiwan seems to be lower (or higher) in summer (or winter). However, abnormal occurrence was happened in winter 2008 where the SLA has a tendency become lower. The lowest state of the SLA was happened in May 2007 and then followed by January 2008 as the second.

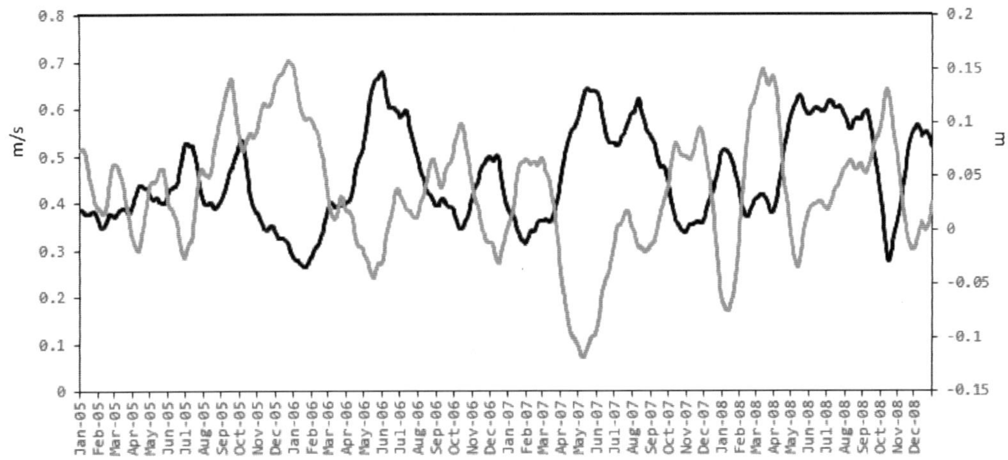

Figure 8. Time series of the Kuroshio speed (black line) and SLA (red line) with 35-days low pass filtered. The Kuroshio speed is the mean of geostrophic velocities from satellite altimetry that across black line (123.375°E) in figure 6, whilst the SLA derived from average value at broken line (122.375°E). Y-axis in the left (or right) indicates value of the Kuroshio speed (or the SLA).

Bottom panel on Fig. 9 displays the Kuroshio speed same as Fig. 8, but data in 2005 have removed for convenient. The boxes from the bottom to the top pose latitude-time plot of the SLA (35-days low passed) along northeast of Taiwan in latitude 25.125°N to 27.25°N range at longitude from 123.875°-122.375°E. The SLA inside the ring is higher (or lower) than surrounding which indicates the appearance of anti-cyclonic (cyclonic) eddy. Those figures indicate that dominant eddy which forms in northeast of Taiwan or at 122.375°E propagate eastward towards the centre of HF radar observation. Furthermore, the eddies reach the maximum state at 122.875°E and then weakened as it is approaching the Kuroshio meander. In the longitude 123.875°E, the eddies would be merged and disappeared gradually by the Kuroshio. Most of northeast of Taiwan eddies have periods of 30-45 days before they disappear. Figure 9 also shows when the Kuroshio speed is faster (slower), the SLA is carried away by the meander is higher (lower), meanwhile the SLA along continental slope tends to be lower (higher).

To investigate the relationship of both parameters, we have constructed a correlation between a time series of the Kuroshio speed and SLA data from 2005-2008, in the upstream areas as Fig. 7(c) to 7(f). Correlation coefficient derived from calculations that performed on each grid point. Calculations are not only done with the SLA on the same date, but also 20 days after and 20 days, 40 days before, as suggested in previous studies [2].

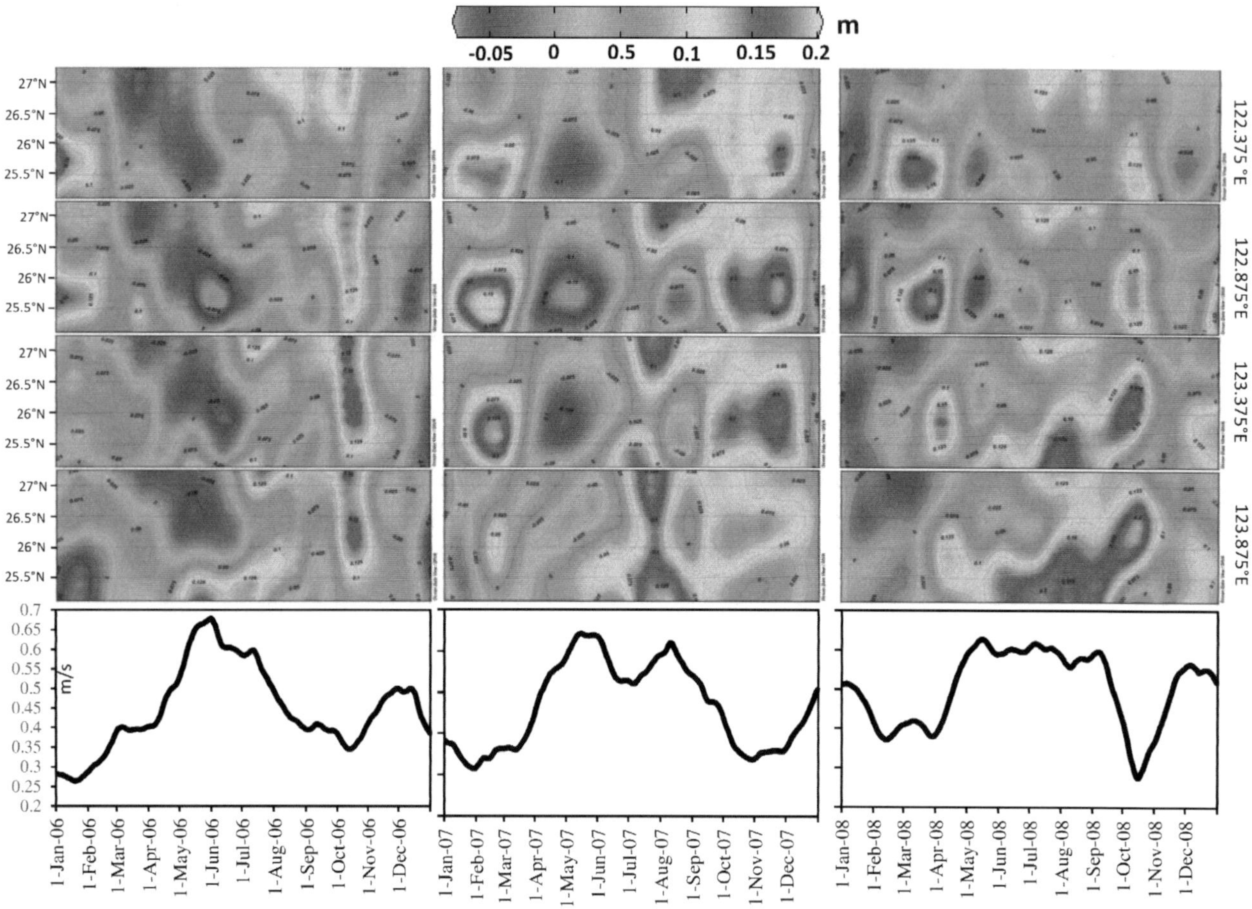

Figure 9. Bottom panel: The Kuroshio speed same as Fig. 7 and Hovmöller (latitude-time) plots of variabilities of the SLA northeast of Taiwan from 2006-2008 (35-days low pass filtered) for longitudes 123.875° to 122.375°E. The contour intervals are 0.05 m.

In Fig.10(c) every positive (or negative) correlation indicates that the SLA at location is higher (lower), while the Kuroshio speed tends to increase (or decrease). Before calculating the correlation, both geostrophic velocity and SLA data were smoothed by 35-days low-passed filtered to reduce small-scale noises. When there are no temporal lags, negative correlation is found in the continental slope. In the northeast of Taiwan also found positive between a pair of negative correlation, this looks similar to Fig. 7(d) in which the positive SLA is surrounded by negative SLA. Positive correlation centred between Yonaguni and Iriomote Island. Negative correlation emerges in the continental slope with the SLA 20 days before (Fig. 10b) and still exists with the SLA 20 days after (Fig.10d), while two small areas of positive correlation are found at 23°N southeast of Taiwan when correlation is accounted with the SLA 40 days before (Fig. 8b). Furthermore, these correlations are seen mixed up and form a larger area western Yonaguni Island with the SLA 20 day before. The positive correlation near east coast of Taiwan still exists, but it disappears at meridional line 123°E. The movement of positive correlation areas with temporal lags implies that anti-cyclonic (cyclonic) eddy southeast of Taiwan through 122.5°-123°N stimulates the Kuroshio speed, which tends to be faster (or slower) at centre of HF radar observation, followed by the emergence of cyclonic (anti-cyclonic) eddy northeast of Taiwan that indicated by negative correlation.

IOP Conf. Series: Earth and Environmental Science **162** (2018) 012045 doi:10.1088/1755-1315/162/1/012045

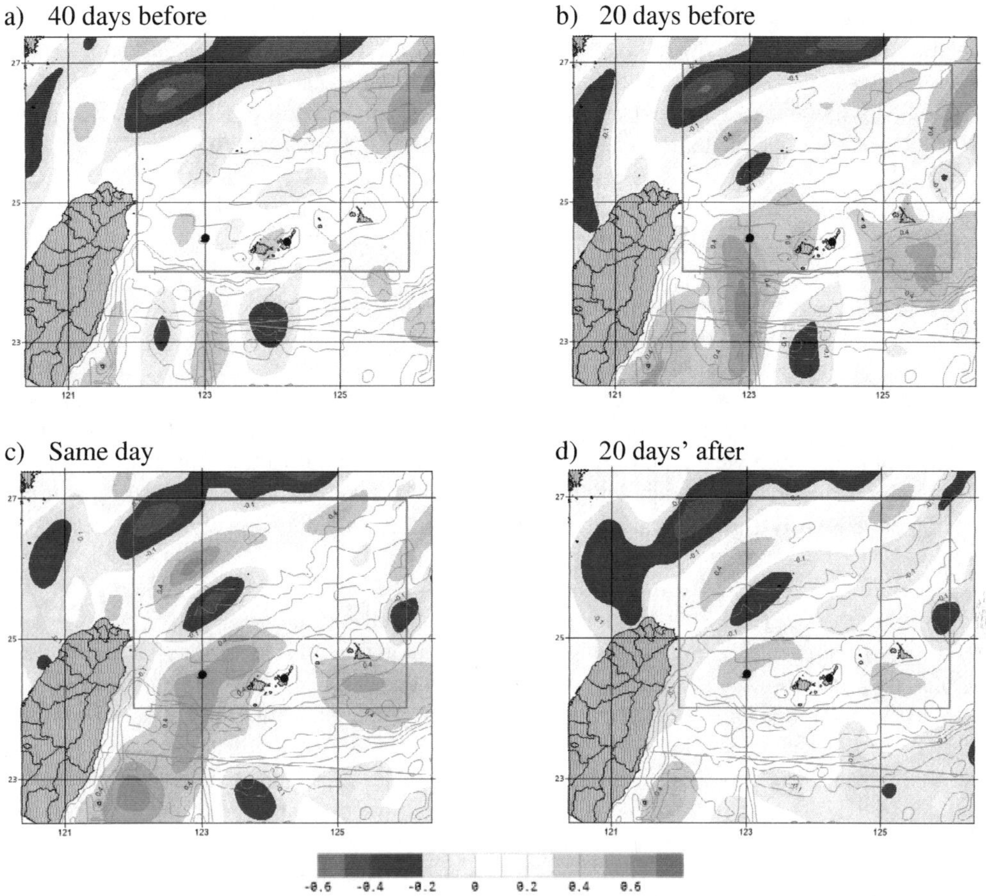

Figure 10. Cross correlation map of 35 days' low pass filtered the Kuroshio speed and SLA 40 days before (a), 20 days before (b), at the same day (c) and 20 days after (d) with 0.1 contour intervals. Dotted marks and bathymetry contours are the same as Fig. 1.

4. Summary and discussion

The combination of HF radar and altimetry data shows seasonal variabilities of the Kuroshio speed in the upstream areas, in particular along 123.375°N. The Kuroshio speed is being faster in summer and slower in winter (Fig. 3 and Fig. 4). Monsoon system becomes the main factor that causes these variabilities. In summer (or winter), the Kuroshio speed may be increased (or decreased) due to southwesterly (or northeasterly) wind effect [17]. However, several prior studies have been recording the strengthening (or weakening) of the Kuroshio speed in summer (or winter) [2,18,19].

The Kuroshio tends to move northward in winter and southward in summer (Fig. 3 and 4), but from the HF radar is not clear the following factor that generates the shift. We may be able to make a preliminary statement as mentioned in the above paragraph, therefore we need accurate evidence to prove it. Variability of the Kuroshio by satellite altimetry describes the shifts of the Kuroshio more clearly (Fig. 5). As expected from Fig. 3 and 4, the southward (or northward) Kuroshio takes place in summer (or winter), especially in July (or February). It seen that the Kuroshio shifts northward when the speed decreased, meanwhile southward shift is followed by increased speed. In fact, southward and northward shifts of the Kuroshio have been reported in prior studies [2,12,16].

In summer (or winter), the SLA northeast of Taiwan has a tendency to be formed cold (or warm) eddy which are depicted by the SLA inside the ring is lower (or higher) than surrounding. This is in line with [10] that found the number of cyclonic eddies are more significant in summer than in winter. In addition to the SLA, cold dome also one indication of the emergence of cyclonic eddy that comes up mainly in summer and associates with upwelling [20]. The SLA northeast of Taiwan seems contrary to the Kuroshio speed (Fig. 8). The SLA is lower (or higher) in summer (or winter), while at the same time the Kuroshio becomes faster (or slower). Even though, there is any exception when the Kuroshio shifted southward and stronger which led to the SLA to be lower, in January 2008. We try to figure out climatological factors that cause these abnormal conditions. Furthermore, El Nino-Southern Oscillation (ENSO) appears as one of the leading factor that estimated to be mastermind of. In 2007-2008, there was a moderate La Nina phenomenon. In the winter when La Nina occurred, surface current speed that crossed near northeast of Taiwan will be faster than usual as suggested in prior studies [5]. However, ENSO is not the only cause of variability of the Kuroshio, so it needs to be investigated further.

Figure 9 depicts eastward propagating of eddy in the northeast of Taiwan. The successive phenomena of the formation to disappearance of the eddy are described as follows: As mentioned above, when the Kuroshio speed is being faster (slower), cold (warm) eddy sets to form at 122.375°E and then propagates eastward towards the Kuroshio meander. On its way, cold (warm) eddy reaches its maximum state at 122.875 and would undergo weakening processes afterwards. As approaching the Kuroshio meander, cold (warm) eddy would merge/interact with the stronger (weaker) Kuroshio. Furthermore, cold (warm) eddy disappears gradually through mixed up with the higher (lower) SLA which is carried by the meander.

By taking into account correlation of the Kuroshio speed and the SLA with temporal lag, the Kuroshio speed in the centre of HF radar location looks to be influenced by eddy southeast of Taiwan, as reported in prior study [2]. Eddy southeast of Taiwan (along 123°E) propagates into the Kuroshio meander and any possibility merges with another eddy along coastline of Taiwan 20 days later, actually eddy southeast of Taiwan is westward propagating eddies from western Pacific and would significantly reduce the Kuroshio transport [3], then the merged eddy propagates together with the meander. Furthermore, the meander arrives at HF radar observation areas 40 days later with decreased (or increased) of the Kuroshio speed, followed by anti-cyclonic (or cyclonic) eddy northeast of Taiwan. When there are no temporal lags, negative correlation is found in the continental slope (Fig.10c). It begins to emerge 20 days before (Fig. 10b) when westward propagating eddy appears in east of Taiwan and still exists 20 days after (Fig. 10d). It indicates that the centre of cyclonic (or anti-cyclonic) eddy appears along continental slope, those events have been recorded in previous studies [6]. In this case, the Kuroshio intrusion in the continental slope plays a main role [21], increased (or decreased) of intrusion coincides with the emergence of cold (or warm) eddies [20]. The appearance of negative correlation from 20 days before to 20 days after implies the dominant eddies that form northeast of Taiwan, have a period of approximately 40 days, in line with Fig. 9 which implies that the eddies have periods of 30-45 days. Along 123°E meridional line is the main path of eddy southeast of Taiwan, towards the centre of the HF radar observation area. It is indicated by positive correlation that appears in the 40 days before (Fig. 10a) then will be vanished 20 days after the same day (Fig. 10d). Positive correlation in the east of Taiwan (Fig. 11c) explains that the appearances of warm (cold) propagating eddy in east of Taiwan, will be increased (decreased) the Kuroshio speed also. Hsu et al., 2016 was mentioned that the Kuroshio speed dropped approximately 84% of the seasonal average under the appearances of cyclonic eddy. The movement of negative correlation north of Taiwan indicates SLA at location not too much influenced by the Kuroshio. In the area is more affected by Taiwan warm current and China coastal current.

The present study only describes and represents specific mesoscale areas, especially predefined reference lines, either longitude or latitude limits. The formation of eddy northeast of Taiwan needs to be investigated further using HF radar, which is capable of produce surface current data with high spatial resolution. It is important to remove tidal components, wind-driven Ekman current component and inertial oscillations [2,8], in order to generate small scale variations of the Kuroshio on the shelf slope northeast of Taiwan.

Acknowledgments
The authors would like to thank National Institute of Information and Communications Technology (NICT) Japan for providing the HF radar data. Satellite altimetry data used in the present study was accessed from Copernicus Marine Environment Monitoring Service (http://marine.copernicus.eu/). Yusuf Jati Wijaya is supported by Ministry of Education and Culture, Indonesia, under grant 54285/A1.4/LN/2016. The travelling support for attending the conferences that held in Bali by MSAT Bandung Institute of Technology and University of the Ryukyus are also acknowledged.

References
[1] Wei L, Qinyu L and Yinglai, J 2004 The Kuroshio Transport East of Taiwan and the Sea Surface Height Anomaly from the Interior Ocean *Journal of Ocean University of China* **3(2)** 135-140
[2] Ichikawa K, Tokeshi R, Kashima M, Sato K, Matsuoka T, Kojima S and Fujii S 2008 Kuroshio variations in the upstream region as seen by HF radar and satellite altimetry data *International Journal of Remote Sensing* **29(21)** 6417-6426
[3] Zhang D X, Lee T N, Johns W E, Liu C -T and Zantopp R 2001 The Kuroshio east of Taiwan: Modes of variability and relationship to interior ocean mesoscale eddies *Journal of Physical Oceanography* **31(4)** 1054–1074
[4] Soeyanto E, Guo X, Uno J and Miyazawa Y 2014 Interannual variations of Kuroshio transport in the East China Sea and its relation to the Pacific Decadal Oscillation and mesoscale eddies *J. Geophys. Res. Ocean* **119(6)** 3595-3616
[5] Liu X, Dong C, Chen D and Su J 2014 The pattern and variability of winter Kuroshio intrusion northeast of Taiwan *J. Geophys. Res. Oceans* **119(8)** 5380–5394
[6] Yuqi Y, Xiaopei L, Yizhen L and Xiangming Z 2014 Seasonal variability of Kuroshio intrusion northeast of Taiwan Island as revealed by self-organizing map *Chinese Journal of Oceanology and Limnology* **32(6)** 1435-1442
[7] Yang Y, Liu C -T, Hu J -H and Koga M 1999 Taiwan current (Kuroshio) and impinging eddies *Journal of Oceanography* **55(5)** 609–617
[8] Takahashi D, Guo X, Morimoto A and Kojima S 2009 Biweekly periodic variation of the Kuroshio axis northeast of Taiwan as revealed by ocean high-frequency radar *Continental Shelf Research* **29(15)** 1896-1907
[9] Chow C H, Liu Q and Xie S -P 2015 Effects of Kuroshio Intrusions on the atmosphere northeast of Taiwan Island *Geophys Res Lett.* **42(5)** 1465–1470
[10] Wu C -R, Lu H -F and Chao S -Y 2008 A numerical study on the formation of upwelling off northeast Taiwan *J. Geophys. Res. Ocean* **113(C8)** 1-12
[11] Hsin Y -C, Chiang T -L and Wu C -R 2011 Fluctuations of the thermal fronts off northeastern Taiwan *J. Geophys. Res. Ocean* **116(C10)** 1-14
[12] Tang T Y, Tai J H and Yang Y J 2000 The flow pattern north of Taiwan and the migration of the Kuroshio *Continental Shelf Research* **20(4-5)** 349-371.
[13] Arbic B K, Scott R B, Chelton D B, Richman J G and Shriver J F 2012 Effects of stencil width on surface ocean geostrophic velocity and vorticity estimation from gridded satellite altimeter data *J. Geophys. Res. Ocean* **117(C3)** 1-18
[14] Hisaki Y 2005 Ocean Wave Directional Spectra Estimation from an HF Ocean Radar with a Single Antenna Array: Methodology *Journal of Atmospheric and Oceanic Technology* **23(2)** 268-286.
[15] CMEMS 2016 Product User Manual for Sea Level SLA products *EU Copernicus Marine Service – Public* p. 19.
[16] Oey L -Y, Hsin Y -C and Wu C -R 2010 Why does the Kuroshio northeast of Taiwan shift shelfward in winter? *Ocean Dynamics* **60(2)** 413-426
[17] Cheng K -H, Lin C -C, Hsu P -C and Huang S -J 2016 Reconstruction and analysis of surface velocity from drifters in the Kuroshio area east of Taiwan *Journal of Marine Science and Technology* **24(6)** 1217-1225

[18] Hsin Y -C, Wu C -R and Shaw P -T 2008 Spatial and temporal variations of the Kuroshio east of Taiwan,1982–2005: A numerical study *J. Geophys. Res. Ocean* **113(C4)** 1-15
[19] Hsu P -C, Lin C -C, Huang S -J and Ho C -R 2016 Effects of Cold Eddy on Kuroshio Meander and Its Surface Properties, East of Taiwan *IEEE Journal of Selected Topics in Applied Earth Observations and Remote Sensing* **9(11)** 5055-5063
[20] Gawarkiewicz G, Jan S, Lermusiaux P F J, McClean J L, Centurioni L, Taylor K, Cornuelle B, Duda T F, Wang J, Yang Y J, Sanford T, Lien R -C, Lee C, Lee M -A, Leslie W, Haley Jr P J, Niiler P P, Gopalakrishnan G, Velez-Belchi P, Lee D -K and Kim Y Y 2011 Circulation and intrusions northeast of Taiwan: chasing and predicting uncertainty in the cold dome *Oceanography* **24(4)** 110–121
[21] Chuang W -S, and Liang W -D 1994 Seasonal variability of intrusion of the Kuroshio water across the continental shelf northeast of Taiwan *Journal of Oceanography* **50(5)** 531-542

IOP Conf. Series: Earth and Environmental Science **162** (2018) 012046 doi:10.1088/1755-1315/162/1/012046

Analysis of dissemination methods for remote sensing information in maritime field to support the local government

G D Yudha and S M Purnama

Remote Sensing Technology and Data, LAPAN, Indonesia

E-mail: gusti.darma@lapan.go.id

Abstract. The demand for remote sensing imagery data has been increasing in recent years in Remote Sensing Technology and Data Center LAPAN, especially the utilization of such data for the maritime field. Data request is recorded increasing trend from 808 scenes in 2015 to become 6773 scenes in 2016. Therefore, Remote Sensing Technology and Data Center should improve its services so that information provided to users could be delivered as fast as possible. The objective of this research was to find the best alternatives dissemination from the current method of data distribution and remote sensing information in terms of speed, scalability, functionality and data maintenance in order to support government program on maritime sectors. Alternative to be examined was based on Web-GIS which means that access to data and remote sensing information could be done online through internet. The used of Web-GIS in this study due to the utilization of the internet already prevalent in the present day. Through this method, users could have obtained data and information from anywhere and anytime. As an example, this paper uses mangrove distribution data in Banten as remote sensing data provided. Results are easily obtained mangrove information; the government can overcome the problems that has emerged as the reduced area of mangrove forest which resulted in causing chain actions such as floods and other disasters. Using Web-GIS can be an effective means of dissemination to speed up service.

1. Introduction

The demand for remote sensing imagery data has been increasing in recent years in Remote Sensing Technology and Data Center, especially the utilization of such data for the maritime field. Data request is recorded increasing trend from 808 scenes in 2015 to become 6773 scenes in 2016. Therefore, Remote Sensing Technology and Data Center should improve its services so that information provided to users can be delivered quickly. Existing condition, when users request remote sensing data and information, they should come to the office to obtain the data and information. Depend on how big the data, if the data size is not big, it can be send through email otherwise copied to other media. This dissemination method takes much time on the operational purpose, so the operational needs some improvement on the dissemination of data and information.

Geographic Information Systems (GIS) are tools for acquiring, managing, analyzing, and presenting spatially related information [1]. GIS convert diverse data into easy-to-read and easy-to-access maps and information [2]. In addition, the advantages of the world wide web are numerous, the two primaries being time and spatial independences. Distributing data over the Internet is more efficient than transmitting data through disks. Both Internet and GIS changed the processes of accessing, sharing, disseminating and analyzing data. Technology to share GIS data, such as Web GIS, Open GIS and distributed GIS on the Internet is rapidly progressing [3]. Spatial data can be processed through Web Map Servers and Web Servers (Figure 1). The difference between these two servers is only on the platform to be used by the client. The client can access the spatial data through the Desktop Application or Web Application. If the client uses the Desktop Application, then the Web Map Server is used. And if using a Web Application, then the Web Server that serves to provide the service. They provide a standard-based solution for serving digital maps, using predefined image tiles [6].

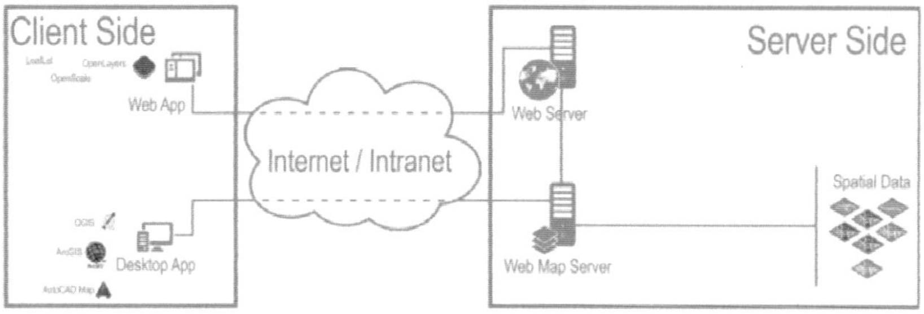

Figure 1. Server and client-side model standard

The objective of this research was to find the best alternatives dissemination from the current method of data distribution and remote sensing information in terms of speed, scalability, functionality and data maintenance in order to support government program on maritime sectors. Alternative to be examined was based on Web-GIS which means that access to data and remote sensing information could be done online through internet. The used of Web-GIS in this study due to the utilization of the internet already prevalent in the present day. Through this method, users could have obtained data and information from anywhere and anytime. As an example, this research use mangrove distribution data in Banten as remote sensing information provided. The results showed mangrove information easily to obtained, and the government could overcome the problems that has emerged as the reduced area of mangrove forest which resulted in causing chain actions such as floods and other disasters.

2. Methods

The development of this application, a WMS is needed that can serve imagery in a variety of coordinate systems, particularly WGS:84 latitude, longitude, and north as well as south polar stereographic. Other coordinate systems would be required in future development. Low latency is an important requirement to support interactive browsing of images on a draggable, zoomable dynamic map component [4]. This study was used Geoserver as a data publisher and remote sensing image information, loading and rendering of maps takes much longer compared to professional mapping services. It was the Java 2 Platform, Enterprise Edition (J2EE) realization of Open GIS Web server [7]. It is an open source platform that supports the OGC standards like Web Map Service (WMS), Web Coverage Service (WCS), Web Feature Service (WFS) and Web Feature Service Transactional protocols. It can work with a large range of data formats like Shapefile, ArcSDE, Oracle Spatial, PostGIS and other spatial formats. It has fully-featured Web administration interface [8]. This can be a result of limitation on powerful hardware or insufficient knowledge in efficient caching techniques, because those standard installations are normally not optimized [5].

There were two sides of activity: user and LAPAN activities (Figure 2). The user requests data or information through the data request mechanism. After the data was requested based on the desired area and level of data, LAPAN attempted to search and prepare according to the area and uploaded to Geoserver. In the Geoserver filter settings, styles and topology according to the method of cartography. System was processed to render the raster and vector data, and converted to image format, for example PNG, JPEG, GIF depends on user preferable. Three types of map service in this system were WCS, WMS, and WFS. This map service was depended on the form of data desired by the user. For example, if user need data only for layer base for digitization, then WMS was chosen. Users who test this system are from college students (2 users/clients) and Banten local government (1 user/client). They just need to set up internet connection, Laptop / PC and Quantum GIS as GIS Desktop software.

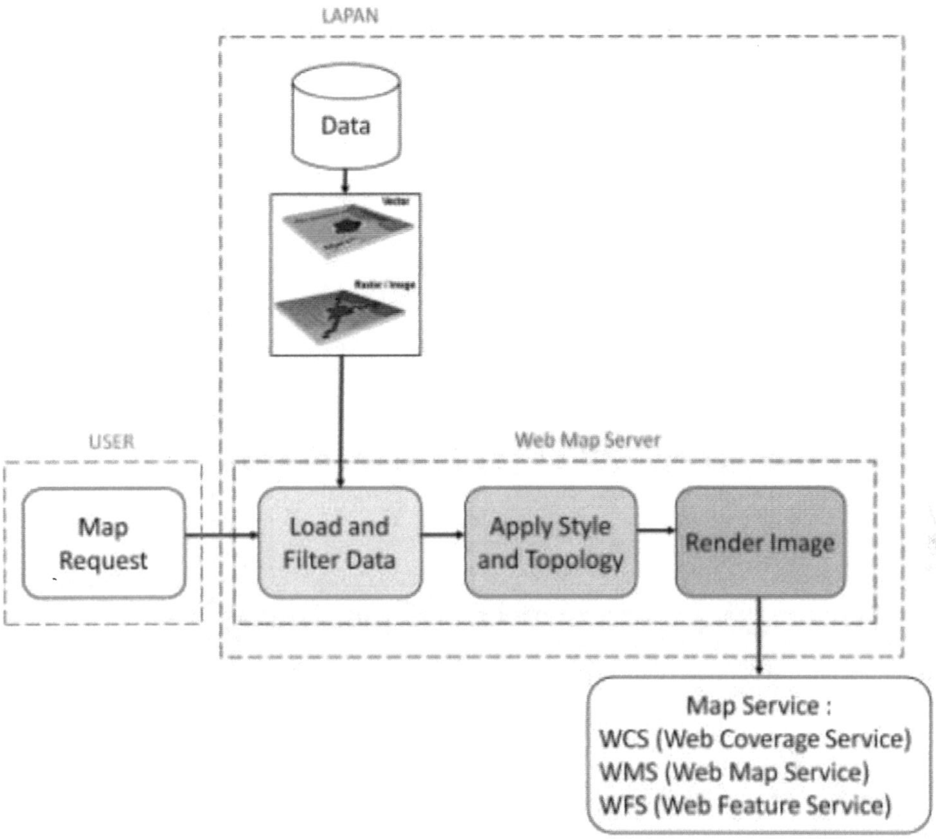

Figure 2. Server and client-side design model

3. Results and Discussion

3.1. Existing condition

Measurement of public satisfaction index on public service in Remote Sensing Technology and Data Center contains 8 elements of questions. That element should be finished when the user received of the data and information. Focus of the element of question to be used in this study such as certainty of service schedule, namely the implementation of service time, in accordance with the stipulated provisions; and the quality and appropriateness of remote sensing satellite data established by Technology Center and Remote Sensing Data. According to user satisfaction, the results was indicated less satisfy (i.e. B option) (Table 1). This condition need to be improve in order to satisfy the users.

Table 1. Public satisfaction index on public service in 2016 and 2017

2016		**2017**	
Certainty of service schedule	*The quality and appropriateness of remote sensing satellite data*	*Certainty of service schedule*	*The quality and appropriateness of remote sensing satellite data*

Choice of answers	*Amount*	*Choice of answers*	*Amount*	*Choice of answers*	*Amount*	*Choice of answers*	*Amount*
A. Not on time	0	A. Not qualified	0	A. Not on time	0	A. Not qualified	0
B. Sometimes on time	2	B. Less qualified	1	B. Sometimes on time	10	B. Less qualified	1
C. On time	45	C. Qualified	44	C. On time	50	C. Qualified	65
D. Always on time	41	D. Highly qualified	43	D. Always on time	56	D. Highly qualified	50

Table 2. Service Speed Average on public service in 2015, 2016 and 2017

Year	Service Speed Average (days)
2017	10
2016	16
2015	17

Table 1 shows that there was some user answered option B on both elements. Users still complain about the service in terms of time and quality of data received. On the other hand, service speed was still more than 2 days. This condition need to be fix and find the best solution. Table 2 shows that average of service speed from 2015 to 2017 was improved. Significant changes occur in 2017, decreasing from 16 days to 10 days, however this need to be improve as much as possible in 2018. Because of this condition, Remote Sensing Technology and Data Center need to improve the services. Improvements focused on speed of service and data quality of remote sensing.

3.2. System development

We tested this service on college student and local government users. The users were requested the data of Banten area. We were used WMS for this system, which the users could access it using the Quantum GIS without having to go back to LAPAN to retrieve the data. At first, we choose the data of Banten and upload to Geoserver. After that, we create WMS link and gave it to users and ask them to open it on Quantum GIS (Figure 3), and the raster data could open normally (Figure 4 and 5). Quantum GIS provides features like map navigation, attribute table, vector symbolization, predefined point symbols, labelling of vector features by attribute, map editing, on-the-fly projection for vector layers, print composer, raster pyramids, spatial index, etc [9]. We ask to try to do some on the fly process like delineation. The result, with the internet connection under 2 Mbps, the image data access is less than 10 seconds and they can use very well. They can do delineation by overlaying the satellite imagery data provided.

The result can improve measurement of public satisfaction index especially element certainty of service schedule and the quality and appropriateness of remote sensing satellite data (Table 1) because if the data is ready, we can provide and publish directly without any waiting time. And also, for Table 2 can accelerate service time up to 50%.

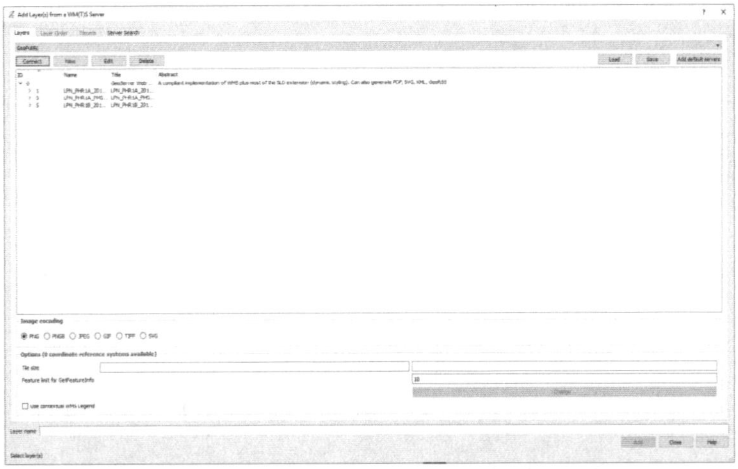

Figure 3. Connect to our WMS on QGIS (Banten data)

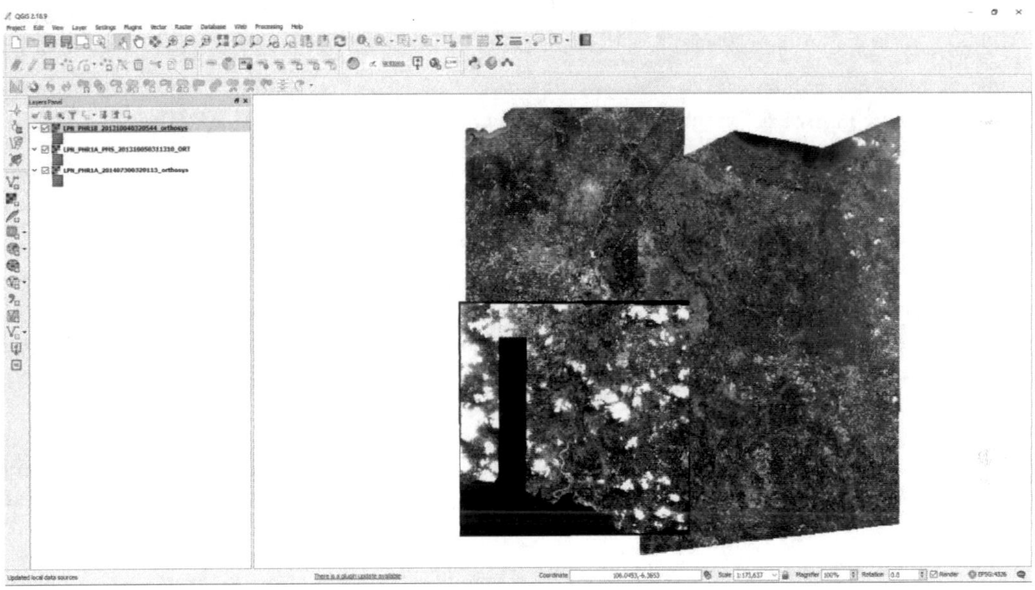

Figure 4. Open data results using WMS (Banten data)

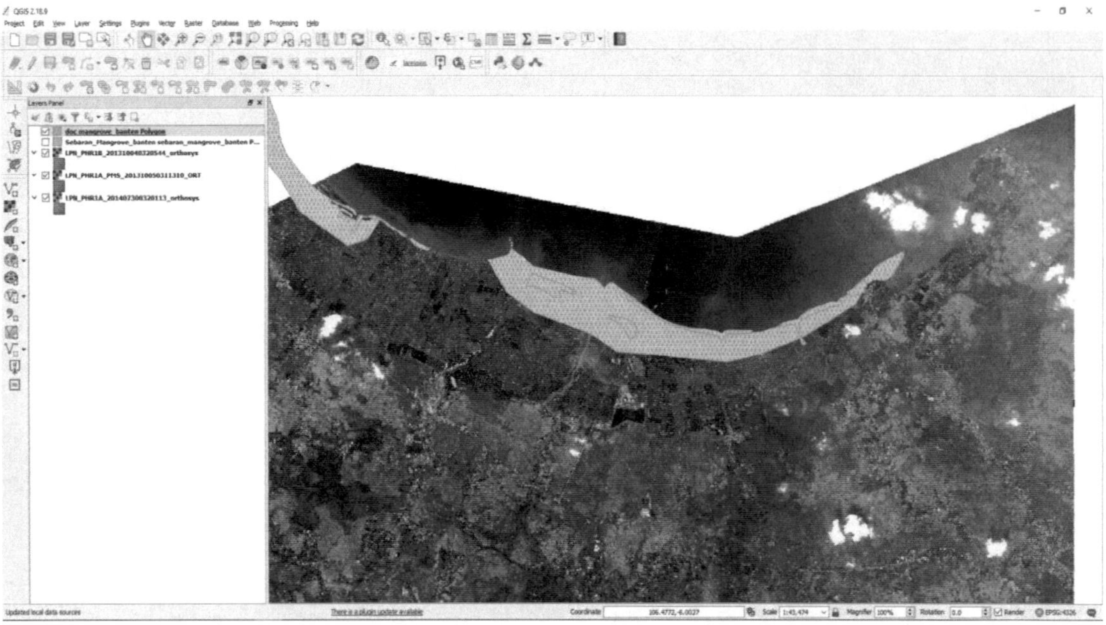

Figure 5. Open Mangrove data results using WMS (Banten data)

4. Conclusion

Utilization of Web-GIS could be an effective way of dissemination to speed up service because it has advantages such as (1) reduce lead times; (2) improve cross-collaboration; (3) low cost barrier to entry; (4) connect multiple structured data source; (5) support real-time spatial analysis; (6) access web maps from any device remotely; and (7) publish and share multi-layer features and points on a single web map. Web-GIS can solve the problems faced like speed of service, A definite schedule of when data is ready to take and data quality in accordance with applicable standards.

Acknowledgments

We thank Mr. Dedi Irawadi, Chief of Remote Sensing Technology and Data Center LAPAN, for his kind support. We would also like to show our gratitude to Mr. Rubini Jusuf for the permission to attend to this Seminar, and to present our study.

References

[1] Coors V 1998 *Interactive Applications of Mobile Computing Using Wearable GIS in outdoor applications* (Germany: CumInCAD)
[2] Mohler JL and Duff J.M 1999 *Designing Interactive Web Sites* (Publishing NY USA: Thomson Learning EMEA, Limited)
[3] Honda K 2003 *Digital ASIA Concept and Activity* (Thailand: Proceedings of the Regional Conference on DIGITAL GMS)
[4] Blower J.D 2010 *GIS in the cloud: implementing a Web Map Service on Google App Engine* (United Kingdom: Reading e-Science Centre, University of Reading, RG6 6AL)
[5] Loechel Alexander J, Stephan Schmid 2012 *Caching techniques for high-performance Web Map Services* (France: Proceedings of the AGILE'2012 International Conference on Geographic Information Science)
[6] Dražen Odobašić et.al 2011 *Web mapping is distributed – overview of open source proxy and processing services* (Croatia: Geoinformatics and Geomedical Engineering in Environmental Protection)

[7] Huang Z and Xu Z 2011 *A Method of Using GeoServer to Publish Economy Geographical Information* (Singapore: Control, Automation and Systems Engineering IEEE. doi:10.1109/ICCASE.2011.5997789) pp. 1–4

[8] Agrawal Sonam and Gupta Rajan Dev 2014 *Development and Comparison of Open Source Based Web GIS Frameworks on Wamp And Apache Tomcat Web Servers* (China: The International Archives of the Photogrammetry, Remote Sensing and Spatial Information Science, ISPRS Technical Commission IV Symposium)

[9] Hugentobler M 2008 *Quantum GIS. In S. Shekar and H. Xiong (Eds.)* (New York: Springer) pp. 935–939

IOP Conf. Series: Earth and Environmental Science **162** (2018) 012047 doi:10.1088/1755-1315/162/1/012047

Predictor-corrector scheme for simulating wave propagation on shallow water region

Rifky Fauzi[a], **L. Hari Wiryanto**[b]

Institut Teknologi Bandung, Jalan Ganesha 10, Bandung, Indonesia

E-mail: [a]`rifkyfauzi9@gmail.com`, [b]`leo@math.itb.ac.id`

Abstract. In this paper, a well-known Saint-Venant's Equation is solved by using Predictor-Corrector method to simulate the propagation of wave in shallow water region with several type of bottom friction effects. The predictor step is chosen to be Adam-Bashforth, meanwhile Adam-Moulton for the corrector step. Numerical stability is also discussed by applying von Neumann stability into the linear equations.

1. Introduction

Shallow water equation or Saint Venant's equation (SVE) is a system of equations that describe fluid flow in relatively long channel. This system though is able to perform simulation for geophysical flow such as ocean lake and atmosphere and even to simulate tsunami waves. The followings are shallow water system of equations

$$\frac{\partial \eta}{\partial t} + \frac{\partial((d_0 + \eta)u)}{\partial x} = 0 \tag{1}$$

$$\frac{\partial(hu)}{\partial t} + \frac{\partial}{\partial x}\left(hu^2 + \frac{1}{2}g\cos(\theta)h^2\right) = g\sin(\theta)h - C_f\frac{u|u|}{h^\alpha} \tag{2}$$

where η surface elevation, $h = d_0 + \eta$ denoting height of water, u is horizontal water velocity, θ is bed inclination, d_0 is water depth and g is acceleration due to gravity. Meanwhile C_f is bottom friction parameter that depends on the choice of constant α. SVE is a limiting case from Navier-Stokes equation by assuming that the horizontal length scale is much larger than vertical length scale. This shallowness assumption leads the SVE to be a hydrostatic model.

Bottom friction constant C_f is non-physical parameter that can be obtained by empirical approach based on mathematical model [1]. However this constant is prescribed without any definite and complete physical concept. Thus in nature the value of this constant cannot be uniform. In [2] different value of C_f define different areas such as smooth area, coral area and scattered tree or building in performing tsunami flooding simulation. One of effort in obtaining constant C_f can be seen in [1]. The determination of C_f is based on the optimization from SVE solution comparing to observational data. This approach shows that there is spatial variation in obtained value.

The bottom friction effect plays a significant role in the tidal phenomenon [3]. It gives rise to shallow area let the waves interact with bottom [4]. In other words, as the waves propagate from deeper area to shallower area those waves start to interact with bottom. Such frictional effect may appear as the existence of pores of the sandy bottom, muddy bottom or bottom irregularities [4]. This effect affects momentum transfer of the propagating waves since one has already familiar from basic mechanics that friction effect acts opposite to the motion.

Wave-bottom interaction rises in the momentum equation modeled as the resistive force fluid over the bottom. The resistance rises from the characteristics of the bottom which can sometimes also be called roughness coefficient that resist the momentum transport. The form of the bottom friction effect or simply the resistance force varies due to the existence of the internal resistance. In this paper, it is sufficient to assume that there is no internal resistance the bottom friction effect term in Equation 4 is valid. Otherwise, this form has to be changed to include the internal resistance effect.

To simulate wave propagation in shallow water region the inclination effect in the SVE need to be vanished. Thus by taking $\theta = 0$ the SVE reads

$$\frac{\partial \eta}{\partial t} + \frac{\partial (hu)}{\partial x} = 0 \tag{3}$$

$$\frac{\partial (hu)}{\partial t} + \frac{\partial}{\partial x}\left(hu^2 + \frac{1}{2}gh^2\right) = -C_f \frac{u|u|}{h^\alpha} \tag{4}$$

The objectives of this paper are for simulating wave propagation in shallow water region with several value of friction coefficient. The inclination effect is neglected due to physical reason that in shallow water region (coastal region to be specific) this effect plays insignificant role than bottom topography. This effect in fact has an important role in the instability of uniform fluid flow which can generate rollwaves or progressing periodic waves see [5] and [6].

The purpose of this study is to simulate wave propagating in shallow water region by taking several value of friction coefficient which obtained from several literature. A good approximation solution of Equation 3 and Equation 4 is extremely needed to obtain best result. The solution of the equations obtained by solving it using predictor-corrector scheme. Additionally, to prevent instability in the numerical iteration, numerical stability is also discussed. Once the numerical scheme has arranged the effect of bottom friction is investigated by varying the value of the parameter.

2. Bottom Friction Effect

There are two choices of bottom friction i.e. analytical approach and empirical approach. The bottom friction effect from empirical approach contains non-physical parameter that can be obtained by trial and error comparing to the experiment [8]. In contrast with the empirical approach, the analytical approach of bottom friction can be obtained from *Poiseuille flow* which hold the analytical meaning that this effect depends on physical parameter such as viscosity and gravitational constant. It is, however, the analytical approach is not the discussion in this study.

The development of bottom friction approach frequently found in hydraulics problem. Nevertheless, the approach can be used in coastal problem [1], [4] and [9] . Since it is derived from hydraulics problem the bottom friction effect varies as the choice of α in the following term

$$(h, u) = C_f \frac{u|u|}{h^\alpha} \tag{5}$$

The friction constant C_f depends on the choice of α. There are three main approach of the choice α that derived by Manning, Chezy and Darcy-Weisbach with their own approach. The differences arise from the experiment that they did in the beginning of the development of bottom friction effect.

For Manning's approach the $\alpha = \frac{4}{3}$ with the corresponding $C_f = n^2$ that can be obtained experimentally depending on the type of the bottom whether the bottom is muddy bottom or there is strong irregularity on the bottom profile [8]. This value also appears in Chezy approach that is for $\alpha = 0$ where $C_f = n^2$ called Chezy's coefficient [6]. Meanwhile for the Darcy-Weisbach that is $\alpha = 1$ the value of bottom friction C_f only depends on the Reynold number (Re) of the flow that is

$$C_f = \begin{cases} \frac{3f}{g} & \text{for } Re < 48 \\ \frac{f}{16g} & \text{for } Re \geq 48 \end{cases} \tag{6}$$

3. Numerical Procedure

Predictor-Corrector scheme has been succeeded simulating coupled dispersive shallow water equation in [10] by using Adam-Bashforth for predictor and Adam-Moulton for corrector. In this paper, similar schemes are also used to simulate wave propagation in shallow water region. Firstly, one need to rewrite the SWE into following form

$$\begin{aligned} \frac{\partial h}{\partial t} &= H\Big(h, u, \frac{\partial h}{\partial x}, \frac{\partial u}{\partial x}\Big) \\ \frac{\partial v}{\partial t} &= U\Big(h, u, \frac{\partial h}{\partial x}, \frac{\partial u}{\partial x}\Big) \end{aligned} \tag{7}$$

Then discretize right hand side Equation 7 by using central difference

$$H_j^n = -h_j^n \frac{u_{j+1}^n - u_{j-1}^n}{2\Delta x} - u_j^n \frac{h_{j+1}^n - h_{j-1}^n}{2\Delta x} \tag{8}$$

$$U_j^n = -u_j^n \frac{u_{j+1}^n - u_{j-1}^n}{2\Delta x} - g \frac{u_{j+1}^n - u_{j-1}^n}{2\Delta x} + C_f \frac{u_j^n |u_j^n|}{(h_j^n)^\alpha} \tag{9}$$

Subsequently, the mass equation i.e. Equation 3 and Equation 4 discretized by using Adam-Bashforth time integration gives

$$\frac{\bar{\eta}_j^{n+1} - \eta_j^n}{\Delta t} = -\frac{1}{2}(3H_j^n - H_j^{n-1}) \tag{10}$$

$$\frac{\bar{u}_j^{n+1} - u_j^n}{\Delta t} = -\frac{1}{2}(3U_j^n - U_j^{n-1}) \tag{11}$$

meanwhile for Adam-Moulton time integration gives

$$\frac{h_j^{n+1} - h_j^n}{\Delta t} = -\frac{1}{2}(\bar{H}_j^{n+1} + H_j^n) \tag{12}$$

$$\frac{u_j^{n+1} - u_j^n}{\Delta t} = -\frac{1}{2}(\bar{U}_j^{n+1} + U_j^n) \tag{13}$$

$\bar{H}_j^{n+1}$ and $\bar{U}_j^{n+1}$ are obtained from predictor scheme.

In these schemes, one need two initial condition. In [10], it is suggested that for such multistep method need special treatment. The initial condition is given by $n = 0$, meanwhile for $n = -1$ is defined as zero.

4. Numerical Stability

Since it has been a problematic matter to investigate the stability of nonlinear equation. A strategic step should be taken to manage the stability analysis of the numerical scheme. The strategy is to perturb the stationary solution of Equation 3 and Equation 4 by giving a small parameter namely $\epsilon \ll 1$ then analyze the numerical solution of $O(\epsilon)$ equations.

The effect of bottom friction only appears in nonlinear form by taking the perturbed uniform solution from Equation 3 and 4. The perturbed solutions read

$$\begin{aligned} u(x,t) &= 0 + \epsilon u_1(x,t) \\ \eta(x,t) &= \eta_0 + \epsilon \eta_1(x,t) \end{aligned} \tag{14}$$

Substituting 14 into Equation 3 and Equation 4 and neglecting $O(\epsilon^2)$ gives

$$\begin{aligned} \frac{\partial \eta_1}{\partial t} + d_0 \frac{\partial u_1}{\partial x} &= 0 \\ \frac{\partial u_1}{\partial t} + g \frac{\partial \eta_1}{\partial x} &= 0 \end{aligned} \tag{15}$$

which are set of linear coupled equation. From 15 one can clearly see that the bottom friction does not appear. In contrast with the case for $\theta > 0$, the bottom friction appears within the linear form [5].

Equation 15 in predictor scheme reads

$$\bar{\eta}_j^{n+1} = \eta_j^n - d_0 \Big(3\frac{u_{j+1}^n - u_{j-1}^n}{2\Delta x} - \frac{u_{j+1}^{n-1} - u_{j-1}^{n-1}}{2\Delta x} \Big) \tag{16}$$

$$\bar{u}_j^{n+1} = u_j^n - g \Big(3\frac{\eta_{j+1}^n - \eta_{j-1}^n}{2\Delta x} - \frac{\eta_{j+1}^{n-1} - \eta_{j-1}^{n-1}}{2\Delta x} \Big) \tag{17}$$

and the corrector scheme

$$\eta_j^{n+1} = \eta_j^n - d_0 \Big(\frac{\bar{u}_{j+1}^{n+1} - \bar{u}_{j-1}^{n+1}}{2\Delta x} + \frac{u_{j+1}^n - u_{j-1}^n}{2\Delta x} \Big) \tag{18}$$

$$u_j^{n+1} = u_j^n - g \Big(\frac{\bar{\eta}_{j+1}^{n+1} - \bar{\eta}_{j-1}^{n+1}}{2\Delta x} + \frac{\eta_{j+1}^n - \eta_{j-1}^n}{2\Delta x} \Big) \tag{19}$$

The stability analysis can be done by using Von Nuemann stability

$$\eta_j^n = \hat{\eta} e^{i\theta \Delta x j + i\beta n \Delta t} \tag{20}$$

$$u_j^n = \hat{u} e^{i\theta \Delta x j + i\beta n \Delta t} \tag{21}$$

By substituting into predictor scheme one gets

$$\begin{pmatrix} \lambda(\lambda-1) & -ir(3\lambda-1) \\ -ir(3\lambda-1) & \lambda(\lambda-1) \end{pmatrix} \begin{pmatrix} \hat{\eta} \\ \hat{u} \end{pmatrix} = 0 \tag{22}$$

Meanwhile for the corrector scheme

$$\begin{pmatrix} \lambda-1 & -ir(\lambda+1) \\ -ir(\lambda+1) & \lambda-1 \end{pmatrix} \begin{pmatrix} \hat{\eta} \\ \hat{u} \end{pmatrix} = 0 \tag{23}$$

where

$$\lambda = e^{i\beta\Delta t} \tag{24}$$

is amplification factor and

$$r = g d_0 \frac{\Delta t}{2\Delta x} \sin\theta \tag{25}$$

IOP Conf. Series: Earth and Environmental Science **162** (2018) 012047 doi:10.1088/1755-1315/162/1/012047

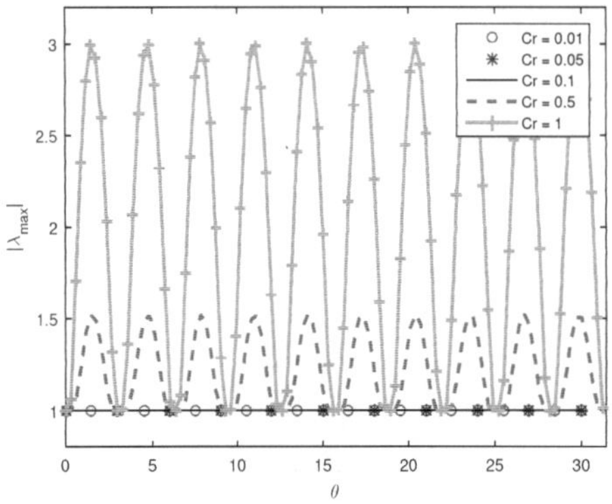

Figure 1. Largest norm of amplification factor

The main idea to analyze the numerical scheme is by designing the amplification so that it is not larger than 1, otherwise the numerical solution would blow up. We know that Equation 22 and 23 give nontrivial solution only if we have singlar matrices. Thus, for the amplification for 22 is given by fourth order polynomial as follows

$$\lambda^2(\lambda - 1)^2 + r^2(3\lambda - 1)^2 = 0 \tag{26}$$

Equation 26's root can be found numerically and is a function of phase angle θ. Here we employ Courant number as the numerical stability criterion

$$Cr = \sqrt{gd_0}\frac{\Delta t}{\Delta x} \tag{27}$$

and by setting $\Delta x = 0.1$ and choose $Cr = \{0.01, 0.05, 0.1, 0.5, 1.0\}$. Figure 4 shows the maximum of norm of the roots of Equation 26 from figure 4, it can be seen that the numerical scheme is stable for predictor if we choose $Cr = 0.1$.

Meanwhile for the corrector scheme, the amplification is given by the roots following polynomial

$$\lambda^2 + 2\Big(\frac{r^2 - 1}{r^2 + 1}\Big)\lambda + 1 = 0 \tag{28}$$

The maximum modulus of the root is given by

$$|\lambda| = 1 \leq 1 \tag{29}$$

Thus the numerical scheme is stable if

$$gd_0\frac{\Delta t}{2\Delta x} \leq 1 \tag{30}$$

5. Results

In the simulation, waves are propagating from left to right. Both spatial and temporal steps are chosen so that satisfy the numerical stability. The effect of bottom friction is investigated by analyzing the propagating waves for a given bottom friction constant. For convenient comparison, waves propagate without bottom friction i.e. $C_f = 0$ is considered. In Figure 2 monochromatic waves propagating from left to right. The waves profile deform due to nonlinear effect in the model. As we can see that the most severe effect from bottom effect is sought in result for $\alpha = 0$. The biggest the value of α gives less damping effect.

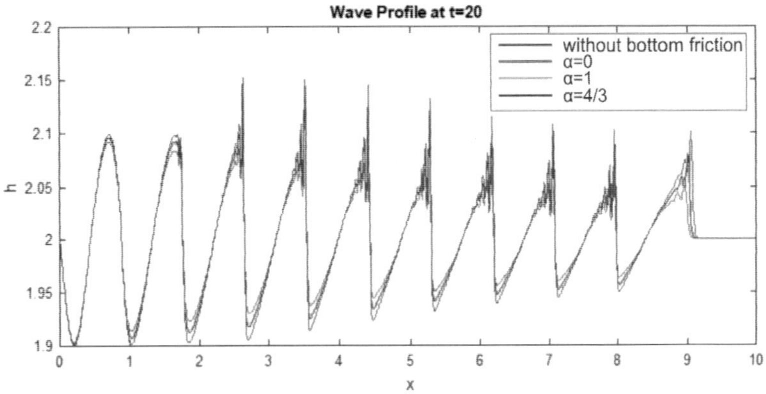

Figure 2. Comparison for several type of bottom friction effect

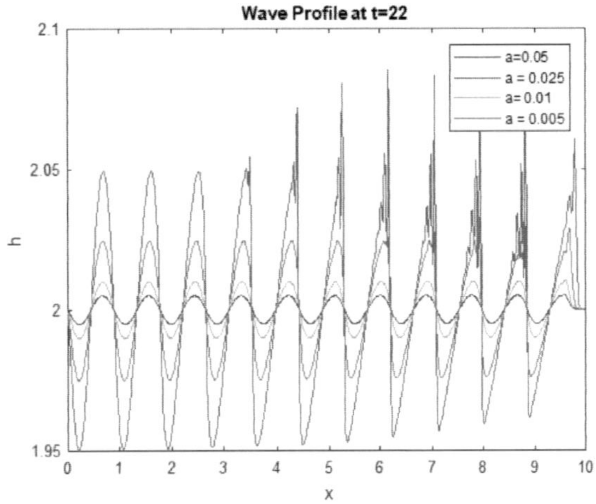

Figure 3. Wave propagation for several value of amplitude

6. Conclusions

Numerical code for waves propagation in shallow water region has been developed by using Adam-Bashforth and Adam-Moulton as Predictor and Corrector respectively. The numerical stability based on von Neumann stability analysis has also been studied to prevent infinite numerical result from finite initial condition. The numerical results show that for any type of bottom friction has decay effect to wave amplitude. This fact shows that rollwaves in shallow water region is different comparing to rollwaves down inclined channel i.e. this phenomenon does not arise from instability of a particular solution. Alternatively, rollwaves can be simulated by taking small but not too small amplitude thus the nonlinear effect still can affect at the same time it can maintain its shape so that it does not break.

References

[1] Ding Y Jia Y and Wang S S Y 2004 Identification of mannings roughness coefficients in shallow water flows *Journal of Hydraulic Engineering* **130** 501-510
[2] Mader C L 2004 *Numerical Modeling of Water Waves*
[3] Wang D, Liu Q and Lv X 2014 *Mathematical Problems in Engineering* **2014** 1-7
[4] Luo W and Monbaliu J 1994 *Journal of Geophysical Research* **99** 18501-18511

[5] Fauzi R and Wiryanto L H 2017 *AIP Conf. Proceedings* **1867** 1-8
[6] Dressler R F 1949 *Communications on Pure and Applied Mathematics* **2** 149-194
[7] Needham D J and Merkin J H 1984 *Proc. R. Soc. Lond. A* **394** 259-278
[8] Kirstetter G, Hu J, Delestre F, Darboux F, Lagree P Y, Popinet S, Fullana J M and Josserand C 2016 *Journal of Hydrology* **536** 1-9
[9] Liu Y, Shi Y, Yuen D A, Sevre E O D, Yuan X and Xing H L 2009 *Acta Geotechnica* **4** 129-137
[10] Wiryanto L H and Mungkasi S 2014 *Applied Mathematical Sciences* **8** 5293-5302

Monitoring remote tidal based on web

Engki A Kisnarti, Suryadhi

Department of Oceanography, University of Hang Tuah Surabaya, Indonesia

E-mail: engkiandri@hangtuah.ac.id

Abstract. Currently oceanographic data can be obtained through overseas websites that provide online and costless, but these online data generally have rough resolution with global or regional space coverage. The purpose of this study is to apply tidal measurements in three different waters (Surabaya, Gresik and Lamongan, East Java-Indonesia) that can transmit data via SMS gateway and store data on the web server to be accessible by people. The first step is to combine the electronic and mechanical circuits and then it was followed by experimental tools and data retrieval conducted in the laboratory. After the trial is successful, it was followed by the actual data retrieval that it tested the transmission of data from the observer station to the data center via modem that uses AT Command, a computer language program. Data will be received and stored in a web server, and then the data can be accessed via internet. The results showed that tidal instrument can work well and gave an estimated error of 3-4%. The choice of location of instrument placement was very important. The most important in this research was to avoid the crowded boat traffic and under the bridge.

1. Introduction

Indonesia as a maritime country is a country that most of its territory is the sea should make marine sector development as the main priority. The development in the marine sector can run well if all aspects related to the development is prepared. One important aspect in development in the marine sector is the existence of oceanographic data.

Some of parameters such as tides, waves, currents, and temperatures can be measured directly using measurement tools [1]. We have designed tidal and current tools [2]. The oceanographic data observed are the speed data, the direction of currents data and the tidal data. The oceanographic data obtained directly from the observed area uses the sensors that are connected to the microcontroller and sent via a modem. In the real time, these data submitted by the microcontroller via the modem which also serves as a gateway SMS, directly to the server. From the server, the community can access those online data using the internet.

The purpose of this study is to apply tidal measurements in three different waters (Surabaya, Gresik and Lamongan, East Java-Indonesia) that can transmit data via SMS gateway and store data on the web server to be accessible by the people wherever they are as long as they have media connected to the internet network.

2. Methods

In this research, we built an oceanographic data processing system using internet network [2], so the person who takes the data no longer has to be at the location during the data retrieval process, whereas in manual usage, the person must be always at the location for a long time. This tool can also send data from several locations at once, so it is very effective and efficient in the process of taking oceanographic data in various places (Figure 1 and 2).

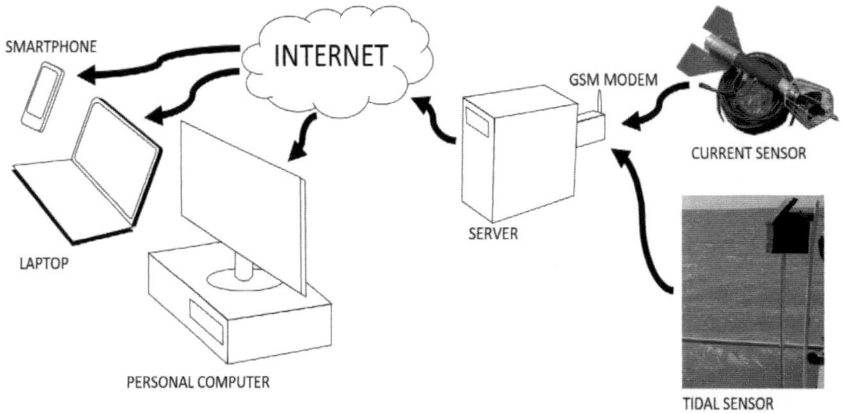

Figure 1. Oceanographic data processing system using internet network.

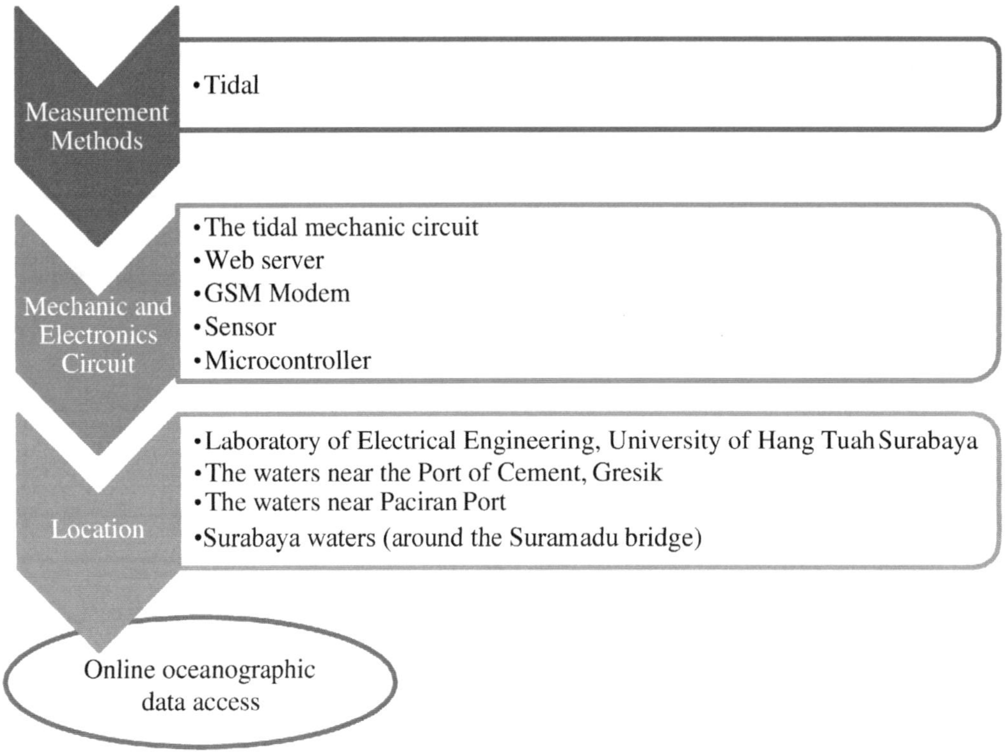

Figure 2. Schematic Diagram

2.1. Location of Research
The research location is in Laboratory of Electrical Engineering, University of Hang Tuah Surabaya, the waters near the Port of cement, Gresik, the waters near Port of Paciran and Surabaya waters (around the Suramadu Bridge).

2.2. Electronic Network Planning
At the stage of designing this electronic circuit and mechanical design is done with steps such as the flow chart in Figure 3 and 4.

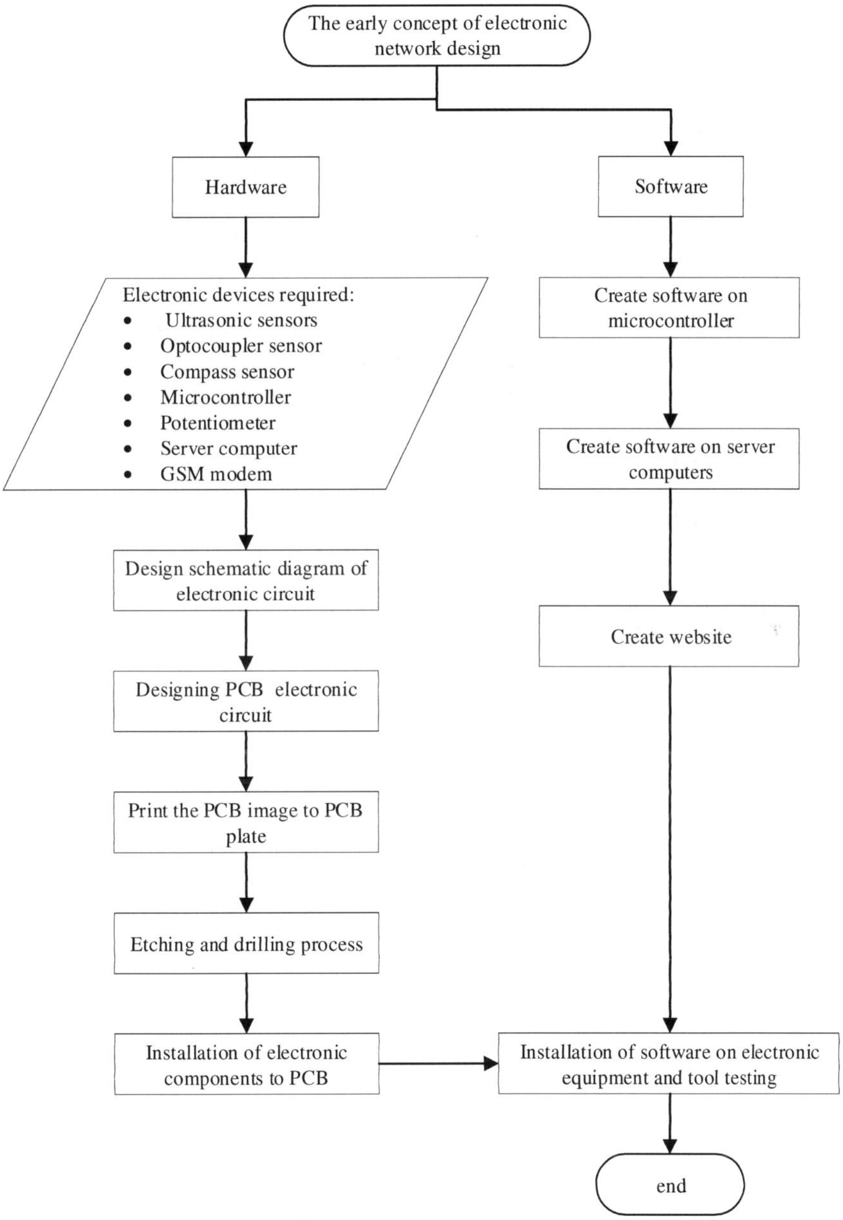

Figure 3. Flow chart of electronic circuit design stage

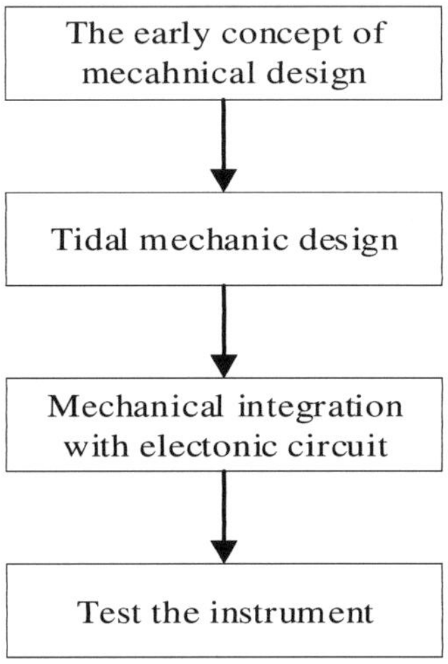

Figure 4. Mechanic planning phase chart

2.3. Testing Instrument and Data Retrieval

Testing instrument and data retrieval will be done first in the laboratory, and then the instrument will be displayed to the actual location of the waters to measure tides of sea water. The measurement method is a direct measurement method that uses tidal measuring instrument [3]. Then, it tested to send data from observer station to data center through SMS gateway. The data will be received and stored in a web server so that data can be accessed via internet.

3. Results and Discussion

3.1. Wavecom Modem Testing

Before using wavecom modem, the users must prepare power supply first. The required voltage for the modem to work is 5.5-12 Volt DC. The voltage is taken from a power bank with a capacity of 1,500 mAh that can last about 2 days. To be able to make this tool work on a longer time, just replace the power bank with a larger capacity. The modem uses AT Command program, so it can receives and sends the data in the form of SMS, or this process can be called as 'SMS Gateway'. The trial needs two different places to prove that the sending-receiving method works well. The first trial was conducted in the laboratory, but the second was undergone in the field.

3.2. Tidal Device Testing

To test the strength and magnitude of errors of the tidal set, tidal tools are placed in three waters location, i.e. the water near Port of Paciran, Lamongan, the water near Port of Cement, Gresik and the waters around Surabaya Bridge. The laying of the tidal appliance is in a conventional way (using human's palm) to determine the magnitude of errors of this tool. To test the robustness of the tool, this tidal set of devices is used for 15 days (half a month cycle around the earth).

To finding out the result of the new tool, the installation of this tidal set is displayed next to the installation of a conventional (human palm), so the observer directly know the efficacy of both tools. The first set of tidal appliances is placed in Lamongan waters, near the Fish Auction Place Dock, Paciran as in Figure 5. The second set of tools is placed in Gresik waters, close to the Semen Gresik Pier (Figure 6). The third set of tidal tools is placed under the Surabaya-Madura Bridge (Figure 7.).

Figure 5. The first set of tidal appliances in Lamongan waters, near the Fish Auction Place Dock, Paciran

Figure 6. Setting location of tidal set circuit in Gresik

Figure 7. Setting location of tidal set circuit in Surabaya

After the tool circuit was installed in the determined location, then the data will be sent using a modem that was equipped with a power supply first. The required voltage for the modem to work is 5.5 - 12 Volt DC. Once the modem was ready, the next equipment that needs to be prepared is the server. The server used in this circuit testing tool was located in University of Hang Tuah, at the Electrical Laboratory.

The three series of devices installed in three locations (Lamongan, Gresik and Surabaya) will transmit data, received and stored into the database from existing sensors. This software was made using the basic programming language of compiler Delphi 7. Delphi 7 software that had been made received SMS and saved SMS data into MySQL database that had been provided.

Data from the set of tidal tools that had been sent in the software were combined and processed with tidal observation data manually to determine the magnitude of errors of each device placed in the waters.

The results of manual measurements and using a set of installed tidal instrument in Lamongan can be seen in Figure 8. The result of the graph in red was the result of measurement using the conventional tide sign (palm), while green color was the result of tidal measurement using the tidal circuit. From both tools, it gives an approximately error of 3%. At this location, there was hardly any problem in sending data from the device to the server that was placed in the Laboratory of Electrical, University of Hang Tuah Surabaya. This was because the location of the placement of the tool was in open water with traffic that was not too crowded.

IOP Conf. Series: Earth and Environmental Science **162** (2018) 012048 doi:10.1088/1755-1315/162/1/012048

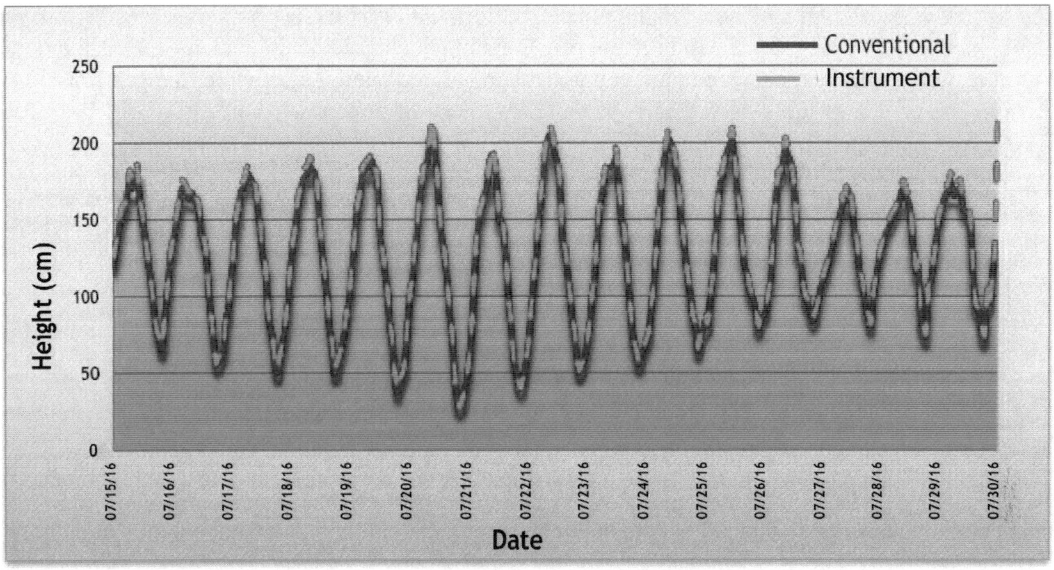

Figure 8. Graph of tidal observation using conventional and automatic in Lamongan

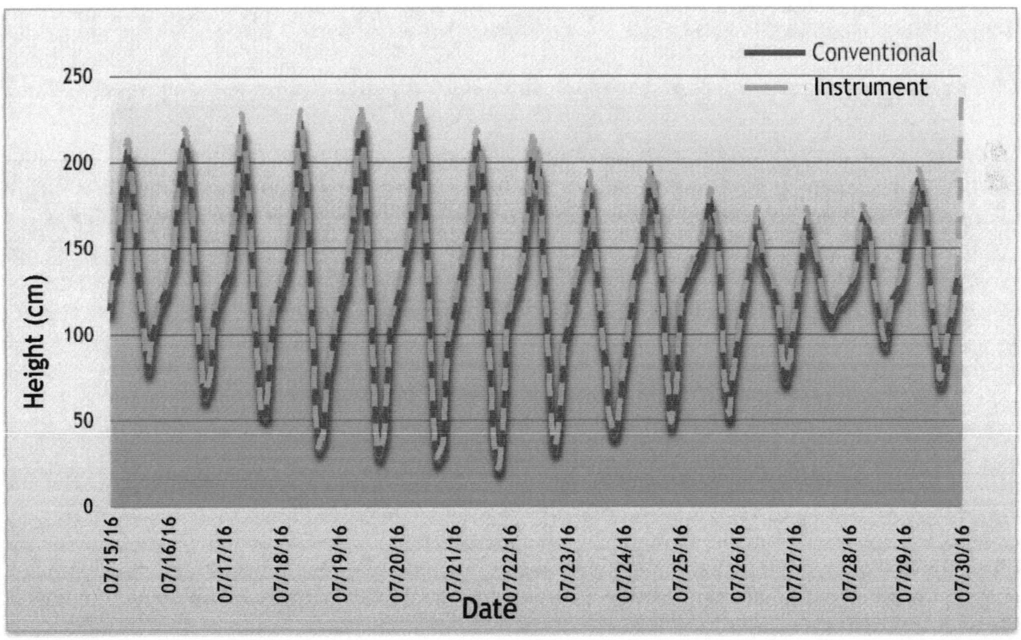

Figure 9. Graph of tidal observation using conventional and automatic in Gresik

The results of manual measurements and using tidal set of instrument in Gresik can be seen in Figure 9. Graph with red is the result of measurement using conventional tide sign (palm), while green color is result of tidal measurement by using tidal circuit. From both these tools, gives a big error of 4%. At this location, there is little problem in the placement of tidal circuit sets. This happens, because the location of the waters is too crowded by boats that go back and forth and is often used by small children to play water. Conditions of water that are only about 50 cm from the surface of the water, causing the device exposed to water splashes, causing some components to be replaced. Meanwhile, the transmission of data from the device to the server placed at the Laboratory of Electrical, University of Hang Tuah, Surabaya, did not experience any problems.

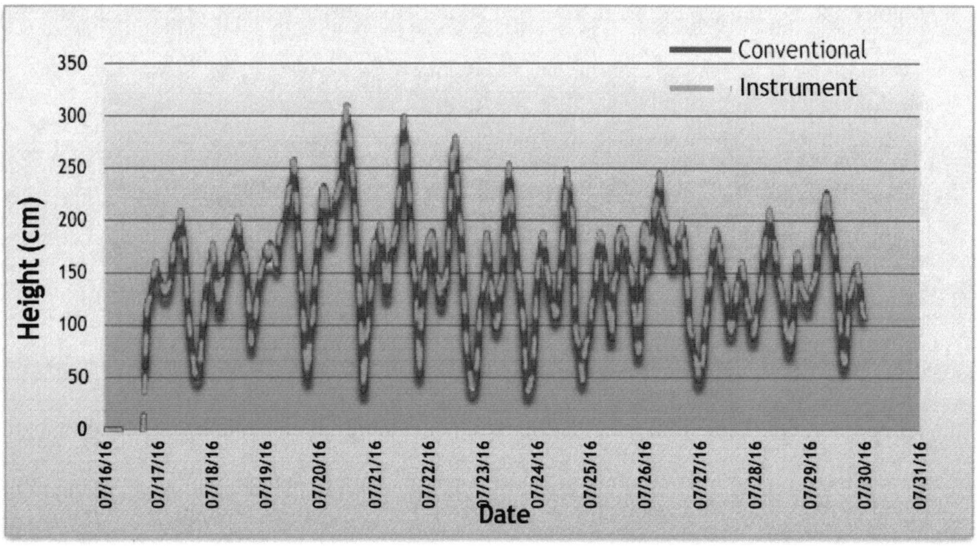

Figure 10. Graph of tidal observation using conventional and automatic in Surabaya

The results of manual measurements and using tidal set of instrument in Surabaya can be seen in Figure 10. Graph with red is the result of measurement using conventional tide sign (palm), while green color is result of tidal measurement by using tidal circuit. From both these tools, gives an approximately error of 3%. Consideration instrument placed under the bridge Suramadu Surabaya for tools protected from rain. However, the problem then arises, namely in the delivery of data from the device to the server that is placed in the University Laboratory Hang Tuah Surabaya. This is possible because Suramadu bridge structure interferes with signal delivery from the device to the server.

4. Conclusions

The conclusion of this research is tidal instrument can work well and give error of 3-4%. The choice of location of instrument placement is very important. Therefore, the research must avoid the crowded boat traffic and under the structure of the building or bridge to get accurate data.

Acknowledgments

The authors would like to thank the Ministry of Research and Technology and University of Hang Tuah for their support in this research.

IOP Conf. Series: Earth and Environmental Science **162** (2018) 012048 doi:10.1088/1755-1315/162/1/012048

References

[1] Setiawan A 2008 *Modul Pengenalan Data Oseanografi* (Jakarta: Pusat Teknologi Lingkungan-BPPT)

[2] Suryadhi and E A Kisnarti 2015 *Oceanography Data Processing Onling using Internet*. Applied Mechanics and Materials. ISSN: 1662-7482 Vol 862 pp 61-66

[3] Davis R A 1972 *Principle of Oceanography* (Philipines: Addison-Wesley Publing Company, Inc.)

IOP Conf. Series: Earth and Environmental Science **162** (2018) 012049 doi:10.1088/1755-1315/162/1/012049

Stress concentration factor distribution of inclined brace in multiplanar offshore tubular double kt joints

Rudi Walujo Prastianto, Yoyok Setyo Hadiwidodo, Ibnu Fasyin Fuadi, Mhd Bonar Yudha Prawira Naibaho and Prasetyo Nugroho Rahardianto

Department of Ocean Engineering, Faculty of Marine Technology, Institut Teknologi Sepuluh Nopember, Surabaya, Indonesia.

E-mail: rudiwp@oe.its.ac.id

Abstract. The majority of tubular joints commonly found in offshore structures are in the form of multi-planar tubular joints, but very few investigations have been reported due to the complexity and high cost involved. The majority of research have been focused on the study of stress distribution at certain position such as saddle and crown points, and the hot spot stress (HSS) at other position along the weld toe of brace/chord intersection have been ignored. In this paper a 60° two-planar double KT (DKT) tubular joint is modelled as a finite element model from an offshore jacket platform. The effect of dimensionless geometrical parameters on the geometrically stress distribution and SCF distribution along the weld toe of inclined brace in axially loaded on the joint are investigated. Non-dimensional parameters that are varied include β, τ, γ and θ. Validation of the finite element model shown a good agreement to the global structural analysis results. The results of parametric studies show that the increase of the β leads to decrease of SCF. While the increase of the τ, γ and θ leads to increase of SCF. The peak SCF mostly occurs at the outer saddle point.The effect of parameters β, γ and θ on the SCF are greater than the effect of parameters τ.

1. Introduction

Jacket platform is an offshore structure often used for oil and gas exploitation of offshore hydrocarbon reservoirs in shallow water areas, which have main structure consists of tubular members welded between braces and chords [1]. Stresses occurring at intersection line brace-chord of the tubular joint is unevenly distributed. The ratio between the maximum stress at the intersection line and the nominal stress in the brace is called Stress Concentration Factor (SCF). The SCF is one of the important parameters in the calculation of fatigue damage [2]. Therefore, more accurate prediction of the SCF is necessary in the calculation of fatigue life of offshore structures.

The majority of tubular joints commonly found in offshore structures are in the form of multi-planar tubular joints. The multi-planar effect plays an important role in the stress distribution at the brace-to-chord intersection areas of the tubular joints as well as stresses in the intersection line are distributed very complex. The fact shows that very few investigations have been reported due to the complexity and high cost involved. For such multi-planar joints, the parametric stress formulae of simple uni-planar tubular joints are not applicable in SCF prediction.

Due to the above situations, the significance of this study is to obtain proper SCF specific for a multi-planar double KT tubular joint. From design point of view this specific SCF would increase accuracy in fatigue life prediction for the joint itself. This research will study the effect of dimensionless geometrical parameters (β, τ, γ and θ) on the stress distribution along the weld toe in the intersection region between brace-to-chord and SCF of a multi-planar double KT tubular joint with several variations of the geometrical parameters subjected to axial loading at the braces. Finite element models of the tubular joint are used for local stress analysis to obtain stress distribution in the intersection region which is the accuracy obtained is strongly influenced by the numerical model to be analyzed [3].

2. Literature review

The understanding of the behaviour of completely multiplanar joints is very limited. During the past thirty years, a significant number of scientific papers and reports have beenpublished on the study of SCFs in various uni-planar tubular joints. As a result, many parametric study in terms of the joint geometrical parameters have been proposed, providing SCF values at certain locations adjacent to the weld for several loading conditions.Two followingparagraphs review the research papers on the determination of SCF distribution along the brace-to-chord intersection of uni-planar joints and on the calculation of SCFs in the multi-planar tubular joints, respectively.

Seng-Keat Yeoh, Ai-Kah Soh and Chee-Kiong Soh [4] presented about behavior of tubular T-joints subjected to combined loadings. Ai-Kah Soh and Chee-Kiong Soh [5] researched about hot spot stress of K tubular joints subjected to combined loadings. Pang and Lee [6] presented three-dimensional finite element analysis of a tubular T-joint under combined axial and bending loading. Chang and Dover [7] predicted of stress distributions along intersection of tubular Y-joints and T-joints. Shao [8] predicted of hot spot stress distribution for tubular K-joint under basic loadings.

Research on the distribution of SCF on multiplanar tubular joints was done by Chiew et al [9], for the multiplanar tubular XX-joints. Karamos et al [10] derived the SCF formula and investegated stress distribution along intersection line on the multiplanar DT-joint due to the bending load. Research was also undertaken by Wingerde et al [11] formulating the SCF formula for uniplanar tubular joints K and multiplanar KK. Woghiren and Brennan [12] developed a setof parametric formulae to predict the values of SCF in multi-planar rack-stiffened tubular KK-joints. Bonar Naibaho [13] presented about SCF parametric study of multiplanar tubular KDT-joint. Arvina Puspaningtyas [14] researched about stress intensity factor parametric study of multiplanar tubular joint with fracture mechanics. Ibnu Fasyin Fuadi [15] derived the SCF formula for multiplanar tubular DKT-joint.

3. Parametric numerical study of weld toe geometric stress distribution on the outer brace

3.1. Finite element modelling, analysis, and validation

The structure to be analyzed in this case is a Multi-Planar Double KT tubular joint on the bracing of jacket platform structure (see Figure 1). Data used in this local analysis are structural geometry data, material properties and loading of the global analysis of structure jacket platform. The loading data is obtained from the global modeling of the structure (see Table 1 and Table 2).

The chord and brace have the same material properties. The steel density is 563,5lb / ft^3. Yield strength and Poisson's ratio are 50 ksi and 0.3, respectively, and with Young modulus of 30,000 ksi.

In this multiplanar double KT tubular joint will be varied with several parameters, namely the parameters of beta (β), tau (τ), gamma (γ), and the angle between chord and brace (θ). The definition of each the tubular joint geometric parameters are: β (= d/D) is ratio between diameter of brace and chord; τ (= t/T) is ratio between wall thickness of brace and chord; γ (= D/2T) ratio between diameter and wall thickness of chord; and θ stands for angle between chord and brace.

IOP Conf. Series: Earth and Environmental Science **162** (2018) 012049 doi:10.1088/1755-1315/162/1/012049

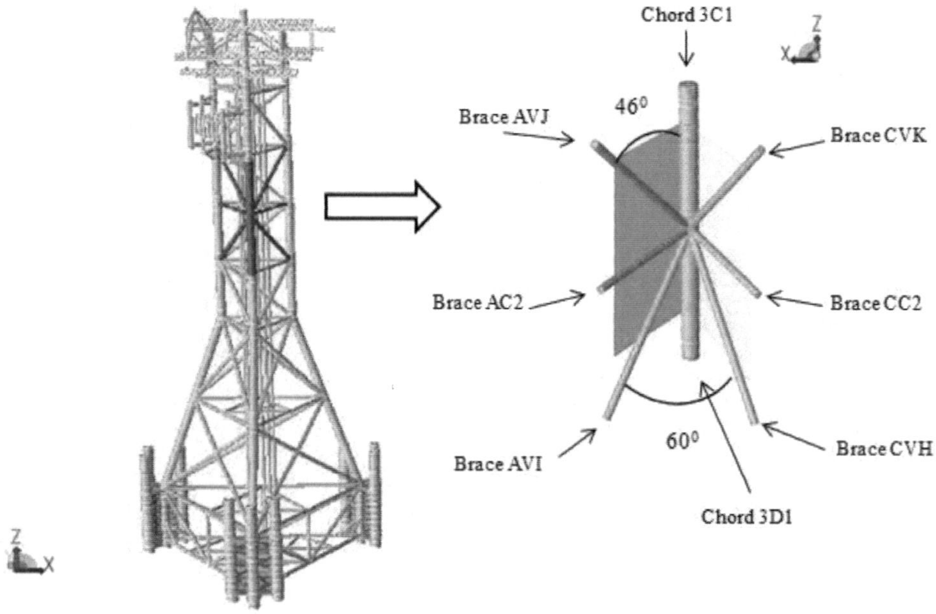

Figure 1. Multiplanar double KT tubular joints taken from global analysis

Table 1. Geometry of multi-planar double KT tubular joint

Joint Member	Thickness (in)	Outside Diameter (in)	Inside Diameter (in)	Length (in)
Chord 3C1	2	54	50	312
Chord 3D1	2	54	50	312
Brace CVK	1	24	22	540
Brace AVJ	1	24	22	540
Brace CC2	1	24	22	384
Brace AC2	1	24	22	384
Brace CVH	1	24	22	540
Brace AVI	1	24	22	540

As mentioned earlier, the aim of the present research is to study the effect of dimensionless geometrical parameters on the SCF distribution along the weld toe (on the chord side) of the main (inclined) brace. Different values assigned to each non-dimensional parameter are as follows: $\beta = 0.35$, 0.4, and 0.45; $\gamma = 10$, 12, and 15; $\tau = 0.35$, 0.45, and 0.55; $\theta = 36°$, 46°, and 56°. These values are selected to cover the practical range of the normalized parameters typically found in multi-planar tubular joints of offshore structures, namely: $0.3 \leq \beta \leq 0.5$; $0.3 \leq \tau \leq 0.6$; $10 \leq \gamma \leq 14$; and $30^{o} \leq \theta \leq 60^{o}$.

Table 2. Moment and forces on multi-planar double KT tubular joint

Joint Member	Moment (kips-in)			Force (kips)		
	Mx	My	Mz	Fx	Fy	Fz
Brace CVK	-0.5	-32.9	-10	-48.6	-0.01	-0.1
Brace AVJ	2.3	-116.6	12.7	-13.5	-0.05	1.4
Brace CC2	213.4	-675.8	-5.4	9.9	0.01	9.8

Brace AC2	-94.6	-384.8	3.7	10.1	-0.01	5.6
Brace CVH	1.2	-42.9	20.5	49.7	0.04	-0.1
Brace AVI	-0.68	-186.3	31.4	-13.8	-0.11	1.6

Multi-planar double KT tubular joint to be analyzed is modeled using preprocessor software of Solid Work software for its geometry (see Figure 2) based on data in Table 1. Then the geometry is exported into finite element method-based software where in the material properties and loading data are inputted according to Table 2. Weld modeling is necessary because the intersection between the two tubular (brace and chord) will form intersection line, where the stress concentration occurs in the area along the intersection line. The welding size along the brace-chord intersection satisfies the AWS specifications [16].

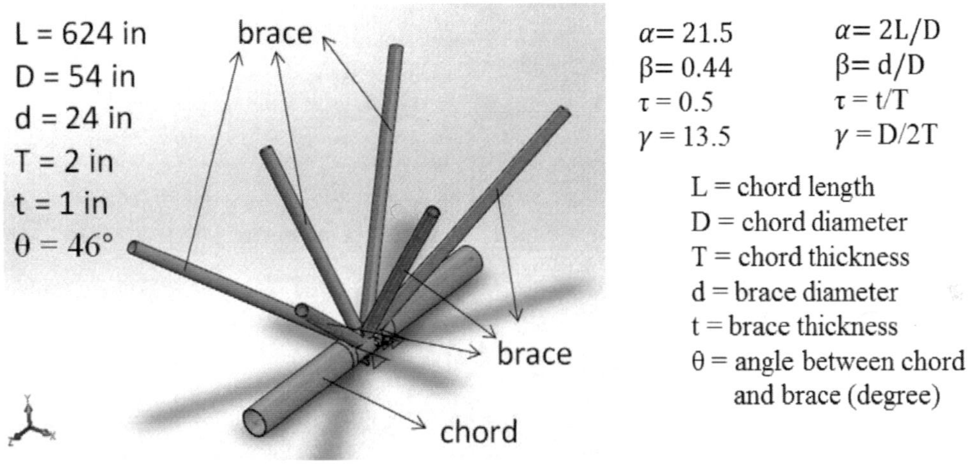

Figure 2. Multiplanar double KT tubular joints for local analysis

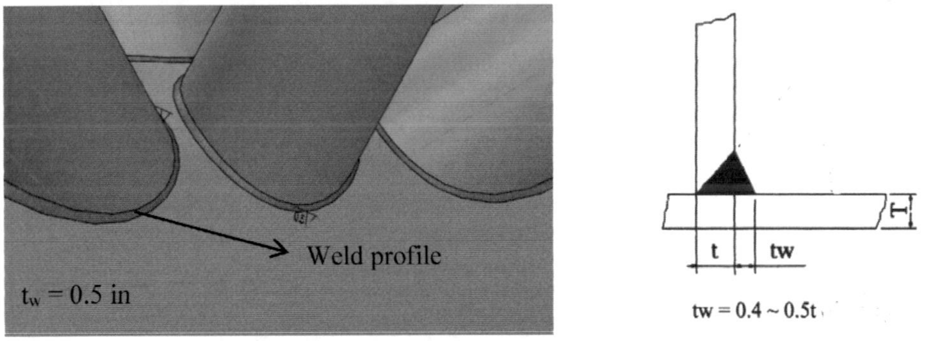

(a) (b)

Figure 3. Weld size along the brace-chord intersection satisfies the AWS specifications (a) Weld profile on the model (b) AWS D 1.1. Specifications for the weld

After modeling the multiplanar DKT tubular joint, then meshing is done. Meshing strategy for the model is performed with meshing size is made smaller (smooth) on regions around the connection between the brace and chord. While in regions far from the connection area as observation areas, meshing size is made larger (coarser). The element used for the model is a solid three-dimensional element with a linear element type of tetrahedron (see Figure 4).

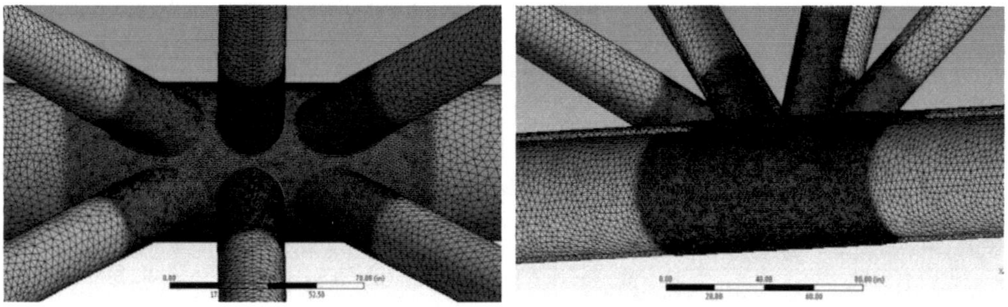

(a)

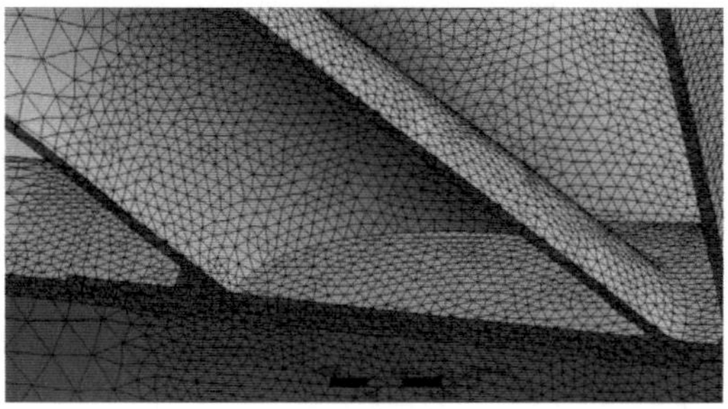

(b)

Figure 4. Meshing on multiplanar double KT tubular joints (a) Global view of the generated mesh (b) Local view of the generated mesh

The chord end fixity conditions of tubular joints in offshore structures may range from "almost fixed" to "almost pinned" with generally being closer to "almost fixed". For analysis in this study, it is assumed that all of the constraints are fixed on both ends of the chord. The tubular joint model with the parameter $\alpha > 12$, the magnitude of the stress at the joint area is not affected by the boundary conditions on the chord. In this study the value of parameter $\alpha = 21.5$ was assigned for all the models. The loading applied to the model are axial forces that the values are obtained from the results of the global modeling as presented in Table 2. Loading directions and boundary condition of the models can be seen at Figure 5.

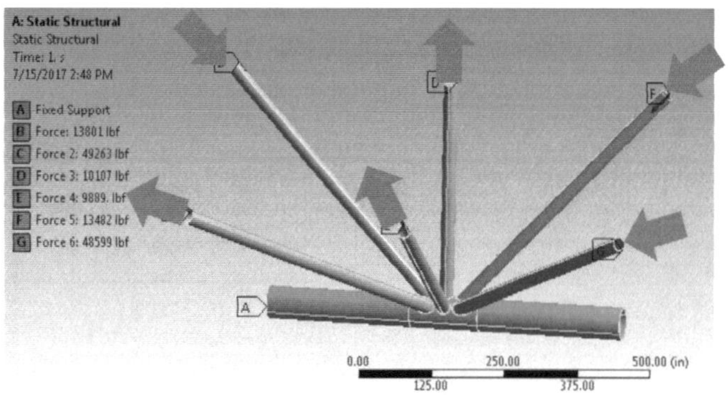

Figure 5. Loading and boundary condition of the model

The meshing sensitivity analysis is performed to ensure the consistency and accuracy of the analysis output from the finite element model as a function of the number of elements used. Size of meshing is iteratively changed, until the stress generated at a particular location becomes constant. From the results of meshing sensitivity test (see Figure 6), it is found that the number of meshing elements whose stress output is near constant at a particular point is meshing with the number of elements as many as 1,000,000 elements.

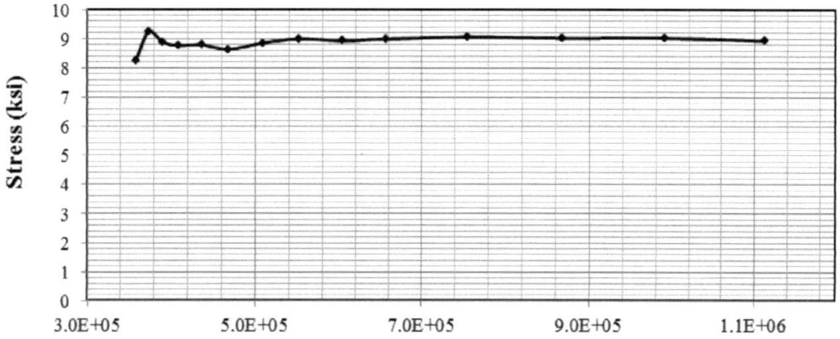

Figure 6. Graph of meshing sensitivity analysis

Table 3. Result of meshing sensitivity test

No	Number of elements	CPU time (minute)	Disk Storage (bytes)	Stress (ksi)	Change (%)
1	358,179	10	1,168,567,338	8,261	
2	373,059	10.2	1,179,997,513	9,235	10,547
3	390,022	10.3	1,195,417,876	8,877	4,037
4	408,545	10.5	1,205,657,858	8,763	1,301
5	435,671	10.8	1,217,451,712	8,788	0,284
6	467,794	11.3	1,220,329,118	8,633	1,795
7	509,145	11.7	1,239,971,901	8,859	2,551
8	551,325	12.1	1,245,735,142	8,992	1,479
9	603,822	12.6	1,250,661,018	8,933	0,660
10	656,583	13	1,260,421,134	8,991	0,645
11	754,722	13.6	1,277,445,914	9,058	0,740
12	867,228	14.1	1,282,657,781	9,025	0,366
13	990,615	14.7	1,295,957,708	9,029	0,044
14	1,111,926	15.5	1,311,581,327	8,942	0,973

Other step to validate the results of the local analysis is to compare it with the global model in term of the nominal stress in the brace using Von Mises stress, as presented at Table 4.

Table 4. Nominal stress validation

$\sigma_{N. \text{ Von Mises}}$ (ksi)		Error (%)
FEA (Global model)	FEA (Local model)	
9,0297	9,0298	0,000916

From Table 4 it is shown that the percentage error is 0.000916% which is below 1%, indicating that the local analysis is close to the global analysis both in terms of structural modeling, loading, and boundary conditions and also the results of the analysis. So that the local analysis is well validated by the global analysis results.

3.2. Hot-spot stress and SCF calculation

The widely accepted conventional approach for fatigue strength assessment of tubular joints is to use the geometric stresses at the weld toe. According to DNVGL-RP-C203 [17] hot-spot stress determined using linear extrapolation. The linear extrapolation method is done by making two points forming an area which will be extrapolated in linear stress. In this study, to determine the distance of two linear extrapolation points used refers to DNVGL-RP-C203 (see Figure 7). The maximum principal stress is used as linear extrapolation to determine the hot-spot stress [18].

After the known nominal stress and hot-spot stress, the formula stress concentration factor is:

$$SCF = \frac{\sigma_{maks}}{\sigma_{n}} \tag{1}$$

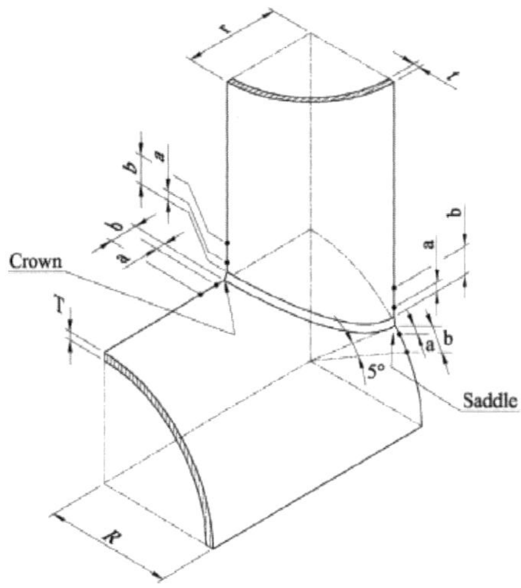

Figure 7. Extrapolation procedure recommended byDNVGL-RP-C203

4. Results and analysis

The result of the parametric study with finite element approach is performed to determine the effect of the non-dimensional parameter variation on the stress distribution along the intersection line between the brace and the chord, which in study is done along the weld toe on the chord side. Stress distribution along the weld toe on the chord side to be investigated only on the joint between the brace with the chord that experienced the greatest stress.

In this study, the joint between the brace and the chord that experienced the greatest stress due to the axial force is in the CVH brace (inclined brace) as can be seen at Figures 8 and 9. The intersection line with oval-shaped approached using angle in degree (see Figure 10). The crown point 1 is assumed to be the angle 0° and the 360° angle indicating that the point of crown 1 is the starting point of the investigation and the end point of the investigation. The saddle outer point is assumed to be an angle of 90°, the crown 2 points assumed to be angle of 180° and the inner saddle is assumed to be angle 270°. Different values assigned to each non-dimensional parameter are as follows: $\beta = 0.35$, $\beta = 0.4$, $\beta = 0.45$; $\gamma = 10$, $\gamma = 12$, $\gamma = 15$; $\tau = 0.35$, $\tau = 0.45$, $\tau = 0.55$; $\theta = 36°$, $\theta = 46°$, $\theta = 56°$.

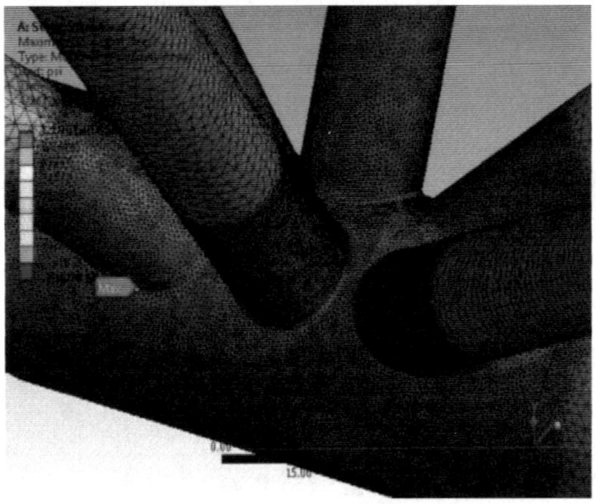

Figure 8. Maximum stress occurs at the inclined brace-CVH

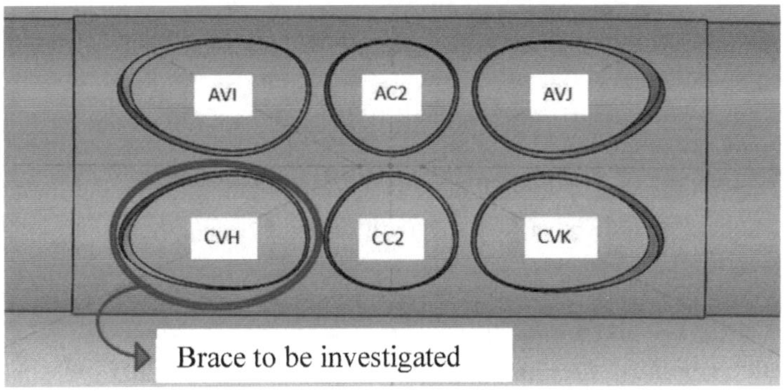

Figure 9. Location of inclined brace to be investigated

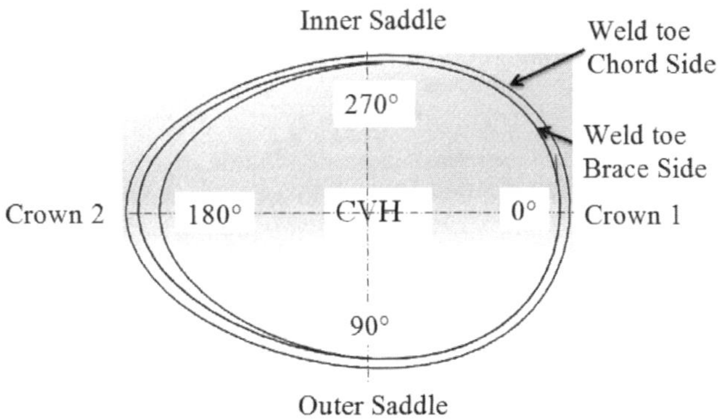

Figure 10. Intersection line of inclined brace to be investigated with angle approach

4.1. Effect of β on the geometric stress and SCF distribution

The parameter β is the ratio of brace diameter to chord diameter. In this study, variation of the β is β = 0.35, β = 0.4 and β = 0.45. Increasing the β in the models having constant value of chord diameter leads to increase of brace diameter. All parameters except the β parameter, i.e., the parameters τ, γ and θ are made equal to the three models (τ = 0.5; γ = 13.5; θ = 46°).

Results of investigation that can be seen in Figure 11 indicate that increase of the β leads to decrease both geometric stress and SCF distribution along the weld toe (on the chord side). According to the three models, the maximum geometric stress and maximum SCF occurs at the same point, i.e. at the outer saddle point (angle 90°).

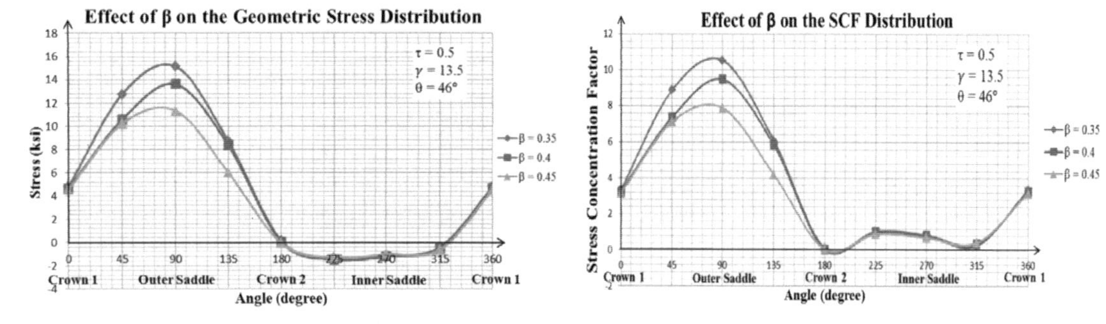

(a) (b)

Figure 11. Effect of β on the (a) geometric stress distribution, (b) SCF distribution

4.2. Effect of γ on the geometric stress and SCF distribution

The parameter γ is the ratio of diameter or radius to thickness of chord. In this study, the γ is varied in three models with different size of parameter γ made for these models, i.e. γ = 10, γ = 12 and γ = 15. Increase of γ in the models having constant value of chord diameter leads to decrease of chord thickness. All parameters except the γ parameter, i.e., the parameters β, τ and θ are made equal to the three models (β = 0.44; τ = 0.5; θ = 46°).

Results of investigation (can be seen in Figure 12) indicate that increase of the γ leads to increase geometric stress and SCF distribution along the weld toe (on the chord side). In the γ = 12 and γ = 15 models, the maximum geometric stress and SCF occurs at the saddle outer point (angle 90°) while for the model γ = 10, the maximum geometric stress and peak SCF occurs at a point with an angle of 45°.

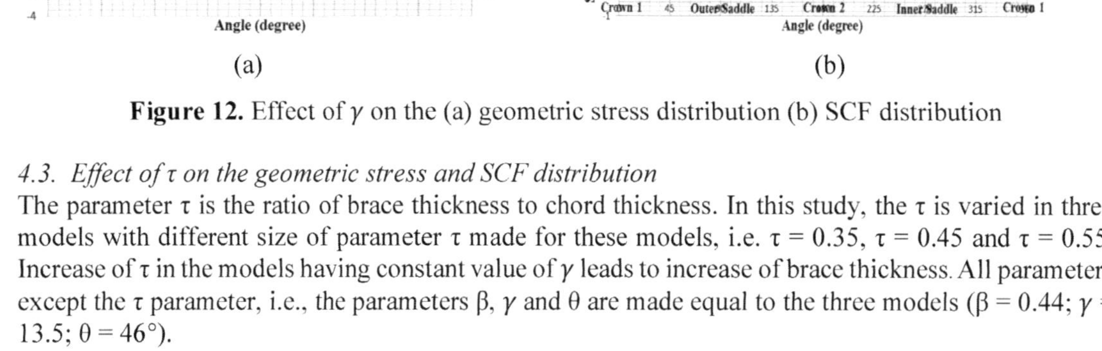

(a) (b)

Figure 12. Effect of γ on the (a) geometric stress distribution (b) SCF distribution

4.3. Effect of τ on the geometric stress and SCF distribution

The parameter τ is the ratio of brace thickness to chord thickness. In this study, the τ is varied in three models with different size of parameter τ made for these models, i.e. τ = 0.35, τ = 0.45 and τ = 0.55. Increase of τ in the models having constant value of γ leads to increase of brace thickness. All parameters except the τ parameter, i.e., the parameters β, γ and θ are made equal to the three models (β = 0.44; γ = 13.5; θ = 46°).

Opposite to the effects of the two previous parameters, β and γ, results of investigation (as presented at Figure 13) indicate that increase of the τ leads to also increase geometric stress and SCF distribution along the weld toe (on the chord side). According to the three models, the maximum geometric stress and peak SCF occurs at the same point which is at an angle of 45°.

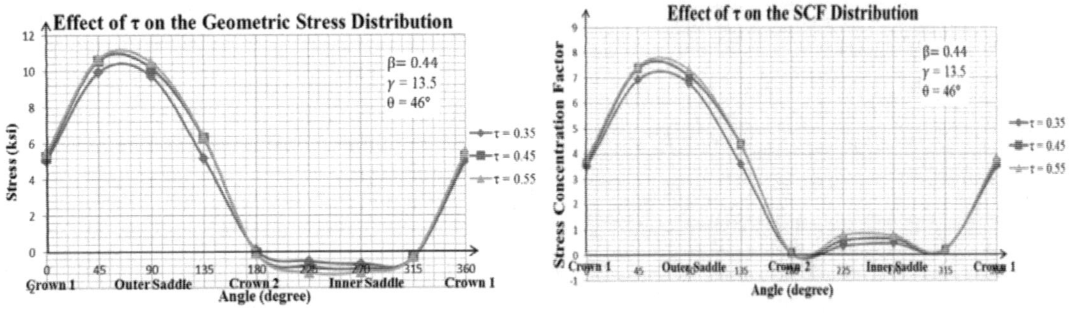

(a) (b)

Figure 13. Effect of τ on the (a) geometric stress distribution, and (b) SCF distribution

4.4. Effect of θ on the geometric stress and SCF distribution

The parameter θ is angle between brace to chord. In this study, the θ is varied in three models with different size of parameter θ made for these models, i.e. θ = 36, θ = 46 and θ = 56. All parameters except the θ parameter, i.e., the parameters τ, γ and τ are made equal to the three models (β = 0.44; τ = 0.5; γ = 13.5). In principle it is free to select values of these β, τ, and γ as long as in the common range as already mentioned previously. For this analysis, the chosen values is in order to enrich data for this research.

Results of investigation (can be seen in Figure 14) indicate that increasing the θ leads to decrease geometric stress and SCF distribution along the weld toe (on the chord side). According to the three models, the maximum geometric stress and peak SCF occurs at the same point, i.e. at the outer saddle point (angle 90°).

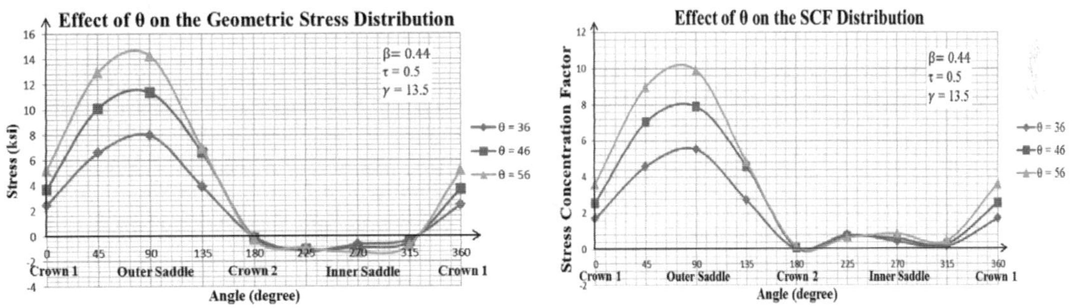

(a) (b)

Figure 14. Effect of θ on the (a) geometric stress distribution (b) SCF distribution

5. Conclusions

The results of parametric studies show that the increase of the β leads to decrease of SCF and geometric stress. While the increase of the τ, γ and θ leads to increase of SCF and geometric stress. The maximum geometric stress and peak SCF occurs at the same point caused by variation of β, i.e. at the outer saddle point (angle 90°). For variation of γ, in the γ = 12 and γ = 15 models, the maximum geometric stress and SCF occurs at the saddle outer point (angle 90°) while for the model γ = 10, the maximum geometric stress and peak SCF occurs at a point with an angle of 45°. The maximum geometric stress and peak SCF occurs at the same pointby variation ofτ, with an angle of 45°.For variation of θ, the maximum geometric stress and peak SCF occurs at the same point, i.e. at the outer saddle point (angle 90°). The effect of parameters β, γ and θ on the SCF are greater than the effect of parameters τ.

References

[1] Ahmadi, Hamid and Esmaeil Zavvar 2015 Stress concentration factors induced by out-of-plane bending loads in ring-stiffened tubular KT-joints of jacket structures, *Thin Walled Structures*, Vol.**91** p 82-95
[2] Hidayat, Urfan, Rudi Walujo P and Wisnu Wardhana 2016 SCF parametric study of multiplanar tubular joint double K *Bachelor Thesis of Ocean Engineering Department FTK-ITS* Surabaya
[3] Murthy D S 2004 An Efficient FE Modelling Strategy for Fracture Analysis of Tubular Joints *Journal Structural Engineering Research Centre* Vol.**85** May
[4] Yeoh, Seng-Keat, Ai-Kah Soh and Chee-Kiong Soh 1995 Behavior of Tubular T-Joints Subjected to Combined Loadings *Journal Construction Steel Research* Vol. 32 p 259-280
[5] Soh, Ai-Kah and Chee-Kiong Soh 1993 Hot Spot Stress of K Tubular Joints Subjected to Combined Loadings *Journal Construction Steel Research* Vol.**26** p 125-140
[6] Pang, H L J and C W Lee 1995 Three-dimensional finite element analysis of a tubular T-joint under combined axial and bending loading *Journal Fatigue* Vol. **17** p 313-320
[7] Chang, E and W D Dover 1999 Predicted of stress distributions along intersection of tubular Y-joints and T-joints *International Journal of Fatigue* Vol. **21** p 361-381
[8] Young-Bo, Shao, Du Zhi-Fu and Lie Seng-Tjhen 2009 Predicted of hot spot stress distribution for tubular K-joint under basic loadings *Journal of Construction Steel Research* Vol. **65** p 2011-2026
[9] Chiew, S P, C K Soh and N W Wu 2000 General SCF design equations for steel multiplanar tubular XX-joints *International Journal of Fatigue* Vol. **22** p 283-293
[10] Karamos, Spyros A, Arie Romejin, Jaap Wardenier 2002 SCF equationsin multi-planar welded tubular DT-joint including bending effects *Marine Structures* Vol. **15** p 157-173
[11] Wingerde, Arno M van, Jeffrey A Packer and Jaap Wardenier 2001 Simplified SCF formulae and graphs for CHS and RHS K- and KK-connections *Journal of Construction Steel Research* Vol. **57** p 221-252
[12] Woghiren, C O and F P Brennan 2009 Weld toe stress concentrations in multi-planar stiffened tubular KK-joints *International Journal of Fatigue* Vol. **31** p 164-172
[13] Naibaho, Bonar, Rudi Walujo P and Wisnu Wardhana 2016 SCF parametric study of multiplanar tubular KDT-joint *Bachelor Thesis of Ocean Engineering Department FTK-ITS* Surabaya
[14] Puspaningtyas, Arvina, Rudi Walujo P and Wisnu Wardhana 2016 Parametric study of stress intensity factor on multiplanar tubular joint with fracture mechanics *Bachelor Thesis of Ocean Engineering Department FTK-ITS* Surabaya
[15] Fuadi, Ibnu fasyin,Rudi Walujo P and Nur Syahroni 2017 Formulation of stress concentration factors for multiplanar tubular DKT-joint using finite element method *Bachelor Thesis of Ocean Engineering Department FTK-ITS* Surabaya
[16] Structural welding code: AWS D 1.1. USA: American Welding Society (AWS): 2002
[17] DNVGL RP-C203 2001 Recommended Practice, Fatigue Strength Analysis of Offshore Steel Structure.
[18] Haghpanahi, M and H Pirali 2006 Hot Spot Stress Determination for a Tubular T-Joint under Combined Axial and Bending Loading *International Journal of Engineering Sciences*, Vol. **17** p 21-28